GEOCHIMICA ET COSMOCHIMICA ACTA

Journal of The Geochemical Society and The Meteoritical Society

SUPPLEMENT 9

MARE CRISIUM: THE VIEW FROM LUNA 24

PROCEEDINGS OF THE CONFERENCE ON LUNA 24

Houston, Texas, December 1-3, 1977

Compiled by

Lunar and Planetary Institute

Houston, Texas

PERGAMON PRESS

NEW YORK · OXFORD · TORONTO · SYDNEY · FRANKFURT

U.K.	Pergamon Press Ltd., Headington Hill Hall, Oxford OX3 0BW, England
U.S.A.	Pergamon Press Inc., Maxwell House, Fairview Park, Elmsford, New York 10523, U.S.A.
CANADA	Pergamon of Canada Ltd., 75 The East Mall, Toronto, Ontario, Canada
AUSTRALIA	Pergamon Press (Aust.) Pty. Ltd., 19a Boundary Street, Rushcutters Bay, N.S.W. 2011, Australia
FRANCE	Pergamon Press SARL, 24 rue des Ecoles, 75240 Paris, Cedex 05, France
FEDERAL REPUBLIC OF GERMANY	Pergamon Press GmbH, 6242 Kronberg-Taunus, Pferdstrasse 1, Federal Republic of Germany

First edition 1978

Library of Congress Cataloging in Publication Data

Conference on Luna 24, Houston, Tex., 1977.
Mare crisium.

(Geochimica et cosmochimica acta: Supplement; 9)
1. Lunar geology--Congresses. I. Lunar and Planetary Institute. II. Title. III. Series.
QB 592.C64 1977 559.9'1 78-10110
ISBN 0-08-022965-4

Contents

Regional Studies

Regolith Studies

Preface

During August, 1976, the Soviet Union's Luna 24 automated spacecraft obtained a 160-cm core sample from Mare Crisium, the third in a series of collecting stations on a traverse across the moon's eastern limb. Under the terms of a joint agreement with NASA, the Soviet Academy of Sciences provided 3 grams of the returned soil for study by U. S. scientists. Allocation and study of this material followed very rapidly upon receipt of the samples, culminating in December, 1977, with a conference to which early results of the sample studies were reported.

The LPI Topical Conference on Luna 24 was held in Houston, December 1-3, 1977. This 2½-day meeting was attended by 105 scientists from the U. S., U. S. S. R., France, Great Britain, and India who presented 67 papers on the following topics:

I. Mare Crisium and the Luna 24 Landing Site: from Kilometers to Micrometers
II. Regolith Studies of the Luna 24 Soil Samples
III. Petrology, Mineral Chemistry, and Cooling Histories of the Lithic Fragments in Luna 24 Samples
IV. Chemistry, Isotopic Studies and Geochronology of Luna 24 Samples

Mare Crisium: The View from Luna 24 contains 45 papers organized in a format similar to that of the conference, in which the authors have reported the results of their sample studies and have attempted to place them in perspective with our knowledge of rocks and soils at other lunar landing sites.

Publication of these proceedings, which is sponsored by the Lunar and Planetary Institute on behalf of Principal Investigators in the Lunar and Planetary Programs, has occurred in parallel with *Proceedings of the Ninth Lunar and Planetary Science Conference*, which include several papers relevant to the Mare Crisium region and to the Luna 24 core sample.

On behalf of the Institute, I thank the authors, editors and reviewers for their cooperation in accomplishing the tight publication schedule. In particular, I thank Jim Papike who, as Science Editor, succeeded admirably in keeping everyone's productivity at a very high level indeed and who, together with the Associate Editors, deserves much of the credit for the scientific quality of this volume. My personal appreciation is extended also to the members of the Institute editorial staff who have shared every stage in the development of this book, and in particular to Ms. Paula Robertson, Technical Editor, who has borne the major responsibility for technical preparation of these *Proceedings*.

RUSSELL B. MERRILL

The Lunar and Planetary Institute
Houston, Texas
May, 1978

1977; Tarasov *et al.*, 1977). High-magnesium materials also include a limited number of lithic fragments (olivine gabbro and olivine vitrophyre). The MgO content of these lithic fragments ranges from 10.4% to 16.9% (Ryder *et al.*, 1977; Tarasov *et al.*, 1977) and their olivines ($\sim Fo_{75}$) are similar in composition to high-magnesium monomineralic fragments from the core. High-magnesium homogeneous green glass (MgO = 10% to 14%) is also present in the Luna 24 soils (Vaniman and Papike, 1977; Ryder *et al.*, 1977; Grove *et al.*, 1977; Norman *et al.*, 1977).

If these high-magnesium materials represent another major basalt unit in Mare Crisium, then any significant surface expression of this unit should be observable in the orbital X-ray fluorescence (XRF) data. These data have been used to map the chemical variations in Mare Crisium in terms of Al/Si, Mg/Si and Mg/Al intensity ratios.

Luna 24 Surface Soils as Characterized by Orbital XRF Data

The Apollo 15 XRF experiment provides broad coverage of southern Mare Crisium, including the Luna 24 landing site. The correlation of these orbital data (refined using the procedures outlined in the "background" section) to Al/Si and Mg/Si concentration ratios from Luna 24 sample analyses is a useful point of reference for interpretation of chemical variations elsewhere in Mare Crisium.

An Al/Si concentration ratio of 0.40 was calculated from XRF data for a 2-degree square area centered at the Luna 24 landing site (62° 12′E; 12° 45′N). This value compares closely to the Al/Si concentration ratio, 0.41, from chemical analysis of the < 74 μm fraction of the drill core (Barsukov *et al.*, 1977). The Mg/Si concentration ratio for the Luna 24 soil is 0.26, both from orbital XRF and analysis of the finest fraction (Fig. 1).

The extraordinary agreement between the orbital XRF and sample analysis concentration ratios for the finest fraction from the Luna 24 site is more than one can reasonably expect. Al/Si and Mg/Si values from the orbital experiment, with a practical resolution of 15–30 km at best are being compared to Al/Si and Mg/Si values from soils in a core 8 mm in diameter. In addition, the 2-degree square area centered on the landing site, for which the concentration ratios were calculated, contains ridges and scarps along the first Crisium basin ring (Olson and Wilhelms, 1974) and may not be chemically homogeneous. Bulk soil analyses have considerably lower Al/Si concentrations (0.32) and Mg/Si values (0.27) are slightly higher than those calculated from XRF data. The Luna 24 core did not sample the surface at the landing site. However, we believe the <74 μm fraction of the core to be most representative of the surface soil composition as measured from orbit. Barsukov *et al.* (1977) reports that this size fraction constitutes 50% of the soil collected. The orbital XRF measurements of elemental abundances are restricted to the top tens of microns of the lunar surface. Bulk soil analysis, however, is skewed toward the chemical composition of large lithic fragments, whereas an essentially surficial analysis of the soil,

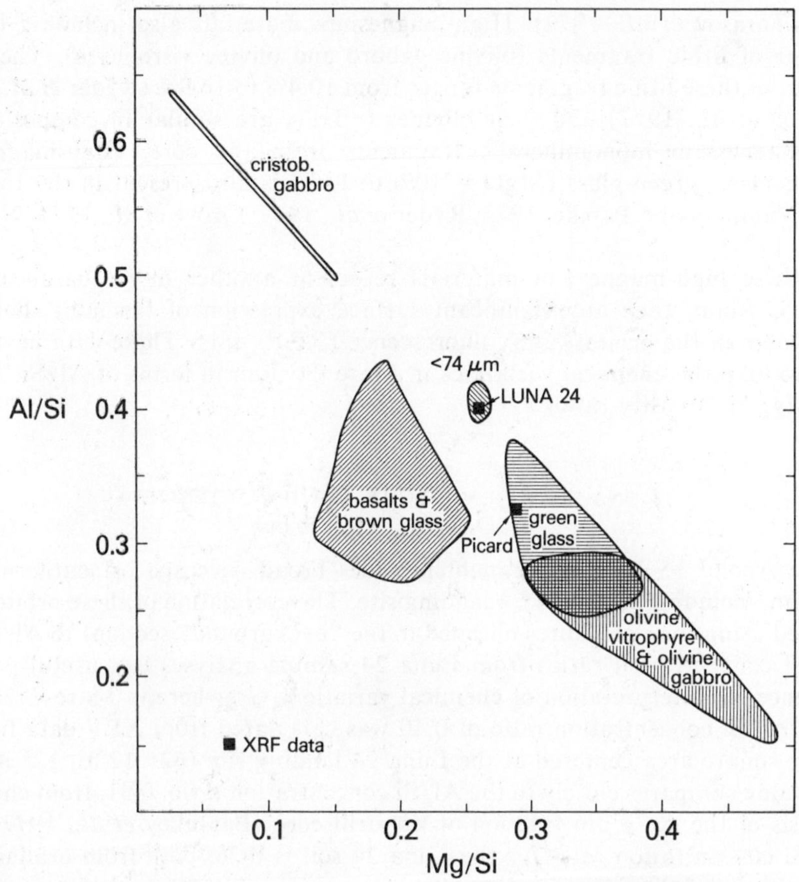

Fig. 1. Orbital XRF and sample Al/Si and Mg/Si concentration values for the Luna 24 soils (Ryder *et al.*, 1977; Vaniman and Papike, 1977; Tarasov *et al.*, 1977).

rather than a volumetric one, is needed for a comparison to orbital data. Thus, one would expect the fines to show the closest correlation to orbital XRF data.

This relationship between orbital and sample data permits further comparisons between the Luna 24 soils and other areas included in the XRF coverage of Mare Crisium.

CHEMICAL VARIATIONS IN MARE CRISIUM SURFACE SOILS

Figures 2, 3 and 4 show respectively Al/Si, Mg/Si and Mg/Al intensity ratios for 0.25° square units, averaged over one degree cells. In each case, the cells have been shaded to represent low (white), medium (cross-lined) and high (black) values for the mare basalts. Cells close to mare/terra borders were left vacant if the signal might include secondary X-rays from terra material beyond the mare

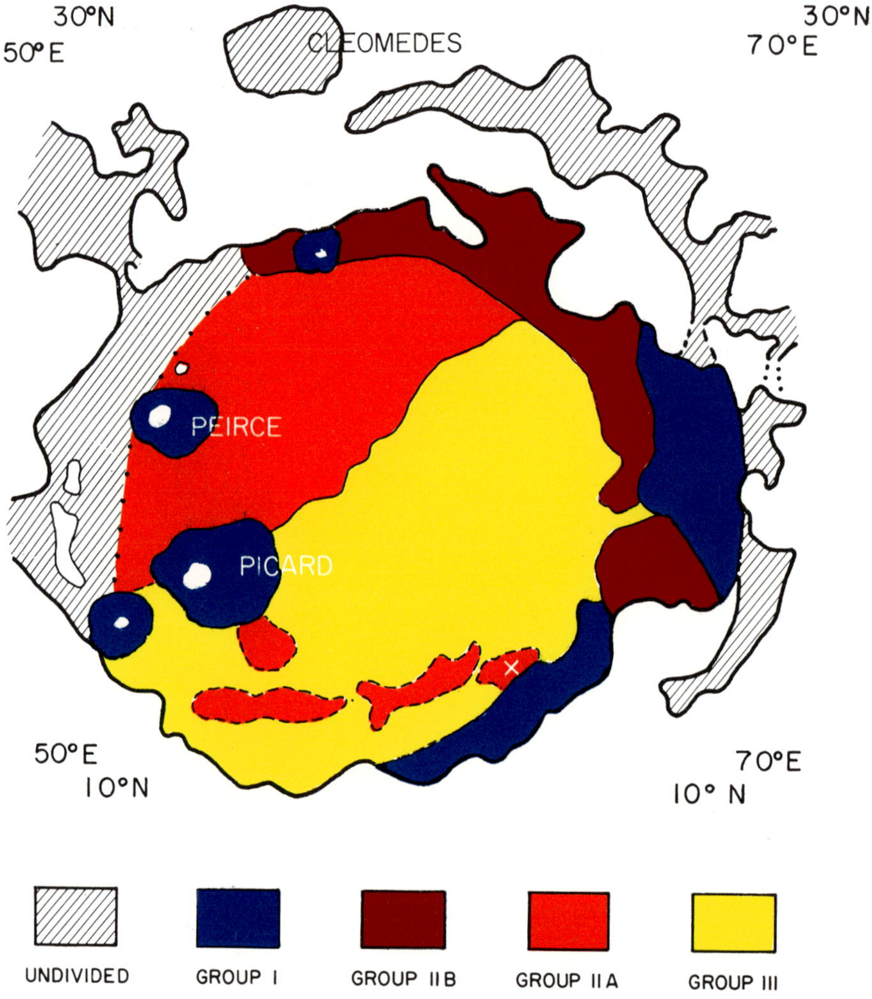

Geologic units of Mare Crisium. Three major basalt groups are recognized: (Group I) a soil derived from Fe- and Mg-rich titaniferous basalt similar to Luna 16 samples; (Group II) soils derived from very low titanium ferrobasalts similar to Luna 24 samples; and (Group III) a soil derived from a low titanium ferrobasalt similar to samples from Apollo 12. Group I basalts were apparently emplaced over the majority of the basin but are presently exposed as a topographic shelf in the southwest part of the basin. The second group of basalts were emplaced subsequently in two stages and now appear as part of the outer shelf and topographic annulus. Subsidence of the basin interior preceded and continued after the emplacement of the third basalt unit. The Luna 24 site is within a patch of Group II and just northwest of Group I, and is marked by an X. Detailed discussion of these units and their relation to Luna 24 and basin history can be found in Head *et al.* (this volume).

Merrill, R. B. and Papike, J. J., editors.
(1978) *Mare Crisium: The View from Luna 24*, Pergamon Press (New York), p. 1–12
Printed in the United States of America

Evidence for a high-magnesium subsurface basalt in Mare Crisium from orbital X-ray fluorescence data

CONSTANCE G. ANDRE*
ROBERT W. WOLFE**
ISIDORE ADLER*

*Chemistry Department University of Maryland, College Park, Maryland 20742
**National Air and Space Museum, Washington, D. C. 20562

Abstract—The primary rock type in the Luna 24 drill core from southeast Mare Crisium is a low-magnesium, low-titanium mare basalt. The soils are also low-titanium, but are enriched in magnesium relative to the basalt lithic fragments. Magnesium-rich exotic particles present in small amounts include olivine gabbro, olivine vitrophyre and green glass. This suggests that the soil is a mixture of local basalt and a more magnesian component, possibly with the composition of the olivine-rich fragments.

Apollo 15 X-ray fluorescence (XRF) data collected in orbit over southern Mare Crisium have been analyzed to characterize chemical variations in the surface soils on the basis of Al/Si, Mg/Si and Mg/Al ratios and to locate any substantial surface expression of the high-Mg component in the core. These orbital data indicate that most of the surface soils in Mare Crisium have magnesium concentrations $\leq$ those at the Luna 24 landing site. Major exceptions are high-magnesium areas northeast of the Luna 24 site and those exposed around the post-mare craters Picard and Peirce. The material around these craters is interpreted to be a magnesium-rich subsurface basalt excavated from depth. Evidence supporting this hypothesis includes: 1) a radar reflector (lunar sounder data) at a depth of 1400 m which most likely represents the interface between the high- and low-magnesium basalt layers below which Picard and Peirce penetrated and 2) spectral reflectance data characterizing the basalts associated with Picard and Peirce as distinctly different from and somewhat higher in TiO_2 content than those at the landing site. On the basis of XRF data alone, the subsurface basalt sampled by Picard and Peirce appears to be representative of that from which the high-Mg components in the Luna 24 soils were derived. However, if the TiO_2 content of this material is indeed significantly higher than that at the Luna 24 site, it is possible that at least three chemically distinct basalt types are present in the Crisium basin: low-Mg, low-Ti basalts typical of lithic fragments in the core; high-Mg, relatively high-Ti basalts excavated from depth at Picard and Peirce, and high-Mg, low-Ti coarse grained basalts exposed on the surface northeast of the landing site which are represented in the core as monomineralic-fragments and a few lithic fragments of olivine gabbro and olivine vitrophyre.

BACKGROUND

The Apollo 15 and Apollo 16 missions to the moon carried an X-ray fluorescence spectrometer which measured from orbit the intensity of the secondary X-ray signals characteristic of aluminum, silicon, and magnesium in the surface soils, using the X-ray emission from the sun as the exciting source (Adler *et al.*, 1972a, 1972b). The primary objective of the experiment was to map chemical variations in the areas overflown in terms of these major rock-forming elements. The elemental measurements are expressed as intensity ratios (Al/Si,

Mg/Si, and Mg/Al) in order to minimize non-chemical effects on the measured signal, such as those caused by differences in sun-moon-spacecraft geometry, shifts in the solar spectrum, and particle size variations on the lunar surface.

The important relationship between orbital X-ray intensity ratios and returned sample analyses is based on the fact that the characteristic secondary X-ray intensity is primarily a function of the element's concentration in surface soils. The conversion factors for intensity to concentration ratios were calculated prior to the Apollo missions. The critical link between concentrations predicted from orbit and those from sample data is the basis for extending detailed chemical information from a few specific landing sites to broad areas of the moon traversed by the Apollo 15 and Apollo 16 spacecrafts.

The X-ray fluorescence data used in this study have been refined in the following manner:

1. The mean level of Al/Si, Mg/Si and Mg/Al values for single orbits was adjusted to compensate for interorbit variations due to solar spectral changes during the 2-hour time lapse between orbits.

2. A weighted-filter procedure (Andre *et al.*, 1977; Bielefeld *et al.*, 1977) was applied to the data to maximize chemical detail. To summarize this procedure, data points are integrated by a series of increasingly large data filters which simulate the radial response of the spectrometer's field of view. The smallest filter represents a 7.5-km^2 area of the lunar surface. The next three filters in the series are concentric about the smallest filter and increase in size up to 52.5 km on an edge. Each filter incorporates data points weighted on the basis of statistical error. Furthermore, each Al/Si value calculated for the four filters is weighted so that the smaller the filter, representing an area closer to the center point, the more influential the value. When this set of concentric filters is moved along the lunar surface in 7.5-km steps, each 0.25° unit value represents the best possible balance between statistical significance and spatial resolution.

INTRODUCTION

The Luna 24 soil is mainly composed of a basalt with a low content of TiO_2 and MgO, and a high Fe:Mg ratio. A puzzling problem which arises from analysis of the Luna 24 drill core samples is the presence of high-magnesium material which is not characteristic of the predominant rock type in the core. The MgO content of bulk soils (~10%) (Blanchard *et al.*, 1977; Simonds *et al.*, 1977) is significantly higher than the 6.5% MgO content of basalt lithic fragments (Taylor *et al.*, 1977; Vaniman and Papike, 1977; Ryder *et al.*, 1977). High-magnesium material is present in three classes of fragments in the Luna 24 drill core. There is an unusually high proportion (~52%) of monomineralic fragments in the soils. Some of these fragments correspond to the composition of minerals found in lithic fragments of the low-magnesium basalt. However, many of the olivine and pyroxene fragments have a high-magnesium content (Mg/Mg + Fe = ~0.75) which is not characteristic of this basalt (Basu *et al.*,

1977; Tarasov *et al.*, 1977). High-magnesium materials also include a limited number of lithic fragments (olivine gabbro and olivine vitrophyre). The MgO content of these lithic fragments ranges from 10.4% to 16.9% (Ryder *et al.*, 1977; Tarasov *et al.*, 1977) and their olivines ($\sim Fo_{75}$) are similar in composition to high-magnesium monomineralic fragments from the core. High-magnesium homogeneous green glass (MgO = 10% to 14%) is also present in the Luna 24 soils (Vaniman and Papike, 1977; Ryder *et al.*, 1977; Grove *et al.*, 1977; Norman *et al.*, 1977).

If these high-magnesium materials represent another major basalt unit in Mare Crisium, then any significant surface expression of this unit should be observable in the orbital X-ray fluorescence (XRF) data. These data have been used to map the chemical variations in Mare Crisium in terms of Al/Si, Mg/Si and Mg/Al intensity ratios.

Luna 24 Surface Soils as Characterized by Orbital XRF Data

The Apollo 15 XRF experiment provides broad coverage of southern Mare Crisium, including the Luna 24 landing site. The correlation of these orbital data (refined using the procedures outlined in the "background" section) to Al/Si and Mg/Si concentration ratios from Luna 24 sample analyses is a useful point of reference for interpretation of chemical variations elsewhere in Mare Crisium.

An Al/Si concentration ratio of 0.40 was calculated from XRF data for a 2-degree square area centered at the Luna 24 landing site (62° 12′E; 12° 45′N). This value compares closely to the Al/Si concentration ratio, 0.41, from chemical analysis of the < 74 μm fraction of the drill core (Barsukov *et al.*, 1977). The Mg/Si concentration ratio for the Luna 24 soil is 0.26, both from orbital XRF and analysis of the finest fraction (Fig. 1).

The extraordinary agreement between the orbital XRF and sample analysis concentration ratios for the finest fraction from the Luna 24 site is more than one can reasonably expect. Al/Si and Mg/Si values from the orbital experiment, with a practical resolution of 15–30 km at best are being compared to Al/Si and Mg/Si values from soils in a core 8 mm in diameter. In addition, the 2-degree square area centered on the landing site, for which the concentration ratios were calculated, contains ridges and scarps along the first Crisium basin ring (Olson and Wilhelms, 1974) and may not be chemically homogeneous. Bulk soil analyses have considerably lower Al/Si concentrations (0.32) and Mg/Si values (0.27) are slightly higher than those calculated from XRF data. The Luna 24 core did not sample the surface at the landing site. However, we believe the <74 μm fraction of the core to be most representative of the surface soil composition as measured from orbit. Barsukov *et al.* (1977) reports that this size fraction constitutes 50% of the soil collected. The orbital XRF measurements of elemental abundances are restricted to the top tens of microns of the lunar surface. Bulk soil analysis, however, is skewed toward the chemical composition of large lithic fragments, whereas an essentially surficial analysis of the soil,

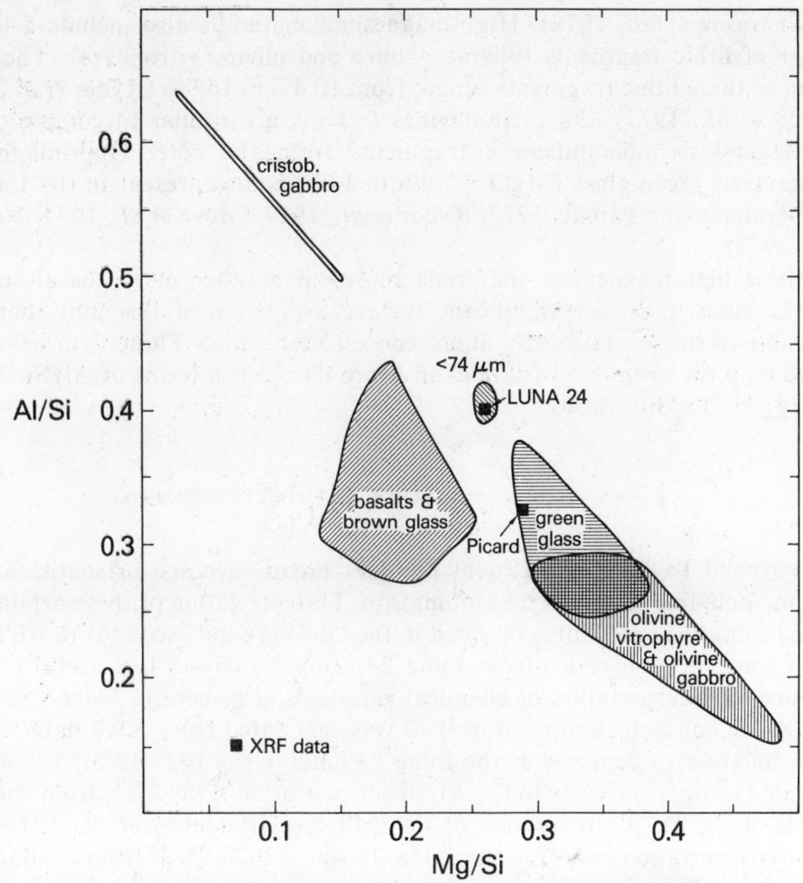

Fig. 1. Orbital XRF and sample Al/Si and Mg/Si concentration values for the Luna 24 soils (Ryder *et al.*, 1977; Vaniman and Papike, 1977; Tarasov *et al.*, 1977).

rather than a volumetric one, is needed for a comparison to orbital data. Thus, one would expect the fines to show the closest correlation to orbital XRF data.

This relationship between orbital and sample data permits further comparisons between the Luna 24 soils and other areas included in the XRF coverage of Mare Crisium.

Chemical Variations in Mare Crisium Surface Soils

Figures 2, 3 and 4 show respectively Al/Si, Mg/Si and Mg/Al intensity ratios for 0.25° square units, averaged over one degree cells. In each case, the cells have been shaded to represent low (white), medium (cross-lined) and high (black) values for the mare basalts. Cells close to mare/terra borders were left vacant if the signal might include secondary X-rays from terra material beyond the mare

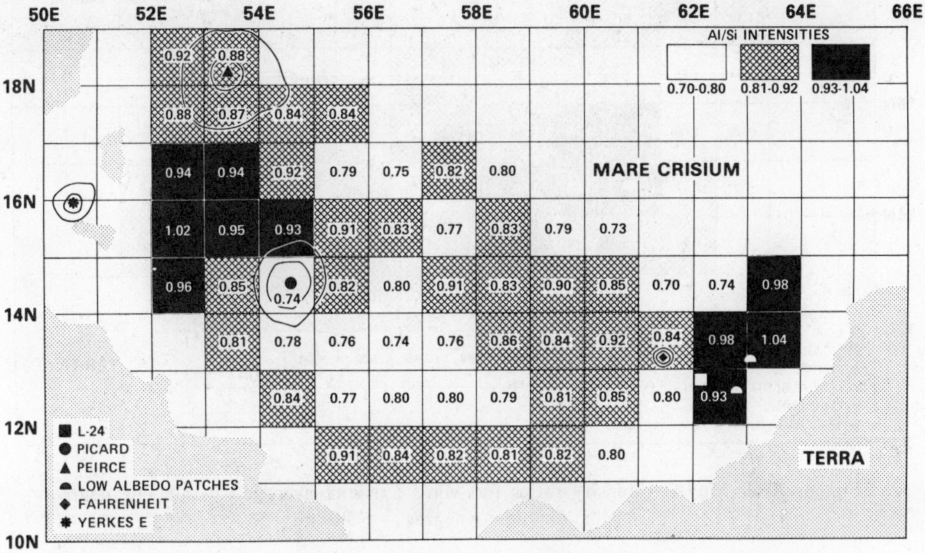

Fig. 2. XRF Al/Si intensity ratios for Mare Crisium, averaged within one degree cells. One lunar degree = ~30 km.

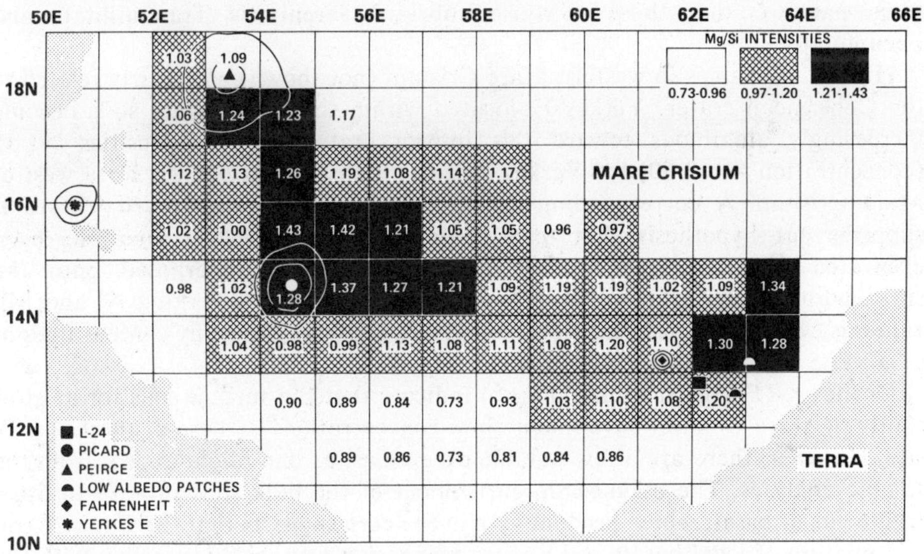

Fig. 3. XRF Mg/Si intensity ratios for Mare Crisium, averaged within one degree cells. One lunar degree = ~30 km.

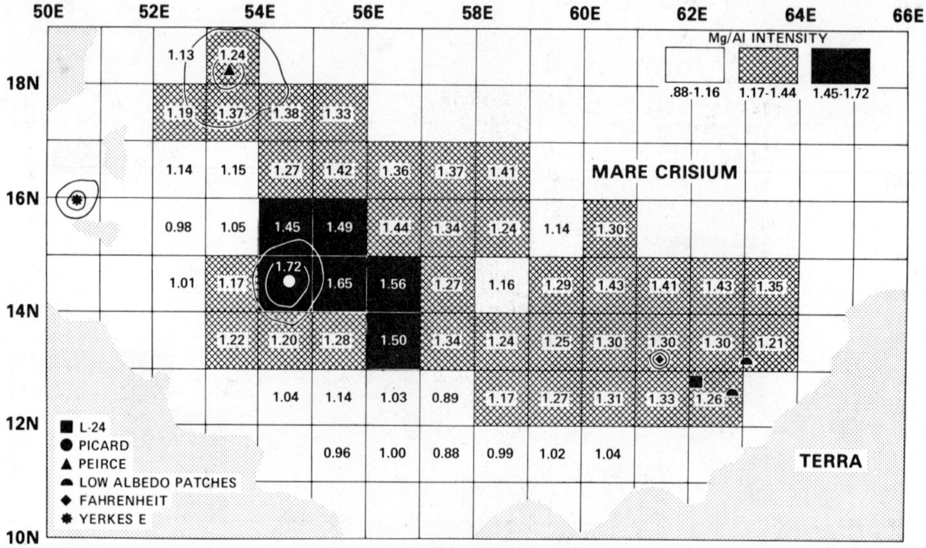

Fig. 4. XRF Mg/Al intensity ratios for Mare Crisium, averaged within one degree cells. One lunar degree = ~30 km.

boundary. Physical features, pertinent to this study, are shown by symbols explained in the keys on each figure.

Al/Si intensity ratios in Mare Crisium (Fig. 2) are, in general, higher and the range narrower than those in Maria Imbrium, Serenitatis, Tranquillitatis and Fecunditatis.

High Al/Si values in western Mare Crisium (not shown) are clearly related to the Copernican crater, Yerkes E, located within the mare. Mare soils become increasingly aluminous toward this impact crater. Al/Si intensities >1.35 (concentration = ~0.60) for Yerkes E are as high as those for the terra west of Mare Crisium. A corresponding decrease in Mg/Si values toward Yerkes E supports our hypothesis that terra material from the basin floor has been excavated from beneath shallow mare basalts and superposed onto the surrounding mare surface. The pronounced effect of the diverging Al and Mg values is evident in a detailed view of Mg/Al variations shown by contour lines of 0.25° (7.5 km) intensity ratio values in Fig. 5.

Orbital XRF Mg/Si values (Fig. 3) indicate that the surface soils for most of southern Mare Crisium have magnesium concentrations ≤ those at the Luna 24 site. However, there are a few notable exceptions, as one might expect from the sample analyses. The magnesium enrichment of the bulk soils relative to lithic fragments of basalt which predominate in the core suggests that the Luna 24 soil is a mixture of local basalt and a more magnesian component, possibly with the composition of the olivine gabbro, olivine vitrophyre or green glass fragments. These fragments have an MgO range of 10% to 16.9% and an Al_2O_3 range of

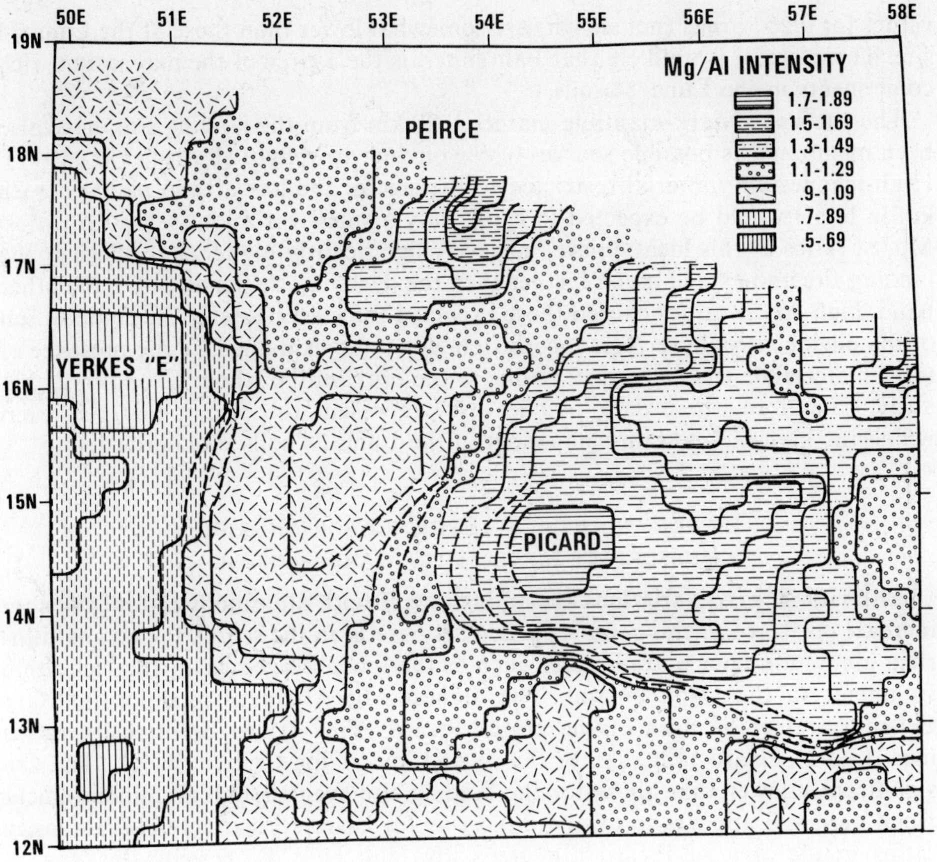

Fig. 5. Mg/Al contour map based on 0.25° × 0.25° area values of XRF Mg/Al intensity ratios.

6.2% to 14.9%. If this magnesium-rich material is exposed on the surface of Mare Crisium with an areal extent commensurate with the field of view of the XRF experiment, the XRF Mg/Si concentration ratio should be greater than 0.285 (intensity = ~1.31) and the Al/Si concentration ratio should be less than 0.372 (intensity = ~0.87).

It has been suggested that the magnesium-rich basalt fragments in the core may have been ejected from the crater Fahrenheit, 18 km from the landing site (McKay *et al.*, 1977; Basu *et al,* 1977). This post-mare crater, ~6.3 km in diameter, is considerably smaller than the spatial resolution of the XRF data. However, the ejecta blanket mapped by Florensky *et al.* (1977) is ~30 km in diameter. If high-magnesium ejecta from this crater reached the landing site 18 km away, a small increase in Mg/Si values would be expected around Fahrenheit even though there is some overlap in the field of view. (This is because signal intensity decreases sharply from the center of the field of view). However, Mg/Si values for 0.25° areas (not shown) are somewhat lower than those at the Luna 24 site. Therefore, it is unlikely that Fahrenheit is the source of the magnesium-rich components in the Luna 24 soils.

The patches of dark-mantling material, 20 km from the landing site, have also been mentioned as possible sources of the olivine vitrophyre (Ryder *et al.*, 1977). High-magnesium material restricted to the size of the low-albedo patches (~28 km in total) would be expected to cause only a small variation in Mg/Si ratios. Mg/Si ratios at this location are, however, significantly higher than those at the landing site and extend much farther than the dark mantle deposits. On the other hand, high Al/Si values exceed those expected for material with the composition of the olivine vitrophyre. This inconsistency may be explained by the presence of terra material on the mare surface due to the proximity of the highlands ~25 km away. Therefore, these dark patches may represent the parent material from which the high-magnesium component in the Luna 24 soils was derived.

PICARD AND PEIRCE—EVIDENCE FOR MG-RICH SUBSURFACE MATERIAL IN MARE CRISIUM

Although Mg/Al variations are ambiguous if both aluminum and magnesium increase or decrease, this ratio is particularly useful in the search for magnesium-rich, aluminum-poor mare basalts, such as the few observed as olivine-rich lithic fragments in the Luna 24 soil. The highest Mg/Al intensity ratios in Mare Crisium are in the area of the crater Picard (Fig. 4). Al/Si intensity ratios measured at Picard (Fig. 2) are the second lowest for Crisium basalts and are significantly lower than those for the landing site even though aluminous ejecta from Yerkes E appears to extend almost as far east as Picard. Mg/Si intensity ratios in the area of Picard are unusually high (Fig 3). Specifically, Mg/Si intensity ratios for 0.25° units within the continuous ejecta blanket of Picard (not shown) are as high as 1.48 (concentration = ~0.32) and Al/Si intensity ratios as low as 0.61 (concentration = ~0.25). These ratios are well within the limits defined by the magnesium-rich grouping of olivine vitrophyre, olivine gabbro and green glass fragments in the Luna 24 soil (Al/Si concentration <0.372 and Mg/Si concentration >0.285). On the elemental composition diagram (Fig. 1) average Al/Si and Mg/Si concentration ratios within the continuous ejecta of Picard lie on a mixing line between the olivine vitrophyre/olivine gabbro and the Luna 24 soil ($<74\ \mu$m). Picard is an Eratosthenian crater, 22.2 km in diameter, which formed after the extrusion of mare basalt lavas in the Crisium basin. We postulate, therefore, that the high-Mg, low-Al measurements associated with Picard represent material excavated from a substratum of magnesium-rich basalt which may typify that from which the Mg-rich components in the Luna 24 soils were derived.

High Mg/Al intensity ratios are also associated with the post-mare crater Peirce (18.4 km in diameter) north of Picard. The gradation of Mg/Al intensity ratios relative to these two craters is shown in Fig. 5. Picard is clearly the focal point of the maximum Mg/Al intensity ratios. The center of Peirce is just outside the northern limit of XRF coverage in Mare Crisium which may explain why Mg/Al values are not as high as those for Picard. However, decreasing values between Picard and Peirce indicate that high Mg/Al values associated with Peirce are not simply ejecta from Picard but that magnesium-rich, aluminum-poor subsurface basalts were samples separately by these two impact events.

Additional evidence for at least two chemically distinct basalt strata in Mare Crisium is a radar-reflecting layer extending across central Crisium at 1400 m beneath the surface (May *et al.,* 1976). This discontinuity detected by the lunar sounder experiment lies above the floor of Picard at ~1750 m and the floor of Peirce at ~1490. Therefore, both Picard and Peirce have sampled material from below this discontinuity. The radar reflector may represent a buried regolith that developed on either the basin floor or on old mare basalts. If Picard and Peirce excavated aluminous basin floor material from depth, there would be a sharp increase in Al/Si values for the ejecta blankets relative to the surrounding mare. At Picard the converse is true and Peirce shows no appreciable increase. Therefore, unlike the crater Yerkes E, Picard and Peirce did not penetrate below mare basalts to sample aluminous material. The radar reflector, then, most likely represents the interface between low-magnesium surface basalts and the earlier Mg-rich basalts that were excavated by Picard and Peirce (Fig. 6).

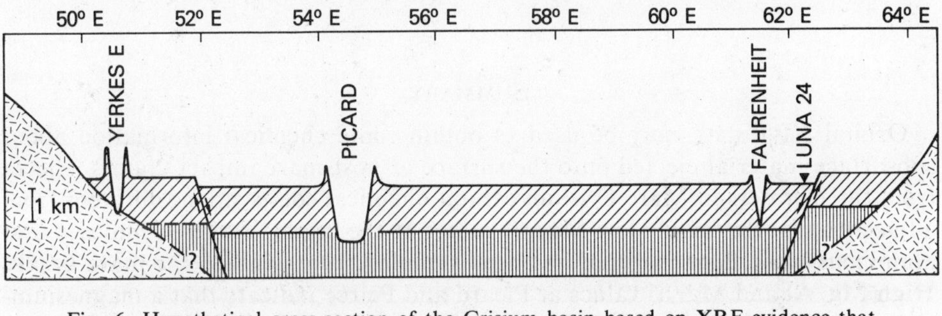

Fig. 6. Hypothetical cross-section of the Crisium basin based on XRF evidence that there is a high-magnesium substratum of basalt (vertical lines) beneath low-magnesium surface basalts (diagonal lines). The interface between these two mare basalts represents the radar reflecting layer at 1400 m from lunar sounder data (May *et al.*, 1976).

The chemical units associated with Picard and Peirce observed in the XRF data are also distinguishable in other data sets. For example, photographs of Mare Crisium display low-albedo material encompassing both Picard and Peirce which coincides with the XRF high-Mg units. In addition, spectral reflectance measurements (Pieters and McCord, 1976; Adams *et al.,* 1978) indicate that ejecta from Picard and Peirce has spectral properties distinctly different from those for central Crisium and the Luna 24 landing site. This supplementary chemical information warrants further discussion.

DISCUSSION

XRF data indicates Mg-rich regolith in three areas of Mare Crisium, northeast of the landing site, at the crater Picard, and at the crater Peirce. These areas may represent the basaltic chemical composition which causes the magnesium enrichment in the bulk soils relative to basalt lithic fragments in the Luna 24 core.

Recent interpretations of spectral reflectance data (Pieters and McCord, 1976; Adams *et al.*, 1978) categorize surface soils at Picard, Peirce and the southeast border of Mare Crisium as one unit which is different from, and higher in TiO_2 content than the soils at Luna 24. XRF data show no corresponding increase in magnesium for this proposed unit southeast of the landing site. The XRF high-Mg unit northeast of Luna 24 is within the lower TiO_2 unit which includes the landing site. On the basis of these two elements, this area northeast of Luna 24 may typify the parent material from which the high-magnesium components in the low-titanium Luna 24 soils are derived. Spectral reflectance studies of TiO_2 content in mare soils are at best semi-quantitative in low-Ti target areas. However, if the magnesium-rich soils at Picard and Peirce are, in fact, significantly higher in titanium than the landing site, it is possible that this material represents a third basalt type in Mare Crisium which is not present in the Luna 24 core.

SUMMARY

Orbital XRF data may be used to obtain some chemical information about subsurface material ejected onto the surface at post-mare impact craters. Ejecta from three craters in Mare Crisium exhibit chemical compositions different from adjacent soils. Extremely low Mg/Al values at the crater Yerkes E indicate that aluminous basin floor material has been sampled from beneath the mare basalt. High Mg/Al and Mg/Si values at Picard and Peirce indicate that a magnesium-rich subsurface basalt has been excavated from depth. There is no evidence from Al/Si data that these craters have exposed basin floor material. Therefore, the radar reflecting layer from lunar sounder data at 1400 m below the surface probably represents the boundary between high-Mg subsurface basalts and low-Mg surface basalts. Spectral reflectance data support our hypothesis that material associated with Picard and Peirce is chemically different from the adjacent soils.

Most of the surface soils in Mare Crisium represent low-magnesium basalts such as those which predominate in the Luna 24 core. Three areas with high Mg/Si values have been identified from XRF data which may typify the parent material from which the magnesium-rich monomineralic fragments and a few olivine-rich lithic fragments in the core are derived: 1) the area northeast of the landing site, a portion of which contains dark mantling material; 2) the area encompassing the crater Picard; 3) the area surrounding the crater Peirce. However, if the basalts at Picard and Peirce are enriched in TiO_2 relative to Luna

24 and the high-Mg area northeast of the landing site, then, we cannot exclude the possibility that there are three basalt types exposed on the surface of Mare Crisium; 1) moderate- to low-Mg basalts with a low-Ti content, typical of the Luna 24 site; 2) high-Mg, relatively high-Ti basalts excavated from depth at Picard and Peirce; and 3) high-Mg, low-Ti basalts northeast of the landing site which are represented in the core as monomineralic fragments and a few lithic fragments of olivine gabbro and olivine vitrophyre.

Acknowledgments—This work was supported by NASA grant NGR-21-002-368.

REFERENCES

Adams J. B., Head J. W. III, McCord T. B., Pieters C., and Zisk S. (1978) Mare Crisium: Regional stratigraphy and geologic history. Submitted to *Geophys. Res. Lett.*

Adler I., Gerard J., Trombka J., Schmadebeck R., Lowman P., Blodget H., Yin L., Eller E., Lamoth R., Gorenstein P., Bjorkholm P., Harris B., and Gursky H. (1972a) X-ray fluorescence experiment. *Proc. Lunar Sci. Conf. 3d,* p. 2157–2178.

Adler I., Trombka J., Lowman P., Schmadebeck R., Blodget H., Eller E., Yin L., Lamothe R., Osswald G., Gorenstein P., Bjorkholm P., Gursky H., and Harris B. (1972) Apollo 16 Geochemical X-ray fluorescence experiment: Preliminary report. *Science* **177,** 256–259.

Andre C. G., Bielefeld M. J., Eliason E., Soderblom L., Adler I. and Philpotts J. A. (1977) Lunar Surface Chemistry: A New Imaging Technique. *Science* **197,** 986-989.

Barsukov V. L., Tarasov L. S., Dmitriev L. V., Kolesov G. M., Shevaleyevsky I. D., and Garanin A. V. (1977) The geochemical and petrochemical features of regolith and rocks from Mare Crisium (preliminary data). *Proc. Lunar Sci. Conf. 8th*, p. 3319–3332.

Basu A., McKay D. S., and Fruland R. M. (1977) Petrography, mineralogy and source rocks of Luna 24 drill core soils (abstract). In *Papers Presented to the Conference on Luna 24*, p. 18–26. The Lunar Science Institute, Houston.

Bielefeld M. J., Andre C. G., Eliason E. M., Clark P. E., Adler I, and Trombka J. I. (1977) Imaging of lunar surface chemistry from orbital X-ray data. *Proc. Lunar Sci. Conf. 8th*, p. 901–908.

Blanchard D. P., Haskin L. A., Brannon J. C., and Aaboe E. (1977) Chemistry of soils and particles from Luna 24 (abstract). In *Papers Presented to the Conference on Luna 24*, p. 37–40. The Lunar Science Institute, Houston.

Florensky C. P., Basilevsky A. T., Ivanov A. V., Pronin A. A. and Rode O. D. (1977) Luna 24; Geologic setting of landing site and characteristics of sample core (Preliminary data). *Proc. Lunar Sci. Conf. 8th*, p. 3257–3279.

Grove T. L. and Bence A. E. (1977) Petrogenesis of gabbros from Mare Crisium (abstract). In *Papers Presented to the Conference in Luna 24*, p. 64–67. The Lunar Science Institute, Houston.

May T. W., Peeples W. J., Maxwell T., Sill W. R., Ward S. H., Phillips R. J., Jordan R. L. and Abbott E. A. (1976). Subsurface layering in Maria Serenitatis and Crisium: Apollo Lunar Sounder results (abstract). In *Lunar Science VII*, p. 540–542. The Lunar Science Institute, Houston.

McKay D. S., Basu A., Waits G., Clanton U., Fuhrman R., and Fruland R. (1977) Grain size and evolution of Luna 24 soils (abstract). In *Papers Presented to the Conference in Luna 24*, p. 114–117. The Lunar Science Institute, Houston.

Norman M., Coish R. A. and Taylor L. A. (1977) Glass chemistry and magma evolution at Mare Crisium (abstract). In *Papers Presented to the Conference on Luna 24*, p. 136–138. The Lunar Science Institute, Houston.

Olson A. B. and Wilhelms D. E. (1974) Geologic map of the Mare Undarum quadrangle. U.S. Geol. Survey, Misc. Investigation Series Map I-837.

Pieters C. and McCord T. B. (1976) Regional basalt types in the Luna 24 landing area as derived from remote observations. *Geophys. Res. Lett.* **3**, 697–700.

Ryder G., McSween H. Y., and Marvin U. B. (1977) Basalts from Mare Crisium. *The Moon.* In press.

Simonds C. H., Warner J. L., McGee P. E., Cochran A., and Brown R. W. (1977) Luna 24: Petrochemical potpourri (abstract). In *Papers Presented to the Conference on Luna 24*, p. 170–173. The Lunar Science Institute, Houston.

Tarasov L. S., Nazarov M. A., Shevaleyvsky I. D., Kudryashova A. F., Gaverdovskaya A. S., and Horina M. I. (1977) Mineralogy and petrography of lunar rocks from Mare Crisium (preliminary data). *Proc. Lunar Sci. Conf. 8th*, p. 3333–3356.

Taylor G. J., Keil K., and Warner R. D. (1977) Very low-Ti mare basalts. *Geophys. Res. Lett.* **4**, 207–210.

Vaniman D. T. and Papike J. (1977) Ferrobasalts from Mare Crisium: Luna 24. *Geophsy. Res. Lett.* In press.

Merrill, R. B. and Papike, J. J., editors.
(1978) *Mare Crisium: The View from Luna 24*, Pergamon Press (New York), p. 13–32
Printed in the United States of America

From Serenity to Langemak: A regional chemical setting for Mare Crisium

NORMAN J. HUBBARD

NASA Johnson Space Center, Houston, Texas 77058

FAITH VILAS

Lockheed Electronics Co. Inc., Houston, Texas 77058

JAMES E. KEITH

NASA Johnson Space Center, Houston, Texas 77058

Abstract—Changes recently made in the reduction of the orbital X-ray fluorescence data from Apollo 15 and 16 produce improved Al/Si and Mg/Si intensity ratios, some of which are used here to describe the chemistry of the region surrounding Mare Crisium. For the Apollo 15 data presented in this paper, the Al/Si and Mg/Si intensity ratio data have been improved by factors of 1.7 and 4.6 respectively. These data are used to show that terra (or nonmare) materials in the area of study have approximately constant Al/Si ratios, and that the Mg/Si ratios vary by a factor of two. This is interpreted to mean that the average Mg/Si ratio of the olivines plus pyroxenes is equally variable, even though the plagioclase content is roughly constant. The existence of these Mg/Si variations in terra regions may suggest either: 1) that the terra regions were once chemically diverse and that the heavy early bombardment has not chemically homogenized them; or 2) that volcanism subsequent to the heavy bombardment (but pre-mare) has produced these Mg/Si variations.

The Al/Si and Mg/Si data for mare areas also show similarly independent variations. Maria Smythii, Crisium and Serenitatis show only small variations in Al/Si ratios, while the Mg/Si ratios for different parts of these maria vary by a factor of at least two. Further, the variations within a single mare are almost equally large. One of the materials showing high Mg/Si intensity ratios is the dark mantling material northwest of the Apollo 17 landing site. Others are the ejecta from the crater Picard and material along the southern edge of Mare Crisium, the material in the most deeply filled part of Mare Smythii, and areas in south central Mare Serenitatis. Where spectral reflectance data are available, they indicate that the materials with high Mg/Si ratios have diverse mineralogical compositions. Material with lower Mg/Si intensity ratios occurs in or near the dark annulus in southeastern Mare Serenitatis and in much of Mare Crisium.

I. INTRODUCTION

Lunar sample analyses have provided an extensive and detailed body of information about the chemical composition and genesis of lunar rocks, and the chronology of lunar differentiation and volcanism that produced the rocks. An equally extensive and detailed body of information about the regional chemical variations of the pre-mare lunar crust, and about the chemical diversity among and within areas of mare volcanism, is required before major improvements can be made in our understanding of the igneous differentiation and volcanism that have occurred on the moon. Apollo 15 and 16 orbital X-ray data for Al/Si and Mg/Si ratios can provide some of this information. Recent improvements in the data reduction procedures for this experiment have improved the Mg/Si intensity

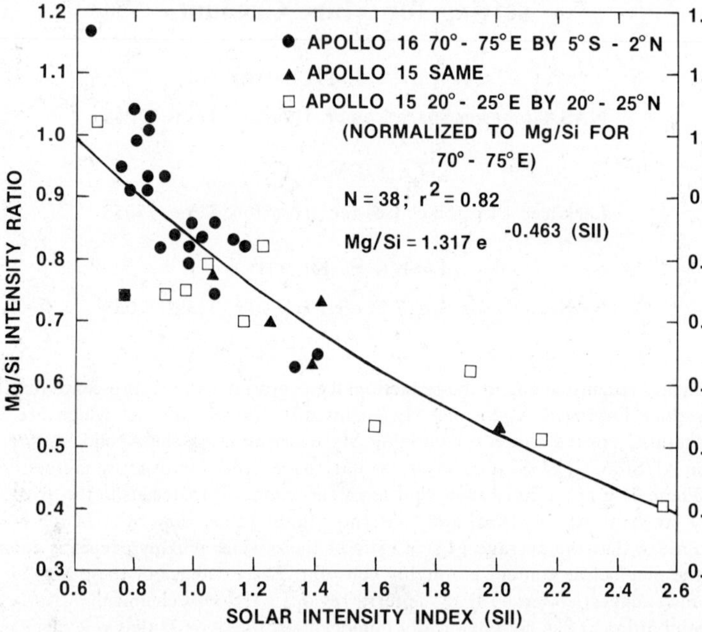

Fig. 1. This figure shows the empirical relationship between the measured intensity ratios for Mg/Si and our measure of the intensity of the X-ray flux emitted by the Sun. Our method of measuring and removing the effects due to variations in the solar X-ray flux is closely analogous to the one proposed by Trombka and coinvestigators for use with the X-ray fluorescence spectrometer that was part of the proposed Lunar Polar Orbiter. That design places a ceramic plate in the sunlight and measures the intensity of Mg and Si X-rays (potentially also those from other elements) emitted from the plate as a consequence of excitation by solar X-rays. This allows a direct correction of the measured Mg/Si and Al/Si intensity ratios for the lunar surface. In our method, the ceramic plate is replaced by a series of "standard" areas on the lunar surface. The essential feature (assumption) of these areas is that each one has a constant chemical composition and physical state. For each area, we construct a graph of measured intensity ratio vs. solar intensity index. The data for different areas are compiled onto one graph by choosing one plot as standard and normalizing the others to it. A curve is fitted to the composite set of data and the resulting equation is then used to remove the solar effects from all data where we have reason to believe that it applies. Our method uses an empirical description of the photometric variations of the X-ray flux measured by the X-ray spectrometers and the assumption that the Si concentration in soils on the surface of the moon is constant. This assumption is true to better than about ±15% of the Si concentration, which in turn will cause less than a ±7.0% relative error in the solar correction for Mg/Si, and less than a ±3.0% relative error in the solar correction for Al/Si. That is, the removal of solar artifacts from the data may introduce systematic errors of up to 7.0% and 3.0%, whereas the solar artifacts removal commonly causes errors of ~20% to ~100% for individual orbits.

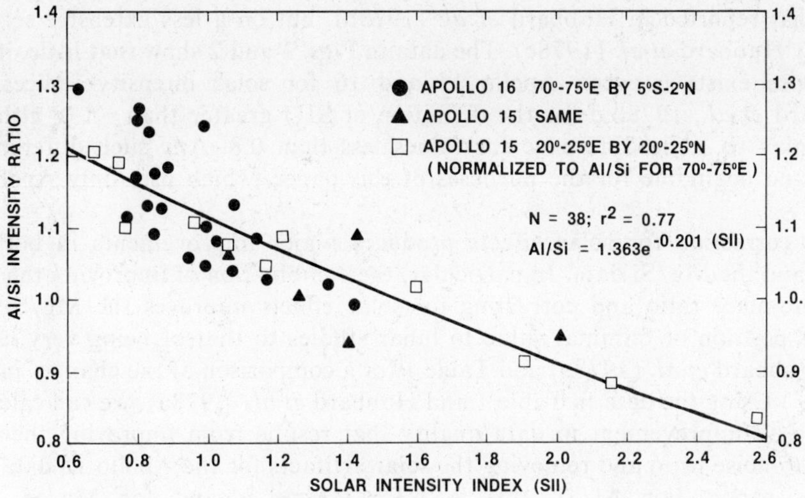

Fig. 2. This is the same kind of graph as Fig. 1 except that it is for Al/Si intensity ratio data.

ratio data so that they can now be used with the Al/Si intensity ratio data to interpret regional geologic patterns with a spatial resolution of better than 60 km. We report here on the quality of these improved data, and their application to the study of mare volcanism and the determination of regional chemical variations in the pre-mare lunar crust. Mare Crisium is the focal point for the initial conclusions that can be drawn from the improved orbital X-ray data.

II. Improvements in Data Reduction

Reduction procedures used for these data are described fully in two papers (Hubbard *et al.*, 1978c,d) and two abstracts (Hubbard *et al.*, 1978a and b). For the convenience of the reader, we provide a brief description of the improvements that we have made and the overall quality of the data used in this paper. First, we increased the basic signal-to-noise ratio by using only those channels of the 8 channel pulse height analyzer that had high signal-to-noise ratios, thus improving the signal-to-noise ratio of the raw data by a factor of three. Al/Si and Mg/Si intensity ratios were then produced from these raw data. These calculated Al/Si and Mg/Si ratios were improved by factors of ~1.3 and ~2.0, respectively (Hubbard *et al.*, 1978a). Next, we smoothed the data along individual orbits using a weighted 5-point sliding average that simulates the field of view of the detectors, reducing the scatter in the individual data points by a factor of about 1.8. Finally, we removed those changes in the Al/Si and Mg/Si intensity ratios that are associated with changes in the intensity of the incident solar X-ray flux. This was done using a simple empirical technique that is described in Hubbard *et al.* (1978b,c,d). The empirical equations and the data used to derive them are shown in Figs. 1 and 2. These equations are based on a more extensive set of data than that reported in Hubbard *et al.* (1978b), but on a less extensive set than used by Hubbard *et al.* (1978c). The data in Figs. 1 and 2 show that little, if any, difference exists between Apollo 15 and 16 for solar intensity indices (see Hubbard *et al.*, 1978b,d for the definition of SII) greater than ~1.0, although there may be some difference for values less than 0.8. Any such difference is small and negligible for the purposes of this paper, which uses only Apollo 15 data.

The correction for solar effects produces major improvements in both the Al/Si and the Mg/Si data. In particular, the combination of improving the basic signal-to-noise ratio and correcting for solar effects improves the Mg/Si data from a position of minimal value to lunar studies to that of being very useful. (See Hubbard *et al.* (1978a) and Table 1 for a comparison of the changes in data quality.) Using the data in Table 1 and Hubbard *et al.* (1978a), we can calculate the overall improvement in data quality that results from improving the basic signal-to-noise ratio and removing the solar artifacts for the Apollo 15 data used in this paper. For Al/Si, this is $1.3 \times 1.3 = 1.7$; and for Mg/Si, it is $2.3 \times 2.0 = 4.6$. Table 1 also shows the systematic errors removed by the procedure for solar corrections, demonstrating further improvements of $1.06 - 1.2 \times$.

The internal precision of the Al/Si and Mg/Si intensity ratios is dependent on the basic quality of the data, which is variable. In Fig. 3, we show how the % standard deviation of the data varies with the magnitude of the solar-corrected intensity ratio. These values are calculated directly from the data shown in this paper for multiple orbits and therefore include interorbit errors. As discussed in the figure caption, they are conservative estimates of the sum of all errors. We hope to have more accurate estimates of the internal precision in subsequent papers. In a given orbit, we reject all Mg/Si data at greater than 50° in longitude from the solar maximum point (see Hubbard *et al.*, 1978b, for a definition of SMP) and all Al/Si data at greater than 60° away from the SMP. The Mg/Si data will always be somewhat inferior to the Al/Si data in absolute quality; however, they are equally useful in lunar studies because of the independent and valuable information that they provide.

III. LIMITATIONS OF THE DATA

Two limitations should be kept in mind when reading the remainder of this paper: 1) the uncertainties in relating intensity ratios to concentration ratios; and 2) the smearing effect of the field of view.

The fidelity with which the Al/Si and Mg/Si intensity ratios reflect the Al/Si and Mg/Si concentration ratios (and therefore Al_2O_3 and MgO concentrations) in the lunar surface soils is very poorly known. Preliminary studies allow one to suppose that the intensity ratios reflect concentration ratios to within approximately 20% of the true value (Maxwell *et al.*, 1977; Bielefeld, 1977). The procedure used by previous authors to derive concentrations from the orbital X-ray data has never been published, so that the quality of those data can only be

Table 1. A comparison of the mean and standard deviation of orbital X-ray fluorescence data with and without removal of errors due to variations in the solar X-ray flux.

Longitude Interval	No. of Data Points	Before Solar Corrections		After Solar Corrections		Improvement in Precision		% Systematic* Error in Non-Corrected Data	
		Al/Si Mean, Std. Dev. % Std. Dev.	Mg/Si Mean, Std. Dev. % Std. Dev.	Al/Si Mean, Std. Dev. % Std. Dev.	Mg/Si Mean, Std. Dev. % Std. Dev.	Al/Si	Mg/Si	Al/Si	Mg/Si
45°–50°	52	1.055 ± .112 ±10.6%	0.694 ± .146 ±21.0%	1.124 ± .081 ±7.2%	0.791 ± .060 ±7.6%	1.5	2.8	−6.1%	−12.3%
70°–75°	66	1.020 ± .091 ±8.9%	0.668 ± .154 ±23.0%	1.136 ± 0.071 ±6.3%	0.841 ± .071 ±8.4%	1.4	2.7	−10.2%	−20.6%
95°–100°	37	1.172 ± .088 ±7.5%	0.536 ± .096 +17.9%	1.263 ± .083 ±6.6%	0.633 ± .077 ±12.2%	1.1	1.5	−7.2%	−15.3%

The data in these longitude intervals are shown in Figs. 8, 9 and 10. The caption to Fig. 3 contains more information about the data in this table.
*Note that the standard deviations in this table are those of the distributions, not the means, so that all the systematic errors quoted here are significant at confidence levels greater than 95%.

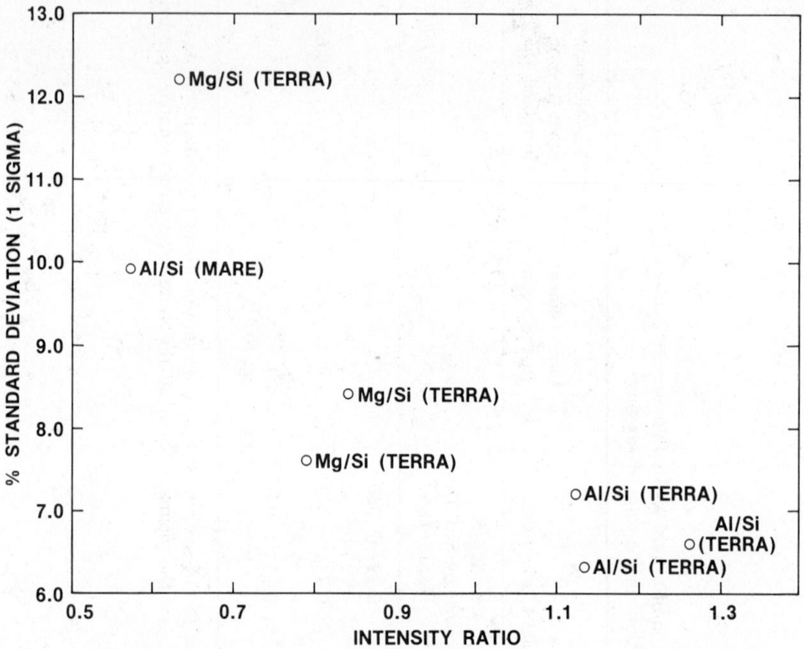

Fig. 3. The results of an error analysis of the data presented in this paper are shown here. We have chosen a few segments (5°–10° longitude in length) of data where the Al/Si and Mg/Si ratios are approximately constant and calculated the mean and standard deviation for the data from 3 to 5 orbits. These error estimates are conservative because all errors associated with the data reduction are included (those due to channel selection, determinations of background, smoothing and the removal of solar artifacts). In addition, there are unknown errors associated with chemical and physical variations in the areas of lunar regolith overflown. We have plotted the relative error estimates (one standard deviation) vs. the value of the intensity ratio to show that there is a general increase in the error as the relative intensity of the Al and Mg X-ray signals decrease (that is, lower Al/Si and Mg/Si ratios).

inferred from comparison with the results of Maxwell *et al.* (1977) and Bielefeld (1977), which give similar values. These authors ignore the effects that differences between mare and terra (here used to refer to nonmare areas on the moon's surface) compositions can have on the accuracy of the concentrations calculated for mare areas relative to terra areas, and use data which lacked corrections for solar effects. In this paper, we will often compare intensity ratios for different mare areas (or terra areas) as if they are directly related to differences in Al or Mg concentrations, and we will calculate approximate MgO values for some areas. We will not make similar direct comparisons of data for mare with data for terra areas.

The smearing effect of the field of view is shown in Fig. 4. We have illustrated a mare-terra edge as a simple, one dimensional contact between terra material with Al/Si = 1.36 and mare material with Al/Si = 0.75, and then calculated

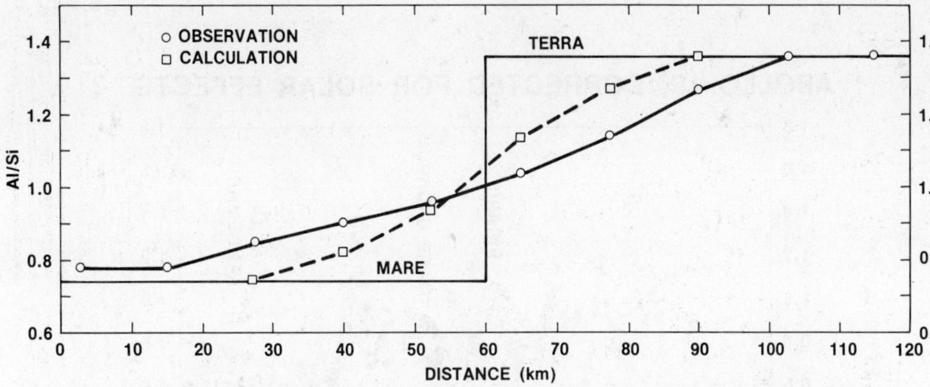

Fig. 4. The smearing of chemical boundaries by the field of view of the detectors is shown here. We show the way in which the measured Al/Si values will charge as the detector is flown at 110 km altitude across a mare-terra boundary. This boundary is illustrated as a simple step function where the mare has Al/Si = 0.75 and the terra has Al/Si = 1.36. The Al/Si intensity ratio data for orbit 16, which crosses the eastern mare-terra boundary of Mare Smythii, are shown for comparison. (The step function values were chosen because the observed Al/Si ratios plotted here attain approximately these values on either side of the boundary.)

what the orbital X-ray detectors would see as the spacecraft flew over this boundary at 110 km altitude. For comparison, the Al/Si data on orbit 16 of Apollo 15 have been plotted where this orbit crossed the eastern edge of Mare Smythii. As shown in Fig. 4, this sharp, one dimensional boundary is a fair approximation of the data for this mare-terra boundary.

IV. Presentation and Discussion of the Data

A. *General Comments*

The major contribution of this paper is the presentation of the improved Mg/Si data in the region including Mare Crisium.

The general features of the data are shown by Figs. 5 and 6 where we have plotted the Apollo 15 data for several orbits that are relevant to the study of major areas of the lunar surface. In Fig. 7, we have plotted the ground tracks of three of these orbits on a sketch showing the major lunar features. Comparison of Fig. 5 with Fig. 6 shows the difference between the Al/Si and Mg/Si data. The predominant general feature of the Mg/Si data is the slow increase from Mg/Si $\simeq$ 0.6 in the region east of Smythii (>90° long.) to Mg/Si $\simeq$ 1.0 in western Mare Serenitatis. Of the maria, only Mare Smythii is clearly distinguishable from the surrounding terra because it has Mg/Si ratios that are higher than those of the adjacent terra. In contrast, the Al/Si data show the well known sharp decreases in all mare areas.

The regional increase in Mg/Si ratios to the west, and the lack of contrast for

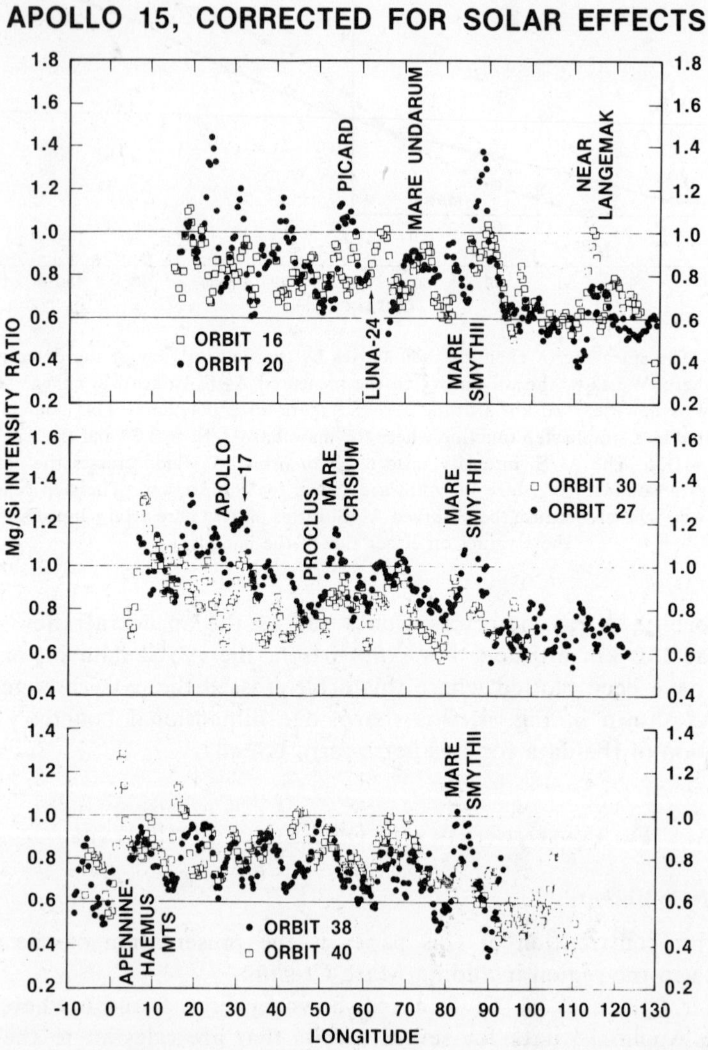

Fig. 5. The large-scale variations in the Mg/Si intensity ratio data are shown in this figure. We have plotted the data for three pairs of orbits in order to show the internal consistency of data for pairs of orbits that are relatively closely spaced, and to show simultaneously the regional differences that are recorded by more widely spaced orbits. The ground tracks of one orbit of each pair are shown later in Fig. 7. The second orbit in each pair is separated by 2.0 degrees or less. We have plotted data on this graph that are further from the solar maximum than our data acceptance criteria. Comparison of data in this figure and Fig. 6 with the data in Fig. 8 thru 11 will show the data that should be treated with extreme caution.

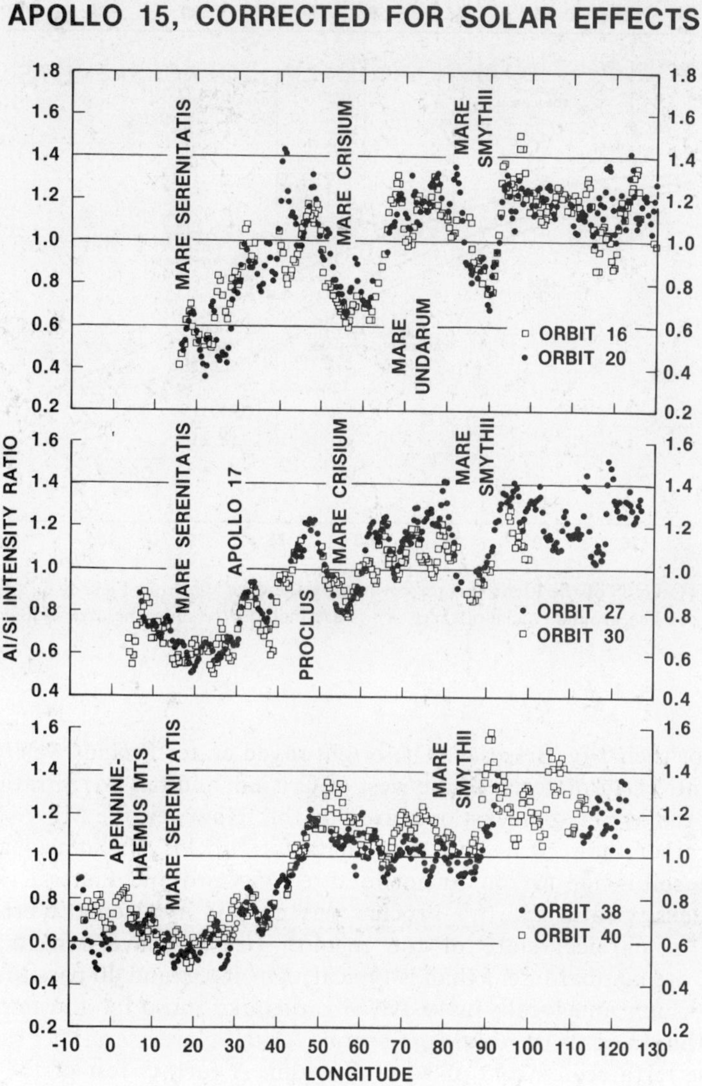

Fig. 6. The large-scale variations in the Al/Si intensity ratio data are shown here. Otherwise, this graph is the same as Fig. 5.

the mare vs. terra regions, suggest that the terra areas west of Mare Smythii have higher Mg/Si ratios than those areas east of Mare Smythii between ~90° and ~113° longitude that Wilhelms and El-Baz (1977) have mapped as terra. Verification of this requires care because of the presence of abundant mare material from Mare Smythii westward. We use two approaches: 1) the data for a large post-mare crater in terra material are examined; and 2) the terra areas west of Mare Smythii that have Al/Si ratios similar to those east of Smythii are

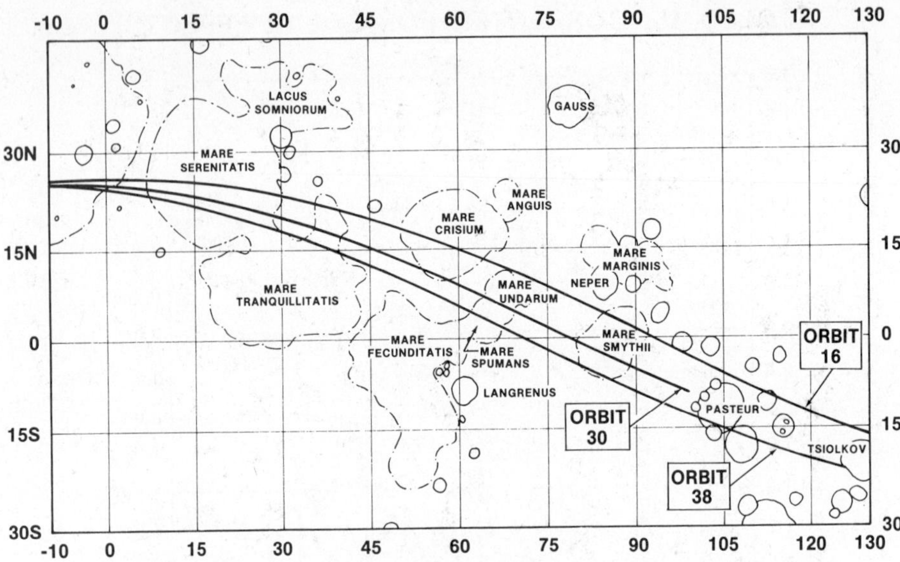

Fig. 7. A sketch of the lunar surface covered by the data shown in Figs. 5, 6, 8, 9, 10 and 11. The ground tracks of three orbits are included to show the area studied.

examined for their Mg/Si ratios. The bright rayed crater Proclus was formed by an impact into terra material to the west of Crisium and has Al/Si ratios similar to those for the terra region east of Mare Smythii. However, the Mg/Si ratios are about 0.8 rather than $\lesssim$0.7. The X-ray data for Proclus are somewhat suspect because the soil inside Proclus is not mature (McCord and Pieters, 1978), and thus the intensity ratio data for Proclus may contain a systematic error due to differences in physical state of the regolith (lesser conversion to glass by micrometeorite bombardment than is typical). Orbits 38 and 40 pass to the south of Mare Crisium and locally have Al/Si ratios like those for the terra east of Mare Smythii. In contrast, the Mg/Si ratios for these areas are $\gtrsim$0.7 rather than $\lesssim$0.7 for the terra region east of Mare Smythii. A further test of the chemical uniformity of the terra material in this area can be made by examining the Mg/Si values for all areas west of Smythii (some not shown here) that have Al/Si values similar to the median Al/Si values east of Smythii. This again yields Mg/Si ratios of 0.8-1.0. From these results we conclude that the terra material throughout the area covered by these data probably has a uniform average Al/Si ratio and hence plagioclase content. The roughly 50% increase in Mg/Si ratio (from $\lesssim$0.7 for the terra east of Smythii to ~0.9 for the terra material west of Smythii), at an almost constant average Al/Si ratio, implies a similar increase in the Mg/Si ratio in the pyroxenes plus olivines of this material. This exercise also shows that the general increase in Mg/Si ratio to the west is not due solely to the presence of mare material, which would decrease the Al/Si ratio. The terra area east of Mare Smythii will be discussed in more detail later.

B. *Regional Areas*

Here we will specifically present and discuss the data for three maria and their immediate environs (Smythii, Crisium and Serenitatis), and the terra region to the east of Mare Smythii.

In Fig. 8, we have plotted data from five orbits that cross Mare Smythii. The boundaries of Mare Smythii are at about 83° and 92° longitude (the exact location is not critical to this discussion). Two orbits (16 and 20) cross the lower albedo portion of the mare fill (see Andre *et al.*, 1977) for a more detailed discussion of the Al/Si data for Mare Smythii) showing Al/Si ratios which decrease rapidly from the terra values at ~93° longitude to about 0.75 at 89°–

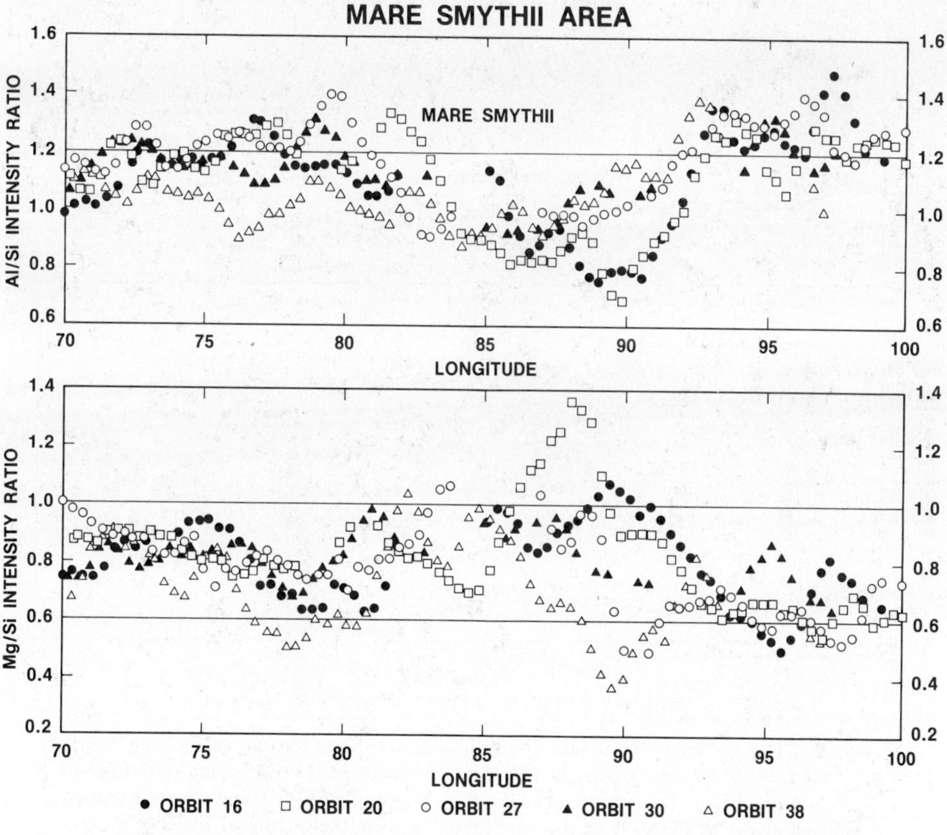

Fig. 8. The Mg/Si and Al/Si intensity ratio data are shown for five orbits of data from Apollo 15 that cross Mare Smythii. The data from 70°–75° and 95°–100° longitude were used to calculate some of the error estimates shown in Fig. 3 and Table 1.

90°. The other orbits (27, 30 and 38) pass to the south of the lower albedo area and do not show the rapid decrease in Al/Si; all show very much the same pattern of variations in Al/Si ratios across Smythii and environs. In contrast, the Mg/Si data show a very different pattern. The Mg/Si ratios on orbits 16 and 20 increase together from about 94° to about 89°, where orbit 20 shows a prominent peak and orbit 16 shows Mg/Si ratios like those of orbits 27 and 30. Orbits 27 and 38 show similarly decreasing Mg/Si ratios from about 90° to 92°, where the values for orbit 27 rise from ~0.5 to ~0.8, while the values for orbit 38 continue to decrease to Mg/Si ~0.4 and then only increase to intersect the values (~0.9) found on other orbits at 85° longitude. The high Mg/Si values from orbit 38 at 86° to 82° correspond to the low albedo mare material in southern Mare Smythii (Stewart *et al.*, 1975). The low Mg/Si values from orbit 38 approximately correlate with the crater Brunner, but may actually be due to ejecta from the large post-mare crater Hirayama (Stewart *et al.*, 1975). These data for Mare

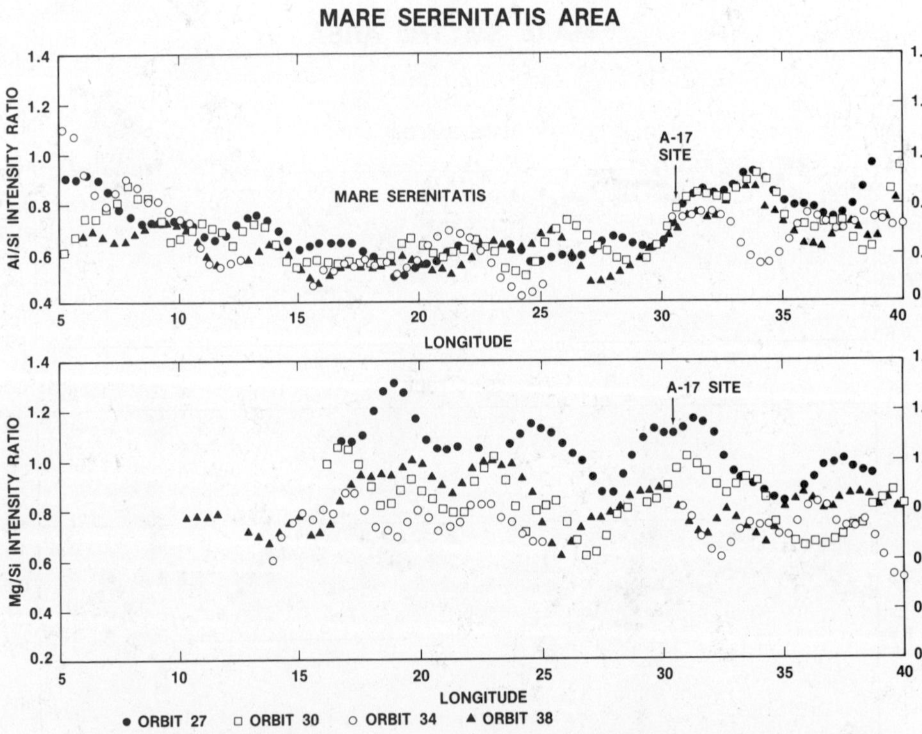

Fig. 9. The Mg/Si and Al/Si intensity ratio data are shown for four orbits from Apollo 15 that cross Mare Serenitatis. The Al/Si data from 15°–25° longitude were used to calculate one of the error estimates shown in Fig. 3. The Mg/Si data have not been extended as far to the west as the Al/Si data because the quality of the Mg/Si data decreases rapidly beyond 40° from the solar maximum point (see Hubbard *et al.*, 1978a, e and d for a definition of the solar maximum point). We reject *all* Mg/Si data that are more than 50° from the solar maximum point.

Smythii clearly show that Al/Si and Mg/Si are independent variables: decreases in Al/Si may be accompanied by *either* increases or decreases in Mg/Si.

If the high Mg/Si ratios from orbit 20 can be directly related to MgO concentrations as presented later in the discussion of data for Mare Crisium, then Mare Smythii contains mare materials with MgO >14.0 wt. %. Data for other orbits suggest maximum MgO values of about 14.0 wt. %.

Moving to Mare Serenitatis (Fig. 9), we see again that Al/Si variations are not a guide to Mg/Si variations. The eastern terra-mare boundary of Mare Serenitatis is crossed at ~29° by orbit 27 and ~27° by orbit 38. The western boundary is at about 10°. Careful study of these data suggests that there may be small variations in Al/Si across Mare Serenitatis, but only the area between 10°–15° may be measurably different (see caption to figure). In contrast, the Mg/Si data for the area show some of the lowest values just inside the eastern boundary of the mare. From ~17° to ~24° longitude there is an area of elevated Mg/Si ratios, which is also seen in the data for orbits that are not used in this paper. The relatively poor reproducibility of these solar-corrected Mg/Si data is related to their nearness to the cutoff point for accepting Mg/Si data. If we assume that these orbits of Mg/Si data should show the same general variations and that the divergences seen in Fig. 9 are due to effects not fully accounted for, then we can normalize the Mg/Si values for orbit 27 to those for 30 and 38. If this normalization is done using the Mg/Si data between 20° and 24°, one finds that Mg/Si intensity ratios for the south-central part of Mare Serenitatis between 16° and 26° are slightly higher than those found on orbit 27 over the dark mantling material to the north of the Apollo 17 landing site at 29.5° longitude (~0.95 vs. ~0.85). The material in south central Mare Serenitatis does not have the spectral reflectance signature (Johnson *et al*., 1977a) or the extremely low albedo of the dark mantling material. However, a correlation of albedo with Mg/Si is not to be expected. The low albedo of the dark mantling material is almost certainly related to its mode of formation or Fe and Ti concentrations, rather than to its Mg/Si ratio.

Assignment of a Mg/Si ratio to the dark mantling material overflown by orbit 27 is somewhat arbitrary, but can be argued as follows: the probable lower limit on the MgO concentration is that of the Apollo 17 valley floor soils (~9.5 wt.% MgO). The upper limit (14.4 wt.% MgO) may be assumed to be given by the Apollo 17 orange soil (Rhodes *et al*., 1974) and its devitrified counterpart (Haggerty, 1974), which is reported to be a major component of the dark mantling material (Adams *et al*., 1974). A technical problem is that the patch of dark mantling material seen on orbit 27 at 29.5° longitude does not completely fill the field of view, although it contributes $\gtrsim$70% of the total signal. Note that the terra material at ~32° on orbits 27 and 30 has similar Mg/Si ratios, but orbit 30 has higher Al/Si ratios (~0.8 vs. ~0.6). Examination of the Al_2O_3 and MgO data for the samples returned by Apollo 17 (Rhodes *et al*., 1974) shows that terra compositions can have Mg/Si ratios that are as high or higher than mare materials. Therefore, the Mg/Si intensity ratio for this area of dark mantling material is probably not affected much by the presence of terra material.

If we proceed with the reasonable assumption that the dark mantling material overflown on orbit 27 at 29.5° longitude has a MgO concentration between 10.0 and 14.0 wt.%, then we can conclude that the material in the southeastern part of Mare Serenitatis has similar or somewhat higher MgO values. In contrast, the low Mg/Si ratios at 26° to 28° longitude imply MgO values of less than about 8.0 wt.%. These low Mg/Si ratios are located near the western edge or on the dark annulus that rings the eastern edge of Mare Serenitatis. Johnson *et al.* (1977a,b) show that the spectral reflectance signature of the dark annulus is similar to that of the high Ti mare areas at the Apollo 11 and 17 sites and that the inferred TiO_2 concentrations of the annulus are lower. Further, the MgO contents of the Apollo 11 and 17 mare soils are 7.8 and 9.6 wt.% respectively, and the average MgO contents of the Apollo 17 and Apollo 11 mare basalts are 9.1 and 6.9 (Low-K) or 7.6 (high-K) wt.% respectively. Interestingly, the MgO contents of the Apollo 11 soil and mare basalts are similar to our approximate MgO value for the annulus and perhaps immediately adjacent mare material. The inferred MgO values for south central Serenitatis may be 10–15% too low because the mare materials in Mare Serenitatis may have more "normal" SiO_2 values (~46.0 wt.%) than the Apollo 17 orange soil and the Apollo 17 valley floor soils (38–40 wt.%). The material in the annulus may have the low SiO_2 values associated with high TiO_2 Apollo 11 materials (~40.0 wt.%). A similar argument using the Al/Si intensity ratio data for orbit 27 at 29.5° longitude is of minimal value because the bounding Al_2O_3 values are 6.3 wt.% for the Apollo 17 orange soil and ~11.0 wt.% for the Apollo 17 valley floor soils—a spread too wide to be useful.

The data shown in Fig. 10 are for the orbits that intersect Mare Crisium (16, 20 and 27) and one that passes to the south over terra material (38). The Al/Si from orbits 16, 20 and 27 show the well known decrease in Al/Si over maria relative to terra. The Mg/Si data seem at first glance to show the inverse relationship. Careful examination shows that this is not true. Orbit 16 crosses the central part of Mare Crisium and only shows increases in Mg/Si for local areas of the mare relative to the terra. Much of orbits 20 and 27 show the same thing. High Mg/Si values are seen in orbit 20 for the post-mare crater Picard (~55° longitude) and just to the east of Picard at 57.5°, and in orbit 27 along the southwestern edge of the Crisium basin near the crater Lick. This latter area of high Mg/Si ratios is broad enough (52°–57° longitude) to suggest a major area of high magnesian material along the southern edge of Mare Crisium.

Picard ejecta and the material to the east of Picard have spectral reflectance signatures that are different from each other and from those of the dark mantling material north of the Apollo 17 site and the areas of high Mg/Si in Mare Serenitatis (Head *et al.*, 1978). The TiO_2 contents inferred from the spectral reflectance data for these areas are also different (Head *et al.*, 1978; Johnson *et al.*, 1977a). Therefore, we conclude that the mare materials showing high Mg/Si intensity ratios are diverse in their overall chemical composition and origin.

The chemical data from the Luna 24 sample can be used as follows to get approximate MgO concentrations for materials in Mare Crisium. If we assume

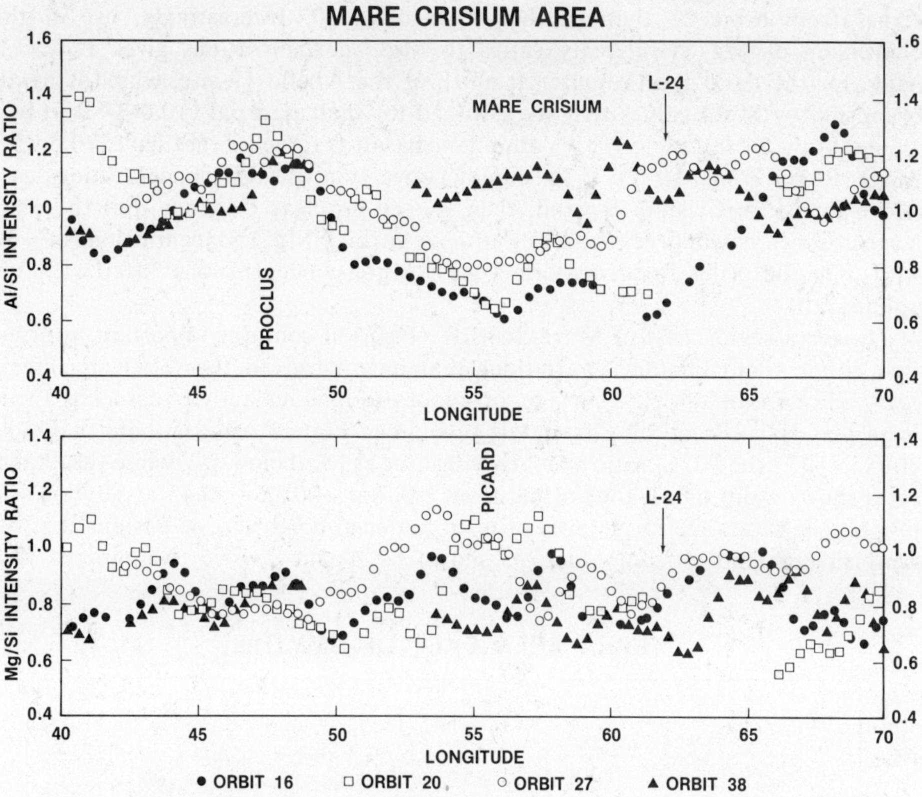

Fig. 10. The Mg/Si and Al/Si intensity ratio data are shown for four orbits of data from Apollo 15 that cross Mare Crisium and environs. The data from 45° to 50° longitude were used to calculate one of the error estimates shown in Fig. 3 and Table 1.

that the MgO concentration of the average (bulk) Luna 24 soil is 10.0% (Blanchard *et al.*, 1977) and that this value also applies to the area to the west of the landing site where the orbital X-ray detectors saw only mare material (~61°), then we calculate $(1.1/0.8) \times 10.0 = 13.8$ wt.% MgO for the ejecta of Picard crater and for the areas of high Mg/Si just to the east of Picard. A similar average MgO value is indicated for the area near Lick crater. At both areas there is reason to believe that the X-ray detectors measured the Mg/Si ratios of high Mg mare materials that contains some admixed terra material. Picard appears to have excavated some small amount of terra material (Head *et al.*, 1978), and there is high albedo material (presumably terra material) in the area of Lick crater. The presence of terra material probably lowers the observed Mg/Si ratios somewhat, but the data presented earlier for the area around the Apollo 17 landing site show that this is not necessarily true. The Al/Si ratios on orbit 27 near Lick crater are somewhat higher (~0.8) than those from orbit 20 over Picard (~0.7), suggesting that terra material contributes more to the high Mg/Si signal from orbit 27 than to that from orbit 20. Interestingly, use of this conversion of Mg/Si intensity ratios to MgO concentrations gives 12.7 wt.% MgO for the dark mantle material north of the Apollo 17 site, which is neatly bracketed by MgO values that we assumed for that material (10.0–14.0 wt.%).

The Luna 24 sample contains abundant basalt fragments that have ~6.5 wt.% MgO (Ryder *et al.*, 1977). If the simple conversion to MgO concentrations used above for Mare Crisium is valid, then we see no clear evidence that there are large areas (a few degrees square) that have surface MgO concentrations of ~6.5 wt.%. On the other hand, materials with MgO concentrations $\gtrsim$10.0 wt.% are abundant.

The terra region east of Mare Smythii (Fig. 11) contains important information either about chemical variations that have survived the relatively intense early bombardment of the surface, or about the igneous activity associated with terra materials. Proceeding from left to right in Fig. 11, we first note that from 100° to 105° the Al/Si ratio is nearly constant at just below 1.2, while the Mg/Si ratio shows a dip on all four orbits from ~0.7 at 100° to ~0.45 at 103°. These low Mg/Si ratios are for an area of terra material northwest of Pasteur that has uniform highland albedo values, as shown by Apollo metric photographs. The

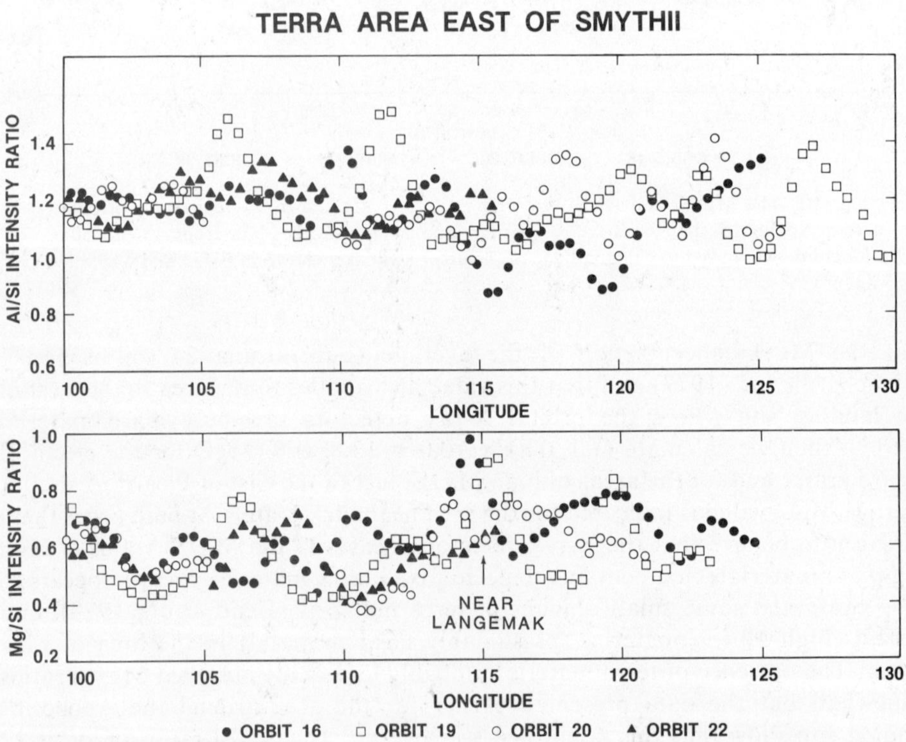

Fig. 11. The Mg/Si and Al/Si intensity ratio data are shown for four orbits of data from Apollo 15 that cross the terra region to the east of Mare Smythii.

area around 115° shows the lowest average Al/Si ratios (~1.0) and the highest average Mg/Si ratios (~0.8) for this area. All four orbits show distinct peaks (16 and 19) or substantial increases (20 and 22) in Mg/Si at this area. [Once again, as for Mare Serenitatis, the Mg/Si data are diverging as the cut off point for accepting data is approached.] The Mg/Si ratios of this area near Langemak are similar to those for terra regions west of Mare Smythii, whereas the Al/Si ratios are somewhat lower (~1.1 vs. ~1.2). The Mg and Al variations are very strikingly shown on a plot of Al/Mg ratios (see Hubbard and Vilas, 1977, for such a graph). This area is mapped by Wilhelms and El Baz (1977) as terra. Overall, the data for terra areas show a considerable diversity of Al/Si and Mg/Si intensity ratios. If this diversity is not related to the physical state of the terra soils or to the extreme topographic roughness of the terra areas, then we have two alternatives: 1) that the terra regions were once chemically diverse and have not been chemically homogenized by the heavy bombardment; or 2) that volcanism subsequent to the heavy bombardment (but pre-mare?) has produced the observed chemical variations.

V. Regional Setting for Mare Crisium and Conclusions

The Mare Crisium basin was excavated in terra material that has Al/Si ratios typical of a broad region today extending from 10–15° east of Mare Smythii to the west and south of Mare Crisium. However, the terra material to the east, west and south of Crisium has higher Mg/Si intensity ratios than the terra east of Smythii, except for the area near Langemak. These higher Mg/Si ratios at an approximately constant Al/Si ratio imply a higher Mg/Fe ratio in the olivine and pyroxenes of the terra material around Mare Crisium. This difference in Mg/Si may be either an ancient regional difference, or it may be due to the excavation of originally deeper located materials by the Crisium and other basin-forming impacts (where it is assumed that the deeper layers have higher Mg/Si ratios).

The Mare Crisium basin, like the Serenitatis and Smythii basins, was filled by low albedo materials (presumed to be mare basalts) that have a wide range of Mg/Si ratios. The orbital X-ray data can be calibrated with sample data to indicate that local average MgO concentrations for mare soils in these three maria range from <8.0 to >14.0 wt.%. The sample data alone show MgO values for pure mare materials (mare basalts, Apollo 17 orange glass, Apollo 15 green glass) that range from ~6.5 wt.% to over 16.0 wt.%. The observed variations in Mg/Si intensity ratios for these mare areas are not accompanied by equally diverse Al/Si intensity ratios. In addition, only Mare Smythii shows large differences in Al/Si ratios for material emplaced in the mare basin, some of which may not have been emplaced by mare volcanism (Stewart *et al.*, 1975 and Andre *et al.*, 1977).

There are small and apparently consistent differences in Al/Si ratio between Mare Serenitatis (0.57), Mare Crisium (~0.7), and Mare Smythii (~0.75 for the low albedo portion). It is difficult to conclude today the extent to which these observed small differences in Al/Si ratios reflect chemical differences among the mare basalts that flooded the different mare basins, as opposed to different ratios of mare/terra material in the soils developed on the surface of mare basalt fills. On the other hand, the large differences in Mg/Si intensity ratios for mare areas, and the general lack of mare-terra contrast for Mg/Si, strongly suggest similar variations in the MgO concentration of mare basalts. Spectral reflectance data also indicate a considerable diversity among the materials that fill mare basins (Johnson *et al.*, 1977a; Head *et al.*, 1978). The spectral reflectance data, like the Mg/Si data, are much less sensitive to the admixture of terra material than the Al/Si data. In addition, it has been pointed out that the albedo and color differences of mare surfaces are commonly very sharp (Johnson *et al.*, 1977a), strongly suggesting that the surface chemical compositions seen from orbit are primarily those of indigenous mare materials. From these arguments it seems safe to conclude that the Mg/Si variations are directly linked to Mg variations in mare materials. Perhaps the much smaller differences in Al/Si should not be too readily dismissed as due to admixture of terra material.

Acknowledgments—We thank Jacob Trombka, Richard Schmadebeck, Constance Andre, and Michael Bielefeld for the assistance, criticism and encouragement that they provided during our development of the improvements in data reduction which produced the data and results presented in this paper.

REFERENCES

Adams J. B., Pieters C. and McCord T. B. (1974) Orange glass: Evidence for regional deposits of pyroclastic origin on the Moon. *Proc. Lunar Sci. Conf. 5th*, p. 171–186.

Andre C. G., Wolf R. W., Adler I., Clark P. E., Weidner J. R. and Philpotts J. A. (1977) Chemical character of the partially flooded Smythii Basin based on Al/Si orbital X-ray data. *Proc. Lunar Sci. Conf. 8th*, p. 925–931.

Bielefeld M. J. (1977) Lunar surface chemistry of regions common to the orbital X-ray and gamma-ray experiments. *Proc. Lunar Sci. Conf. 8th*, p. 1131–1147.

Blanchard D. P., Haskin L. A., Brannon J. C. and Aaboe E. (1977) Chemistry of soils and particles from Luna 24 (abstract). In *Papers Presented to the Conference on Luna 24*, p. 37–40. The Lunar Science Institute, Houston.

Haggerty S. E. (1974) Apollo 17 orange glass: Textural and morphological characteristics of devitrification. *Proc. Lunar Sci. Conf. 5th*, p. 193–205.

Head J. W., Adams J. B., McCord T. B., Pieters C. and Zisk S. (1978) Regional stratigraphy and geologic history of Mare Crisium. In *Mare Crisium: The View from Luna 24*. (R. B. Merrill and J. J. Papike, eds.) This volume.

Hubbard N. J. and Vilas F. (1977) From Serenity to Langemak: A regional chemical setting for Crisium (abstract). In *Papers Presented to the Conference on Luna 24*, p. 85–87. The Lunar Science Institute, Houston.

Hubbard N. J., Vilas F. and Keith J. E. (1978a) Improvements in orbital X-ray fluorescence data reduction procedures (abstract). In *Lunar and Planetary Science IX*, p. 555–557. Lunar and Planetary Institute, Houston.

Hubbard N. J., Vilas F. and Keith J. E. (1978b) Solar corrections for the Apollo orbital X-ray fluorescence data (abstract). In *Lunar and Planetary Science IX*, p. 558–560. Lunar and Planetary Institute, Houston.

Hubbard N. J., Vilas F. and Keith J. E. (1978c) An improved procedure for reducing the orbital X-ray fluorescence data collected by Apollo 15 and 16. In preparation.

Hubbard N. J., Vilas F. and Keith J. E. (1978d) Ot Yasnosti do Langemaka: Regionalnoye polozheniye dlya Morya Krizisov. Luna-24: sbornik statyei. V pechati.

Johnson T. V., Mosher J. A. and Matson D. L. (1977a) Lunar spectral units: A northern hemispheric mosaic. *Proc. Lunar Sci. Conf. 8th*, p 1013–1028.

Johnson T. V., Saunders R. S., Matson D. L. and Mosher J. A. (1977b) A TiO_2 abundance map for the northern maria. *Proc. Lunar Sci. Conf. 8th*, p. 1029–1036.

Maxwell T. A., Strain D. L. and El-Baz F. (1977) Mare Crisium: Compositional inferences from low altitude X-ray fluorescence data. *Proc. Lunar Sci. Conf. 8th*, p. 933–944.

McCord T. and Pieters C. (1978) Atlas of reflectance data. Unpublished data, Remote Sensing Laboratory, Univ. of Hawaii.

Rhodes J. M., Rodgers K. V., Shih C., Bansal B. M., Nyquist L. E., Wiesmann H. and Hubbard N. J. (1974) The relationship between geology and soil chemistry at the Apollo 17 landing site. *Proc. Lunar Sci. Conf. 5th*, p. 1097–1117.

Ryder G., McSween H. Y. and Marvin U. (1977) Luna 24 VLT basalt: Character and origin (abstract). In *Papers Presented to the Conference on Luna 24*, p. 160–163. Lunar Science Institute, Houston.

Stewart H. E., Waskom J. D. and DeHon R. A. (1975) Photogeology and basin configuration of Mare Smythii. *Proc. Lunar Sci. Conf. 6th*, p. 2541–2551.

Wilhelms D. E. and El-Baz F. (1977) Geological Map of the East Side of the Moon. Geological Atlas of the Moon, Map I-948.

Merrill, R. B. and Papike, J. J., editors.
(1978) *Mare Crisium: The View from Luna 24*, Pergamon Press (New York), p. 33–42
Printed in the United States of America

Correlation of chemistry with normal albedo in the Crisium region

MICHAEL J. BIELEFELD

Computer Sciences Corporation, Silver Spring, Maryland 20910

ROBERT L. WILDEY

Department of Physics and Astronomy, Northern Arizona University and U. S. Geological Survey
Flagstaff, Arizona 86001

JACOB I. TROMBKA

Goddard Space Flight Center, Greenbelt, Maryland 20771

Abstract—The region in the vicinity of the LUNA 24 handling site (40–70° E; 0–20° N) is studied by means of the Apollo orbital fluorescent X-rays Al/Si and Mg/Al data and earth-based normal albedo measurements. By means of unsupervised non-parametric cluster analysis this region is divided into eight surface units. One of these units centered in the Proclus region deviates from the usual trend of increasing normal albedo with increasing Al/Si values and decreasing Mg/Al values.

INTRODUCTION

The portion of the lunar surface which includes Mare Crisium, Mare Fecunditatis and the non-mare regions between is well suited for the intercomparison of orbital X-ray data and earth-based normal albedo measurements. The Al/Si and Mg/Al concentration ratios and the normal albedo of the region 40°–70° E, 0°–20° N show nearly the full range of their measured values for the whole moon. Although coverage is limited, the comparison should be valid for larger segments of the moon. In addition, the remote sensing data in this study area demonstrate the representativeness of the Soviet Luna 24 (62.2° E, 12.7° N) and Luna 20 (56.3° E, 3.3° N) sample chemistries for the surface soils.

The correlation between normal albedo and the Apollo 15 and 16 orbital fluorescent X-ray Al/Si intensity ratios was pointed out by Adler *et al.* (1972, 1973). Clark *et al.* (1976) treated the same correlation in more detail for the Palus Somni just north of M. Tranquilitatis. The laboratory work of Gold *et al.* (1974, 1977) on returned samples shows that the albedo of the lunar surface is more directly related to the Fe concentration in the outermost layers of the lunar soil particles than to Al concentration. Nevertheless, the link between the Al/Si and Mg/Al X-ray data and the normal albedo measurements can be indirectly made through the correlations pointed out by Bielefeld (1977) between the X-ray ratio data and surface iron concentrations as measured by the Apollo orbital gamma-ray experiment.

Recent advances in the standardization of multiple lunar data sets by the La

Jolla Consortium on remote sensing of the moon have facilitated the comparison among geochemical and geophysical parameters (Frontispiece, 1977). All data sets were properly registered and digitized in a 0.25° × 0.25° format. The preparation of the normal albedo file is outlined by Wildey (1977). The details of the imaging of the X-ray data sets are given by Bielefeld *et al.* (1977a) and Andre *et al.* (1977).

This paper will present enlargements of these geochemical and geophysical images which appeared as false color plates in the frontispiece of the *Proceedings of the Eighth Lunar Science Conference* and which are similar to the false color presentations in abstract form by Bielefeld *et al.* (1977b). Correlations between the individual X-ray chemical ratios and the normal albedo will be graphically demonstrated. Finally, a computerized automatic classification of the study area will be performed using all three measurements as the discriminating functions.

PRESENTATION OF RESULTS

Figure 1 depicts a shaded map of the study area (40–70° E, 0–20° N) and is included for orientation purposes only. The area is representative of the full dynamic range of each of the three variables for the observed frontside of the

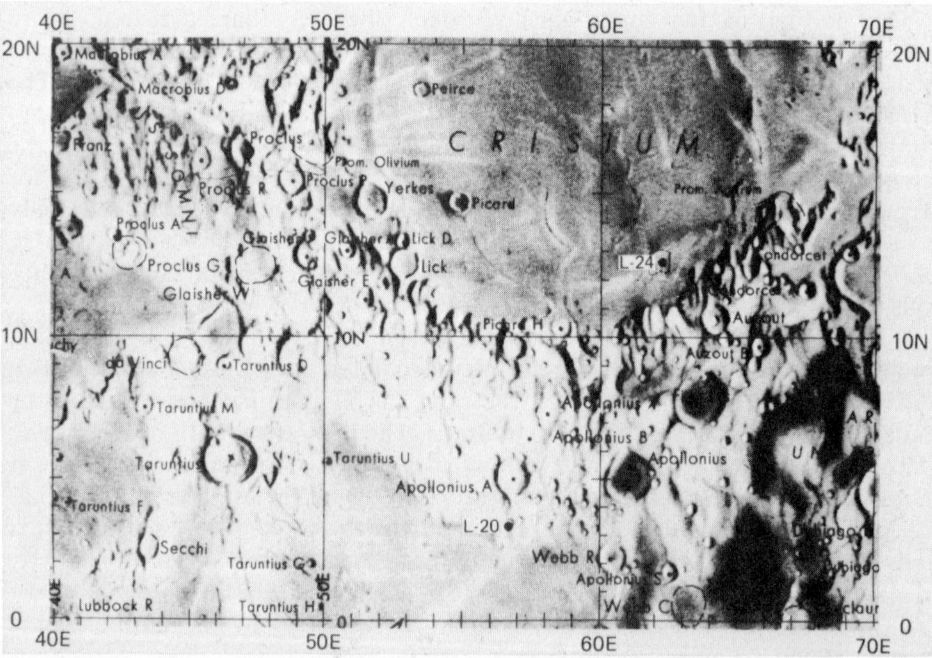

Fig. 1. Map of the M. Crisium study area.

moon (see Frontispiece, 1977 plates 4–6). The area includes the landing sites of Luna 24 (62.2° E, 12.7° N) and Luna 20 (56.3° E, 3.3° N).

Classification of the study area by a process of density slicing in a single dimension for the images of normal albedo, Al/Si and Mg/Al concentration ratios by weight are given in Figs. 2–4. Within the constraints of reasonable graphical presentation and statistical reliability, we have chosen the widths and positions of the density slices specified in the legend of each figure to reflect the natural grouping in the one-dimensional histogram of each data set which can be found along the appropriate axes in Fig. 5. Certain areas of the study region are not covered by orbital X-ray data due to trajectory limitations. The reader should be aware that this type of presentation can overemphasize the contrast between small variations in grey levels *at* the boundaries while ignoring small changes in grey levels *within* the boundaries.

The information contained in the three images of Figs. 2–4 is presented in Fig. 5 as a pair of two-dimensional histograms of albedo versus chemistry. These histograms are two-component frequency distributions of the grey level values within the area of common coverage. Some 7000 $0.25° \times 0.25°$ regional units, which comprise the images of Figs. 2–4, are positioned on the arrays of Fig. 5 at the coordinates specified by the intensity value of each component. The frequency of occurrence of the same values is reflected in the height of the plot. The plots along the axes are frequency histograms of all the values of each component considered separately. The correlation between normal albedo and Al/Si and Mg/Al is readily seen in this form of presentation.

An automatic classification scheme has been used to integrate all of the above information. Many methods for classification rely on "training areas" of known composition. Since direct chemical analysis of the moon is restricted to a few landing sites, the classifier chosen for this study, known as HINDU, does not rely on training areas but locates the natural density clusters in the data itself. HINDU is an unsupervised non-parametric classifier well suited for lunar classification since neither training areas of known composition nor information about the type of distributions underlying the measurements is available. Presently, HINDU can classify a regional area using from two to ten different geochemical and geophysical parameters simultaneously. Information about the reliability and strategy of HINDU is presented by Jayroe *et al.* (1976) and Bielefeld *et al.* (1978). Factors which increase chances for misclassification are: statistical uncertainty in the data; incompatibility in the areal resolution of the different detectors; poor separation of the clusters; and proximity of data units to both the geographical borders between classes and the borders separating the clusters in the parameter density space.

Figure 6 shows the results of a three-component classification scheme in which the information in the images of normal albedo, Al/Si ratio and Mg/Al ratio are simultaneously used to find the natural clustering of these three parameters. Eight major classes were determined by HINDU and the characteristics of each are given in Table 1. The legend in Fig. 6 maps the eight clusters into the geographical study area.

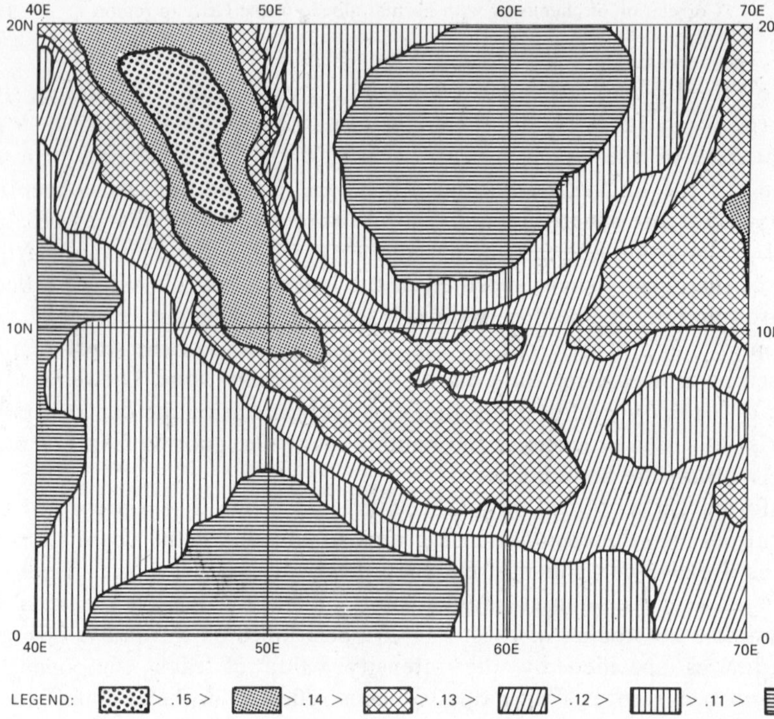
40E
50E
60E
70E
20N
10N
0
LEGEND: > .15 > > .14 > > .13 > > .12 > > .11 >

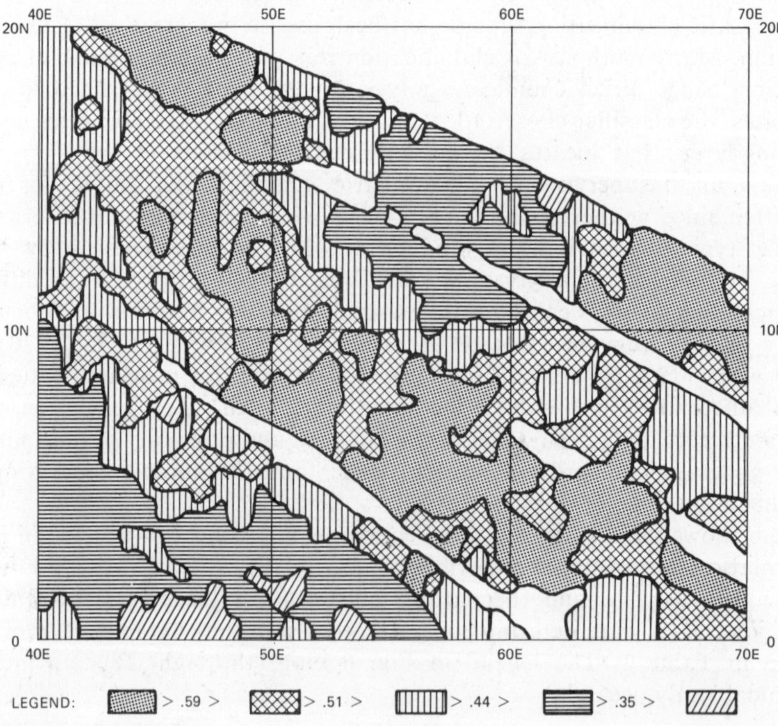
40E
50E
60E
70E
20N
10N
0
LEGEND: > .59 > > .51 > > .44 > > .35 >

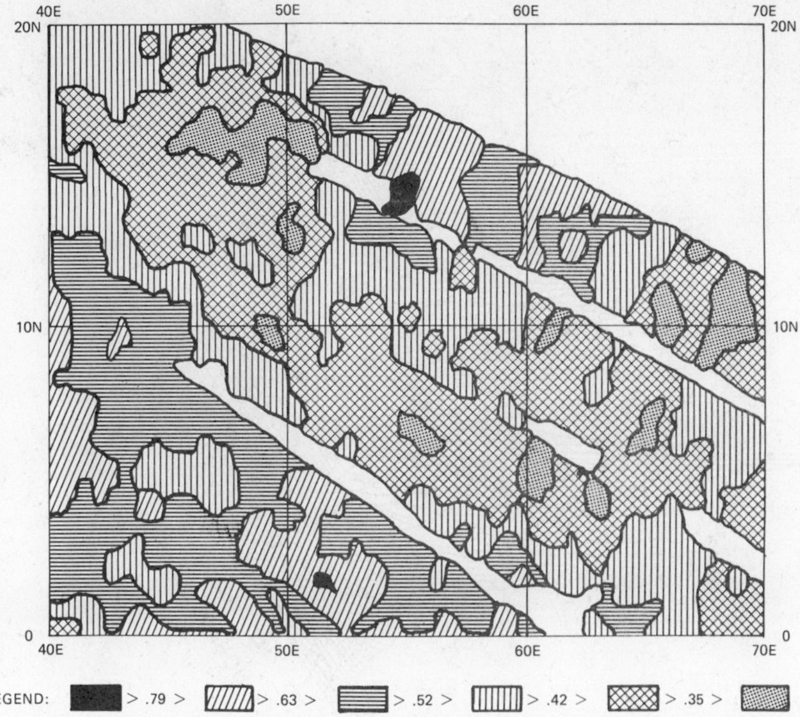

Fig. 2. Variations in the normal albedo of the M. Crisium study area as prepared for the La Jolla Consortium by Wildey (1977). Enlargement of Frontispiece (1977) plate 6.

Fig. 3. Variations in the Al/Si concentration ratio by weight for the M. Crisium study area as prepared for the La Jolla Consortium by Bielefeld *et al.* (1977) from the Apollo orbital fluorescent X-ray data. Enlargement of Frontispiece (1977) plate 4.

Fig. 4. Variations in the Mg/Al concentration ratio by weight for the Crisium study area as prepared for the La Jolla Consortium by Bielefeld *et al.* (1977) from the Apollo orbital fluorescent X-ray data. Enlargement of Frontispiece (1977) plate 5.

Discussion

The low albedo surface material, which in general terms is associated with the mare material of the inner Crisium basin and M. Fecundatatis (see Fig. 2), has a natural clustering at the albedo value of 0.102. This is most clearly seen in the frequency histogram along the albedo axis of Figure 5a. The associated value of the Al/Si concentration ratio by weight for this low albedo material is approximately 0.41 as can be seen in the histogram along the Al/Si axis. The two-dimensional histogram of Fig. 5a shows a pronounced clustering at these coordinates. However, when the Mg/Al concentration ratio information is added to the discrimination scheme, this general class is subdivided into at least three

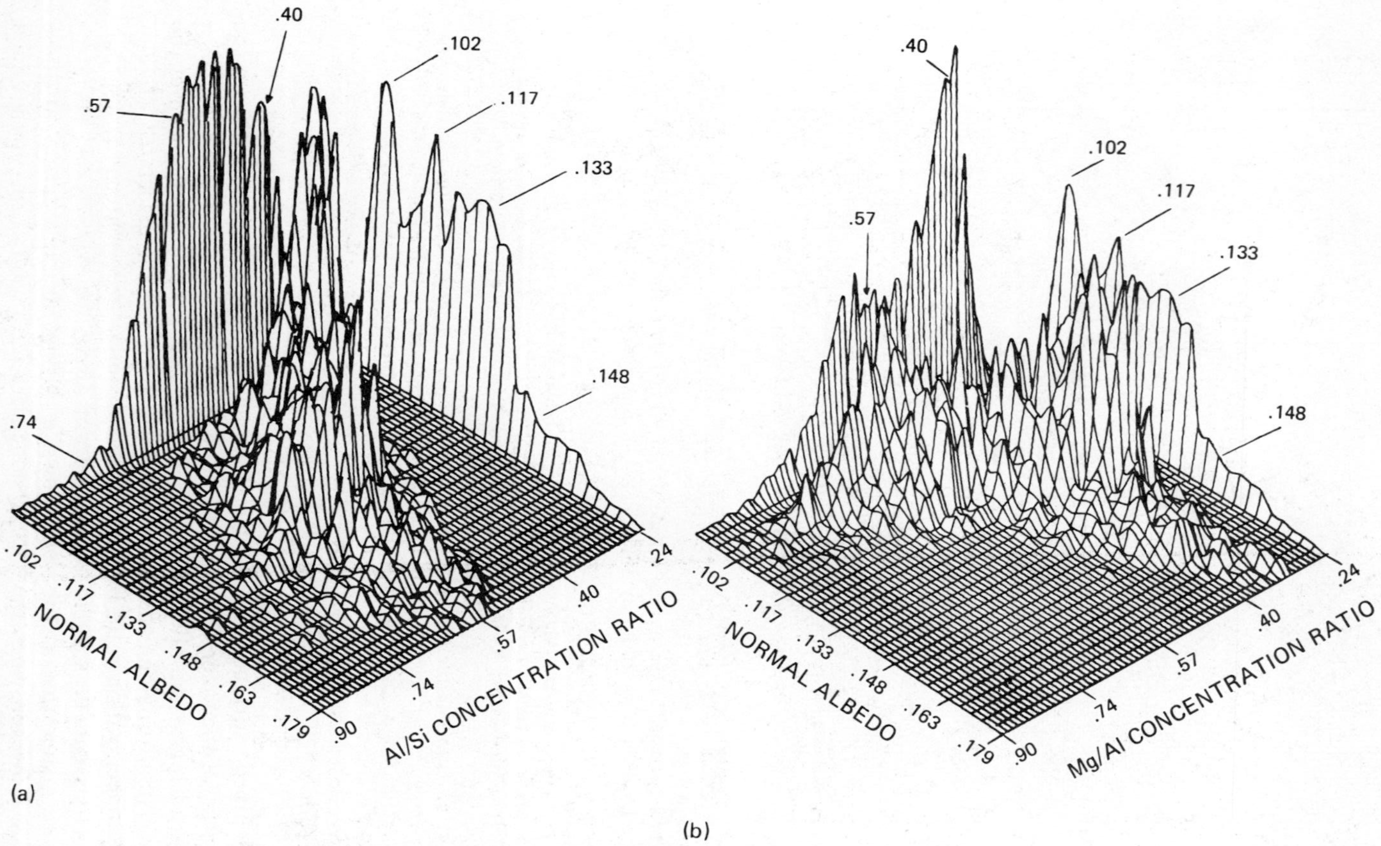

Fig. 5. (a) Frequency distribution of Al/Si vs. albedo and (b) frequency distribution of Mg/Al vs. albedo for the area of common coverage in the Crisium study region. There are approximately 7000 0.25° × 0.25° areal units present in each plot. The plots along the axes are not projections but frequency histograms of all the values of each component considered separately.

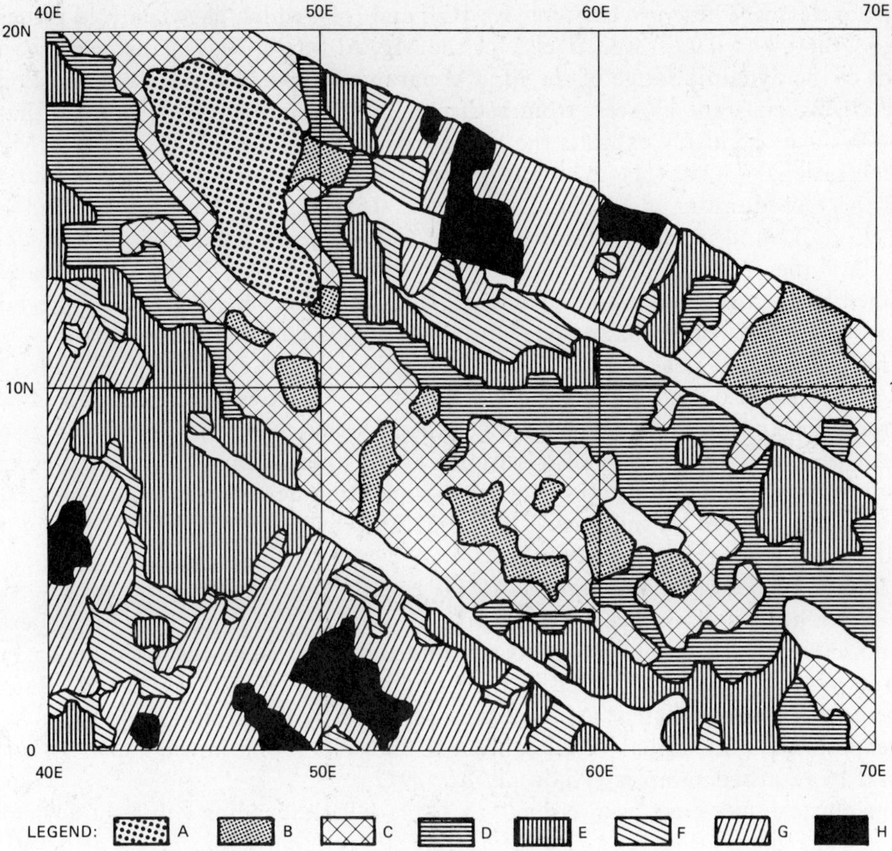

Fig. 6. Classification map of the Crisium study area using the natural clustering of the albedo vs. Al/Si *vs*. Mg/Al data as determined by the HINDU linear unsupervised non-parametric classifier.

Table 1. Parameters of the clusters

CLUSTER	POPULATION	CENTROID OF CLUSTER Al/Si*	Mg/Al*	Albedo
A	373	.57	.37	.157
B	468	.62	.35	.136
C	1348	.57	.40	.132
D	1287	.50	.42	.125
E	1285	.48	.50	.114
F	678	.41	.53	.107
G	1220	.40	.60	.101
H	333	.36	.73	.101

*Values of the fluorescent X-ray intensity ratios have been converted to concentration ratios by weight according to the methodology outlined by Bielefeld (1977).

separate classes. These are clusters [F], [G], and [H], which have centroid values of 0.53, 0.60, and 0.73, respectively, for the Mg/Al ratios. This spans nearly two thirds of the dynamic range of the Mg/Al parameter. Since the Si concentration is relatively constant as seen from returned sample analysis, this implies that mare surface chemistry exhibits the highest two thirds of orbitally observed Mg concentrations. Cluster [H] which identifies the regions of highest Mg is found near the Picard crater in M. Crisium (55° E, 15° N), near Taruntius E and F (40° E, 5°N) and near Taruntius G and H (50° E, 2° N).

Within the Crisium basin the mare material of cluster [E] forms a band outlined by the first and second geological rings of the basin. This same material flooded the crater Firmicus (63° E, 8° N), M. Undarum and M. Spumans.

The pronounced peak at 0.40 in the one-dimensional histogram of Mg/Al concentration ratio seen along the Mg/Al axis in Fig. 5b is broken into the two clusters [C] and [D] with two distinct albedo values of 0.132 and 0.125 also clearly seen in Fig. 5. The distribution around these centroid coordinates is closely packed and is reflected in the high population density of the two classes. They are composed of non-mare material and in general ring M. Crisium separating it from the other maria.

Cluster [B] possesses the highest Al/Si centroid value at 0.62 and the lowest Mg/Al value at 0.35 concentration ratio by weight; yet it has only a modest albedo value of 0.136. This high Al soil intermittently rings the Crisium basin with its largest expanse centered in the Condorcet region (67° E, 11° N). It is representative of the highest Al/Si values observed by the orbital fluorescent X-ray experiment and similar in Al content to the soils found in the Luna 20 and Apollo 16 returned samples (Andre, *et al.*, 1977).

The cluster units have been ordered in Table 1 in descending values of normal albedo. The populations of the clusters reflect the identifiable peaks at 0.160, 0.137, 0.132, 0.125, 0.117, and 0.102 seen in the histogram on the albedo axis of Fig. 5. The descending order of albedo values generally corresponds to the descending order of Al/Si ratios and the ascending order of Mg/Al. The regression curve for normal albedo versus either of the X-ray geochemical parameters is not perfectly linear as seen in Fig. 5 and in Bielefeld *et al.* (1978). Gold *et al.* (1974, 1977) showed that the logarithm of the albedo can be linearly related to surface Fe concentrations. Bielefeld (1977) observed a strong linear relationship between orbital gamma-ray Fe concentrations and orbital X-ray Al/Si and Mg/Al concentration ratios for large integrated areas of the lunar surface. Therefore, the regression curves in this study should also be logarithmic.

When the logarithm of albedo centroid values are plotted against the chemistry values, cluster [A] clearly breaks from the regression curve exposing an anomalous soil type centered in the Proclus region (47° E, 12° N). The breadth of the area argues against the anomaly arising from simple misclassification. This is an example of the ability of the classification procedure to identify regional deviations from simple correlation.

Chemistry is not the only factor determining normal albedo. Soil maturity and other factors also influence albedo causing scatter and deviation from the major correlation with chemistry. Cluster [A] possesses the highest albedo seen in the study area with a centroid value of 0.157 which is comparable in brightness to the Tycho region in the southern hemisphere of the moon (Frontispiece, 1977). On the grounds that the very high albedo values of the study area cluster about the less extreme Al/Si values of 0.57 and Mg/Al values of 0.35 concentration ratios by weight, the suggestion can be made that the other very high albedo regions of the moon, such as Tycho, have chemistries similar to those of cluster [A].

The ultimate classification achieved through the present cluster analysis might be expected to correspond to classification by geologic formation. Preliminary comparison of Fig. 6 with the geologic map of the east side of the moon (Wilhelms and El Baz, 1977) shows heartening similarities. The present study yields additional separation as compared to the geologic map within the maria where albedo variations and surficial expression is greatly subdued. A few formations on the geologic map are not found in Fig. 6; for example, the crater Apollonius (61° E, 5° N) shows no sign of mare flooding.

The discussion here was limited as much as possible to the correlation of the data sets themselves with the application and interpretation left to the various disciplines of lunar and planetary science. Clustering of the data from X-ray chemistry and normal albedo is not random but follows the general regression curve of low-to-high albedo, low-to-high Al/Si and high-to-low Mg/Al. Within this generally continuous correlation there is clustering of the data which is indicative of natural classes of the geochemical and geophysical properties of the lunar surface soils. Clustering also occurs off the general correlation indicating distinctive soil types. The methods employed here allow one to locate areas of common properties geographically. Multiparameter classification (see Fig. 6) is more discerning in finding soil types than single parameter classification (see Figs. 2–4). The use of additional parameters, such as orbital gamma-ray chemical data, visible and near-infrared reflectivity data, altimetry data and crustal thickness data is the next logical step in this investigation, and should help in the understanding of the processes of lunar evolution and differentiation.

Acknowledgments—We would like to acknowledge several helpful discussions with Walter C. Shoup concerning data presentation. The SMIPS/VICAR image processing computer software which is maintained by Code 930 of Goddard Space Flight Center was used extensively in this study. This work was entirely supported by NASA.

References

Adler I., Gerard J., Trombka J., Schmadebeck R., Lowman P., Blodget H., Yin L., Eller E., Lamothe R., Gorenstein P., Bjorkholm P., Harris B., and Gursky H. (1972) The Apollo 15 X-ray fluorescence experiment. *Proc. Lunar Sci. Conf. 3rd*, p. 2157–2178.

Adler I., Trombka J. I., Schmadebeck R., Lowman P., Blodget H., Yin L., Eller E., Podwysocki M., Weidner J. R., Bickel A. L., Lum R. K. L., Gerard J., Gorenstein P., Bjorkholm P., and Harris B. (1973) Results of the Apollo 15 and 16 X-ray experiment. *Proc. Lunar Sci. Conf. 4th*, p. 2783–2791.

Andre C. G., Bielefeld M. J., Eliason E., Soderblom L. A., Adler I. and Philpotts J. A. (1977) Lunar surface chemistry: a new imaging technique. *Science* **197**, 986–989.

Bielefeld M. J. (1977) Lunar surface chemistry of regions common to the orbital X-ray and gamma-ray experiments, *Proc. Lunar Sci. Conf. 8th*, p. 1131–1147.

Bielefeld M. J., Andre C. G., Eliason E. M., Clark P. E., Adler I., and Trombka J. I. (1977a) Imaging of the lunar surface chemistry from orbital X-ray data. *Proc. Lunar Sci. Conf. 8th*, p. 901–908.

Bielefeld M. J., Shoup W. C., Evans L. G., and Trombka J. I. (1978) Classification of the lunar surface using remote sensing data (abstract). In *Lunar and Planetary Science IX*, p. 85–87. Lunar and Planetary Institute, Houston.

Bielefeld M. J., Wildey R. L. and Trombka J. I. (1977b) Correlation of chemistry with normal albedo in the Crisium region. In *Papers Presented to the Conference in Luna 24*, p. 28–33. The Lunar Science Institute, Houston.

Clark P. E., Andre C. G., Adler I., Podwysocki M. and Weidner J. (1976) Palus Somni: anomalies in the correlation of Al/Si X-ray fluorescence intensity ratios and broad-spectrum visible albedos. *Geophys. Res. Lett.* **3**, 421–424.

Frontispiece (1977) *Proc. Lunar Sci. Conf. 8th*, Vol. I, plates 4–6.

Gold T., Bilson E. and Baron R. L. (1974) Observation of iron-rich coating on lunar grains and a relation to low albedo. *Proc. Lunar Sci. Conf. 5th*, p. 2413–2422.

Gold T., Bilson E. and Baron R. L. (1977) The relationship of surface chemistry and albedo of lunar soil samples. *Philos. Trans. R. Soc. London* **A 285**, 427–431.

Jayroe, Jr. R. R., Atkinson R., Dasarathy B. V., Lybanon M. and Ramapryian H. K. (1976) *Classification Software Technique Assessment*. NASA TN D-8240.

Wildey R. L. (1977) A digital file of the lunar normal albedo. *The Moon* **16**, 231–277.

Wilhelms D. E. and El-Baz F. (1977) Geologic map of the east side of the moon. U. S. Geol. Survey, Misc. Investgation Series MAP I-948.

Merrill, R. B. and Papike, J. J., editors.
(1978) *Mare Crisium: The View from Luna 24*, Pergamon Press (New York), p. 43–74
Printed in the United States of America

Regional stratigraphy and geologic history of Mare Crisium

JAMES W. HEAD, III

Department of Geological Sciences, Brown University, Providence, Rhode Island 02912

JOHN B. ADAMS

Department of Geological Sciences, University of Washington, Seattle, Washington 98195

THOMAS B. MCCORD

Institute for Astronomy, University of Hawaii, Honolulu, Hawaii 96822

CARLE PIETERS

NASA Johnson Space Center, Houston, Texas 77058

STANLEY ZISK

Haystack Observatory, MIT, Westford, Massachusetts 01886

Abstract—A variety of remote sensing and Luna 24 sample data are used to develop a summary of the regional stratigraphy and geologic history of Mare Crisium and to place the Luna 24 site in a regional context. The spectral reflectance curves of returned Luna 24 soils have a low 0.40 μm/0.56 μm ratio and a strong 1 μm band depth different from the mare soils of the Apollo missions. These unique spectral properties, attributed to the combination of high iron and low titanium concentration, are also seen in reflectance spectra and spectral images of the Luna 24 site and other areas of Mare Crisium. Three major basalt groups are recognized: (Group I) a soil derived from Fe- and Mg-rich titaniferous basalts similar to Luna 16 samples; (Group II) soils derived from very low titanium ferrobasalts similar to Luna 24 samples; and (Group III) a soil derived from a low titanium ferrobasalt similar to samples from Apollo 12. Group I basalts were apparently emplaced over the majority of the basin but are presently exposed as a topographic shelf in the southwest part of the basin. The second group of basalts were emplaced subsequently in two stages and now appear as part of the outer shelf and topographic annulus. Subsidence of the basin interior preceded and continued after the emplacement of the third basalt unit. The Luna 24 site is within a patch of Group II and just northwest of Group I basalts. Each Mare Crisium basalt unit can be found elsewhere in the nearside maria.

I. INTRODUCTION

The Soviet spacecraft Luna 24 landed in southeastern Mare Crisium in August 1976 and returned a core sample from the site (12°45′N; 60°12′E) (Barsukov, 1977) (Figs. 1,2). Shortly after this mission, Pieters *et al.* (1976) used multispectral images and a single telescopic reflectance spectrum from central Crisium to identify distinct geochemical surface units in the Mare Crisium region. The soil of the area for which the spectrum was available was inferred to have from 2–4% TiO_2 based on the .40/.56 μm ratio (Charette *et al.*, 1974). The area lies 70 km

Fig. 1. Photographic mosaic of the front side of the Moon.

northwest of the Luna 24 landing site and, prior to analysis of the samples, was thought likely to be representative of material returned in the Luna 24 core. Luna 24 samples, however, have ~1% TiO_2 and the laboratory spectra (Adams

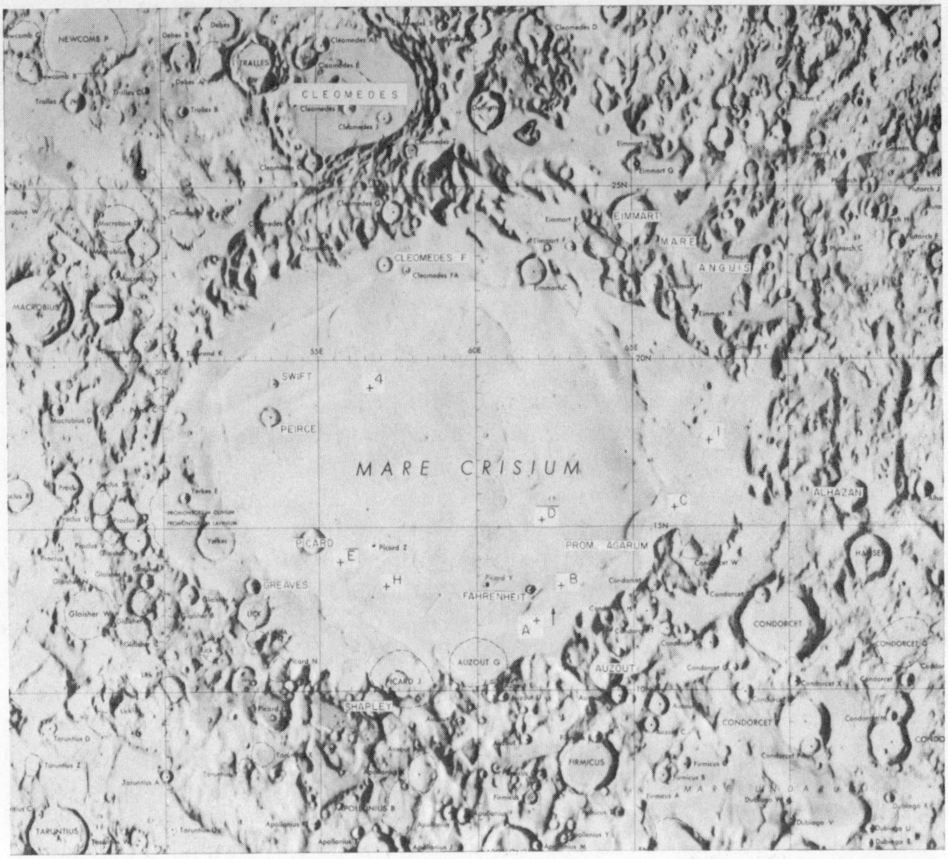

Fig. 2. Crisium basin geography. Crosses indicate location of spectra (see Fig. 10) and small arrow indicates location of Luna 24 landing site. Base map is NASA Lunar Planning Chart LOC-3.

and Ralph, 1977) differ from the original telescopic spectra. It became clear that the geology of the Luna 24 landing site was complex and a more detailed study of the region was necessary.

In a subsequent paper (Adams *et al.*, 1978) we combined evidence from Luna 24 samples, new telescopic reflectance data, and radar and other topographic data to produce a preliminary regional geologic map of the Crisium basin. Spectral reflectance measurements of five Luna 24 samples and new telescopic reflectance spectra of 10–20 km areas of seven sites in Mare Crisium were used to calibrate multispectral images of mare units in the Crisium basin. Three major mare groups were defined.

The present article builds on the Adams *et al.* (1978) study and presents a more detailed and complete analysis of the regional stratigraphic relationships and the geologic history of Mare Crisium. The Luna 24 site and the returned samples are placed in a regional context in order to relate them to other events in lunar history. The approach is to use remote sensing data to define and characterize surface units. The local and regional geology is then reconstructed to derive insight into the global, compositional, thermal, and geologic evolution. The primary information used is derived from Apollo, Lunar Orbiter and earth-based telescopic photographs; ground-based reflectance spectra; multispectral, radar and visible albedo maps; lunar samples; and Apollo geochemical remote sensing studies. The reader is directed to Head *et al.* (1978) for a discussion of our approach and of the nature and usefulness of various data bases.

II. LUNA 24 SAMPLES

In previous work (e.g., Adams and McCord, 1972) laboratory spectra of Apollo soil samples (on the millimeter scale) have been successfully compared with telescopic spectra of the Apollo landing sites (10–20 kilometer scale). In this section, we review the properties of the Luna 24 soil, and discuss the laboratory spectral reflectance data on the samples from the upper part of the core. We show that the Luna 24 samples have unique spectral reflectance properties. In a subsequent section it is shown that these unique properties are seen in telescopic spectral curves and spectral images of the Luna 24 landing site and of other areas in Mare Crisium.

It is not known to what extent the original lunar surface is preserved in the Luna 24 core. The top few centimeters of the core may have been lost during drilling, or mixed with stratigraphically lower material in the process of extracting or later handling of the core. An unknown thickness of soil may have been jetted from the landing area during firing of the descent engine. Bogard and Hirsch (1977) considered extreme models for the core and concluded that sample 24077, the uppermost sample available to United States scientists, could have come from the lunar surface or from as far as 70 cm below the surface. They calculated cosmic ray exposure ages using the assumption that 24077 was a near-surface sample.

Noble gas data (Bogard and Hirsch, 1977), and grain-size data (McKay *et al.*, 1977) indicate that the Luna 24 soils are mixtures of two soils that differ in maturity. Ferromagnetic resonance data (Morris, 1977) show that the bulk soil samples are all immature to submature, with the most mature soil being 24077. Florensky *et al.* (1977) and Butler and Morrison (1977) suggest that immature basalt (gabbro) was derived from ejecta from the 6 km diameter crater Fahrenheit which is 18 km from the landing site.

The bulk of the soils consist of mineral and lithic fragments and glasses, agglutinates and breccias that were derived from very low titanium (VLT) ferrobasalt (e.g., Basu *et al.*, 1977; Hu and Taylor, 1977; McKay *et al.*, 1977; Taylor *et al.*, 1977; Blanchard *et al.*, 1977; Simonds *et al.*, 1977). Particles derived from the highlands are rare in the Luna 24 samples (Bence and Grove, 1977). A variety of less abundant or even exotic particles is discussed in the above papers. Although of considerable petrologic interest, the rare components of the soil are of less importance for remote sensing unless they represent known units, and will not be discussed here.

Laboratory reflectance measurements were made in the wavelength region 0.35 μm to 2.5 μm on five samples of Luna 24 soils. Each soil had been sieved to <90 μm and each sample weighed <50 mg. Details of the laboratory measurements are given by Adams and Ralph (1977). Of particular interest to remote sensing applications are the spectra of the near surface samples 24077 and 24109 (Fig. 3).

The reflectance curves of the Luna 24 soils show the typical steeply sloping continuum and low integral reflectance (albedo) of other Apollo and Luna mare

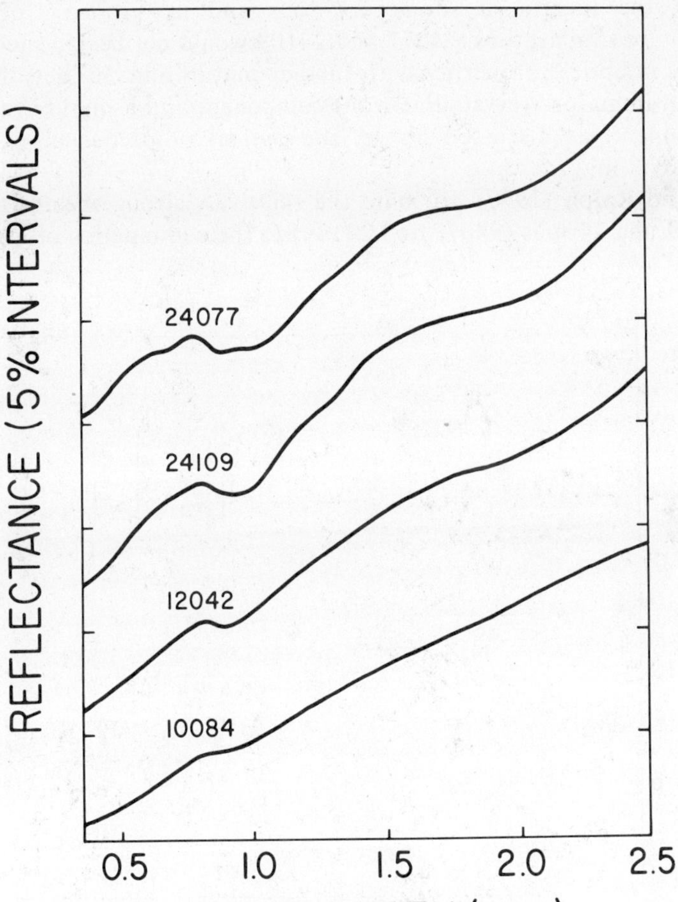

Fig. 3. Laboratory diffuse reflectance spectra of near-surface Luna 24 samples compared with Apollo 11 and 12 soils. Spectra are offset vertically for clarity of illustration. Note relatively deep absorption bands in spectra of Luna 24 samples.

soils. The spectra of rocks, breccias, and glasses are distinctly different (Charette and Adams, 1975). The Luna 24 spectra have characteristic absorption bands near 1 μm and 2 μm that arise mainly from Fe^{2+} in pyroxene and in glass. Absorption in the UV and blue is caused by multiple bands including strong bands that involve iron and titanium charge-transfers. A weak band at 1.25 μm is caused by Fe^{2+} in plagioclase.

Absorption bands in the reflectance spectra of Luna 24 soils are unusually deep as compared with the spectra of other lunar samples. In the lower part of the core the strong spectral contrast of the samples is in part an expression of the immaturity of the soils, as can be seen in the deep band structure of the immature soil 24182 (Adams and Ralph, 1977). However, submature samples 24077 and 24109 also have relatively deep bands (Fig. 4). Adams and Ralph (1978) have shown that the spectra of submature soils and mature soils of the same composition are similar; thus, 24077 and 24109 would not be expected to change significantly in optical properties with further maturation. In fact, if these soils consist of an immature (crystalline-rich) component and a mature (agglutinate-rich) component, as discussed above, the optical properties may already be dominated by a mature soil.

Adams and Ralph (1977) attribute the relatively strong spectral contrast of the surface Luna 24 soils (24077 and 24109) to the combination of high iron and

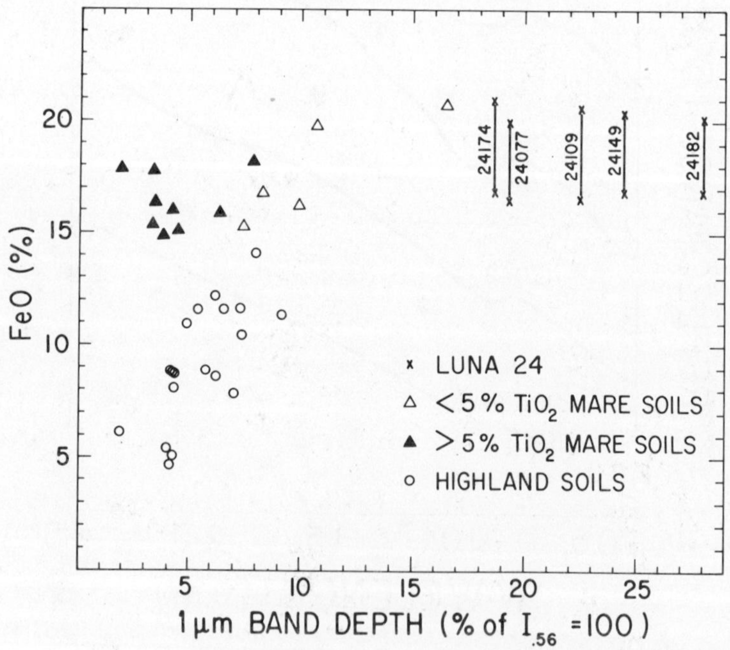

Fig. 4. Depth of 1 μm band in diffuse reflectance spectra of lunar samples vs. weight percent FeO. Luna 24 samples show a deep band and have a very low TiO_2 content.

low titanium. The FeO content of bulk Luna 24 soils is 16–20%, and TiO_2 is about 1% (Blanchard *et al.*, 1977). The concentration of Fe^{2+} in agglutinitic glass is approximately that in the bulk soil (Taylor *et al.*, 1977). The depths of all of the observed absorption bands are proportional to the Fe^{2+} concentration, an example being the 1 μm band shown in Fig. 4. The addition of titanium to the soil, and especially to the agglutinitic glass, affords the opportunity for $Fe^{2+} \rightarrow Ti^{4+}$ and other charge-transfers that produce intense absorption in the UV and visible portions of the spectrum. In iron-rich mare soils TiO_2 at the 1% level has the effect of deepening the UV-blue absorption band. Higher concentrations of titanium, however, spread the UV-blue "band" to longer wavelengths and increase the opacity of the soil through opaque phases, thereby obscuring the other bands contributed by Fe^{2+} alone. Increasing the concentration of titanium, therefore, decreases the spectral contrast in the UV and visible regions and flattens the slope of the spectral curve. This is the origin of the TiO_2-slope relationship for remote sensing of lunar soil that was first presented by Charette *et al.* (1974). As TiO_2 increases above the 1–2% level the UV-visible slope in reflected light becomes flatter, i.e., less red, bluer, higher ratio .40 μm/.56 μm (see Fig. 5).

The UV-visible region of the Luna 24 curves has the lowest .40 μm/.56 μm ratio yet observed in the laboratory for lunar mare soils, which is consistent with $<2\%$ TiO_2 (Charette *et al.*, 1974; Adams and Ralph, 1978).

As TiO_2 increases, the 1 μm band becomes shallower for a given FeO content, as is seen in Fig. 4. The 1 μm band for Luna 24 soils is deeper than other mare soil samples measured, again consistent with the very low concentration of TiO_2.

Therefore, two properties of the spectral reflectance curves of Luna 24 soils (low .40 μm/.56 μm ratio and strong 1 μm band depth) can be identified which differ from the mare soils of the Apollo missions, and which we attribute to the combination of a high concentration of iron and a very low concentration of titanium. These unique spectral properties of the Luna 24 samples are observed in two types of telescopic spectral data for Mare Crisium: 1) spectral reflectance curves, and 2) spectral vidicon and photographic images.

III. Unit Definition and Characterization

A variety of remotely sensed data is available for Mare Crisium and is used here in the manner outlined by Head *et al.* (1978) for unit definition and characterization. The techniques that provided the most useful data for unit mapping are multispectral images, earth-based and orbital photography (relative albedo measurements), and radar backscatter maps. Each unit was further characterized by spectral reflectance measurements for small regions in the units. Some of the recently processed orbital X-ray spectroscopy data was of sufficient resolution to provide further characterization of the Mare Crisium units. Supporting techniques include orbital γ-ray spectroscopy, UV spectroscopy, and α-particle measurements.

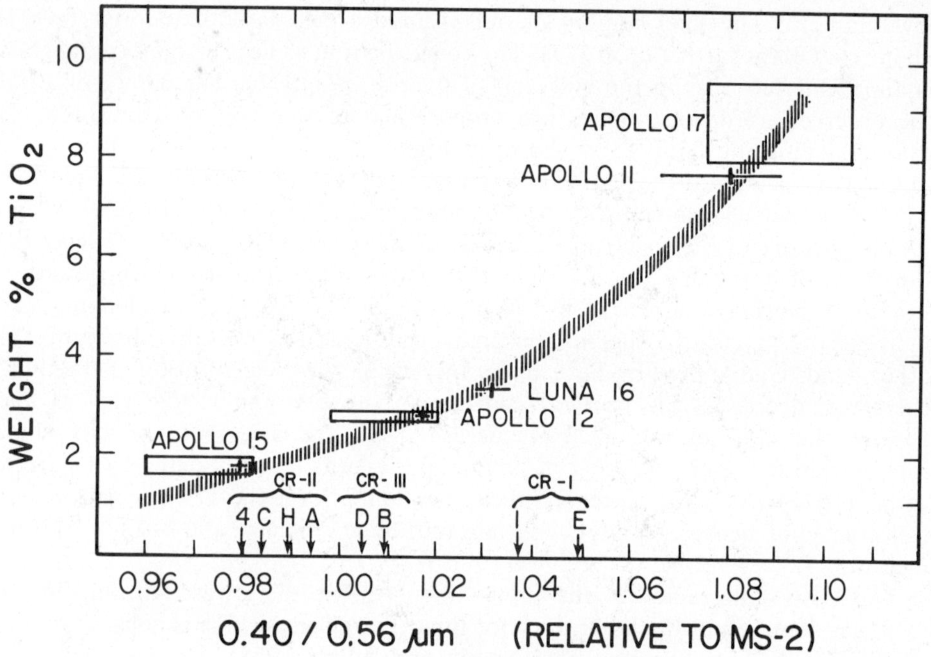

Fig. 5. Ratio .40/.56 μm from relative lunar telescopic reflectance spectra vs. weight percent of TiO_2 of representative soil samples from the sites. Figure is adapted from Charette *et al.* (1974) and Pieters *et al.* (1976). Measurements of areas in Mare Crisium are plotted on the horizontal axis and are keyed to Figs. 2 and 11 and Table I.

A. *Description of defining techniques*

An observable characteristic that easily distinguishes among many basalt types is the general slope of the lunar spectrum in the ultraviolet and visible (Fig. 3, 6). This slope, as measured by reflectance ratios, varies only a few percent ($<10\%$) between .40 and .56 μm for different maria. For unit definition we have relied heavily on two types of multispectral images that provide regional information on this spectral parameter: 1) vidicon spectral ratio mosaics (McCord *et al.*, 1976, 1978), and 2) color difference photographs (Whitaker, 1972, 1976). The spectral band passes used in producing these multispectral images (McCord *et al.*, 1976; Zisk *et al.*, 1977) are shown in Fig. 6. The vidicon ratios (Fig. 7a, b, c) provide high precision data in a single frame that can be strongly enhanced to bring out fine detail, It is, however, difficult to calibrate adjacent frames with each other. The color difference photograph of Whitaker (Fig. 8) provides information on the regional color relationships but does not provide precise quantitative data. Allowing for the differences in spectral coverage, we have used both sources of data (calibrated with spectra) to map individual units. Three general unit groups can be distinguished in the UV/VIS spectral images based on whether the

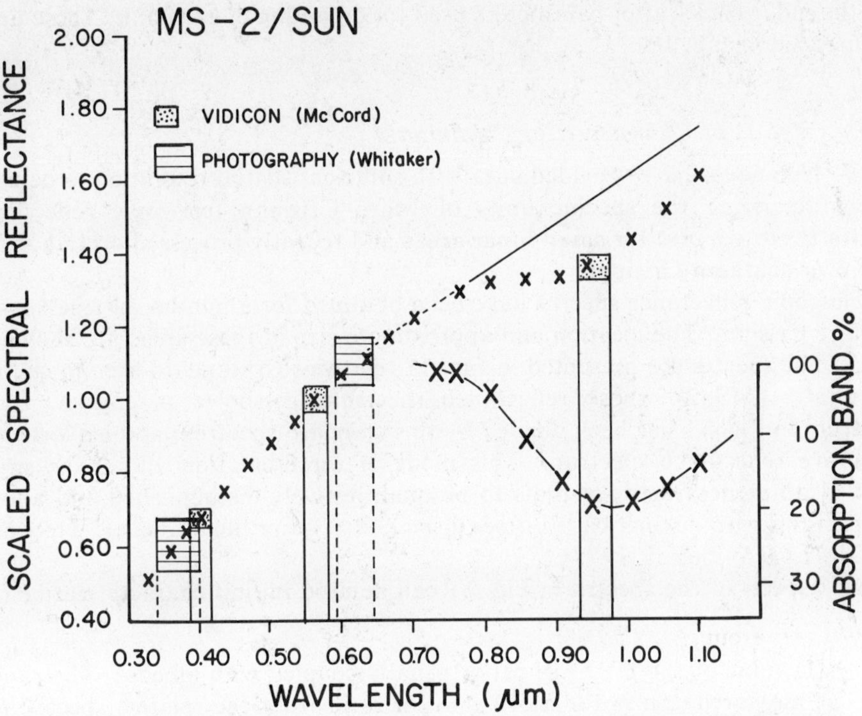

Fig. 6. Reflectance spectrum of the MS-2 standard in central Serenitatis. Superimposed on the spectrum are the band passes for filters used in vidicon multispectral imagery (McCord) and recent color difference photography (Whitaker).

surface has a relatively high, intermediate, or low UV/VIS reflectance ratio (see Table I). In central Crisium the boundaries of these units are often indistinct, creating a mottled appearance of the units.

Two other types of defining techniques were used to a lesser extent to distinguish the Mare Crisium units. In various albedo maps (Pohn and Wildey, 1970; Evsyukov and Barabashov, 1973a, b) Mare Crisium has irregular high and low albedo areas, the boundaries of which are dependent on the resolution of the mapping technique coupled with the abundance of ray material on the surface. We have used, therefore, the higher resolution relative albedo that can be distinguished in earth-based and orbital photographs as a supplementary defining parameter. Some areas have a lower albedo and are distinctly dark relative to their surroundings (see Table I). It was also noted that many of the well defined unit boundaries closely follow boundaries seen in a depolarized radar (3.8 cm) backscatter map (Fig. 9). For some areas where the boundaries on the spectral images were indistinct, the radar data were used as supplementary information for unit definition.

Summarized in Table I are the relative values of the UV/VIS ratio, albedo, and the radar backscatter parameters used for definition of each unit. These units are mapped in Fig. 10.

B. *Description of Characterizing Techniques*

Two techniques have provided data with sufficient spatial resolution to be used in characterizing the specific units of Mare Crisium: telescopic reflectance spectra (.3 to 1.1 μm) for small lunar areas and recently processed orbital X-ray data over southern Crisium.

Telescopic reflectance spectra have been obtained for a number of small areas in Mare Crisium. The location and approximate size of these areas are shown in Fig. 2. The spectra are presented in Fig. 11 relative to a standard area in central Serenitatis (MS–2) whose reflectance spectrum is shown in Fig. 6. Each spectrum in Fig. 11 has been divided by this standard spectrum and is plotted as a relative reflectance spectrum. This mode of representation allows the small spectral differences between units to be quantitatively distinguished and allows comparisons with distinct basalt types discerned at other lunar areas (Pieters and McCord, 1976; Pieters, 1978).

Two aspects of the spectra in Fig. 11 can be used in unit characterization.

1. The value of the UV/VIS ratio (usually coupled with albedo). The range of measured values for a .40/.56 μm ratio from the relative spectra for three groups are (I) 1.025–1.06, (II) less than 1.00, and (III) 1.00–1.025.
2. The nature of the absorption band near 1 μm (this band can be seen in the MS-2 spectrum of Fig. 6). If the area observed has a weaker band than the MS-2 standard, the relative reflectance spectrum of Fig. 11 has a rise in the curve near .95 μm (e.g., spectrum I). If the area observed has a stronger band than the standard, the relative reflectance curve is low at 1 μm (e.g., spectrum 4).

An estimate of the strength of the 1 μm band can also be derived from the .95/.56 μm vidicon ratio images (Fig. 7c). In this mosaic, dark areas indicate a strong absorption at .95 μm relative to .56 μm indicative of abundant lithic fragments in the soil. Fresh craters, for example, have a strong 1 μm band and appear very dark. On the other hand, the strength of the 1 μm band for mature mare surfaces is controlled largely by the chemistry of the agglutinates. It can be seen that the entire eastern region of Mare Crisium, which contains the areas for spectra C and I, has a relatively weak absorption (appears as relatively bright in Fig. 7c) compared to the mare to the west. Both the spectra and the vidicon multispectral imagery were used to estimate the nature of the 1 μm absorption band for each of the Crisium units (indicated in Table I).

The spectra of Fig. 11 have put some important constraints on the units identified from the defining techniques. Group II units, with a low UV/VIS

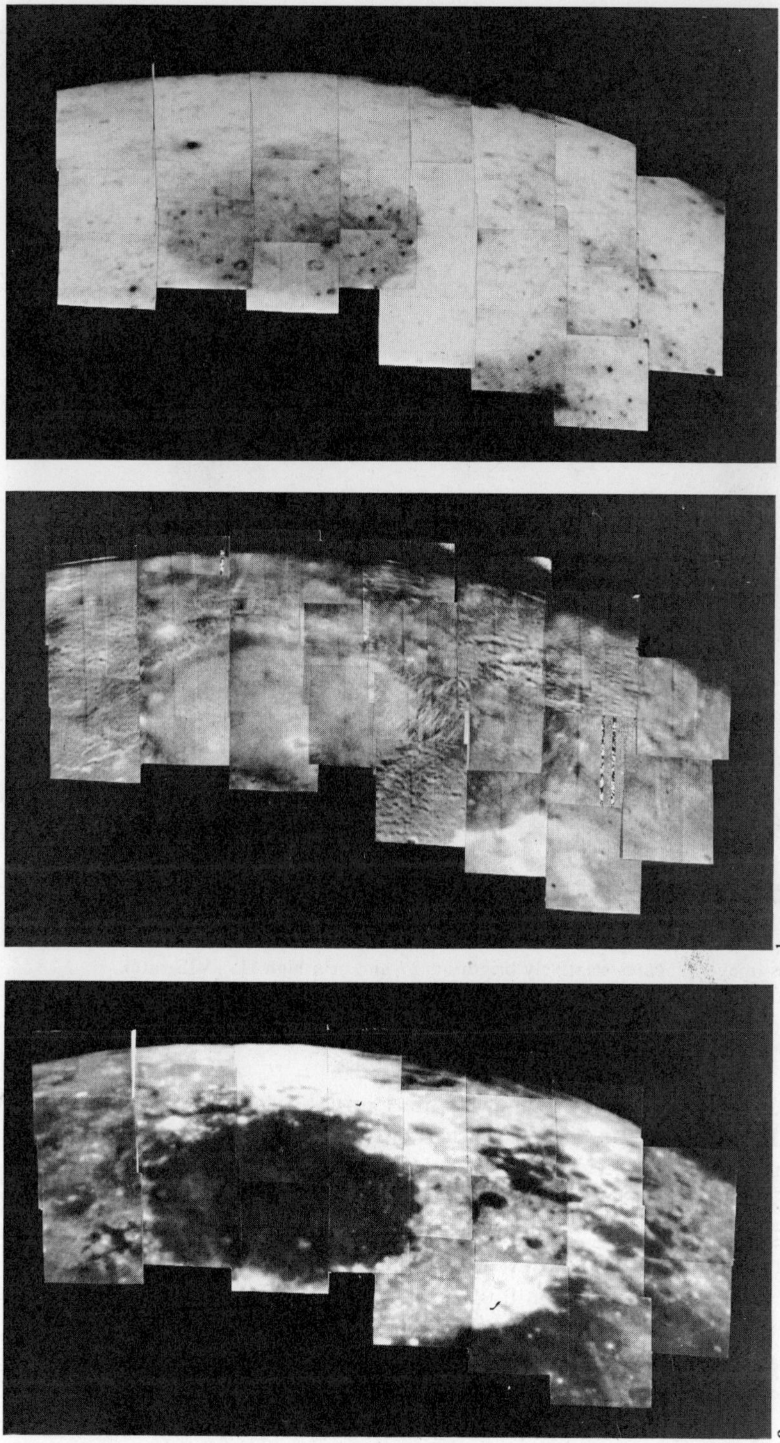

a b c

Fig. 7. Multispectral maps of the Mare Crisium region of the Moon. These maps are mosaics of digital images acquired using ground-based telescopes and a silicon vidicon imaging system (McCord *et al.*, 1976). Fig. 7a is an image made through a 0.56 μm filter; gray tones correspond to albedo. Fig. 7b is a gray tone representation of the areal distribution of the reflectance ratio 0.40 μm/0.56 μm and is made by dividing images made through filters centered at these two wavelengths. Bright indicates a high UV/VIS ratio. Fig. 7c is like Fig. 7b, but is for the reflectance ratio 0.95 μm/0.56 μm. Bright indicates a weak 1 μm band.

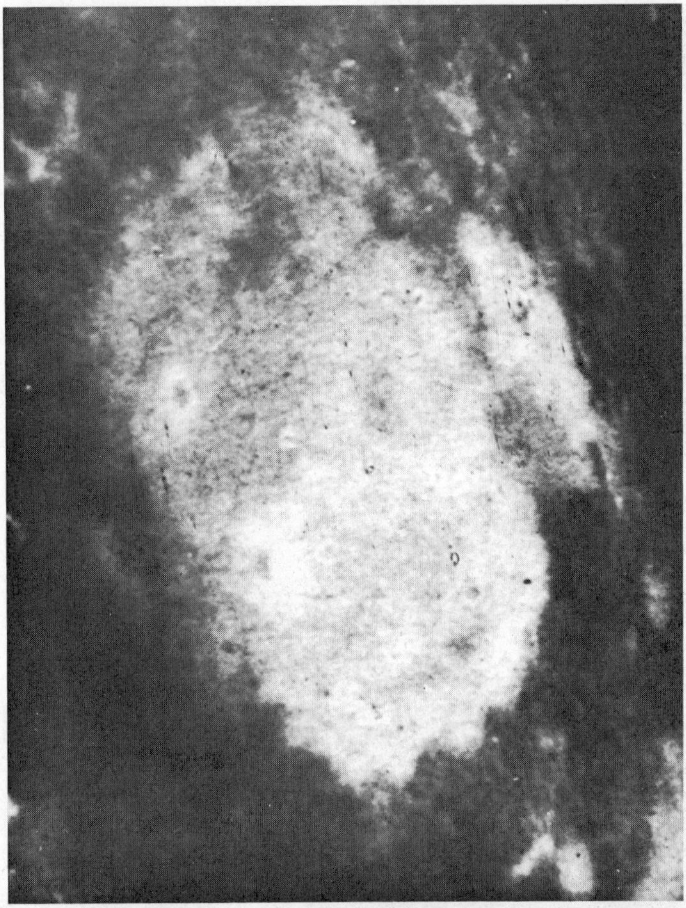

Fig. 8. Color difference photograph of Mare Crisium (.37–.61 μm) (Whitaker, 1976). Bright in this image indicates relatively bright at .37 μm, or a high UV/VIS ratio.

ratio, can be subdivided according to whether the 1 μm absorption band is strong (IIa) or weak (IIb). This distinction is critical in linking the remote measurements to the returned samples. Furthermore, the spectra allow the nature of the mottled area in southern Crisium to be understood in terms of these two unit subdivisions. Regions H, A, and B occur in this mottled central area and, with the spatial resolution of the spectra, apparently include more than one unit. The spectra for areas A and H have clear affinities to the spectrum for area 4 and are interpreted as being dominated by spectral properties of unit IIa. Similarly, the spectrum for area B seems to be dominated by material of unit III.

The orbital X-ray experiment on Apollo 15 provides some additional information for some of the units. In particular, Andre *et al.* (1977) have shown that there is a clear enrichment of MgO in the surface material in and around the crater Picard (and perhaps also Peirce).

C. *Definition of Mare Crisium Units*

The distinguishing characteristics of Mare Crisium units (Table I) are as follows: (Group I) a relatively dark unit with a relatively high ultraviolet reflectance and a weak 1 μm band; (Group IIa) a unit of intermediate albedo with a low ultraviolet reflectance and a strong 1 μm absorption; (Group IIb) a similar unit with weaker 1 μm absorption; and (Group III) a unit of intermediate albedo, intermediate ultraviolet reflectance, and average 1 μm absorption.

We consider the surface of *Group I* to be a mature soil, based on the low albedo, the characteristic spectral properties, and the low radar backscatter. The TiO_2 content as estimated from Fig. 5 is 3.5–5.5%. The data of Pieters and McCord (1976) show that Group I is spectrally similar to the Luna 16 landing site based on a classification of the overall shapes of relative telescopic spectra into "spectral types." The TiO_2 content of returned Luna 16 soil is 3.4%, only slightly less than our telescopic estimate for Group I.

The relatively high magnesium value for the area around Picard (Andre *et al.*, 1977; Hubbard and Vilas, 1977) indicates that Group I is possibly richer in Mg than other Crisium units. The low radar backscatter around Picard is consistent with a relatively mafic soil. Hubbard and Vilas (1977; pers. comm.) have presented evidence that an increase in Mg in the orbital XRF data also occurs for the area immediately east of the Luna 24 site, which may correspond to our Group I material (Auzout) in that area. We interpret Group I, therefore, as a mature soil developed on a relatively Mg-rich ferrobasalt having a TiO_2 content in the 3–5% range, and having a close similarity to the samples from Luna 16.

Group II basalts are subdivided into two distinctly different units based on the 1 μm characteristics. The combination of strong UV absorption and a deep 1 μm band of Group IIa indicates high FeO and very low TiO_2. Since this combination of characteristics is unique to the laboratory spectra of returned Luna 24 soils, we have designated the basalts of Group IIa as those most likely to be representative of the material sampled at Luna 24. The characteristics of group IIb basalts are similar to IIa except that Group IIb spectra contains a weaker 1 μm absorption. Although further spectral studies are needed to quantify this distinction, one possible interpretation is that Group IIb basalts contain less FeO. The general spectral characteristics of Group IIb have strong similarities to Apollo 15 soils.

An estimate of the TiO_2 content from Fig. 5 for Group II basalts would be 1.5–2.5%. The discrepancy between these values and the measured TiO_2 content of ~1% for Luna 24 soils (Blanchard *et al.*, 1977) has several possible sources. One possibility is that at low TiO_2 concentrations ($\lesssim$2%) the .40/.56 μm ratio cannot be used to measure to a percent TiO_2. This is the most likely explanation for soils with a UV/VIS ratio between .98 and 1.01 and is due to the interaction of iron and titanium in controlling spectral contrast as discussed in section II. Another

Table I. Properties of Geologic Units in Mare Crisium.

Group	Spectrum Code	Unit Name	Location	Defining and Characterizing Techniques: Albedo	UN-VIS ratio	1 μm band depth	Radar Back-scatter	Orbital Mg/Al	Interpretation
I	I	Alhazen	eastern edge	low	high	shallow	low	—	Soil derived from Fe- and Mg-rich titaniferous basalt similar to Luna 16 sample. TiO_2 content of Luna 16 soil is 3.4%. TiO_2 range estimated from telescopic data is 3.5–5.5%.
		Auzout	southeast ring	low	high	shallow	intermed	high?	
	E	Picard ejecta	around Picard	low	high	intermed	low	high	
		Peirce ejecta	around Peirce	low	high	—	low	high	
		Cleomedes F ejecta	around Cleomedes F	low	high	—	low	—	
		Greaves ejecta	around Greaves	low	high	—	intermed	—	
IIa	4	Swift	north, northwest part of basin	intermed	low	deep	intermed	—	Soil derived from very low titanium ferrobasalt similar to samples from Luna 24 core.
	A, H	Swift mottled	small patches in south part of basin	intermed	low	deep	intermed	intermed	
IIb	C	Agarum	east ring	intermed	low	shallow	high	—	Similar to IIa (possibly less FeO)
		Eimmart	northeast ring	intermed	low	shallow	high	—	

III	B, D	Shapley	east central	intermed	intermed	intermed	high and low	intermed	Soil derived from low titanium ferrobasalt similar to samples from Apollo 12. (Apollo 12 soils have 2.7% TiO_2. Telescopic data suggest 2–3% TiO_2 for Crisium III.)
		Picard crater floor	Picard	high	intermed	intermed	intermed	not resolved?	Mature to submature soil developed on basement crustal material. Interpretation is tentative. Reflectance spectra and/or improved resolution of orbital XRF data for Al/Si are needed.
		Peirce crater floor	Peirce	high	intermed	intermed	intermed	—	

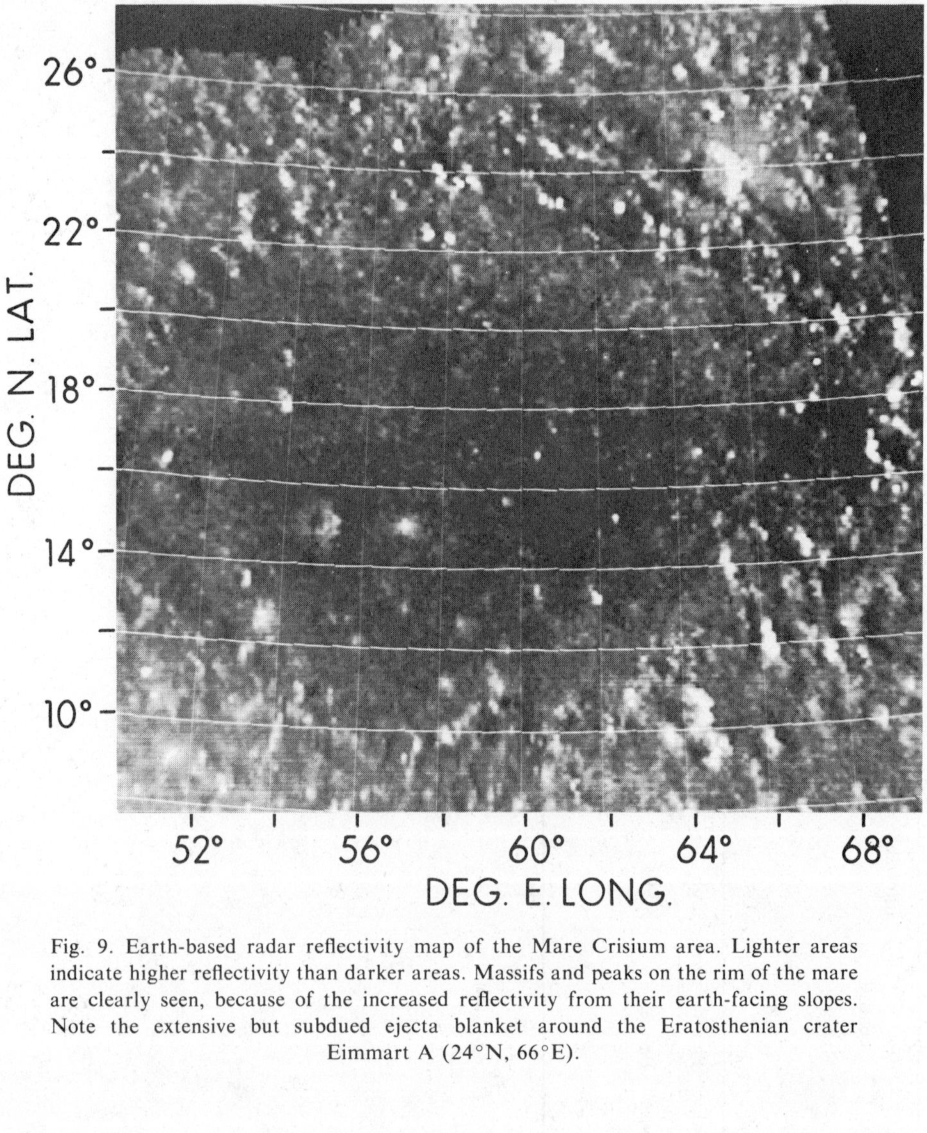

Fig. 9. Earth-based radar reflectivity map of the Mare Crisium area. Lighter areas indicate higher reflectivity than darker areas. Massifs and peaks on the rim of the mare are clearly seen, because of the increased reflectivity from their earth-facing slopes. Note the extensive but subdued ejecta blanket around the Eratosthenian crater Eimmart A (24°N, 66°E).

possibility is that unit II contains basalts that range in TiO_2 from less than 1% to about 2%, and that on a scale of kilometers these variations are averaged in the telescopic observations. In this second model the Luna 24 samples would represent the low TiO_2 end member. If the curve shown in Fig. 5 is accurate for low TiO_2 soils, a 1% TiO_2 intercept would call for a .40/.56 μm ratio of 0.96. Ratios this low and lower occur for spectra of several mare areas elsewhere on the moon (Pieters and McCord, 1976; Pieters, 1978) and are most likely due to very low titanium soils, but current data do not indicate such areas occur extensively in Mare Crisium.

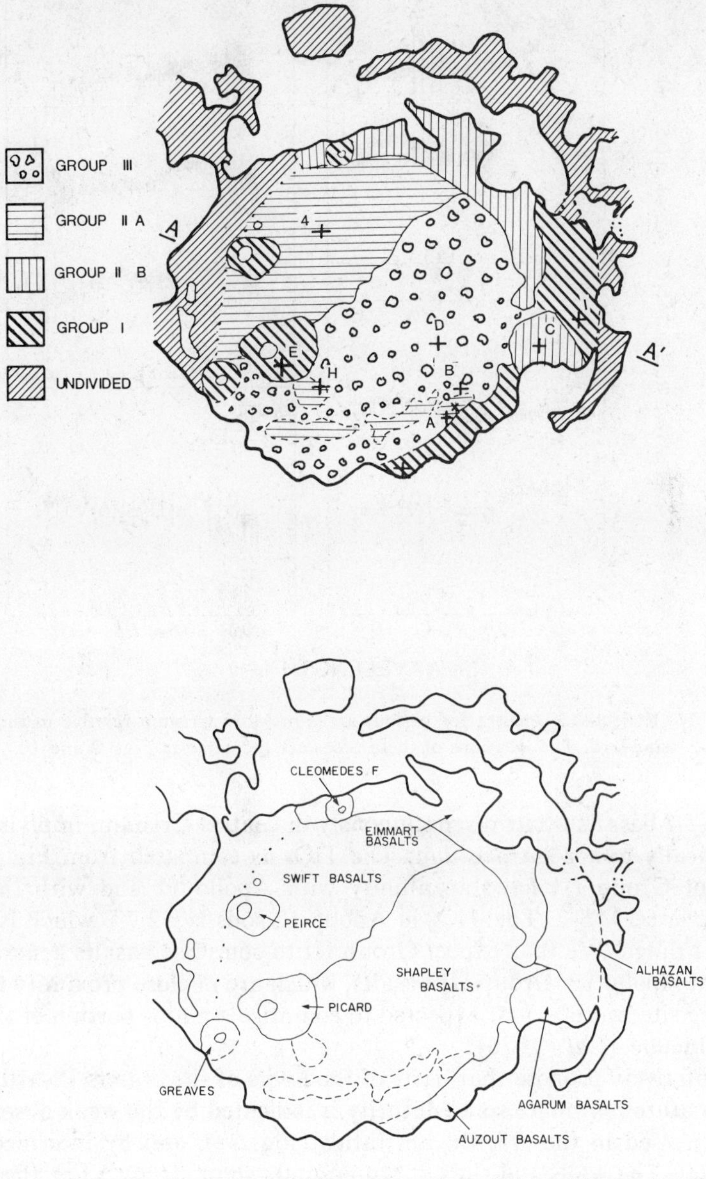

Fig. 10. (a) Geologic units of Mare Crisium. Luna 24 site indicated by small x. Small circle above site is Fahrenheit crater. Crosses indicate locations of telescopic spectra, each of which covers an area of 15 km diameter (see Fig. 2). Units are listed approximately in order of age, with oldest at bottom (Fig. 15 shows more detailed age relationships). (b) Informal unit nomenclature.

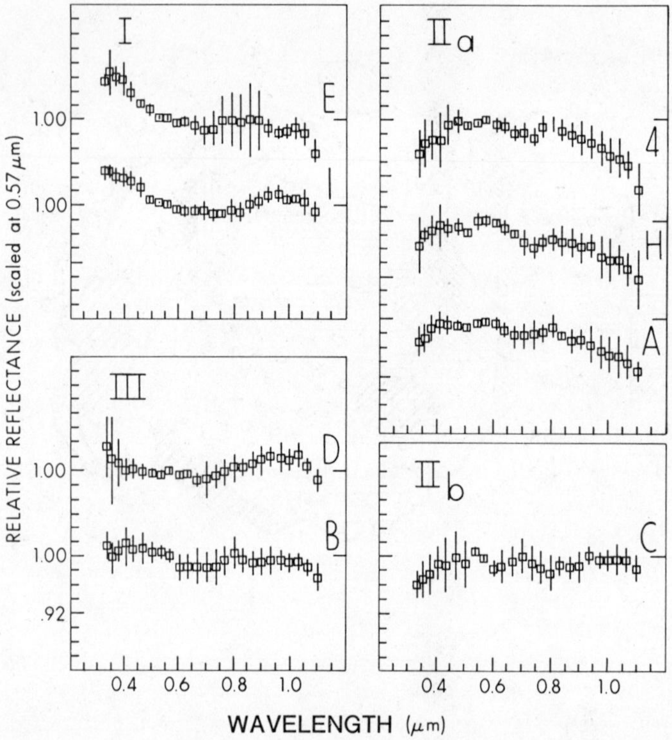

Fig. 11. Reflectance spectra for various areas in Mare Crisium relative to the MS-2 standard. The locations of these areas are indicated in Figs. 2 and 10.

Group III basalts occur discontinuously in central Crisium, implying a thin or volumetrically minor surface unit. The TiO_2 as estimated from Fig. 5 is 2–3%. Spectra of Group III have an affinity with Apollo 12 and with the standard telescopic area, MS-2. The TiO_2 in Apollo 12 soils is ~2.7% which is within the Group III range. We thus expect Group III to comprise basalts generally similar to those at Apollo 12. Group III basalts, which are in close proximity to the Luna 24 landing site, were at first expected to comprise a major portion of the returned sample (Pieters *et al.*, 1976).

We tentatively propose that parts of the floors of the craters Picard and Peirce contain mature non-mare soil. Maturity is indicated by the weak absorption near 1 μm expressed in the .95/.56 μm ratios (Fig. 7c), and by intermediate radar backscatter. The walls and rims of both craters show strong 1 μm absorption and high radar backscatter, as is expected from steep blocky slopes. If the floors of Picard and Peirce are covered with mature soil, they should have a low albedo similar to the surrounding mare regions. Instead, both craters show bright floors in orbital and telescopic photographs. We assign bright, mature soils to material with highland affinities; however, independent evidence from spectra or orbital XRF data are needed to test this interpretation.

IV. Present Topography as Related to Geologic Evolution

The present topography of the Crisium basin and its mare fill, as well as its relationship to the mare fill in nearby areas (e.g., Mare Anguis and Cleomedes crater, Fig. 2), yield clues to the detailed sequence of mare emplacement and surface subsidence. Accurate topographic information is available from three sources: 1) the Lunar Topographic Orthophotomaps (Kinsler, 1976), based on Apollo Mapping Camera data; 2) Earth-based Radar Topography, obtained by radar interferometer measurements for more than 60% of the lunar earthside hemisphere (Zisk, 1972a,b); and 3) Apollo Laser Altimeter (Kaula *et al.*, 1973) and Lunar Sounder (Brown *et al.*, 1974) data for those parts of the area along the appropriate Apollo ground tracks. Radar topography provides the only available Crisium area data north of 20° latitude, and since it agrees with the other data to better than 100 m rms (Zisk, 1978), the following analysis is based largely on the radar topography map of Fig. 12.

Examination of Fig. 12 shows the following major characteristics of Mare Crisium topography: 1) There is a distinctly annular appearance to the mare, with some exceptions as noted below. The mare surface is about 4 km below the basin rim, with an outer shelf about 50 km wide around most of the basin circumference. The shelf is about 400 m above the neighboring mare surface, and its inner edge is marked by a series of scarps and mare ridges (Figs. 13,14). 2) There is a central elevation about 400 m above the surrounding mare. 3) Between the shelf and the central elevation, there is an annular depression, relatively flat-bottomed, about 80 to 120 km in width. The central elevation rises smoothly from the inner edge of this depression. However, several exceptions to the annular geometry are noted. 4) The southwest part of the basin rim is distinctly linear and the shelf is absent along this linear edge. 5) The central elevation actually slopes away from a central, 150 km long north-south ridge which is at about the elevation of the shelf. The region of highest elevation, at the north end of the ridge, is about 40 km in diameter and, at its highest point, is about 300 m higher than the rest of the ridge and displaced 60 km northwest of the center of the mare. Another lesser ridge extends from this point about 80 km toward the west-southwest. 6) There is an extension of the mare beyond the annular shelf to the east.

Even more interesting in terms of the mare filling process is the elevation of the floor of Mare Anguis, part of a concentric ring structure northeast of the rim of Crisium (Fig. 2); Anguis is about 800 m higher than Crisium (Fig. 12). Another mare area, on the same concentric ring to the northwest of Crisium (about 23°N, 51°E; Fig. 12), is at the same elevation as Anguis within the accuracy of the map. The floor of Cleomedes crater, located between these two mare patches and also partly mare-filled, is at a lower topographic level typical of Crisium. For both Mare Anguis and the western mare-patch there are apparently channels leading down to the floor of Crisium (Wilhelms, 1973), at (21°N, 67.2°E) and (20.2°N, 69.2°E) for Anguis and at (22°N, 52°E) for the western patch. Although it is not certain that these channels did in fact conduct lava during episodes of mare filling, there appears in both Figs. 12 and 13 a strong

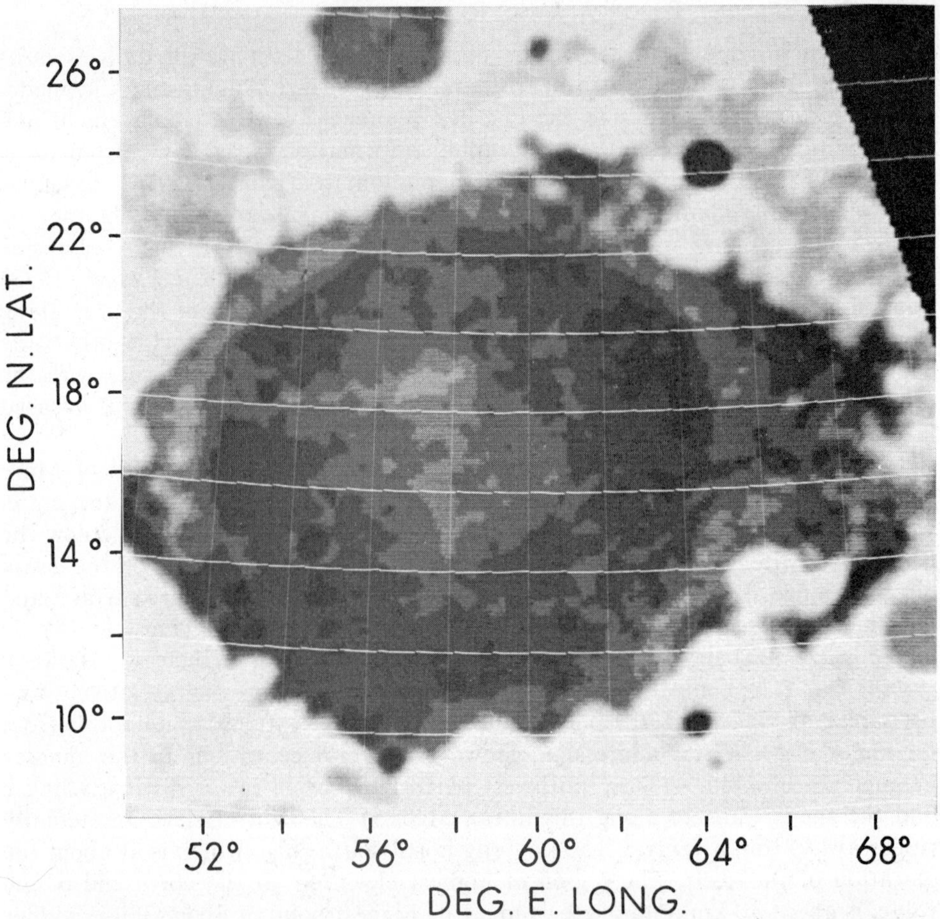

Fig. 12. Topography of Mare Crisium. Contour intervals are 200 m; white is higher. The elevation of Mare Anguis (at 25°N, 66°E) is 800 m higher than Mare Crisium (at 16°N, 62°E). The map is based on earth-based radar interferometer data (Zisk, 1972a).

suggestion of a flow channel extending from the western patch across the Crisium shelf, at (20.8°N, 52.6°E).

V. STRATIGRAPHY AND GEOLOGIC HISTORY

A. *Previous work*

Previous stratigraphic studies of the Crisium region concentrated on distinguishing and subdividing broadly defined highland and mare units. Wilhelms and McCauley (1971) mapped the mare material within Crisium as a single unit of

Fig. 13. Topographic relief in the western half of Mare Crisium. Low sun-angle photograph showing the shelf and circumferential mare ridge and scarp structure, and the central ridge. Note the possible flow channel crossing the mare shelf from the northwest rim, perhaps an outlet from the higher-elevation mare area in the northwest concentric ring. From the Consolidated Lunar Atlas, plate C3 (USAF, 1968).

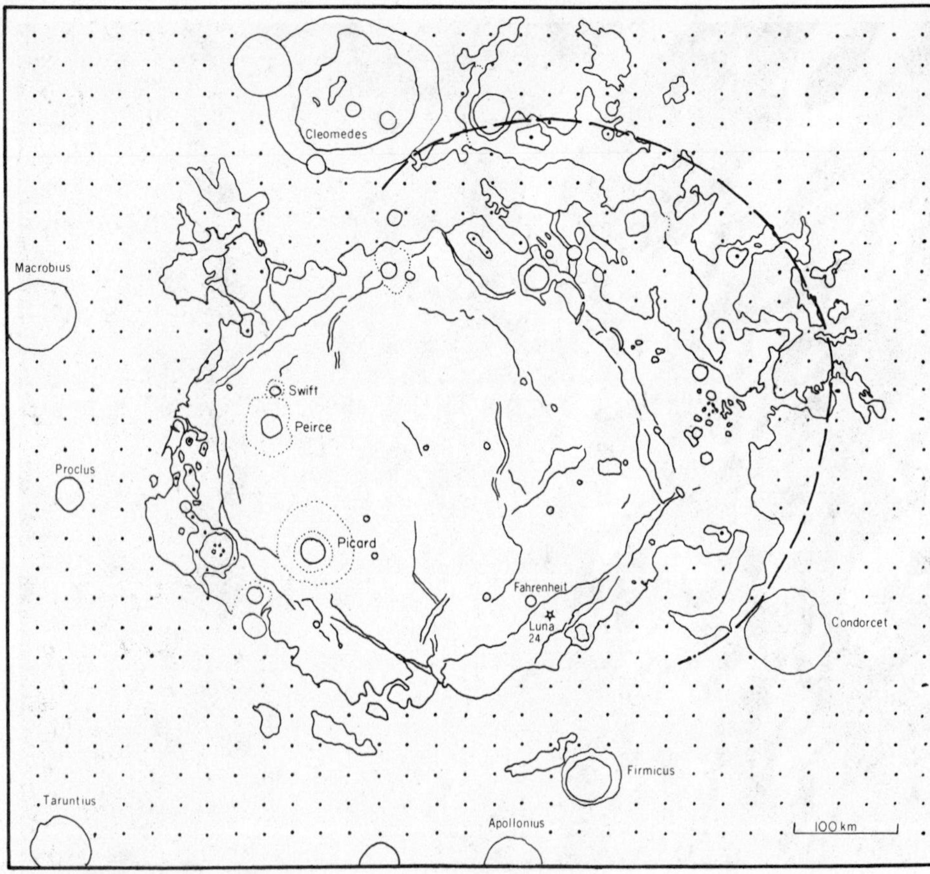

Fig. 14. Major mare ridge and scarp systems in Mare Crisium. Peripheral mare areas are also indicated. Distinct dark line indicates possible rim of pre-Crisium basin.

Eratosthenian age and the mare deposits of the broadly concentric ring (Spumens, Undarum, Anguis) as Imbrian. Casella and Binder (1972), mapping the northern half of Crisium, designated the central mare as a single unit of Imbrian age, but mapped the area along the eastern margin as Eratosthenian-Imbrian and subdivided the concentric mare into these two units. The two mare units were distinguished on the basis of albedo (darker equals younger) and estimates of crater density. Dark mantling material was also mapped in several locations along the basin margins and as units superposed on numerous Eratosthenian-aged mare craters such as Peirce, Swift and Cleomedes F. Olson and Wilhelms (1974), mapping in the southern Crisium region, distinguished an older Imbrian-aged mare unit along the eastern edge of the basin north of Promintorium Agarum, and a younger Imbrian unit in the central and southern basin. Both of these units occurred in the concentric mare (Undarum, Spumens) to the south. No Eratosthenian mare was mapped near Crisium but two patches of dark mantle were outlined just east of the Luna 24 landing site. Wilhelms (1973) utilized Apollo 17 metric photography to update the geologic map of northern Crisium. On the basis of data on morphology and size-frequency distribution of superposed craters reported, but not included, Wilhelms concluded that there was only one major unit in northern Crisium and associated concentric maria and that it was of Imbrian age. This convention was followed by Wilhelms and El Baz (1977) who mapped all maria in the Crisium region as a single unit of middle to late Imbrian age.

Attempts to date the Crisium mare as a whole have suggested that it is young, relative to other maria (Hartmann and Wood, 1971; Offield and Pohn, 1970). Additonal information on ages of various areas in Crisium has been obtained through crater size-frequency distribution studies (Neukum *et al.*, 1975; Boyce *et al.*, 1977; Boyce and Johnson, 1977). These studies show evidence for a wide range of ages for Crisium basalts, ranging from as old as 3.75 billion years to perhaps as young as 2.5 billion years (Boyce and Johnson, 1977).

B. *Stratigraphic Settings of Basin Units*

The three major groupings of Crisium basalts described in the previous sections have distinctive geologic characteristics. In general, each group represents units emplaced in a different, but perhaps overlapping time period. Groups can be divided into informal subunits on the basis of physical differences and geographic discontinuities. Such subdivisions are useful in documenting stratigraphic correlations and geologic history.

Group I basalts

Basalt near Auzout—This unit occurs in a narrow strip along the southeastern edge of Crisium (Fig. 10) and forms a bench bordered on one side by the basin rim and the other by a series of mare ridges just southeast of the Luna 24 landing site (Fig. 14). The bench, which is one of the highest mare topographic features in the area, stands several hundred meters above basinward mare and tilts slightly back toward the basin rim (Fig. 12). Two patches of dark mantling material have been mapped within this unit just southeast of the Luna 24 landing site (Olson and Wilhelms, 1974). An albedo and color boundary can be seen along the northeastern boundary of this unit in Apollo images (e.g., AS17-22787). This unit is assigned an early age on the basis of the onlap of Group III (Shapley) basalts and its preservation and present exposure due to its topographic elevation. Similar relationships are seen for units along basin margins in southwest Serenitatis (Howard *et al.*, 1973) where portions of early basalt deposits did not downwarp as much as adjacent rocks, were left as topographic benches along the basin margin, and were partly embayed by later flows.

Basalts exposed in ejecta deposits of Greaves, Picard, Peirce and Cleomedes F—These Eratosthenian-aged craters have penetrated the basalt of Crisium in different areas (Figs. 2,10) and to different depths. Each of these craters exposes a distinctive unit in its ejecta deposit which, on the basis of its characteristics (Table I), is placed in Group I. Since these basalts, as presently exposed, are parts of ejecta units emplaced on Eratosthenian units, their age of emplacement as flows is not known. However, they overlie several other units (Group II, Eimmart, Swift basalts; and Group III, Shapley basalts; see Fig. 11). Since they are exposed adjacent to and surrounding the rim crest of each crater, these occurrences indicate that the unit excavated by the craters was emplaced early in the history of mare filling, perhaps contemporaneous with the similar Auzout (Group I) basalt. If the correlation is correct, these occurrences indicate that this early stage of filling extended at least throughout much of the eastern part of the basin, including the presently topographically high annulus along the northwest and west (Fig. 12).

The Group I *Alhazan basalts* are exposed in a low-lying area in the eastern part of the basin between the annulus ridge and the basin margin (Figs. 10, 12). In Lunar Orbiter and Apollo images, this unit appears less cratered than adjacent maria, suggesting a younger age. In addition, Boyce and Johnson (1977) map a region of relatively sparsely cratered terrain in this area. However, similar relationships in other areas of low albedo units are often misleading (Howard *et al.*, 1973; Lucchita and Sanchez, 1975) such that the low albedo unit may actually be older than apparent from visual observation of crater densities. Examples are known in the Serenitatis basin (Howard *et al.*, 1973) where units on the outer edge of the topographic annulus are actually older than the units exposed in the annulus (e.g., the Apollo 17 region). Superposition and onlap relationships cannot be determined conclusively for these Group I basalts. The Group I Alhazen basalts are thus correlated with the Auzout basalts and those basalts exposed in several Eratosthenian crater deposits in terms of characteristics outlined in Table I, but their exact age relationships are uncertain (Fig. 15).

Group II basalts

An annulus of distinctive basalts coincides in its distribution with the topographic annulus extending from Promontorium Agarum (Fig. 2) around the northeast part of the basin to Cleomedes F. The Goup IIb *Agarum basalts* are exposed in a broad patch just north of the promontory. The western and northern boundaries in part coincide with a mare ridge system (Fig. 14). Group III (Shapley) basalts appear to embay this area (Fig. 10) and separate the Agarum unit from the Eimmart basalts, a similar unit to the north. The Group IIb *Eimmart basalts* extend to the north around the mare margin to near Cleomedes F. Ray materials associated with Proclus obscure the characteristics of maria along the northwest margin (Fig. 10) so the affinity of basalts in this area is uncertain. The inner boundary of the Eimmart basalts is marked by a distinctive topographic drop toward the basin of several hundred meters (Fig. 12). This boundary coincides with a similar mare ridge system in the northwest which is less prominent in the north (Fig. 14). Cleomedes F apparently excavates Group I

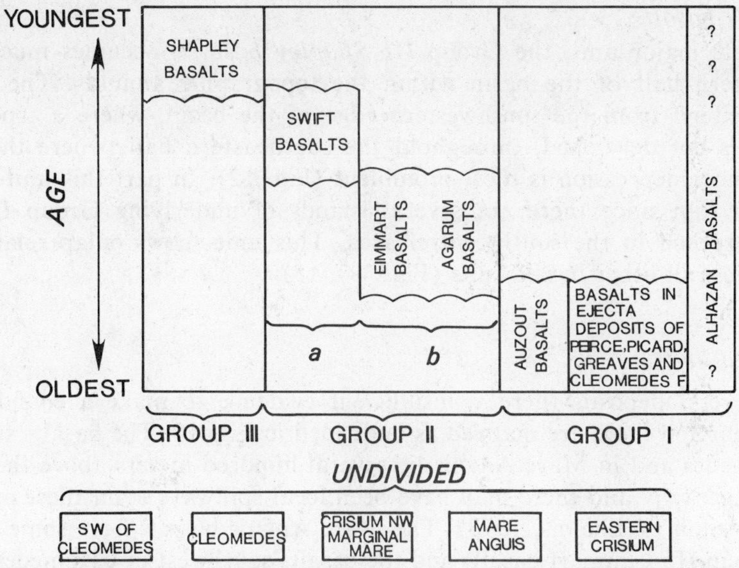

Fig. 15. Stratigraphic correlation chart of Mare Crisium.

basalts from beneath this IIb unit. Group IIa (Swift) and Group III (Shapley) basalts appear to onlap the Eimmart basalts and embay them, suggesting an intermediate age (Fig. 15).

The Group IIa *Swift basalts* which are distinct from the annulus IIb basalts in IR spectral properties, occur in central Crisium in two regions. The largest exposure occupies the northwest part of the central basin area. In general, this region of the basin is topographically lower than the marginal annulus but relatively higher than the southeast half of the basin interior (Fig. 12). From the spectral data discussed previously, units believed correlative with the Swift basalt (Table I) also occur as patches in south-central Crisium, including the Luna 24 landing site (Fig. 10). The presence and shape of these irregular patches suggest that the Group IIa Swift basalts extend over essentially the whole mare inside the annulus, but have been subsequently covered to varying degrees by the Group III Shapley basalts in the southeast. Onlap and topographic relationships suggest that the Group IIa Swift basalts postdate the Group IIb Eimmart basalts (Fig. 15). Crater counts in an area centered on the Luna 24 landing site (corresponding approximately to one of the patches of Group IIa Swift basalt), give an age of 3.5 ± 0.1 BY (Boyce *et al.*, 1977).

Fahrenheit crater, apparently lying within Group III basalt near the contact with Group IIa (about 18 km from the site, Fig. 10), may have delivered an important contribution to the Luna 24 site (Barsukov *et al.*, 1977; Butler and Morrison, 1977). Settle *et al.* (1978) have presented evidence that any Fahrenheit ejecta contribution to the Luna 24 site is likely to be derived from the same unit on which the site is located.

Group III basalts

A single major unit, the Group III *Shapley basalts*, occupies much of the southeastern half of the basin within the topographic annulus. The Shapley basalts extend from the southwestern edge of the basin, where a topographic annulus is not developed, throughout the southeastern half, where the donut-shaped inner depression is most prominent (Fig. 12). In part this unit must be relatively thin since there are several islands of underlying Group IIa Swift basalts exposed in the south-central area. This unit shows onlap relationships with almost all other basalt units (Figs. 10, 15).

Undivided units

For several deposits there is insufficient evidence to make a confident unit assignment and these are mapped as undivided in Fig. 10. The basalts southwest of Cleomedes and in Mare Anguis lie several hundred meters above the present mare topography and there may have been local spillways from these areas into Mare Crisium (Wilhelms, 1973). The Mare Anguis basalts have some affinities with Group IIb Eimmart basalts and the basalts southwest of Cleomedes may be related to Group I basalts. Additional information is required before an accurate assignment can be made.

C. *Structure and basin evolution*

Present topography (Fig. 12) and the outcrop pattern of the various geologic units demonstrate that the Crisium basin underwent topographic evolution (predominantly relative subsidence of the interior) during and after the emplacement of basalts (Head *et al.*, 1977; Zisk *et al.*, 1978). Many basalt surfaces have slopes of several degrees and the donut-shaped depression within Crisium (Fig. 12) is several hundreds of meters below both the marginal shelf and the basin center. Mare ridges are dominantly compressional features (Howard and Muehlberger, 1973) concentrated near areas of maximum topographic change. The major characteristics and stages of evolution of Crisium are strikingly parallel to the sequence of events documented in the Serenitatis basin (Howard *et al.*, 1973; Soloman and Head, 1978), although some variations are evident. In Serenitatis deformation was in part responsible for the patterns of emplacement of later lavas. Earliest basalt deposits of Serenitatis occur around the southern rim; subsequent to their emplacement they were downwarped and faulted by a series of basin-concentric graben. A second stage of filling is represented by basalts presently exposed as part of an annulus along the eastern and southwestern parts of Serenitatis. These units have been subsequently downwarped toward the basin center and the central part of Serenitatis was filled with a third stage of lavas. Mare ridges are developed extensively in this unit and appear to indicate continued downwarping well past emplacement of the youngest lavas in Serenitatis (Howard *et al.*, 1973; Soloman and Head, 1978).

There is no evidence for concentric graben development in Mare Crisium although downwarping and downfaulting of the basin obviously occurred throughout mare emplacement. Group I basalts were apparently emplaced over the majority of the basin. Their preservation in part as a topographic shelf (Auzout basalts) in the southeast part of the basin indicates that some downwarping of this unit occurred. Group II basalts were emplaced subsequently in two stages. The Agarum (IIb) and Eimmart (IIb) basalts, which today occupy part of the outer shelf and topographic annulus (Figs. 10, 12) were presumably emplaced over a larger area. Preferential subsidence of the central part of the basin occurred although there may have been little actual movement of the annulus. The coincidence of the unit boundaries and mare ridge systems strongly suggests that subsidence was less of a broad flexure and more concentrated as vertical movement with emphasis on downdropping of the central mare side. Subsequent flooding (IIa Swift and III Shapley basalts) was concentrated in the inner basin primarily within the major concentric mare ridge system. Central subsidence had continued to such a degree (several hundred meters) that this stage of flooding did not cover the earlier units along most of the basin edge (Fig. 10). The central mare ridges and variable topography of the youngest units provide evidence that subsidence continued past their emplacement as it did in Serenitatis.

The asymmetrical development of the annulus of topography and unit boundaries (generally absent in the southwest) appears to be correlated with a major linear segment of the basin rim (Fig. 12). This linear segment reflects a regional lineament that may have facilitated downward movement of the mare portion of the basin edge, causing younger units to flood up to the base of the highlands.

The Apollo Lunar Sounder Experiment provides important information on the present configuration of units in the subsurface and the relative amounts of vertical movement accompanying mare unit emplacement. May *et al.* (1976) present evidence for subsurface layering in Mare Crisium; one distinctive reflector was detected at an average depth of 1400 m (assuming an 8.7 dielectric constant). Reflectors were also found on the shelf regions along the eastern and western margins of the basin at shallower depths. On the basis of our geologic reconstruction we interpret the reflector to be approximately the boundary between basalt Groups I and II (Fig. 16). The evidence for this is as follows: 1) Group I basalts occur at depth within the basin as evidenced by their presence in numerous crater ejecta deposits; 2) all such craters have *present* depths below the mare surface of greater than about 1400 m (Table II). Excavation depths were greater than the present observed depths (Dence, 1968; Settle and Head, 1978), reaching well below 1400 m depth; 3) the reflector is discontinuous in the vicinity of Peirce (Fig. 16); 4) the reflector is closer to the surface in an area coincident with the eastern annulus of Agarum basalt (Group II), consistent with the relative downdropping of the Agarum basalt toward the basin center as deduced from geologic relationships; 5) the reflector disappears along the eastern basin edge where it encounters a surface exposure of Group I basalts (Alhazan basalts). This interpretation agrees with more recent analyses of sounder data by Phillips *et al.* (1978).

Table II. Depths of Craters below Surrounding Mare Surface (determined from Lunar Topographic Orthophotomaps). See Fig. 2 for crater locations.

Crater Name	Diameter (km)	Present Depth Below Mare Surface (m)
Cleomedes K	12	1370
Fahrenheit	6.2	1269
Greaves	13.5 ave.	1805
Peirce	19	1490
Picard	22.5	1788
Swift	10.2	1633

The reconstruction presented in Figs. 10, 15, and 16 and the present topography (Fig. 12) strongly suggest that the most significant subsidence of the mare occurred basinward of the major mare ridge system. The position of this ridge system coincides approximately with the inner or first basin ring which may have been an active fault throughout and subsequent to mare emplacement as previously suggested by Butler and Morrison (1977).

VI. Correlation with Other Maria

A global classification of basalt types based on spectral reflectance and albedo

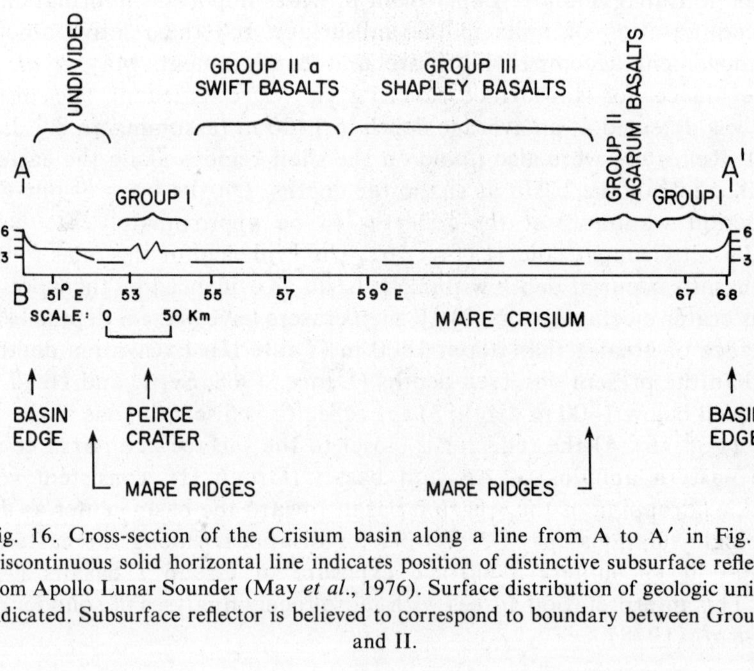

Fig. 16. Cross-section of the Crisium basin along a line from A to A′ in Fig. 10. Discontinuous solid horizontal line indicates position of distinctive subsurface reflector from Apollo Lunar Sounder (May *et al.*, 1976). Surface distribution of geologic units is indicated. Subsurface reflector is believed to correspond to boundary between Groups I and II.

data (Pieters and McCord, 1976; Pieters, 1978) indicates that each Mare Crisium basalt type can also be found elsewhere in the nearside maria. Basalt regions similar to the medium-high titanium basalts of Group I are found almost entirely in the eastern maria: NE Fecunditatis (Luna 16), south central Tranquillitatis, the northeastern and western edge of Serenitatis and most of Mare Vaporum. Basalts similar to the subgroups of the low titanium Group II basalts are more difficult to distinguish from each other. From the currently available data, regions similar to Group IIa (most likely sampled by Luna 24) occur in western Mare Frigoris and northern Sinus Roris. Regions of Group IIb material occur predominantly around basin margins of Mare Serenitatis and Imbrium. Basalts with characteristics similar to the low titanium Group III occur extensively in most of the major mare areas of the lunar frontside.

The Luna 24 very low titanium samples have contributed substantially to our knowledge of lunar basalt types and have provided a key piece of ground truth information for global remote sensing studies. From the currently available data, it is apparent that there remain major additional unsampled groups of basalt types, particularly in the western maria.

Acknowledgments—Of great aid in carrying on this research was the unpublished multispectral map provided by E. Whitaker. In addition, we thank Lucy McFadden for acquiring some of the telescopic spectra, Donna MacLaskey and Duncan Chesley for helping to prepare the multispectral maps and Russell Ralph for assisting with the laboratory reflectance measurements. James Whitford-Stark provided helpful discussion of the manuscript in preparation. Zeny Cocson, Lianne Uno, Nancy Christy and Laurie Raymond aided in manuscript preparation. This research was supported by NASA Grants NSG 7312, NSG 7048, NGL-22-009-790, NSG 7197, NSG 7173 and NGR-40-002-116.

REFERENCES

Adams J. B., Head J., McCord T. B., Pieters C. and Zisk S. H. (1978) Mare Crisium: Regional stratigraphy and geologic history. Submitted to *Geophys. Res. Lett.*

Adams J. B. and McCord T. B. (1972) Electronic spectra of pyroxenes and interpretation of telescopic spectral reflectivity curves of the moon. *Proc. Lunar Science Conference 3rd*, p. 3021–3034.

Adams J. B. and Ralph R. L. (1977) Spectral reflectance of Luna 24 soils (abstract). In *Papers Presented to the Conference on Luna 24*, p. 1–4. The Lunar Science Institute, Houston.

Adams J. B. and Ralph R. L. (1978) Diffuse reflectance spectra of Luna 24 soils. In *Mare Crisium: The View from Luna 24* (R. B. Merrill and J. J. Papike, eds.). This volume.

Andre C. G., Wolfe R. W. and Adler I. (1977) Evidence for a high-magnesium subsurface basalt in Mare Crisium from orbital x-ray fluorescence data (abstract). In *Papers Presented to the Conference on Luna 24*, p. 8–11. The Lunar Science Institute, Houston.

Barsukov V. L. (1977) Preliminary data for the regolith core brought to earth by the automatic lunar station Luna 24. *Proc. Lunar Sci. Conf. 8th*, p. 3303–3318.

Barsukov V. L., Tarasov L. S., Dmitriev L. V., Kolesov G. M., Shevaleevsky I. D. and Garanin A. V. (1977) The geochemical and petrochemical features of regolith and rocks from Mare Crisium (preliminary data). *Proc. Lunar Sci. Conf. 8th*, p. 3319–3332.

Basu A., McKay D. S., Clanton U., Waits G. and Fruland R. M. (1977) Glass particles in Luna 24 drill core soils (abstract). In *Papers Presented to the Conference on Luna 24*, p. 14–17. The Lunar Science Institute, Houston.

Bence A. E. and Grove T. L. (1977) The highland component in the Luna 24 core (abstract). In *Papers Presented to the Conference on Luna 24*, p. 22–24. The Lunar Science Institute, Houston.

Blanchard D. P., Haskin L. A., Brannon J. C. and Aaboe E. (1977) Chemistry of soils and particles from Luna 24 (abstract). In *Papers Presented to the Conference On Luna 24*, p. 37–40. The Lunar Science Institute, Houston.

Bogard D. D. and Hirsch W. C. (1977) Noble gases in Luna 24 core soils (abstract). In *Papers Presented to the Conference on Luna 24*, p. 44–47. The Lunar Science Institute, Houston.

Boyce J. M. and Johnson D. A. (1977) Ages of flow units in Mare Crisium based on crater density. *Proc. Lunar Sci. Conf.* 8th, p. 3495–3502.

Boyce J. M., Schaker G. and Dial A. (1977) Age of Luna 24 mare basalts based on crater studies. *Nature* **265**, 38–39.

Brown W. E., Jr., Adams G. F., Eggleton R. E., Jackson P., Jordan R., Kobrick M., Peeples W. J., Phillips R. J., Porcello L. J., Schaber G., Sill W. R., Thompson T. W., Ward S. H. and Zelenka J. S. (1974) Elevation profiles of the moon. *Proc. Lunar Sci. Conf. 5th*, p. 3037–3048.

Butler P. and Morrison D. A. (1977) Geology of the Luna 24 landing site. *Proc. Lunar Sci. Conf. 8th*, p. 3281–3301.

Casella C. J. and Binder A. B. (1972) Geologic map of the Cleomedes quadrangle of the Moon. U.S. Geol. Survey Map I-707.

Charette M. P. and Adams A. B. (1975) Mare basalts: Characterization of compositional parameters by spectral reflectance (abstract). In *Papers Presented to the Conference on Origins of Mare Basalts and Their Implications For Lunar Evolution*, p. 25–28. Lunar Science Institute, Houston.

Charette M. P., McCord T. B., Pieters C. and Adams J. B. (1974) Application of remote spectral reflectance measurements to lunar geology classification and determination of titanium content of lunar soils. *J. Geophys. Res.* **79**, 1605–1613.

Dence M. R. (1968) Shock zoning at Canadian craters: Petrography and structural implications. In *Shock Metamorphism of Natural Materials*, (B. M. French and N. M. Short eds.), p. 169–184. The Mono Book Corp., Baltimore.

Evsyukov N. N. and Barabashov N. P. (1973a) Photometric map of The Visible Hemisphere of the Moon. Academy of Science, U.S.S.R.

Evsyukov N. N. and Barabashov N. P. (1973b) Albedo map of the Visible Hemisphere of the Moon. Academy of Science, U.S.S.R.

Florensky C. P., Basilevsky A. T., Ivanov A. V., Pronin A. A. and Rode O. D. (1977) Luna 24: Geologic setting of landing site and characteristics of sample core (preliminary data). *Proc. Lunar Sci. Conf. 8th*, p. 3257–3279.

Hartmann W. K. and Wood C. A. (1971) Moon: Origin and evolution of multi-ringed basins. *The Moon* **3**, 3–78.

Head J. W., Pieters C., McCord T. B., Adams J. B. and Zisk S. H. (1978) Definition and detailed characterization of lunar surface units using remote observations. *Icarus,* **33**, 145–172.

Head J. W., Zisk S. H., Adams J. B., McCord T. B. and Pieters C. (1977) Geologic setting and topography of the Crisium basin and the Luna 24 landing region (abstract). In *Papers Presented to the Conference on Luna 24*, p. 78–81. The Lunar Science Institute, Houston.

Howard K., Carr M. and Muehlberger W. (1973) Basalt stratigraphy of southern Mare Serenitatis. NASA SP-330, p. 29-1 to 29-25.

Howard K. and Muehlberger W. (1973) Lunar thrust faults in the Taurus-Littrow region. NASA SP-330, p. 31-22 to 31-25.

Hu H.-N. and Taylor L. A. (1977) Luna 24 soils from Mare Crisium: Agglutinate chemistry (abstract). In *Papers Presented to the Conference on Luna 24*, p. 82–84. The Lunar Science Institute, Houston.

Hubbard N. J. and Vilas F. (1977) From Serenity to Langemak: A regional chemical setting for Crisium (abstract). In *Papers Presented to the Conference on Luna 24*, p. 85–88. The Lunar Science Institute, Houston.

Kaula W. M., Schubert G. and Lingenfelter R. E. (1973) Lunar topography from Apollo 15 and 16 laser altimetry. *Proc. Lunar Sci. Conf. 4th*, p. 2811–2819.

Kinsler D. C. (1976) New lunar cartographic products. *Proc. Lunar Sci. Conf. 7th*, p. i–x.

Lucchitta B. and Sanchez A. (1975) Crater studies in the Apollo 17 region. *Proc. Lunar Sci. Conf. 6th*, p. 2427–2441.

May T., Peeples W., Maxwell T., Sill W., Ward S., Phillips R., Yordan R. and Abbott E. (1976) Subsurface layering in Maria Serenitatis and Crisium: Apollo lunar sounder results. *Proc. Lunar Sci. Conf. 7th*, p. 540–542.

McCord T. B., MacLaskey D., Feierberg M. and Pieters C. (1978) More multispectral maps of the lunar frontside. In preparation.

McCord T. B., Pieters C. and Feierberg M. A. (1976) Multispectral mapping of the lunar surface using groundbased telescopes. *Icarus* **29**, 1–34.

McKay D. S., Basu A., Waits G., Clanton U., Fuhrman R. and Fruland R. (1977) Grain size and evolution of Luna 24 soils (abstract). In *Papers Presented to the Conference on Luna 24*, p. 114–117. The Lunar Science Institute, Houston.

Morris R. V. (1977) FMR and magnetic properties of Luna 24 soils and 1 mm soil particles (abstract). In *Papers Presented to the Conference on Luna 24*, p. 121-124. The Lunar Science Institute, Houston.

Neukum G., Konig B. and Arkani-Hamed J. (1975) A study of lunar impact crater size-distribution. *The Moon*, **12**, 201–229.

Offield T. and Pohn H. (1970) Lunar crater morphology and relative age determination of lunar geologic units. Part 2: Applications. *U.S.G.S. Prof. Paper 700 C*, p. C163–C169.

Olson A. B. and Wilhelms D. E. (1974) Geologic map of the Mare Undarum quadrangle of the Moon. *U.S.G.S. Map I-837*.

Phillips R. J. and Maxwell T. A. (1978) Lunar sounder revisited: Stratigraphic correlations and structural inferences (abstract). In *Lunar and Planetary Science IX*, p. 890–891. The Lunar and Planetary Institute, Houston.

Pieters C. (1978) Distribution of basalt types on the frontside of the moon. *Proc. Lunar and Planet. Sci. Conf. 9th*, in press.

Pieters C. and McCord T. B. (1976) Characterization of lunar mare basalt types: I. A remote sensing study using reflection spectroscopy of surface soils. *Proc. Lunar Sci. Conf. 7th*, p. 2677–2690.

Pieters C., McCord T. B. and Adams J. B. (1976) Regional basalt types in the Luna 24 landing area as derived from remote observation. *Geophys. Res. Lett.*, **3**, 697–700.

Pohn H. and Wildey R. L. (1970) A photoelectric-photographic study of the normal albedo of the Moon. *U.S.G.S. Prof. Paper 599-E*.

Settle M., Cintala M. and Head J. (1978) Mode of emplacement of Fahrenheit ejecta at the Luna 24 site. *Proc. Lunar and Planet. Sci. Conf. 9th*, in press.

Settle M. and Head J. (1978) Rim slumping modification of lunar craters: Implications for the depth of excavation of basin-sized impacts. *Proc. Lunar and Planet. Sci. Conf. 9th*, in press.

Simonds C. H., Warner J. L., McGee P. E., Cochran A. and Brown R. W. (1977) Luna 24: Petrochemical potpourri (abstract). In *Papers Presented to the Conference on Luna 24*, p. 170–173. The Lunar Science Institute, Houston.

Solomon S. and Head J. W. (1978) Vertical movement in mare emplacement, basin tectonics, and lunar thermal history. *Proc. Lunar and Planet. Sci. Conf. 9th*, in press.

Taylor G. S., Warner R. D., Wentworth S. and Keil K. (1977) The Luna 24 regolith: Lithologic abundances in the 250–500 μm size fraction, and compositions of agglutinates and nonmare lithic fragments and glasses (abstract). In *Papers Presented to the Conference on Luna 24*, p. 186–188. The Lunar Science Institute, Houston.

Tarasov L. S., Nazarov M. A., Shevaleevsky I. D., Kudryashova A. F., Gaverdovskaya A. S. and Korina M. I. (1977) Mineralogy and petrography of Luna rocks from Mare Crisium (preliminary data). *Proc. Lunar Sci. Conf. 8th*, p. 3333–3356.

Whitaker E. A. (1972) Lunar color boundaries and their relationship to topographic features. *The Moon* **4**, 348–355.

Whitaker E. A. (1976) Color-difference photograph of the Moon. Unpublished manuscript.

Wilhelms D. E. (1973) Geologic map of the northern Crisium region. *Apollo 17 Preliminary Sci. Report, NASA SP-330*, p. 29-29–p. 29-35.

Wilhelms D. E. and El Baz F. (1977) Geologic map of the east side of the Moon. U.S. Geol. Survey Map I-948.

Wilhelms D. E. and McCauley J. (1971) Geologic map of the near side of the moon. U.S. Geol. Survey Map I-703.

Zisk S. H. (1972a) A new, earth-based radar technique for the measurement of lunar topography. *The Moon* **4**, 296–306.

Zisk S. H. (1972b) Lunar topography: First radar-interferometer measurements of the Alphonsus-Ptolemaeus-Arzachel region. *Science* **178**, 977–980.

Zisk S. H. (1978) Mare Crisium area topography: A comparison of earth-based radar and Apollo mapping camera results. In *Mare Crisium: The View from Luna 24* (R. B. Merrill and J. J. Papike, eds.). This volume.

Zisk S. H., Adams J. B., Head J., McCord T. B. and Pieters C. (1978) Topographic evolution of Mare Crisium and its relation to mare filling. *Proc. Lunar and Planet. Sci. Conf. 9th*. In press.

Zisk S. H., Hodges C. A., Moore H. J., Shorthill R. W., Thompson T. W., Whitaker E. A. and Wilhelms D. E. (1977) The Aristarchus-Harbinger region of the Moon: Surface geology and history from recent remote sensing observations. *The Moon* **17**, 59–99.

Merrill, R. B. and Papike, J. J., editors.
(1978) *Mare Crisium: The View from Luna 24*, Pergamon Press (New York), p. 75–80
Printed in the United States of America

Mare Crisium area topography: A comparison of Earth-based radar and Apollo mapping camera results

STANLEY ZISK

Haystack Observatory, Westford, Massachusetts 01886

Abstract—An earth-based radar topography (ERT) map has been constructed of the Mare Crisium area. Systematic and random sources of error are discussed. A comparison between the ERT map and Lunar Topographic Orthophotomaps shows a random mean discrepancy of less than 100 m between the two maps, except for small-scale (20 km or less in diameter) features, where systematic smoothing reduces the ERT elevation contrast.

INTRODUCTION

In a companion paper (Head *et al.*, 1978) there is a new map of the topography of the floor of Mare Crisium and vicinity based on the earth-based radar topography (ERT) measurements of Zisk (1972a). The complex nature of the interaction between mare filling and subsidence (or elevation) episodes makes the detailed topography of the present mare surface an important source of information in developing the stratigraphic history of the mare and mare basalt flows in Mare Crisium.

It is evident that an estimation of the accuracy of the ERT map will be of interest in evaluating the conclusions of Head *et al.* (1978). However, no detailed estimate of ERT accuracy has been published until now. Thus, this paper will present a detailed comparison between the ERT map for Mare Crisium and the Lunar Topographic Orthophotomap series, which is based on photogrammetry of Apollo Mapping Camera photographs. A more complete error analysis of the radar interferometry technique would be beyond the scope and interest of this volume, but will appear elsewhere.

The topographic map

Figure 1 is a contour map which presents in considerably more detail the ERT data used in Head *et al.* (1978). Because of time limitations during the preparation of this paper, the reference sphere used for Fig. 1 has a radius of 1731.60 km, compared to 1730.000 km for the LTO map series. Hence, a constant 1600 m. must be added to the elevations of Fig. 1 for comparison with LTO elevations.

Discrepancies between the two maps can arise from several factors, not all of which represent possibly erroneous data.

1) The scale of Fig. 1 is large enough that small-size features are smoothed because of the fixed number of grid points in the map-processing computer

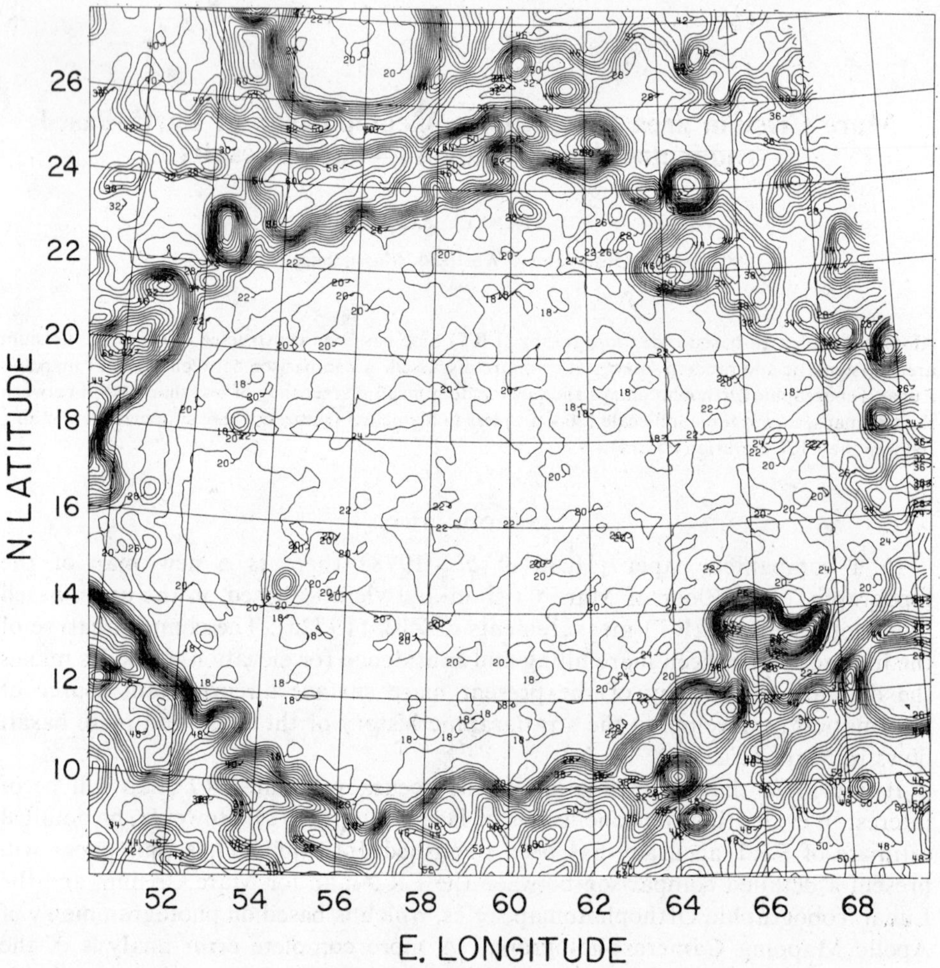

Fig. 1a. Contour map of Mare Crisium Topography based on earth-based radar interferometer measurements. Contour interval is 200 meters. Contours are in 100's of meters, referred to a datum sphere of 1731.6 km radius (e.g. 18 = 1800 meter above datum = 1733.4 km).

program. Craters or peaks smaller than about 5 to 8 km in diameter would not appear or would be greatly reduced in relief. This effect would not be as pronounced if a different smaller-scale map were prepared from the same data.

2) Narrow peaks and craters are reduced in amplitude by the ERT observing process. A surface area of about 1 × 2 km is measured as a single datum point by the radar interferometer. Within that cell, the measured elevation is weighted by the radar reflectivity of different units within the cell. Hence, the elevation of flat or gently undulating terrain would be

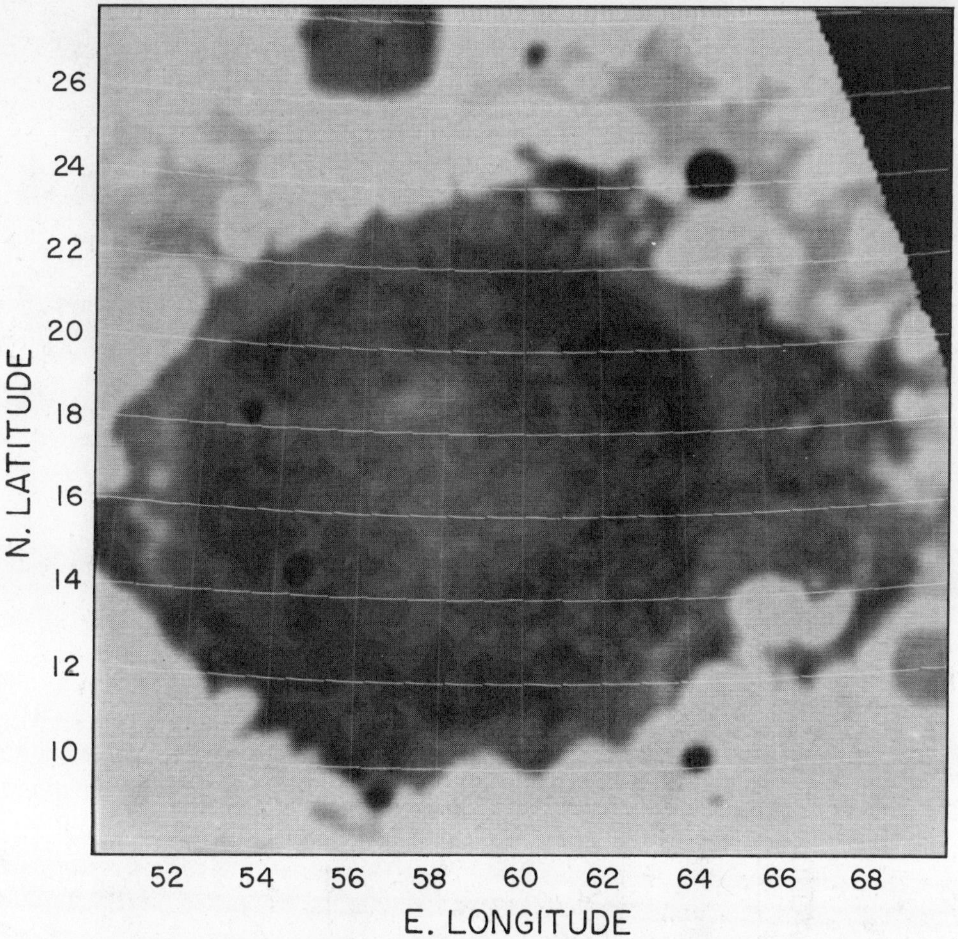

Fig. 1b. Alternate presentation of the map of Fig. 1a, showing more clearly the annular shelf and central ridge system. Contour interval = 100 m.

measured quite accurately whereas an area containing steep slopes within the 1 × 2 km cell boundaries would measure preferentially the (average) elevation of earth-facing slopes. The result is that sharp peaks or sharp-bottomed craters are often measured at a lesser displacement from the surrounding lunar surface than the true value.

3) The latitude/longitude grid is occasionally misplaced by as much as 0.5 deg, the result of errors in the lunar ephemeris used in the radar measurements. A number of small features can always be recognized in the co-aligned backscatter-type map which is a by-product of the topography measurement. An accurate grid can be applied, if necessary, by relocating radar map features to features in a standard lunar map.

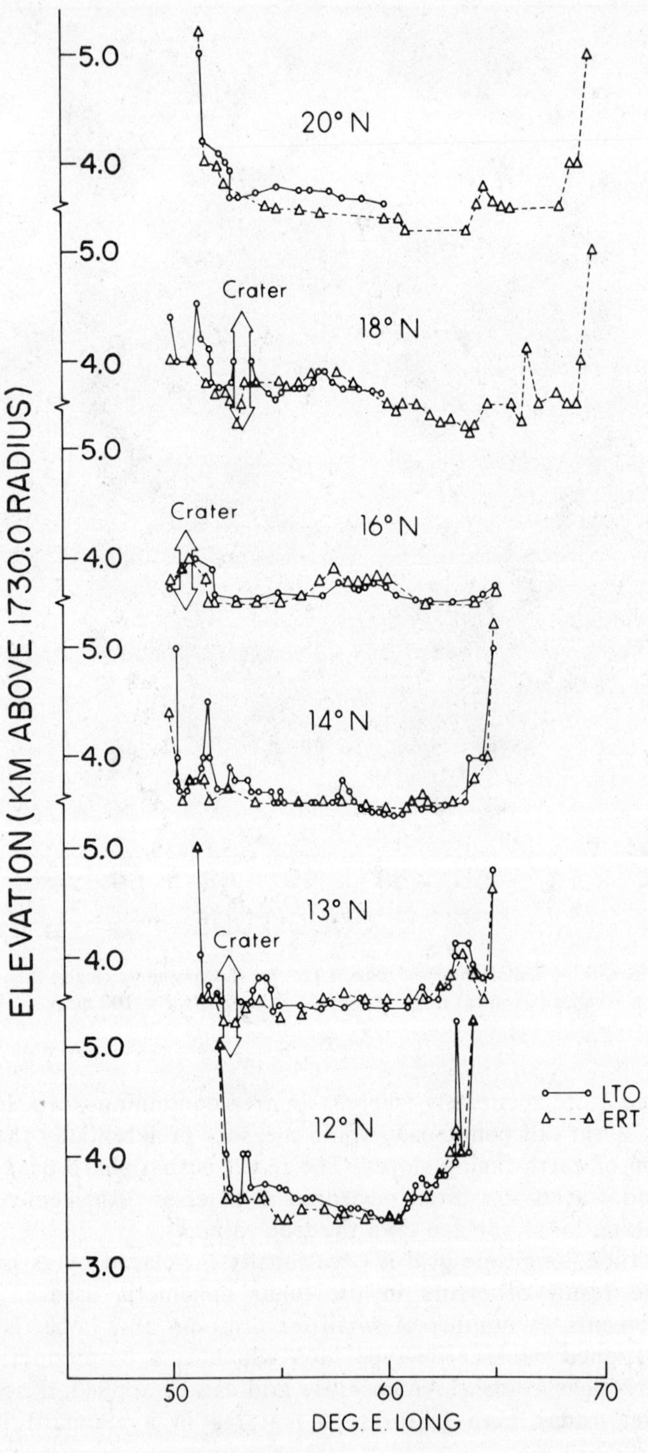
ELEVATION (KM ABOVE 1730.0 RADIUS)
5.0
4.0
20° N
5.0
Crater
18° N
4.0
5.0
Crater
16° N
4.0
5.0
14° N
4.0
5.0
13° N
Crater
4.0
5.0
LTO
ERT
12° N
4.0
3.0
50
60
70
DEG. E. LONG

Fig. 2. Elevation profiles of ERT and LTO maps along lines of constant latitude. Areas marked "crater" show too much high-relief topography on the LTO map to be presented at this scale. The data of the entire ERT map has been adjusted upward by 1600 m. to facilitate this comparison. Where the LTO profiles are incomplete, no data was available because of the limited Apollo ground-track coverage. Source: LTO maps 62A (1–4), 62B (1–4), and 44D (3,4) (NASA/DMA, 1976).

However, in the current investigations, features are usually compared directly with other data, e.g. multispectral vidicon images, without reference to a grid.

4.) A shallow slope may remain in the ERT maps artificially as a result of the same ephemeris error. The largest such slope noted to date was 0.1 deg, although all other maps appear to have much less than 0.3 times this value when compared with other data.

5) The datum sphere for the ERT topography is arbitrary. In the map of Fig. 1, the datum level was chosen for convenience of data presentation, before an accurate datum level was obtained from the LTO maps. The inability to measure the radar datum independently is largely a result of the ephemeris errors. The generation of accurate ephemerides is still being considered, but the local topography data is sufficiently useful in its own right that current efforts have concentrated on utilizing this local topography only.

The interpretation of the topographic data in Fig. 1 is discussed in detail in Head *et al.* (1978) and will not be repeated here. Briefly, the annular nature of the topography is evident, including the 400 m high mare shelf and two rings of highlands (the second only partly shown in the figure) at about 6 and 4.8 km elevation.

There are additional patches of mare fill to the northeast and northwest, beyond the first ring of highlands. These are essentially parallel to the surface of Mare Crisium, but are at an elevation of 800 m above Crisium. An erosion channel about 100–200 m deep appears on the Crisium shelf at about 20.6N/52.5E, where material from the northwestern patch flowed down a short 20% grade out onto Crisium. The external mare fill was therefore at least simultaneous to, if not later than, the Crisium shelf—possibly coincident with the later flows in the annular depression in Crisium (Groups II-A and III, Head *et al*, 1978).

Craters of 20 to 30 km in size are clearly seen, for example, Peirce at 18.3 deg N/53 deg E, and Picard at 14.5 deg N/54.5 deg E. The ERT depth of Peirce is 600 m and of Picard 1 km, compared with the LTO depths of 1300 m and 1600 m. Note that the artificial smoothing (i.e. reduction in relief; item 2, above) for Peirce, which is only 16 km in diameter, is 55% compared to only 35% for Picard which is 23 km in diameter. These are typical values for features of this size and slope. A flat-bottomed crater with at least 20 km diameter floor will exhibit little or no smoothing. Smaller craters can be measured accurately from the unsmoothed radar data, at a cost of considerably more low-scale randomness (i.e. "noise") on large, flat surfaces.

Comparison of ERT and mapping camera topography

Figure 2 is a series of profiles of the ERT and LTO maps along lines of constant latitude, from 12 deg to 20 deg N. A fixed 1600 m datum shift was added to all ERT values and the result plotted in pairs with the corresponding LTO profile.

At first glance, the two maps appear to be in excellent agreement. The mean discrepancy across the floor of Mare Crisium is less than 100 m rms, whereas in this region the published mean accuracy for the LTO maps varies from 50 to 80 m relative to Apollo 15 datum. However, several small features in the LTO maps are either heavily subdued or absent on the ERT maps. Typical instances appear on the 14 deg N profile, at 52 deg and 53 deg, for example.

Another feature of LTO and ERT profiles is a 200 m discrepancy at 20 deg N. ERT appears 200 m lower than LTO, and suggests a slope in the ERT data. However, it should be noted that the 18 deg profiles do not show such a discrepancy, and that the ERT map at 20 deg N actually indicates a rise in elevation in the center of the mare (Fig. 1a). It is possible that an unknown error has appeared in the ERT data along this one profile. Alternately, since the LTO map in this region is based on the extreme edge of the Apollo 15 Mapping Camera band, it may itself be in error; or the discrepancy may be a combination from both data sets.

CONCLUSIONS

The profiles of Fig. 2 suggest strongly that the ERT data is a reliable source of lunar topography, comparable in accuracy to photogrammetry of Mapping Camera photos over moderate scales of 600 km or less but lacking in fine detail for scales below 20 km. Since such detail is often quite accurately provided by shadow-measuring techniques in areas of no Apollo coverage, it is not likely to prove a drawback to the use of the radar topography data.

Acknowledgments—I wish to express my appreciation for the many discussions and the continuing and fruitful cooperation with my colleagues J. Adams, J. Head, T. McCord and C. Pieters. Research was supported by NASA Grant NSG-7197. Haystack Observatory is operated by MIT for the Northeast Radio Observatory Corporation (NEROC) under NSF Grant AST 75-14701.

REFERENCES

Head J. W., Adams J. B., McCord T. B., Pieters C. and Zisk S. H. (1978) Regional stratigraphy and geologic history of Mare Crisium. In *Mare Crisium: The View From Luna 24* (R. B. Merrill and J. J. Papike, eds.). This volume.

Lunar Topographic Orthophotomaps, NASA (1973-76), Defense Mapping Agency.

Zisk S. H. (1972a) A New Earth-based Radar Technique for the Measurement of Lunar Topography. *The Moon* **4**, 296–306.

Zisk S. H. (1972b) Lunar Topography: First Radar-Interferometer Measurements of the Alphonsus-Ptolemaeus-Arzachel Region. *Science* **178**, 977–980.

Merrill, R. B. and Papike, J. J., editors.
(1978) *Mare Crisium: The View from Luna 24*, Pergamon Press (New York), p. 81–87
Printed in the United States of America

Diffuse reflectance spectra of Luna 24 soils

JOHN B. ADAMS and RUSSELL L. RALPH

Department of Geological Sciences, University of Washington, Seattle, Washington 98195

Abstract—Visible and near-infrared (0.35 μm to 2.5 μm) diffuse reflectance spectra were measured of five samples from different depths in the Luna 24 core. Absorption bands in the spectra of the near-surface, relatively mature soils are deeper than previously measured for Apollo or other Luna samples. The unusual depths of the UV-visible band and the 1 μm band are an expression of the combination of high FeO and low TiO_2 in the agglutinates and thus in the bulk soil. The Luna 24 soil, therefore, has a characteristic spectral signature that can be utilized for identifying and mapping similar soils by remote spectral reflectance methods.

INTRODUCTION

We report here the results of laboratory spectral reflectance analyses of five samples of soils from the core returned by the Soviet unmanned spacecraft Luna 24 (Barsukov, 1977). The purpose of our study is to provide surface calibration of telescopic spectral reflectance and spectral imaging measurements in Mare Crisium. Laboratory spectra measure areas of only about 1 cm^2, whereas telescopic spectra cover areas 10–20 km in diameter. Comparisons of data that differ so widely in scale were successfully made early in the Apollo program (e.g., Adams and McCord, 1972) and rely on the sampled soil being representative of the surface on the kilometer scale.

In separate papers (Adams *et al.*, 1978, and Head *et al.*, 1978) the laboratory spectra of Luna 24 samples are shown to correspond closely to a telescopic spectrum of the Luna 24 landing site and to the spectral properties of a major geologic unit in the Crisium basin. Spectral mapping of soils on basalt units in Mare Crisium is then used along with other remote sensing data to interpret the stratigraphic and geologic history of the mare basin. The laboratory spectra thus are an important link in the chain of evidence connecting the wealth of detailed chemical, petrologic and other analyses on the samples to the regional geology.

EXPERIMENTAL PROCEDURE

Diffuse reflectance spectra were obtained of five samples of Luna 24 soil: 24077,27; 24109,21; 24149,23; 24174,18 and 24182,27. The samples had been sieved to <90 μm prior to receipt from the NASA curatorial facility. Samples ranged in weight from 29 to 44 mg.

Samples were spread on an aluminum substrate over an area sufficient to fill the spectrophotometer beam, and covered with an optical quality silica coverglass. Samples were mounted on the integrating sphere of a Beckman DK-2A ratio-recording spectroreflectometer and measured with reference to MgO (and to Halon G-80) that had been covered with a matching coverglass. The MgO and the Halon each were calibrated using Pd and Au standards for monitoring instrumental drift and long term changes in MgO or Halon.

DISCUSSION

Spectra of the Luna 24 samples are shown in Fig. 1. Curves for Apollo 11 and Apollo 12 surface soils are shown in the same figure for comparison. All of the Luna 24 soils show the same general spectral characteristics of other soils from Apollo and Luna sites, namely a steeply sloping continuum and subdued absorption bands. Lunar rocks and breccias by comparison have flatter continuum slopes and deeper absorption bands.

The origin of the steep continuum in the lunar soils is not precisely known, however it is correlated with the agglutinate fraction. Charge-transfer absorptions in the ultraviolet and visible portions of the spectrum that involve iron and titanium may contribute to the continuum slope; however, other factors such as scattering by submicroscopic inclusions of metallic iron in agglutinates (Hapke *et al.*, 1975) may be the main cause.

Absorption bands are superposed on the continuum, and can be fitted by Gaussian functions that define the wavelength, band depth and half-width. Overlapping bands can be separated by sequential fitting and subtraction of bands. Thus, for example, we can characterize the spectrum (normalized to 100% at 560 nm) of sample 24077 by a linear continuum having a slope of 0.049%/nm and absorption bands with the following parameters:

band center	band half-width	band depth
280 nm	320 nm	16.2%
675	100	3.9
872	80	5.4
1040	300	20.8
1325	200	6.2
1813	110	3.9
2096	450	11.6
2321	140	3.1

where band depth is measured as a percent of I_{560}.

In general, similar absorption bands occur in each of the Luna 24 samples, although their depths vary depending on the abundances of the absorbing mineral and glass phases, differences in particle size, and other factors. Broad absorption in the ultraviolet-visible (UV-VIS) end of the spectrum is the most difficult to characterize as it probably is a combination of scattering, charge transitions in Fe^{2+} and in Ti^{3+}, and several charge transfers, such as oxygen → metal, $Fe^{2+} \rightarrow Ti^{4+}$ and $Ti^{3+} \rightarrow Ti^{4+}$.

Charge-transfers involving iron and titanium appear to be the most important cause of the UV-VIS optical absorption (Loeffler *et al.*, 1974, 1975) in mature (agglutinate-rich) mare soils. These absorptions are so intense that they can extend to longer wavelengths thereby masking other absorption bands (lowering

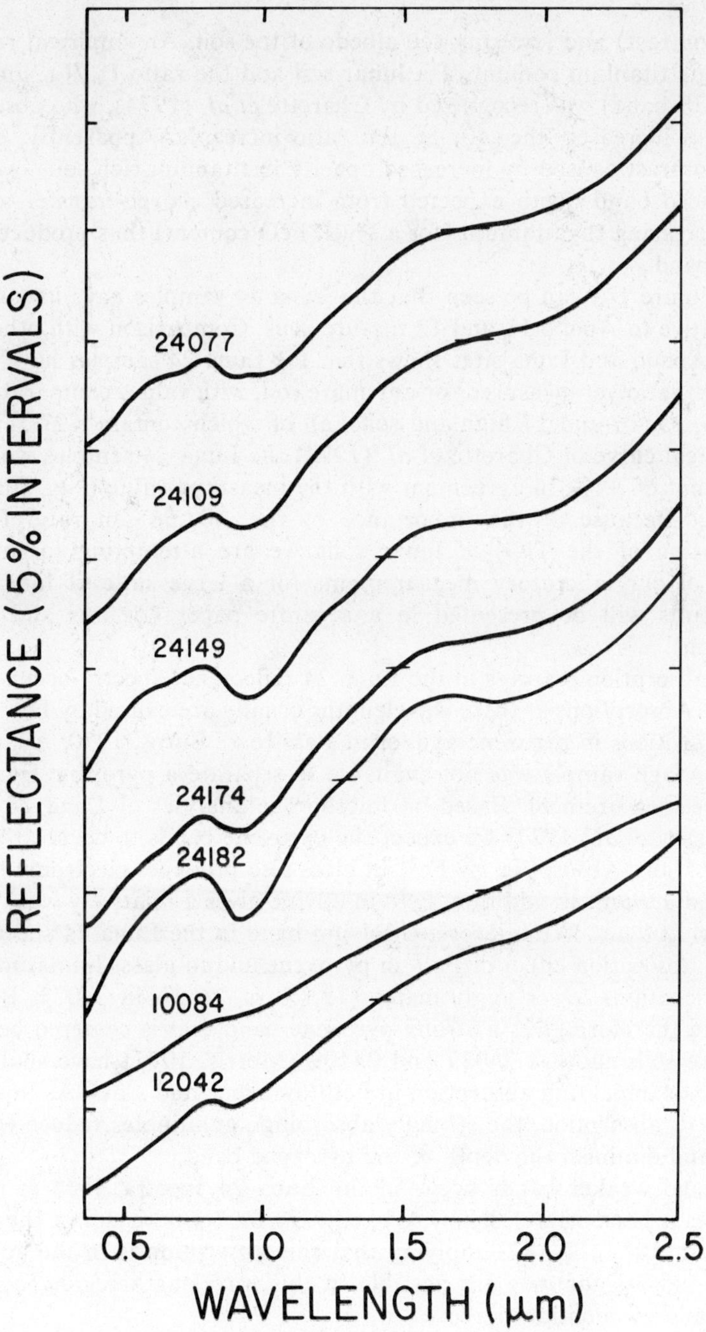

Fig. 1. Diffuse reflectance spectra of Luna 24 soils and representative Apollo soils relative to MgO in the wavelength interval 0.35 μm to 2.50 μm measured in percent. Successive spectra are offset 5% vertically for clarity. Note the relatively deep absorptions near 1.0 μm and 2.0 μm on the Luna 24 spectra. Horizontal tic marks indicate 20% reflectance.

spectral contrast) and lowering the albedo of the soil. An empirical relationship between the titanium content of a lunar soil and the ratio $I_{.40}/I_{.56}$ μm (slope of the UV-VIS band) was recognized by Charette *et al.* (1974), who observed that, as titanium increases, the .40/.56 μm ratio increases. Apparently, the loss of spectral contrast caused by increased opacity in titanium-rich soils overbalances the enhanced band depth expected from increased charge-transfer absorptions alone. Decreasing the titanium (for a given FeO content) thus produces a deeper UV-VIS band.

From Figure 1 it can be seen that the Luna 24 samples have low .40/.56 μm ratios relative to Apollo 11 and 12 mature soils. Comparison with other samples from the Apollo and Luna sites shows that the Luna 24 samples have the lowest .40/.56 μm ratio yet measured for any mare soil, with values comparable to some Apollo 14, 15, 16 and 17 highland soils, all of which contain <2% TiO_2. Using the empirical curve of Charette *et al.* (1974) the Luna 24 samples would have a TiO_2 content of ~1%, in agreement with the measured value (e.g., Blanchard *et al.*, 1977). Because of the importance of the .40/.56 μm ratio for remote determination of the TiO_2 of lunar soils we are attempting to improve the accuracy of our laboratory measurements for a large suite of lunar samples. These results will be presented in a separate paper (Adams and Ralph, in preparation).

Major absorption features in the Luna 24 reflectance spectra occur near 1 μm and 2 μm. Absorptions at these wavelengths usually are caused by Fe^{2+} electronic charge transitions in pyroxene and/or in glass (e.g., Burns, 1970, Adams, 1975). A large enough sample was not available to separate a pyroxene fraction for a reflectance measurement. Based on microprobe analyses of Luna 24 pyroxenes (e.g., Tarasov *et al.*, 1977) we expect the pyroxene bands to occur at about 0.95 μm and 2.1 μm. Absorption by Fe^{2+} in glass also produces electronic bands near 1.0 μm and 2.0 μm. In addition, Fe^{2+} in olivine gives a relatively weak electronic band near 1.05 μm. We interpret the 1 μm band in the Luna 24 spectra to be a combined absorption effect of Fe^{2+} in pyroxene and in glass. Immature soil such as 24182 contains fewer agglutinates (12.7% vs. 26.0% in 24077, Basu *et al.*, 1977), and therefore gives a strong pyroxene band that is centered below 1 μm. Submature soils such as 24077 and 24109 (Morris, 1977) have shallow asymmetric bands indicating absorption in both pyroxene and glass. As in the case of the UV-VIS absorption, the strongly absorbing agglutinates reduce the spectral contrast and diminish the depth of the pyroxene band.

Additional weaker bands occur in the Luna 24 spectra. Fe^{2+} in plagioclase contributes a band near 1.3 μm. A band near 0.67 μm occurs in the spectra of samples other than 24182, implying that this absorption is in the agglutinates. Several band assignments are possible in this wavelength region including the charge-transfers mentioned above.

Charette and Adams (1975) and Charette *et al.* (1975) noted that in mature (agglutinate-rich) lunar soils the depth of the band near 1 μm was proportional to the FeO content of the bulk soil. This relationship holds because the 1 μm band in mature soils is in large part caused by the Fe^{2+} electronic absorption in the

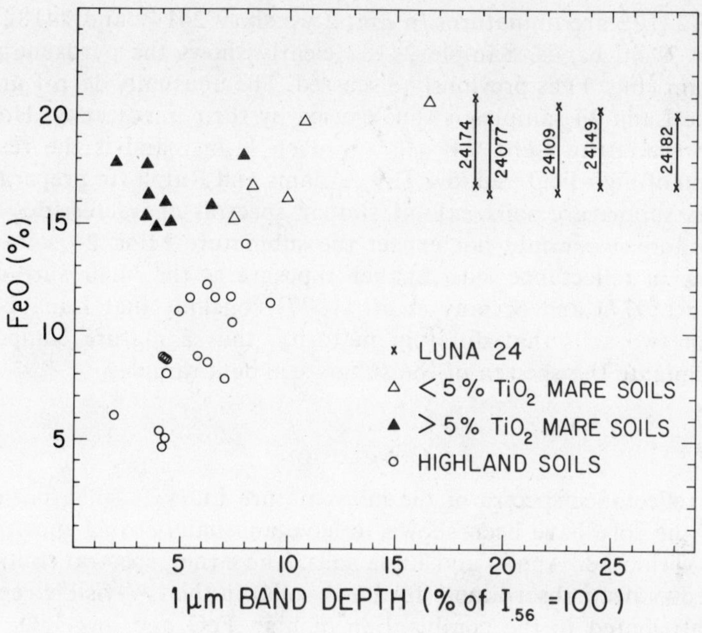

Fig. 2. Depth in percent of the absorption band near 1 μm *vs* weight percent FeO. Band depths are measured on curves that have intensity values scaled to 1 at 0.56 μm. FeO values for Luna 24 soils are from Blanchard *et al.*, (1977). The higher Luna 24 FeO values are for <250 μm sieve fractions, whereas the lower values are for <20 μm sieve fractions. Luna 24 samples can be distinguished from other FeO-rich mare soils on the basis of band depth.

agglutinitic glass, where the band depth is proportional to the Fe^{2+} concentration (Bell and Mao, 1972). Via and Taylor (1976) and Hu and Taylor (1977) have shown that the FeO in the agglutinate fraction of soils is close to that in the bulk soil analysis. Taylor *et al.* (1977) confirmed that FeO in Luna 24 agglutinates also is approximately the same as in the bulk soil.

Figure 2 shows the increase in the 1 μm band depth in the spectra of highland and mare soils. Note, however, that the presence of titanium in mare soils influences the FeO-band depth relationship. In the titanium-rich mare soils the 1 μm band depth is shallow because $Fe^{2+} \rightarrow Ti^{4+}$ and $Ti^{3+} \rightarrow Ti^{4+}$ charge-transfer absorptions centered in the UV-VIS have lowered the overall albedo and the spectral contrast. The Luna 24 samples, being unusually low in TiO_2 for mare soils, have deep 1 μm Fe^{2+} bands.

The FeO-band depth relationship holds only for mature and submature soils, for the reason that in immature soils there may be strong absorption from Fe^{2+} in pyroxene near 1 μm, and the pyroxene band depth is not a simple function of Fe^{2+} concentration. Morris (1977) classifies 24077 and 24109 as being submature on the basis of ferromagnetic reasonance data. Sample 24174 is considered by Morris as immature; however, it is close to his submature definition. Samples

24149 and 24182 are immature. In Fig. 2 we show 24149 and 24182 as having the deepest 1 μm bands. Sample 24182 clearly shows the pyroxene absorption short of 1 μm (Fig. 1) as previously discussed. The unusually deep 1 μm band for some of the Luna 24 samples is thus caused by their immaturity. However, for the relatively mature Luna 24 soils the deep 1 μm band is the result of the combination of high FeO and low TiO_2. Adams and Ralph (in preparation) have shown that submature soils exhibit similar spectral characteristics to mature soils. Therefore, we would not expect the submature Luna 24 soils to change significantly in reflectance with further exposure at the lunar surface. Bogard and Hirsch (1977) and McKay *et al.* (1977) conclude that Luna 24 soils are mixtures of two soils that differ in maturity, thus a mature component may already dominate the spectra of the submature bulk samples.

CONCLUSIONS

Diffuse reflectance spectra of the more mature Luna 24 soils that occur near the top of the core have been shown to have unusually strong spectral contrast compared with other Apollo and Luna soils. The strong spectral contrast, which is expressed as deep absorptions (for lunar soils) in the UV-visible region and at 1 μm, is attributed to the combination of high FeO and low TiO_2. The ratio $I_{.40\,\mu m}/I_{.56\,\mu m}$ is an index of the combined effect of spectral contrast and of charge-transfer absorptions in the UV-visible region, and has been related to the TiO_2 content of lunar soils. The low $I_{.40\,\mu m}/I_{.56\,\mu m}$ ratio for the Luna 24 soils is consistent with 1% TiO_2, and is a parameter that can be measured from telescopic reflectance spectra of the moon or spectral vidicon images. Electronic absorption at 1 μm by Fe^{2+} in agglutinitic glass in shown to be unusually strong in the Luna 24 spectra due to the low TiO_2. The relatively deep 1 μm band is therefore a second spectral characteristic that can be measured telescopically on the moon for remote sensing of very low titanium basalts.

REFERENCES

Adams J. B. (1975) Interpretation of visible and near-infrared diffuse reflectance spectra of pyroxenes and other rock-forming minerals. In *Infrared and Raman Spectroscopy of Lunar and Terrestrial Minerals* (C. Karr, Jr., ed), p. 91–116.

Adams J. B. and McCord T. B. (1972) Electronic spectra of pyroxenes and interpretation of telescopic spectral reflectivity curves of the moon. *Proc. Lunar Sci. Conf. 3rd*, p. 3021–3034.

Adams J. B., Head J. W., McCord T. B., Pieters C. and Zisk S. (1978) Mare Crisium: regional stratigraphy and geologic history. *Geophys. Res. Lett.* In press.

Barsukov V. L. (1977) Preliminary data for the regolith core brought to earth by the automatic lunar station Luna 24. *Proc. Lunar Sci. Conf. 8th*, p. 3303–3318.

Basu A., McKay D. S., Clanton V., Waits G. and Fruland R. M. (1977) Glass particles in Luna 24 drill core soils (abstract). In *Papers Presented to the Conference on Luna 24*, p. 18–21. The Lunar Science Institute, Houston.

Bell P. M. and Mao H. K. (1972) Crystal-field effects of iron and titanium in selected grains of Apollo 12, 14 and 15 rocks, glasses, and fine fractions. *Proc. Lunar Sci. Conf. 3rd*, p. 545–554.

Blanchard D. P., Haskin L. A., Brannon J. C. and Aaboe E. (1977) Chemistry of soils and particles from Luna 24 (abstract). In *Papers Presented to the Conference on Luna 24*, p. 37–40. The Lunar Science Institute, Houston.

Bogard D. D. and Hirsch W. C. (1977) Noble gases in Luna 24 core soils (abstract). In *Papers Presented to the Conference on Luna 24*, p. 44–47. The Lunar Science Institute, Houston.

Burns R. G. (1970) *Mineralogical applications of crystal field theory*. Cambridge University Press. 224 pp.

Charette M. P. and Adams J. B. (1975) Mare basalts: Characterization of compositional parameters by spectral reflectance (abstract). In *Papers Presented to the Conference on Origins of Mare Basalts and Their Implications for Lunar Evolution*, p. 25–28. The Lunar Science Institute, Houston.

Charette M. P., Adams J. B., Gaffey M. G. and McCord T. B. (1975) Determination of the iron content of lunar soils using remote spectral reflectance measurements (abstract). *EOS (Trans. Amer. Geophys. Union)* **56**, 1013.

Charette M. P., McCord T. B., Pieters C. and Adams J. B. (1974) Application of remote spectral reflectance measurements to lunar geology classification and determination of titanium content of lunar soils. *J. Geophys. Res.* **79**, 1605–1613.

Hapke B., Cassidy W. and Wells E. (1975) Effects of vapor-phase deposition processes on the optical, chemical, and magnetic properties of the lunar regolith. *The Moon* **13**, 339–353.

Head J. W., Adams J. B., McCord T. B., Pieters C. and Zisk S. (1978) Mare Crisium: regional stratigraphy and geologic history. In *Mare Crisium: The View from Luna 24* (R. B. Merrill and J. J. Papike, eds). This volume.

Hu H. and Taylor L. A. (1977) Lack of chemical fractionation in major and minor elements during agglutinate formation. *Proc. Lunar Sci. Conf. 8th*, p. 3645–3656.

Loeffler B. M., Burns R. G. and Tossell J. A. (1975) Metal-metal charge transfer transitions: Interpretation of visible-region spectra of the moon and lunar materials. *Proc. Lunar Sci. Conf. 6th*, p. 2663–2676.

Loeffler B. M., Burns R. G., Tossell J. A., Vaughan D. J. and Johnson K. H. (1974) Charge transfer in lunar materials: Interpretation of ultraviolet-visible spectral properties of the moon. *Proc. Lunar Sci. Conf. 5th*, p. 3007–3016.

McKay D. S., Basu A., Waits G., Clanton V., Fuhrman R. and Fruland R. (1977) Grain size and evolution of Luna 24 soils (abstract). In *Papers Presented to the Conference on Luna 24*, p. 114–117. The Lunar Science Institute, Houston.

Morris R. V. (1977) FMR and magnetic properties of Luna 24 soils and 1 mm soil particles (abstract). In *Papers Presented to the Conference on Luna 24*, p. 121–124. The Lunar Science Institute, Houston.

Tarasov L. S., Nazarov I. D., Shevaleevsky I. D., Kudryashova A. S., Gaverdovskaya A. S. and Korina M. I. (1977) Mineralogy and petrography of lunar rocks from Mare Crisium (preliminary data). *Proc. Lunar Sci. Conf. 8th*, 333–356.

Taylor G. J., Warner R., Wentworth S., Keil K. and Sayeed V. (1977) Petrology and chemistry of Luna 24 mare basaltic glasses (abstract). In *Papers Presented to the Conference on Luna 24*, p. 189–192. The Lunar Science Institute, Houston.

Via W. N. and Taylor L. A. (1976) Chemical aspects of agglutinate formation: Relationships between agglutinate composition and composition of the bulk soil. *Proc. Lunar Sci. Conf. 7th*, p. 393–402.

Merrill, R. B. and Papike, J. J., editors.
(1978) *Mare Crisium: The View from Luna 24*, Pergamon Press (New York), p. 89–103
Printed in the United States of America

The nature of rays and sources of highland material in Mare Crisium

TED A. MAXWELL and FAROUK EL-BAZ
National Air and Space Museum, Smithsonian Institution, Washington, D.C. 20560

Abstract—Although highland fragments are rare in the Luna 24 core, they should provide an important source of information on the sequence of formation and lithology of lunar east-limb basins. The numerous rays that cross the Crisium basin suggest several sources of highland materials. Proclus and Giordano Bruno are responsible for the most conspicuous rays in the basin, but discontinuous rays and secondary craters of Langrenus and Taruntius are found in the central and eastern parts of the basin. Individual bright components of Giordano Bruno rays contain several small fresh craters (probable secondaries) that contribute to the overall brightness of the ray. However, rays of Proclus have very gradational boundaries with the surrounding mare and there is no apparent contrast between the rays and surrounding mare, with respect to small (>100 m) crater frequencies. Both deposition of primary ejecta and reworking of local material were most likely responsible for the formation of ray systems in Mare Crisium. In addition to the fact that Luna 24 landed near the uprange end of a Giordano Bruno ray, incomplete sampling of the uppermost layer and formation of the ray by reworking of *in situ* regolith suggest that its materials may be scarce or not present in the core. Highland fragments present are most likely to be representative of several highland sources.

INTRODUCTION

Highland components of the Luna 24 samples are volumetrically small (Florensky *et al.*, 1977). In the 0.15 to 0.5 mm size fraction of samples studied by U. S. investigators, highland material comprises only 2% (Bence and Grove, 1977). However, the possible sources of highland material are extremely important for: 1) understanding the time sequence of formation of the eastern lunar basins, 2) correlating orbital and earth-based data to the mare material of the Crisium basin, and 3) testing impact mixing and ejecta distribution models. Interpretation and identification of the Luna 24 highland components have been hampered by the fact that the local mare basalts are low in TiO_2 and high in Al_2O_3 (Bence and Grove, 1977; Vaniman and Papike, 1977). Low values of TiO_2 for Mare Crisium have also been noted in earth-based spectral reflectance measurements (Pieters *et al.*, 1976; Adams *et al.*, 1978), and relatively high values of Al_2O_3 have been reported from the Apollo 15 X-ray fluorescence data (Adler *et al.*, 1973; Maxwell *et al.*, 1977). Consequently, it is possible that some highland materials present have not yet been recognized.

The purpose of this paper is to evaluate the probable sources and possible amounts of the external contribution (originating from outside the basin) to Crisium mare regolith. The intent is not to point out the exact sources for individual fragments returned by Luna 24, but rather to suggest the relative amounts of highland material that should be present in the mare regolith.

The distribution pattern of rays and secondary craters within the basin may aid interpretation of remotely-sensed information. With the exception of spectrally determined units (Adams *et al.*, 1978), there are no major clear-cut intramare boundaries that can be distinguished on the basis of visual albedo (Wilhelms, 1973; Frontispiece, 1977). Consequently, one of the causes for different color shades may be the numerous highland crater rays that cross the basin. If the rays are indeed the result of reworking of *in situ* material (Oberbeck, 1971), they still may affect remote-sensing methods that rely on the maturity of the regolith.

Despite the difficulties imposed by viewing angle geometry, earth-based full-moon photographs provide the best source of data for mapping Crisium ray systems. In addition, high sun angle photographs taken on Apollo 15, and trans-earth photographs from Apollo 8, 11 and 17 provide valuable coverage of the lunar east limb. A complete index of photographic coverage is given by Wilhelms and El-Baz (1977a). In addition, visual observations from Apollo 15 and 17 command module pilots A. M. Worden and R. E. Evans have been reported by El-Baz *et al.* (1972) and Evans and El-Baz (1973), and summarized by Butler and Morrison (1977).

GEOLOGIC SETTING

According to age relations summarized by Wilhelms and El-Baz (1977a), older basins that provided the substrate for the Crisium impact are Smythii, Marginis, Humboldtianum, Fecunditatis, and probably Nectaris (Fig. 1). Because of the scarcity of radial lineations and crater chains, it was originally thought that Crisium was a pre-Nectarian basin, and that Serenitatis was younger than Crisium (Wilhelms and El-Baz, 1977b). However, on the basis of superposition relations among interpreted basin secondaries, and crater counts excluding basin secondaries, Wilhelms (1976) has proposed that Crisium is younger than Nectaris, and that both the Crisium and south Serenitatis basins are therefore Nectarian in age.

The relative sequence of formation of the Crisium and Serenitatis basins is not distinct. The region between Crisium and Serenitatis where superposition relations would be observable is too complex for a simple interpretation. Lineations and basin secondaries from both north and south Serenitatis, Crisium, Humboldtianum, Tranquillitatis and Imbrium all affect this area, confusing the stratigraphic relations (Wilhelms, 1976). Interpretation of the relative ages of these two basins directly influences the proper interpretation of the circum-Crisium highlands, since considerable amounts of Serenitatis ejecta could be present in the Crisium region (McGetchin *et al.*, 1973). Generally speaking, Crisium appears to be the fresher of the two basins, but this criterion may not be valid since Serenitatis was subject to a great deal of degradation by the Imbrium event. At present, therefore, the exact origin (and source) of circum-Crisium highland material is subject to debate. It is possible that further results from analyses of Luna 20 samples may help to resolve this problem.

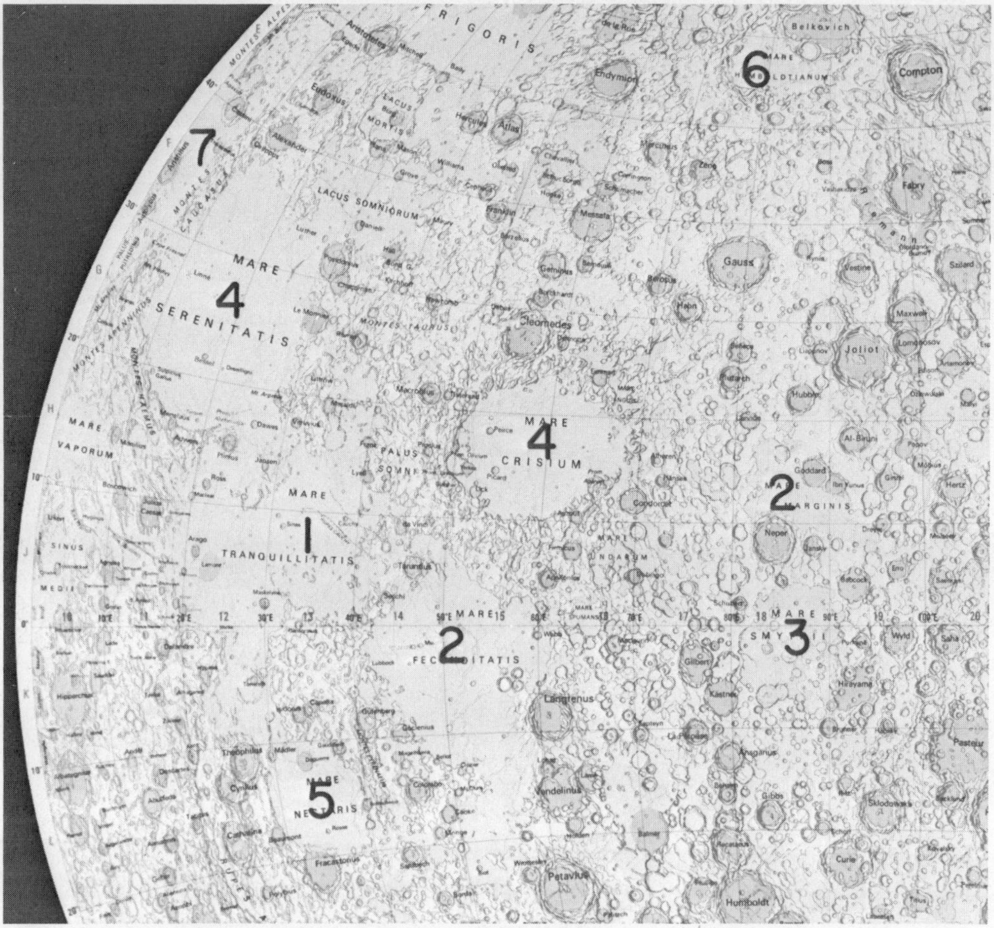

Fig. 1. Sequence of formation of major east-limb basins. Age relations (1 = oldest; 7 = youngest) are from Wilhelms and McCauley (1971) and Wilhelms and El-Baz (1977a).

On the basis of spectral reflectance measurements, Adams *et al.* (1978) have mapped three mare units in the Crisium basin. These units cannot be distinguished on the basis of visual albedo. Instead, the mare material appears to be largely homogeneous with respect to both color and visual albedo (Fig. 2). In areas unaffected by rays from Proclus, mare color is similar to that of Fecunditatis and northern Mare Tranquillitatis. Correlation of discrete mare units with those determined spectrally is made difficult by the multitude of both fresh and subdued rays that cross the basin. One of the most vivid descriptions of the Crisium mare surface (particularly near Proclus) was given by Apollo 15 command module pilot Worden during debriefing: "It is almost like flying above a haze layer and looking down through the haze at the surface. Ejecta from that

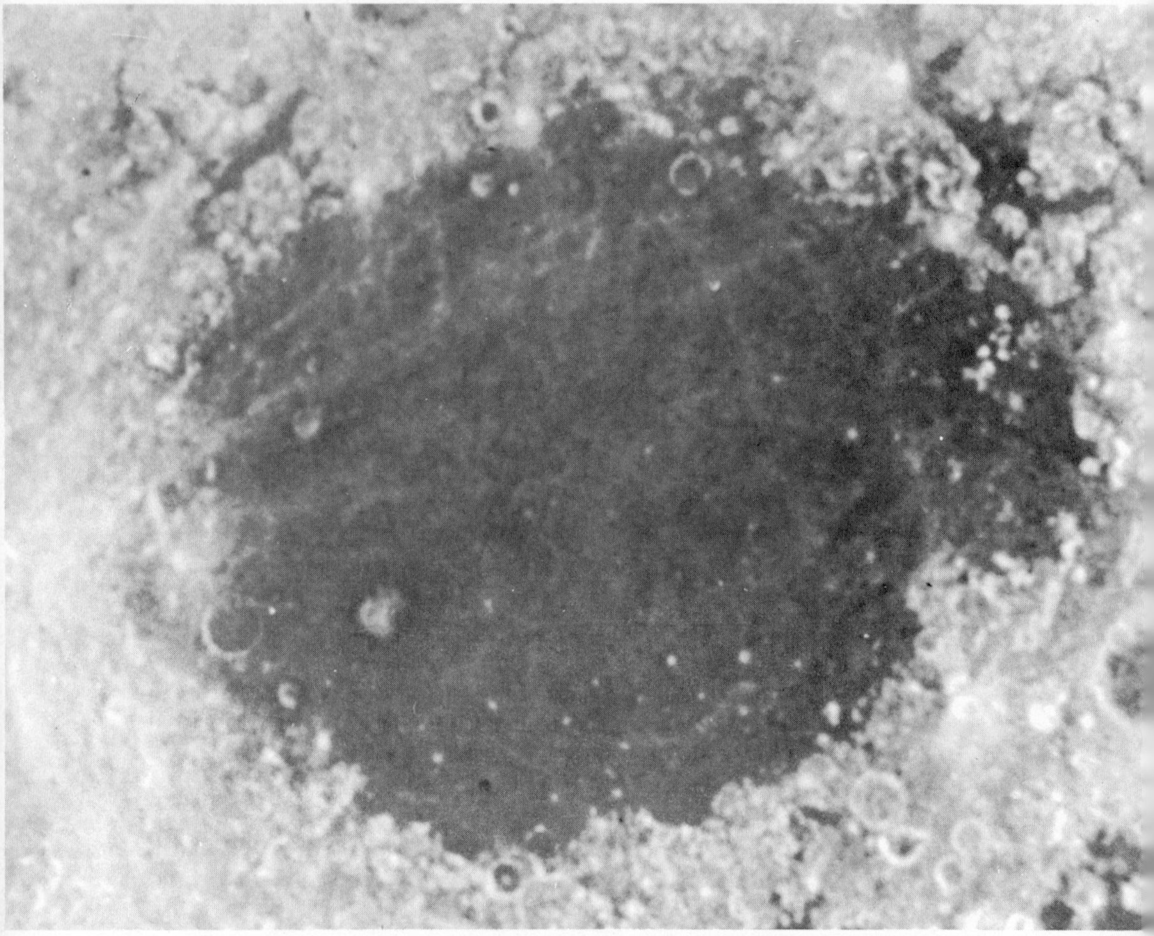

Fig. 2. A) High sun angle trans-earth photograph of the Crisium basin (Apollo 11 Fr. 6664). B) Edge-enhanced version of the Crisium basin. Note the extreme contrast of the Proclus rays, boundary of the low-albedo ejecta blanket of Picard, and more subdued, generally north-trending rays.

crater [Proclus] does not look like it is resting on top of Crisium. It looks like it is suspended over it. It gives a very filmy, very gauzy appearance to the whole thing," (El-Baz *et al.*, 1972; p. 95).

DISTRIBUTION OF RAYS AND SECONDARY CRATERS

As in other regions of the moon, rays in Mare Crisium are very gradational and indistinct. Depending on which photographs are used, rays may range from very bright with clear-cut boundaries to virtually indistinguishable from the

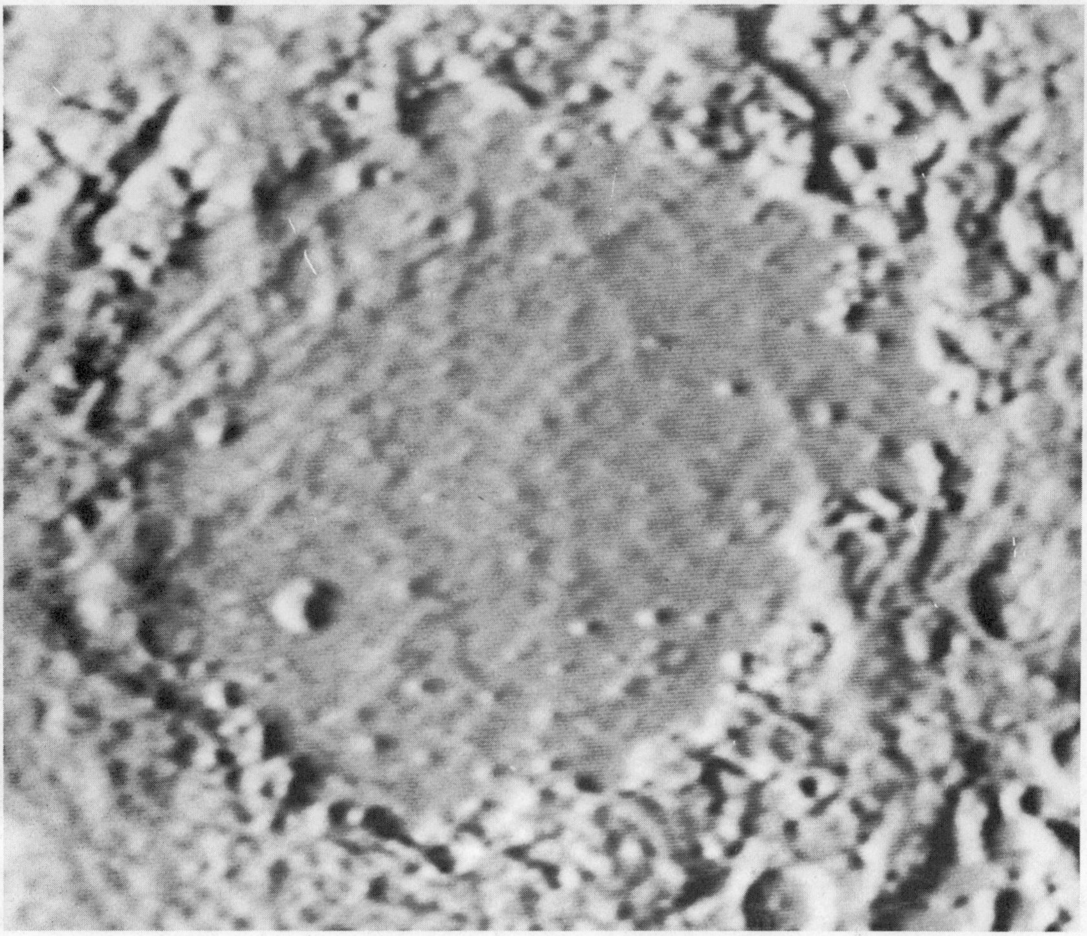

surrounding mare. For this reason, only those portions of rays or secondary clusters with linear trends and distinct boundaries were mapped (Fig. 3). Consequently, the map does not indicate all ray components (especially the subtle north-trending patches), but it does include deposits of the major sources of highland material. The source craters of these major rays were determined from orientation of the rays and of secondary crater ejecta (Fig. 4). As is evident in earth-based photographs, the brightest rays are from Proclus, Giordano Bruno, and possibly Thales and Stevinus.

On the basis of brightness on high sun photographs, Butler and Morrison (1977) have shown that Giordano Bruno is most likely younger than Proclus. However, there is a further distinction in the type of rays. In northwestern Crisium, ray material from Proclus does not appear to have any distinct associated secondary craters, but instead appears as a "haze" of light colored

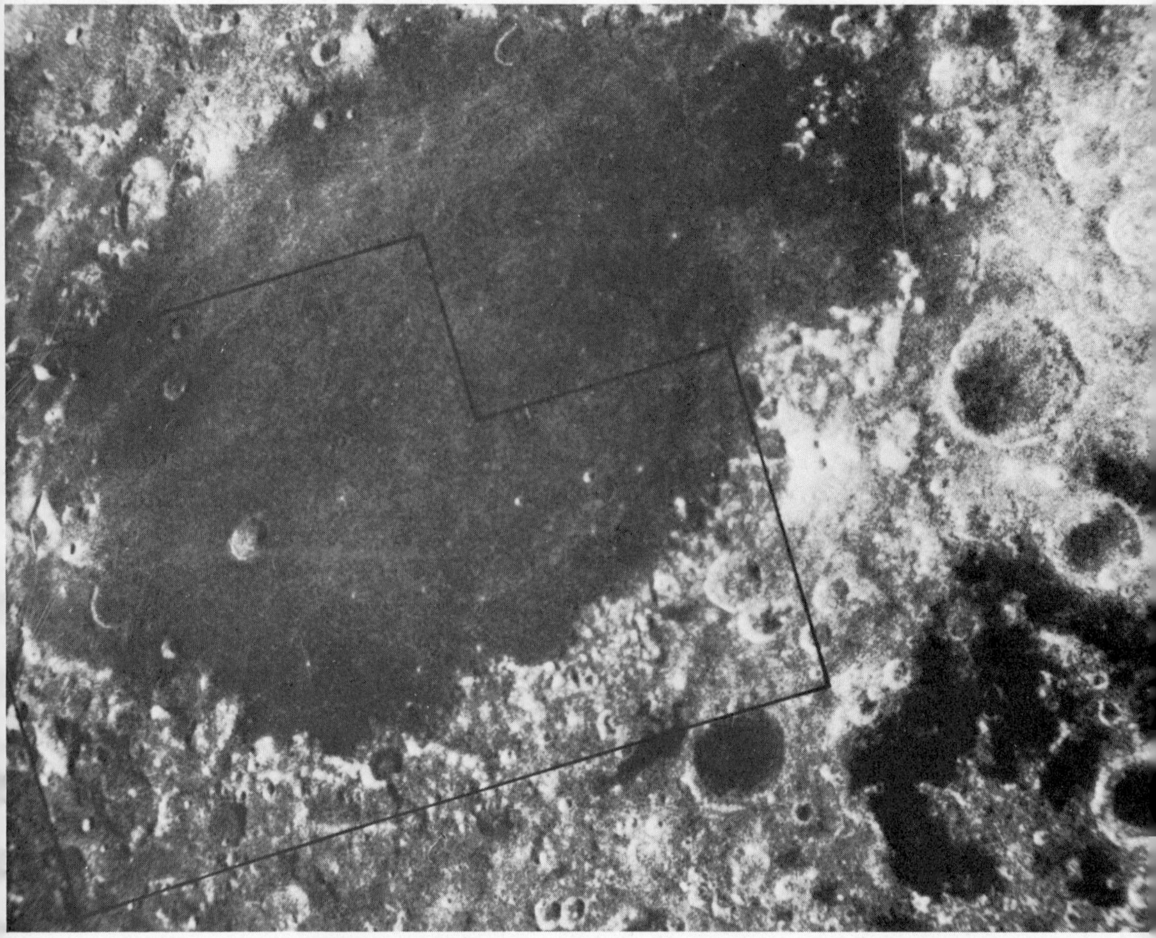

Fig. 3. A) Location of mapped region in Mare Crisium (Apollo 8 Fr. 2864). Mapping was restricted to areas covered by existing Lunar Topographic Orthophotomaps. B) Map of major rays and secondary crater clusters in Mare Crisium. Boundaries in

mare. Rays from Giordano Bruno in the northeastern and southeastern parts of the basin have sharp-rimmed 50–100 m craters at their uprange end (Fig. 5). Viewed on enlarged panoramic photographs, these Bruno secondaries are neither elongate nor do they have asymmetric ejecta. This morphology is consistent with a high ejection angle which would be needed for material to travel such a great distance (1300 km). In addition to differences in craters on the ray, there is also a difference in ray pattern. Rays from Proclus have no distinct uprange contact with darker mare, and deviate from a true radial direction to the east of Swift. Individual components of Giordano Bruno rays, however, are lineated radially to the crater, but have sharp uprange boundaries that are approximately perpendicular to the source.

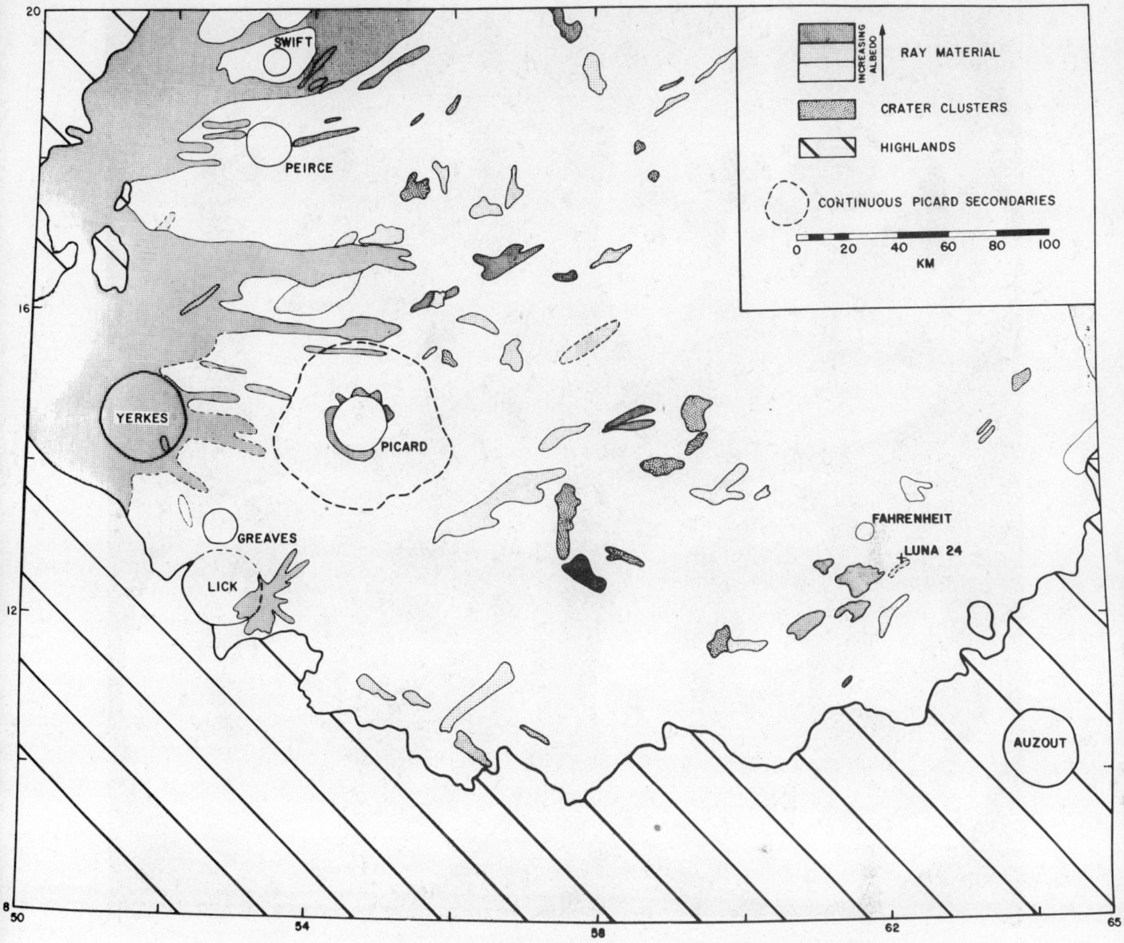

northwestern part of the basin are schematic due to the variability of Proclus rays on different high-sun photographs.

Less distinct, north-northwest-trending rays between Peirce and Picard are likely to have originated from Taruntius. Since they are subdued by Proclus rays, identification of discrete components is difficult, but age relations (rays radial to Taruntius are not as bright as Proclus rays) are consistent (Fig. 6). Only a few scattered secondary crater clusters in the central and eastern part of the basin have the correct orientations for a Langrenus origin. Based on ejecta thickness calculations (McGetchin *et al.*, 1973), Langrenus should have contributed the greatest percentage of highland material to southeastern Mare Crisium. It is likely, therefore, that many of the subdued, generally north-trending rays originate from Langrenus.

By restricting the map area to Lunar Topographic Orthophotomap coverage,

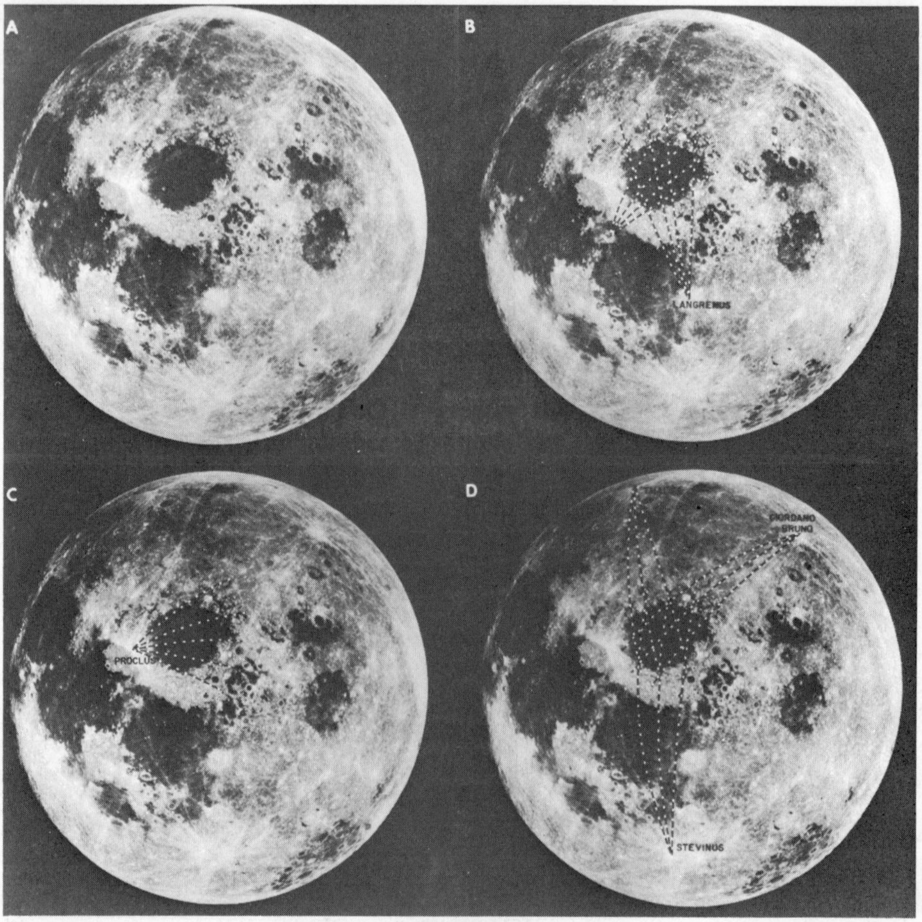

Fig. 4. Orientations of rays and secondary craters originating from Copernican craters surrounding Mare Crisium. A) Apollo 11 Fr. 6664 showing ray orientations; B–D) orientations of crater ejecta plotted with a stereo net to account for spherical distortion.

not all the rays have been considered, especially in the northwestern part of the basin where rays from Proclus obscure older north and northwest-trending rays aligned with Thales and smaller craters on the northwest rim of the basin. However, these north and northwest-trending rays are not prominent in the central and southeastern portions of the basin. In the eastern part of the basin, a prominent north-northwest ray originating from Condorcet T obscures much of unit 2 of Adams *et al.* (1978). This crater, however, has an elongate floor and a very asymmetric ejecta pattern to the north. Its rays are not prominent in the direction of the Luna 24 site (Florensky *et al.*, 1977).

Discussion

Ray origin

The origin of lunar ray systems has been attributed to either 1) finely comminuted debris originating from the primary crater and deposited as a very thin, discontinuous ejecta layer (Baldwin, 1963; Allen, 1977), or 2) dominantly local material overturned by numerous secondary impacts that increase the blockiness and hence the albedo of the surface (Oberbeck, 1971). Both processes have support in photogeologic and analog studies; rays customarily have a larger number of small fresh craters within their boundaries than the surrounding terrain. However, small-scale cratering studies have reproduced characteristic ray patterns, and suggest that a competent substrate beneath the primary crater may contribute to their formation (Andrews, 1976).

The major unresolved problem, therefore, is how much material ejected from the primary crater contributes to a ray. Based on crater counting results on Lunar Orbiter V frames, Oberbeck (1971) attributed rays to the excavation of substrate by secondary or tertiary craters. In that study, a 3–5% larger concentration of small concentric craters on a Copernicus ray matches well with the increase in visual albedo from the surrounding mare. In Crisium, however, the frequency of craters (visually estimated from craters >100 m) on rays from Proclus does not appear different from the surrounding mare, thus suggesting deposition of primary ejecta. The presence of secondary swarms on Giordano Bruno rays, however, favors a "secondary" origin. Small concentric craters on the Copernicus ray (Fig. 4 in Oberbeck, 1971) are randomly dispersed throughout the ray, whereas the fresh craters on Bruno rays (probable secondaries) are concentrated at the uprange end.

Assuming that Proclus and Giordano Bruno target material is similar, this comparison of rays indicates that both mechanisms are valid. It is still possible, however, that secondary and tertiary craters below the limit of resolution may significantly contribute to the rays. It is suggested that distance from the primary crater is the controlling factor as to which process will dominate (Oberbeck *et al.*, 1975). As is the case with ejecta, rays nearest the crater may be composed of primary ejecta, whereas the distal ends are more heavily influenced by secondary craters and reworking of local regolith.

Implications for source of Luna 24 highland material

It is doubtful that highland material immediately surrounding the Crisium basin will be significantly represented in the Luna 24 core. The most likely contributions from the Crisium rim are from the Copernican crater Condorcet T, and the 14 km Eratosthenian crater Condorcet TA (Olson and Wilhelms, 1974), both of which are 106 km east of the landing site. Continuous reworking of the regolith has obscured any evidence for deposition of Condorcet TA material near the site, and the rays from Condorcet T extend only to the north. Consequently,

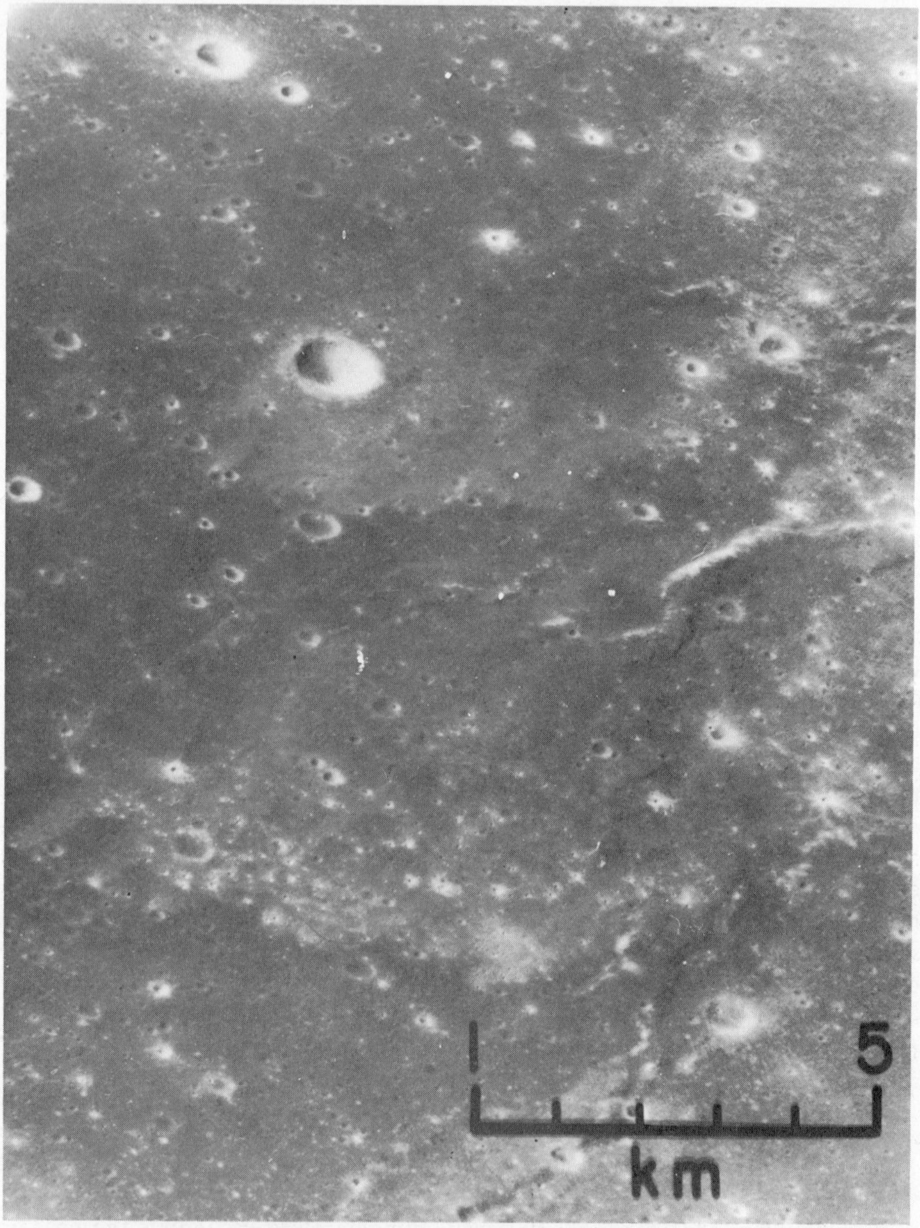

Fig. 5. A) Rays from Giordano Bruno near the Luna 24 landing site (Apollo 17 pan Fr. 2223). Numerous small bright craters (about 100 m dia.) are prevalent at uprange boundary of ray (Butler and Morrison, 1977). B) Rays originating from the crater

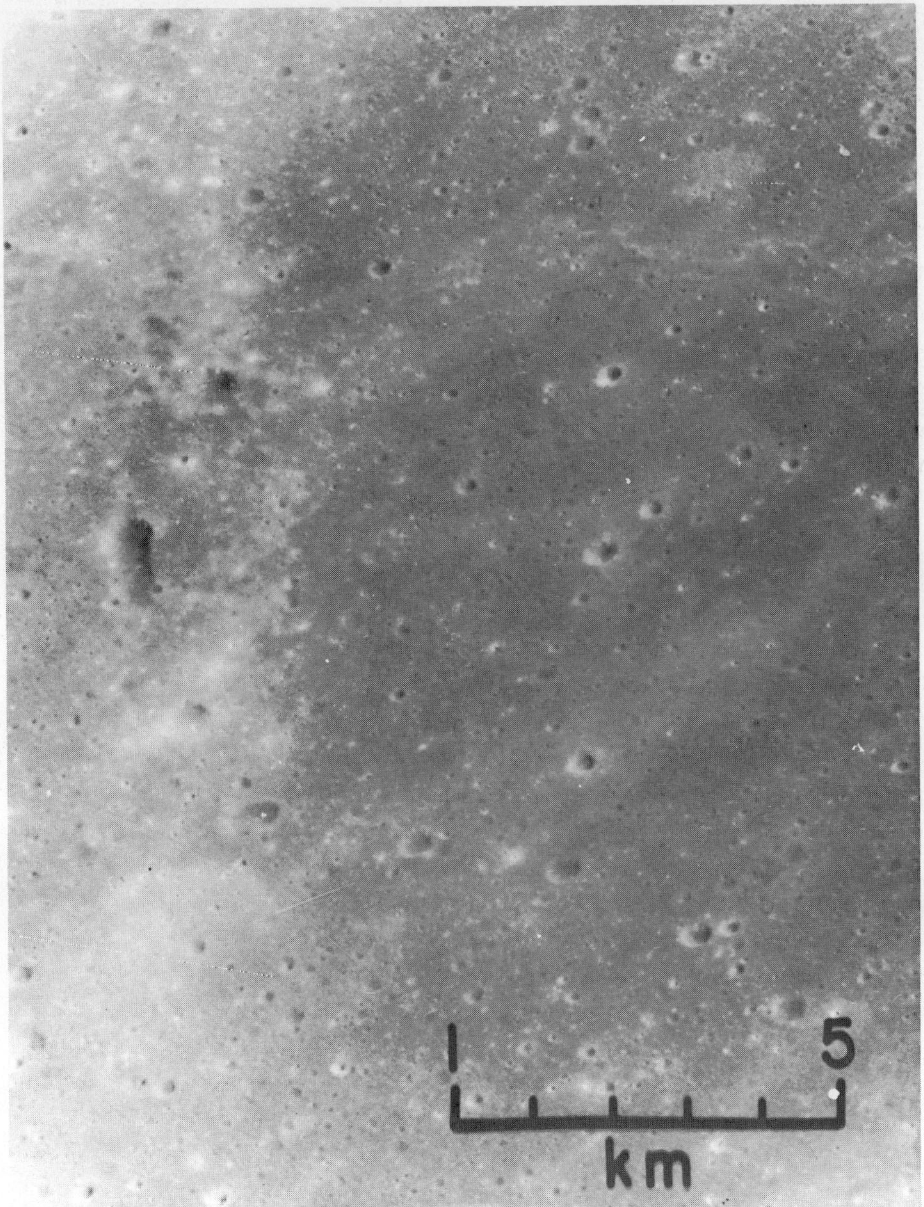

Proclus (trending from lower left to upper right). No similar small craters are present, and ray has a hazy appearance (Apollo 17 pan Fr. 2233).

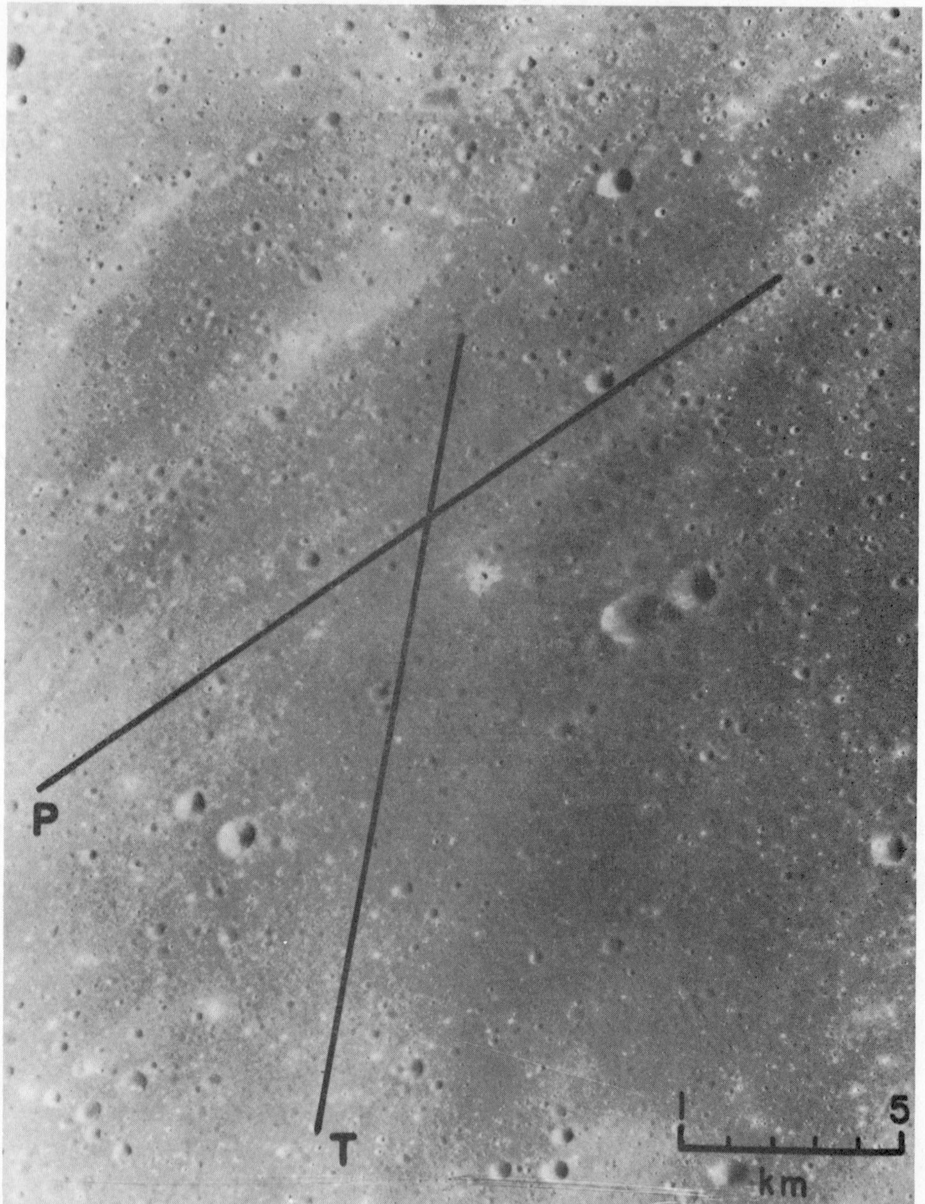

Fig. 6. Complex field of secondaries originating from Proclus (P) and Taruntius (T). Taruntius rays and secondaries are obscured by effects of the younger crater Proclus (Apollo 17 pan Fr. 2247).

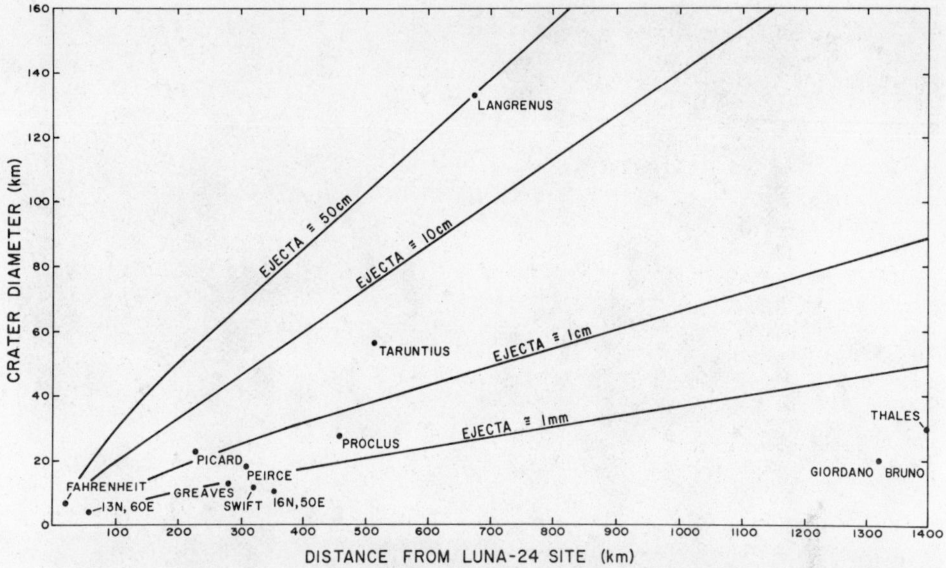

Fig. 7. Diameter of major Coperican craters as a function of distance from the Luna 24 site. Ejecta thickness lines are calculated from hypothetical extension of empirical relationships of McGetchin *et al.* (1973). Both diameters and ejecta thicknesses are not corrected for crater enlargement through slumping or terrace formation.

the amount of highland material from the immediate rim may be very small, if present at all.

Based on the size of Copernican craters and their distance from the Luna 24 site (Fig. 7), Langrenus may have contributed the largest percentage of highland material. Although the Langrenus impact occurred at the highland-mare contact (Fig. 8), X-ray fluorescence data suggest that the primary ejecta is dominantly composed of highland material (Wilhelms and El-Baz, 1977a). Reid *et al.* (1972) have suggested that a substantial part of the cindery glass found at the Luna 16 site may have been derived from Langrenus ejecta. A comparison of this regolith type to Luna 24 regolith would provide one test (although not necessarily conclusive) for the presence of Langrenus ejecta. It is also possible, however, that the shallow mare component of Langrenus ejecta may be represented, although this may be much more difficult to identify. In addition, theoretical calculations of continuous deposits from Taruntius (McGetchin *et al.*, 1973) suggest that there should be about twice as much material from Taruntius than from Proclus.

Despite this evidence for highland material in Mare Crisium, present investigations indicate extremely small amounts of highland fragments in the Luna 24 core. This scarcity is not unexpected, and is consistent with the indistinct, patchy nature of the rays, an origin of the distal ends of rays by reworking of *in situ*

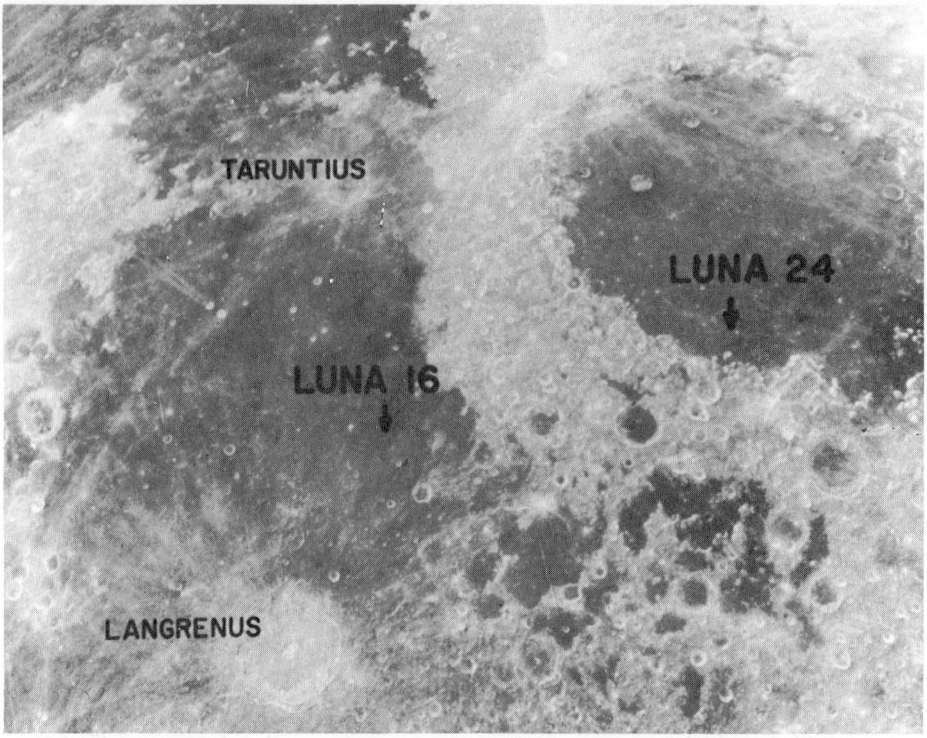

Fig. 8. Soviet Luna landing sites in mare regions on the east limb of the moon. The presence of ray systems of Langrenus and Taruntius in Mare Crisium is suggested by discontinuous north and northeast trends (Apollo 17 Fr. 23316).

regolith, and the possibility that the uppermost 40 cm of regolith were incompletely sampled (Florensky *et al.*, 1977).

This study of ray systems in Mare Crisium indicates that no single source will be responsible for highland fragments of the Luna 24 sample. The position of the landing site at the uprange end of a Giordano Bruno ray suggests that material from that crater will be minimal, if sampled at all. The most likely candidates as sources of highland material are the craters Langrenus and Taruntius which impacted on probable Fecunditatis ejecta. Consequently, highland fragments of the Luna 24 core are not likely to be representative of a single suite with a distinctive lithology.

Acknowledgments—We thank Don Wilhelms for providing earth-based photographs, and Ann Gifford for helping with the illustrations. M. Cintala and D. Morrison provided constructive reviews of the manuscript. Some of the Apollo photography was provided by the National Space Science Data Center. This research was supported by NASA grant NSG-7188.

REFERENCES

Adams J. B., Head J. W., McCord T. B., Pieters C., and Zisk S. (1978) Mare Crisium: Regional stratigraphy and geologic history. *Geophys. Res. Lett.* **5**, 313–316.

Adler I., Trombka J. I., Lowman P., Schmadebeck R., Blodget H., Eller E., Yin L., Lamothe R., Osswald G., Gerard J., Gorenstein P., Bjorkholm P., Gursky H., Harris B., Arnold J., Metzger A., and Reedy R. (1973) Apollo 15 and 16 results of the integrated geochemical experiment. *The Moon* **7**, 487–504.

Allen C. C. (1977) Rayed craters on the Moon and Mercury. *Phys. Earth Planet. Inter.* **15**, 179–188.

Andrews R. J. (1976) Characteristics of debris from small-scale cratering experiments (abstract). In *Papers Presented to the Symposium on Planetary Cratering Mechanics*, p. 9–11. The Lunar Science Institute, Houston.

Baldwin R. B. (1963) *The Measure of the Moon.* Univ. Chicago Press, Chicago. 488 pp.

Bence A. E. and Grove T. L. (1977) The highland component in the Luna 24 core (abstract). In *Papers Presented to the Conference on Luna 24*, p. 22–24. The Lunar Science Institute, Houston.

Butler P. Jr. and Morrison D. A. (1977) Geology of the Luna 24 landing site. *Proc. Lunar Sci. Conf. 8th*, p. 3281–3301.

El-Baz F., Worden A. M. and Brand V. D. (1972) Astronaut observations from lunar orbit and their geologic significance. *Proc. Lunar Sci. Conf. 3rd*, p. 85–104.

Evans R. E. and El-Baz F. (1973) Geological observations from lunar orbit. In *Apollo 17 Prelim. Sci. Rep.* NASA SP-330, p. 28-1–28-32.

Florensky C. P., Basilevsky A. T., Ivanov A. V., Pronin A. A., and Rode O. D. (1977) Luna 24: Geologic setting of landing site and characteristics of sample core (preliminary data). *Proc. Lunar Sci. Conf. 8th*, 3257–3279.

Frontispiece (1977) *Proc. Lunar Sci. Conf. 8th.* Vol. 1, Plate 6.

Maxwell T. A., Strain P. L., and El-Baz F. (1977) Mare Crisium: Compositional inferences from low altitude X-ray fluorescence data. *Proc. Lunar Sci. Conf. 8th*, p. 933–944.

McGetchin T. R., Settle M., and Head J. W. (1973) Radial thickness variation in impact crater ejecta: Implications from lunar basin deposits. *Earth Planet. Sci. Lett.* **20**, 226–236.

Oberbeck V. R. (1971) A mechanism for the production of lunar crater rays. *The Moon* **2**, 263–278.

Oberbeck V. R., Hörz F., Morrison R. H., Quaide W. L., and Gault D. E. (1975) On the origin of the lunar smooth-plains. *The Moon* **12**, 19–54.

Olson A. B. and Wilhelms D. E. (1974) Geologic map of the Mare Undarum quadrangle of the Moon. U. S. Geol. Survey Map I-837.

Pieters C., McCord T. B., and Adams J. B. (1976) Regional basalt types in the Luna 24 landing site as derived from remote observations. *Geophys. Res. Lett.* **3**, 697–700.

Reid J. B. Jr., Taylor G. J., Marvin U. B., and Wood J. A. (1972) Luna 16: Relative proportions and petrologic significance of particles in the soil from Mare Fecunditatis. *Earth Planet. Sci. Lett.* **13**, 286–298.

Vaniman D. T. and Papike J. J. (1977) Ferrobasalts from Mare Crisium: Luna 24. *Geophys. Res. Lett.* **4**, 497–500.

Wilhelms D. E. (1973) Geologic map of the Northern Crisium Region. In *Apollo 17 Prelim. Sci. Rep.* NASA SP-330, p. 29-29–29-35.

Wilhelms D. E. (1976) Secondary impact craters of lunar basins. *Proc. Lunar Sci. Conf. 7th*, p. 2883–2901.

Wilhelms D. E. and El-Baz F. (1977a) Geologic map of the east side of the Moon. U. S. Geol. Survey Map I-948.

Wilhelms D. E. and El-Baz F. (1977b) The Crisium basin and adjacent terra (abstract). In *Papers Presented to the Conference on Luna 24*, p. 200–203. The Lunar Science Institute, Houston.

Wilhelms D. E. and McCauley J. F. (1971) Geologic map of the near side of the Moon. U. S. Geol. Survey Map I-703.

Merrill, R. B. and Papike, J. J., editors.
(1978) *Mare Crisium: The View from Luna 24*, Pergamon Press (New York), p. 105–116
Printed in the United States of America

Noble gases in Luna 24 core soils

D. D. BOGARD and W. C. HIRSCH[1]

NASA Johnson Space Center, Houston, Texas 77058
([1]Also, Northrop Services, Inc.)

Abstract—Isotopic abundances of the noble gases have been determined in grain size separates of six soils from different depths in the Luna 24 core. The compositions of trapped solar gases in these Luna 24 soils are generally typical of those measured in mare soils. Concentrations of cosmogenic ^{21}Ne in these soils show a complex pattern as a function of soil depth, and are inconsistent with a single stage, *in situ* production in a static regolith or with a simple model of slow accumulation of previously unirradiated soils followed by a static period. All soils have apparently had a complex pre-irradiation history. Some soils show significant differences in $[^{21}\mathrm{Ne}]_c$ between different grain sizes, which is consistent with suggestions made by other investigators that they formed as a mixture of soils with different maturities. The exact subsurface depth relationships of the soils are not known, and several possibilities are discussed. Because of this factor and the complex nature of the soils, the cosmogenic Ne data do not define the depositional-irradiational history of the core. Luna 24 core soils may have been deposited any time during the past $\sim 3 \times 10^8$ yr.

INTRODUCTION AND EXPERIMENTAL

Our collaborative group for lunar core studies has extensively analyzed common <250 μm samples from six depths in the Luna 24 core. Each ~50 mg sample was carefully sieved (McKay *et al.*, 1978) into five grain sizes of <20, 20–45, 45–90, 90–150, and 150–250 μm. For these separates soil petrography and grain size distribution are reported by McKay *et al.* (1978) and Basu *et al.* (1978), magnetic properties and relative surface maturity by Morris (1978), chemical compositions by Blanchard *et al.* (1978), and particle track densities by Blanford and Wood (1977). Here we report isotopic abundances of the noble gas elements on <20μm and 45–90 μm sizes of all six soils and 150–250 μm sizes of five soils. Noble gas data were obtained mass spectrometrically by standard techniques (Bogard & Hirsch, 1975).

The surface maturity of these Luna 24 soils, as defined by the FMR maturity index, I_s/FeO, range from submature for the two soils nearest the lunar surface to immature for the four lowermost soils (Morris, 1978). In contrast, particle track densities on plagioclases from the 45–90 μm sizes are rather high (generally $>10^9$ cm^{-2}) and indicate substantial irradiation at the lunar surface (Blanford and Wood, 1977). Grain-size data (McKay *et al.*, 1978) indicate that the soils consist of two components of different maturities which have not been strongly blended by reworking. The mean grain size of the finer component generally increases with core depth. Elemental abundances (Blanchard *et al.*, 1978) show appreciable differences between the 90–150 μm and <20 μm grain size fractions of each soil (but not among bulk soils), which are interpreted as consistent with either mixing of soil components with different chemical compositions or differential comminution. Several investigators (e.g., Barsukov *et al.*, 1977; Blanchard *et al.*, 1978; McKay *et al.*, 1978; Basu *et al.*, 1978; Hu and Taylor, 1977) have pointed out that substantial differences in mineralogical and chemical properties exist between the bulk soils and many coarse (>0.25 mm) fragments.

RESULTS

Measured isotopic abundances of the noble gases are presented in Tables 1–3. Various possible models of subsurface depths of our soils in the Luna 24 core are shown in Fig. 1 and will be discussed in detail later. Noble gas data for grain sizes of each soil were subjected to ordinate-intercept analyses (Eberhardt *et al.*, 1970) to determine the isotopic composition of the surface-correlated, trapped component and abundances of the volume-correlated component produced by

Table 1.

Sample	Wt. (mg)	^{3}He $\times10^{-6}$	^{4}He $\times10^{-2}$	^{22}Ne $\times10^{-5}$	$^{20}Ne/^{22}Ne$	$^{22}Ne/^{21}Ne$	^{36}Ar $\times10^{-5}$	$^{36}Ar/^{38}Ar$	$^{40}Ar/^{36}Ar$	^{84}Kr $\times10^{-8}$	^{132}Xe $\times10^{-8}$
24077,9											
45– 90 μm	3.66	14.6	3.35	6.72	12.28 ± .28	22.83 ± .24	14.4	5.11 ± .01	0.79 ± .02	10.1	1.33
<20 μm	1.55	58.1	16.5	29.4	12.65 ± .08	29.64 ± .07	68.9	5.23 ± .01	0.67 ± .02	38.5	5.69
24109,13											
150–250 μm	3.16	8.29	1.66	3.66	12.29 ± .04	20.24 ± .21	9.53	5.09 ± .01	0.76 ± .04	7.03	0.90
45– 90 μm	6.04	14.3	3.32	6.64	12.41 ± .04	23.34 ± .15	14.6	5.12 ± .01	0.71 ± .02	10.2	1.25
<20 μm	1.04	70.6	18.7	30.2	12.64 ± .03	29.37 ± .28	68.3	5.21 ± .01	0.72 ± .02	41.6	5.26
24149,15											
150–250 μm	3.50	10.7	2.19	3.96	12.44 ± .03	21.23 ± .28	7.07	5.09 ± .01	0.78 ± .04	4.65	0.52
45– 90 μm	4.91	12.3	2.76	5.68	12.45 ± .05	23.89 ± .18	10.3	5.14 ± .05	0.81 ± .03	5.03	0.93
<20 μm	1.33	52.3	14.1	23.6	12.57 ± .04	29.37 ± .42	48.1	5.21 ± .01	0.74 ± .03	30.5	3.68
24174,10											
150–250 μm	1.98	6.18	1.22	2.65	12.36 ± .04	22.10 ± .58	5.54	5.11 ± .01	1.03 ± .06	4.38	0.64
45– 90 μm	6.45	12.7	3.03	6.18	12.49 ± .03	24.86 ± .19	11.7	5.14 ± .01	0.72 ± .02	7.18	0.93
<20 μm	1.46	76.8	20.5	32.7	12.62 ± .03	30.62 ± .42	65.6	5.22 ± .01	0.66 ± .02	43.4	5.41
24182,15											
150–250 μm	2.35	3.61	0.57	1.37	12.09 ± .05	16.78 ± .85	3.79	5.02 ± .01	1.34 ± .12	2.70	0.40
45– 90 μm	4.63	7.83	1.49	3.35	12.38 ± .03	20.34 ± .19	6.10	5.07 ± .01	1.11 ± .05	4.61	0.78
<20 μm	1.38	42.3	11.3	20.3	12.53 ± .07	29.28 ± .36	40.8	5.20 ± .01	1.03 ± .02	25.8	3.41
24210,9											
150–250 μm	3.33	6.91	1.51	3.13	12.34 ± .10	20.94 ± .30	6.45	5.10 ± .01	1.02 ± .07	4.96	0.69
45– 90 μm	4.71	11.1	2.50	5.15	12.42 ± .04	22.96 ± .28	10.9	5.15 ± .07	0.84 ± .03	7.52	1.06
<20 μm	1.38	43.6	11.7	18.1	12.65 ± .03	30.03 ± .29	31.1	5.28 ± .01	0.90 ± .02	13.1	1.05

Table 1. Isotopic concentrations (cm^3STP/g) of noble gases in grain size separates of Luna 24 soils. Abundance uncertainties are estimated as ±5–10% for He, Ne, Ar, and Xe and ±10% for Kr. Relative uncertainties for isotopic ratios are one sigma of the mean of multiple measurements of individual ratios plus one-half the applied blank corrections. Absolute isotopic ratios have an additional uncertainty of ±0.1%/mass unit. Blank corrections were nearly always <1%.

Table 2.

Sample	$^{84}Kr/^{86}Kr$	$^{83}Kr/^{86}Kr$	$^{82}Kr/^{86}Kr$	$^{80}Kr/^{86}Kr$	$^{78}Kr/^{86}Kr$
24077,9					
45– 90 μm	3.256 ± .014	.662 ± .004	.678 ± .005	.1815 ± .0047	.0268 ± .0008
<20 μm	3.271 ± .007	.659 ± .003	.705 ± .008	.2095 ± .0047	.0319 ± .0054
24109,13					
150–250 μm	3.258 ± .013	.665 ± .003	.667 ± .003	.1388 ± .0065	.0248 ± .0016
45– 90 μm	3.263 ± .012	.664 ± .004	.666 ± .003	.1352 ± .0010	.0211 ± .0005
<20 μm	3.270 ± .014	.660 ± .005	.662 ± .005	.1347 ± .0028	.0205 ± .0004
24149,15					
150–250 μm	3.275 ± .006	.667 ± .006	.672 ± .005	.1436 ± .0058	.0215 ± .0014
45– 90 μm	3.253 ± .016	.663 ± .004	.688 ± .005	.1926 ± .0094	.0294 ± .0013
<20 μm	3.302 ± .034	.668 ± .008	.680 ± .010	.1635 ± .0074	.0253 ± .0010
24174,10					
150–250 μm	3.265 ± .022	.664 ± .008	.669 ± .008	.1436 ± .0043	.0254 ± .0032
45– 90 μm	3.269 ± .015	.665 ± .004	.666 ± .004	.1326 ± .0007	.0227 ± .0010
<20 μm	3.272 ± .008	.659 ± .002	.660 ± .004	.1304 ± .0007	.0212 ± .0007
24182,15					
150–250 μm	3.291 ± .035	.673 ± .006	.672 ± .010	.1422 ± .0080	.0296 ± .0046
45– 90 μm	3.284 ± .022	.672 ± .003	.676 ± .007	.1390 ± .0020	.0243 ± .0016
<20 μm	3.274 ± .011	.663 ± .004	.665 ± .004	.1326 ± .0015	.0220 ± .0008
24210,9					
150–250 μm	3.272 ± .022	.669 ± .008	.669 ± .006	.1427 ± .0051	.0273 ± .0018
45– 90 μm	3.277 ± .014	.668 ± .004	.671 ± .008	.1354 ± .0021	.0251 ± .0016
<20 μm	3.300 ± .011	.666 ± .005	.675 ± .005	.1374 ± .0023	.0275 ± .0019

Table 2. Isotopic ratios of Kr in grain size fractions of Luna 24 soils. Small corrections for $(^{40}Ar)_2^+$ have been applied to ^{80}Kr. Relative uncertainties given are one sigma of the mean of multiple measurements, plus one-half the magnitude of the above corrections for ^{80}Kr. Absolute ratios have an additional uncertainy of ±0.1%/mass unit in the applied correction for instrument discrimination.

cosmic ray interactions (Table 4). These data are not as precise as we have usually obtained because of the small sample weights and relatively few grain sizes analyzed for each soil. The trapped $^4He/^3He$ and $^{36}Ar/^{38}Ar$ ratios and the concentrations of cosmogenic ^3He and ^{38}Ar were obtained from such plots using data from all three grain sizes of each soil (except 24077). The trapped $^{22}Ne/^{21}Ne$ and $^{20}Ne/^{22}Ne$ ratios were obtained from only the two finest grain sizes of each soil. However, using all three available grain sizes for each soil would only change the $^{20}Ne/^{22}Ne$ and $^{22}Ne/^{21}Ne$ ratios by amounts comparable to the variations among soils shown in Table 4. Ordinate-intercept plots for $^{40}Ar/^{36}Ar$ were non-linear, and the upper limits to this ratio given in Table 4 are the lowest measured value for each soil.

Abundances and isotopic compositions of the trapped gas in these Luna 24 samples are generally typical of mare soils. Within their relative uncertainties

Table 3.

Sample	$^{134}Xe/^{136}Xe$	$^{132}Xe/^{136}Xe$	$^{131}Xe/^{136}Xe$	$^{130}Xe/^{136}Xe$	$^{129}Xe/^{136}Xe$	$^{128}Xe/^{136}Xe$	$^{126}Xe/^{136}Xe$	$^{124}Xe/^{136}Xe$
24077,9								
45– 90 μm	1.218 ± .013	3.263 ± .017	2.693 ± .023	.536 ± .008	3.377 ± .034	.288 ± .006	.0180 ± .0035	.0121 ± .0042
<20 μm	1.227 ± .011	3.310 ± .026	2.726 ± .016	.542 ± .006	3.417 ± .029	.284 ± .006	.0155 ± .0027	.0124 ± .0036
24109,13								
150–250 μm	1.218 ± .027	3.251 ± .048	2.674 ± .034	.533 ± .011	3.354 ± .043	.274 ± .006	.0183 ± .0030	.0162 ± .0024
45– 90 μm	1.229 ± .013	3.300 ± .038	2.721 ± .033	.545 ± .009	3.423 ± .038	.281 ± .005	.0175 ± .0026	.0162 ± .0036
<20 μm	1.218 ± .020	3.266 ± .035	2.688 ± .027	.533 ± .011	3.365 ± .027	.271 ± .006	.0146 ± .0022	.0152 ± .0024
24149,15								
150–250 μm	1.241 ± .031	3.327 ± .054	2.756 ± .046	.539 ± .012	3.442 ± .052	.282 ± .012	.0307 ± .0127	.0240 ± .0024
45– 90 μm	1.231 ± .017	3.298 ± .060	2.717 ± .040	.535 ± .011	3.406 ± .066	.289 ± .009	.0251 ± .0037	.0189 ± .0049
<20 μm	1.220 ± .018	3.326 ± .078	2.723 ± .061	.541 ± .015	3.406 ± .067	.283 ± .006	..0138 ± .0030	.0149 ± .0030
24174,10								
150–250 μm	1.228 ± .033	3.208 ± .064	2.616 ± .049	.513 ± .021	3.268 ± .063	.262 ± .012	.0223 ± .0073	.0178 ± .0092
45– 90 μm	1.219 ± .012	3.277 ± .013	2.709 ± .022	.541 ± .005	3.410 ± .017	.275 ± .006	.0200 ± .0025	.0150 ± .0032
<20 μm	1.225 ± .013	3.301 ± .026	2.702 ± .021	.540 ± .008	3.417 ± .021	.274 ± .004	.0145 ± .0013	.0132 ± .0031
24182,15								
150–250 μm	1.213 ± .035	3.203 ± .072	2.637 ± .053	.506 ± .025	3.267 ± .069	.271 ± .009	.0259 ± .0088	.0257 ± .0128
45– 90 μm	1.226 ± .010	3.237 ± .028	2.651 ± .033	.523 ± .011	3.323 ± .035	.270 ± .005	.0186 ± .0023	.0171 ± .0045
<20 μm	1.225 ± .015	3.305 ± .044	2.714 ± .035	.541 ± .005	3.441 ± .046	.273 ± .007	.0177 ± .0033	.0159 ± .0044
24210,9								
150–250 μm	1.234 ± .036	3.301 ± .090	2.724 ± .069	.549 ± .025	3.400 ± .090	.274 ± .015	.0178 ± .0055	.0166 ± .0086
45– 90 μm	1.216 ± .011	3.249 ± .028	2.669 ± .024	.527 ± .004	3.340 ± .036	.272 ± .008	.0195 ± .0021	.0166 ± .0019
<20 mm	1.221 ± .022	3.279 ± .032	2.684 ± .031	.534 ± .020	3.391 ± .033	.278 ± .017	—	—

Table 3. Isotopic ratios of Xe in grain size fractions of Luna 24 soils. Relative uncertainties given are one sigma of the mean of multiple measurements. Absolute ratios have an additional uncertainty of ±0.1%/mass unit in the applied correction for instrument discrimination.

$^4He/^3He$ ratios are essentially the same for all Luna 24 soils. Trapped $^{36}Ar/^{38}Ar$ are also the same within uncertainties except for 24210,9, which is higher. Trapped $^{20}Ne/^{22}Ne$ and $^{22}Ne/^{21}Ne$ show only a narrow spread among soils. The Xe and Kr isotopic data for the <20 μm grain sizes plot within the mass fractionation trends found for a large number of Apollo 15, 16, and 17 soils by Bogard *et al.* 1974). We interpret this to indicate that trapped Kr and Xe in Luna 24 soils have mass fractionated isotopic compositions relative to each other and to other lunar soils and were probably derived from a common solar wind composition. Trapped gas abundances for a given grain size vary among Luna 24 soils in relation to their surface maturity, as shown by a plot of ^{36}Ar and percent agglutinates (Fig. 2). Such a relationship has been attributed to incorporation of trapped gases into agglutinates as a soil matures at the lunar surface (e.g., Schultz *et al.*, 1977; Bogard, 1977)

Whereas $[^3He]_c$ and $[^{38}Ar]_c$ were obtained from ordinate-intercept analyses, concentrations of cosmogenic ^{80}Kr and ^{126}Xe were calculated from the measured $^{80}Kr/^{86}Kr$ and $^{126}Xe/^{136}Xe$ ratios in the 150–250 μm fractions and the trapped solar wind ratios reported by Eberhardt *et al.* (1972). Concentrations of cosmogenic ^{21}Ne can be obtained more precisely than the other cosmogenic

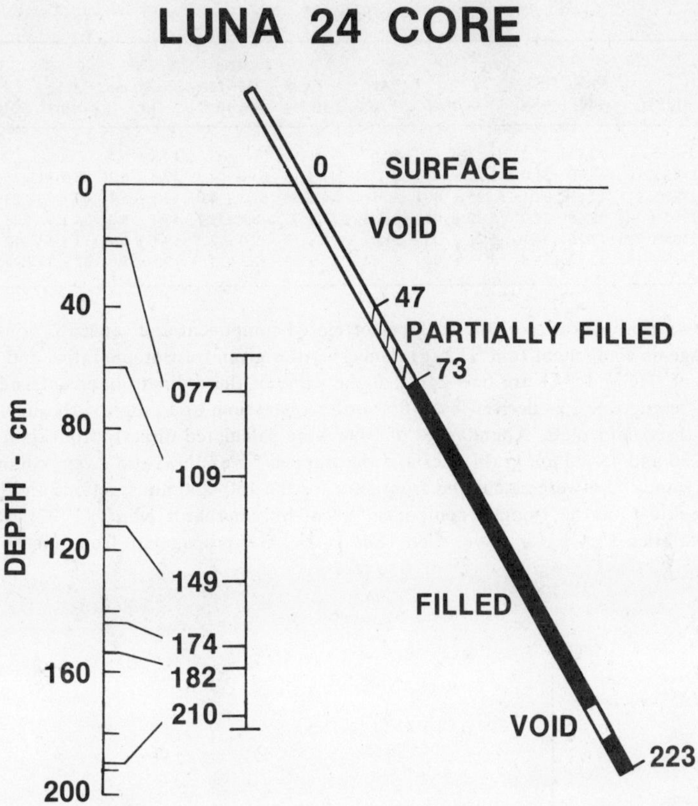

Fig. 1. Schematic model of possible depth relationships of Luna 24 soils. On the right the coring tube is shown to have entered the lunar regolith at a 30° angle and to have been incompletely filled with soil (Florensky *et al.*, 1977). On the left two extreme models of possible subsurface depth relationships of our samples are indicated, depending on explanations used for void spaces in the core (see text). Sample numbers are indicated.

nuclides, and consequently $[^{21}Ne]_c$ were calculated independently for the 45–90 μm and 150–250 μm size fractions using the trapped $^{22}Ne/^{21}Ne$ values given in Table 4.. The relative uncertainties given for the $[^{21}Ne]_c$ values do not consider the unkown uncertainty in the trapped values used. If we assume as an extreme case that the uncertainties in trapped $^{22}Ne/^{21}Ne$ are comparable to the total spread of values (i.e., ±1), additional uncertainties would be produced in $[^{21}Ne]_c$ of 9–14% for the 45–90 μm grain sizes and 7–10% for the 150–250 μm grain sizes. An uncertainty of ±1 unit in trapped $^{22}Ne/^{21}Ne$ is the maximum we would expect because trapped $^{22}Ne/^{21}Ne$ determined from ordinate-intercept analyses of a large number of lunar soils typically give $^{22}Ne/^{21}Ne$ values of 31–33.

To compare concentrations of cosmogenic gases among soils we applied corrections for differences in chemical composition using the bulk soil chemical

Table 4.

Sample	$^{4}He/^{3}He$	$^{20}Ne/^{22}Ne$	$^{22}Ne/^{21}Ne$	$^{36}Ar/^{38}Ar$	$^{40}Ar/^{36}Ar$	^{3}He 10^{-8}	^{21}Ne 150–250 μm 10^{-8}	^{21}Ne 45–90 μm 10^{-8}	^{38}Ar 10^{-8}	^{80}Kr 10^{-11}	^{126}Xe 10^{-11}
24077,9	~2975	12.75	31.6	—	≤ .67	—	—	84.7 ± 6.5	—	—	—
24109,13	2745 ± 10	12.70	31.0	5.22 ± .02	≤ .71	247 ± 13	65.0 ± 4.3	72.9 ± 6.2	58 ± 11	25 ± 14	1.1 ± .8
24149,15	2880 ± 115	12.61	31.1	5.225 ± .004	≤ .74	316 ± 56	61.3 ± 4.0	57.1 ± 5.1	40 ± 7	23 ± 8	2.6 ± 1.0
24174,9	2750 ± 30	12.65	32.0	5.22 ± .03	≤ .66	171 ± 45	38.4 ± 2.7	57.4 ± 5.2	28 ± 9	22 ± 6	1.6 ± 1.5
24182,15	2565 ± 295	12.56	31.1	5.21 ± .03	≤1.03	197 ± 18	39.0 ± 2.2	59.1 ± 3.9	33 ± 6	12 ± 7	1.5 ± 1.1
24210,9	2715 ± 155	12.72	31.6	5.31 ± .04	≤ .84	172 ± 53	52.3 ± 3.5	63.6 ± 5.0	62 ± 14	23 ± 8	0.7 ± 1.1

Table 4. Isotopic ratios of the surface-correlated component and concentrations of the cosmogenic component (cm^3STP/g). Uncertainties given for isotopic ratios and abundances of ^{3}He and ^{38}Ar are one sigma of the value of the ordinate intercept and of the slope, respectively, as derived by a first-order regression, and describe the linearity of these three-data plots. Abundances of ^{21}Ne were calculated directly from data on the 150–250 and 45–90 μm grain sizes and the trapped $^{22}Ne/^{21}Ne$ ratio given. Abundances of ^{80}Kr and ^{126}Xe were calculated from data for the 150–250 μm fractions and isotopic compositions of the trapped component given by Eberhardt *et al.* (1972). Relative uncertainties shown for ^{21}Ne, ^{80}Kr, and ^{126}Xe are propagated from measurement errors.

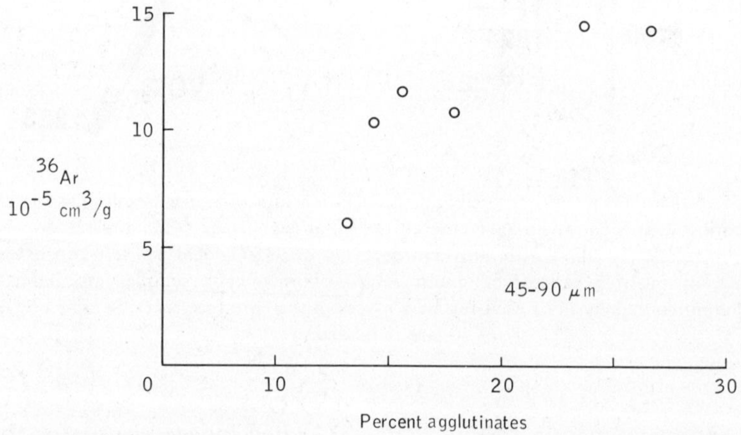

Fig. 2. Variation of ^{36}Ar concentrations with percent agglutinates in the 45–90 μm size fraction of Luna 24 soils.

data given by Blanchard *et al.* (1978) and the technique discussed by Bogard and Hirsch (1975). For $[^{21}Ne]_c$ these corrections were <2%, except for soil 24210,9, for which the correction was 6%. Analogous corrections to $[^{38}Ar]_c$ are <3%. The relevant trace element concentrations for making chemical corrections to $[^{80}Kr]_c$ and $[^{126}Xe]_c$ have been reported for four of our soils (Nyquist *et al.*, 1978), but the large experimental uncertainties associated with the $[^{80}Kr]_c$ and $[^{126}Xe]_c$ data dominate over applied corrections. Corrected data are used in the figures, whereas Table 4 gives uncorrected data. Blanchard *et al.* (1978) found considerable differences in concentrations of some elements between the <20 μm sizes on one hand and the 90–150 μm sizes and bulk soil on the other hand. Because the 45–90 μm size fraction includes the mean grain size of the bulk soil below 250 μm (McKay *et al.*, 1978), we expect that chemical compositions, and therefore our calculated correction factors, would be similar for the 45–90 μm and 150–250 μm fractions and the bulk soil.

Sample Depths

Cosmic ray-produced gases as a function of core depth, along with such parameters as particle tracks and I_s/FeO, can often be used to determine the depositional and irradiational history of a regolith section. Unfortunately, the subsurface depths of our Luna 24 samples are not precisely known. From descriptions of the core given by Florensky *et al.* (1977) and from discussions with C. Simonds (pers. comm., 1977), who observed the core under dissection in Moscow, we have attempted to determine probable depth relationships for our samples (Fig. 1). The right hand side of the Fig. represents the 260 cm core tube which penetrated the surface at an angle of 30 degrees. Soil numbers assigned by the Soviets represent distances measured along the core from the point of surface penetration, which the Soviets assume to be that point on the core where traces of soil are first found. Regolith filled the core below the 73 cm level indicated, except for an ~11 cm void near the bottom. A few grams of soil and coarse fines incompletely filled the 47–73 cm levels and apparently correspond to ~11 cm of filled core. Only traces of soil were observed between the 47 cm level and the indicated surface. If all material in the core above the 73 cm level were packed down to fill the total cross sectional area of the core, the first occurrence of material as measured along the core would be at the ~63 cm level (with an uncertainty of a few cm). To explain the greater depth of core tube penetration compared to recovered core length (approx. 220 cm and 150 cm, respectively), the Soviets (Florensky *et al.*, 1977) offer several possibilities: 1) uppermost parts of the regolith could have been incompletely sampled; 2) the soil column may have shortened after entering the core due to compaction; or 3) the soil may have expanded to fill the greater diameter of the core (~10–12 mm) compared to the bit (8 mm). Considering the small bit diameter, temporary blockage by coarse fragments conceivably could have occurred at several points. Rolling of the core and its cloth container into a drum for return conceivably could have distorted depth relations as well.

On the left hand side of Fig. 1 we have rotated the core into a vertical orientation and have indicated two extreme, but possible models of subsurface depth relationships among our Luna 24 soils. One interpretation (extreme left) assumes that all void spaces in the ~223 cm of core tube which presumably penetrated the regolith were caused by expansion of the core in cross-section to fill the tube. This interpretation indicates that ~193 cm of regolith were sampled and that sample 24077 and the coarse fragments located above it were located within ~18 cm of the lunar surface. The second interpretation shown in Fig. 1 is that no expansion of the core occurred and that the void spaces are due to incomplete recovery during coring. This interpretation results in a vertical section only ~121 cm deep, which in Fig. 1 we indicate to have been taken from subsurface depths of ~55 cm to ~178 cm. This model would indicate that sample 24077 was recovered from a subsurface depth of ~67 cm. Other depth models of the core might also be made. For example, it is conceivable that the ~121 cm long core section just discussed was taken from various levels along the cored section or from the lunar surface to a depth of ~121 cm. Radionuclide data reported by Surkov (1978) appear to favor the core model which sampled material just below the lunar surface, whereas radionuclide data reported by Rao (1978) appear to favor loss of appreciable material from the top of the core. Therefore, we do not know accurately the subsurface depths of our samples.

IRRADIATION HISTORY

In an effort to determine the depositional and irradiational history of the core we have examined the concentrations of cosmogenic gases as a function of core depth (Figs. 3 and 4). We use $[^{21}Ne]_c$ because we have more data of better accuracy for this nuclide, although $[^{3}He]_c$ and $[^{38}Ar]_c$ show generally the same trends. Fig. 3 shows $[^{21}Ne]_c$ plotted against depth for a core model which assumes a cored section ~193 cm deep (i.e., model 1 above). The curve indicates the approximate $[^{21}Ne]_c$ production profile with depth (Reedy and Arnold, 1972), assuming an average regolith density of 1.7 g/cm^3, and has been arbitrarily normalized to the 45–90 μm datum for 24109 (60 cm). The collective data are grossly inconsistent with a single stage, *in situ* production in a static regolith, as most $[^{21}Ne]_c$ data from soils below 80 cm are too high relative to the normalized production curve. Neither our choice of the datum to which the production curve is normalized, nor consideration of additional uncertainties in cosmogenic gas contents due to possible variations in the composition of trapped Ne would alter this conclusion. Fig. 4 shows the same $[^{21}Ne]_c$ data plotted against two different models of core depth which assume a core section ~121 cm deep. The lower depth scale in the figure assumes that the cored section was taken from the surface down to a depth of ~121 cm, whereas the upper depth scale assumes that the cored section was taken from ~55 cm depth down to the presumed maximum penetration of the core, ~193 cm. Subsurface depths for a given soil differ by ~55 cm between the two depth models. The theoretical production curves (Reedy and Arnold, 1972), again arbitrarily normalized to the 45–90 μm datum for 24109, are shown as dashed and solid lines for the lower and upper depth scales, respectively. As for Fig. 3, the collective data are inconsistent with a single stage, *in situ* irradiation for all soils. Neither are the collective data consistent with a simple model of slow accumulation of previously unirradiated soils over a long period of time followed by a long static period. The lowermost three soils are required to have been substantially pre-irradiated with cosmic rays before

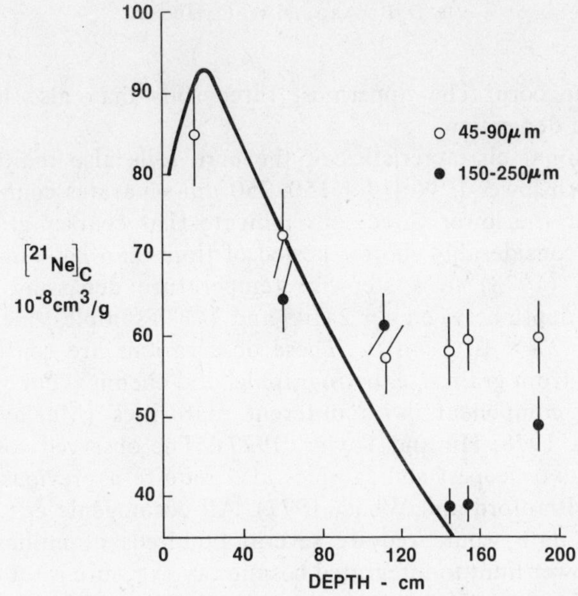

Fig. 3. Concentrations of cosmogenic ^{21}Ne as a function of soil depth for the model of a core taken from the lunar surface to a depth of ~193 cm. $[^{21}Ne]_c$ have been chemistry-corrected to data for 24149. The curve indicates the approximate production rate profile (Reedy and Arnold, 1972) of $[^{21}Ne]_c$ in a static regolith (assuming $\rho = 1.7$ g/cm^3) and has been arbitrarily normalized to the data for the 45–90 μm size fraction of 24109.

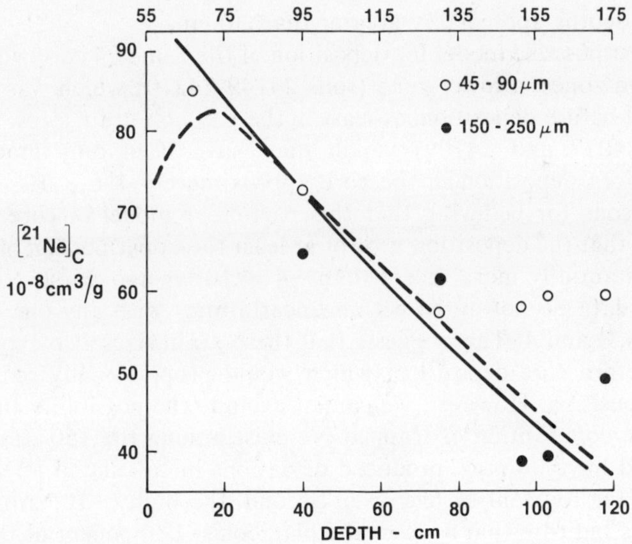

Fig. 4. Concentrations of cosmogenic ^{21}Ne (chemistry-corrected) as a function of soil depth for two models of a core which sampled a regolith section ~121 cm in vertical depth. The curves indicate the approximate production rate profile of $[^{21}Ne]_c$ in a static regolith (Reedy and Arnold, 1972) and have been arbitrarily normalized to the data for the 45–90 μm size fraction of 24109. The lower depth scale and dashed curve represent a core taken from the lunar surface to a depth of ~121 cm. The upper depth scale and solid curve represent a core taken between depths of ~55 cm and ~176 cm.

deposition in the core. The uppermost three soils may also have been pre-irradiated before deposition.

Several additional characteristics of the core soils also indicate a complex history. The much lower $[^{21}Ne]_c$ for 150–250 μm separates compared to 45–90 μm separates for the lower three soils indicate that coarser grains have been irradiated for a considerably shorter period of time than have finer grains. Rao and Venkatesan (1978) in a stepwise temperature degassing of soil 24163 (intermediate in depth between our 24149 and 24174 samples) measured an even lower $[^{21}Ne]_c$ of 27×10^{-8} cm^3/g. These observations are consistent with the conclusion made from grain size, petrographic, and chemical data that these soils are mixtures of components with different maturities (McKay *et al.*, 1978; Blanchard *et al.*, 1978; Hu and Taylor, 1977). The observed cosmic ray track densities in the two deepest soil samples also require a previous irradiation at shallow depths (Blanford and Wood, 1977). All cosmogenic gas concentrations observed in our data would require several hundreds of millions of years to accumulate. A lower limit to integrated cosmic ray exposure is set by the data for 150–250 μm sizes of 24174 and 24182 and by the data of Rao and Venkatesan (1978), which under the assumption of irradiation at the maximum in the production curve (i.e., ~20 cm depth), would require $\sim 3 \times 10^8$ yr and $\sim 2 \times 10^8$ yr of irradiation, respectively. On the other hand, the integrated exposure time of soil 24077 could be no shorter than $\sim 6 \times 10^8$ yr if it were irradiated within ~30 cm of the surface, and even longer exposure times are indicated if this soil were irradiated at depths appreciably greater than 30 cm.

The simplest possible model for deposition of the Luna 24 core would be that it consists of two zones: a lower zone (soils 24149-24210) which was substantially pre-irradiated before deposition, at least in the <90 μm grain sizes, and an upper zone (soils 24077 and 24109) which may have been only moderately pre-irradiated before deposition in the core approximately $4\text{–}6 \times 10^8$ yr. However, there are reasons for believing that this particular model of core deposition is incorrect and that the deposition time of at least the upper portion of the Luna 24 core was substantially more recent than $\sim 4 \times 10^8$ yr ago. First, $[^{21}Ne]_c$ for our 150–250 μm data do not agree within uncertainties with any one of the model curves in Figs. 3 and 4. This suggests that these grain sizes also experienced pre-irradiation before core deposition, which would proportionally reduce the time since core deposition. However, we cannot exclude the possibility that variations in the isotopic composition of trapped Ne exist among the 150–250 μm sizes of these soils and have, in part, produced deviations in calculated $[^{21}Ne]_c$ from the model curves for four out of five soils. Second, the high ($\sim 10^9$ cm^{-2}) solar flare track densities indicate that at least the plagioclase component of these soils has been extensively irradiated within ~1 cm of the surface (Blanford and Wood, 1977). Third, the I_s/FeO and agglutinate maturity indices (Morris, 1978; McKay *et al.*, 1978) are too low for any of these soils to have lain within ~20 cm of the surface for period of ~300 my. Morris (1978) suggests that the decrease in I_s/FeO from 24077 to 24149 and the bimodal grain size distribution for these soils are not consistent with an *in situ* reworking profile but were probably produced at the time of core deposition.

Last, and perhaps most important, the isotopic changes which have been produced in Sm by neutron capture are a factor of two higher in soil 24109 than that observed in any other lunar soil, whereas, the neutron fluence seen by coarse-grain soil 24170 is a factor of two lower than that in 24109 (Wasserburg *et al.*, 1978). This indicates that both soils have experienced a long irradiation at depths greater than the recovered depth of 24109, and that for much of this pre-irradiation the two soils could not have existed for the same time in the same relative depth position as that in which they were recovered. If we accept the arguments given above, we can rule out any combination of models of core deposition with models of core depth which require the Luna 24 soils to have received a major portion of their irradiation by cosmic rays since their deposition in the Luna 24 core. This would suggest that all soils which we analyzed were substantially pre-irradiated before deposition in the core, and that this deposition could be relatively recent.

In conclusion, cosmogenic noble gas data, even when combined with currently available data on soil maturity, particle tracks, radionuclides, and neutron fluence do not suggest any definitive model of the depositional and irradiational history of the Luna 24 core. This ambiguity has a variety of causes, including the uncertainty in the subsurface depths of the soils, the apparent multistage history of irradiation by cosmic rays, the evidence for soil origins by a mixture of components with different maturities and irradiation histories, and the seemingly contradictory suggestions that the core soils are old and strongly irradiated with cosmic rays and solar flares (noble gases and tracks) and at the same time have relatively low surface maturities (agglutinates and I_s/FeO). Additional data on neutron fluence, radionuclides, and particle tracks in ferromafic minerals may help resolve these problems. Our preferred model for the core is that it was deposited over some relatively short period during the past 3×10^8 yr. as a mixture of soils having variable irradiation histories. The question of whether an appreciable amount of material was lost from the top of the core during acquisition also remains ambiguous, as deposition of the core a relatively short time ago would permit sample 24077 to have lain within a few cm of the lunar surface since deposition.

Acknowledgments—Grain-size separations of these Luna 24 samples were made by D. S. McKay. We acknowledge informative discussions with D. S. McKay, R. V. Morris, D. P. Blanchard and G. Blanford.

References

Barsukov V. L., Tarasov L. S., Dmitriev L. V., Kolesov G. M., Shevaleevsky I. D. and Garanin A. V. (1977) The geochemical and petrochemical features of regolith and rocks from Mare Crisium (preliminary data). *Proc. Lunar Sci. Conf. 8th*, p. 3319–3332.

Basu A., McKay D. S. and Fruland R. M. (1978) Origin and modal petrology of Luna 24 soils. In *Mare Crisium: The View from Luna 24* (R. B. Merrill and J. J. Papike, eds.). This volume.

Blanchard D. P., Brannon J. C., Aaboe E. and Budahn J. R. (1978) Major and trace element chemistry of Luna 24 samples from Mare Crisium. In *Mare Crisium: The View from Luna 24* (R. B. Merrill and J. J. Papike, eds.). This volume.

Blanford G. E. and Wood G. C. (1977) Particle track densities in the Luna 24 core (abstract). In *Papers Presented to the Conference on Lune 24*, p. 41. The Lunar Science Institute, Houston.

Bogard D. D. (1977) Effects of soil maturation on grain size-dependence of trapped solar gases. *Proc. Lunar Sci. Conf. 8th*, p. 3705–3718.

Bogard D. D. and Hirsch W. C. (1975) Noble gas studies on grain size separates of Apollo 15 and 16 deep drill cores. *Proc. Lunar Sci. Conf. 6th*, p. 2057–2083.

Bogard D. D., Hirsch W. C. and Nyquist L. E. (1974) Noble gases in Apollo 17 fines: Mass fractionation effects in trapped Xe and Kr. *Proc. Lunar Sci. Conf. 5th*, p. 175–2003.

Eberhardt P., Geiss J., Graf H., Grogler N., Krahenbuhl U., Schwaller H., Schwarzmuller J. and Stettler A. (1970) Trapped solar wind noble gases, exposure age and K/Ar-age in Apollo 11 fine material. *Proc. Apollo 11 Lunar Sci. Conf.*, p. 1037–1070.

Eberhardt P., Geiss J., Graf H., Grogler M. D., Mendia M., Morgeli M., Schwaller H. and Stettler A. (1972) Trapped solar wind gases in Apollo 12 lunar fines 12001 and Apollo 11 breccia 10046. *Proc. Lunar Sci. Conf. 3rd*, p. 1821–1856.

Florensky C. P., Basilevsky A. T., Ivanov A. V., Pronin A. A. and Rode O. D. (1977) Luna 24: Geologic setting of landing site and characteristics of sample core (preliminary data). *Proc. Lunar Sci. Conf. 8th*, p. 3257–3280.

Hu H. N. and Taylor L. A. (1977) Luna 24 soils from Mare Crisium: Agglutinate chemistry (abstract). In *Papers Presented to the Conference on Luna 24*, p. 82. The Lunar Science Institute, Houston.

McKay D. S., Basu A. and Waits G. (1978) Grain size and the evolution of Luna 24 soils. In *Mare Crisium: The View from Luna 24* (R. B. Merrill and J. J. Papike, eds.). This volume.

Morris R. V. (1978) FMR and magnetic studies of Luna 24 soils and >1 mm soil particles. In *Mare Crisium: The View from Luna 24* (R. B. Merrill and J. J. Papike, eds.). This volume.

Nyquist L., Wiesmann H., Bansal B., Wooden J. and McKay G. (1978) Chemical and Sr-isotopic characteristics of the Luna 24 samples. In *Mare Crisium: The View from Luna 24* (R. B. Merrill and J. J. Papike, eds.). This volume.

Rao M. N. (1978) Data presented at the IX Lunar and Planetary Science Conference, (March 13–17) Houston.

Rao M. N. and Venkatesan T. R. (1978) Noble gas studies of Luna 24 drill core samples (abstract). In *Lunar and Planetary Science IX*, p. 934–936. Lunar and Planetary Institute, Houston.

Reedy R. C. and Arnold J. R. (1972) Interaction of solar and galactic cosmic ray particles with the Moon. *J Geophys. Res.* **77**, 537–555.

Schultz L., Weber H. W., Spettel B., Hintenberger H., and Wänke H. (1977) Noble gas and element distribution in agglutinate and bulk grain size fractions of soil 15601. *Proc. Lunar Sci. Conf. 8th*, p. 2799–2816.

Surkov Yu. A. (1978) Data presented at the IX Lunar and Planetary Science Conference, (March 13–17) Houston.

Wasserburg G. J., Papanastassiou D. A., McCulloch M. T., Dymek R. F., DePaolo D. J., Chodos A. A. and Albee A. L. (1978) Petrology, chemistry, age, and irradiation history of Luna 24 samples. In *Mare Crisium: The View from Luna 24* (R. B. Merrill and J. J. Papike, eds.). This volume.

Merrill, R. B. and Papike, J. J., editors.
(1978) *Mare Crisium: The View from Luna 24*, Pergamon Press (New York), p. 117-123
Printed in the United States of America

FMR and magnetic studies of Luna 24 soils and >1 mm soil particles

RICHARD V. MORRIS

Code SN7, Geochemistry Branch, NASA Johnson Space Center, Houston, Texas 77058

Abstract—Six Luna 24 soils (24077, 24109, 24149, 24174, 24182, 24210), their grain-size separates, and twenty-five >1 mm soil particles were analyzed by ferromagnetic resonance (FMR) and static magnetic techniques. The values of the FMR maturity index (I_s/FeO) for the soils ranged from 19.0 to 39.0 units; soils 24077 and 24109 are submature and soils 24149, 24174, 24182, and 24210 are immature. The metallic iron contents of the soils ranged from 0.30 to 0.51 wt.% and are commensurate with their maturities. The "best estimate" value for the concentration of metallic iron of Luna 24 mare basalts is 0.038 wt.%.

INTRODUCTION

The Luna 24 core was drilled into the regolith in the southeastern part of Mare Crisium. The regional geologic setting is discussed by Florensky *et al.* (1977) and Butler and Morrison (1977). Six samples of soil (24077, 24109, 24149, 24174, 24182, and 24210) and one sample of crushed rock (24170) were presented by the Soviet Union to NASA at the Eighth Lunar Science Conference. The sample locations along the core in units of centimeters are given by the last three digits of the sample identification. This paper pertains to ferromagnetic resonance (FMR) and magnetic studies of the bulk soil (<250 μm), grain-size separates thereof, and >1 mm soil particles. Surface maturity data for the soils is determined as a part of these studies. Grain-size and petrographic, rare gas, particle track, and chemical studies of the same allocation of bulk soil and its grain-size separates are reported elsewhere in this volume (McKay *et al.*, 1978; Basu *et al.*, 1978; Bogard and Hirsch, 1978; Blanford and Wood, 1978; Blanchard *et al.*, 1978). The best estimate of the metallic iron concentration of Luna 24 mare basalts is obtained from the studies of the >1 mm particles. The experimental procedures are discussed elsewhere (Morris, 1976, 1978a,b; Morris and Gose, 1976).

SOIL (<250 μM) STUDIES

Results and discussion

The FMR- and magnetically-derived parameters for the <250 μm fraction of soil are summarized in Table 1. The concentration of magnetically-determined FeO is quite uniform about an average value of 20.6 ± .5 wt.%. The error quoted is the standard deviation of the average; the accuracy of the FeO determinations for these soils is about ±0.8 wt.% at the one standard deviation level (Morris *et*

Table 1. FMR- and magnetically-derived parameters for the Luna 24 soils. The maturity classification is defined by Morris (1976).

Soil	FeO* (wt.%)	I_s/FeO <250 μm (Arb)	ΔH (Gauss)	Fe°_{VSM}[+] (wt.%)	n	Comment
24077,7	21.0	39.0	787	0.51	0.45	Submature
24109,14	21.0	31.0	783	0.46	0.47	Submature
24149,16	20.4	21.0	798	0.33	0.56	Immature
24174,17	21.0	27.0	790	0.37	0.78	Immature
24182,12	20.1	19.0	793	0.30	0.74	Immature
24210,10	19.8	19.0	790	0.33	0.48	Immature

*Calculated from the paramagnetic susceptibility.
[+]Concentration of metal particles having diameters ≳20–40 A.

al.,1978). The values of FeO reported here are in excellent agreement with the values reported by Blanchard *et al.* (1978). The values of the FMR linewidth average 790 ± 5 Gauss and are in the range expected from the FeO content of the soils (Morris, 1976). The values of the FMR maturity index, I_s/FeO, range from 19.0 to 39.0 units. Since the boundaries between immature and submature soils and between submature and mature soils are at 30.0 and 60.0 units of I_s/FeO, respectively (Morris, 1976), the Luna 24 soils are immature to submature. Soils having comparable maturities were returned from every Apollo landing site except Apollo 11 (Morris, 1976, 1978b). The values of Fe°_{VSM} (the concentration of all metal particles ≳20–40 A in diameter as measured on a vibrating sample magnetometer) average 0.38 ± 0.08 wt.%. Fe°_{VSM} covaries with I_s/FeO because a large fraction of the metal in these soils (~60%) is acquired during exposure at the lunar surface (Morris, 1978a).

There is an apparent conflict between the I_s/FeO and agglutinate (McKay *et al.*, 1978) maturity data and the solar-flare track maturity data. According to Blanford and Wood (1978), solar-flare track densities for plagioclase grains from the six Luna 24 soils are mostly $>10^9$ cm^{-2}. From a comparison of I_s/FeO and particle track data for soils from 60009/10 (Blanford *et al.*, 1977), such a high proportion of grains with $>10^9$ cm^{-2} indicates mature soils and not immature to submature soils as is indicated by I_s/FeO and agglutinates. This apparent contradiction can be resolved by allowing that the Luna 24 soils are mixtures in which the major plagioclase-bearing component is mature. However, at least some pyroxene grains also have high track densities (Crozaz, 1978; Blanford and Wood, 1978). Mass balance considerations then make the I_s/FeO and agglutinate data and the track data harder to reconcile by the above argument; that is, there may not be enough of an immature component to balance the mature component so that the bulk maturity as determined by I_s/FeO or agglutinates is immature to submature. If so, the conflict between the I_s/FeO and agglutinate data and the track data remains unresolved.

The grain-size dependence of I_s (relative concentration of fine-grained metal) for the six soils is shown in Fig. 1 as a log-log plot of I_s versus soil particle diameter (d). For each soil the data were fit to the following equation:

$$\log (I_s) = -n \log (d) + \text{constant} \quad (1)$$

The calculated values of *n* range between 0.45 and 0.78 and are listed for the individual soils in Table 1. Although the 0.78 value is among the largest observed for lunar soils, on a plot of I_s/FeO versus *n* the Luna 24 data fall in the range defined by the great majority of other lunar soils (Morris, 1977).

Evolution of the Luna 24 core

As discussed in detail by Bogard and Hirsch (1978), there is considerable ambiguity in the depth of the samples with respect to the lunar surface. This circumstance, especially for exposure parameters (e.g., $^{21}Ne_c$) which are produced over a large range in depth, greatly complicates interpretation of exposure-related data. In Fig. 2, the depth variation of I_s/FeO is shown for two of the models proposed by Bogard and Hirsch (1978). These models were chosen because they maximize and minimize the depth of the uppermost soil 24077. It is

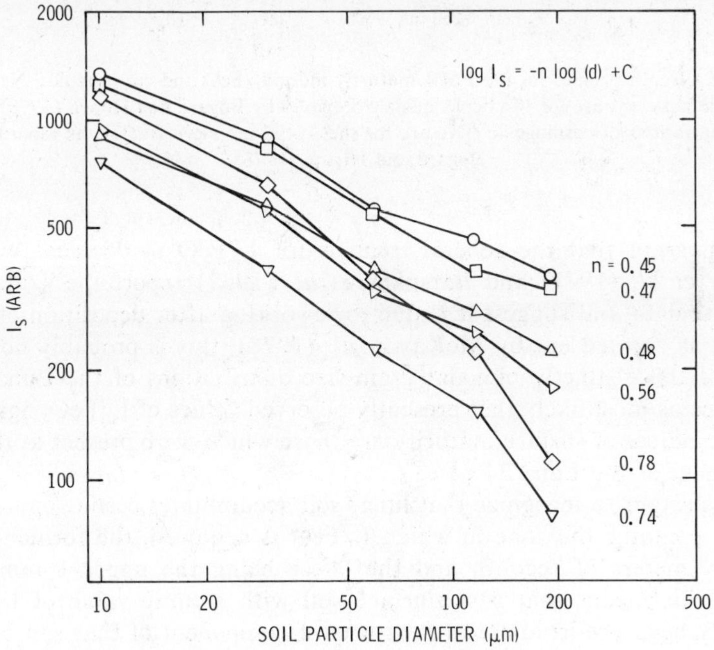

Fig. 1. Dependence of I_s (relative concentration of fine-grained metal) on soil-particle diameter. For each soil, the data points represent the values of I_s for the <20, 20–45, 45–90, 90–150, and 150–250 μm sieve fractions. The symbol for soil 24077 is a circle; 24109, square; 24149, flag; 24174, diamond; 24182, inverted triangle; and 24210, triangle.

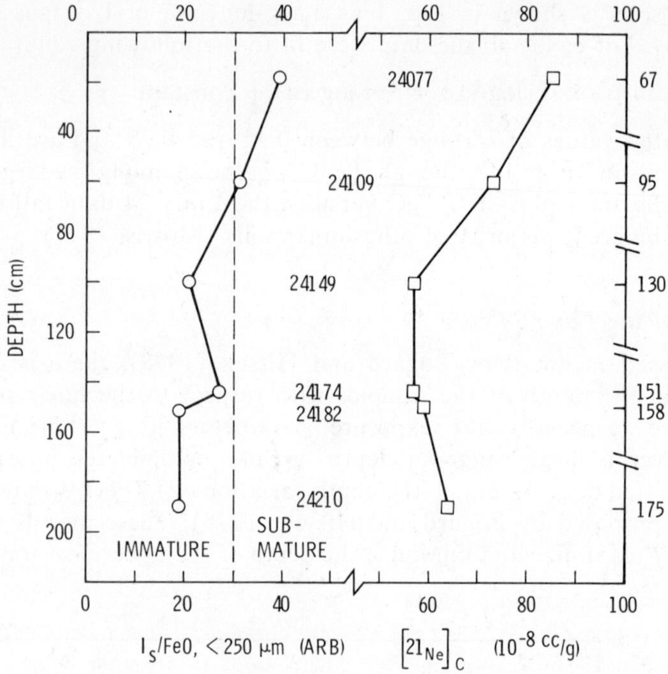

Fig. 2. Depth profiles for the FMR maturity index I_s/FeO and cosmogenic ^{21}Ne. The sample locations are for two depth models proposed by Bogard and Hirsch (1978). The concentrations of cosmogenic ^{21}Ne are for the 45–90 μm sieve fraction as reported by Bogard and Hirsch (1978).

readily apparent that the general trend is for I_s/FeO to decrease with depth. Florensky *et al.* (1977) and Barsukov *et al.* (1977) report the same trend in agglutinate data and suggest it is due to reworking after deposition of the core. However, as pointed out by McKay *et al.* (1978), this is probably not the case because of the distinctly bimodal grain-size distributions of the Luna 24 soils. Thus, it seems most likely that presently observed values of I_s/FeO, agglutinates, and other indices of surface maturity are those which were present at the time of emplacement of the Luna 24 core.

It is important to recognize that lunar soil accumulates cosmogenic gases in a zone that contains the zone in which I_s/FeO is acquired, the former being the upper few meters of regolith and the latter being the upper 1 mm or so of regolith. This means that a subsurface soil with a finite value of I_s/FeO has necessarily been pre-irradiated (or at least a component of that soil has) in the upper 1 mm of the lunar surface. Since I_s/FeO varies from 19.0 to 39.0 units for the Luna 24 soils, it is not surprising that Bogard and Hirsch (1978) were not able to fit their cosmogenic ^{21}Ne data to an *in situ* irradiation profile for any depth model. In fact, Fig. 2 shows that the concentration of cosmogenic ^{21}Ne tends to covary with I_s/FeO, implying that depth variations in cosmogenic ^{21}Ne

may simply be related to differences in maturity and not to differences in the production of cosmogenic ^{21}Ne with depth. Thus, the cosmogenic ^{21}Ne data do not tightly constrain the time of emplacement of the core. The final emplacement could have occurred recently in association with one of the Giordano Bruno secondaries proximate to the nominal Luna 24 landing site (e.g., Butler and Morrison, 1977). In this case, virtually all the cosmogenic ^{21}Ne is due to pre-irradiation. The final emplacement could also have occurred long ago, and several limiting cases are discussed by Bogard and Hirsch (1978). The longest emplacement time they give is ~600 m.y., but this is an upper limit for the time of final emplacement because, as discussed above, at least some of the cosmogenic ^{21}Ne for all the soils is due to pre-irradiation.

>1 mm Particle Studies

Metallic iron concentrations in two ranges of metal grain diameters were obtained for >1 mm soil particles in order to distinguish particle types. The results for 25 particles are given in Table 2. The parameter Fe°_{VSM} denotes the concentration of metal grains ≳20–40 A in diameter; the parameter Fe°_{FMR} denotes the concentration of metal grains in the diameter range between ~40 Å and somewhat greater than 330 Å. As shown in Fig. 3, the data cluster into three groups on a plot of Fe°_{VSM} versus Fe°_{FMR}. Previous FMR and magnetic studies of lunar rocks (e.g., Housley *et al.*, 1976; Pearce *et al.*, 1974; Cisowski *et al.*, 1974) have shown that (1) basalts tend to have relatively low concentrations (≲0.2 wt.%) of metal and it is coarse-grained, (2) high-grade breccias tend to have relatively high concentrations (≳0.3 wt.%) of metal and it is coarse-grained, and (3) soil or low-grade breccias are distinctive in having an appreciable proportion of fine-grained metal. It then follows that the individual Group 1 particles are Luna 24 basalts, mineral grains or lithic fragments from basalts, or perhaps impact melts of basalts. In the last instance, the impact melt would not have incorporated meteoritic metal. The Group 2 particles are most likely impact melts incorporating meteoritic metal or high-grade breccias. The Group 3 particles appear to be constructional particles of soil, such as soil breccias or agglutinate particles, which have undergone insufficient heating to coalesce the fine-grained metal. The Luna 24 soils should and do plot in the vicinity of the Group 3 particles, as shown in Fig. 3.

A "best estimate" for the metallic iron concentration of Luna 24 mare basalts was calculated by averaging the values of Fe°_{VSM} for the Group 1 particles. The value obtained is 0.038 wt.%, and it is on the low end of the range (~0.02 to ~0.21 wt.%) observed for mare basalts.

Summary

(1) The values of the FMR maturity index (I_s/FeO) range from 19.0 to 39.0 units. The upper two soils (24077 and 24109) are submature, and the lower four soils (24149, 24174, 24182, and 24210) are immature.

Table 2. Values of Fe°_{VSM} and Fe°_{FMR} for >1 mm particles from Luna 24 soils. The parameter Fe°_{VSM} denotes the concentration of metal having diameters ≳20–40 A; the parameter Fe°_{FMR} denotes the concentration of metal in the diameter range between ~40 A and somewhat greater than 330 A. The >1 mm particles were classified into groups according to Fig. 3. The error in Fe°_{VSM} is about ±.015 wt.% and that of Fe°_{FMR} about ±10% relative.

Particle	Fe°_{VSM} (wt.%)	Fe°_{FMR} (wt.%)	Group	Particle	Fe°_{VSM} (wt.%)	Fe°_{FMR} (wt.%)	Group
24077,4	0.018	0.009	1	24109,8	0.065	0.010	1
,11	0.23	0.011	2	,9	0.58	0.14	3
,12	0.031	0.003	1	24149,5	0.48	0.16	3
,13	0.019	0.001	1	24174,5	0.023	0.006	1
,14	0.021	0.003	1	,6	0.057	0.012	1
,16	1.41	0.21	3	,7	0.064	0.016	1
,17	0.31	0.003	2	24182,5	0.27	0.096	3
,18	0.72	0.24	3	,8	0.044	0.005	1
,19	0.027	0.004	1	,12	0.042	0.001	1
,21	0.70	0.26	3	24210,5	0.41	0.013	2
24109,5	—	0.17	3	,6	0.33	0.006	2
,6	0.044	0.008	1	,50	0.47	0.009	2
,7	0.032	0.007	1				

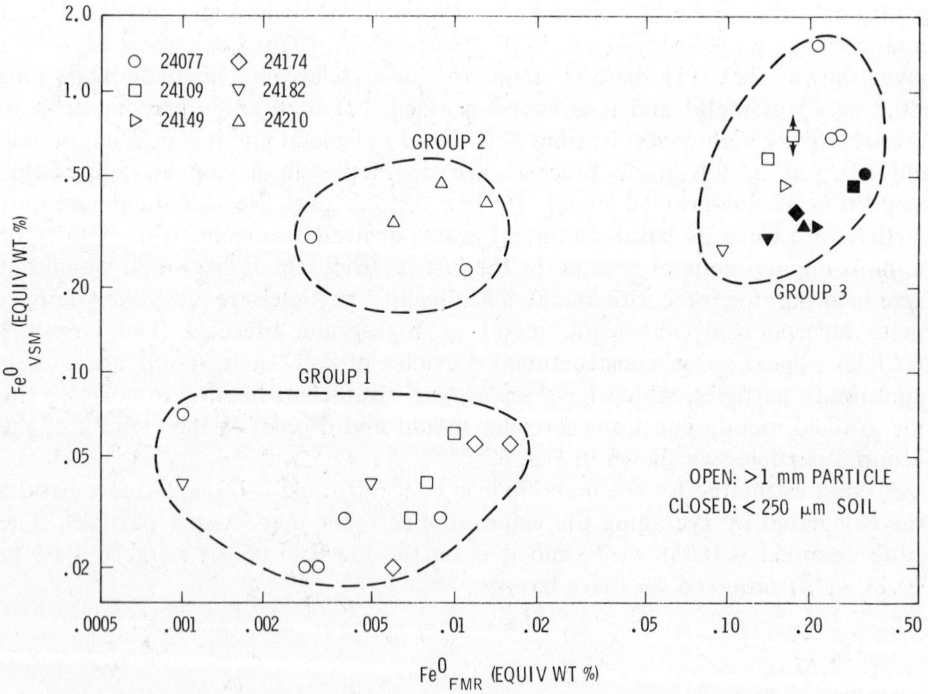

Fig. 3. This figure is a plot of Fe°_{VSM} versus Fe°_{FMR} for >1 mm particles from Luna 24 soils and <250 μm Luna 24 soil. The values for the soils were taken from Table 1 and Morris (1978a).

(2) The values of I_s/FeO imply that the Luna 24 soils are pre-irradiated with respect to cosmogenic ^{21}Ne. In part for this reason, the time of final emplacement of the Luna 24 core is not well constrained.

(3) The "best estimate" for the metallic iron concentration of Luna 24 mare basalts is 0.038 wt.%.

Acknowledgments—This paper has benefited from discussions with Drs. D. D. Bogard and D. S. McKay and the reviews of Drs. G. Crozaz and R. Housley. Miss C. Hardy is thanked for typing the manuscript.

REFERENCES

Barsukov V. L. (1977) Preliminary data for the regolith core brought to Earth by the automatic lunar station Luna 24. *Proc. Lunar Sci. Conf. 8th*, p. 3303–3318.

Basu A., McKay D. S., and Fruland R. M. (1978) Petrography, mineralogy, and source rocks of Luna 24 soils. In *Mare Crisium: The View From Luna 24* (R. B. Merrill and J. J. Papike, eds.). This volume.

Blanchard D. P., Brannon J. C., Aaboe E., and Budahn J. (1978) Major and trace element chemistry of Luna 24 samples from Mare Crisium. In *Mare Crisium: The View From Luna 24* (R. B. Merrill and J. J. Papike, eds.). This volume.

Blanford G. E., McKay D. S., and Wood G. C. (1977) Particle track densities in double drive tube 60009/10. *Proc. Lunar Sci. Conf. 8th*, p. 3017–3025.

Blanford G. E. and Wood G. C. (1978) Particle track densities in the Luna 24 core. In *Mare Crisium: The View From Luna 24* (R. B. Merrill and J. J. Papike, eds.). This volume.

Bogard D. D. and Hirsch W. C. (1978) Noble gases in Luna 24 core soils. In *Mare Crisium: The View From Luna 24* (R. B. Merrill and J. J. Papike, eds.). This volume.

Butler P. Jr. and Morrison D. A. (1977) Geology of the Luna 24 landing site. *Proc. Lunar Sci. Conf. 8th*, p. 3281–3301.

Cisowski C. S., Dunn J. R., Fuller M., Ross M. F., and Wasilewaki P. J. (1974) Impact processes and lunar magnetism. *Proc. Lunar Sci. Conf. 5th*, p. 2841–2858.

Crozaz G. (1978) Nuclear particle tracks and the regolith at the Luna 24 site. In *Mare Crisium: The View from Luna 24* (R. B. Merrill and J. J. Papike, eds.). This volume.

Florensky C. P., Basilevsky A. T., Ivanov A. V., Pronin A. A., and Rode O. D. (1977) Luna 24: Geologic setting of landing site and characteristics of sample core (preliminary data). *Proc. Lunar Sci. Conf. 8th*, p. 3257–3279.

Housley R. M., Cirlin E. H., Goldberg I. B., and Crow H. (1976) Ferromagnetic resonance studies of lunar core stratigraphy. *Proc. Lunar Sci. Conf. 7th*, p. 13–26.

McKay D. S., Basu A., and Waits G. (1978) Grain size and evolution of Luna 24 soils. In *Mare Crisium: The View from Luna 24* (R. B. Merrill and J. J. Papike, eds.). This volume.

Morris R. V. (1976) Surface exposure indices of lunar soils: A comparative FMR study. *Proc. Lunar Sci. Conf. 7th*, p. 315–335.

Morris R. V. (1977) Origin and evolution of the grain-size dependence of the concentration of fine-grained metal in lunar soils: The maturation of lunar soils to a steady-state stage. *Proc. Lunar Sci. Conf. 8th*, p. 3719–3747.

Morris R. V. (1978a) Reduction excess, meteoritic excess, and initial metals in lunar soils: Origins of the excess metal. Unpublished manuscript.

Morris R. V. (1978b) The maturity of lunar soils: Concepts and more values of I_s/FeO. In *Lunar and Planetary Science IX*, p. 760–762. Lunar and Planetary Institute, Houston.

Morris R. V. and Gose W. A. (1976) Ferromagnetic resonance and magnetic studies of cores 60009/60010 and 60003: Compositional and surface exposure stratigraphy. *Proc. Lunar Sci. Conf. 7th*, p. 1–11.

Morris R. V. Gose W. A. and Lauer H. V. (1978) Depositional and surface exposure history of the Shorty Crater core 74001/2: FMR and magnetic studies. *Proc. Lunar Sci. Conf. 9th*. In press.

Pearce G. W., Strangway D. W., and Gose W. A. (1974) Magnetic properties of Apollo samples and implications for regolith formation. *Proc. Lunar Sci. Conf. 5th*, p. 2815–2826.

Merrill, R. B. and Papike, J. J., editors.
(1978) *Mare Crisium: The View from Luna 24*, Pergamon Press (New York), p. 125–136
Printed in the United States of America

Grain size and the evolution of Luna 24 soils

D. S. McKay

NASA Johnson Space Center, Houston, Texas 77058

A. Basu

Dept. of Geology, Indiana Univ., Bloomington, Indiana 47401

G. Waits

Lockheed Electronics Co., Houston, Texas 77058

Abstract—We have determined the grain size distribution of six Luna 24 samples. These samples are characterized by a bimodal distribution which is indicative of soils formed primarily by mixing rather than by reworking. Although agglutinate content decreases with depth, it is not likely that the Luna 24 soils have undergone appreciable *in situ* reworking.

We have also petrographically determined particle types and abundances in each of four size fractions. Mineral fragments are very abundant in all analyzed size fractions. Pyroxene and plagioclase increase slightly in abundance at finer grain sizes but olivine decreases significantly. Compared to typical mare soils, the Luna 24 trends are anomalous. They are compatible with the hypothesis that many, if not most, of the mineral grains in the 20–250 μm fractions come from coarse-grain rocks having average mineral grain sizes >250 μm. The mineralogy and chemistry of the coarse-grained rocks has not been well characterized but there is evidence that at least some of them are higher in MgO than the analyzed finer-grained basalts.

Introduction

Grain size relationships combined with petrographic modal data have been useful in unraveling some of the complex history of the lunar regolith. Such data have been particularly useful in determining whether a set of regolith samples have undergone significant *in situ* reworking or whether they were mainly the product of mixing events (McKay *et al.*, 1977). We have investigated six samples from the Luna 24 core by grain size analysis and by petrographic techniques to determine the grain size distributions, the particle types present, and their relative abundances in each size fraction. Detailed petrographic descriptions and electron probe analyses of soil particles are reported separately (Basu *et al.*, 1978). Portions of our sieved sample were also analyzed for major and trace elements (Blanchard *et al.*, 1978), rare gases (Bogard and Hirsch, 1978), FMR and magnetic properties (Morris, 1978), and particle tracks (Blanford and Wood, 1978).

METHOD AND RESULTS

We sieved our initial allocation of 40–50 mg of each soil through 150 μm, 90 μm, 45 μm and 20 μm sieves using liquid freon. Samples had been previously sieved in the JSC Curatorial Laboratory at 1 mm, 500 μm and 250 μm (Nagle and Walton, 1977). In addition to sieving we analyzed the <20 μm fraction with a Coulter counter to determine grain size distribution down to about 2 μm. Coulter counter data on the samples showed that the grain size distribution is approximately log normal down to between 5 and 10 μm depending on the sample. Below this point a slight negative deviation (fines deficiency) develops and continues down to our limit of size resolution 2 μm. The shapes of the curves suggest that between 0.5 and 1 percent of the samples are finer than 2 μm. Sieve results are presented in Table 1. We constructed cumulative size curves by combining our sieve data and Coulter counter data with sieve data on the coarser fractions from the Luna 24 catalog (Nagle and Walton, 1977). Summary size parameters Mz (graphic mean grain size) and σ_I (graphic standard deviation) were determined from the cumulative curves and are presented in Table 2 both for the entire sample (subcentimeter) and for the <250 μm fraction which we sieved. Grain size histograms derived from the cumulative curves are shown in Fig. 1.

The standard deviation of these samples shows a negative correlation with the mean grain size (Fig. 2). For comparison the least-squares lines for the Apollo 16 drive tube and Apollo 17 soil samples are also shown in Fig. 2. We have found that this negative correlation holds for all related groups of lunar samples which we have analyzed and which have been analyzed by other investigators provided that data from the entire size range below 1 cm is used in the grain size calculations.

Using both transmitted and reflected light we petrographically identified approximately 300 particles in each of the size fractions above 20 μm. Table 3 presents major particle categories and their relative abundances.

The bimodal grain size distribution

The most prominent feature shown in the grain size histograms (Fig. 1) is the bimodal nature of the samples. They have a coarse mode at about 1 mm for most samples and a finer mode in the vicinity of 100 μm. This bimodal property is particularly strong in the uppermost sample 24077. For four of the six samples the minimum part of the histogram separating the two modes occurs between 250 and 500 μm. It is important to note that 250 μm is the size at which all of the samples were separated prior to analyses by various investigator groups. Consequently, mineralogical and chemical data on the finer mode (<250 μm) may not be representative of the coarser mode or of the entire sample.

The bimodal nature of these soils strongly suggests that they are mixtures of a coarse soil with a finer soil and that these mixtures have not been well blended by meteorite reworking. Bimodal soils are likely to be products of path 2 soil

Table 1. Sieving weights from Luna 24 samples. All values in milligrams.

	24077	24109	24149	24174	24182	24210
150–250 μm	4.70	7.82	8.37	6.63	7.11	7.58
90–150 μm	6.05	8.97	9.45	8.78	9.68	8.32
45– 75 μm	10.22	12.53	12.75	12.20	13.55	11.56
20– 45 μm	8.87	11.59	11.56	10.54	11.95	9.66
<20 μm	10.75	10.09	8.66	9.84	12.63	6.82
Total wt.	40.59	51.00	50.79	47.99	54.92	43.94

Table 2. Summary grain size parameters of Luna 24 samples.

	Subcentimeter Grain Size Parameters			<250 μm	
	Mean Grain Size (M_Z)		Standard Deviation (σ_I)	Mean Grain Size (M_Z)	
	Φ	μm	Φ	Φ	μm
24077	2.85	139	2.38	4.59	42
24109	3.21	108	2.35	4.27	52
24149	3.41	94	2.12	4.17	56
24174	3.74	75	1.96	4.32	50
24182	3.59	83	2.24	4.34	49
24210	3.79	72	1.70	4.09	59

evolution (mixing) rather than path 1 soil evolution (reworking) (McKay *et al.*, 1974). As the uppermost sample is the most strongly bimodal it is the least likely to have undergone significant *in situ* reworking (McKay *et al.*, 1977). If 24077 was not reworked after deposition, the samples beneath it in the core could not have been reworked after deposition of 24077; they would have been protected by the overlying soil. However, the lower samples could have been reworked to some extent prior to deposition of 24077 or before deposition in their present position.

Depth variation of mean grain size

Variation with depth of the mean grain is shown in Fig. 3. True depth below the lunar surface is not known with certainty for any of these samples (Bogard and Hirsch, 1978) but relative depth is represented by distance down the core tube. For the subcentimeter data which includes the entire sample, the analyzed samples are coarsest at the top and, except for a small reversal, become finer toward the bottom. The depth relationships for the mean grain size of just the <250 μm size fraction shows an opposite trend to that shown for the subcentimeter data. This mean grain size generally becomes coarser with depth.

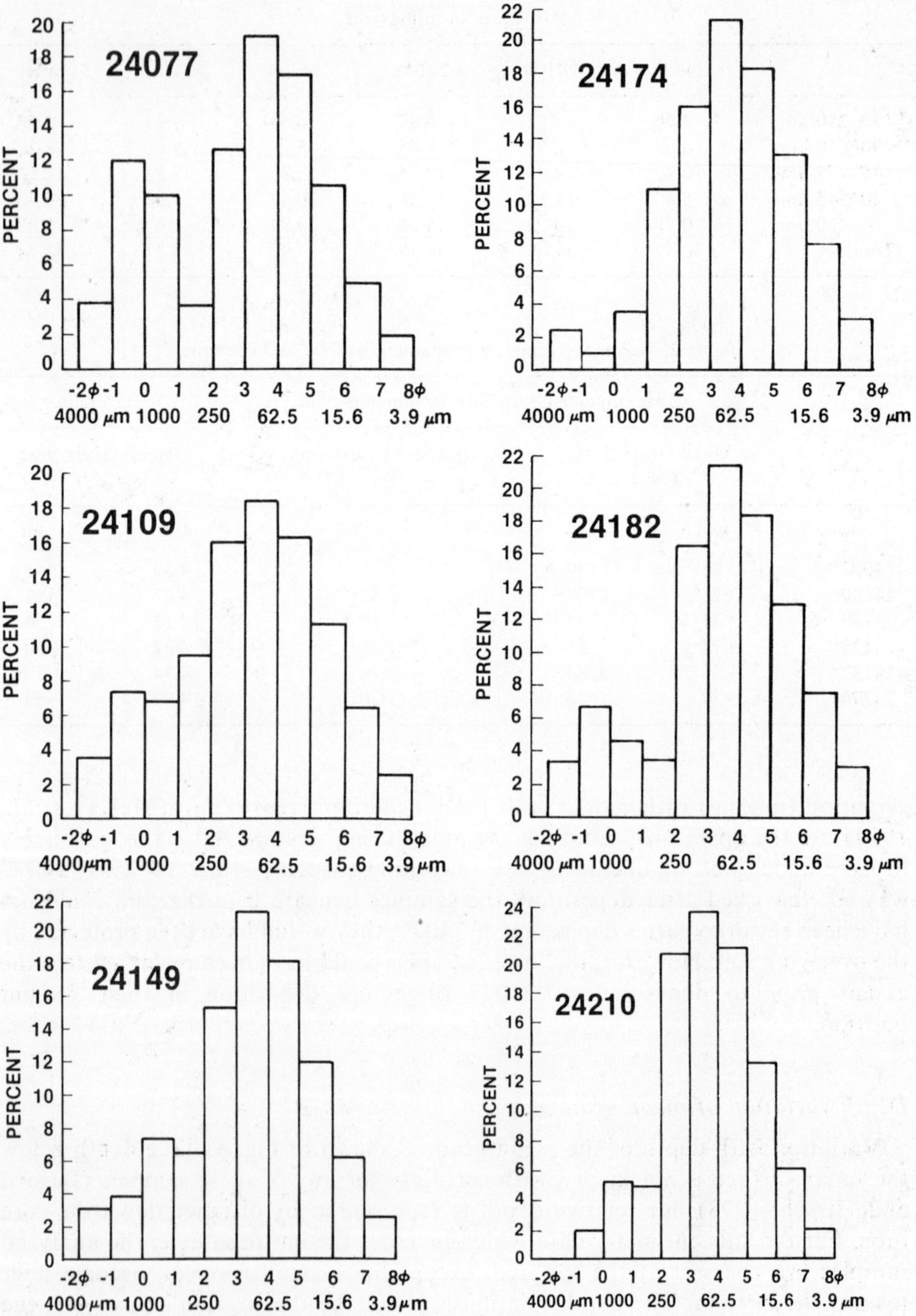

Fig. 1. Histograms of grain size distributions of Luna 24 samples.

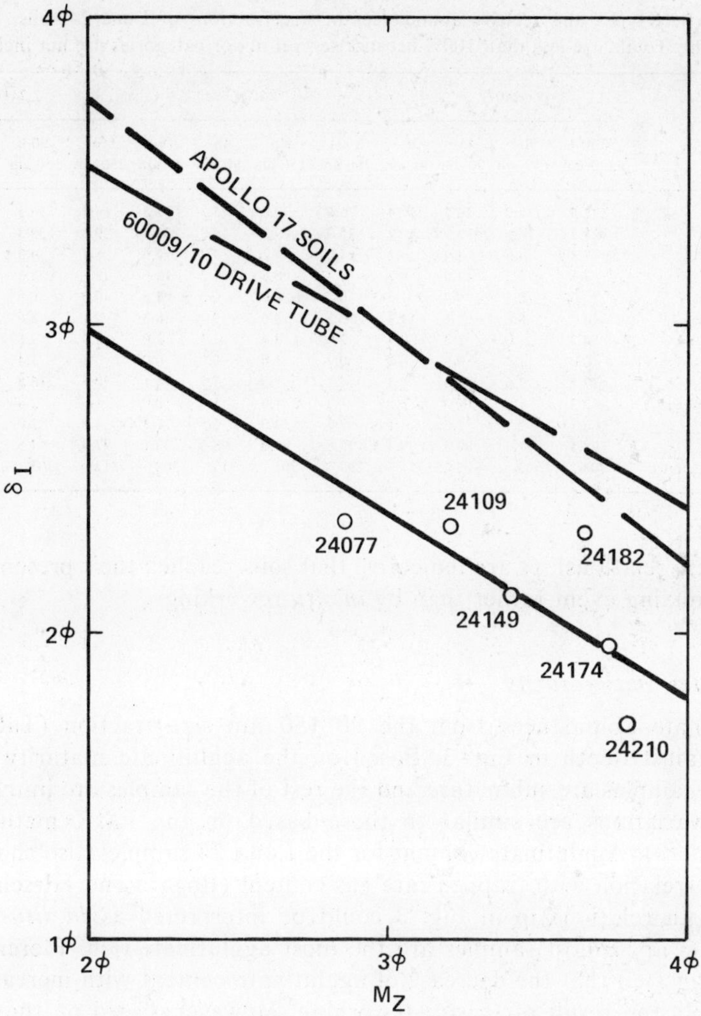

Fig. 2. Mean grain size (Mz) plotted against graphic standard deviation (σ_I) for Luna 24 samples. Least-squares lines for other samples are also shown.

If the subcentimeter mean grain size is used as a maturity index, the lowest samples would be considered the most mature and the uppermost samples would be the least mature. However, if only the <250 μm mean grain size is used as a maturity index the opposite is true. This relationship suggests the fine-grained mode, as represented by the <250 μm mean grain size, differs in maturity from the coarser mode; these soils have fractional rather than homogeneous maturity (McKay *et al.*, 1974). Furthermore, the uppermost samples are the most fractionated. That is, the grain size maturity index of the fine-grained mode is relatively the farthest separated from the maturity index for the entire sample.

Table 3. Particle types and relative abundances in size fractions of Luna 24 soils. All values in percent. Totals are less than 100% because several minor categories are not included.

	24077				24109				24149			
	150–250 μm	90–150 μm	45–90 μm	20–45 μm	150–250 μm	90–150 μm	45–90 μm	20–45 μm	150–250 μm	90–150 μm	45–90 μm	20–45 μm
Agglutinates	32.5	32.7	26.7	17.4	25.6	30.6	23.3	17.5	15.4	12.5	14.2	13.6
Pyroxene	22.7	26.4	33.9	40.2	35.0	29.2	35.8	33.8	29.2	39.3	34.8	40.1
Olivine	7.8	4.6	3.9	3.2	4.0	4.7	4.2	2.6	6.4	4.0	1.9	2.5
Plagioclase	7.1	6.3	14.0	8.7	9.4	6.6	9.3	13.6	8.7	8.6	13.9	7.6
Opaque Minerals	0.0	0.3	0.7	1.3	0.7	0.0	0.6	1.8	0.6	0.0	0.0	3.2
Olivine-Free Basalt	8.7	8.6	6.1	11.9	10.1	8.6	9.3	8.3	7.1	8.6	11.4	7.3
Olivine Basalt	4.2	3.0	1.3	0.3	2.0	3.4	1.3	2.0	3.8	2.6	2.2	0.6
Highland Derived ANT	0.0	1.7	0.3	0.0	0.7	1.0	0.0	0.0	0.3	1.4	0.3	0.0
Vitric Breccia	7.1	7.6	8.8	9.3	4.1	7.3	7.7	8.1	20.5	16.5	13.0	12.7
Melt-Matrix Breccia	2.3	1.3	1.0	0.0	1.3	1.6	1.3	0.0	1.6	2.0	0.0	0.3
Glass	6.1	5.3	2.6	7.1	2.7	4.0	5.5	11.3	4.0	2.0	5.0	10.5
Total	98.5	97.8	99.3	99.4	95.6	97.0	98.3	99.1	97.6	97.5	96.7	98.4
Total Grains Counted	308	303	307	311	297	301	313	308	312	303	316	314

Again, these relationships are indicative that soils reached their present composition by a mixing event rather than by *in situ* reworking.

Agglutinates and maturity

Agglutinate abundances from the 90–150 μm size fraction (Table 3) are plotted against depth in Fig. 3. Based on the agglutinate maturity index the upper two samples are submature and the rest of the samples are immature. The maturity variations are similar to those based on the FMR maturity index (Morris, 1978). Agglutinate content for the Luna 24 samples also shows a good positive correlation with trapped rare gas content (Bogard and Hirsch, 1978).

The depth relationship in Fig. 3 could be interpreted as *in situ* reworking because the uppermost samples are the most agglutinate-rich. Florensky *et al.* (1977) suggested that the decrease of agglutinate content with increasing depth was possibly the result of *in situ* reworking. However, based on the grain size arguments previously discussed, we doubt that appreciable *in situ* reworking has occurred in these samples. If *in situ* reworking has not occurred, agglutinate maturity reflects surface reworking and mixing history prior to deposition of the core section.

Does Luna 24 soil contain two distinct populations?

The grain size data indicate that Luna 24 soils are mixtures of at least two populations which differ in mean grain size. Although exceptions exist, previous studies of lunar soils have shown that coarse-grained soils are likely to be relatively immature and fine-grained soils are likely to be relatively mature. Consequently, based on grain size alone we can assume that the coarse mode of the bimodal Luna 24 soils is likely to be less mature than the fine-grained mode.

Table 3. (*Continued*).

	24174				24182				240210			
	150–250 μm	90–150 μm	45–90 μm	20–45 μm	150–250 μm	90–150 μm	45–90 μm	20–45 μm	150–250 μm	90–150 μm	45–90 μm	20–45 μm
Agglutinates	13.9	20.8	15.3	11.60	7.6	15.9	13.2	12.6	15.9	19.6	17.7	15.6
Pyroxene	40.1	34.6	36.7	43.89	42.8	29.8	33.1	41.0	30.0	26.6	31.9	37.8
Olivine	4.8	2.6	4.2	4.39	7.2	5.5	4.8	3.5	7.9	7.7	2.6	1.3
Plagioclase	12.6	9.9	13.6	11.60	13.3	19.1	14.8	15.8	8.6	9.0	16.8	14.3
Opaque Minerals	0.0	0.3	0.9	3.13	0.3	0.6	1.0	0.6	0.0	0.9	0.3	0.6
Olivine-Free Basalt	9.4	13.1	11.1	14.11	9.4	10.3	13.8	9.2	8.2	9.6	9.9	14.0
Olivine Basalt	2.7	1.9	0.6	1.25	3.6	2.5	2.0	0.9	4.5	1.6	0.6	0.6
Highland Derived ANT	2.0	2.2	0.9	0.00	1.0	1.6	0.0	0.0	1.7	2.2	0.3	0.0
Vitric Breccia	7.1	7.7	11.0	6.27	5.5	6.7	8.0	5.8	14.8	16.0	9.5	6.3
Melt-Matrix Breccia	1.4	0.3	0.6	0.31	2.3	1.0	1.3	0.0	2.8	1.9	1.1	0.0
Glass	3.1	5.7	2.5	2.19	4.2	4.9	5.8	8.4	5.0	3.6	6.9	8.0
Total	97.1	99.1	97.4	98.7	97.1	97.9	97.8	97.8	99.4	98.7	97.6	98.5
Total Grains Counted	294	312	308	300	306	309	311	310	290	312	304	314

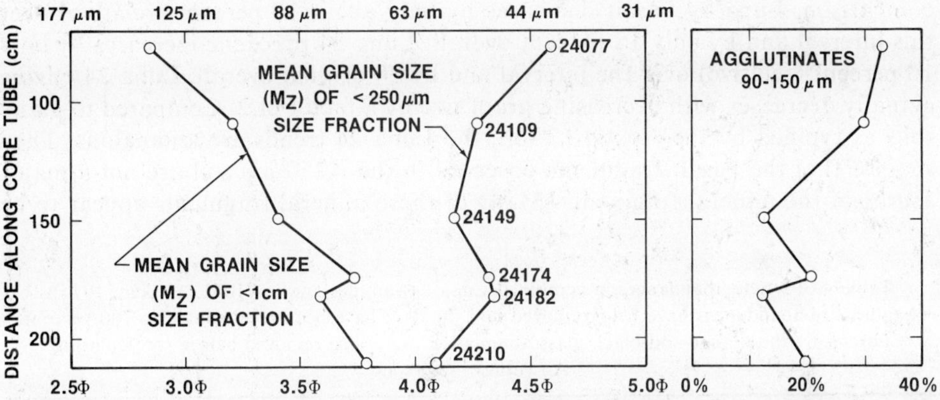

Fig. 3. Variation with distance down the core of subcentimeter mean grain size, <250 μm mean grain size and agglutinate content of the 90–150 μm grain size fraction.

Is there other evidence that this maturity difference exists? Blanford and Wood (1978) found bimodal track distributions for two of the samples (24182 and 24210) which they analyzed. These bimodal distributions were in an intermediate size range (45-90 μm) and may reflect a distribution overlap between a coarser, less mature component and a finer, more mature component. Goswami (1978) also found a bimodal distribution for 24190 and suggests that this sample represents a mixture between a mature and immature component. Bogard and Hirsch (1978) found that coarser grain size fractions for some soils contained less cosmogenic ^{21}Ne and suggested that this relationship indicates the mixing of components of differing pre-irradiation histories, the coarser material being less irradiated.

If the two grain size populations differ in maturity, do they perhaps differ in other properties such as mineralogy and chemical composition? While we have not made detailed studies of material coarser than 250 μm, we can perhaps make inferences about the mineralogy of this coarser material based on grain size trends of particle types in the finer material which we have analyzed. Table 4 presents particle abundances for major particle types in the Luna 24 samples. Data from the 6 Luna 24 samples have been combined and each size fraction represents over 1800 counted particles. Additionally, the data have been normalized to agglutinate-free soils in order to remove the effects of differing maturities. For comparison, normalized data from a typical mare soil from Apollo 17 is included (Heiken, 1975).

In most lunar soils, lithic fragments decrease and mineral fragments increase with decreasing grain size. This results as lithic fragments are progressively comminuted into their constituent minerals at finer and finer grain sizes. The Apollo 17 mare soil nicely demonstrates this relationship (Fig. 4). With decreasing grain size from 150-250 μm to 45-90 μm, basalt abundance decreases by more than a factor of 2 and pyroxene increases by a factor of 1.6. By comparison, Luna 24 basalt decreases by only about 10 percent (relative) over this interval and by only 15 percent overall. Luna 24 pyroxene increases by only 10 percent (relative) over the interval and by 20 percent overall. Luna 24 olivine actually decreases with decreasing grain size by a factor of 2. Compared to many soils as typified by the Apollo 17 soil, the Luna 24 trends are anomalous. They suggest that the basalt fragments observed in the <250 μm soil are not a major source of the mineral fragments. Many of these mineral fragments appear to be

Table 4. Particle abundances in combined Luna 24 samples and in 71061 (Heiken, 1975). Agglutinate abundances have been removed and the data have been recalculated to 100 percent. For 71061 the orange and black glass abundances were also removed before recalculating. All values in percent.

	150–250 μm	90–150 μm	45–90 μm	20–45 μm
Pyroxene	38.6	39.7	42.3	46.1
Olivine	7.9	6.3	4.5	3.4
Plagioclase	12.1	12.5	16.8	14.0
Olivine-Free Basalts	10.8	12.5	12.2	11.8
Olivine Basalts	4.3	3.3	1.4	1.1
Highland-Derived ANT	1.1	2.2	0.4	0.0
Vitric Breccia	12.0	13.0	11.8	9.5
Melt Matrix Breccia	2.4	1.8	1.1	0.1
Glass	5.3	5.6	5.8	9.0
MARE REGOLITH SAMPLE 71061				
Pyroxene	22.1	28.1	36.1	—
Plagioclase	11.5	23.1	22.5	—
Basalt	43.6	26.2	17.9	—
Vitric Breccia	6.5	5.6	3.5	—

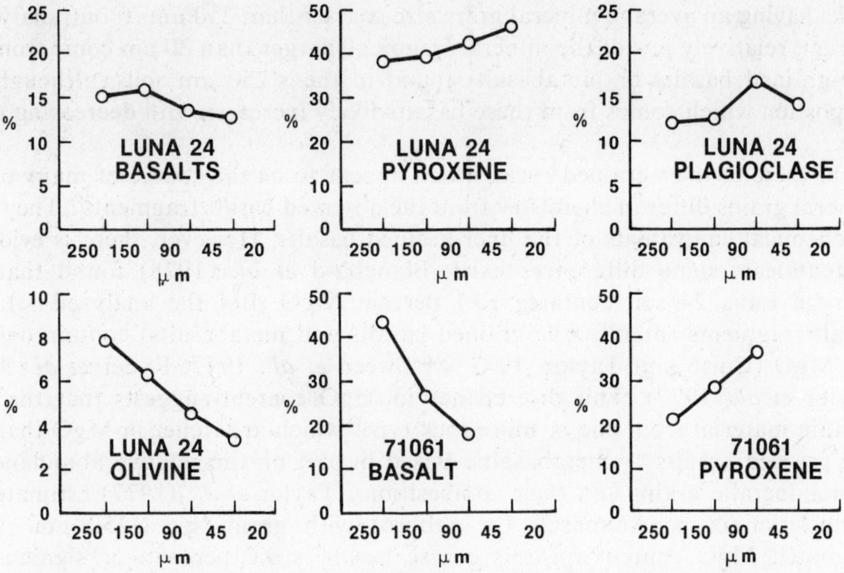

Fig. 4. Grain size variation of basalts and mineral fragments from Luna 24 and from an Apollo 17 mare soil. Abundances have been recalculated to remove agglutinates.

coming from rock types which are much coarser grained than the lithic fragments in the <250 μm soils. Mineral fragments from a very coarse-grained rock, for example a rock with average mineral size >1 mm, would be completely liberated from the rock by the time the rock fragments were comminuted to 250 μm. As comminution proceeded below 250 μm, these mineral fragments would be pulverized to smaller sizes but would not increase appreciably in abundance relative to other components in the soil except as the other components were destroyed by comminution. This describes the Luna 24 trends for pyroxene and plagioclase (Fig. 4). If comminution of the coarse-grained rock has only slightly proceeded, the grain size of many of the liberated minerals will be close to their original grain size and at finer grain sizes mineral abundance decreases because not many of the coarser mineral fragments have been comminuted to finer grain sizes. This describes the trend observed for olivine in Luna 24. By this analysis the olivine was either much coarser initially than the pyroxene and plagioclase or else came from a rock which had not been comminuted as long.

By contrast, if the average grain size of minerals in a rock are in the range of 45 to 250 μm, comminution over that size interval efficiently liberates those minerals from their parent rock fragments and causes the abundance of mineral fragments relative to rock fragments to sharply increase with decreasing grain size. This describes the trends for 71061 (Fig. 4).

This analysis based on Fig. 4 suggests that many, if not most, of the mineral fragments found at all grain sizes larger than 20 μm in Luna 24 soils come from rocks having an average mineral grain size coarser than 250 μm. If our analysis is correct, relatively few of the mineral fragments larger than 20 μm come from the fine-grained basalts or metabasalts found in the <250 μm soils, although the proportion which comes from these basalts likely increases with decreasing grain size.

Do these coarser-grained rocks which appear to be the source of many of the mineral grains differ in chemistry from the observed basalt fragments? They may be chemical equivalents of the finer-grained basalts. However, there is evidence that at least some differences exist. Blanchard *et al.* (1978) found that the average Luna 24 soil contains 10.1 percent MgO. But the analyzed <1 mm basalt fragments (mainly fine-grained basalts and metabasalts) contain only 5–8% MgO (Coish and Taylor, 1977; McSween *et al.*, 1977; Ryder *et al.*, 1977; Taylor *et al.*, 1977). This discrepancy in MgO content suggests that the soils contain material from one or more rock types which are richer in MgO than the fine-grained basalts or metabasalts. Based in part on the relative abundance of monomineralic grains and their compositions, Taylor *et al.* (1977) estimate the composition of coarse basalt (or gabbro) with grain size >250 μm. Their estimated MgO content for this coarse basalt is 9.0 percent, or significantly higher than that of the analyzed fine-grained basalts. This analysis assumes that the coarse mineral fragments came from a single rock type. Two of the gabbro fragments analyzed by Barsukov (1977) had MgO greater than 10%. However, it is unlikely that a small fragment of a coarse-grained rock would be representative of the overall bulk chemical composition. A good representative analysis of the coarse-grained (>250 μm) gabbro may therefore not be possible for the Luna 24 samples.

Basu *et al.* (1978) and Taylor *et al.* (1977) hypothesize, based on the abundance of high Mg olivines (>Fo70), that a rock type more mafic than any of the observed basalts has contributed significantly to the Luna 24 soils. If this suggestion is correct then the grain size behavior of olivine (Fig. 4) indicates that this mafic rock is very coarse-grained. Alternatively, the high Mg olivines could be associated with the observed coarse gabbro. However, none has been reported in the analyzed fragments of this gabbro; these olivines are always much more iron-rich (Taylor *et al.*, 1977; Vaniman *et al.*, 1977; Barsukov, 1977; Albee *et al.*, 1977).

Regardless of the source of the olivines, our grain size data indicate that they are concentrated in the coarser size range and are likely to belong primarily to the coarser grain size population revealed by the bimodal grain size distribution.

Another indication that a significant fraction of olivine is associated with the more immature, coarser population comes from the track data of Goswami (1978). His combined data on 6 different Luna 24 core samples show that 70 percent of the 28 analyzed olivine grains and less than 10^8 tracks/cm^2 but only 36 percent of the 108 analyzed plagioclase grains had less than 10^8 tracks/cm^2. This difference may result from a difference in track retention properties or etching properties between the two minerals. However, if the track densities reflect real exposure times, these data suggest that the olivine population is less mature than the plagioclase population and may therefore be more closely associated with the coarse-grained, immature mode of the samples. Pyroxene is intermediate with 51% of the 49 analyzed grains having less than 10^8 tracks/cm^2.

CONCLUSIONS

1. The bimodal grain size distribution of the Luna 24 samples supports the hypothesis that Luna 24 core soils are mixtures of a coarse, immature component and a finer, more mature component.
2. Decrease of agglutinate content with depth is compatible with *in situ* reworking; however, the strongly bimodal nature of the soils appears to preclude appreciable *in situ* reworking of any of the samples.
3. The source of many, if not most, of the mineral fragments in the 20–250 μm size range is probably one or more coarse-grained rock types not well represented in the <250 μm lithic fragments. There is evidence that some of these coarse-grained rocks have a higher MgO content than the fine-grained basalt.

Acknowledgments—Part of this research was conducted by one of us (AB) as a Visiting Scientist at the Lunar and Planetary Institute which is operated by the Universities Space Research Association under contract No. NSR-09-051-001 with the National Aeronautics and Space Administration. This paper constitutes the Lunar and Planetary Institute Contribution No. 319.

One of us (AB) is grateful to T. R. McGetchin and the staff of the LPI for many personal favors, and to J. Mitra of N.Y.U. for unlimited encouragement.

We also thank R. Fuhrman for operating the Coulter counter.

REFERENCES

Albee A. L., Chodos A. A. and Dymek R. F. (1977) Petrography of Luna 24 sample 24170 (abstract). In *Papers Presented to the Conference on Luna 24*, p. 5–7. The Lunar Science Institute, Houston.

Barsukov V. L. (1977) Preliminary data for the regolith core brought to earth by the automatic lunar station Luna 24. *Proc. Lunar Sci. Conf. 8th*, p. 3303–3318.

Basu A., McKay D. and Fruland R. (1978) Origin and modal petrography of Luna 24 soils. In *Mare Crisium: The View From Luna 24* (R. B. Merrill and J. J. Papike, eds.). This volume.

Blanchard D. P., Brannon J. C., Aaboe E. and Budahn J. R. (1978) Major and trace element chemistry of Luna 24 samples from Mare Crisium. In *Mare Crisium: The View From Luna 24* (R. B. Merrill and J. J. Papike, eds.). This volume.

Blanford G. E. and Wood G. C. (1978) Particle track densities in the Luna 24 core. In *Mare Crisium: The View From Luna 24* (R. B. Merrill and J. J. Papike, eds.). This volume.

Bogard D. D. and Hirsch W. C. (1978) Noble gases in Luna 24 core soils. In *Mare Crisium: The View From Luna 24* (R. B. Merrill and J. J. Papike, eds.). This volume.

Coish R. A. and Taylor L. A. (1977) Igneous rocks from Mare Crisium: Mineralogy and petrology (abstract). In *Papers Presented to the Conference on Luna 24*, p. 48–50. The Lunar Science Institute, Houston.

Goswami J. N. (1978) Particle track based irradiation and accretion history of the Luna 24 regolith column. *Geophys. Res. Lett.* In press.

Florensky C. P., Basilevsky A. T., Ivanov A. V., Pronin A. A. and Rode D. D. (1977) Luna 24: Geologic setting of landing site and characteristics of sample core (preliminary data). *Proc. Lunar Sci. Conf. 8th*, p. 3257–3279.

Heiken G. (1975) Petrology of lunar soils. *Rev. Geophys. Space Phys.* **13**, 567–587.

McKay D. S., Dungan M. A., Morris R. V. and Fruland R. M. (1977) Grain size, petrographic, and FMR studies of the double core 60009/10: A study of soil evolution. *Proc. Lunar Sci. Conf. 8th*, p. 2929–2952.

McKay D. S., Fruland R. M. and Heiken G. H. (1974) Grain size and the evolution of lunar soils. *Proc. Lunar Sci, Conf. 5th*, p. 887–906.

McSween H. Y. Jr., Taylor L. A. and Clark J. C. (1977) Luna 24 metabasalts: Possible evidence for assimilation of highlands materials? (abstract). In *Papers Presented to the Conference on Luna 24*, p. 118–120. The Lunar Science Institute, Houston.

Morris R. V. (1978) FMR and magnetic studies of Luna 24 soils and >1 mm soil particles. In *Mare Crisium: The View From Luna 24* (R. B. Merrill and J. J. Papike, eds.). This volume.

Nagle J. S. and Walton W. J. A. (1977) Luna 24 catalog and preliminary description of samples presented to the U.S.A. Curatorial office, NASA Johnson Space Center, Houston, 88 pp.

Ryder G., McSween H. Y. and Marvin U. B. (1977) Luna 24 VLT basalt: Character and origin (abstract). In *Papers Presented to the Conference on Luna 24*, p. 160–163. The Lunar Science Institute, Houston.

Taylor G. S., Warner R., Wentworth S., Keil K. and Sayeed U. (1977) Petrology and chemistry of Luna 24 mare basalts and basaltic glasses (abstract). In *Papers Presented to the Conference on Luna 24*, p. 189–192. The Lunar Science Institute, Houston.

Vaniman D. T. and Papike J. J. (1977) Luna 24 ferrobasalt: Petrology and comparison with other mare basalt types (abstract). In *Papers Presented to the Conference on Luna 24*, p. 196–199. The Lunar Science Institute, Houston.

Merrill, R. B. and Papike, J. J., editors.
(1978) *Mare Crisium: The View from Luna 24*, Pergamon Press (New York), p. 137–155
Printed in the United States of America

Microcrater and solar-flare track maturation of the lunar regolith

G. POUPEAU[+°], M. CHRISTOPHE-MICHEL-LÉVY[||], J. C. MANDEVILLE[§], J. JOHNSON[°] and PH. ROMARY[+]

[+]Centre des Faibles Radioactivités, Laboratoire mixte CNRS/CEA, 91190 Gif-sur-Yvette, France
[°]McDonnell Center for the Space Sciences, Department of Physics, Washington University, St. Louis, MO 63130
[||]Laboratoire de Minéralogie, Université Paris VI, 75005 Paris
[§]ONERA, CERT/DERTS, Toulouse

Abstract—We have been studying the microcrater, pancake and cosmic ray track record in the 100–200 microns feldspar fraction of a variety of surface soils and Apollo and Luna 16–24 cores.

In Luna 24, a study of the glassy agglutinates + breccias in 9 levels revealed this core was deposited as discrete layers, most of which (below level 150), are immature, the upper part of the core being submature. Our Luna 16 and 20 cores were found respectively mature and immature.

Glassy agglutinates and impact microcraters, absent to rare in immature soils such as 12033 and 14141, are present in similar abundances on crystals from submature to mature soils. In submature and mature soils, pancakes are present on 60 to 100% of crystals and craters on 50 to 70%.

Microcrater, track and solar wind gas data suggest feldspar crystals in the grain size range considered, reach an equilibrium maturity grade faster than bulk soils. This is interpreted as due to the limited lifetime of these crystals in the lunar regolith.

I. INTRODUCTION

Although the existence of impact microcraters on lunar fines material has been known since the earliest Apollo sample studies (Carter and MacGregor, 1970; McKay *et al.*, 1970; Crozaz *et al.*, 1970), they did not receive much attention up to the work of Blanford *et al.* (1974), who showed that crystals with such features are common in mature soils. Since 1975 (Poupeau *et al.*, 1975), we have pursued a systematic study of the cosmic-ray track and microcrater record in lunar and meteoritic crystals, as part of a general approach to the problem of solar and micrometeorite effects on extraterrestrial regoliths. The ultimate goal of this work (in conjunction with the measurement of solar wind effects) is to determine the constancy of our interplanetary environment during the past 4 billion years (Poupeau and Johnson, 1976; Poupeau *et al.*, 1978a,b; Etique *et al.*, 1978).

We present a series of results on microcraters and cosmic ray tracks in various levels of the Luna 16, 20 and 24 cores, the Apollo 16 deep drill core and a number of surface soils from Apollo 12 to 17 sites.

The latter were chosen among soils with widely varying degrees of maturity to follow the buildup of microcraters and track densities with soil "aging". In order to place our Luna samples in this soil maturity sequence, we also measured their abundance of glassy agglutinates and regolith breccia fragments in the relevant grain size fractions.

II. EXPERIMENTAL PROCEDURES

Our study of the microcrater/solar flare track record in single grains follows basically the techniques described earlier by Poupeau *et al.* (1975). We therefore refer below only to modifications of the routine procedures previously described.

The present work has been done exclusively on feldspar crystals taken in the 100–200 μm size fraction. Clear, relatively unfractured crystals were selected by handpicking under a stereomicroscope.

A systematic search for microcraters and glassy deposits on the crystals was done with a scanning electron microscope (SEM) at a magnification of 7,500× (corresponding to a smallest observable crater of 0.1 to 0.2 μm in diameter). Following the discovery of microcrater(s) on a given crystal, local surveys were performed at higher magnifications, up to 30,000×. In those cases where the total number of microcraters per crystal was ≳30, a photomosaic was made at high magnification and the cumulative size distribution of microcraters measured. In practice, this occurred only for microcrater densities $>10^6/cm^2$.

Following this, the crystals were mounted in epoxy, polished and etched in a boiling 6.25N NaOH solution. (For those crystals having microcrater densities $>10^6/cm^2$, the tracks were studied on a polished section perpendicular to the most cratered surface). Track density counts were done on photographs taken with a SEM standard magnification of 28,000×.

In order to evaluate the degree of maturation of our Luna samples, their glassy agglutinate content in various grain size fractions was measured. For the Luna 24 samples, counts were performed by direct observation of soil particles under a polarizing microscope. Seven samples taken from the top to the bottom of the core were studied on the 40–500 μm fraction, obtained after dry sieving of ~10 mg of bulk soil. Two other samples, 24096 (particles larger than 80 μm) and 24184 (fraction 200–370 μm) were received already sieved.

For Luna 16 and 20, we used previously prepared polished sections for the determination of glassy agglutinates in the grain size range 50–100 μm.

III. RESULTS

Surface features

More than 300 crystals have been studied, coming from 16 different soil samples. Our results are reported in Table 1.

In most soils, from 30% to 60% of the crystals have measurable microcraters. Lower abundances were found only in two soils, about 14% for soil 14141 and none in soil 12033. The abundance of microcratered crystals in soil 10084 has been revised upward from our earlier paper (Poupeau *et al.*, 1975) by a factor of 20% following the re-examination of a suite of crystals with our present SEM procedure (i.e. using a magnification of 7,500× rather than 3,000× to 5,000×). The abundances of microcratered crystals previously reported in the Apollo 15 core and soil 64421 (Poupeau *et al.*, 1975) were correspondingly corrected in Table 1.

The fraction of crystals with a large number of microcraters >0.2 μm diameter (corresponding to crater densities in excess of about $10^6/cm^2$) decreases with the overall percentage of microcratered crystals in a given soil. For example, soil 14141 has only about 5% of crystals with microcrater densities $>10^6/cm^2$, as compared to about 10–25% for more heavily microcratered soils.

Whatever the care brought to the search of microcraters in individual soil

Table 1. Microcraters and glassy discs on lunar soil feldspars.

Sample	Depth	Number of crystals	Fraction of crystals with: Glassy discs <2 μm dia.	Microcraters >0.2 μm dia.	Microcrater density[§] $\geq 10^6/cm^2$
LUNA					
Luna 16	20–22 cm	31	0.94	0.31	0.09
Luna 20		30		0.20	0
Luna 24	96 cm	17	0.85	0.50	0.17
	172 cm	24		0.37	
APOLLO					
12033		14	0.07	0	0
14141		19	0.74	0.14	0.05
67481, 67601	North Ray Crater rim	41	0.95	0.58	0.12
64421		13	1	0.46	0.15
10084		131	0.6	0.30	0.11
60007+		42	0.90	0.66	0.25
core 15000°		49	0.61	0.43	0.10

§For craters >0.1 microns dia.
+Grouped results from two levels at 2.5 cm and 9.5 cm depth, respectively (Poupeau and Johnson, 1976).
°Grouped from four levels at 37,124,240 and 242 cm respectively (Poupeau *et al.*1975).

crystals, it should be kept in mind that the abundances of microcratered crystals reported in Table 1 are lower limits to the actual values: only about 70% to 80% of the crystal surface is accessible to observation; also, most crystals have only a very small number of microcraters, making it easy to miss them. Typically no more than 1 to 5 microcraters were found per crystal, even after a careful examination at high magnification following the discovery of a first crater. For crystals having densities of microcraters greater than about $10^6/cm^2$, the microcratered area is usually very small compared to the total grain surface. Although precise estimates are difficult, the heavily cratered area generally represents <10% of the area accessible to observation (see also Poupeau *et al.*, 1975). Only on three crystals was a major crystal face found which was more or less regularly cratered.

So far, a total of about 40 crystals have been found to contain locally high densities ($>10^6/cm^2$) of microcraters. In all crystals but one, the microcrater (>0.1 μm diameter) density falls between about 10^6 and $6 \times 10^7/cm^2$; only one (see Zinner and Walker, 1975, Fig. 9) had a higher density of about $10^8/cm^2$.

For crystals with a local microcrater density of $\geq 10^6/cm^2$ and a total number of microcrater ≥ 30, the cumulative frequency of the microcrater diameters was measured. In all cases the cumulative frequency is represented by a straight line with a -2 slope in a log-log diagram (Fig. 1). For diameters smaller than 0.1 μm and down to about 600Å the slope may approach a value of 1, depending mostly on the quality of the cratered surface (and not on the microcrater density as would be expected for an erosion effect).

Glassy discs ≤2 μm in diameter have also been found on crystals from all soils. They were notably less abundant in soil 12033 than in any other soil with only one crystal (out of a total of 14 examined) having glassy discs. In the other soils, these "pancakes" (Blanford *et al.*, 1974) were present on more than 60% of the crystals. Usually, they are present on all major faces of the crystals, although their densities vary from face to face and even on a given face of a crystal. Usually, the pancake density is much below $10^6/cm^2$. Their abundance on a given crystal surface is *not* correlated with the density of microcraters, as shown on

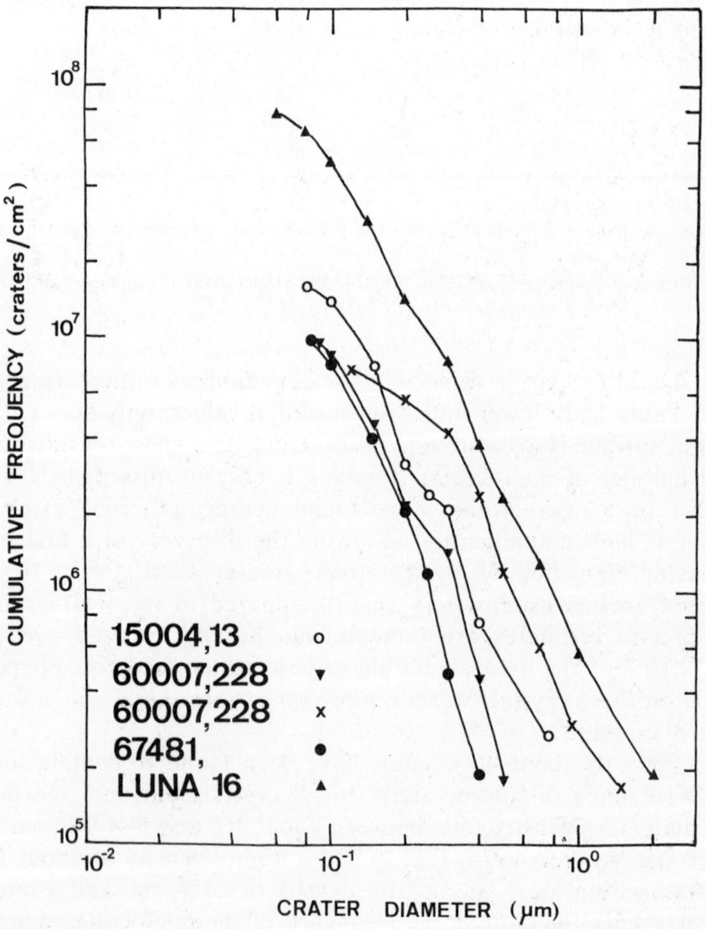

Fig. 1. Cumulative size-frequency (craters per unit area) distribution of microcraters in Apollo and Luna soils.

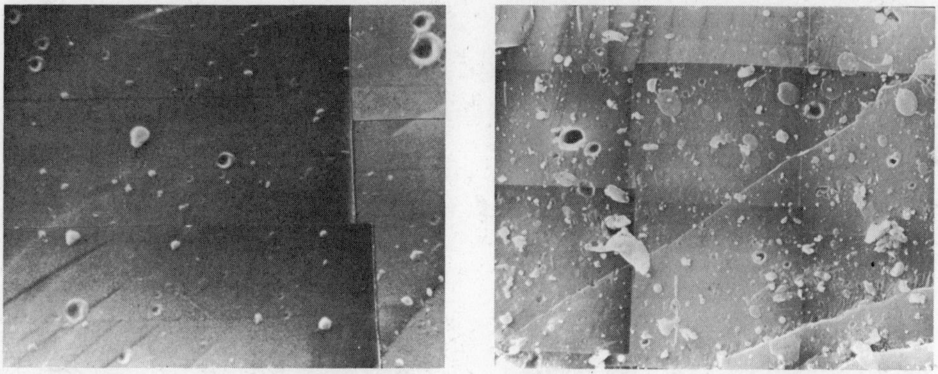

Fig. 2. Scanning electron micrographs of two crystals with high microcrater densities. The largest craters in both photos have diameters of 1 micron. Note that whereas the crater densities are about the same ($\sim 10^7/cm^2$), the abundances of pancakes and other accretionary particles on these two crystals differ widely. (G 10,000×).

Fig. 2, where crystals with about the same densities of microcraters have considerably different abundances of glassy discs.

Although not specified in Table 1, micron-sized accretionary particles and large, thick glassy splashes are quite ubiquitous on feldspars of all soils. They were especially abundant on some crystals in soils 12033. In this soil, the crystal surfaces were literally covered by adhering micron-sized particles, apparently glued on a more or less glassy surface (Fig. 3). These unique features gave these crystal surfaces a very peculiar appearance, quite different from crystals in any other soil. In soil 12033, crystals with such a glassy cover were excluded from the data reported in Table 1.

Nuclear tracks

Usually, more than 80% of the crystals in which surface features were studied could be conveniently etched for track counting. In all soils except 14141 and 12033, the track densities fell in the same range, generally between $(0.1\text{–}4) \times 10^9$ tracks/cm^2. Track densities were about one order of magnitude lower in soil 14141 and two orders of magnitude lower in soil 12033. In these last two soils, track density gradients between the center and edges of the crystals were generally absent, in contrast to the center and edges of the crystals were generally absent, in contrast to the other soils where most grains had track density gradients near their edges. In no soil was the track density distribution inside single crystals found to be correlated to the density of microcraters on the same grains. However, no microcraters were found on crystals with central track densities lower than about $2.5 \times 10^8/cm^2$. In contrast, glassy discs were present, although with low frequency, on grains with low uniform track densities ($<10^8/cm^2$) e.g. in soils 12033 and 14141.

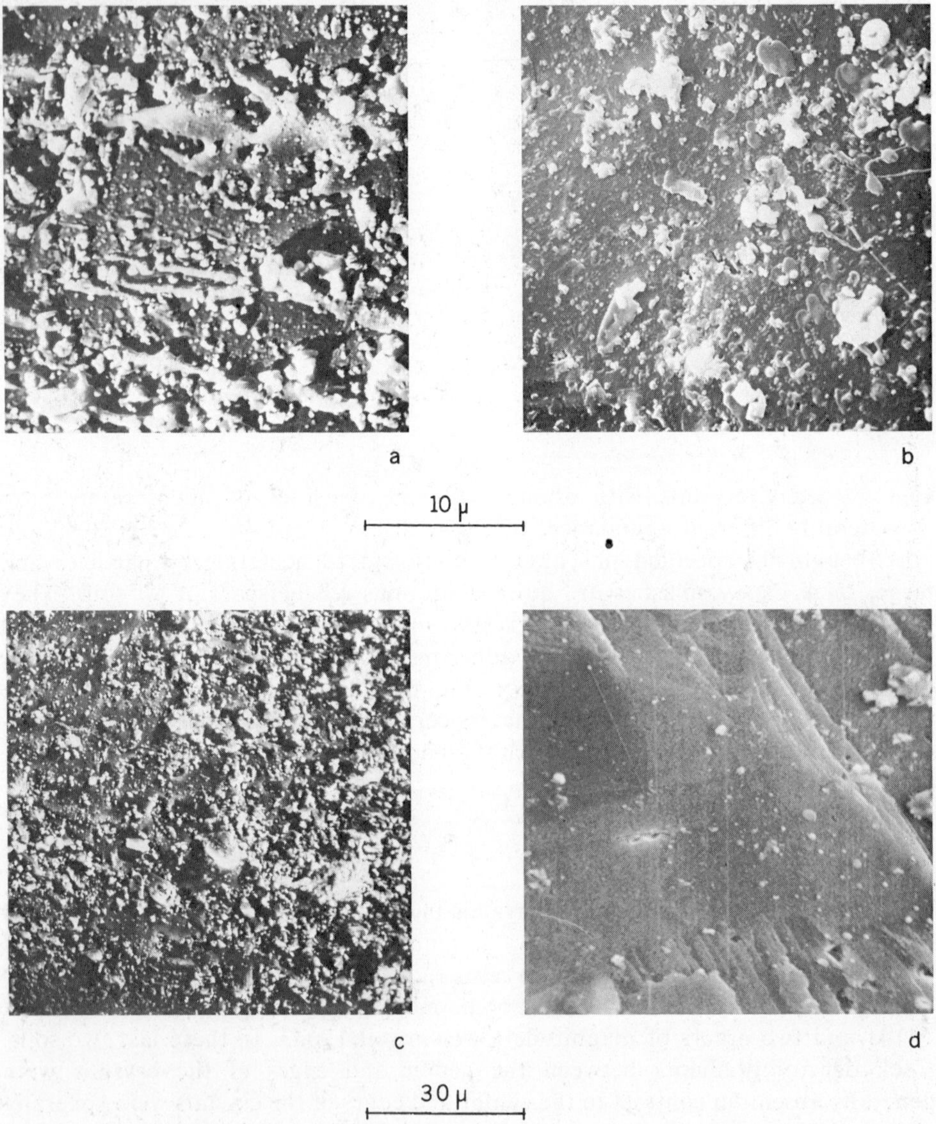

Fig. 3. Scanning electron micrographs of typical glassy blanketing in soil 12033. They show glassy mantling more or less oriented (a,c) with glued particles, abundant glassy deposits and accretionary particles (b), as compared to an ordinary regolith crystal surface (d).

Glassy agglutinates in Luna samples

Glassy agglutinates and breccias have been measured in several grain size fractions between 40 and 370 μm. The results are presented in Fig. 4. In these histograms, we have not always distinguished glassy agglutinates from soil breccias. This results from our determination procedure: observing soil particles directly under a microscope has the advantage of not altering them in any way but may make the distinction between petrographic species difficult. However, as shown in Fig. 5, our results do agree quite well with those of Hu and Taylor (1977), who examined thin sections. It may be noted that both series of results differ somewhat, for various levels, from those of McKay *et al.* (1977). This is probably due to different observation conditions and selected criteria for the definition of glassy particles, although differences in the mode of sieving (dry seiving in our case, wet sieving in McKay's *et al.*) might also have introduced some bias. When considered altogether however, all these results are consistent with the more detailed data given by Rode *et al.* (1977) for the whole core (Fig.

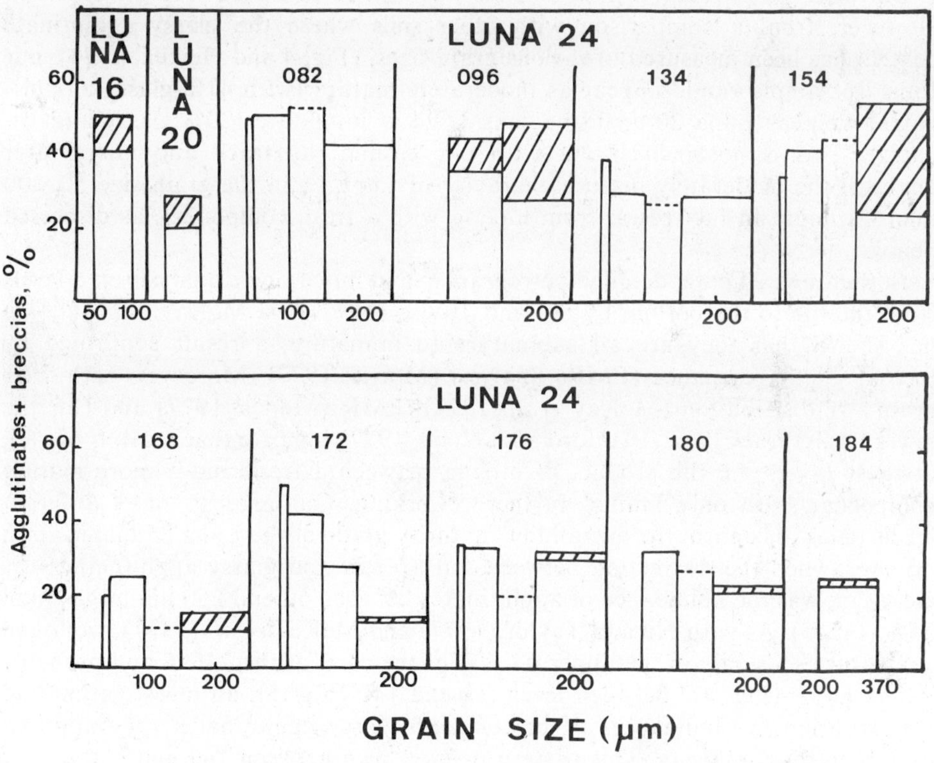

Fig. 4. Abundances of glassy agglutinates and breccias in various grain sizes (between 50 and 370 microns) of Luna 16, 20 and 24 core samples.

Measurements were made under a polarizing microscope on unpolished particles. When breccias could be differentiated from agglutinates, they are reported under shadowed areas. Broken lines indicate measurements made on less than 100 particles.

5). All indicate a general tendency for the soil to be more immature in the deepest half of the core, as indicated by the lowest glassy agglutinate contents.

IV. DISCUSSION

Luna soil maturity

The maturity of Luna soils was estimated from their glassy agglutinate content. This method was in effect the most practicable because of the small size of our samples. When doing this we have to keep in mind that, strictly speaking, this analysis applies to the grain size for which the percent glassy agglutinate content is measured. It qualifies the degree of maturity of bulk soils only when reworking is dominant over mixing (McKay *et al.*, 1974; Poupeau, 1974). As a first approximation, therefore, we considered immature soils as those having less than 30% glassy agglutinates in the 100–150 μm fraction, mature as those with more than 40% glassy agglutinates and submature as soils in between.

Our analysis of Luna 16 and 20 soils was limited to the 50–100 μm fraction. However, from a comparison with other soils where the glassy agglutinate content has been measured in various grain sizes, (Fig. 4 and Heiken, 1974), our Luna 16 sample would appear as moderately mature (with 41% glassy agglutinates) whereas Luna 20 could be considered at immature (20% glassy agglutinates). This is not inconsistent with our cosmic ray track and microcrater analysis; the moderately mature character of Luna 16 in the grain size 50–200 microns might in fact result from mixing with a fresh component, as discussed below.

In the Luna 24 core, detailed petrographic and mineralogic analysis on 6 levels from the top to the bottom by Hu and Taylor (1977) and McKay *et al.* (1977) has shown that they are all submature to immature, a result confirmed by ferromagnetic resonance (FMR) analysis (Morris, 1977). Moreover, both their grain size distribution (McKay *et al*, 1977), FMR (Morris, 1977) and (for the lower levels) track data (Blanford and Wood, 1977) indicate that most, if not all, of these layers are the product of mixing between a fresh and a more mature component, with only limited further reworking. Our analysis on 9 different levels tends to confirm the overall low maturity grade of the Luna 24 material. In no case, when the distinction between soil breccia and glassy agglutinates was clear-cut, was the abundance of agglutinates between 50 and 250 μm larger than 35% (Fig. 5). As with Hu and Taylor (1977) and McKay *et al.* (1977), we found submature soils only in the upper portion of the core, at the 24096 and probably 24082 levels (Fig. 4). Between levels 77 and 140 (Fig. 5), all investigators find the variation in abundances of breccia and glassy agglutinates vary approximately in the same way. This pattern breaks down between 140 and 170 cm, as appears in the French-American data (Fig. 5, right side). This corresponds, as shown by the more detailed sampling of Rode *et al.* (1977) to an anticorrelation between the abundances of these two types of particles (Fig. 5, left side), with a low percentage of glassy agglutinates (as well as gabbros) and a large (up to

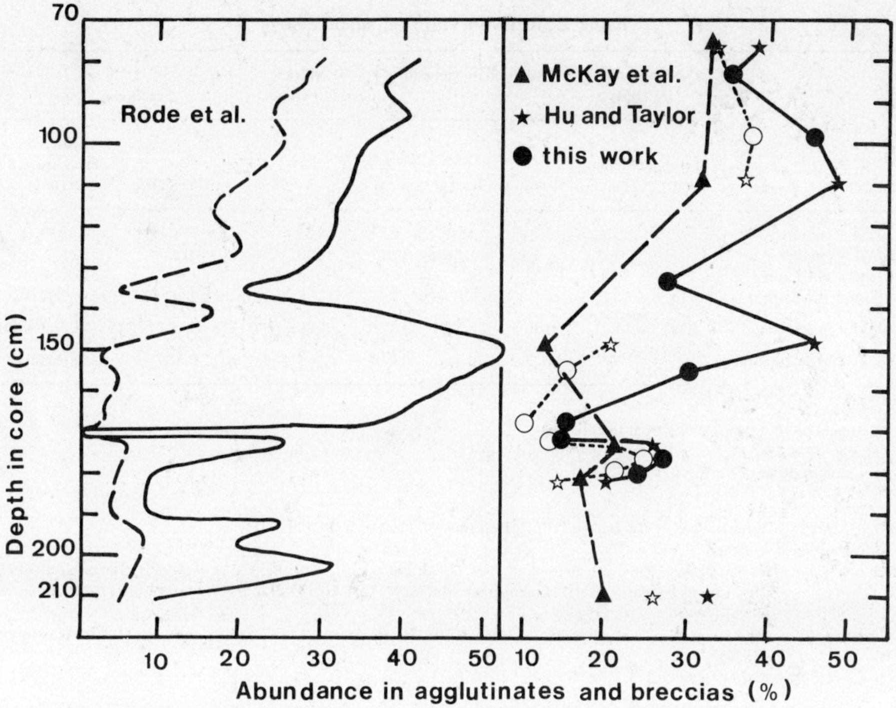

Fig. 5. Abundances of glassy agglutinate and breccia fragments in the Luna 24 core, for the 200 microns size fraction (left) and the 90–150 microns fraction (right). Close symbols in our own and Hu and Taylor's (1977) data are used for the percentage of breccia + agglutinates; open symbols for the percentage of agglutinates only.

30%) percentage of breccias in the >200 microns fraction. This anomalous relationship between agglutinates and breccias supports the idea that this layer (140–170 cm) was formed at once from fresh and brecciated materials and quickly covered and protected against reworking.

Levels 168 to 184 are composed of rather immature material. They are marked by a sharp decrease both in breccias and agglutinate around 170 cm. This spectacular depletion at 170 cm, peculiarly visible in Rode's data, corresponds to the presence of a gabbroic fragment apparently broken within the core during sampling into several pieces, one of which, at level 171, was dated by Birck *et al.* (1977) at 3.74 ± 0.28 b.y. Layer 172, just below this gabbro, presents a very peculiar glassy agglutinate distribution shape (Fig. 4), with an abundance continuously decreasing from 50% in the 40–50 μm fraction to 13% in the interval 150–250 μm. In this layer, breccia fragments were found to be especially rare; in contrast, feldspars in the 150–250 μm fraction were more abundant than usual, (35% of the particle population). Possibly these characteristics are related to the proximity of the above gabbro: either the material of layer 172 was deposited at the same time as the gabbro or it has been contaminated by the

Table 2. Relevant data on surface soils

		Median grain size (Md)		Agglutinate content (%)		Track-rich crystals content	Fraction <1 mm	
Soil number	Maturity degree	Total soil	Fraction <1 mm	90–150 μm	62–125 μm	%	^{36}Ar $\times 10^{-4}$ccSTP/g	Spallation exposure age $\times 10^{6}$ yr
10084	mature		52			≥85	3.3 → 4.36††	≥440 (^{126}Xe)
64421	mature		75†		54	100	4.83; 5.14	350 (^{21}Ne)
67481	submature	272	114	23	24	86	1.66	53 (^{21}Ne)
67601	submature	110	82		20	98	2.26	55 (^{21}Ne)
14141	immature	735		36		30		45^{+} 75^{++} (^{126}Xe)
12033	very immature		97	5.2		3	0.22	>250 (^{21}Ne)

†Graphic mean Mz.
††Extremes of 10 measurements from the literature.
$^{+}$Fraction >300 μm.
$^{++}$Fraction <150 μm.

Data sources:
Md. Butler *et al.*(1973); Heiken (1974); Heiken *et al.* (1973); LSPET. (1971).
Agglutinate content. Heiken (1974).
Track-rich grains. Berdot *et al.* (1972 a, b); Behrmann *et al.* (1973; Lal *et al.* (1970); Phakey *et al.* (1972); Storzer *et al.* (1973).
^{36}Ar. Bogard and Nyquist (1973); Hintenberger *et al.* (1971); Kirsten *et al.* (1973); Marti *et al.* (1970); Pepin *et al.* (1972); Walton *et al.* (1973).
Exposure ages. Burnett *et al.* (1972); Eberhardt *et al.* (1970); Heymann *et al.* (1972); Kirsten *et al.* (1973); Walton *et al.* (1973).

gabbroic material by later reworking or during sampling and transport.

Levels 180 and 184 belong to the most immature unit of the core found by Rode *et al.* (Fig. 5). The relatively large content of agglutinates + breccias in the smallest grain sizes of 180 may be due to its proximity to the less immature overlaying unit. These two levels are particularly rich in gabbroic fragments which account for 80% of the particles in soil 184.

The Luna 24 core is therefore a unit composed of predominantly submature levels in the upper part turning to immature material downwards. It is quite inhomogeneous, and is marked by layers with widely varying characteristics. Mixing must then be predominant over reworking in the constitution of these layers, although limited surface exposure is indicated in crystals from all levels as indicated by cosmic-ray track data (Blanford and Wood, 1977; Poupeau, unpublished data; Pellas, pers. comm.).

Surface features and soil maturation

The degree of maturity of a lunar soil as defined by its bulk properties (Table 2) such as grain size distribution parameters and agglutinate content, solar wind species content, etc., is a function of the residence time of its components within the top few millimeters of the regolith surface (Gault *et al.*, 1974). It was therefore expected that the abundance of microcraters and small glassy discs ("pancakes" of Blanford *et al.*, 1974) on single feldspar crystals in a given soil would be correlated to the degree of maturity of this soil. Our results will be discussed in this context.

Microcraters

The two most immature soils of our collection, soil 12033 and 14141, have by far the lowest percentage of cratered grains. In soil 12033 for instance, not one crystal among 14 studied was found to have microcraters, suggesting that none or very few grains in this soil were ever exposed atop the regolith. This is in agreement with the very low percentage of track-rich crystals (i.e. with track densities $>10^8/cm^2$) in this soil, as well as its very low glassy agglutinates and solar wind ^{36}Ar content (Table 2). In the more mature soils 14141 and Luna 20, microcraters were found to be present in 14 to 20% of the crystals. This is still notably less than in mature soils such as soil 10084, 64421 and Luna 16, where the abundance of cratered crystals may reach—as for Apollo 15 and 16 and Luna 24 cores—values of ~30% to ~60% of the population scanned. Therefore, apparently there is some correlation between the abundance of cratered crystals and the degree of maturity of lunar soils in the grain size range considered. We think the variations in the percentages of cratered crystals among mature soils reflects both our limited statistics and the variable near-surface exposure history of these soils, which may consist of components with various degrees of maturity. This last point may be illustrated by our Luna 16 sample, which can be considered mature from glassy agglutinates measurements (Fig. 4) as well as high solar wind gas content (Schultz *et al.*, 1973) and high percentage of micron-sized grains with an amorphous coating layer (Maurette and Price, 1975). However, we found only 31% of the 100–200 μm feldspars with impact microcraters, which suggests some fresh component is present in this grain size. This in turn is in agreement with earlier suggestions based on track data (Dran *et al.*, 1972; Poupeau *et al.*, 1973) that a young component might be present in the very same grain size range.

Most interesting in this context is the fact that the submature soils 67481 and 67601 have the same percentage of cratered crystals with crater densities $>10^6/cm^2$ as the mature soils in Table 2 and may have among the highest percentage of microcratered grains (Table 1). Both soil 67481 and 67601 were collected on the rim of the 50 m.y. old North Ray Crater (Arvidson *et al.*, 1975). Extensive contamination of these soils by nearby older regolithic material is excluded by the fact that their rare gas exposure ages are also of the order of 50 m.y. (Kirsten *et al.*, 1973; Walton *et al.*, 1973). This time therefore appears as an upper limit to the time necessary to build up an equilibrium microcrater record in soil grains of the grain size considered.

Glassy deposits

In a previous paper (Poupeau *et al.*, 1975), we considered pancakes as indicators of surface exposure. In effect, the morphology of these impact glasses suggests they were formed from very fluid droplets. This implies very high temperatures (>1300°C) for the droplets (Blanford *et al.*, 1974) and therefore a very nearby origin for this material, i.e. generation from microimpacts at the regolith surface. This receives some support from the fact that the most immature soil, 12033, has very few crystals with pancakes, whereas soil 14141 which is more mature, has already as many pancakes as fully mature soils, in which almost all crystals have such features. It may also be noted that crystal surfaces saturated with pancakes were encountered only in the submature soils 67481 and 67601 (see also Zinner *et al.*, 1977) and more mature soils, as reported by Blanford *et al.* (1974) and Poupeau *et al.* (1975). In soils 12033 and 14141, some pancakes were found on crystals whose low cosmic ray track densities indicated they had never been exposed within the top millimeter of the regolith. Thus, pancake production can also occur as the result of impacts penetrating and reorganizing the regolith over a depth of a few millimeters.

The large glassy splashes occasionally found on regolith crystals and whose abundances are not correlated with the degree of maturity of soils, must be produced by impacts penetrating at least several centimeters into the regolith. Oriented glassy covers as observed frequently on feldspars of soil 12033 (Fig. 2), which seem to correspond to those observed on ropy glasses of this soil by McKay *et al.* (1971), could have originated in large scale events such as base surges as already advocated by these authors.

Cosmic-ray tracks and soil maturation

The cosmic-ray track production rate varies strongly with depth, decreasing in the first millimeter of the regolith (where the solar flare component is predominant) by several orders of magnitude. Indeed, cosmic-ray track densities at the center of crystals in the 100 microns size range have been found to vary, among soils from the Apollo and Luna sites, from $10^6/cm^2$ to $5 \times 10^9/cm^2$. Because of the complexity of the surface processes controlled by the micrometeorite rain, the deciphering of this track record is not yet completely understood (compare, e.g., the differences in interpretation of track data from the Monte Carlo calculation models of Bibring *et al.*, 1975 and Comstock, 1977). However, there is some agreement that the residence time within the top millimeter of 100 μm diameter crystals (with track densities at their center $<10^8/cm^2$), was at best brief and with limited changes in orientation, as indicated by their largely anisotropic track density gradients (when present). Also, it is generally agreed that crystals with central track densities $>10^8/cm^2$, as they more frequently show track density gradients decreasing from the edges to the center, may have had a longer exposure time near the regolith surface. All Monte Carlo models also predict that from immature to mature soils, the population of crystals with low ($<10^8/cm^2$) central track densities would decrease to some equilibrium value, due to the continuous influx of fresh regolith material, whereas in the fraction of crystals with high track densities ($>10^9/cm^2$), the mean value of the central track density would increase with maturation.

Some of these trends were found for the suite of soils in Table 2, as shown in Fig. 6, in which we compare the histograms of central track densities in feldspars as measured by us in soils 67481, 67601, 64421, Luna 20 (Berdot *et al.*, 1972a,b), and by Crozaz *et al.* (1972) for soil 14141 and Arrhenius *et al.* (1972) for 12033. While our own measurements on 12033 and 14141 on a limited number of crystals do not appear in this figure, they are in the same range as found by other authors. One may observe (Fig. 6) that in soil 12033, where no cratered crystals were found, very few crystals have track densities above

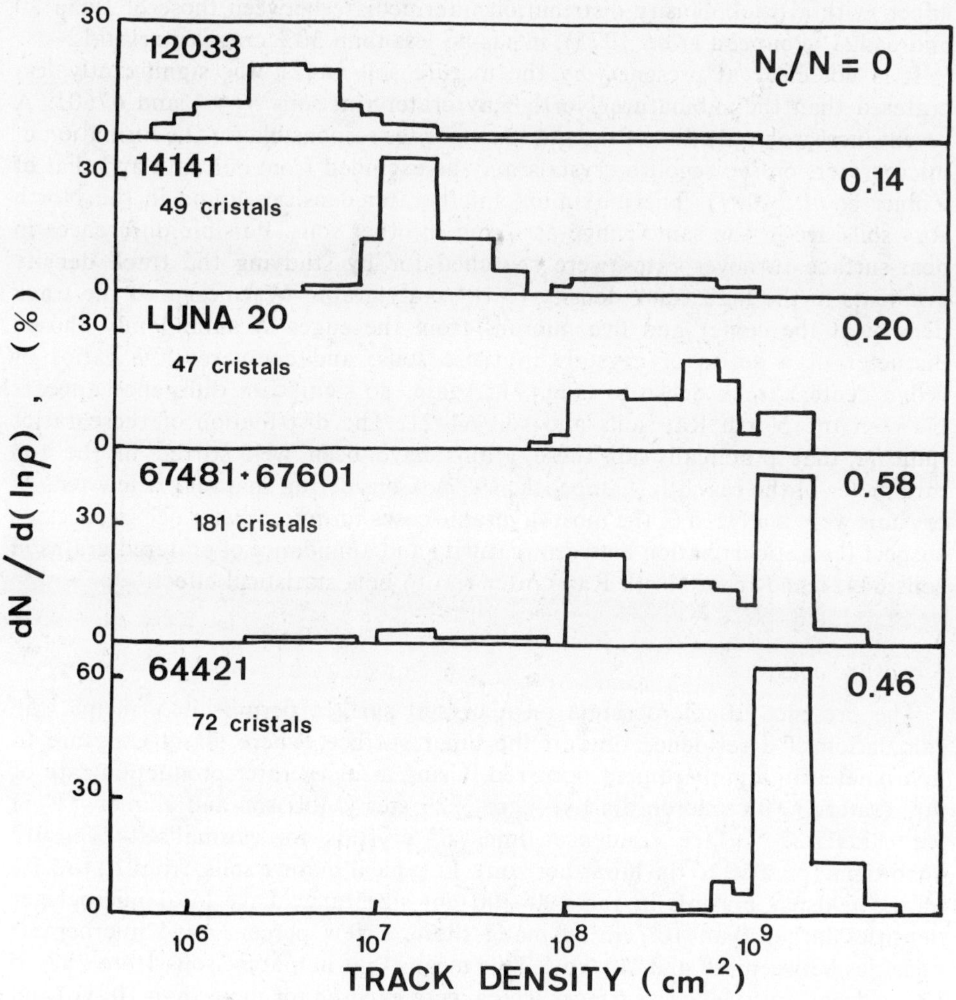

Fig. 6. Histograms of track density distribution in a suite of immature (12033, 14141), submature (Luna 20, 67481, 67601) and mature (64421) soils. N_c/N represents the fraction of cratered crystals found in each of these soils.

$10^8/cm^2$; while in soils 14141 and Luna 20, with respectively 20 and nearly 100% crystals with central track densities $>10^8/cm^2$, from 14 to 20% cratered crystals were found. More mature soils, such as the North Ray crater soils 67481 and 67601, and soil 64421 in which a large fraction of crystals have central track densities larger than $10^9/cm^2$, have between respectively 58% and 46% cratered crystals. There is therefore a net qualitative correlation between the occurrence of microcratered crystals and crystals with high track densities. The results shown in Fig. 6 also suggest that in lunar soils a notable fraction of crystals with track densities $>10^8/cm^2$ must have spent some time on the lunar surface. Our Luna 16 sample, not shown in Fig. 6, shows the same behavior as other soils; in effect, with a track density distribution intermediate between those of Luna 20 and 64421 (Poupeau *et al.* 1973), it has no less than 30% cratered crystals.

It is not clear at present why the mature soil 64421 was significantly less cratered than the submature North Ray crater rim soils 67481 and 67601. A recent increase in the flux of the micrometeorites responsible for the formation of microcraters on the regolith crystals may be excluded from our data and that of Zinner *et al.* (1977). The maximum microcrater densities found in the North Ray soils are in the same range as found in other soils. Possible differences in near-surface turnover rates were searched for by studying the track density gradients in the high track density ($>10^9/cm^2$) grains. We measured the track density at the center and five microns from the edges of a randomly chosen diameter of a series of crystals in these soils, and compared the ratios of (edge/center) track densities (Fig. 7). Again, no significant difference appears between the North Ray soils and soil 64421. The distribution of these ratios indicate that practically all these grains have been well stirred in the top millimeter of the regolith (Comstock, 1977). Considering that only a few tens of crystals were analyzed in the most favorable cases for microcrater occurrence, we suspect the anticorrelation between maturity and abundance of cratered grains in soils 64421 and from North Ray crater rim to be a statistical effect.

Exposure ages

The presence of microcraters on a crystal surface permits in principle the calculation of a residence time at the lunar surface, where direct exposure to micrometeorite bombardment occurred. Using a microcrater production rate of 900 craters (>0.1 micron dia.).$yr^{-1}.cm^{-2}/2\pi$ ster (Morrison and Zinner, 1975) we calculated surface residence times of crystals for normalised exposure conditions (parallel to the lunar horizon). In typical mature soils, from 12 to 25% of the feldspar crystals in the 100–200 μm size range have local microcrater densities larger than $10^6/cm^2$. Among these, a few percent have microcrater densities between 10^7 and $10^8/cm^2$. This means that in mature soils 10 to 25% of the feldspar crystals have surfaces which were exposed for more than 10^3 yr (and up to 10^4–10^5 yr for a few percent grains), whereas in the remaining crystals another 20 to 40% indicate an exposure of less than 10^3 yr; in others, from absence of microcraters, there is no evidence of micrometeoritic bombardment. It

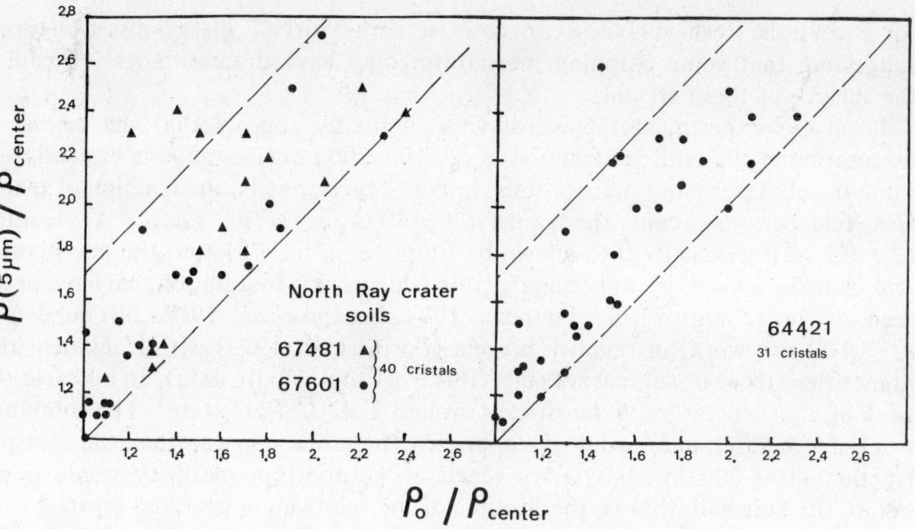

Fig. 7. Correlation plot between the ratios of track densities measured at five microns from the edges of a crystal at the two ends of a randomly chosen diameter to the density at center.

must be recalled at this point that all inferences discussed above rely on the data from Table 1, which give the results of our observations. Taking into account the fact that all surfaces were not accessible to observation during our SEM survey of individual grains, the percentage of crystals exposed for various periods of time to micrometeorite bombardment should be considered as lower limits by about 20 to 30%.

The figure of exposure times at the lunar surface deduced from our crater measurements correlates well with those calculated by Signer *et al.* (1977) from solar wind gas in feldspar crystals in the same grain size range. In effect, in mature soils these authors estimate an average surface exposure of $\sim 10^4$ yr necessary to account for solar ^{36}Ar content in feldspars (150–200 μm diameter). That seems to be somewhat higher than microcrater-derived times, but it must be taken into account that many accretionary particles and glassy splashes present at the crystal surfaces may themselves bear some solar wind gases. All in all, we consider microcrater and solar wind estimates to be compatible.

The surface exposure times for 100 μm feldspar crystals obtained from microcrater and solar wind ^{36}Ar data are much shorter than those derived by Monte Carlo calculations from central track densities (Bibring *et al.*, 1975). Several explanations have been offered to account for this discrepancy: a dust cover effect (Poupeau *et al.*, 1975; Comstock, 1977) and a discontinuous erosion by flaking (Bibring *et al.*, 1975), since solar wind erosion seems to be inefficient for grains in this size range (Zinner *et al.*, 1977). Although no overwhelming evidence in favor of the flaking effect was found by us, it should be noted that on some crystals fresh surfaces can co-exist with "dirty", glassy disc-rich ones, suggesting that some chipping mechanism may have discontinuously modified the surface of these crystals.

From our experimental point of view, it finally appears that the degree of maturation of the feldspar fraction in the 100–200 μm size range is basically the same in submature and mature soils: in both types of soils, the fraction of track-rich feldspars is about the same (70–100%) as is its solar ^{36}Ar loading ($2 \pm 0.2 \times 10^{-5}$ccSTP/g) as shown by Etique *et al.* (1978); and the microcrater and pancake record are indistinguishable (this work). In addition, we have never seen any lunar mature soil (Poupeau, 1974; Berdot *et al.*, 1972a,b; Poupeau *et al.*, 1973; this work) or regolith breccia (Lorin *et al.*, 1973) with track densities higher than those of submature soils (this work and unpub. data). In all cases the peak of high central track densities is around 2×10^9 tracks/cm^2. The combined track-microcrater-solar wind gas approach therefore suggests that the feldspar fraction (100–200 microns) may reach an equilibrium maturity grade much before the bulk soil; that is, the lifetime of the relatively unshocked crystals such as those used in these studies may be quite limited in the lunar regolith. In this respect it may be noted that the feldspar component still immature in the 25×10^6yr Cone Crater ejecta blanket (soil 14141), is already mature in the 50×10^6yr North Ray Crater rim soils 67481 and 67601. This time span could then mark a critical period for transition of a several centimer layer from a noticeably immature feldspar component to a mature one as measured by tracks, micrometeorites (Poupeau and Johnson, 1976) and solar wind gases. In mature soils the difference in the high density peak value between our data (Fig. 6 and references cited above), of $(1–2) \times 10^9$/cm^2 and the one predicted by Monte Carlo calculations, up to 5×10^9/cm^2 (Bibring *et al.*, 1975), although partly due to track "aging" (Price *et al.*, 1973; Dran *et al.*, 1973), could be mostly a result of the continuously renewed population of feldspar (i.e. equilibrium between addition of fresh material and elimination by breakening and glassy agglutinates creation).

V. SUMMARY AND CONCLUSIONS

(1) From the content of agglutinate plus breccia, we found the maturity of the Luna 24 core to vary mostly from submature in the upper part to immature downwards. Widely varying relative abundances of agglutinates plus breccias at various levels indicate this material was deposited as discrete units and that reworking of the core was limited to shallow depth.

(2) The abundances of pancakes and impact microcraters in feldspar crystals (100–200 μm range) is not strongly correlated to the maturity of bulk soils. They are absent to sparse in immature soils such as 12033 and 14141, they may be present in similar abundances in the submature soils

67481 and 67601 (from the rim of the North Ray Crater), and more mature soils (e.g. 64421).

Typically, in submature and mature soils, pancakes are present on at least 60% and up to 100% of the crystals and microcraters on 50 to 70%. From a comparison with the abundance of track-rich crystals (with central track density $>10^8/cm^2$), one can deduce that a noticeable fraction of the latter was once exposed atop the regolith.

(3) Microcrater, track and rare gas data on 100–200 μm feldspars suggest that they reach an equilibrium maturity state at the submature stage of bulk soils. This result, obtained from a limited number of data, is interpreted as an equilibrium between the rate of addition and destruction of grains in the size range considered.

Acknowledgments—We would like to acknowledge many helpful discussions with Dr. R. M. Walker at various stages of this work and encouragement and interest from Dr. J. Labeyrie. We appreciated the expert technical assistance of L. Ross and P. Swann. This work was supported by NASA grant No. NGL 26 008-065 and the French Centre National de la Recherche Scientifique and Commissariat à l' Energie Atomique.

References

Arrhenius G., Liang S., MacDougall D., Wilkening L., Bhandari N., Bhat S., Lal D., Rajagopalan G., Tamhane A. S. and Venkatavaradan V. S. (1972) The exposure history of the Apollo 12 regolith. *Proc. Lunar Sci. Conf. 3rd*, p. 2583–2598.

Arvidson R., Crozaz G., Drozd R., Hohenberg C. and Morgan C. (1975) Cosmic ray exposure ages of features and events at the Apollo landing sites. *The Moon* **13**, 259–276.

Berdot J. L., Chetrit G. C., Lorin J. C., Pellas P. and Poupeau G. (1972a) Track studies in Apollo 14 rocks and Apollo 14, Apollo 15 and Luna 16 soils. *Proc. Lunar Sci. Conf. 3rd*, p. 2867–2881.

Berdot J. L., Chetrit G. C., Lorin J. C., Pellas P. and Poupeau G. (1972b) Irradiation studies of lunar soils: 15100, Luna 20 and compacted soil from breccia 14307. In *The Apollo 15 Lunar Samples* (J. W. Chamberlain and C. Watkins, eds.), p. 333–335. The Lunar Science Institute, Houston.

Berhmann C., Crozaz G., Drozd R., Hohenberg C., Ralston C., Walker R. M. and Yuhas D. (1973) Cosmic ray exposure history of North Ray and South Ray materials. *Proc. Lunar Sci. Conf. 4th*, p. 1957–1974.

Bibring J. P., Borg J., Burlingame A. L., Langevin Y., Maurette M. and Vassent B. (1975) Solar wind and solar flare maturation of the lunar regolith. *Proc. Lunar Sci. Conf. 6th*, p. 3471–3493.

Birck J. L., Manhes G., Richard P., Joron J. L., Treuil M. and Allègre C. J. (1977) ^{87}Rb-^{87}Sr age of Luna 24 microgabbros, and isotopic and trace element study of soil 24096 (abstract). In *Papers Presented to the Conference on Luna 24*, p. 34–36. The Lunar Science Institute, Houston.

Blanford G. E., Fruland R. M., McKay D. S. and Morrison D. A. (1974) Lunar surface phenomena: solar flare track gradients, microcraters and accretionary particles. *Proc. Lunar Sci. Conf. 5th*, p. 2501–2526.

Blanford G. E., McKay D. S. and Wood (1977) Particle track density in double drive tube 60009, 10. *Proc. Lunar Sci. Conf. 8th*, p. 3017–3025.

Blanford G. and Wood G. C. (1977) Particle track densities in the Luna 24 core. In *Papers Presented to the Conference on Luna 24*, p. 41–43. The Lunar Science Institute, Houston.

Bogard D. D. and Nyquist L. E. (1973) $^{40}Ar/^{36}Ar$ variations in Apollo 15 and 16 regolith. *Proc. Lunar Sci. Conf. 4th*, p. 1975–1985.

Burnett D. S., Huneke J. C., Podosek F. A., Russ III G. P., Turner G. and Wasserburg G. J. (1972) The irradiation history of lunar samples. In *Lunar Science III*, p. 97–100. The Lunar Science Institute, Houston.

Butler J. C., Greene G. M. and King E. A. (1973) Grain size frequency distributions and modal analyses of Apollo 16 fines. *Proc. Lunar Sci. Conf. 4th*, p. 267–278.

Carter J. L. and MacGregor I. D. (1970) Mineralogy, petrology and surface features of some Apollo 11 samples. *Proc. Apollo 11 Lunar Sci. Conf.*, p. 247–265.

Comstock G. M. (1977) On deciphering the particle-track record of lunar regolith history. *J. Geophys. Res.* **82**, 357–367.

Crozaz G., Drozd R., Hohenberg C. M., Hoyt H. P., Ragan D., Walker R. M. and Yuhas D. (1972) Solar flare and galactic cosmic ray studies of Apollo 14 and 15 samples. *Proc. Lunar Sci. Conf. 3rd*, p. 2917–2931.

Crozaz F., Haack U., Hair M., Maurette M., Walker R. M. and Woolum D. (1970) Nuclear tracks studies of ancient solar radiations and dynamic lunar surface processes. *Proc. Apollo 11 Lunar Sci. Conf.*, p. 2051–2080.

Dran J. C., Duraud J. P., Maurette M., Durrieu L., Jouret C. and Legressus C. (1972) Track metamorphism in extra terrestrial breccias. *Proc. Lunar Sci. Conf. 3rd*, p. 2883–2903.

Dran J. C., Maurette M., Dumail M. and Duraud J. P. (1973) Aging of fossil tracks in extra terrestrial material. *Meteoritics* **8**, 347–355.

Eberhardt P., Geiss J., Graf H., Grögler M., Krähenbühl U., Schwaller H., Schwarsmüller J. and Stettler (1970) Trapped solar wind noble gases, exposure age and K-Ar ages in Apollo 11 lunar fine material. *Proc. Apollo 11 Lunar Sci. Conf.*, p. 1037–1070.

Etique Ph., Funk H., Poupeau G., Romary Ph., Signer P. and Wieler R. (1978) Solar wind gas and solar flare track correlations in lunar soils; revisited (abstract). In *Lunar and Planetary Science IX*, p. 300–302. The Lunar Science Institute, Houston.

Gault D. E., Hörz F., Brownlee D. E. and Hartung J. B. (1974) Mixing of the lunar regolith. *Proc. Lunar Sci. Conf. 5th*, p. 2365–2386.

Heiken G. H. (1974) A catalog of lunar soils. Preprint.

Heiken G. H., McKay D. S. and Fruland R. M. (1973) Apollo 16 soils: grain size analyses and petrography. *Proc. Lunar Sci. Conf. 4th*, p. 251–265.

Heymann D., Yaniv A. and Lakatos S. (1972) Inert gases from Apollo 12, 14 and 15 fines. *Proc. Lunar Sci. Conf. 3rd*, p. 1857–1863.

Hintenberger H., Weber H. W. and Takaoka N. (1971) Concentrations and isotopic abundances of the rare gases in lunar matter. *Proc. Lunar Sci. Conf. 2nd*, p. 1607–1625.

Hu N.-H. and Taylor L. A. (1977) Luna 24 soils from Mare Crisium: agglutinate chemistry. In *Papers Presented to the Conference on Luna 24*, p. 82–85. The Lunar Science Institute, Houston.

Kirsten T., Horn P., and Kiko J. (1973) ^{39}Ar–^{40}Ar dating and rare gas analysis of Apollo 16. *Proc. Lunar Sci. Conf. 4th*, p. 1757–1784.

Lal D., McDougall D., Wilkening L. and Arrhenius G. (1970) Mixing of the lunar regolith and cosmic ray spectra: evidence from particle track studies. *Proc. Apollo 11 Lunar Sci. Conf.*, p. 2295–2303.

Lorin J. C., Pellas P., Poupeau G., Schultz L. and Signer P. (1973) Irradiation et préirradiation de la brèche 14307. *The Moon* **8**, 149–175.

LSPET (Lunar Sample Preliminary Examination Team) (1971) Preliminary examination of lunar samples from Apollo 14. *Science* **173**, 681–693.

Marti K., Lugmair C. W. and Urey H. C. (1970) Solar wind gases, cosmic ray spallation products and the irradiation history of Apollo 11 samples. *Proc. Apollo 11 Lunar Sci. Conf.*, p. 1357–1367.

Maurette M. and Price P. B. (1975) Electron microscopy of irradiation effects in space. *Science* **187**, 121–129.

McKay D. S., Fruland R. M. and Heiken G. (1974) Grain size and the evolution of lunar soils. *Proc. Lunar Sci. Conf. 5th*, p. 887–906.

McKay D. S., Basu A., Waits G., Clanton U., Furhman R. and Fruland R. (1977) Grain size and evolution of Luna 24 soils (abstract). In *Papers Presented to the Conference on Luna 24*, p. 114–117. The Lunar Science Institute, Houston.

McKay D. S., Greenwood W. A. and Morrison D. A. (1970) Origin of small lunar particles and breccias from the Apollo 11 site. *Proc. Apollo 11 Lunar Sci. Conf.*, p. 673–694.

McKay D. S., Morrison D. A., Clanton U. S., Ladle G. H. and Lindsay J. F. (1971) Apollo 12 soils and breccias. *Proc. Lunar Sci. Conf. 2nd*, p. 755–773.

Morris R. (1977) FMR and magnetic properties of Luna 24 soils and 1 mm soil particles (abstract). In *Papers Presented to the Conference on Luna 24*, p. 121–124. The Lunar Science Institute, Houston.

Morrison D. A. and Zinner E. (1975) Studies of solar flares and impact craters in partially protected crystals. *Proc. Lunar Sci. Conf. 6th*, p. 3373–3390.

Pepin R. O., Bradley I. C., Dragon J. C. and Nyquist L. E. (1972) K-Ar dating of lunar fines: Apollo 12, Apollo 14 and Luna 16. *Proc. Lunar Sci. Conf. 3rd*, p. 1569–1588.

Phakey P. P., Hutcheon I. D., Rajan R. S. and Price P. B. (1972) Radiation effects in soils from five lunar missions. *Proc. Lunar Sci. Conf. 3rd*, p. 2905–2915.

Poupeau G. (1974) Etude de l'irradiation par le vent solaire et le rayonnement cosmique solaire et galactique des météorites riches en gaz et des sols lunaires. Doctoral thesis, University of Paris VI.

Poupeau G., Chetrit G. C., Berdot J. L. and Pellas P. (1973) Etude par la méthode des traces nucléaires du sol de la Mer de la Fécondité. *Geochim. Cosmochim. Acta* **37**, 2005–2016.

Poupeau G. and Johnson J. (1976) Micrometeorites and the solar flare maturation of lunar soils (abstract). In *Lunar Science VIII*, p. 709–711. The Lunar Science Institute, Houston.

Poupeau G., Mandeville J. C. and Christophe-Michel-Lévy M. (1976) Impact features in Luna 16, 20 and 24 soils and the maturation of the Lunar regolith (abstract). In *Papers Presented to the Conference on Luna 24*, p. 148–151. The Lunar Science Institute, Houston.

Poupeau G., Mandeville J. C., Christophe-Michel-Lévy M. and Romary Ph. (1978a) Microcrater, cosmic-ray tracks and petrographic studies of Luna 16, 20 and 24 soils and a comparison with Apollo soils. In Luna 24 Publication, Moscow (in Russian). In press.

Poupeau G., Rajan R. S., Walker R. M. and Zinner E. (1978b) The modern and ancient flux of solar wind particles, solar flare particles and micrometeoroids. *Proc. 9th COSPAR Meeting*, 1976, Philadelphia. In press.

Poupeau G., Walker R. M., Zinner E. and Morrison D. A. (1975) Surface exposure history of individual crystals in the lunar regolith. *Proc. Lunar Sci. Conf. 6th*, p. 3433–3448.

Price P. B., Lal D., Tamhane A. S., and Perelygin (1973) Characteristics of tracks of ions of 14 Z 36 in common rock silicates. *Earth Planet. Sci. Lett.* **9**, 377–395.

Rode O. D., Ivanov A. V., Tarasov L. S. and Korina M. I. (1977) General lithologic-morphologic characteristics of Luna 24 regolith core. *Geokhymiya* **n°10**, 1465–1476. (In Russian).

Schultz L., Phinney D. and Signer P. (1973) Helium, neon and argon in level C of Luna 16 fines. *Geochim. Cosmochim. Acta* **37**, 2021–2023.

Signer P., Baur H., Derksen U., Etique Ph., Funk H., Horn P. and Rainer W. (1977) Helium, neon and argon records of lunar soil evolution. *Proc. Lunar Sci. Conf. 8th*, p. 3657–3683.

Storzer D., Poupeau G. and Kratschmer W. (1973) Track exposure and formation ages of some lunar samples. *Proc. Lunar Sci. Conf. 4th*, p. 2363–2377.

Walton J. R., Lakatos S. and Heymann D. (1973) Distribution of inert gases in fines from the Cayley Descartes region. *Proc. Lunar Sci. Conf. 4th*, p. 2079–2095.

Zinner E. and Walker R. M. (1975) Ion probe studies of artificially implanted ions in lunar samples. *Proc. Lunar Sci. Conf. 6th*, p. 3601–3617.

Zinner E., Walker R. M., Chaumont J. and Dran J. C. (1977) Ion probe surface concentration measurements of Mg and Fe and microcraters in crystals from lunar rock and soil samples. *Proc. Lunar Sci. Conf. 8th*, p. 3859–3883.

Merrill, R. B. and Papike, J. J., editors.
(1978) *Mare Crisium: The View from Luna 24*, Pergamon Press (New York), p. 157–163
Printed in the United States of America

Particle track densities in the Luna 24 core

GEORGE E. BLANFORD and GORDON C. WOOD
University of Houston at Clear Lake City, Houston, Texas 77058

Abstract—We report track density measurements in soil grains at six locations in the Luna 24 core. Irradiation levels are generally high precluding information on the depositional history, but indicating that the soils were exposed at the lunar surface at sometime in the past. Maturity levels determined from track density parameters disagree with those determined from ferromagnetic resonance and agglutinate contents. This discrepancy is not understood, but it could possibly indicate an ancient epoch of increased solar-flare activity.

We are reporting on measurements of particle track densities in soil grains at six locations in the Luna 24 core (24077,9; 24109,13; 24149,15; 24174,10; 24182,15; 24210,9). These grains are from the 45–90 μm sieve fraction prepared by McKay *et al.* (1978). Other measurements from the same allocation of bulk soils are being reported by Basu *et al.* (1977 & 1978), Blanchard *et al.* (1978), Bogard and Hirsch (1978), McKay *et al.* (1978) and Morris (1978 a) concerning petrographic modal analyses, chemical analyses, rare-gas contents, grain-size distributions and magnetic and ferromagnetic-resonance measurements.

EXPERIMENTAL METHODS

We etched polished grain mounts in boiling 1 N NaOH (~118 °C) for 3 h or until the average track diameter in plagioclase was ~60 nm. We located plagioclase mineral fragments or grains containing plagioclase inclusions by using X-ray energy dispersive analysis. We measured track densities in thirty plagioclase grains per sample by counting tracks on scanning electron micrographs. Figure 1 displays histograms showing the distribution of track densities in the plagioclase soil grains. We etched two samples (24077, 24182) for long periods in 1 N NaOH and also for several hours in 6.25 N NaOH to enlarge particle tracks in pyroxene grains to ~40 nm. Figure 1 also displays histograms based on measurements in thirty grains per sample.

DISCUSSION

Particle tracks in lunar samples have two sources, solar-flare particles and galactic cosmic rays. Because galactic cosmic rays are modulated by the magnetic field of the solar cavity, solar flare particles dominate the population of track producing particles at low energies and therefore at very small penetration depths. Galactic cosmic rays are more energetic and dominate the population of track producing particles at depths greater than 1 cm, but their rate of track production decreases as a power law below ~10 cm. Walker and Yuhas (1973), Hutcheon *et al.* (1974), Yuhas (1974), and Blanford *et al.* (1975) have presented measurements of track production profiles in lunar materials. These measurements indicate that solar-flare particles produce tracks at a depth of ~100 μm at

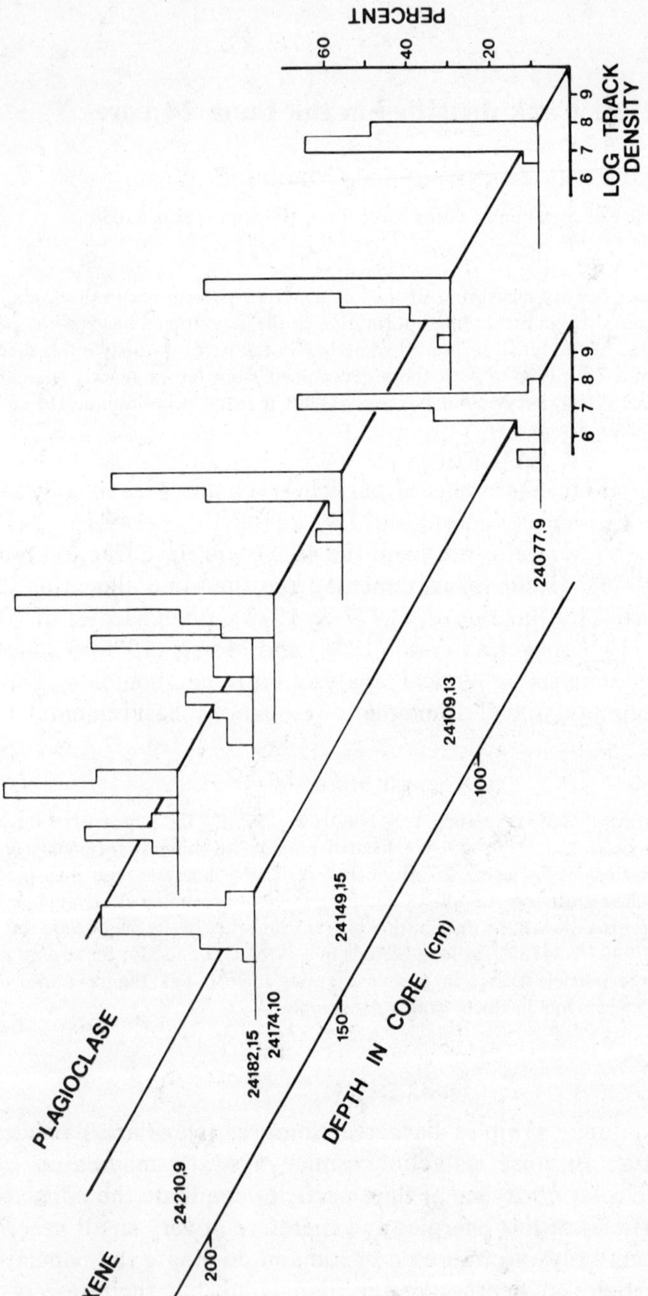

Fig. 1. Histograms of particle track densities in plagioclase and pyroxene crystals in the 45–90 μm fraction of Luna 24 core samples. Each histogram is based on the measurement of 30 soil grains. Correlations of core depths with depths from the lunar surface are discussed by Bogard and Hirsch (1978).

a rate of $\sim 3 \times 10^9$ cm^{-2} $(10^6$ y$)^{-1}$ and galactic cosmic rays produce tracks at a depth of ~1 cm at a rate of $\sim 1 \times 10^6$ cm^{-2} $(10^6$ y$)^{-1}$. From these rates we can conclude that soil grains with track densities less than 1×10^8 cm^{-2} most probably have never been exposed at the immediate lunar surface (i.e. $\lesssim 1$ mm) and that grains with track densities $\gtrsim 1 \times 10^9$ cm^{-2} have had exposure to solar flares at the immediate lunar surface.

Track density distributions displayed in Fig. 1 indicate that most of the soil grains from the Luna 24 core have been exposed at the lunar surface. We classify the plagioclase fraction for the upper four soils (24077, 24109, 24149, and 24174) as mature because almost all of the grains have track densities $\geq 10^9$ cm^{-2} (Fig. 2). The lowest two soils (24182, 24210) have a significant component of plagioclase grains that were not exposed at the lunar surface and this component appears to be sharply divided from the mature component. We would favor interpreting this type of distribution to imply a mixture of two soils with different predeposition histories. Results from Monte Carlo calculations by Bibring *et al.* (1975) suggest an alternative interpretation, namely that this soil may exhibit partial reworking near the lunar surface; the final depth from which the soil is extracted would depend on the lifetime of this ancient surface-exposure epoch. Because plagioclase is a small soil component at the Luna 24 site (~22%; Basu *et al.*, 1978), we measured track density distributions in pyroxene grains at two locations (24077, 24182). There is a shift to lower track densities in these distributions relative to the plagioclase distributions, a feature that we would attribute to the fact that average track diameters were ~40 nm as compared to ~60 nm in plagioclase. Because the resolution of the microscope is on the order of 10 nm, if the tracks are smaller, some can be overlooked in counting which results in smaller densities. Considering this interpretation of the shift, we conclude that the pyroxene component of the soil (~66%; Basu *et al.*, 1978) is mature without an admixture of immature material. Therefore, with the exception of the low track density plagioclase, the soil in the Luna 24 core had had sufficient near surface exposure to be classified as mature. Experimental results of Crozaz (1978) and Goswami and Lal (1978) agree with those presented here.

Our conclusions concerning soil maturity contradict conclusions made from ferromagnetic resonance measurements (Morris, 1978a) and agglutinate contents (McKay, *et al.*, 1978). These data indicate that the upper four soils are submature and the lower two soils are immature. We can think of two possible explanations for this discrepancy, one based on the uncertain character of maturity indices, and the other based on an ancient epoch of increased solar-flare activity.

Agglutinate contents, mean grain-size, rare gas contents, ferromagnetic resonance intensity (Is/FeO) and the percentage of track-rich grains are all useful maturity indices in a lunar soil and, although their relationships are becoming better understood, some uncertainity in our understanding of these relationships may be the source of discrepancy between the percentage of grains with track densities greater than or equal to 10^9 cm^{-2} (% $\geq 10^9$) and Is/FeO and agglutinate contents. Morris (1978b) argues that Is/FeO is the most generally applicable maturity index because agglutinate contents and mean grain-size saturate at high maturity values. He also discredits the use of the percentage of track-rich grains as a maturity index because they are restricted to plagioclase grains. Etique, *et al.* (1978) also show that the restriction of track density measurements to plagioclase biases results relative to rare-gas contents. Blanford, *et al.* (1977) show that % $\geq 10^9$ tends to saturate relative to Is/FeO. Correlation of a track maturity index with Is/FeO is qualitatively based on experimental results for the Apollo 16 cores 60009/60010 (Morris and Gose, 1976; Blanford *et al.*, 1977) and 60002 (Blanford and Morrison, 1976; Gose and Morris, 1977; Blanford and Wood, 1978). From these earlier experiments, if % $\geq 10^9$ is greater than ~70% and if Is/FeO is greater than ~60%, the soil is classed as mature. If % $\geq 10^9$ is greater than ~30% and Is/FeO is greater than ~30, the soil is classed as submature. However, the number of sample sets compared is not large and we may not yet know how Is/FeO and track densities should correlate. For Luna 24, Is/FeO is less than 60 (submature) for the upper two samples and less than 30 (immature) for the lowest four samples (Morris, 1978a) (Fig. 2); % $\geq 10^9$ is greater than 70% for the upper four soils (mature) and between 70–30% for the lower two soils (submature) (Fig. 2). Agglutinate contents indicate maturity generally in agreement with Is/FeO (Basu *et al.*, 1978) (Fig. 2) whereas rare gas contents in the <20 μm fraction generally agree with the track maturity index (Bogard and Hirsch, 1978) (Fig. 2). Goswami and Lal (1978) use a track maturity index N_H/N, which is the fraction of grains with track densities greater than 10^8 cm^{-2}, which agrees with % $\geq 10^9$ (Fig. 2). There is no simple explanation for the disagreement in these maturity parameters for the Luna 24 core. Because track data indicates maturity where Is/FeO and agglutinate contents indicate less maturity, the saturation effect for % $\geq 10^9$ relative to Is/FeO is not relevant. The bias that track measurements suffer by being restricted to plagioclase and pyroxene should lead to an underestimate of maturity and is therefore not applicable to this case. Since agglutinate contents and Is/FeO are physically interrelated but are not directly related to the percentage of track-rich grains, the nature of the discrepancy probably indicates some change in this relationship that is not understood.

An ancient epoch of increased solar-flare activity can explain the differences between the percentage of track-rich grains and agglutinate production. Agglutinate contents and Is/FeO depend on the same formation process, namely melting by meteoroid impact. Solar-flare tracks depend on the meteoroid flux and size-frequency distribution as well as the flux of solar-flare particles because it is necessary to stir the soil grains to within sufficiently near the surface in order for them to receive exposure. Furthermore, it is clear from the level of irradiation in even the immature soil grains that we are concerned with soil irradiated at some epoch in the past. Therefore, we can ask whether a change in the meteoroid flux and size-frequency distribution or the solar-flare particle flux could be the cause of the maturity index discrepancy. A decline in the meteoroid flux or a change in the size-frequency distribution could lead to reduced Is/FeO and agglutinate

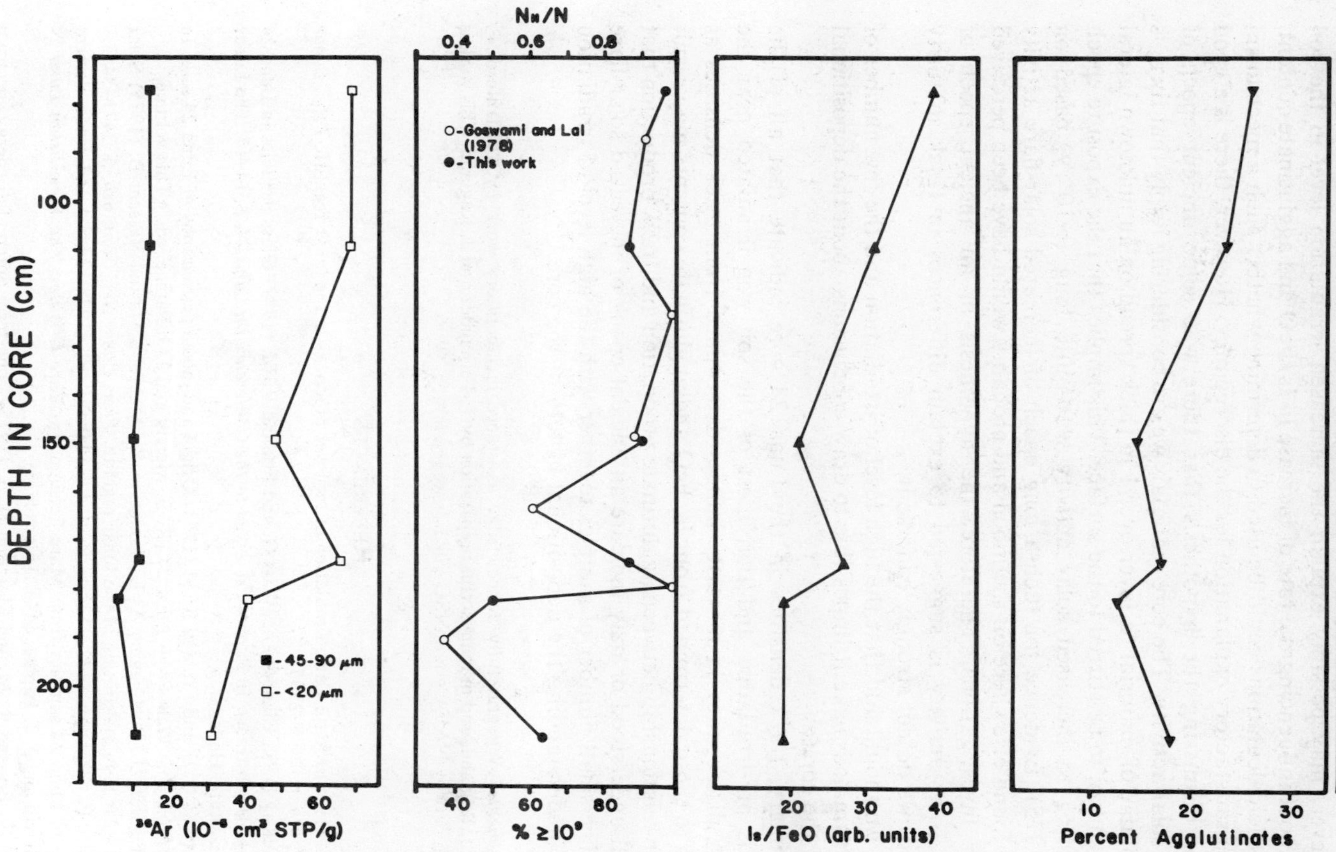

Fig. 2. Depth in the Luna 24 core versus various lunar soil maturity indices. Trapped ^{36}Ar is from Bogard and Hirsch (1978); % $\geq 10^9$, the percentage of grains with track densities $\geq 10^9$ cm^{-2}, is from this work; N_H/N, the fraction of grains with track densities $\geq 10^8$ cm^{-2}, is from Goswami and Lal (1978); Is/FeO, the normalized ferromagnetic resonance intensity, is from Morris (1978a); the agglutinate contents are from Basu *et al.* (1978).

formation, but it would also lead to reduced stirring and, therefore, to reduced exposure of grains to the lunar surface. An increase in the solar-flare particle flux, however, may possibly explain the increased irradiation level in the soil grains without influencing the rate of increase in Is/FeO and agglutinate content. Monte Carlo calculations could be used to determine whether such a mechanism would be a satisfactory explanation for the discrepancy. However, there is a good reason for dismissing the hypothesis that there was some ancient epoch of increased solar activity. The core materials we are considering imply that there is a large volume of material (a layer over 1 m thick spread on an unknown lateral area) which had to be stirred to the surface. This implies that the exposure epoch corresponding to increased solar activity was quite long ($\sim 10^9$ y, based on Arnold, 1975). Evidence for such a long epoch of increased solar-flare activity should be found elsewhere on the moon and probably would have been perceived by earlier investigations. Therefore, the hypothesis of an ancient epoch of increased solar activity is suggested to explain differences in track maturity indices, but we do not strongly endorse it.

Finally we should add that the high level of irradiation and the few number of samples examined make it impossible to draw conclusions about the depositional history of the core.

In summary, track densities in the Luna 24 core indicate that all of the material is pre-irradiated and that most of the soil was irradiated near the surface. A discrepancy in maturity indices determined from track densities as compared to those determined from Is/FeO and agglutinate contents may result from some relationship between agglutinate production and track production that is not well understood or may indicate an ancient epoch of increased solar flare activity. The small number of samples together with the high level of irradiation precludes understanding the depositional history of the core.

Acknowledgments—We gratefully acknowledge assistance in laboratory work by J. Blanford, P. Grover and J. Ragland, and in manuscript preparation by T. Fernandez and J. Ragland. This work is supported in part by NASA grant NSG 9043.

REFERENCES

Arnold J. R. (1975) Monte Carlo simulation of turnover processes in the lunar regolith. *Proc. Lunar Sci. Conf. 6th*, p. 2375–2395.

Basu A., McKay D. S., Clanton U., Waits G. and Fruland R. M. (1977) Glass particles in Luna 24 drill core soils (abstract). In *Papers Presented to the Conference on Luna 24*, p. 14–17. The Lunar Science Institute, Houston.

Basu A., McKay D. S. and Fruland R. M (1978) Origin and modal petrography of Luna 24 soils. In *Mare Crisium: The View From Luna 24* (R. B. Merrill and J. J. Papike, eds.) This volume.

Bibring J. P., Borg J., Burlingame A. L., Langevin Y., Maurette M. and Vassent B. (1975) Solar wind and solar flare maturation of the lunar regolith. *Proc. Lunar Sci. Conf. 6th*, p. 3471–3493.

Blanchard D. P., Brannon J. C., Aaboe E. and Budahn J. R. (1978) Major and trace element chemistry of Luna 24 samples from Mare Crisium. In *Mare Crisium: The View From Luna 24* (R. B. Merrill and J. J. Papike, eds.). This volume.

Blanford G. E., Fruland R. M. and Morrison D. A. (1975) Long-term differential energy spectrum for solar-flare iron-group particles. *Proc. Lunar Sci. Conf. 6th*, p. 3557–3576.

Blanford G. E., McKay D. S. and Wood G. C. (1977) Particle track densities in double drive tube 60009/10. *Proc. Lunar Sci. Conf. 8th*, p. 3017–3025.

Blanford G. E. and Morrison D. A. (1976) Stratigraphy in Apollo 16 drill section 60002. *Proc. Lunar Sci. Conf. 7th*, p. 141–154.

Blanford G. E. and Wood G. C. (1978) Irradiation stratigraphy in Apollo 16 cores (abstract). In *Lunar and Planetary Science IX*, p. 106–108. Lunar and Planetary Institute, Houston.

Bogard D. D. and Hirsch W. C. (1978) Noble gases in Luna 24 core soils. *Mare Crisium: The View From Luna 24* (R. B. Merrill and J. J. Papike, eds.). This volume.

Crozaz G. (1978) Nuclear particle tracks and the regolith at the Luna 24 site. *Mare Crisium: The View From Luna 24* (R. B. Merrill and J. J. Papike, eds.). This volume.

Etique Ph., Funk H., Poupeau G., Romary Ph., Signer P. and Wieler R. (1978) Solar-wind gas and solar flare track correlations in lunar soils; revisited (abstract). In *Lunar and Planetary Science IX*, p. 300–302. Lunar and Planetary Institute, Houston.

Gose W. A. and Morris R. V. (1977) Depositional history of the Apollo 16 deep drill core. *Proc. Lunar Sci. Conf. 8th*, p. 2909–2928.

Goswami J. N. and Lal D. (1978) Particle tracks and microcraters in Luna 24 drill core soil samples (abstract). In *Lunar and Planetary Science IX*, p. 400–402. Lunar and Planetary Institute, Houston.

Hutcheon I. D., Macdougall D. and Price P. B. (1974) Improved determination of the long-term average Fe spectrum from 1 to 460 MeV/a.m.u. *Proc. Lunar Sci. Conf. 5th*, p. 2561–2576.

McKay D. S., Basu A. and Waits G. (1978) Grain size and the evolution of Luna 24 soils. In *Mare Crisium: The View From Luna 24* (R. B. Merrill and J. J. Papike, eds.). This volume.

Morris R. V. (1976) Surface exposure indices of lunar soils: A comparative FMR study. *Proc. Lunar Sci. Conf. 7th*, p. 315–335.

Morris R. V. (1978a) FMR and magnetic studies of Luna 24 soils and >1 mm soil particles. In *Mare Crisium: The View From Luna 24* (R. B. Merrill and J. J. Papike, eds.). This volume.

Morris R. V. (1978b) The maturity of lunar soils: Concepts and more values of Is/FeO (abstract). In *Lunar and Planetary Science IX*, p. 760–762. Lunar and Planetary Institute, Houston.

Morris R. V. and Gose W. A. (1976) Ferromagnetic resonance and magnetic studies of cores 60009/10 and 60003: Compositional and surface-exposure stratigraphy. *Proc. Lunar Sci. Conf. 7th*, p. 1–11.

Walker R. and Yuhas D. (1973) Cosmic ray production rates in lunar materials. *Proc. Lunar Sci. Conf. 4th*, p. 2379–2389.

Yuhas D. E. (1974) The particle track record in lunar silicates: Long-term behavior of solar and galactic VH nuclei and lunar surface dynamics. Ph. D. thesis, Washington University, St. Louis.

Merrill, R. B. and Papike, J. J., editors.
(1978) *Mare Crisium: The View from Luna 24*, Pergamon Press (New York), p. 165–169
Printed in the United States of America

Nuclear particle tracks and the regolith at the Luna 24 site

GHISLAINE CROZAZ
Department of Earth and Planetary Sciences and McDonnell Center for the Space Sciences, Washington University, St. Louis, Missouri 63130

Abstract—Track investigations of feldspars and pyroxenes in Luna 24 soils indicate that most of the grains resided within millimeters of the lunar surface. The two deepest samples appear to be less mature than the upper ones. Lack of correlation between tracks, agglutinates and I_s/FeO is puzzling and unique.

INTRODUCTION

The following is a report of a study of nuclear particles tracks in soils from the Luna 24 core.

Tracks, produced by both solar flare and galactic heavy cosmic ray nuclei, are found in abundance in lunar soils and have been used to try to decipher the depositional histories of lunar cores. The main advantage of this method resides in the fact that, in favorable cases, the irradiation history of individual grains can be traced whereas most other types of studies of the lunar regolith measure integrated effects on large numbers of grains. In addition, the track method can provide time constraints.

It is now well established that the irradiation and depositional histories of lunar cores vary greatly (see, for example, Crozaz and Dust, 1977). Two extreme modes of accumulation seem to be operating; on one hand, accumulation of layers generally a few cm thick but occasionally as thick as 60 cm (like the coarse grained layer between ~20 and 80 cm in the Apollo 17 deep drill stem) and, on the other hand, accumulation of smaller increments of heavily irradiated material. Layers of the first type are characterized by a decrease in the minimum track densities with depth in agreement with the expected track production from galactic cosmic rays. However, in order to recognize such cm units and fully exploit the potentials of the track method, it is necessary to study the cores in some detail, i.e. analyze samples at ~cm-intervals or, better even, use impregnated sections.

Conclusions from the present study are somewhat limited as only 7 samples from a 160 cm section were available. Six soil samples were analyzed as well as feldspars from the microgabbro layer at the 170 level. This layer has been assumed to be formed by a single rock which may have been crushed during the drilling operations (Barsukov, 1977).

EXPERIMENTAL

Individual feldspar grains >50 μm in size were mounted in epoxy resin, polished and etched to reveal cosmic ray heavy nuclei tracks following techniques described earlier (Crozaz and Dust, 1977). In addition, 20 pyroxene grains were also analyzed. Track densities in excess of $\sim 3 \times 10^7/cm^2$ were measured with a stereoscan electron microscope (SEM). SEM counts were not, as usually done in our laboratory, divided by two to account for track density differences obtained while using both the optical microscope and the SEM in order to compare our results with those of other authors. Application of this correction factor does not in any way alter the interpretations of the data and the conclusions.

RESULTS AND DISCUSSION

Track densities in individual feldspar and pyroxene grains are shown in Fig. 1. If these crystals are representative of the bulk soils, we are led to conclude that all samples analyzed can be considered as mature: only ~10% of the crystals have track densities lower than $10^8/cm^2$ indicating that most of the grains analyzed

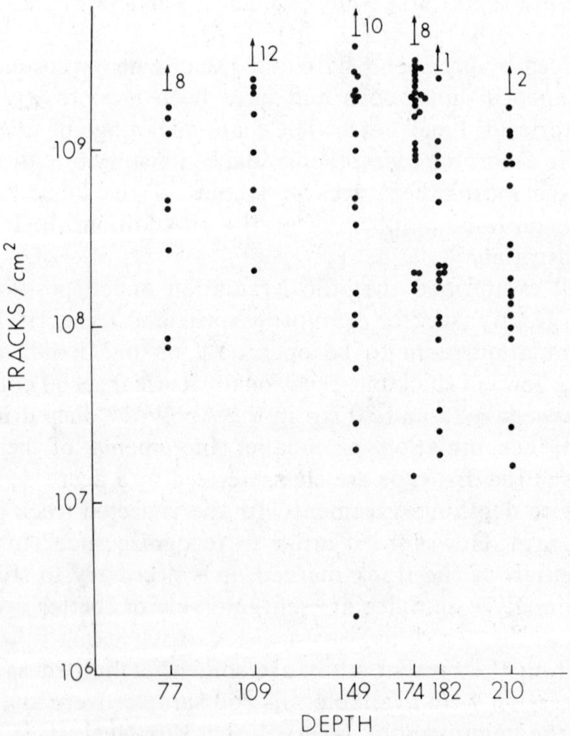

Fig. 1. Track densities in individual feldspar and pyroxene grains at 6 levels in the Luna 24 core.

acquired their tracks while residing within millimeters of the lunar surface. Although all the samples appear mature, the lower two (24182 and 24210) are less so (this is best seen in Fig. 2 where the quartile track density—density above which 75% of the grains are found—is shown as a function of the depth). Sample levels are indicated on Figs. 1 and 2. Because considerable ambiguity exists as to the depths of the samples, a problem which Bogard and Hirsch (1977) addressed at the Conference on Luna 24, depth estimates have not been included in the figures. Results similar to the ones presented here have also been obtained by Blanford and Wood (1977).

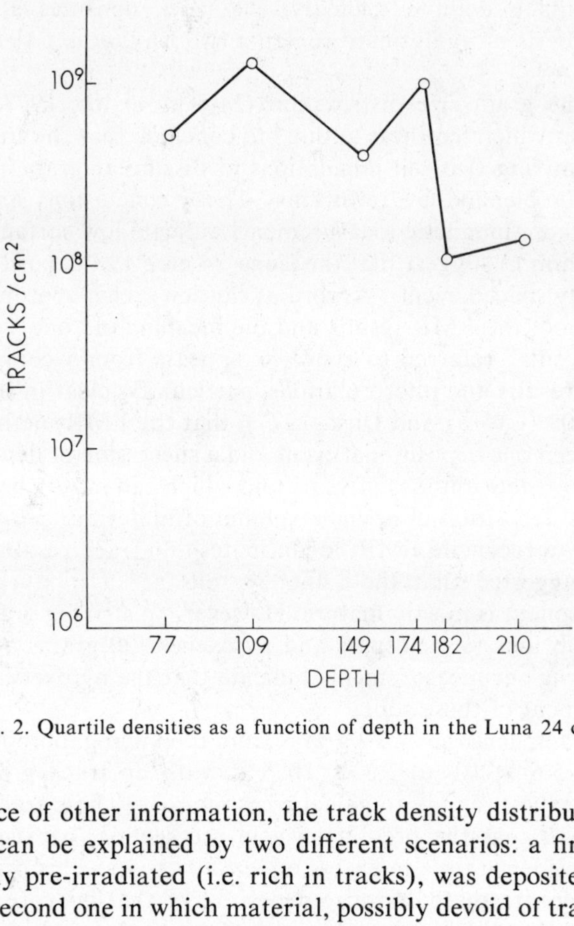

Fig. 2. Quartile densities as a function of depth in the Luna 24 core.

In the absence of other information, the track density distribution observed in these six soils can be explained by two different scenarios: a first one in which material, largely pre-irradiated (i.e. rich in tracks), was deposited, possibly very rapidly, and a second one in which material, possibly devoid of tracks, accreted so slowly that the reworking was intense enough to bring most of the grains close to the surface where they were exposed to solar flares and accumulated their high track densities.

Indication that the first scenario may be more valid is provided by the study of noble gas and isotopic effects. Bogard and Hirsch (1977), using their cosmogenic

^{21}Ne data, deduced that probably all the soils were substantially pre-irradiated before deposition in the core. Papanastassiou *et al.* (1977) reached a similar conclusion from their samarium isotopic data. However, it should be remarked that this does not automatically imply that the grains were track rich when they were deposited at the sampling site as solar flare tracks are acquired only within the upper few millimeters of the lunar regolith, whereas cosmogenic noble gases and isotopic effects extend to depth measured in meters.

Previous studies (see, for example, Blanford, *et al.*, 1977) have shown that solar flare tracks, agglutinate contents and I_s/FeO, all indicators of residence very close to the surface, correlate very well. Up to the return and analysis of the Luna 24 samples, there was no exception to this correlation. However, the Luna 24 core represents a definite anomaly: the track densities are higher than expected on the basis of agglutinate contents (McKay, *et al.*, 1977) and I_s/FeO (Morris, 1977).

In addition, the grain size distribution (McKay *et al.*, 1977) has a strong bimodal character which led these authors to conclude that the Luna 24 soils are the products of mixing two soil populations of dissimilar grain size which have not been strongly blended by reworking. These conclusions are endorsed by Morris (1977) whose magnetic measurements indicate low surface soil residence times which led him to suggest that the Luna 24 core was deposited in one event or series of closely spaced events. A word of caution seems appropriate here as to the interpretation of the FMR results and the meaning of "one event or series of closely spaced events" referred to above. It appears from a comparison of both track and FMR results and interpretations, particularly clear in the case of drive tube section 60009 (Crozaz and Dust, 1977), that the FMR method is unable to distinguish between one depositional event and a succession of depositional events separated each by a few millions of years and which can hardly be identified with the simultaneous deposition of a whole column of material.

In an attempt to reconcile FMR, agglutinates and track results, Blanford and Wood (1977) suggested that the Luna 24 soils are a mixture in which the feldspathic component is mostly mature. However, no striking difference in track distributions is obvious for feldspars and pyroxenes (20 grains analyzed) in the present work. Thus our measurements indicate that the pyroxenes also represent a mature component of these soils.

In addition, track densities in 22 grains from the microgabbro layer at the 170 level vary from 5.6×10^6 to $1.3 \times 10^7/\text{cm}^2$ with no track gradient observed indicative of an exposure only to galactic cosmic rays. These tracks cannot have accumulated *in situ*, at the present level in the section, because their density would imply a residence time in excess of 3×10^{11} years. This proves that either the sample was pre-irradiated prior to deposition or that the core was built up in a series of events that left the sample much closer to the surface in the past.

In summary, it appears that track, agglutinate and FMR studies lead to conclusions that are not always mutually consistent. As pointed out above, the Luna 24 core represents a definite anomaly in this respect. Possibly, it was formed at a time when the irradiation-gardening conditions were different than for most lunar samples. Finally, it should be stressed again that the limited number of samples available for study (only six) severely constrains the potential of, or the conclusions derived from, the different types of investigations. A more detailed investigation should lead to a better understanding of the depositional history at the Luna 24 site.

Acknowledgments—The expert help of Lou Ross is gratefully acknowledged. This research was supported by NASA grant 59220B.

References

Barsukov V. L. (1977) Preliminary data for the regolith core brought to earth by the automatic lunar station Luna 24. *Proc. Lunar Sci. Conf. 8th*, p. 3303-3318.

Blanford G. E., McKay D. S. and Wood G. C. (1977) Particle track densities in double drive tube 60009/10. *Proc. Lunar Sci. Conf. 8th*, p. 3017-3025.

Blanford G. E. and Wood G. C. (1977) Particle track densities in the Luna 24 core (abstract). In *Papers Presented to the Conference on Luna 24*, p. 41-43. Lunar Science Institute, Houston.

Bogard D. D. and Hirsch W. C. (1977) Noble gases in Luna 24 core soils (abstract). In *Papers Presented to the Conference on Luna 24*, p. 44-47. Lunar Science Institute, Houston.

Crozaz G. and Dust S. (1977) Irradiation history of lunar cores and the development of the regolith. *Proc. Lunar Sci. Conf. 8th*, p. 3001-3016.

McKay D. S., Basu A., Waits G., Clanton U., Fuhrman R. and Fruland R. (1977) Grain size and evolution of Luna 24 soils (abstract). In *Papers Presented to the Conference on Luna 24*, p. 114-117. Lunar Science Institute, Houston.

Morris R. V. (1977) FMR and magnetic properties of Luna 24 soils and >1 mm soil particles (abstract). In *Papers Presented to the Conference on Luna 24*, p. 121-124. Lunar Science Institute, Houston.

Papanastassiou D. A. (1977) Conference on Luna 24, oral presentation.

Merrill, R. B. and Papike, J. J., editors.
(1978) *Mare Crisium: The View from Luna 24*, Pergamon Press (New York), p. 171–177
Printed in the United States of America

Cosmic ray track record and maturity of the Luna 24 regolith

D. CHAILLOU[1], P. PELLAS[1], M. CHRISTOPHE-MICHEL LÉVY[2] and D. STORZER[1]

[1]Laboratoire de Minéralogie du Muséum, 75005 Paris
[2]Laboratoire de Minéralogie-Cristallographie, Université Paris VI

Abstract—The cosmic ray track record is investigated in feldspars, pyroxenes and olivines from 10 soil levels of the Luna 24 core. The results show that all layers contain a major fraction of crystals with a high level of solar flare irradiation. For the gabbro clast (24171) an upper limit of 4 Myr can be estimated for its surface residence time. A correlation plot of agglutinate content vs. percentage of highly solar-flare irradiated crystals indicates that most of the Luna 24 soils follow a normal maturation sequence. Two soils (24172,-174) show, in this frame, a totally abnormal behaviour.

INTRODUCTION

This note reports cosmic-ray track densities measured in olivines, pyroxenes and feldspars from ten different levels of the Luna 24 core, including the leucocratic microgabbro (24171). Through the courtesy of our Russian colleagues, the samples studied could be selected according to their different soil maturity parameters (variations in contents of agglutinates, breccias, etc. . .) reported by Florensky *et al.* (1977) and Barsukov (1977). This selection seemed to be important as it is well known that solar-flare track densities correlate quite well with soil maturity parameters (e.g. Blanford *et al.*, 1977). From each level about 30 mg of bulk soils were available. The grain-size fractions $<40\ \mu m$ (400 mesh) are being investigated by the Orsay group as one part of a cooperative study. We investigated the size fractions $>40\ \mu m$ for their agglutinate contents and their cosmic-ray track records. Cosmic-ray tracks were also studied in 200–370 μm sized feldspars from Sample 24184 for comparison with the data obtained by Kashkarov *et al.* (1978).

EXPERIMENTAL PROCEDURES

The percentages of agglutinates, breccias, igneous fragments and monomineralic grains in the different grain sizes between 40 and 250 μm were defined by means of a polarizing stereomicroscope.

For cosmic-ray track investigations, 80–150 μm-sized crystals of olivines, pyroxenes, and feldspars were mounted separately in epoxy, polished and etched using the following conditions: olivines (1 hour in boiling WN solution, slightly modified by adding 4 g of oxalic acid per 100 cc of solution; the final pH being adjusted to 8.00 ± 0.05) (Krishnaswami *et al.*, 1971); pyroxenes (5 minutes at 23°C in a mixture of 2 vol.48% HF + 1 vol.96% H_2SO_4 + 1 vol.65% HNO_3 + 6 vol. H_2O); feldspars (boiling solution of 6 g NaOH + 16 cc H_2O for time intervals of 6, 9, 15, 20, 30, and 60 minutes, in order to account for differences in plagioclase composition). It should be noted that the feldspars from soil 24180 in the size range 80–120 μm were etched by the Orsay group in a boiling solution of 12 g

NaOH + 16 cc H_2O for 1 to 2 minutes, resulting in a much stronger etching than with our conditions. All track densities were determined with a stereoscan electron microscope. The track counts correspond to the track densities measured in the grain centers, or, in the case of crystals with track density gradients, to the lowest density.

RESULTS AND DISCUSSION

The cosmic-ray track densities in individual feldspar-pyroxene-olivine crystals from the 10 levels studied are reported in Fig. 1. Except for the crystals from the gabbroic layer 24171 and for those of the grain-size fraction 200–370 μm from level 24184, all crystals show track densities in excess of 10^8 cm^{-2}, and mostly even in excess of 10^9 cm^{-2}. This indicates an intense solar-flare irradiation of these soils, and consequently, a rather high level of maturity. Despite the quite large dispersion in track densities among individual crystals from a given level, most of the soils show no indication of a bimodal track density distribution, with the exception of levels 24171, 24184 (size fraction 200–370 μm), and possibly level 24168. Observed bimodal track density distributions are believed to indicate mixing of mature and immature components (Blanford *et al.*, 1977). The absence of such bimodal distributions in most of our Luna 24 soils suggests, therefore, a rather high degree of reworking, hence of maturity. In all levels (except 24134) there is a general tendency for the track densities to decrease from feldspar through pyroxene to olivine (Figs. 1 and 2). For an identical irradiation history in a given level, this is in fact to be expected, as the etchable ranges of cosmic-ray iron-group nuclei decrease in the same crystal sequence. In this respect, soil 24134 is exceptional since the track densities in feldspars are on the average lower than those measured in pyroxenes and olivines. This might suggest a mixing of two components differing in maturity as well as in petrology. However, the unimodal track density distribution does not seem to endorse such a mixing event.

Crystals from the gabbroic layer 24171 show a bimodal track density distribution for pyroxene and olivine. All feldspars and the pyroxenes and olivines with track densities below 10^8 cm^{-2} belong to the microgabbro. In these crystals galactic cosmic-ray and solar-flare tracks are observable. This proves that the original gabbroic clast was once exposed at the very top of the lunar surface. A maximum surface residence time of about 4×10^6yr can be estimated from the lowest track density observed in feldspar (4×10^6/cm^2) and assuming a corresponding depth of 2 cm in the gabbro clast (track production rate of Blanford *et al.*, 1975). In addition to the microgabbro crystals, a population of solar-flare irradiated pyroxenes and olivines is present with track densities ranging between 4×10^8 and $\sim 10^9$ cm^{-2}. This contamination is probably due to the drilling process, as suggested by Florensky *et al.* (1977).

In the level 24168, directly overlying the gabbroic layer, there seems to be an indication of a bimodal track density distribution in the feldspars which might be due to a contamination from the microgabbro. On the other hand, level 24172 does not show any clear contamination from the adjacent gabbro.

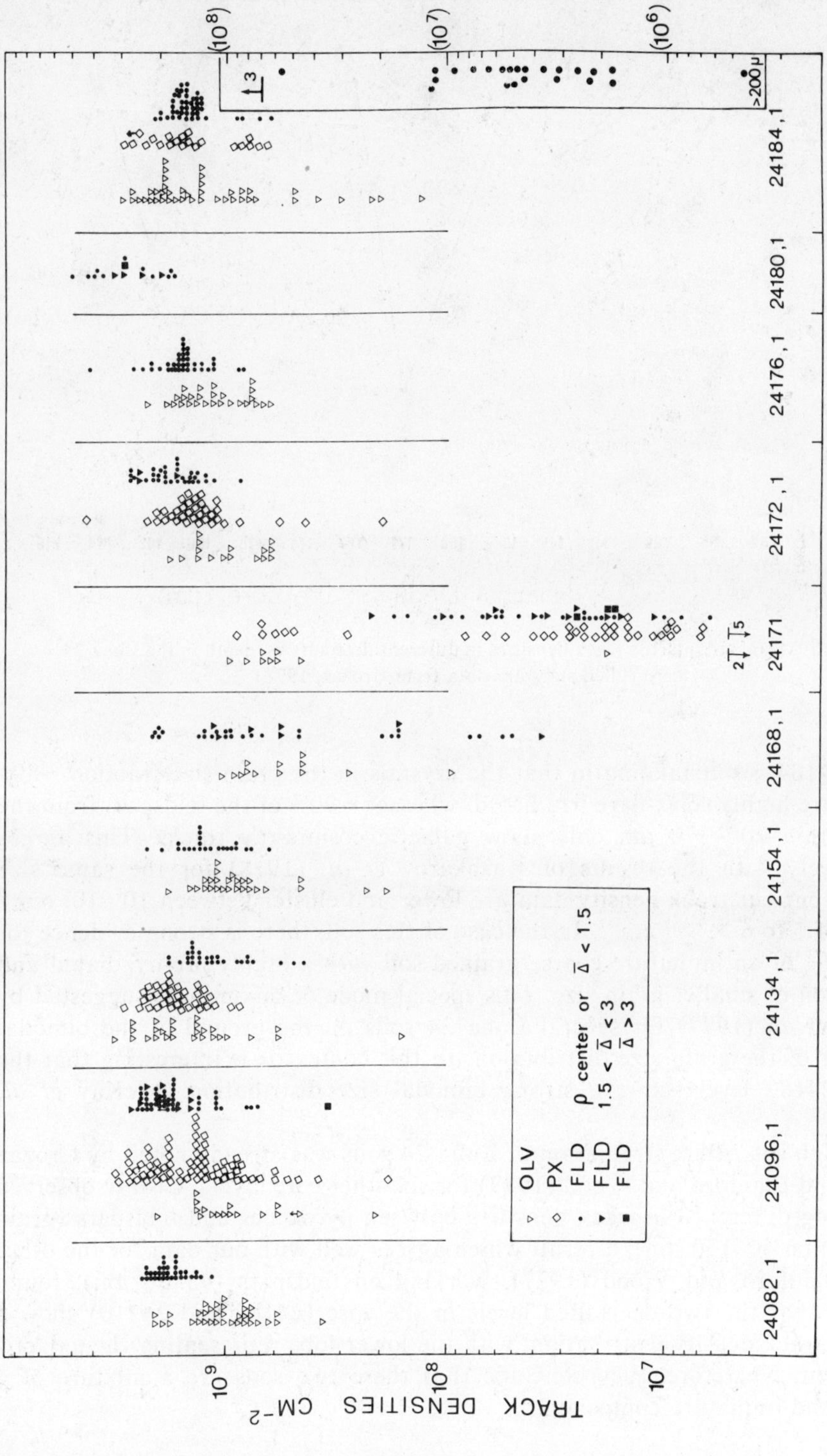

Fig. 1. Track densities in individual feldspar, pyroxene and olivine crystals (80–150 μm) at 10 levels of the Luna 24 core (large filled circles on the right hand side correspond to feldspars (200–370 μm) from level 24184).

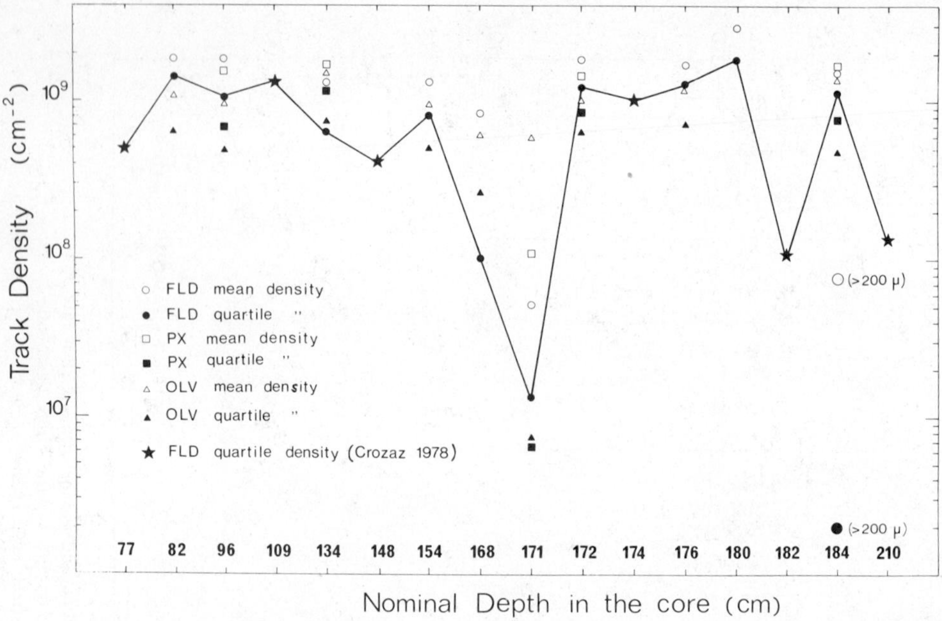

Fig. 2. Mean and quartile track densities in different detectors vs. depth in the Luna 24 core (filled stars are data from Crozaz, 1978).

Soil 24184 is outstanding in that the crystals in the grain size fraction ~80–150 μm are highly solar-flare irradiated, whereas ~90% of the feldspars from the size fraction 200–370 μm only show galactic cosmic-ray tracks. This agrees qualitatively with the results of Kashkarov *et al.* (1978) for the same size fraction, but our track density data are lower and cluster between 10^6–10^7 cm^{-2} instead of 1 to 6 $\times$ 10^7 cm^{-2}. In the case of this soil, there is strong evidence for the mixing of an immature coarse grained soil with a highly pre-irradiated and mature soil of smaller grain size. This special mode of mixing was suggested by McKay *et al.* (1977) for several Luna 24 soils on the ground of the bimodal character of the grain size distribution. In this context it is interesting that the nearby 24182 level shows a strong bimodal size distribution (McKay *et al.* 1977).

The high solar-flare irradiation of Luna 24 soils was already noted by Crozaz (1978) and Blanford and Wood (1977) for six other core layers. Crozaz observed no striking difference in track densities between pyroxenes and feldspars (grain size fraction 50–150 μm), a result which agrees well with our data for the other levels. Blanford and Wood (1977), working on feldspars (45–90 μm), found evidence that the two deep-sited levels in the core (24182 and 24210) show a bimodal track density distribution, with the lower lobe representing the galactic component. Therefore, they conclude that these two soils are a mixture of a mature and immature component.

The mean and the quartile track densities of the various minerals from the ten levels studied are reported in Fig. 2, together with the results obtained by Crozaz (1978) for the six additional levels. In the frame of the track data, it is obvious that all "soils" (excluding of course the gabbro clast) display a high degree of maturity down to the level 24180. Below this level, varying admixtures of immature components seem to be present.

There is a general consensus that in Luna 24 soils there exists an inconsistency between solar-flare track data (Crozaz 1978; Blanford and Wood, 1977) and other maturity parameters, e.g. agglutinate contents or FMR maturity index (McKay *et al.*, 1977; Morris, 1977). The agglutinate contents of the soils studied in this work are reported in Table 1 for the different grain size fractions. In Fig. 3, the percentage of agglutinates in the 150–250 μm size fraction is plotted against the percentage of feldspars with track densities exceeding 10^9 cm^{-2} in the size fraction 80–150 μm, in order to compare the Luna 24 soils with those of the double drive tube 60009/10 (Blanford *et al.*, 1977). For completeness, the track data for Crozaz (1978) and the agglutinate fractions (150–250 μm) determined by Nielsen and Drake (1977) are also plotted. It is apparent that most Luna 24 soils (9 out of 14) clearly follow the normal maturation sequence. In this sequence 3 soils (24182, -210, -168) are quite immature, whereas the soils 24077, -134, -096, -176, -154, and -082 show an increasingly higher maturity. Soils 24149 and 24109 could well be within the maturation sequence. Soil 24184 might also fall into this normal correlation, because the agglutinate content reported in Fig. 3 corresponds to a larger size fraction (200–370 μm, Table 1). Undoubtedly,

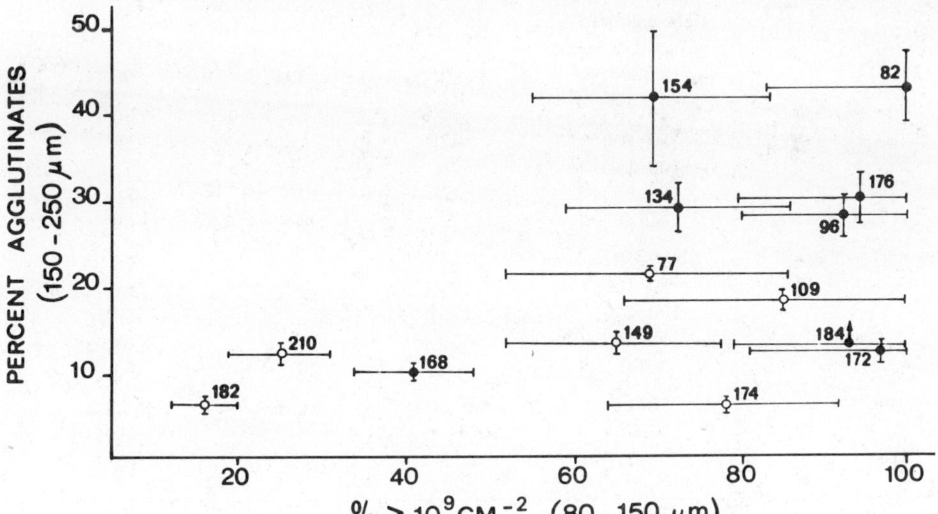

Fig. 3. Percentage of agglutinates in the 150–250 μm size fraction vs. percentage of feldspars with track densities $>10^9$ cm^{-2} in the size fraction 80–150 μm. Open circles are data from Crozaz (1978) and Nielsen and Drake (1977).

Table 1. Particle types (%) as a function of grain sizes in 9 levels of the Luna 24 core.

Sampling depth in the core	Grain sizes (μm)	Agglutinates	Breccias	Igneous rocks and monomineralic grains
	40–100	~46		54
	100–150	(52)[a]		(48)[a]
	150–250	43	2	55
24082,1	>200[1]	28	12	50
	80–150	35	10	55
	150–250	28	22	50
24096,1	>200[1]	25	20	55
	40–50	40		60
	50–150	30		70
	150–250	29		71
24134,1	>200[1]	~10	20	50 total (80%)
	40–50	35		65
	50–150	43		
	150–250	42	20	38
24154,1	>200[1]	6	50	34
	40–50	20		80
	50–100	25		75
	100–150	(10)[a]		(90)[a]
	150–250	10	5	85
24168,1 a	>200[1]	5	38	57
	light-coloured microgabbro layer			
24171	>200[1]	2	3	95
	40–50	50		50
	50–100	42		58
	100–150	28		72
	150–250	12	2	86
24172,1	>200[1]	8	23	69
	40–50	34		66
	50–100	33		67
	100–150	(20)[a]		(80)[a]
	150–250	30	2	68
24176,1	>200[1]	8	20	63
	50–150	no data	no data	no data
	>200[1]	6	5	80 total (91%)
24184,1	200–370	13	2	85

[1](Florensky *et al.*, 1977)

[a]Large statistical uncertainties: less than 100 grains studied.

only two soils (24172, -174) fall off the sequence and show an abnormal behaviour. So far we do not understand the special process responsible for the formation of these two soils.

Acknowledgments—We are indebted to the Academy of Sciences of the U.S.S.R. for making samples returned by the Luna 24 probe available for scientific studies. Especially we acknowledge the cooperation of Dr. A. V. Ivanov during the sample selection. The generous help of G. Crozaz for making her unpublished manuscript available and for critically reading and correcting our manuscript is highly appreciated. We are also grateful to Dr. J. C. Lorin for his critical comments.

References

Barsukov V. L. (1977) Preliminary data for the regolith core brought to earth by the automatic lunar station Luna 24. *Proc. Lunar Sci. Conf. 8th*, p. 3303–3318.

Blanford G. E., Fruland R. M. and Morrison D. A. (1975) Long term differential energy spectrum for solar-flare iron-group particles. *Proc. Lunar Sci. Conf. 6th*, p. 3557–3576.

Blanford G. E., McKay D. S. and Wood G. C. (1977) Particle track densities in double drive tube 60009/10. *Proc. Lunar Sci. Conf. 8th*, p. 3017–3025.

Blanford G. E. and Wood G. C. (1977) Particle track densities in the Luna 24 core (abstract). In *Papers Presented to the Conference on Luna 24*, p. 41–43. The Lunar Science Institute, Houston.

Crozaz G. (1978) Nuclear particle tracks and the regolith at the Luna 24 site. In *Mare Crisium: The View from Luna 24* (R. B. Merrill and J. J. Papike, eds.). This volume.

Florensky C. P. (1977) Luna 24: Geologic setting of landing site and characteristics of sample core (preliminary data). *Proc. Lunar Sci. Conf. 8th*, p. 3257–3279.

Kashkarov L. L., Genaeva L. I. and Lavrukhina A. K. (1978) Track studies in four samples of Luna 24 core (abstract). In *Lunar and Planetary Science IX*, p. 612–614. Lunar and Planetary Institute, Houston.

Krishnaswami S., Lal D., Prabhu N. and Tamhane A. S. (1971) Olivines: revelation of tracks of charged particles. *Science* **174**, 287–291.

McKay D. S., Basu A., Waits G., Clanton U., Fuhrman R. and Fruland R. (1977) Grain size and evolution of Luna 24 soils (abstract). In *Papers Presented to the Conference on Luna 24*, p. 114–117. The Lunar Science Institute, Houston.

Morris R. V. (1977) FRM and magnetic properties of Luna 24 soils and >1 mm soil particles (abstract). In *Papers Presented to the Conference on Luna 24*, p. 121–124. The Lunar Science Institute, Houston.

Nielsen R. L. and Drake M. J. (1977) Magma types at the Luna 24 site in Mare Crisium (abstract). In *Papers Presented to the Conference on Luna 24*, p. 132–134. The Lunar Science Institute, Houston.

Merrill, R. B. and Papike, J. J., editors.
(1978) *Mare Crisium: The View from Luna 24*, Pergamon Press (New York), p. 179–194
Printed in the United States of America

Radiation and thermal histories of Luna 24 drill-core samples

S. A. Durrani, R. K. Bull and S. W. S. McKeever

Department of Physics, University of Birmingham, Birmingham B15 2TT, England

Abstract—Luna 24 drill-core samples from three different depths (~90 cm, 125 cm, and 196 cm) have been studied, using charged-particle track analysis and thermoluminescence methods. Track-density data as a function of depth suggest that the most mature component of the Luna 24 soils exists among the fine-grained fraction and that the overall maturity of the soil decreases with increasing depth. A high proportion (~30%) of mineral grains from all levels exhibit anisotropic track-density gradients, which points either to a low degree of stirring at the lunar surface or to pre-burial irradiation of these grains on a rock-face. TL studies suggest the possibility of using the TL sensitivity of samples, measured by inducing artificial TL in them, as an indicator of their past radiation history. The low-temperature TL (below ~160°C) in the initially cold samples from all three levels is found to have been lost during their 12-month storage at room temperature. We recommend refrigeration of a part of the drill-core material from future Luna missions.

1. Introduction

We have carried out charged-particle track analysis and thermoluminescence (TL) studies of Luna 24 samples provided to us by the U.S.S.R. Academy of Sciences through the Royal Society, London. The main purpose of these studies is to interpret the radiation and thermal histories of these samples in their natural environment on the moon. As the Luna 24 samples come from successive depths below the lunar surface (down to ~2 m), it is of particular interest to study the radiation and temperature effects as a function of depth.

Since only very small amounts of sample were made available to us (~3.5 mg for track analysis and ~16 mg for TL work), our data are necessarily limited, and one cannot be sure that they are fully representative. They, however, supplement the results of other Luna 24 workers, and hence we have attempted to construct an integrated picture of relevant effects wherever possible.

1.1 *Description of samples*

The samples for both track and TL studies came from three positions within the Luna 24 deep drill core, with the following designations and depths below the core surface given by the U.S.S.R. Academy of Sciences: 24090,1 (from a depth of 89–90 cm in the core tube); 24125,1 (124–125 cm); and 24196,1 (195–196 cm). According to Barsukov and Florensky (1977), the actual length of the recovered core material was less than the tube penetration depth, so that the real sample depths must be considered to be somewhat uncertain (see Bogard and Hirsch, 1977, for discussion). In view of these uncertainties, we shall always refer in this paper to depths in the core tube, which is the convention generally followed by most Luna 24 workers.

After preliminary sample processing carried out by the British curatorial facility at Cambridge (see Pillinger and Fabian, 1977, for details), the following samples were allocated to us. For track analysis, we were given approximately 20 mineral grains of size 150–250 μm from each depth. These had been hand-picked out of the 'heavy fractions' (>2.96 $g.cm^{-3}$) of 24090,1,9303; 24125,1,9303; and 24196,1,9303; we subsequently also received ~1 mg of the 40–53 μm fraction from *each* of the three locations. When received, none of the grains had been divided into mineral types. For TL work, about 4.5 mg of the sieved fraction below 75 μm from *each* location was made available to us, bearing the numbers: 24090,1,9601; 24125,1,9601; and 24196,1,9601; we were subsequently issued with another ~2 mg of the 75–106 μm fraction from sample 24125,1,9503. The first three TL samples (<75 μm) had been subjected to heat-drying in an oven at ~100°C for 10–15 min following the wet-sieving and washing in acetone during sample processing at Cambridge; the fourth TL sample (75–106 μm), in common with all other fractions above 75 μm, had received no heat treatment, having been dried at room temperature.

2. Charged-Particle Track Studies

Implicit in the grains of lunar minerals is a record of cosmic rays, solar flares and micrometeorite activity on the moon extending back to several aeons (10^9 years). We have studied the distribution of track densities exhibited by grains from various depths of the drill core in order to gain an indication of the 'maturity' (i.e. degree of exposure at or near the lunar surface to low-energy nuclear particles and micrometeorites) of the corresponding soil layers as well as to investigate the soil-mixing characteristics of the lunar regolith.

2.1 *Experimental procedure*

Grains from the 150–250 μm size-fraction were mounted individually in epoxy resin, allowed to set, and then ground and polished. The final polishing stage was carried out with 0.3 μm diamond dust. Minerals were identified on the basis of low-power microscope examination and, on occasion, checked by using the non-dispersive X-ray analyser attached to our Cambridge Mk IV scanning electron microscope (SEM). The grains were then etched as appropriate for the mineral type. Thus feldspars and glasses were etched in a boiling solution containing 1 g NaOH to 2 g H_2O (or occasionally in the proportion 2 g:3 g), usually for 2–5 min. Pyroxenes were etched in a 3:2 NaOH solution for 5–20 min. Olivines were etched in boiling WN solution (Krishnaswami *et al.*, 1971) at pH 8, for 20–60 min.

The 40–53 μm grains were mounted in batches of 20–30 per sample holder and were rather lightly etched. In this case preliminary selection of etched grains was made under our Cambridge Stereoscan 600 SEM at magnifications of ~1000×, followed by more detailed investigation at 5–10 K. The analysis of this size fraction is not yet complete.

Track densities were generally determined from SEM (Stereoscan 600) photographs at 5000–10,000×, although lower track densities ($\lesssim 10^7$ tracks cm^{-2}) were measured directly under an optical microscope at 600× magnification. Track densities in the central regions of a given plane under examination (termed the 'central track densities') were recorded, and where marked track-density gradients were noticed, more detailed photomosaics were constructed. From these observations, profiles of track-density versus depth within the grain were built up.

2.2 *Results and discussion*

Figure 1 shows the distribution of central track densities for the three sample depths, ~90, 125, and 196 cm below the top surface of the core. Solid symbols refer to the 150–250 μm fraction, and open symbols to the 40–53 μm fraction. Track densities in olivine grains have been multiplied by 2 to allow for the

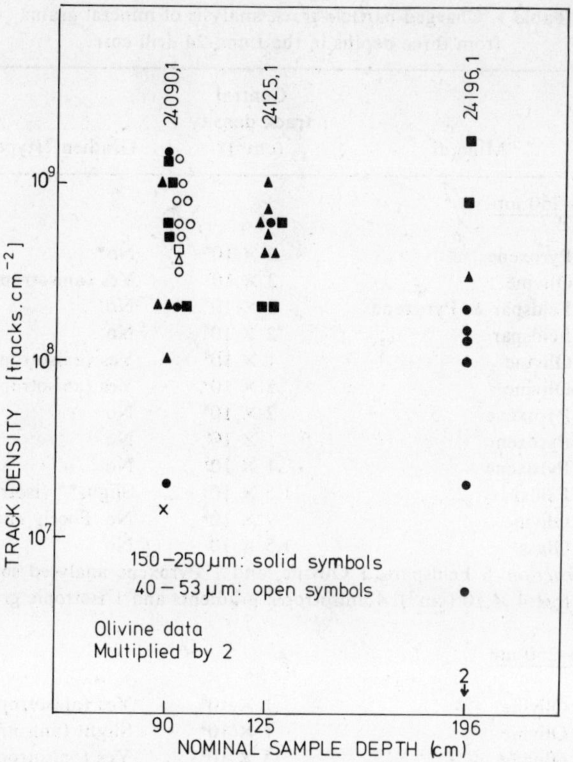

Fig. 1. Distributions of central track density for the three Luna 24 samples (24090,1,9303; 24125,1,9303; 24196,1,9303). Two size fractions of feldspar, pyroxene and olivine grains have been studied: 150–250 μm (solid symbols) and 40–53 μm (open symbols). The symbols used are: feldspar (●,○); pyroxene (▲,△); olivine (■,□). For high track densities SEM photography (5–10 K), and for lower track densities ($\lesssim 10^7$ cm^{-2}) optical microscopy (600×), was employed. Track densities in olivines have been multiplied by 2 for normalization (see text). The one point marked × (at the 90 cm depth) refers to a glass fragment; this value has not been corrected, as the correction factor (up to ~10×) is uncertain.

generally accepted low registration efficiency of this mineral compared to feldspars and pyroxenes (Bhandari *et al.*, 1972). The value for the glass fragment is uncorrected, since the degree of adjustment necessary is uncertain at present; but it may be as large as ~10×.

Table 1 records a detailed break-down of the mineral type, central track density, and the nature of any track-density gradient for each grain from the 150–250 μm fraction. Also shown is a summary of the information available at this stage for the 40–53 μm grains from sample 24090,1. The track-density data in Table 1 are *uncorrected* for registration efficiency.

It may be noted that at each location a majority of the grains have track densities $\gtrsim 10^8$.cm^{-2}, and that a fairly high proportion (~30%) show track-density

Table 1. Charged-particle track analysis of mineral grains from three depths in the Luna-24 drill core.

Sample; Size-fraction Grain No.	Mineral	Central track density (cm^{-2})†	Gradient?(type)	Comments††
24090,1,9303; 150–250 μm				
Grain No.				
1	Pyroxene	2×10^8	No*	
3	Olivine	3×10^8	Yes (anisotropic)*†	
4	Feldspar & Pyroxene	2×10^7	No	
5	Feldspar	2×10^8	No	
6	Olivine	1×10^8	Yes (anisotropic)	
7	Olivine	5×10^8	Yes (anisotropic)	
8	Pyroxene	2×10^8	No	
9	Pyroxene	1×10^8	No	
10	Pyroxene	1×10^9	No	
11	Feldspar	1.5×10^9	Slight** (isotropic)	
12	Olivine	7×10^8	No. Poorly etched, shocked?	
13	Glass	1.5×10^7	No	
Also, *40–53* μm *fraction*: 8 Feldspars, 1 Olivine, and 1 Pyroxene analysed so far. Central track densities: 4×10^8 to 1.4×10^9(cm^{-2}). 4 anisotropic gradients and 1 isotropic gradient found.				
24125,1,9303; 150–250 μm				
Grain No.				
1	Olivine	1×10^8	Yes (anisotropic)	
2	Olivine	1×10^8	Slight (anisotropic)	
3	Olivine	3×10^8	Yes (anisotropic)	
5	Pyroxene	4×10^8	No	
6	Pyroxene	7×10^8	No	
7	Pyroxene	4×10^8	Yes (isotropic)	
9	Pyroxene	5×10^8	No. Fractured surface	
10	Pyroxene	6×10^8	No	
11	Feldspar	6×10^8	No	
12	Pyroxene	1×10^9	Slight (anisotropic)	

gradients. Figure 2 shows a typical crystal with a track-density gradient in one corner on the periphery of the grain. Profiles of track-density versus depth within a grain were also built up for a number of grains from each location, but are not shown.

It seems reasonable to conclude from the above observations that these irradiation characteristics (viz. broadly similar central track densities irrespective of the sample depth below the lunar surface, and frequent track gradients within grains of mere 150–250 μm in size) could not have been produced if the grains had always remained *in situ* where they were found.

A series of studies on Luna 24 drill-core material has recently been carried out by a consortium in the U.S.A., using samples 24077,9; 24109,13; 24149,15; 24174,10; 24182,15; and 24210,9 (from nominal tube depths of 77 cm, 109 cm,

Table 1. (*cont.*)

Sample; Size-fraction Grain No.	Mineral	Central track density (cm^{-2})†	Gradient?(type)	Comments††
24196,1,9303; 150–250 μm				
Grain No.				
1	Olivine	4×10^8	Yes (anisotropic)	
2	Olivine	9×10^8	No	
3	Feldspar	2×10^7	No	
4	Feldspar	1×10^8	No	
7	Feldspar	2×10^8	Yes (anisotropic)	
8	Feldspar	3×10^5	No	
10	Feldspar	1.5×10^8	Yes (anisotropic)	
11	Feldspar	1.4×10^8	No	
12	Pyroxene	5×10^5	No	
13	Pyroxene	3×10^8	No	
14	Feldspar	5×10^6	No	

†Track densities are *uncorrected* for different registration efficiencies of the various minerals. See text for description of etching procedures. Track densities of $\lesssim 1 \times 10^7$ cm^{-2} are measured by optical microscopy; all the others by SEM. Random errors arising from counting statistics are ~10–20%.

*The statement 'No' throughout this column means that the central track density (well away from the edges) and the track density measured at ~5 μm from the edge are equal within experimental uncertainties.

*†The gradient is taken to be 'anisotropic' throughout this table if, while being present at some edges, it is absent or significantly less marked on at least one edge of the grain.

**'Slight' gradient signifies that track densities at the edge and at the centre of a grain differ by ≲50%.

††Taking the whole 150–250 μm fraction (32 mineral grains in all) from all three levels together, we find that 20 grains (i.e. ~63%) have no detectable track-density gradients; 10 grains (~31%) have anisotropic gradients; and 2 grains (~6%) have isotropic gradients. The proportions in grains from each individual level are fairly close to these mean values.

etc.) The picture which has emerged from some of these studies (Blanford and Wood, 1977; McKay *et al.*, 1977; and Morris, 1977) is of a column in which the soil generally consists of a mixture of materials which had apparently been deposited by rather few events. The evidence seems to argue against a large amount of reworking of this mixture.

McKay *et al.* (1977) propose that the soils generally consist of a mixture of two components: one fine-grained (centering at ~100 μm) and one coarser-grained (around 1 mm), the former being of greater maturity. They show that if only the fine-grained mode (say, the <250 μm fraction) is considered, the mean-grain-size parallels the agglutinate content more closely, with more mature soils being finer-grained and richer in agglutinate content, whereas there is a lack of correlation if the soils as a whole are taken into account, owing to the diluting

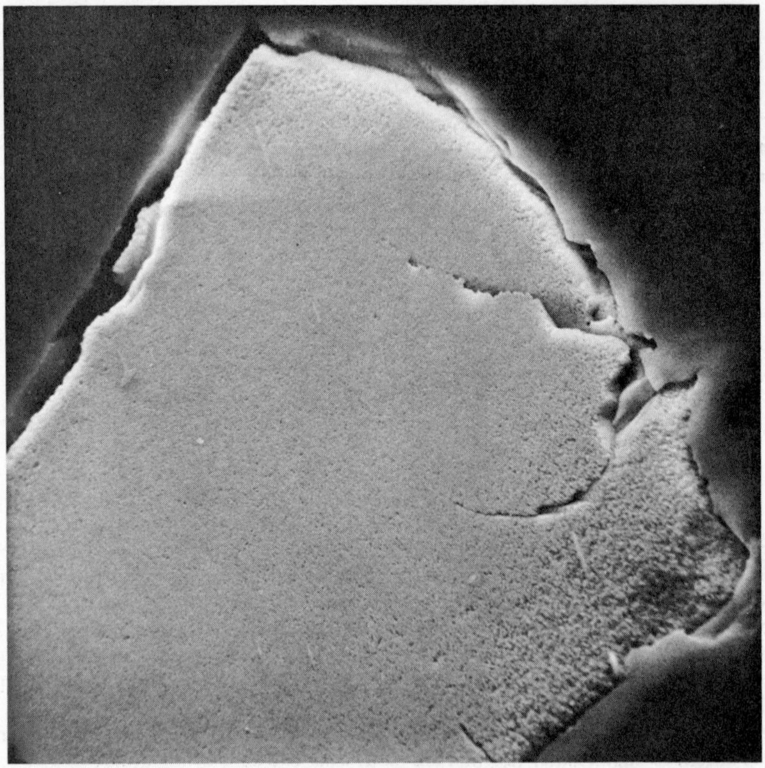

Fig. 2. Olivine grain number 3 from 24090,1,9303, clearly showing a track-density gradient from the lower right-hand corner. The grain was etched in WN solution for 30 min; the magnification is 550×.

effect of the immature coarse soil. They conclude that the Luna 24 samples have not been appreciably reworked after the mixing event which created the bimodal distributions. This conclusion is at variance with that of Florensky *et al.* (1977), who have suggested that the upper part of the core had reworked progressively downward from the surface.

Blanford and Wood (1977) have found that feldspars from the 45–90 μm fraction show track densities predominantly of $\gtrsim 10^9$ tracks.cm^{-2} (caused primarily by solar flares at the very top surface of the moon), with a significant admixture of lightly irradiated grains in the bottom two layers examined by them (182 cm and 210 cm), these latter low track densities being attributed to cosmogenic particles. They conclude that, since the two groups of track densities are sharply divided, the soil is a mixture of a mature and an immature component.

Let us now consider our own track results and their relationship to the preceding observations, even though the fact that there is no overlap between the sample locations for the material issued to the British and the American workers makes direct comparison impossible.

The bulk of our track densities lie between $\sim 10^8$ and 10^9 tracks.cm^{-2}. These track densities do not indicate a significant component of immature material being present in the soil fraction analysed. Although we have found fewer grains with $\gtrsim 10^9$ tracks.cm^{-2} than did Blanford and Wood (1977), it must be remembered that our SEM magnifications (5–10 K) were lower than those used by these authors (>20 K), so that the difference may be observational rather than real. Our magnification makes track densities of greater than $1\text{–}2 \times 10^9$ tracks.cm^{-2} difficult to measure. There is, thus, evidence of fairly high track densities in all minerals from the 150–250 μm fraction of our three samples as well as from the 40–53 μm fraction of 24090,1 so far analysed. This observation should be considered in conjunction with that of Blanford and Wood (1977), namely that highly irradiated feldspars are present in the 45–90 μm fraction in the soils from all six depths studied by the U.S. consortium.

It is thus clear that, in order to account for the low overall maturity of the core material, there must exist some major component of the soil which is less highly irradiated. The most likely source of this immature component is the coarse, >250 μm, fraction as postulated by McKay *et al.* (1977). Further support of this suggestion may be adduced from the reported observation of Barber (1977), who found track densities $\lesssim 10^8$.cm^{-2} in the 250–1000 μm grains from various depths in the Luna 24 core. However, against the above evidence must be weighed the results of chemical studies by Blanchard *et al.* (1977), which suggest that the 90–150 μm fractions are dominated by relatively fresh material and are typical of immature to sub-mature bulk soils. These authors, too, nevertheless conclude that the Luna 24 soils are mixes of two or more components: an old, mature component having been diluted by fresh, immature material. The point at issue is, at what size fraction does the immature soil begin to predominate? It is to be hoped that further track studies as a function of grain size will throw more light on this complex and currently unresolved question.

The maturity of the deepest of our three samples (viz. at 196 cm) seems to be significantly lower than that of the two upper ones. This is in agreement with the pattern of decreasing maturity with increasing depth observed by Barsukov *et al.* (1977). Of particular interest in this context are the results of Gardiner *et al.* (1977), who, on the basis of contents of iron metal and hydrolysable carbon, find a maturity sequence 240901,1 >24125,1 >24196,1. Whilst it is difficult to separate samples 24090,1 and 24125,1 on the basis of the track data, we can certainly confirm the lower maturity of sample 24196,1.

Finally, as noted above, it is significant that a high proportion (~30%) of the grains at all levels show anisotropic track-density gradients (cf. Table 1). This is particularly evident amongst the olivines (see Fig. 2). The presence of these anisotropic gradients suggests that such grains have been cycled on rather few occasions through the top few 100 μm of the lunar soil or, alternatively, that pre-irradiation on the surface of larger rocks has taken place, as indicated by the observations of Barber (1977).

3. THERMOLUMINESCENCE (TL) STUDIES

The TL glow exhibited by a sample as it is heated has implicit in it information on the processes that have been operating on the sample in the past. These processes include the filling and creation of 'traps' of charged carriers by radiation, the effects of temperatures of irradiation as well as of subsequent storage, and the dose rate to which the sample has been subjected. It is our purpose in the following sections to attempt to retrieve some of this information on the environmental history of the Luna 24 samples as a function of their depth. The results obtained are discussed in the light of recent TL work on lunar, meteoritical and terrestrial materials both by ourselves and by other workers.

3.1 *Experimental procedure*

The apparatus used consists essentially of a heating chamber in which samples of ~1–2 mg are heated on a tantalum strip in an oxygen-free nitrogen atmosphere. The heating rate of $3.6°C.s^{-1}$ is automatically controlled by a comparative amplifier. The TL output, normally transmitted through an Ilford 'Broad Spectrum Blue' filter No. 622 (375–530 nm transmission band) is seen by a specially selected quartz-window photomultiplier (EMI 9804 QB) in a thermoelectrically cooled housing. The PM output, measured by a sensitive electrometer, is registered on an X(t),Y chart recorder. The details of grain size and sample description have been given in section 1.1.

3.2 RESULTS

3.2.1 *Natural glow curves*

All four samples produced similar natural TL glow curves (Fig. 3). A noteworthy feature in these curves is the fact that there is no natural glow below ~160°C. Normally one would expect that, in view of the low temperatures prevailing at the depths below the lunar surface from which these samples were retrieved, a substantial component of natural TL would have survived below 160°C (Durrani and Hwang, 1974). Figure 4 shows the glow curve from sample 24196,1,9601 after irradiation with a small dose (50 krad) of β-particles from a ^{90}Sr source; and here it can be seen that there is a substantial component starting at a read-out temperature of ~65°C and peaking at ~135°C. The implications in terms of room-temperature storage over the last one year are discussed in section 3.3.1.

3.2.2. *Normalized TL versus depth*

In order to draw any meaningful conclusions from the variation of TL with depth below the lunar surface it is necessary to 'normalize' the glow for the TL sensitivity of each sample. This is because the lunar soil is not homogeneous in 'TL phosphors', i.e. in minerals specially endowed with TL-emitting properties, e.g. plagioclase (see discussion by Hoyt *et al.*, 1972). To overcome this problem it is necessary to determine the 'TL sensitivity' per unit weight of each sample by giving it a standard test dose and to use this factor for normalizing purposes.

Figure 5 shows the normalized TL versus depth profile for the Luna 24

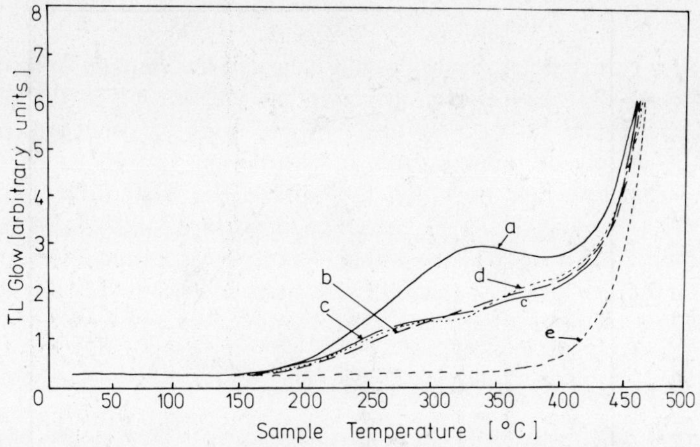

Fig. 3. Natural TL glow curves from samples (a) 24090,1,9606; (b) 24125,1,9601; (c) 24196,1,9601, and (d) 24125,1,9503. The first three samples consisted of grains <75 μm in diameter, and had received a heat treatment during preliminary separation; the fourth sample (d) had grains in the size range 75–106 μm, and had not received such heat treatment. Little TL glow is given out below ~160°C by any of the four samples, indicating drainage of this part of the glow curve (cf. Fig. 4). The heating rate for these and all other TL glow curves was 3.6°C.s.$^{-1}$; (e) is the signal due to the black-body radiation.

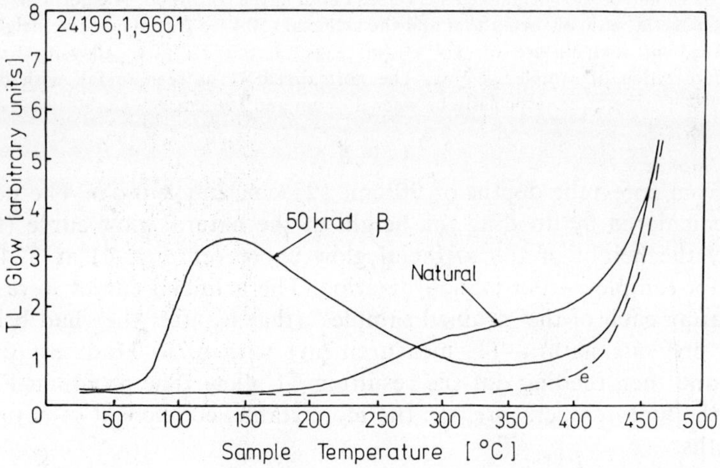

Fig. 4. Glow curve from 24196,1,9601, obtained by artificially irradiating the sample with 50 krad of ^{90}Sr β-particles after its natural TL had already been measured (cf. Fig.3). The natural glow curve and the black-body signal (curve e) are also shown. Notice the prominent low-temperature peak (centering at ~135°C) induced by artificial irradiation (cf. Fig. 3).

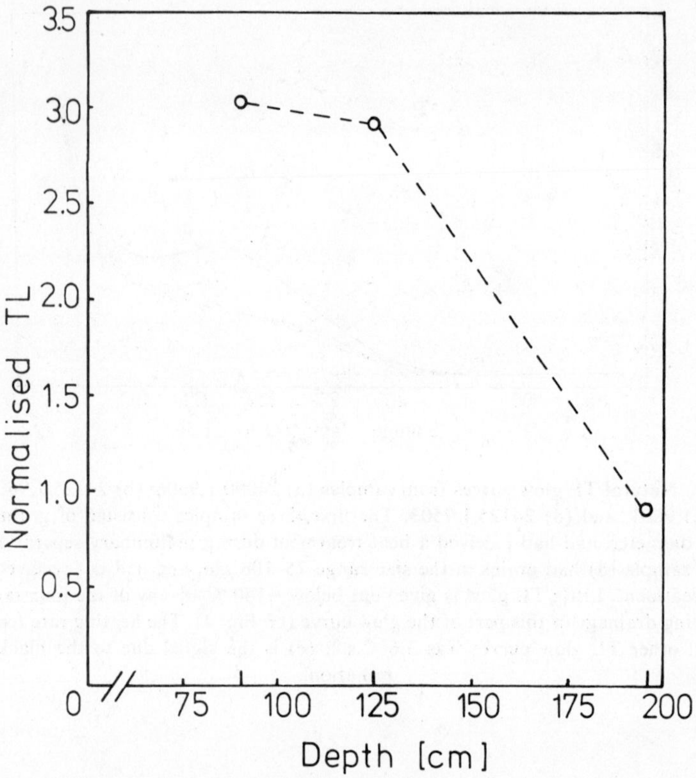

Fig. 5. Profile of the normalized TL versus depth in the core-tube. The normalized TL is taken as the ratio of the natural and the artificial (50 krad β's) glow-curve heights at the read-out temperature of 324°C (cf. Figs. 3 and 4), after allowing for the standardization of sample weights. The normalized TL is seen to fall with depth, especially between 125 cm and 196 cm.

samples from core-tube depths of 90 cm, 125 cm, and 196 cm. The normalized TL was calculated by dividing the height of the natural glow curve (Fig. 3) at 324°C by the height of the artificial glow curve (cf. Fig. 4) at 324°C, after allowing for sample-weight standardization. The artificial curves were produced by irradiating each of the 'drained samples' (that is, after they had been heated during a previous natural-TL measurement) with a 50 krad test dose of β-particles and then reading out the resulting TL glow. The profile in Fig. 5 thus refers to the high-temperature, i.e. the more stable, component of natural TL at these depths.

3.2.3. *Repeated gamma irradiations*

Since the amount of Luna 24 materials made available to us was small (a few mg from each location), it was necessary for us to use the same sample over and over again in our investigations. In our previous lunar and meteoritical studies, we have always avoided doing so in view of the sensitization effects of 'pre-dose' and 'thermal cycling' (Durrani *et al.*, 1973). The effects of the present procedure may be gauged from the following observations.

Figure 6 shows the growth curves resulting from the repeated γ-irradiation of sample 24125,1,9601, using a ~3000 Ci ^{60}Co source. The two curves are obtained by measuring the height of the TL glow curve at (a) 150°C and (b) 378°C. The points on the abscissa represent increasing individual γ-ray doses given to the same sample after all preceding read-outs. Figure 7 shows the increasing intensity of the actual glow curve in the region of 150°C with increasing dose. From the figures it is clear that saturation of the sample has not been reached even after a γ-ray dose of 2.27 Mrad has been imparted to the sample, and that both curves in Fig. 6 exhibit supralinear growth. This is particularly so for the 378°C read-out region, which, unlike the 150°C peak, shows no trend towards saturation. When the TL sensitivity of the sample was tested by giving it a standard β-dose of 50 krad subsequent to its total γ-ray pre-dose of 8.27 Mrd (which is the sum of all the individual doses shown in Fig. 7), it was found that the sensitivity in the 378°C region had increased by a factor of 18 in comparison with the value before the γ-irradiation (figure not shown). This result indicates a severe sensitization of the TL response due to the combined effect of thermal cycling and pre-dose.

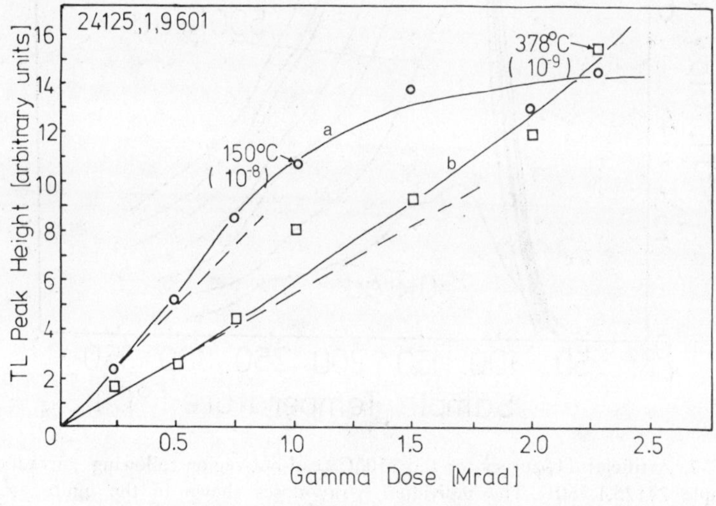

Fig. 6. TL growth curves for sample 24125,1,9601 as increasing doses of γ-rays are imparted to the sample. The same sample is used repeatedly, and the abscissa represents individual γ-ray doses at each successive step. The TL intensity is given in terms of the peak height at a read-out temperature of (a) 150°C (——○——) and (b) 378°C (——□——); curve (b) has been magnified 10 times in this plot. Both curves exhibit 'supralinear' growth (the nominal linear growth curves are indicated by dashed lines), which signifies increases in the TL sensitivity of the sample as a result of repeated γ-ray irradiation and read-out thermal cycling.

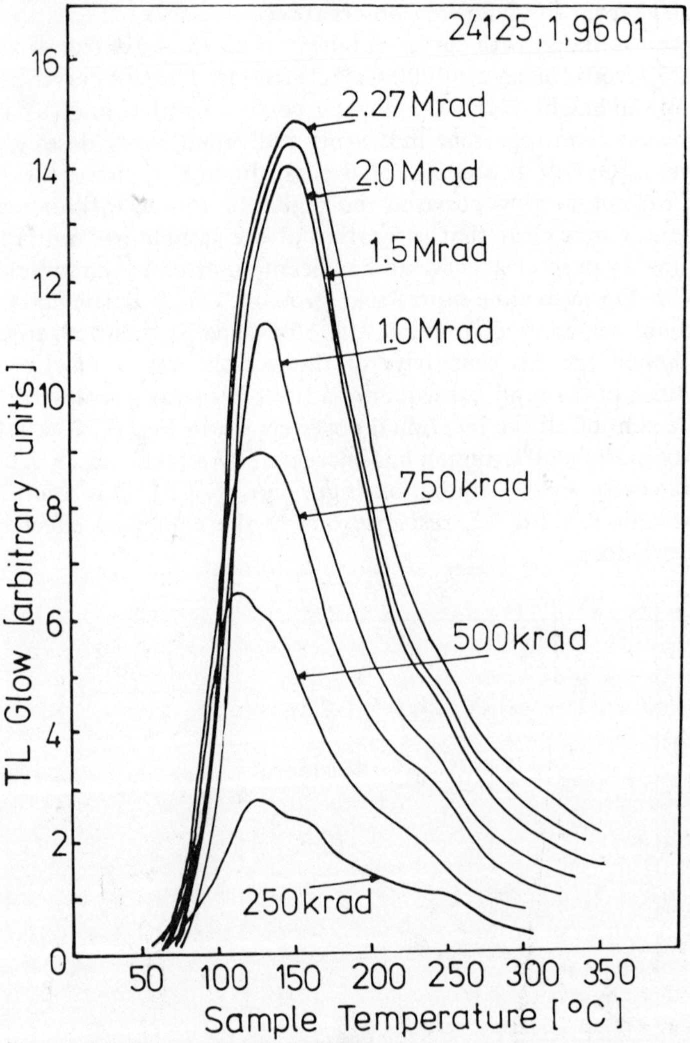

Fig. 7. Artificial TL curves for the 150°C read-out region following γ-irradiation of sample 24125,1,9601. The individual γ-ray doses shown in the figure are given successively to the same sample after each read-out. Note that at least three TL peaks are present in this region of read-out, and that the height of the glow curve at ~150°C is still increasing slightly even after the final γ-ray dose of 2.27 Mrad (i.e. a total cumulative dose of 8.27 Mrad) has been imparted to the sample. The heating rate is 3.6°C.s^{-1}.

3.3 Discussion

3.3.1 *Desirability of sample refrigeration*

One would expect, *a priori*, to find well-preserved TL at low read-out temperatures from deep-core samples in view of the low ambient temperatures at the depths from which they had been recovered (Durrani and Hwang, 1974). The absence of natural TL glow below ~160°C in all four samples, including the one that had *not* been heat-treated during processing at Cambridge, therefore unequivocally establishes the fact that the drainage of the natural TL in this region of the glow curve (compare Figs. 3 and 4) is due to their storage at room temperature for one full year (August 1976 to September 1977) both in the U.S.S.R. and in the U.K. This reinforces the plea made by one of us earlier (Durrani, 1972) that a part, at least, of the lunar material recovered from cooler regions should be kept refrigerated. The efficacy of such a precaution was fully demonstrated in the case of the Apollo 17 material collected from the shadows of certain boulders (Durrani *et al.*, 1976). It is strongly recommended that part or all of any future deep-core material from Luna missions should be placed in deep freeze as soon after retrieval as possible.

3.3.2 *Induced TL as an indicator of radiation history*

An interesting result to emerge from our TL measurements is the observation of changes in TL sensitivity caused by repeated heating (during TL read-outs) and/or repeated irradiation of the same samples. Unpublished work in this laboratory on meterorites (Potter and Estacado) as well as on natural quartz suggests that thermal cycling is not a major cause of increased TL sensitivity. There is strong evidence, however, that large doses of radiation do cause a high degree of sensitization. Thus in connection with experiments on the Oklo phenomenon (prehistoric natural reactor), we have observed (Durrani *et. al.*, 1975) that a γ-dose of 2.5 Mrad increased the TL sensitivity of natural quartz (as tested by a standard β-dose of 10 krad) by a factor of ~600—although subsequent intense radiation damage by 10^5 Mrad of 10 MeV protons counteracted this sensitization. The observation of radiation-induced changes in sensitivity is in keeping with the results of TL studies of meteorites (see, e.g., Vaz and Sears, 1977).

The foregoing observations suggest that there ought to be some correlation between artificially induced TL (which gives a direct measure of the TL sensitivity) and radiation exposure age, such that the most mature samples have the highest TL sensitivity. As track densities in crystalline lunar grains also give an indication of maturity (section 2.2 above), one might reasonably expect some correlation between track density and artificial TL in a sample. One must, however, allow for differences in the 'intrinsic' TL sensitivity of different minerals (Hoyt *et al.*, 1972). If one can be sure of examining the same amounts of the same TL phosphors in different samples, then a correlation between induced TL and sample maturity may be possible for lunar material. The results could then be judged in the light of track density data for the same samples.

3.3.3 *Depth-dependence of natural TL*

Finally, it is interesting to examine the question of normalized natural TL as a function of depth below the lunar surface. There are two opposing main processes in operation, as mentioned at the beginning of section 3, in bringing about a thermodynamic equilibrium in the TL content of a sample. The first is the radiation-filling (and creation) of charge traps in a sample. The second is the thermal emptying of these traps at the relevant ambient temperature. The external radiation dose as well as the temperature varies with depth in a complex way. The net result is that the normalized TL (i.e. natural TL/artificial TL) at first increases rapidly with depth down to ~10 cm below the lunar surface, chiefly because of the rapid attenuation of the diurnal heat-wave; it then decreases slowly with increasing depth, owing mainly to the attenuation of the external radiation flux but also to the outward heat-flow. The deepest that these studies have previously been extended is to ~40 cm below the lunar surface (Crozaz and Plachy, 1976; Hoyt *et al.*, 1972). Our TL versus depth results shown in Fig. 5 are thus of interest, extending as they do to a core depth of ~2 m. They continue the general trend of the data reported by Crozaz and Plachy (1976) and Hoyt *et al.* (1972). The trend shown in Fig. 5 for normalized TL also broadly parallels our track-density versus depth data in Fig. 1.

It must be pointed out, however, that the exact shape of the TL versus depth curve is dependent, amongst other things, on the region of the glow curve used in the measurements. In Fig. 5, the normalized glow refers to the glow curve heights (both natural and those resulting from 50 krad β's) at 324°C. Crozaz and Plachy (1976) and Hoyt *et al.* (1972) used relatively low-temperature TL regions (~200°C), which were more strongly influenced by ambient temperatures, for mapping their TL versus depth profiles. Nash and Conel (1971) as well as Hoyt *et al.* (1972) found no observable depth-dependence in the glow temperatures above ~500°C at depths beyond ~1 cm below the lunar surface, and relatively little depth-dependence even in the read-out region 300–400°C. This is partly explained by the greater thermal stability of the traps corresponding to the higher read-out intervals, so that the temperature variation with depth has less effect on them than on the traps responsible for the earlier regions of the glow curve.

In comparing the track and the TL data one must also bear in mind the fact that the total track-accumulation ages are very much longer (being up to several times 10^9 yr, although the majority of tracks would have been accumulated while the grains were fairly close to the top of the lunar surface, viz. ~10^6–10^7 yr) than those reflected by the TL values. This is because, as new soil layers are laid on top of the old lunar surface, the TL profile re-equilibrates to a new profile which now peaks at ~10 cm below the *new* surface. The time required for this new thermodynamic equilibrium to be established between the radiative filling and the thermal drainage of the traps depends on the region of the glow curve being monitored, but is usually of the order of 10^4–10^5 yr. During this time-period the number of tracks/cm^2 acquired by the grains would be quite negligible at depths greater than a few 100 μm. Both TL data and track-density versus depth profiles are relevant parameters in any consideration of residence times and soil-mixing frequencies of the near-surface layers of the lunar regolith. What light they can throw, in conjunction with each other, on these complex questions merits fuller investigation.

4. Conclusions

From our track data we conclude that the most mature component of the Luna 24 soils exists among the fine-grain fractions ($\lesssim$250 μm). This is in accord with the observations of workers such as Blanford and Wood (1977). We also find evidence for the conclusion that the overall maturity of the lunar soil decreases with increasing depth within the Luna 24 deep-core. The presence of anisotropic track-density gradients within a high proportion of the grains at all levels indicates a low degree of soil churning at the very top of the lunar regolith, unless it can be explained in terms of pre-burial exposure of these grains on solid rock surfaces.

Our TL studies indicate that the TL sensitivity of samples, as determined by inducing artificial TL in them, may prove to be a useful indicator of the radiation history of those samples in their natural environment on the moon. The profiles of both track density and normalized natural TL (at ~324°C) as a function of depth within the core-tube are broadly similar (c.f. Figs. 1 and 5).

Finally, the observation that most of the low-temperature natural TL in the Luna 24 samples from all three deep-core levels has faded during their 12-month storage at room temperature prompts us to urge the U.S.S.R. Academy of Sciences that at least a part of the drill-core material from future Luna missions should be deep-frozen as soon after retrieval as practicable.

Acknowledgments—We wish to thank the U.S.S.R. Academy of Sciences and the Royal Society, London, for making the Luna 24 samples available to us. We would also like to record our appreciation of Dr. Colin Pillinger for his curatorial help. RKB and SWSM wish to thank the Science Research Council (SRC) for Fellowships. The financial support of the SRC and the Royal Society for our lunar and meteoritic research projects is gratefully acknowledged.

References

Barber D. J. (1977) Electron microscopy of radiation damage in Luna 24 soil (abstract). In *Papers Presented to the Conference on Luna 24*, p. 12–13. The Lunar Science Institute, Houston.

Barsukov V. L. and Florensky C. P. (1977) The Lunar Soil from Mare Crisium: Preliminary data (abstract). In *Lunar Science VIII*, p. 61–63. The Lunar Science Institute, Houston.

Barsukov V. L., Ivanov A. V., Nazarov M. A., Rode O. D., Stakeev Yu. I., Tarasov L. S., Tobelko K. I. and Florensky C. P. (1977) Preliminary description of the regolith core from Mare Crisium (abstract). In *Lunar Science VIII*, p. 67–69. The Lunar Science Institute, Houston.

Bhandari N., Goswami J. N., Lal D., Macdougall D. and Tamhane A. S. (1972) A study of the vestigial records of cosmic rays in lunar rocks using a thick section technique. *Proc. Indian Acad. Sci.* **LXXVI** (No.1, Sec.A.), 27–50.

Blanchard D. P., Haskin L. A., Brannon J. C. and Aaboe, E. (1977) Chemistry of soils and particles from Luna 24 (abstract). In *Papers Presented to the Conference on Luna 24*, p. 37–40. The Lunar Science Institute, Houston.

Blanford G. E. and Wood G. C. (1977) Particle track densities in the Luna 24 core (abstract). In *Papers Presented to the Conference on Luna 24*, p. 41–43. The Lunar Science Institute, Houston.

Bogard D. D. and Hirsch W. C. (1977) Noble gases in Luna 24 core soils (abstract). In *Papers Presented to the Conference on Luna 24*, p. 44–47. The Lunar Science Institute, Houston.

Crozaz G. and Plachy A. L. (1976) Origin of the Apollo 17 deep drill coarse-grained layer. *Proc. Lunar Sci. Conf. 7th*, p. 123–131.

Durrani S. A. (1972) Refrigeration of lunar samples destined for thermoluminescence studies. *Nature* **240**, 96–97.

Durrani S. A., Prachyabrued W., Hwang F. S. W., Edgington J. A. and Blair I. M. (1973) Thermoluminescence of some Apollo 14 and 16 fines and rock samples. *Proc. Lunar Sci. Conf. 4th*, p. 2465–2479.

Durrani S. A. and Hwang F. S. W. (1974) Thermoluminescence and thermal environment of some Apollo 17 fines. *Proc. Lunar Sci. Conf. 5th*, p. 2689–2702.

Durrani S. A., Khazal K. A. R., Malik S. R., Fremlin J. H. and Hendry G. L. (1975) Thermoluminescence and fission-track studies of the Oklo fossil reactor materials. In *Le Phenomene d'Oklo*, p. 207–222. International Atomic Energy Agency, Vienna.

Durrani S. A., Khazal K. A. R. and Ali A. (1976) Temperature and duration of some Apollo 17 boulder shadows. *Proc. Lunar Sci. Conf. 7th*, p. 1157–1177.

Florensky C. P., Basilensky A. T., Ivanov A. V., Pronin A. A. and Rode O. D. (1977) Luna 24: Geologic setting of landing site and characteristics of sample core (preliminary data) *Proc. Lunar Sci. Conf. 8th*, p. 3257–3279.

Gardiner L. R., Jull A. J. T., Woodcock M. R., Pillinger C. T. and Stephenson A. (1977) Carbon chemistry and metallic iron content of some Luna 24 core samples (abstract). In *Papers Presented to the Conference on Luna 24*, p. 60–63. The Lunar Science Institute, Houston.

Hoyt H. P. Jr., Walker R. M., Zimmerman D. W. and Zimmerman J. (1972) Thermoluminescence of individual grains and bulk samples of lunar fines. *Proc. Lunar Sci. Conf. 3rd*, p. 2997–3007.

Krishnaswami S., Lal D., Prabhu N. and Tamhane A. S. (1971) Olivines: Revelation of tracks of charged particles. *Science* **174**, 287–292.

McKay D. S., Basu A., Waits G., Clanton U., Fuhrman R. and Fruland R. (1977) Grain size and evolution of Luna 24 soils (abstract). In *Papers Presented to the Conference on Luna 24*, p. 114–117. The Lunar Science Institute, Houston.

Morris R. V. (1977) FMR and magnetic properties of Luna 24 soils and >1 mm soil particles (abstract). In *Papers Presented to the Conference on Luna 24*, p. 121–124. The Lunar Science Institute, Houston.

Nash D. B. and Conel J. E. (1971) Luminescence and reflectance of Apollo 12 samples. *Proc. Lunar Sci. Conf. 2nd*, p. 2235–2244.

Pillinger C. T. and Fabian D. M. (1977) Preliminary separation and allocation of Luna 24 samples presented to the Royal Society by the Soviet Academy of Sciences. To be published.

Vaz J. E. and Sears D. W. (1977) Artificially-induced thermoluminescence gradients in stony meteorites. *Meteoritics* **12**, 47–60.

Merrill, R. B. and Papike, J. J., editors.
(1978) *Mare Crisium: The View from Luna 24*, Pergamon Press (New York), p. 195–206
Printed in the United States of America

Electron microscopy of radiation damage and microstructure in soils from the Luna 24 core

D. J. BARBER
Physics Department, University of Essex, Colchester, Essex, England.

Abstract—Micron-sized and hand-picked sub-millimetre-sized grains from the Luna 24 core have been examined by HVEM and SEM methods to assess levels of radiation damage and internal microstructure. Both sizes of soil grains come from the 90, 125 and 196 cm depths in the core.

The micron-sized grains from all three depths contain heavily-irradiated grains with amorphous coatings and track densities $>10^{10}$ cm^{-2}, but there are also significant numbers of lightly-irradiated grains. Comparison studies on Luna 16 and Luna 20 soils show that the levels of radiation damage in Luna 24 soil are less than those in Luna 16 soil, but much greater than those in Luna 20 soil. It is also concluded that the fine-grained fraction of Luna 24 soil is a mixture of at least two components, one of which is fairly mature, the other being submature.

The sub-millimetre-sized grains mostly exhibit uniform track densities of $\sim 3 \times 10^8$ to 5×10^9 cm^{-2}, although one pyroxene grain shows a track gradient, apparently acquired while it formed part of a whole rock surface for a time of $\lesssim 10^6$ years. The internal microstructures of the grains which have been examined are described; when coupled with X-ray microprobe data, they are consistent with the view that the grains are fragments of local mare basaltic rock.

INTRODUCTION

The formation and maturation of the lunar regolith have been shown (Arrhenius *et al.*, 1971; Phakey *et al.*, 1972; Bibring *et al.*, 1975) to be a very complex process. Hence the deduction from particle track data of meaningful estimates of soil maturity, accumulation rates, or regolith stirring are fraught with difficulty. The very high track densities ($\sim 10^{10}$ cm^{-2}) which occur in micron-sized grains (Borg *et al.*, 1970; Barber *et al.*, 1971b) necessitate the use of transmission electron microscopy, preferably at high voltage (HVEM). This method also provides information about the pre-irradiation condition of each grain examined, so that it is possible to deduce, for example, whether the microstructures are totally consistent with a mare origin. Larger sub-millimetre-sized grains normally have track densities $\lesssim 10^9$ cm^{-2}, which are best measured by scanning electron microscopy (SEM) in conjunction with track etching methods.

In the examination of micro-sized grains by HVEM, measurements should ideally be made on a large number of similarly-sized grains of the same mineral type, which is a lengthy process. This is necessary to eliminate problems arising from the fact that minerals require different radiation doses to produce saturation effects (e.g., loss of crystallinity). Nonetheless, Phakey and Price (1973) have demonstrated that qualitative data on the levels of radiation damage in soil samples can be provided by HVEM fairly easily via selected area diffraction. In reporting these preliminary results on soil from the Luna 24 site, Mare Crisium, the latter approach has been used.

EXPERIMENTAL

Samples consisted of approximately 1 mg of the "smoke-sized fraction" of grains from the 90, 125 and 196 cm depths in the Luna 24 core, together with 34 hand-picked grains (in total) from the same depths having diameters of 250–1000 μm. Smoke-sized grains were signified by adding the code 9990 to the sample number. Grains with diameters of 500–1000 μm have the code 9102 added, while those with diameters of 250–500 μm have the code 9101. In addition, samples of the "smoke-sized" grains from the Luna 16 and Luna 20 landing sites were obtained for the purpose of comparison.

Grains from the "smoke-sized fractions" were distributed on glass slides, which were then carbon coated and stripped by flotation onto water. The carbon-supported grains were then examined directly by HVEM at 1 MV and the nature of each electron-transparent grain was examined with the aid of selected area diffraction. High resolution dark field imaging was usually employed to search for tracks because of their low visibility in bright field. The large hand-picked grains were treated quite differently: after optical examination they were mounted in epoxy cement and the mounts were ground to expose a little of the interior of each grain. The grains were then polished to ¼ μm diamond finish and most of them were analysed for major elements by electron probe X-ray microanalysis. After repolishing, the grains were then etched in appropriate etchants to reveal the density and distribution of particle tracks. Etchant WN (Krishnaswami *et al.*, 1971) was used for olivines and 6N NaOH solution was used for feldspars and most of the pyroxenes. Etched grains were photographed by SEM and then repolished to remove the etched layer. Subsequently the mount was ground from the opposite side (i.e. from the back of the grain) until the grain was reduced to a thin section, approximately 30 μm thick. The section was then mounted on a washer-type copper electron microscope grid and ion-thinned (Barber, 1970) until thin enough for examination by HVEM.

RESULTS: MICRON-SIZED GRAINS

The HVEM observations of the micron-sized particles amongst as-received "smoke-sized fractions" show that the general characteristics of the Mare Crisium soil do not differ from those of soils returned by earlier lunar missions. The soil contains both angular and more rounded crystalline grains and lithic fragments, microspherules, glassy fragments and a complex range of agglutinates, etc.

Irradiation effects

(i) Virtually all of the crystalline grains exhibit some degree of surface amorphization as a result of exposure to low energy solar wind ions. However, grains show wide variations in the thicknesses of the amorphous layers. This observation, coupled with the correlation between thin amorphous coatings with low track densities, indicates that some of the grains resided at the lunar surface for short times. At the other extreme, grains with the thickest amorphous coatings almost invariably possess a rounded habit. Some of these grains have received sufficient irradiation that they have undergone partial loss of internal crystallinity, indicated by arcing of electron diffraction spots and by diffuse scattering. Selected area diffraction also suggests that the majority of the heavily-coated and rounded grains are feldspars. They represent a mature component in the soil, as follows from the work of Borg *et al.* (1976).

(ii) Moderate ($\sim 10^9$ cm^{-2}) or high (from 10^{10} cm^{-2} to $>10^{11}$ cm^{-2}) track densities are apparent in the majority of the crystalline grains from all three levels in the core. Many of the heavily-damaged grains also exhibit signs of gas implantation and precipitation, noted previously (Phakey *et al.*, 1972). Occasionally, irradiation in the electron beam of the microscope induces changes such as the appearance of small gas bubbles and even recrystallization. These observations all serve to reinforce the conclusion, drawn in the previous section, that the soil contains a mature component, i.e. grains which endured long enough in the top stratum of the regolith to acquire near-saturation levels of irradiation from solar wind and solar flare ions.

The presence of grains whose appearance is consistent with low irradiation doses must also be significant, however. Two examples of grains of this type are shown in Figs. 1a and 1b, while Fig. 1c shows an example of the more common heavily-irradiated grains. The grains in Figs. 1a and 1b are angular and possess relatively thin (~300 A) amorphous coatings. The grain in Fig. 1c was apparently originally angular, but it has become rounded off and the amorphous coating is up to 1500 A thick in places. No tracks were visible when the grains of Figs. 1a and 1b were examined in dark field, although a slight mottle is apparent in Fig. 1b, indicative of some irradiation, possibly caused or enhanced by electron-beam-induced damage. The absence of tracks in grains of this size implies residence times within the first 10 μm of the top of regolith of $\lesssim 10^2$ years, as judged by the track accumulation rate of $\sim 3 \times 10^6/(\text{cm}^2\ 2\pi$ ster yr) deduced by Hutcheon *et al.* (1974). Such grains certainly are submature and possibly even immature in terms of the relative gradings of soils used by Morris (1976).

Agglutinates and microbreccias

Approximately 25% of the micron-sized particles from the 90 cm level are either agglutinates or microbreccias. This percentage seems to decrease with depth in the core, as judged from the other two samples available in this study; the exact figures are not of much significance because there is some difficulty in distinguishing between particles which are merely in contact from those which are truly cemented together by glass. The presence of glass is obvious on many particles. The range of characteristics of impact-melted and shock-produced lunar glasses has already been well documented by Christie *et al.* (1973). The features seen in the Luna 24 soil are not notably different.

High densities of tracks (i.e. $>10^{10}$ cm^{-2}) are seen to be retained in crystalline clasts embedded within melt glass, as shown in Fig. 2. The largest grains in this particle from the 90 cm level are approximately 1 μm in diameter and the tracks within them are still clearly visible, although apparently slightly annealed. This is consistent with the findings of Hutcheon *et al.* (1972) that tracks were still present in grains <5 μm in diameter in Apollo 14 breccias.

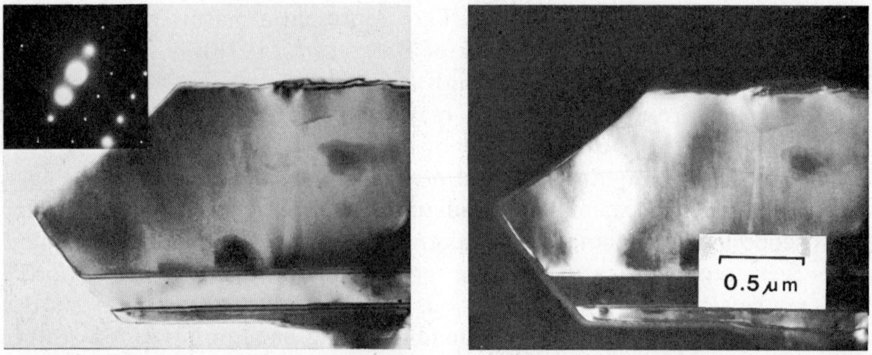

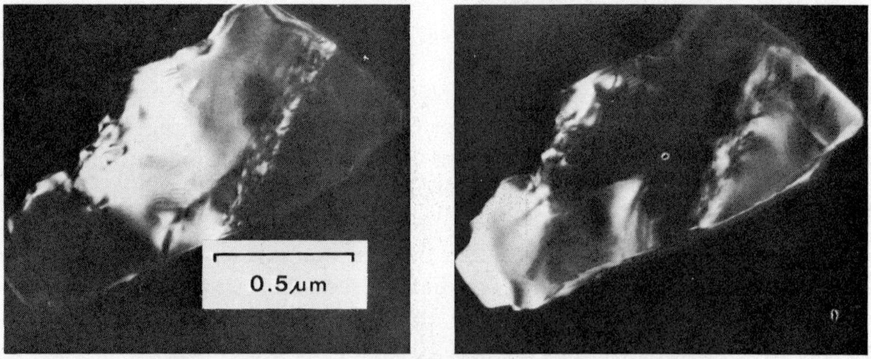

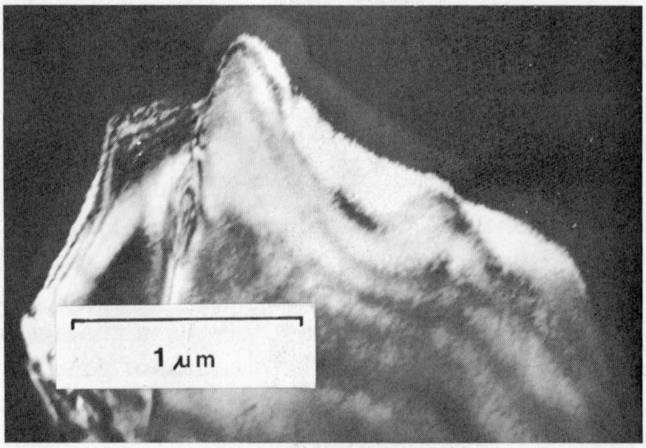

Fig. 1a. Bright and dark field micrographs of lightly-irradiated feldspar grain from 196 cm depth (24196,1,9990).

Fig. 1b. Dark field images (slightly different angles of tilt) of lightly-irradiated olivine grain from 125 cm depth (24125,1,9990).

Fig. 1c. Dark field image of heavily-irradiated grain from 125 cm depth (24125,1,9990).

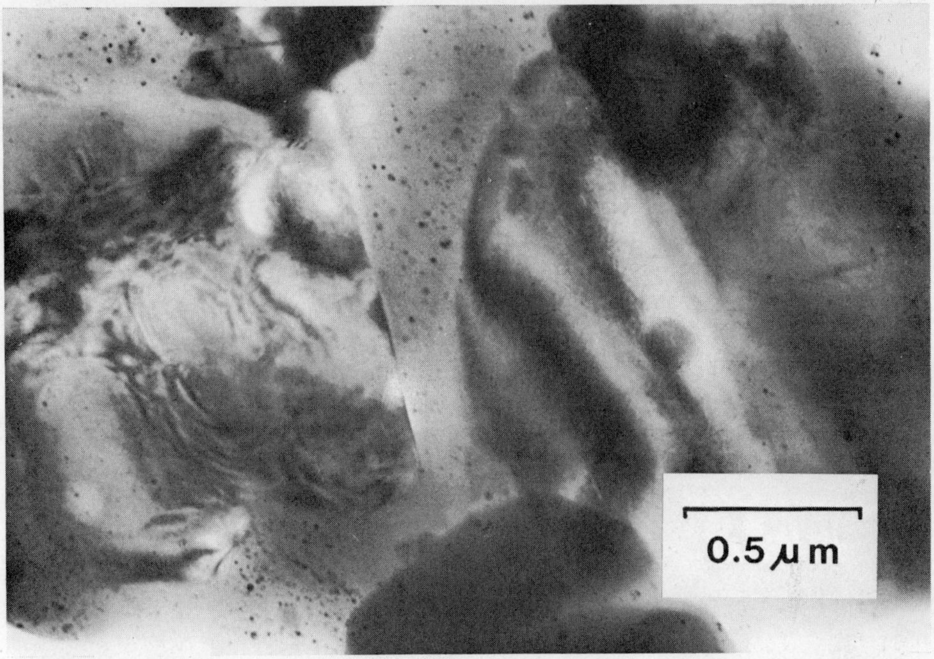

Fig. 2. Bright field micrograph of breccia from 90 cm depth (24090,1,9990) showing track densities $\sim 10^{10}$ cm^{-2} in the crystalline clasts embedded in the impact-melted glass.

Comparison of irradiation effects in Luna 16, −20 and −24 soils

In order to try and relate the history of the Luna 24 soil to that of other lunar soils, comparative electron microscopy was carried out on similar size fractions from Luna 16 and Luna 20 soils. The criteria given by Phakey and Price (1973) to classify electron diffraction effects (and hence the degree of radiation damage) were used. For those who are unfamiliar with the field it is somewhat difficult to illustrate these criteria, which depend upon the relative strengths of diffuse and Bragg scattering. However, Fig. 3 shows a category 1 grain from Luna 16 soil, imaged in dark field using a strong Bragg diffraction spot. In Luna 16 micron-sized soil only about 10% of grains fall in this "least heavily-damaged" category and there are no grains with even lower levels of radiation damage, as are found in Luna 24 soil and illustrated in Figs. 1a and 1b. Category 2 grains typically have track densities $>5 \times 10^{11}$ cm^{-2} and exhibit weak diffraction spots, while category 3 grains are amorphous, or nearly so, and give only diffuse scattering. Table 1 shows the percentage of grains from sample 24090,1,9990 assigned to each category in comparison with the values obtained by the author for Luna 16 and Luna 20 soils. The Luna 24 sample was too small to obtain good results, but there is general agreement with the earlier figures of Phakey and Price (1973).

Table 1. Comparative Radiation Damage of Luna Soils

Category†	Luna 16	Luna 20	Luna 24*
1	~10%	~90%	57%
2	32%	~ 3%	26%
3	60%	—	<20%

†Phakey and Price (1973)
*from 90 cm depth

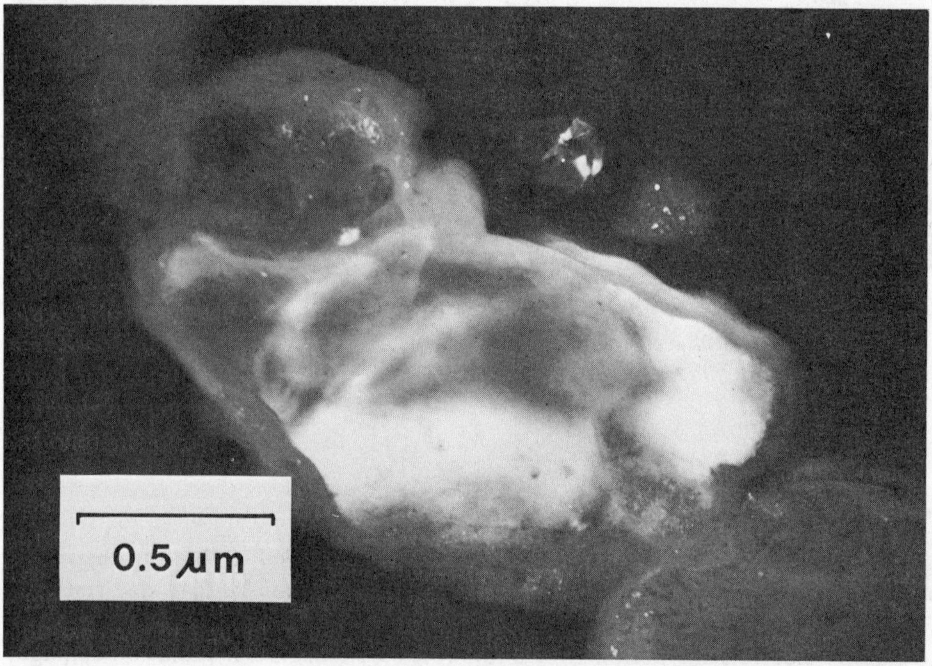

Fig. 3. Dark field micrograph of heavily-irradiated grain from the Luna 16 core soil. The interior is still crystalline.

RESULTS: SUB-MILLIMETRE-SIZED GRAINS

All of the 34 grains of Luna 24 supplied have not yet been fully examined. Apart from three, they are monomineralic; these three are basaltic lithic clasts. Pyroxenes predominate amongst the grains and of feldspars and olivines there are apparently only about five each. The grains were handpicked from the 90, 125 and 196 cm levels of the core material before being supplied to Essex University.

Solar flare tracks

Most of the grains exhibit an essentially uniform density ρ of particle tracks, in the range $\sim 3 \times 10^8$ to 5×10^9 cm^{-2}, determined by etching and SEM methods. One olivine grain (Fo_{56}) from the 196 cm level (specimen number 24196,1,9209KA) has an anomalously low track density of 2×10^6 cm^{-2}, which was measured by SEM and optical microscopy. A few grains showed slight track gradients from one edge, but one pyroxene grain 24196,1,9102X from the 196 cm level had a fairly steep gradient. Such a gradient could only be acquired in an unshielded environment at the lunar surface. Subsidiary track gradients, which might have indicated irradiation at the top of the regolith, were absent. Moreover, a shallow vug-like surface feature had been noted before encapsulating the grain and this was later found to correspond roughly to the position where ρ was at a maximum. It was therefore considered whether the grain could have received its main irradiation as part of a whole rock surface. The observed track gradient, measured along two adjacent paths, is too flat to agree with that simple hypothesis. The track gradients which occur under such conditions have been determined by Barber *et al.* (1971a) and Hutcheon *et al.* (1974). The remaining possibility is that some of the original whole rock surface was removed by impact shortly before the chip itself was formed and became a grain in the regolith. This idea seems very probable, as shown by the plots of measured track density versus depth shown in Fig. 4, together with the common line onto which the same results fall when they are displaced by an amount equal to assumed thicknesses of impact-removed material. By trial and error it was found that if the two sets of values for ρ versus depth were displaced by depth increments of 38 and 26 μm, the adjusted points fell on a line with gradient identical with that found for rock 72315 (Hutcheon *et al.*, 1974) from which an improved energy spectrum for Fe nuclei in solar flares was calculated. It is likely that a few tens of microns of rock surface could be removed by subsidiary ejecta impacts following a nearby primary impact. If this interpretation of the track gradient in specimen 24196,1,9102X is correct, we can calculate an approximate exposure time before the chip was added to the regolith. The result depends upon the solid angle through which solar flares were able to contribute to the track gradients, which cannot be known exactly; however, the exposure is unlikely to have exceeded 10^6 years, which is compatible with the time of formation of the mare rocks and the rate of accumulation of the regolith.

Microstructure

HVEM examination of ion-thinned sections from the handpicked grains has produced the following results. However, the time-consuming nature of the work has so far delayed examination of all the grains allocated.

The pyroxenes are exsolved into profuse fine lamellae, about 0.5 μm thick, of calcium-rich augite and calcium-poor pigeonite; antiphase domains also occur in the pigeonite. Most of the features which are observed have been described previously by Nord *et al.* (1973) and they are characteristic of lunar basaltic

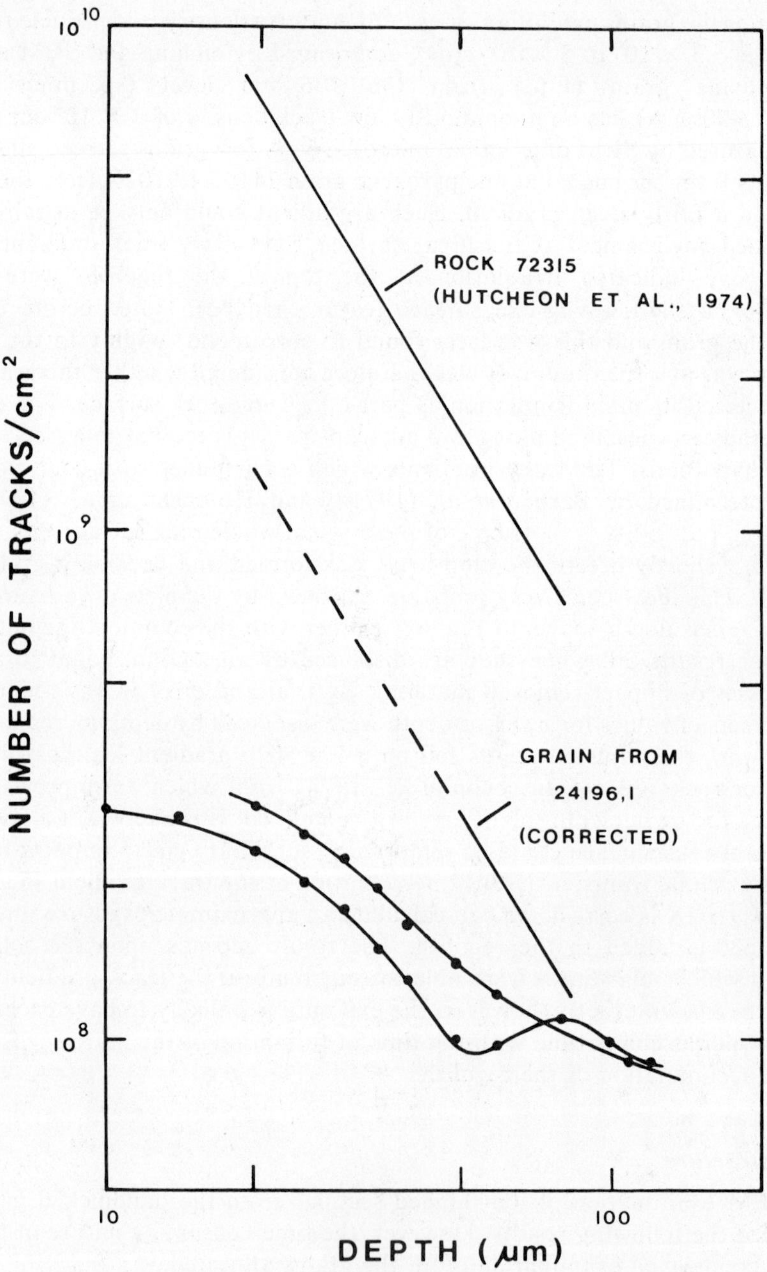

Fig. 4. Plots of track density versus depth, together with depth-corrected track gradient for sub-millimetre grain (24196,1,9102X) and track gradient in rock 72315 (Hutcheon *et al.*, 1974).

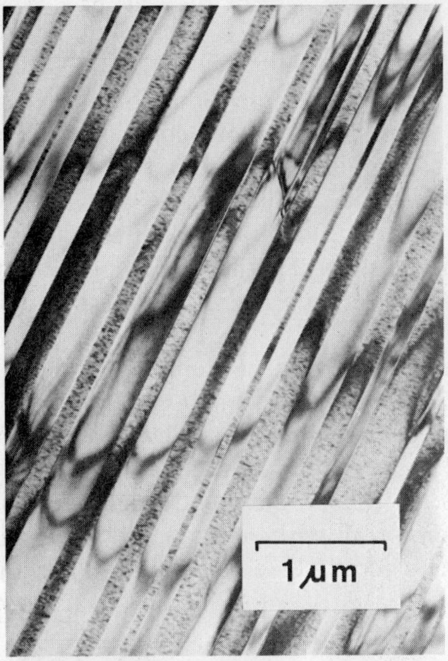

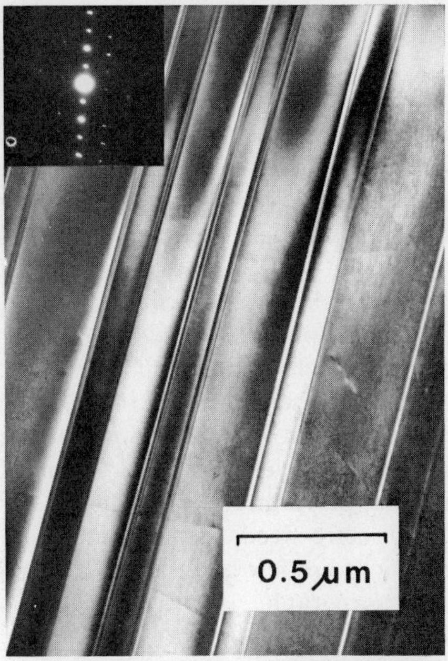

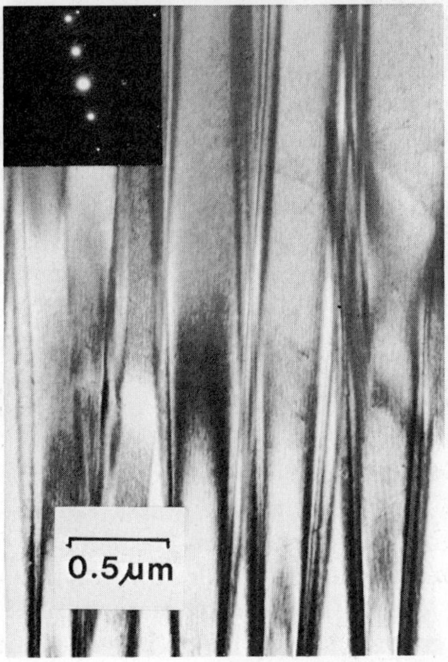

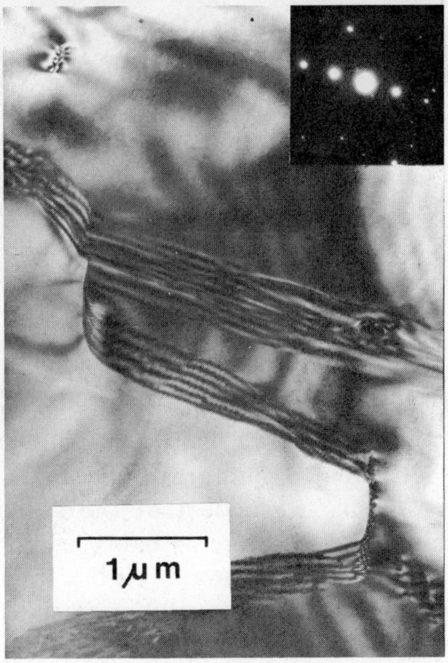

Fig. 5a. Exsolution lamellae of pigeonite in augite in ion-thinned sub-millimetre pyroxene grain from 196 cm depth (24196,1,9102X).

Fig. 5b. Another region of the grain illustrated in Fig. 5a, showing tracks within the exsolution lamellae.

Fig. 5c. Sub-millimetre pyroxene grain from 90 cm depth (24090,1,9102A) showing exsolution lamellae, tweed structure and tracks.

Fig. 5d. Microstructure of feldspar grain within basaltic fragment from the 90 cm depth (24090,1,9102G).

clinopyroxenes. The dislocation densities are low in these Luna 24 pyroxenes, shock effects are largely absent and the microstructures are consistent with a history of slow-cooling. Figure 5a shows exsolution lamellae of pigeonite in augite at the centre of the grain possessing the track gradient, specimen number 24196,1,9102X, discussed in the previous section. Fine-scale antiphase domains are imaged in the pigeonite lamellae. Figure 5b shows another region of the same grain, where the track density is higher and tracks are imaged in both the augite and pigeonite phases.

Exsolution effects and tracks are illustrated in a pyroxene from the 90 cm level (specimen number 24090,1,9102A) in Fig. 5c. Antiphase domains do not occur in this pyroxene but there is widespread occurrence of a tweed structure, indicative of fine scale incipient unmixing. Figure 5d shows the microstructure in a feldspar grain in a polymineralic fragment, 24090,1,9102G, which again indicates that the millimetre-sized soil particles are consistent with a mare basalt origin. Dislocations are few but antiphase boundaries occur, some being in faint contrast in this micrograph.

DISCUSSION AND CONCLUSIONS

At first sight, the foregoing results on the micron-sized grains of Mare Crisium soil are somewhat puzzling because grains which have suffered high doses of radiation coexist with a significant proportion of lightly-irradiated grains. In the highly radiation-damaged grains with thick amorphous coatings, track densities commonly exceed 10^{10} cm^{-2}, as has been previously observed in soil grains from other sites on the moon. (The density of tracks, ρ, determined by HVEM, is a volume track density and should be divided by a factor ~20 when comparing with values of ρ determined on surfaces, as pointed out by Bibring *et al.*, 1975). Such grains in Mare Crisium soil must be regarded as a fairly mature component, even though very heavily-irradiated grains, which have consequently lost internal crystallinity (category 3), are virtually absent. These very mature grains constitute the majority of grains in Luna 16 soil from Mare Fecunditatis, a fact which can be attributed to its greater age and a low rate of accumulation or disturbance. We can thus conclude that the regolith at Mare Crisium is less radiation-damaged and therefore probably less mature than that at Mare Fecunditatis. (A number of workers now take the fraction of agglutinates, i.e. the fraction of grains produced by impact events, as a proper measure of maturity). This conclusion has not been influenced by the presence of the lightly-irradiated grains in the Luna 24 soil, but the latter point must be taken into account. The explanation must lie in the sources from which the soil was formed and the way in which it was mixed; it must contain at least two components, one mature and one less mature. The presence of a mature component has support from the data of other workers: Dran *et al.* (pers. comm., 1977) have found that many grains show heavy radiation damage: Blanford and Wood (1977) have reported that a plagioclase component in the soil is saturated with particle tracks. The absence of tracks in micron-sized grains only implies that the track density, ρ is $\lesssim 10^8$ cm^{-2}

and does not give a lower limit for ρ. It is therefore not possible to say from the present track data whether additional components are submature or immature, in terms of the FMR-derived classification of Morris (1976). In reaching conclusions about the mix of Mare Crisium soil, Blanchard *et al.* (1977) note that surface exposure indices are typical of an immature to submature soil, as also indicated by the data of Bogard and Hirsch (1977), McKay *et al.* (1977) and Morris (1977). The results of Gardner *et al.* (1977) and Morris (1977) suggest that maturity decreases with depth. The results on micron-sized agglutinates reported in this paper are consistent with this depth dependence.

The examination of sub-millimetre-sized grains is not yet complete. The results obtained to date, which include X-ray microprobe analyses not presented here, are consistent with the grains originating from the comminution of local mare basalts. Dislocation microstructures are of low density and exsolution effects are pervasive, both of which indicate a slow cooling history. Track densities in these sub-millimetre grains are mostly in the range of 3×10^8 to 5×10^9 cm^{-2}, which are in line with the volume track densities measured in the micron-sized grains by the HVEM method.

Acknowledgments—My thanks are due to Dr. N. Charnley of Cambridge University for X-ray microanalyses, to the Natural Environment Research Council for financial support (grant GR3/1668a), and to the Science Research Council and Imperial College for making available the facilities of the EM7 1MeV electron microscope.

References

Arrhenius G., Liang S., MacDougall D., Wilkening L., Bhandari N., Bhat S., Lal D., Rajagopalan G., Tamhane A. S. and Venkatavaradan V. S. (1971) The exposure history of the lunar regolith. *Proc. Lunar Sci. Conf. 3rd,* p. 2583–2598.

Barber D. J. (1970) Thin foils of non-metals made for electron microscopy by sputter-etching. *J. Mat. Sci.* **5**, 1–8.

Barber D. J., Cowsik R., Hutcheon I. D., Price P. B. and Rajan R. S. (1971a) Solar flares, the lunar surface, and gas-rich meteorites. *Proc. Lunar Sci. Conf. 2nd*, p. 2705–2714.

Barber D. J., Hutcheon I. and Price P. B. (1971b) Extralunar dust in Apollo cores? *Science* **171**, 372–374.

Bibring J. P., Borg J., Burlingame A. L., Langevin Y., Maurette M. and Vassent B. (1975) Solar-wind and solar-flare maturation of the lunar regolith. *Proc. Lunar Sci. Conf. 6th*, p. 3471–3493.

Blanchard D. P., Haskin L. A., Brannon J. C. and Aaboe E. (1977) Chemistry of soils and particles from Luna 24 (abstract). In *Papers Presented to the Conference on Luna 24*, p. 37–40. The Lunar Science Institute, Houston.

Blanford G. E. and Wood G. C. (1977) Particle track densities in the Luna 24 core (abstract). In *Papers Presented to the Conference on Luna 24*, p. 41-43. The Lunar Science Institute, Houston.

Bogard D. D. and Hirsch W. C. (1977) Noble gases in Luna 24 core soils (abstract). In *Papers Presented to the Conference on Luna 24*, p. 44–47. The Lunar Science Institute, Houston.

Borg J., Comstock G. M., Langevin Y., Maurette M., Jouffrey B. and Jouret C. (1976) A Monte Carlo model for the exposure history of lunar dust grains in the ancient solar wind. *Earth Planet. Sci. Lett.* **29**, 161–174.

Borg J., Dran J. C., Durrieu L., Jouret C. and Maurette M. (1970) High voltage electron microscope studies of fossil nuclear particle tracks in extraterrestial matter. *Earth Planet. Sci. Lett.* **8**, 379–386.

Christie J. M., Heuer A. H., Lally J. S., Nord G. L., Radcliffe S. V., Griggs D. T. and Fisher R. M. (1973) Electron petrography of Apollo 14 and 15 breccias and shock-produced analogs. *Proc. Lunar Sci. Conf. 4th*, p. 365–382.

Gardner L. R., Jull A. J. T., Woodcock M. R. and Pillinger C. T. (1977) Carbon chemistry and metallic iron content of some Luna 24 core samples (abstract). In *Papers Presented to the Conference on Luna 24*, p. 60–63. The Lunar Science Institute, Houston.

Hutcheon I. D., MacDougall D. and Price P. B. (1974) Improved determination of the long-term average Fe spectrum from ~1 to ~460 MeV/amu. *Proc. Lunar Sci. Conf. 5th*, p. 2561–2576.

Hutcheon I. D., Phakey P. P. and Price P. B. (1972) Studies bearing on the history of lunar breccias. *Proc. Lunar Sci. Conf. 3rd*, p. 2845–2865.

Krishnaswami S., Lal D., Prabhu N. and Tamhane A. S. (1971) Olivines: revelation of tracks of charged particles. *Science* **174**, 287–291.

McKay D. S., Basu A., Waits G., Clanton U., Fuhrman R. and Fruland R. (1977) Grain size and evolution of Luna 24 soils (abstract). In *Papers Presented to the Conference on Luna 24*, p. 114–117. The Lunar Science Institute, Houston.

Morris R. V. (1976) Surface exposure indices of lunar soils: A comparative FMR study. *Proc. Lunar Sci. Conf. 7th*, p. 315–335.

Morris R. V. (1977) FMR and magnetic properties of Luna 24 soils and >1 mm particles (abstract). In *Papers Presented to the Conference on Luna 24*, p. 121–124. The Lunar Science Institute, Houston.

Nord G. L., Lally J. S., Heuer A. H., Christie J. M., Radcliffe S. V., Griggs D. T. and Fisher R. M. (1973) Petrologic study of igneous and metaigneous rocks from Apollo 15 and 16 using high voltage transmission electron microscopy. *Proc. Lunar Sci. Conf. 4th*, p. 953–970.

Phakey P. P., Hutcheon I. D., Rajan R. S. and Price P. B. (1972) Radiation effects in soils from five lunar missions. *Proc. Lunar Sci. Conf. 3rd*, p. 2905–2915.

Phakey P. P. and Price P. B. (1973) Radiation damage in Luna 20 soil. *Geochim. Cosmochim. Acta* **37**, 975–977.

Merrill, R. B. and Papike, J. J., editors.
(1978) *Mare Crisium: The View from Luna 24*, Pergamon Press (New York), p. 207–216
Printed in the United States of America

Luna 24 clast populations

J. STEWART NAGLE

Lunar Curatorial Laboratory (Northrop Services, Inc.)
Johnson Space Center, Houston, Texas 77058

Abstract—Most of the Luna 24 soil samples which were presented to the United States are bimodal in size distribution, and are immature to submature with clearly-definable fractional maturity. Because samples 24077, 24109 and 24210 contain two clast populations with differing maturity, they are interpreted as being mixtures of at least two different soils. In samples 24174 and 24182, size distribution of all particle types is similar, indicating only one clast population. This immature population is relatively coarse and unimodal, and there is an abundance of gabbro and basalt fragments. The size and composition pattern points to a major, bedrock-penetrating event as the origin of 24174 and 24182. In 24109, 24149 and 24210, the more mature clast population contains mare basalt material derived from bedrock and probably reworked in the near-by regolith to attain the maturity now seen. Because the coarse-grained, less mature population consists of low-grade breccias which originated within the regolith, it is inferred that these fine grained soils were derived from within the regolith. Sample 24077 is unusually coarse, bimodal, and texturally immature, but has relatively high ferromagnetic resonance and agglutinate values; this anomalous combination of characters could result from mixing by natural lunar processes, or it could arise by stirring by the drill or deflation by rocket exhaust.

INTRODUCTION

It is the purpose of this study to identify and group clast populations in Luna 24 samples on the basis of the maturation theory. In the Luna 24 samples this allows interpretation of which populations are matured soil components and which are newly introduced into the soil column. In this study, a clast population includes all grain sizes of one particle type, and a clast population group includes all clast populations with similar grain size distribution.

METHODS OF STUDY

Particle types were described and petrographically defined during preliminary examination, and the results were published in the Luna 24 catalog (Nagle and Walton, 1977). From data collected during particle description, size histograms were prepared and compared. Particles larger than .5 mm were weighed. Particles under .5 mm were counted, not weighed, and approximate weights were calculated by multiplying point-count percentage by the weight of total samples in each size fraction. Error in relating point-count percentages to measured weights is assessed by comparing numbers and weights in all .5–1 mm samples. In that size fraction, the average variation in particle weight is ±15%. In general, soil breccia particles weigh less than denser basalt fragments, but the difference was not enough to significantly alter the histogram pattern; hence, a correction factor was not applied. Finer particles are assumed to show a similar weight variation, although weighing such particles was beyond the scope of this work.

Size histograms which include all common particle types are plotted for each soil sample in the Luna 24 catalog. Data from Luna 24 samples presented to the United Kingdom (Pillinger and Fabian, 1977) were obtained by similar procedures and are presented for comparison. All gabbro, basalt, and monomineralic igneous fragments are lumped into the "crystalline component" because it was not possible to petrographically separate all particles into discrete, non-intergrading rock types. For example, some large (>1 mm) gabbro fragments contained abundant reddish-brown pyroxene, and minor amounts of yellowish and clove-brown pyroxene. Other large gabbro fragments contained abundant clove-brown pyroxene and minor amounts of reddish-brown and yellowish pyroxene, and yet other large fragments contained equal quantities of different colored pyroxenes. Petrographically, it was not possible to assign small fragments with one or few mineral grains to one of the coarser particle types. If small particles were proportioned according to the abundance of larger particles, the resulting histograms were similar to those of the combined crystalline component. Hence, the crystalline component was not subdivided. On the other hand, most breccia types could be identified in fine as well as coarse size classes. Soil breccia, as recognized here, is a crumbly rock with the appearance of lithified soil. Melt breccia has a dark, tough, irregular fracturing matrix and is differentiated from devitrified glass on matrix fracture characteristics—the devitrified glass shows a relict conchoidal fracture, but is also dark and relatively dull in contrast to fresh glass, also dark, but with vitreous luster. These breccia types intergrade with each other and with impact glasses of regolithic origin, and may be collectively referred to as low grade breccia. Aphanitic breccia is relatively tough tan to brownish rock that is aphanitic under the binocular microscope; presence of a variety of clasts enables identification of this rock as breccia. Black-and-white breccia consists of white clasts of shocked anorthosite in a black glass matrix. Agglutinates are dark, vesicular glass, with or without soil. After size histograms of all particle types were prepared, (as in Nagle and Walton, 1977) it was found that either all histograms were very similar (i.e. comprised one clast population group) or fell into two groups which were very similar (i.e. two clast population groups). All populations with most mass in the coarsest fractions were placed in the less mature mode (Fig.1) and those with maxima in finer sizes were placed in the more mature mode.

BRIEF DESCRIPTION OF CLAST POPULATIONS IN EACH SAMPLE

24210

Sample 24210, located at the base of the core, is relatively fine-grained, bimodal, and contains many breccia but few coarse-sized crystalline fragments. The fine-grained mode is composed of crystalline rock fragments (particles of

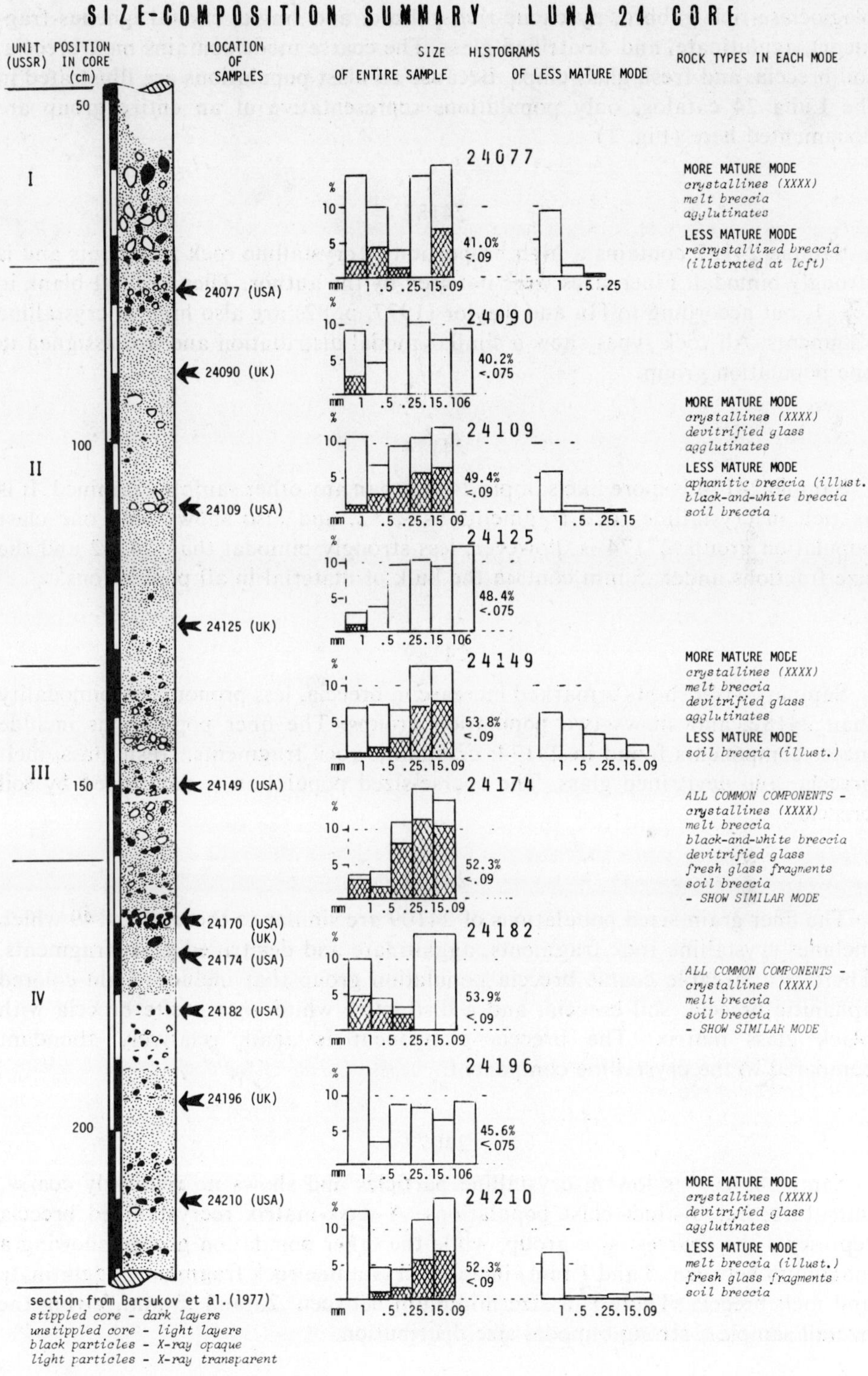
SIZE-COMPOSITION SUMMARY LUNA 24 CORE
UNIT (USSR)
POSITION IN CORE (cm)
LOCATION OF SAMPLES
SIZE HISTOGRAMS OF ENTIRE SAMPLE
OF LESS MATURE MODE
ROCK TYPES IN EACH MODE
I
II
III
IV
50
100
150
200
24077 (USA)
24090 (UK)
24109 (USA)
24125 (UK)
24149 (USA)
24170 (USA)
(UK)
24174 (USA)
24182 (USA)
24196 (UK)
24210 (USA)
24077
41.0%
<.09
24090
40.2%
<.075
24109
49.4%
<.09
24125
48.4%
<.075
24149
53.8%
<.09
24174
52.3%
<.09
24182
53.9%
<.09
24196
45.6%
<.075
24210
52.3%
<.09
mm 1 .5 .25 .15 .09
mm 1 .5 .25 .15 .106
MORE MATURE MODE
crystallines (XXXX)
melt breccia
agglutinates
LESS MATURE MODE
recrystallized breccia (illustrated at left)
MORE MATURE MODE
crystallines (XXXX)
devitrified glass
agglutinates
LESS MATURE MODE
aphanitic breccia (illust.)
black-and-white breccia
soil breccia
MORE MATURE MODE
crystallines (XXXX)
melt breccia
devitrified glass
agglutinates
LESS MATURE MODE
soil breccia (illust.)
ALL COMMON COMPONENTS -
crystallines (XXXX)
melt breccia
black-and-white breccia
devitrified glass
fresh glass fragments
soil breccia
- SHOW SIMILAR MODE
ALL COMMON COMPONENTS -
crystallines (XXXX)
melt breccia
soil breccia
- SHOW SIMILAR MODE
MORE MATURE MODE
crystallines (XXXX)
devitrified glass
agglutinates
LESS MATURE MODE
melt breccia (illust.)
fresh glass fragments
soil breccia
section from Barsukov et al.(1977)
stippled core - dark layers
unstippled core - light layers
black particles - X-ray opaque
light particles - X-ray transparent

plagioclase-rich gabbro, pyroxene-rich gabbro, and single, crystal igneous fragment), agglutinate, and devitrified glass. The coarse mode contains melt breccia, soil breccia, and fresh glass chips. Because all clast populations are illustrated in the Luna 24 catalog, only populations representative of an entire group are documented here (Fig. 1).

24182

Sample 24182 contains a high proportion of crystalline rock fragments and is strongly bimodal. Finer sizes were not seen by the author. They are left blank in Fig. 1, but according to Hu and Taylor (1977, p. 82) are also high in crystalline fragments. All rock types show a similar modal distribution and are assigned to one population group.

24174

Sample 24174 is more like sample 24182 than any other sample examined. It is as rich in crystalline rock fragments as -,182, and also shows only one clast population group. 27174 is, however, less strongly bimodal than 24182 and the size fractions under .5 mm contain the bulk of material in all populations.

24149

Sample 24149 shows a marked increase in breccia, less pronounced bimodality than 24182, and shows two population groups. The finer populations include major components found in 24174: crystalline rock fragments, fresh glass, melt breccia, and devitrified glass. The coarse-sized population is dominated by soil breccia.

24109

The finer grain-sized populations of 24109 are similar to those of 24149 which includes crystalline rock fragments, agglutinate and devitrified glass fragments, There is a sizeable coarse breccia population group that includes light-colored aphanitic breccia, soil breccia, and a distinctive white anorthosite breccia with black glass matrix. The breccia component is again relatively abundant compared to the crystalline component.

24077

Sample 24077 is low in crystalline particles and shows no relatively coarse, immature, but distinct clast populations. A dark-matrix recrystallized breccia represents the coarsest size group, while the other population group (showing a maximum between .5 and 1 mm) includes crystalline rock fragments, agglutinate and melt breccia. There is a size minimum between .25 and .5 mm, giving the overall sample a strong bimodal size distribution.

Comparison with British Samples 24090, 24125, 12196

Samples received by the British are similar to American samples from near-by locations in size distribution and coarse fraction composition. 24196, located halfway between American samples 24182 and 24210, is similar to the American samples in having a bimodal size distribution but is different in that the coarse fraction consists largely of breccia (Pillinger and Fabian, 1977, p. 14) whereas the coarse fraction of 24210 is rich in breccia and the coarse fraction of 24182 is rich in crystalline grains.

Although 24125 is closer to 24109 than 24149, it is very similar to 24149 in size distribution and composition, being unimodal, fine-grained and rich in agglutinate and weakly-sintered soil. However, the abundance of large agglutinate may be comparable to the abundance of agglutinate in 24109 as reported by McKay *et al.* (1977, p. 116). 24090 is much like 24077 in being bimodal with a complex size distribution and a predominance of breccia fragments over crystalline grains.

Discussion

Basis for interpretation—origin and maturation of clast populations

The soil maturation theory applied here is derived from the paper by McKay *et al.* (1974) on textural evolution of lunar soils. Immediately after a rock is impacted and fractured (as are most rocks on the lunar surface), the resulting immature fragment population contains most of its mass in the coarsest size fractions. A size histogram of such a population shows a steady decrease in size classes from coarsest to finest. Such populations (with others noted below) appear in soils with low surface maturity. With increasing surface exposure, bombardment, and comminution, clast populations should mature by becoming progressively finer-grained until constructional processes of agglutination balance the destructive process of comminution. Thus, mature soils should show a great abundance of agglutinates, i.e. a strong agglutinate mode, between 0.31 and 0.62 mm (McKay *et al.*, 1974, p. 893). Because the minimum particle size examined here is .090 mm, the principal phase detected is the comminution phase. Soils which are mixtures of two components at different stages of maturation are termed fractionally mature (McKay *et al.*, 1974, p. 896), and are recognized by a polymodal size distribution.

In this study, in which size distributions of all major particle types are plotted, fractionally mature samples are inferred by establishing the presence of two distinct clast populations. A coarse-grained population is interpreted to be immature, relatively newly-introduced into the regolith and not subject to extensive reworking. Although fragmentation of a friable rock such as white-matrix anorthosite breccia could produce a fine-grained population with a low track density or FMR maturity, a sample of low textural maturity and coarse size must be fragmented from something larger. Therefore it cannot reflect low surface maturity. Higher in a soil column, if local maturation has taken place, the coarse-grained, texturally immature population should become less abundant and finer. This would be the result of reworking and near-surface intraregolithic redistribution of particles. However, a major limitation in interpreting local maturation is that a locally-matured population of a given component cannot be distinguished from an externally-derived mature population of the same component. On the other hand, the appearance of a new coarse-grained "lower maturity" non-agglutinous component does indicate introduction of the new component, because the comminution part of the maturation process does not reverse itself (McKay *et al.*, 1974, p. 894).

Some inferences on size and penetration of impact can be made from analysis of clast populations. If, in a soil column, an extensive immature coarse population of relatively fresh unbrecciated igneous rock types appear above a relatively mature sample, it can be inferred that a regolith-penetrating event deposited the immature stratum. Although an impact into a local boulder could produce an immature fresh-appearing population consisting of several igneous rock types, such an impact would not produce a thick or extensive deposit. On the other hand, if the immature populations are soil breccias and are partially melted breccias that originated within the regolith, it is likely that the impact penetrated only the regolith and redistributed only regolithic-origin particles. Nevertheless, the possibility of a distal deposit from a major event cannot be ruled out.

General Aspects of Luna 24 Soils

Most bulk Luna 24 soils show definable maxima in the coarser grain sizes. On this basis, the samples could be interpreted as immature by the criteria of McKay *et al.* (1974, p. 893). However, samples 24077, 24149 and 24210 also contain more mature populations with most particles in the finer grain sizes. Presence of populations with differing degrees of maturity is termed fractional maturity, and is an indication of mixing of distinct soils.

Furthermore, the relatively mature clast population groups in samples 24077, 24109, 24149 and 24210 contain populations of disparate origin such as mare basalts and low-grade regolithic breccias. These relatively mature, but disparate-origin populations may have had a complex history before they became the precursors of the soils we see today.

There is further evidence for a complex history. Many other authors (Blanchard *et al.*, 1977, p. 27; Blanford and Wood, 1977, p. 41; Bogard and Hirsch, 1977, p. 45; Hu and Taylor, 1977, p. 83; McKay *et al.*, 1977, p. 115, and Simonds *et al.*, 1977, p. 172) also find evidence that the soils are mixed and some (Blanford and Wood) find evidence that the soils contain two distinct components of differing maturity. McKay *et al.* (1977, p. 116) find petrographic evidence that this primary starting material was coarse-grained high Mg olivine and pyroxene from a coarse-crystalline high-Mg igneous rock, and a finer-grained lower-Mg basalt. Both Hu and Taylor (1977, p. 82) and Simonds *et al.*

(1977, p. 172) find evidence in glass chemistry for a mixture of highland and mare glass types. This paper concerns itself with possible mixtures of bedrock-derived basalt and relatively low-grade breccias, probably of regolithic origin. The fact that data in support of all studies are carefully collected and well analyzed indicates that all of the above mixing possibilities can be correct, and further suggests that the Luna 24 soils are very complex mixtures from a number of sources.

Despite the chemical evidence for highland components in fine-grained soil (Blanchard *et al.*, 1977) and glasses (Simonds *et al.*, 1977) there are no coarse-grained petrographically-identified components that are unequivocally highland material. Not surprisingly, most dark-matrix vitric breccia particles that belong to immature clast populations show a high Fe content (Morris, 1977) that is similar to the mare basalts. The most spectacular light-colored breccias (24109,6 -,7 and -,8) are chemically similar to mare basalts in that they contain 22% Fe (Morris, 1977), a value much too high for them to be highland breccias. One low-Fe sample, 24182,8 shows a diabasic texture identifiable with more mafic mare gabbros. This is more likely to be a happenstance fragment of mare basalt that is plagioclase-rich than it is to be a highland rock. 24077,19, a brecciated plagioclase-rich fragment, is also low in Fe content but it contains dull greenish pyroxenes that are similar in appearance and morphology to those in less brecciated mare gabbros (Nagle and Walton, 1977, p. 16). Hence, samples which show some highland-like aspects are likely to show other non-highland characters and are not safely interpreted as highland materials. Although most samples contain two petrographic clast populations, both the mature and immature populations examined in this study are best interpreted as of ultimate mare origin and thus derived from relatively local sources.

Interpretation of the Section

24210, at the base of the section, contains a relatively mature component of fragmented mare basalt, agglutinate, and devitrified glass and a coarser component of less-mature melt breccia, soil breccia and chips of dark-colored fresh glass. All soil breccia fragments, agglutinate and associated igneous fragments, show the same relatively mature size distribution and have probably not been freshly introduced into the regolith. The less mature soil breccia and fresh glass are probably regolithic impact melts that originated within the regolith. The low grade melt breccia particles have a high iron content (Morris, 1977, p. 124) similar to that of mare basalts. These breccias appear to contain basalt fragments and may be derived from brecciated basalts. Because the immature materials are regolithic and are clearly not bedrock in origin, it is evident that 24210 resulted from mixture of mature and immature material that was already in the regolith. In 24182 and 24174 an influx of great amounts of fresh-looking basalts and gabbros and the presence of only one relatively immature clast population suggests an event that penetrated through the regolith and excavated a variety of bedrock basalts. This event homogenized and lowered the maturity of the entire sample. If the regolith in Mare Crisism is 30–50 m thick, a 300 m crater could penetrate to bedrock. Any crater between 300 m and the size of Fahrenheit could have produced these samples. We suggest that this horizon with its high bedrock component and similarity to other texturally and compositionally immature lunar samples is a better candidate for Fahrenheit ejecta than the other, breccia-containing Luna 24 samples. Evidence relating a specific ejecta deposit to a specific crater is rarely very strong, because the distinctive aspects of soil lithology can rarely be related back to specific craters. This sample is no exception.

Components that were immature and very abundant in 24174 and 24182—crystallines, fresh glasses and melt breccia—become less abundant and more mature in 24149 and 24109. This pattern of decreased abundance and increased maturity of crystallines and associated breccias could signify maturation in place, but it could also arise from an influx of more mature materials that superficially resemble prevalent components in 24174 and 24182. The change in maturation between 24149 and 24174 does indicate that different events deposited the two samples. Superimposed on the pattern of increased maturity are populations of low maturity that do represent an influx of material not seen lower in the section. In 24149, soil breccia forms the immature coarse component, while in 24109, a variety of low and medium-grade breccias (including light-colored aphanitic breccia, soil breccia, and black-and-white breccia) form an immature component. The FeO content and ferromagnetic resonance signature of these breccias are close to those of mare basalts. This indicates a relatively local, non-highland source for even the light-matrix breccia.

24077 shows two distinct clast populations, neither of which is very mature, and the entire sample is unusually coarse and texturally immature. There is a very pronounced minimum at .25–.5 mm. Crystalline basalt and associated mineral fragments, fresh glass and melt breccia fragments (as in the lower samples) form the more mature component. The abundance of moderately coarse material in these clast populations is indicative of a lower maturity than -,109, therefore the uppermost soil cannot be derived by reworking or maturation in place of underlying material. Dark vitric breccias form the principal immature component. Both microscope petrography and iron content point to derivation from local mare basaltic terrain. Bulk soil agglutinate maturity (McKay *et al.*, 1977) and FMR maturity (Morris, 1977) are only at the submature level, although they are higher than those of samples lower in the core. Furthermore, mean size of fines is much lower than elsewhere in the section (McKay *et al.*, 1977) and agrees with agglutinate and FMR profiles, and contrasts strongly with clast population maturity. Although this unusually coarse and bimodal soil occurs near the presumed lunar surface, comparison to Apollo mission cores shows that such gravelly material is very unusual at the lunar surface. The discordant agglutinate grain size FMR and clast maturities, coupled with general internal dis-equilibrium (McKay *et al.*, 1977, p. 119) could be an artifact of surface puddling by the drilling action or deflation by rocket exhaust, or it could reflect mixing by natural lunar processes. For example, the unusual coarseness could be due to the sample being on a ray of Giordano Bruno, but such an occurrence is unlikely (Morrison and Butler, 1977).

Conclusions

1. Luna 24 soils 24077, 24109, 24149 and 24210 show two clast populations and are mixtures of at least two precursor soils.
2. Soils 24174 and 24182 contain a much higher percentage of crystalline fragments than other soils. These soils show only one clast population which is relatively immature in a textural sense. A major impact disturbance than penetrated through the regolith to bedrock is suggested for the origin of these soils.
3. Immature clast population groups in 24077, 24109, 24149 and 24210 are dominated by low-grade breccias of mare composition. Such immature populations of intra-regolithic origin were supplied to the Luna 24 site by small or shallow-penetrating impacts.

References

Barsukov B. L., Ivanov A. V., Nazarov M. A., Rode O. D., Stakeev Yu. I., Tarasov L. S., Tobelko K. I., and Florensky C. P. (1977) Preliminary description of the regolith core from Mare Crisium (abstract). In *Lunar Science VIII*, p. 67-69. The Lunar Science Institute, Houston.

Blanchard D. P., Haskin L. A., Brannon J. C., Aaboe E. (1977) Chemistry of soils and particles from Luna 24 (abstract). In *Papers Presented to the Conference on Luna 24*, p. 37-40. The Lunar Science Institute, Houston.

Blanford G. E., and Wood G. C. (1977) Particle track densities in the Luna 24 core (abstract). In *Papers Presented to the Conference on Luna 24*, p 41-43. The Lunar Science Institute, Houston.

Bogard D. D. and Hirsch W. C. (1977) Noble gases in Luna 24 core soils (abstract). In *Papers Presented to the Conference on Luna 24*, p. 44-47. The Lunar Science Institute, Houston.

Hu H.-N. and Taylor L. A. (1977) Luna 24 soils from Mare Crisium: Agglutinate chemistry (abstract). In *Papers Presented to the Conference on Luna 24*, p. 82-84. The Lunar Science Institute, Houston.

McKay D. S., Fruland R. M., and Heiken G. H. (1974), Grain size and the evolution of lunar soils. *Proc. Lunar Sci. Conf. 5th*, p. 887-906.

McKay D. S., Basu A., Waits G., Clanton U., Furhman R., and Fruland R. (1977) Grain size and evolution of Luna 24 soils (abstract). In *Papers Presented to the Conference on Luna 24*, p. 114-117. The Lunar Science Institute, Houston.

Morris R. B. (1977) FMR and magnetic properties of Luna 24 soils and >1 mm soil particles (abstract). In *Papers Presented to the Conference on Luna 24*, p. 121-129. The Lunar Science Institute, Houston.

Morrison D. A. and Butler P., Jr. (1977) Geologic map of the Luna 24 site (abstract). In *Papers Presented to the Conference on Luna 24*, p. 125-128. The Lunar Science Institute, Houston.

Nagle J. S. and Walton W. J. A. (1977) Luna 24: Catalog and preliminary description of samples presented to the U.S.A. Curatorial Office, NASA Johnson Space Center, Houston. 86 pp.

Pillinger C. T. and Fabian D. M. (1977) Luna 24: Preliminary separation and allocation, samples presented to the Royal Society. Publ. of Planetary Sciences Unit, Univ. of Cambridge, Cambridge, England. 21 pp.

Simonds C. H., Warner J. L., McGee P. E., Cochran A., and Brown R. W. (1977) Luna 24: Petrochemical potpourri (abstract). In *Papers Presented to the Conference on Luna 24*, p. 170–173. The Lunar Science Institute, Houston.

could be due to the sample being on a tail of [illegible] trend, but such an occurrence is unlikely ([illegible] and [illegible], 1977).

CONCLUSIONS

1. Luna 24 soils 24077, 24109, 24149 and 24210 show two glass populations and are mixtures of at least two precursor soils.
2. Soils 24174 and 24182 contain a much higher percentage of crystalline fragments than other soils. These soils show only one glass population which is relatively immature in a textural sense. A single impact of [illegible] that penetrated through the regolith to bedrock is suggested as the origin of these soils.
3. Immature glass population groups in 24077, 24109, 24149 and 24210 are dominated by low-grade breccias of mare composition. Such immature populations of autoregolithic origin were supplied to the Luna 24 site by small or shallow-penetrating impacts.

REFERENCES

[illegible] (1977) Preliminary description of the regolith core from Mare Crisium. (Abstract) In *Lunar Science VIII*, [illegible]. The Lunar Science Institute, Houston.

Blanchard D. P., [illegible] (1977) [illegible] In *Papers Presented to the Conference on Luna 24*, [illegible]. The Lunar Science Institute, Houston.

[illegible] (1977) [illegible] in the Luna 24 core [illegible] In *Papers Presented to the Conference on Luna 24*, [illegible]. The Lunar Science Institute, Houston.

[illegible] (1977) [illegible] In *Papers Presented to the Conference on Luna 24*, [illegible]. The Lunar Science Institute, Houston.

[illegible] (1977) [illegible] (abstract). In *Papers Presented to the Conference on Luna 24*, [illegible]. The Lunar Science Institute, Houston.

McKay D. S., [illegible] (1974) [illegible] *Proc. Lunar Sci. Conf.* [illegible], p. 887–906.

McKay D. S., [illegible] (1977) [illegible] In *Papers Presented to the Conference on Luna 24*, [illegible]. The Lunar Science Institute, Houston.

Morris R. V. (1977) FMR and magnetic properties of Luna 24 soils [illegible] (abstract). In *Papers Presented to the Conference on Luna 24*, [illegible]. The Lunar Science Institute, Houston.

Morrison [illegible] (1977) [illegible] In *Papers Presented to the Conference on Luna 24*, [illegible]. The Lunar Science Institute, Houston.

[illegible] (1977) [illegible] NASA Johnson Space Center, Houston. [illegible]

[illegible] (1977) [illegible]

[illegible] (1977) [illegible] (abstract). In *Papers Presented to the Conference on Luna 24*, [illegible]. The Lunar Science Institute, Houston.

Merrill, R. B. and Papike, J. J., editors.
(1978) *Mare Crisium: The View from Luna 24*, Pergamon Press (New York), p. 217–228
Printed in the United States of America

Magnetic properties and carbon chemistry studies pertinent to the evolution of the regolith at the Luna 24 site

C. T. PILLINGER, L. R. GARDINER, A. J. T. JULL and M. R. WOODCOCK.

Planetary Sciences Unit, Department of Mineralogy and Petrology, University of Cambridge, England

and

A. STEPHENSON

Institute of Lunar and Planetary Physics, School of Physics, University of Newcastle-upon-Tyne, England.

Abstract—Magnetic susceptibility, isothermal remanent magnetisation, hydrolysable carbon, ^{20}Ne and major element data are presented for sieved fractions and density / magnetic separates from the 90, 125, and 196 cm levels of the Luna 24 core. The analyses indicate that sample maturity decreases with depth and support the hypothesis that the regolith sampled by Luna 24 is a two component mixture which was deposited either rapidly or as a one step process. The less mature component in the mixture is enriched in FeO, and MgO and depleted in Al_2O_3 relative to the mature fraction.

INTRODUCTION

As part of our investigations into the mechanism of formation of fine-grained iron in lunar soils, we are currently examining three Luna 24 samples 24090, 24125 and 24196 for magnetic properties and carbon chemistry. Although our studies are as yet incomplete, some information relevant to the evolution of the regolith at the Luna 24 site has been revealed. In this paper we report magnetic susceptibility (χ, a measure of superparamagnetic Fe° <130 Å in size) and isothermal remanent magnetisation (IRM_s, Fe° including stable single domain grains >130 Å) data obtained non-destructively from a wide range of sieved, hand-picked and density/magnetic separates. A variety of fractions have been treated with deuterated acid (DCl, 38% in D_2O) to release CD_4 as a measure of hydrolysable carbon (C_{hyd}) and ^{20}Ne as a representative trapped solar wind rare gas. Since our preliminary report on Luna 24 samples (Gardiner *et al.*, 1977a), we have discovered a calibration error for CD_4; the data given herein have been corrected to allow for this discrepancy. In order to use χ and CD_4 as indicators of relative lunar soil maturity these parameters like the ferromagnetic resonance measure I_s, must be normalised for bulk iron (FeO) content (Pillinger *et al.*, 1974; Morris, 1976). Consequently we have carried out partial major element analysis by a thin film XRF technique on some sieved fractions.

This work is part of a collaborative programme being undertaken by ten British laboratories on the Luna 24 core soils.

SAMPLE HANDLING

All three soils have been wet sieved through stainless steel meshes (500, 250, 150, 106, 75, 53, 40 and 30 μm) using spectrographic grade methanol as the liquid phase. The >500 μm fraction was further sieved through a 1 mm nylon screen. Some preliminary details of this procedure and the soil distribution obtained have been reported elsewhere (Pillinger and Fabian, 1977); a complete account is in preparation. In the size range above 75 μm, all three soils exhibit a bimodal distribution similar to the six Luna 24 soils studied by the American Consortium (McKay *et al.*, 1977). Below 75 μm, the samples are markedly depleted in particles in the 75–53 μm range. Comparative data for the other Luna 24 core samples are not available. Material in the size range 150–250 μm and 106–150 μm has been subjected to a density separation in tetrabromoethane (e.g. 2.96 g/cm^3). The ρ <2.96 fractions have been further sub-divided into highly magnetic (moving over 2 mm to the soft iron pole piece of a magnet, HM), medium magnetic (requiring a separation distance of 2 mm to contact, MT) and non-magnetic (not collected under any circumstances, NM) material using our improved magnetic separator and revised procedure (Pillinger and Jull, 1978). The distribution of ρ <2.96 HM, MT, NM and ρ >2.96 particles is given in Table 1. There is an increase in the abundance of material ρ >2.96 g/cm^3 and a decrease of highly magnetic particles in the lower levels of the core. Since the three samples under study have a similar chemistry (see below) we interpret these trends to indicate a greater abundance of vesicular glassy agglutinates near the surface of the regolith and hence an increased maturity. This interpretation is amply borne out by binocular microscope examination and by the data of others (Morris, 1977; McKay *et al.*, 1977).

The techniques used for CD_4 determination by static mass spectrometry, magnetic studies and chemical analysis have been described elsewhere (Gardiner *et al.*, 1977b; Stephenson and de Sa, 1970; Woodcock and Pillinger, 1978).

RESULTS AND DISCUSSION

Bulk chemistry

Partial major element analyses for three sieved fractions from each soil are given in Table 2. There is relatively good agreement between our data and those reported for other levels of the core. The XRF values for FeO are intermediate between those obtained by atomic absorption (Blanchard *et al.*, 1977) or magnetically (Morris, 1977), on the one hand, and the fused bead technique (Simonds *et al.*, 1977) on the other. The data suggest that all three soils are derived from mare type material predominantly the low titanium Crisium basalts recognised by Ryder *et al.* (1977) and Barsukov *et al.* (1977). It appears that the deepest sample 24196 is slightly enriched in FeO and depleted in Al_2O_3. Blanchard *et al.* (1977) report a similar compositional variation for 24210 although it was not seen by Ma and Schmitt (1977). For the three samples in

Table 1. Density/magnetic separation of sieved Luna 24 soils.

	Particle type distribution as % of total size fraction					
	24090		24125		24196	
Separate	250–150 μm	150–106 μm	250–150 μm	150–106 μm	250–150 μm	150–106 μm
ρ <2.96 g/cm^3						
HM	12.3	12.1	6.5	7.9	3.0	3.2
MT	8.6	9.0	9.2	8.8	6.1	3.9
NM	9.6	8.9	9.7	10.6	13.6	13.8
ρ >2.96 g/cm^3	69.4	70.0	74.5	72.7	77.3	79.1

Table 2. Partial major element analysis of Luna 24 soils.

Depth in core (cm)	Size fraction (μm)	Composition % by wt. FeO	TiO_2	MgO	Al_2O_3	CaO
	250–150	19.2	1.24	9.1	12.8	14.2
090	150–106	18.0	0.63	10.7	11.7	14.3
	106–75	16.1	0.46	8.6	10.9	11.5
	250–150	17.7	0.53	9.7	11.0	12.7
125	150–106	18.8	0.61	9.0	11.5	11.9
	106–75	18.2	0.70	9.5	11.4	11.6
	250–150	20.7	0.55	14.3	8.2	10.1
196	150–106	20.7	1.99	9.9	8.9	10.6
	106–75	19.9	1.03	10.2	9.4	10.6
	smoke	13.5	0.91	8.2	18.7	14.6

Errors: FeO, CaO ± 0.3%; TiO_2 ± 0.1%; Al_2O_3, MgO ± 0.5%.

Table 2 there is a slight tendency for FeO to decrease as particle radius decreases. This effect is most markedly discerned from the analysis of the 24196 smoke sized fraction (material <1 μm in size, collected as a suspension in the methanol used for sieving) which is very depleted in FeO, slightly depleted in MgO and strongly enriched in Al_2O_3. Again, Blanchard *et al.* (1977) observed a depletion of FeO in <20 μm fractions from six Luna 24 soils. These data are in accord with the hypothesis (Morris, 1977; McKay *et al.*, 1977) that the Luna 24 core is a mixture of two components. The coarse grained immature fraction is high in FeO, slightly enriched in MgO and low in Al_2O_3 and hence could contain a proportion of the magnesium rich olivine suggested by Basu *et al.* (1977).

Magnetic measurements

All the magnetic measurements made on the various samples are summarised in Table 3. The χ values (and CD_4 values, see below) suggest that 24090 is submature, 24196 is immature and 24125 lies on the border line between the two. The concept of maturity used here is that of Morris (1976); a plot of χ versus I_s for 49 bulk soils (Jull *et al.*, 1978) show that these two parameters are directly related and hence may be interconverted.

For seven size fractions in the range 30–500 μm, each of the three samples of Luna 24 exhibit a linear dependence between χ and the reciprocal of particle diameter (d) (Fig. 1). It can be seen that for any particular sample the susceptibility may be expressed as: $\chi = \chi_o + c/d$ where χ_o and c are constants. It is also apparent from Fig. 1 that c, which represents the slope of the graph, increases as the intercept χ_o (d = ∞) increases. An empirical equation relating c and χ_o may be deduced as follows:

$$\chi = \chi_o + \gamma \chi_o^{1/3}/d$$

where $\gamma = 5.7 \times 10^{-5}$ G cm^4 g^{-1} Oe^{-1}. Using this value of γ give values for χ_o of 0.78, 0.44 and 0.21 $\times 10^{-3}$ G cm^3 g^{-1} Oe^{-1} respectively for the lines drawn through 24090, 24125 and 24196 data points. The χ data can be explained if the three samples comprise mixtures of mature and immature material in variable

Table 3. Magnetic susceptibility and isothermal remanent magnetisation measurements for Luna 24 size fractions and density/magnetic separates.

Sample description*		24090		24125		24196	
		χ	IRM_s	χ	IRM_s	χ	IRM_s
Size fraction							
>1 mm+		2.00	102	1.68	102	0.95	20.1
500–1 mm		0.84	39.9	0.46	23.2	0.21	9.2
250–500		1.46	42.0	0.47	23.4	0.41	21.5
150–250		1.06	43.4	0.62	26.2	0.37	18.2
106–150		1.16	46.1	0.71	29.8	0.45	21.2
75–106		1.32	46.9	0.96	37.3	0.57	26.5
53–75		1.68	52.7	1.14	40.3	0.71	28.5
40–53		1.78	57.4	1.38	47.7	0.90	33.0
30–40		2.29	80.6	1.89	69.4	1.23	51.9
Separates							
150–250							
ρ <2.96	HM	3.59	115.4	3.37	75.3	2.94	61.2
	MT	1.08	46.1	1.09	27.5	0.53	32.5
ρ >2.96		0.41	17.3	0.31	14.4	0.19	8.2
106–150							
ρ <2.96	HM	4.26	128.5	3.83	118.3	3.95	64.0
	MT	0.89	38.9	0.99	35.5	0.88	39.6
ρ >2.96		0.57	21.0	0.43	15.5	0.30	9.0

* sizes in μm except where stated + single grains
Units: χ 10^{-3} G cm^3 g^{-1} Oe^{-1}; IRM_s, 10^{-3} G cm^3 g^{-1}

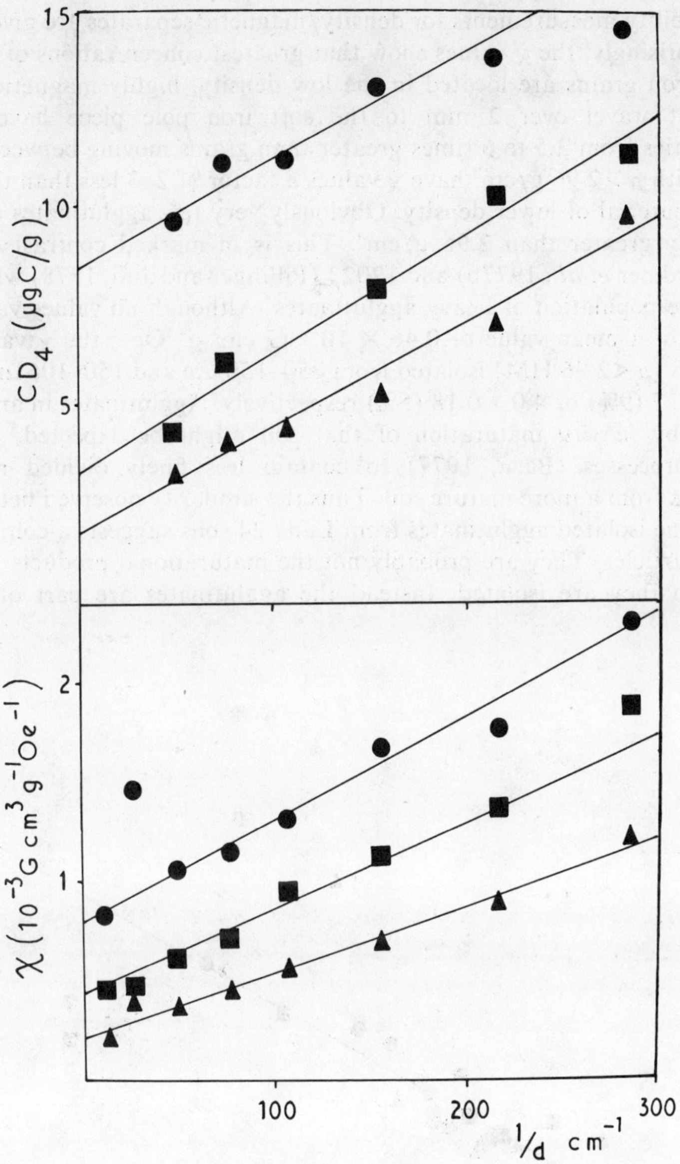

Fig. 1. CD_4 and magnetic susceptibility (χ) versus reciprocal of particle diameter for Luna soils 24090 (●) 24125 (■) and 24196 (▲)

proportions, with an increased abundance of the greater maturity end member at the top of the core. The empirical relationship between the surface component $\gamma\chi_o^{1/3}$ and the volume component χ_o is not fully understood and further investigations are in progress.

Susceptibility measurements for density/magnetic separates are given in Table 3. Not surprisingly, the χ values show that greatest concentrations of superparamagnetic iron grains are located in the low density, highly magnetic fractions. Grains that travel over 2 mm to the soft iron pole piece have magnetic susceptibilities from 3.5 to 6 times greater than grains moving between 0–2 mm. Particles with $\rho > 2.96$ g/cm^3 have χ values a factor of 2–3 less than the medium magnetic material of lower density. Obviously very few agglutinates are present with density greater than 2.96 g/cm^3. This is in marked contrast to samples 15601 (Gardiner *et al.*, 1977b) and 12023 (Pillinger and Jull, 1978) which have a considerable population of heavy agglutinates. Although no values vary by 50% either side of a mean value of 0.46×10^{-3} G cm^3 g^{-1}Oe^{-1} the χ values of the agglutinates ($\rho < 2.96$ HM) isolated from 250–150 μm and 150–106 μm fractions are 3.0 ± 0.27 (9%) or 4.0 ± 0.18 (5%) respectively. Agglutinates in an immature soil made by *in situ* maturation of that soil might be expected, because of recycling processes (Basu, 1977) to contain less finely divided metal than agglutinates from a more mature soil. Thus the similarity observed between the χ values for the isolated agglutinates from Luna 24 soils suggests a common origin for these particles. They are probably not the maturational products of the soils from which they are isolated. Instead the agglutinates are part of a mature

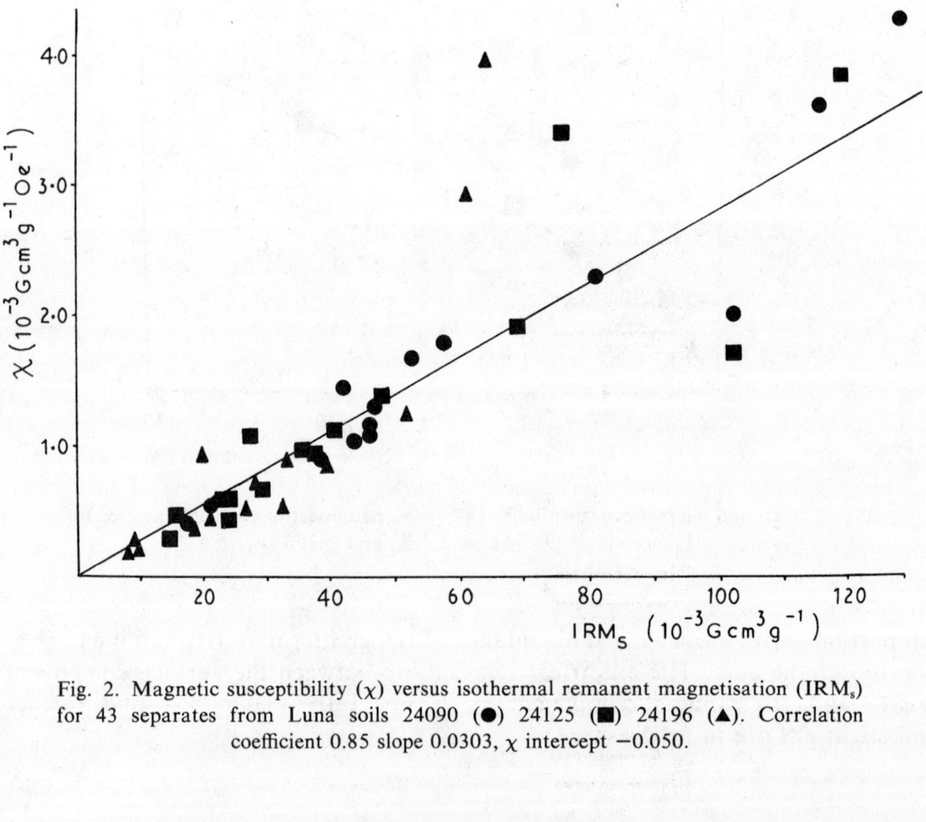

Fig. 2. Magnetic susceptibility (χ) versus isothermal remanent magnetisation (IRM_s) for 43 separates from Luna soils 24090 (●) 24125 (■) 24196 (▲). Correlation coefficient 0.85 slope 0.0303, χ intercept -0.050.

component which has been admixed into the soil from another source. Thus the χ values for sieve fractions within the various soils reflect the proportion of mature agglutinates in the mix. Material balance calculations based on the data in Table 1 give calculated χ values which agree to within 10% of the measured quantities. The slight decrease observed in the χ value $\rho < 2.96$ HM fractions in going from 24090 to 24196 suggests that some local maturation has occurred since mixing took place and that immature unexposed material has been cycled in the agglutinates to a small extent. The fact that the $\rho > 2.96$ g/cm^3 fractions are depleted in agglutinates suggests that this material may represent the immature component. The relatively high density supports our earlier conclusion that this component is rich in ferromagnesian minerals.

Figure 2 shows a plot of χ versus IRM_s for all the fractions (43) examined magnetically. The correlation coefficient is 0.85, the slope of the line 0.030 ± 0.004 and the χ intercept −0.05 ± 0.52. The data suggest that by and large there is a relatively constant size distribution for the iron droplets in Luna 24 materials. The notable exceptions are three fractions of low density ($\rho < 2.96$ g/cm^3) highly magnetic agglutinates (two from 24196 and the other from the 150–250 μm fraction of 24125) and two single grains hand-picked from the >1 mm fractions of 24090 and 24125. The highly magnetic agglutinates appear to be enriched in a fine-grained component of metal whereas the hand-picked fractions contain a component of coarser single domain Fe°. At present we are not sure what significance, if any, may be attributed to this observation.

Hydrolysable carbon

In our preliminary report (Gardiner *et al.*, 1977a) on the Luna 24 samples, we suggested that these materials released considerable quantities of CD_4 on acid dissolution. We have since found that these apparently abnormally high values are attributable to an air leak in our calibration system. The values previously reported have now been adjusted by an appropriate correction factor and are given together with subsequent determinations in Table 4. Like χ, CD_4 released by deuterated acid dissolution increases with the reciprocal of particle diameter (Fig. 1). Neither parameter shows any tendency to exhibit any excess in coarse fractions (Gardiner *et al.*, 1977b). Although the amounts of CD_4 released by bulk sieved fractions of the three Luna 24 soils show considerable variation, the quantities measured from low density, highly magnetic particles are not appreciably different. This observation further supports the view presented earlier that 24090, 24125 and 24196 are two component mixtures of mature and immature material and that most of the agglutinates present in the samples derive from the mature end member.

In Fig. 3 we have plotted χ versus CD_4. A least squares analysis of the data suggests that the two parameters have a correlation coefficient of 0.94. The slope of the line through the data points is 0.109 ± 0.008 and the χ intercept 0.13 ± 0.3. From the slope, it is possible to estimate that the concentration of carbon in iron for samples from the Luna 24 site is 0.26 ± 0.09 wt%. This value is similar to one

Table 4. Deuteromethane and neon concentrations released by DCl treatment of Luna 24 samples.

Sample (μm)	24090 CD_4	24090 ^{20}Ne	24125 CD_4	24125 ^{20}Ne	24196 CD_4	24196 ^{20}Ne
>1 mm	18.7	640	18.9	586	9.5	492
150–250	9.7	198	4.1	99	3.4	100
106–150	11.2	274	6.1	198	4.1	135
75–106	11.2	222	11.5	307	4.4	210
53–75	13.0	505	7.8	450	5.2	321
40–53	13.7	576	10.2	576	7.0	416
30–40	14.2	793	11.3	822	9.7	570
150–250 ρ <2.96 HM	27.0	607	27.3	638	31.1	717

Units: CD_4 – μm/g, ^{20}Ne—arbitrary units relative to CD_4 standard.

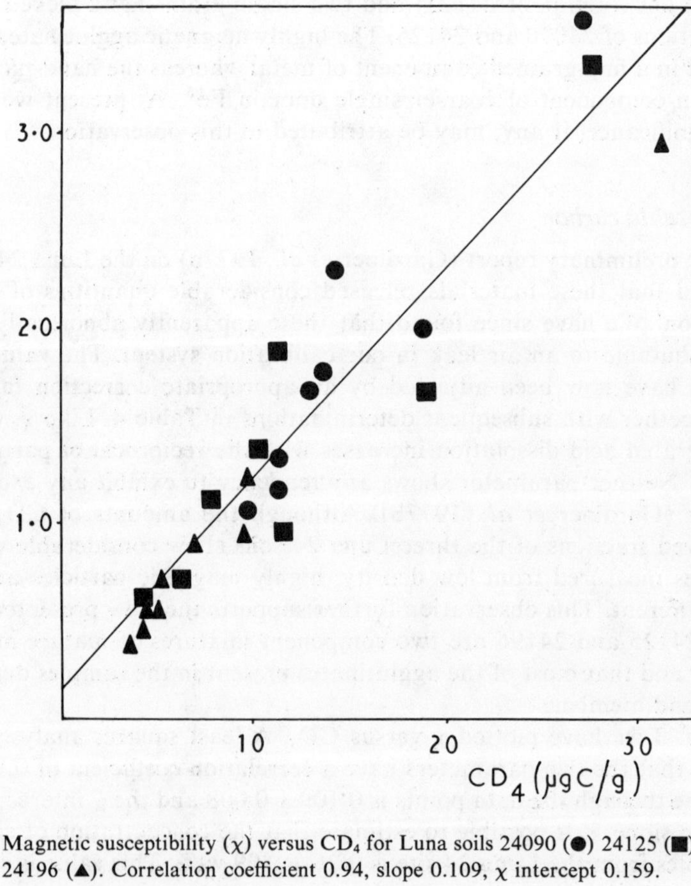

Fig. 3. Magnetic susceptibility (χ) versus CD_4 for Luna soils 24090 (●) 24125 (■) and 24196 (▲). Correlation coefficient 0.94, slope 0.109, χ intercept 0.159.

of 0.24 ± 0.08 wt.% obtained from a plot of χ versus CD_4 for 62 bulk soils and regolith breccias (Jull *et al.*, 1978) suggesting that Luna 24 material has a relatively typical carbon in iron concentration. The value is, however, very different from the 0.12 ± 0.04 wt.% C established for 15601 (Gardiner *et al.*, 1977b) which is similar to Luna 24 in both maturity and bulk iron content. We are currently unable to explain this variation but further samples are being examined in detail in an attempt to determine more precisely the carbon in iron concentration for other bulk soils (Jull *et al.*, 1978). There is some indication from χ versus IRM_s plots that the Luna 24 samples (see above), Luna 16 and 20 material and a number of other bulk lunar soils (Stephenson *et al.*, 1977) contain a higher proportion of stable single domain iron grains than 15601, however, further investigation is required.

Solar wind neon

In addition to releasing CD_4 by hydrolysis, DC1 acid dissolution liberates trapped solar wind rare gases. Helium, neon and argon may be measured concurrently with carbon containing species by our static mass spectrometric method, although argon data require correction to remove the background contribution from atmospheric gas present in the acid. Table 4 gives values for ^{20}Ne (in arbitrary units relative to CD_4 standard) as representative of solar wind rare gases studied. Again the data demonstrate the decreasing maturity of the soils with depth in the core and the surface and volume related components of solar wind derived species. The ^{20}Ne values are also shown plotted against χ in Fig. 4. Low density, highly magnetic samples from the 250–150 μm sieved fractions have lost substantial quantities of ^{20}Ne relative to superparamagnetic iron. This occurrence was also noticed for 15601 (Gardiner *et al.*, 1977b) and is presumably due to fractionation of ^{20}Ne during agglutination. From Fig. 4 it can be seen that for size fractions (including the single >1 mm grains) there is a separate relationship between χ and ^{20}Ne for each soil. 24090 is most depleted and 24196 least depleted in neon. The variable relationships clearly reflect known proportions of agglutinates occurring within each soil. As the three samples of agglutinates (ρ <2.96 HM) are not identical, some reworking of the mature and immature components has taken place since the mixing occurred.

Helium concentrations measured for the density and magnetic separates and sieved fractions, like neon, show the fractionation associated with agglutination and reflect the proportion of agglutinates within the three soils. However we have been unable to detect loss of ^{36}Ar relative to χ in the Luna 24 soils; fractionation of this rare gas relative to Fe° during agglutination was indicated from the analysis of 15601 separates (Gardiner *et al.*, 1977b).

CONCLUSIONS

The processing of samples from the 90, 125 and 196 cm levels of the Luna 24 drill core and the analysis of a variety of separates from each horizon for

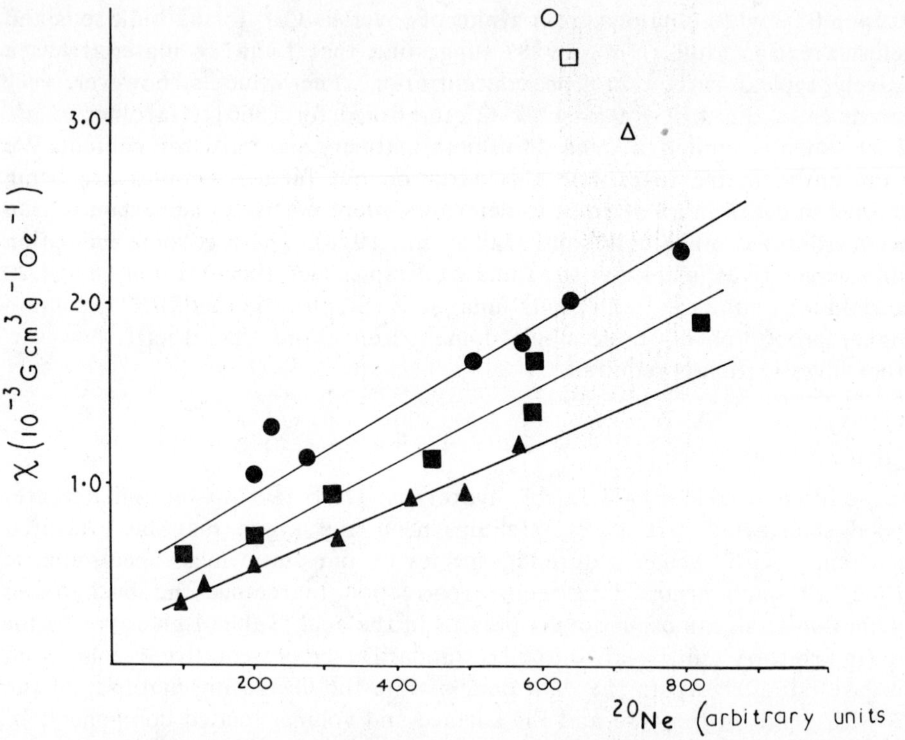

Fig. 4. Magnetic susceptibility (χ) versus ^{20}Ne for sieve fractions (solid symbols) and low density, highly magnetic separates (ρ <2.96, HM, open symbols) from Luna soils 24090 (●) 24125 (■) and 24196 (▲). Correlation coefficients 0.98, 0.98 and 0.99, slopes 0.00190, 0.00186 and 0.00173, χ intercepts 0.751, 0.387 and 0.178 respectively.

magnetic properties, hydrolysable carbon and the rare gas ^{20}Ne suggests that soil maturity decreases with distance from the regolith surface at the Luna 24 site. The range of maturities encountered, however, varies only from submature to immature. Our data corroborate conclusions reached by numerous other investigators (e.g., Morris 1977, McKay *et al.*, 1977, Blanford *et al.*, 1977) using a larger suite of samples. The variation of maturity in 24090, 24125 and 24196 can readily be demonstrated to be associated with the proportion of agglutinates found at any level. However, analytical data obtained from agglutinate separates from each layer are not widely different and hence these particles are thought to be representative of a single mature component mixed into the Luna 24 site rather than distinct products from the *in situ* reworking of local material. The three layers of core examined are thus interpreted as being two component mixtures (cf. Morris, 1977, McKay *et al.*, 1977). The data best fit a model in which the regolith was deposited in a series of closely related events (Morris, 1977) with sufficient time in between each step to allow some reworking of the mature and immature components and hence give rise to the slight variations in agglutinate χ, C_{hyd} and ^{20}Ne content. Alternatively, deposition could have been a one step process with the immature component already having a small fraction of agglutinates due to previous reworking. The nature of the event(s) which allow the two end members to be mixed in the steadily changing proportions indicated by the analysis, are unknown to us. Our chemical analyses indicate that the mature component may have appreciably less FeO and MgO and more Al_2O_3 than the immature end member. Hence, the two components may have derived from some considerable distance apart rather than from the immediate vicinity of the sampling area.

Acknowledgments—We thank the Soviet Academy of Sciences for the provision of Luna 24 samples This work is supported by grants from the Science Research Council and the Royal Society. One of us (AJTJ) is grateful to the National Research Council of Canada for a fellowship; two of us (LRG and MRW) would like to thank the SRC for research studentship maintenance awards.

REFERENCES

Barsukov V. L., Tarasov L. S., Dmitriev L. V., Kolesov G. M., Shevaleesky I. D., and Garanin A. V. (1977) The geochemical and petrochemical features of regolith and rocks from the Mare Crisium (preliminary data). *Proc. Lunar. Sci. Conf. 8th*, p. 3319–3332.

Basu A. (1977) Steady state, exposure age, and growth of agglutinates in lunar soils. *Proc. Lunar Sci. Conf. 8th*, p. 3617–3632.

Basu A., McKay D. S., and Fruland R. M. (1977) Petrography, mineralogy and source rocks of Luna 24 drill core soils (abstract). In *Papers Presented to the Conference on Luna 24*, p. 18–21. The Lunar Science Institute, Houston.

Blanchard D. P., Haskin L. A., Brannon J. C., and Aaboe E. (1977) Chemistry of soils and particles from Luna 24 (abstract). In *Papers Presented to the Conference on Luna 24*, p. 37–40. The Lunar Science Institute, Houston.

Blanford G. E., and Wood C. (1977) Particle track densities in the Luna 24 core (abstract). In *Papers Presented to the Conference on Luna 24*, p. 41–43. The Lunar Science Institute, Houston.

Gardiner L. R., Jull A. J. T., Woodcock M. R., Pillinger C. T., and Stephenson A. (1977a) Carbon chemistry and metallic iron content of some Luna 24 core samples (abstract). In *Papers Presented to the Conference on Luna 24*, p. 60–63. The Lunar Science Institute, Houston.

Gardiner L. R., Woodcock M. R., Pillinger C. T., and Stephenson A. (1977b) Carbon chemistry and magnetic properties of bulk and agglutinate size fractions from soil 15601. *Proc. Lunar Sci. Conf. 8th.*, p. 2817–2839.

Jull A. J. T., Pillinger C. T., and Stephenson A., (1978) The carbon content of finely-divided lunar iron (abstract). In *Lunar and Planetary Science IX*, p. 609–611. Lunar and Planetary Institute, Houston.

Nagle J. S. and Walton W. J. A. (1977) Luna 24: Catalog and preliminary description of samples presented to the U.S.A. Curatorial Office, NASA Johnson Space Center, Houston. 88 pp.

Ma M.-S., and Schmitt R. A. (1977) Luna 24 soils: a chemical study (abstract). In *Papers Presented to the Conference on Luna 24*, p. 98–101. The Lunar Science Institute, Houston.

McKay D. S., Basu A., Waits G., Clanton U., Fuhrman R., and Fruland R. (1977) Grain size and evolution of Luna 24 soils (abstract). In *Papers Presented to the Conference on Luna 24*, p. 114–117. The Lunar Science Institute, Houston.

Morris R. V. (1976) Surface exposure indices of lunar soils: a comparative FMR study. *Proc. Lunar Sci. Conf. 7th*, p. 315–335.

Morris R. V. (1977) FMR and magnetic properties of Luna 24 soils and greater than 1mm soil particles (abstract). In *Papers Presented to the Conference on Luna 24*, p. 121–124. The Lunar Science Institute, Houston.

Pillinger C. T., Davis P. R., Eglinton G., Gowar A. P., Jull A. J. T. and Maxwell J. R. (1974) The association between carbide and finely-divided iron in lunar fines. *Proc. Lunar Sci. Conf. 5th*, p. 1949–1961.

Pillinger C. T. and Fabian D. M. (1977) Preliminary separation of Luna 24 samples presented to the Royal Society by the Soviet Academy of Sciences (copies available from the authors on request).

Pillinger C. T., and Jull A. J. T. (1978) Carbon chemistry studies of agglutinate separates from soil 12023 (abstract). In *Lunar and Planetary Science IX*, p. 910–912. The Lunar and Planetary Institute, Houston.

Ryder G., McSween H. Y., and Marvin U. B. (1977) Basalts from Mare Crisium. *The Moon* **17**, 263–287.

Simonds C. H., Warner J. L., McGee P. E., Cochran A., and Brown R. W. (1977) Luna 24: Petrochemical potpourri (abstract). In *Papers Presented to the Conference on Luna 24*, p. 170–173. The Lunar Science Institute, Houston.

Stephenson A. and de Sa A. (1970) A simple method for the measurement of the temperature variation of the initial magnetic susceptibility between 77 and 1000°K, *J. Sci. Instrum.* **3**, 59–61.

Stephenson A., Collison D. W. and Runcorn S. K. (1977) Magnetic characteristics of Luna 16 and 20 samples. *Phil. Trans. Roy. Soc. London* **A284**, 151–156.

Woodcock M. R. and Pillinger C. T. (1978) The chemistry of glassy agglutinates (abstract). In *Lunar and Planetary Science IX*, p. 1276–1278. Lunar and Planetary Institute, Houston.

Merrill, R. B. and Papike, J. J., editors.
(1978) *Mare Crisium: The View from Luna 24*, Pergamon Press (New York), p. 229–244
Printed in the United States of America

Luna 24: Lateral heterogeneity in the lunar crust

J. L. Warner

NASA Johnson Space Center, Houston, TX 77058

C. H. Simonds

Northrop Services, Inc., Houston, TX 77034

P. E. McGee and A. Cochran

Lockheed Electronics Co., Inc., Houston, TX 77058

Abstract—Petrologic investigations of six samples of Luna 24 soil suggest:

1. Potassium, and by inference all lithophile trace elements, are depleted in the surficial rocks of the moon's eastern limb relative to the moon's central nearside. This suggests a similar depletion in the underlying crust and mantle.
2. Basalts in southern Mare Crisium are derived from two K and Ti depleted magmas; one almost a factor of two richer in MgO than the other.
3. The soil throughout the Luna 24 core may be modeled as a mixture of glass poor mare material (compositionally dominated by the higher MgO basalts) and glass rich highland material.

Introduction

Samples from the Soviet spacecraft Luna 24 allowed us not only to continue the tradition started by the Apollo Soil Survey (for example, Reid *et al.*, 1973) of performing electron microprobe analyses of regolith glass fragments and interpreting the provenance of those glasses, but also to gain insight about a region of the moon not previously investigated. Specifically, the Luna 24 landing site in Mare Crisium had the potential to shed light on another mare and another region of the moon's mantle.

By making comparisons between Luna 24 data and the data obtained from all other Apollo and Luna missions, we have observed that the entire eastern limb of the moon is systematically depleted in lithophile trace elements compared to the more central nearside of the moon. We interpret the observed depletion as not only a characteristic of the surface material, but also as a characteristic of the underlying lunar crust and mantle. Such a systematic lateral compositional heterogeneity must surely be related to the largest scale processes that affected the moon.

Data Sets

Modal analyses of the 90–150 μm fraction of each of six samples were performed by counting every grain in the 11 thin sections of grain mounts

Table 1. Luna 24 modal data for 90–150 μm size fraction

	24077, 49&50	24109, 44&45	24149, 53&54	24174, 42&43	24182, 41&42	24210, 47
Component			*Volume Percent*			
Glass spheres	.71	.54	.32	.19	0	.46
Glass fragments	3.26	3.90	2.16	1.66	2.47	1.86
Devitrified glasses	2.34	2.64	1.19	1.07	1.76	3.96
Heterogeneous glasses	1.53	1.64	1.40	1.17	.70	1.63
Agglutinates	24.71	17.59	10.17	8.90	6.00	10.72
Plagioclase	10.62	11.33	9.09	18.88	20.84	12.58
Pyroxene plus Olivine	21.44	24.93	28.67	26.99	30.14	23.07
Basalts	11.03	10.94	13.96	14.28	14.13	17.48
Crystalline matrix breccias	5.40	2.81	3.02	3.51	5.18	5.58
Vitric matrix breccias	17.97	23.14	28.67	22.99	18.49	22.14
Others	.91	.46	.86	.29	.23	.46
Total	99.92	99.92	99.51	99.93	99.94	99.94
Plagioclase/Mafic mineral	.50	.45	.32	.70	.69	.55
Total grains	979	1279	924	1022	849	429

available to our group; results are presented in Table 1. Each sample was derived from a different depth in the Luna 24 core. Modal analyses performed by Basu *et al.* (1977b) yield similar results.

All glass spheres and fragments, and glass in many agglutinate fragments, were analyzed for 12 elements using an ARL electron microprobe with a 15 μm spot. Analyses were corrected using a Bence-Albee correction procedure. An exception is the sulfur data that are not matrix corrected. About 300 analyses were attempted. After rejecting analyses with totals under 98.5 or over 101.5 we obtained 226 usable analyses that are plotted in Figs. 1–3. Most of the rejected analyses are either agglutinates or heterogeneous impact glass. Two cylindrical fragments that yield high Si analyses are rejected and are considered contamination; perhaps they are a fiber-like glass that was introduced during the sampling procedure on the lunar surface. A table of the 226 analyses is available from the authors.

Agglutinate fragments range from mare to highland compositions, do not form well-defined clusters, and are biased toward mare compositions relative to the glass spheres and fragments. The average agglutinate glass composition is given in Table 2. This composition is within experimental uncertainty of the average agglutinate value given by Hu and Taylor (1977). In fact, our overall distribution of agglutinate compositions is similar to that obtained by Hu and Taylor, even to the extent that the two populations distinguished by Hu and Taylor are observable in our data.

Glass spheres and fragments (other than agglutinates) were sorted into 3 major (many members) and 2 minor (few members) compositional clusters (Table 2) utilizing a new algorithm that initially considers each data point as a cluster and during each iteration combines the two clusters that are closest together in 12 dimensions. This procedure is similar to the dendogram technique as used by Higuchi and Morgan (1975). Clusters have been named following the terminology of previous soil surveys (Reid *et al.*, 1973).

Our glass fragment data are significantly different from the equivalent data obtained by Norman *et al.* (1977) and Basu *et al.* (1977a). Examination of the figures and consideration of the populations of each compositional cluster set out in Table 2 shows that our glasses are split about half and half between mare and highland compositions. This contrasts with Norman *et al.* who found that 70% of their glasses were of mare composition and Basu *et al.* who found that 58% of their glasses were of mare composition. We offer two explanations for this discrepancy. First, we performed our analyses on glasses from the 90–150 μm fraction whereas both Basu *et al.* and Norman *et al.* performed their analyses on glasses from the 150–250 μm fraction. Second, we included the brown glasses that are commonly heterogeneous with the agglutinates; both Basu *et al.* and Norman *et al.* considered the brown glasses as glass fragments. We lumped the brown glasses with the agglutinates because we wanted to segregate the obviously impact-produced glasses early in our data analysis scheme.

The differences in number and composition of glass clusters between this study and the work of Norman *et al.* and Basu *et al.* is due to the different methods each study used in defining compositional clusters.

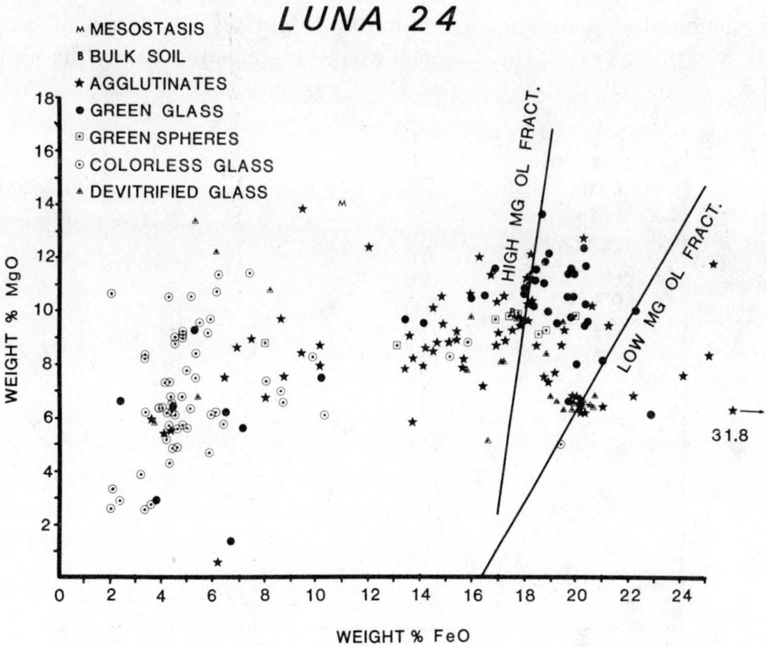

Fig. 1. MgO-FeO plot for glasses in Luna 24 soil. The two lines represent the trend of near surface olivine fractionation for the High and Low Mg basalts. We have used olivine compositions as appropriate from the work of Ryder *et al.* (1977).

Preliminary major element analyses of the <250 μm fraction of six Luna 24 samples were obtained on the MAC automated electron microprobe using the fused bead procedure of Brown (1977). Due to experimental difficulties this experiment was not completed. Our partial results demonstrated that there is no major element chemical stratigraphy in the Luna 24 core. Similar results were obtained by Blanchard *et al.* (1978) who also performed bulk major element analyses. For comparison we have included the average bulk analyses of Blanchard *et al.* in Table 2.

MODAL MAKEUP OF LUNA 24 SOIL

The Luna 24 soil does not have a simple history of *in situ* comminution of a mare basalt terrain as evidenced by differences among the bulk soil analysis, the average agglutinate analysis, and the average glass fragment analysis (Table 2). The nature of the differences are such that there is a greater abundance of highland materials in the average glass with respect to the average agglutinate, and also in the average agglutinate with respect to the bulk soil. The modal data presented below is necessary, but not sufficient, to help understand Luna 24 regolith development.

Glass Spheres

These chemically homogeneous spheroidal glasses are free of schlieren, inclusions, and phenocrysts: the majority are pale green and define the high MgO

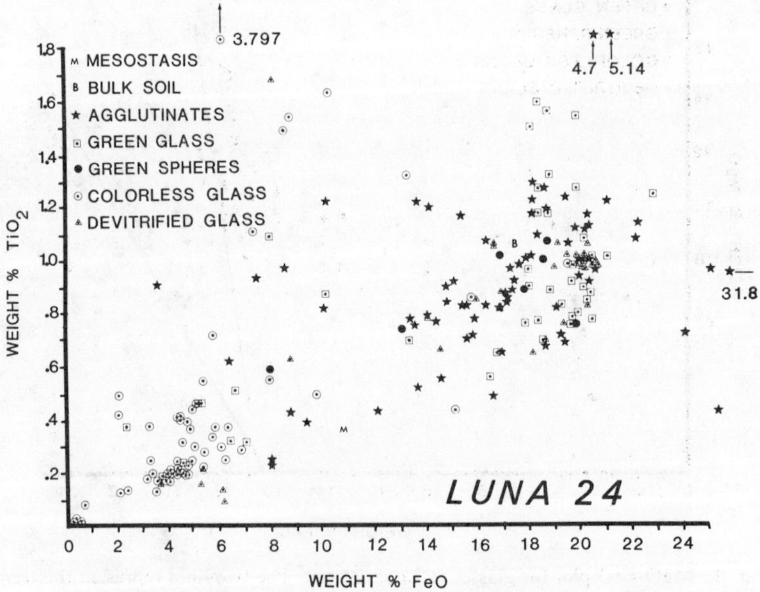

Fig. 2. TiO_2-FeO plot for glasses in Luna 24 soil.

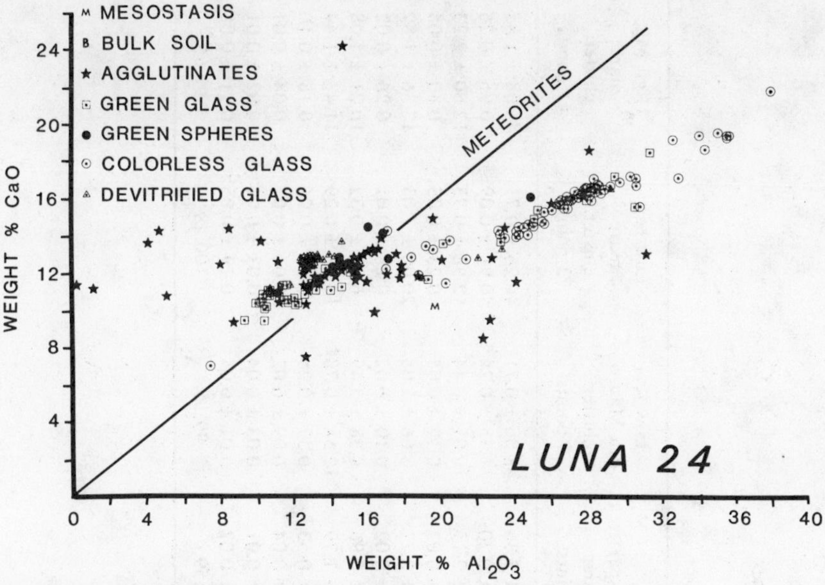

Fig. 3. CaO-Al_2O_3 for glasses in Luna 24 soil. The line labeled meteorite shows the constant CaO:Al_2O_3 ratio in chondritic meteorites.

mare basalt cluster. A few spheres are colorless and are part of the Highland Basalt cluster.

Glass Fragments

These angular, glass fragments are free of inclusions and phenocrysts: pale green fragments are schlieren-free, contain sub-μm troilite blebs, are chemically homogeneous under our 15 μm probe beam, and are part of the high MgO mare basalt cluster. Colorless fragments display some chemical heterogeneities under the probe beam although they do not contain optically identifiable schlieren. Colorless glasses define the feldspar and Highland Basalt cluster in Table 2.

Devitrified Glasses

These angular, brownish devitrified fragments have a variolitic texture consisting of olivine, oxides, clinopyroxene, and plagioclase in spherulitic to dendritic growth forms: these fragments define the tight (low MgO) mare basalt cluster.

Heterogeneous Glasses

These angular, schlieren-filled fragments have been grouped with the agglutinates in Figs. 1–3 and in the clustering procedures.

Table 2. Luna 24 chemical data.

	Average Bulk Soil*	Average Glass 226 points	Average agglutinate 71 points	Highland basalt cluster 67 points	Feldspar Cluster 8 points	Low K Fra Mauro cluster 5 points	Low Mg mare cluster 23 points	High Mg mare cluster 42 points
SiO_2	45.4	45.46 ± 1.73†	45.62 ± 1.44	45.02 ± 1.41	44.44 ± 3.36	48.36 ± 0.71	46.26 ± 0.71	45.08 ± 1.85
TiO_2	1.1	0.68 ± 0.39	0.83 ± 0.27	0.29 ± 0.15	0.13 ± 0.20	1.38 ± 0.24	0.98 ± 0.09	0.96 ± 0.26
Al_2O_3	11.1	18.93 ± 7.55	15.62 ± 5.68	26.44 ± 2.67	34.41 ± 2.17	19.07 ± 0.44	12.98 ± 0.73	12.60 ± 2.23
Cr_2O_3	0.52	0.24 ± 0.18	0.32 ± 0.16	0.11 ± 0.04	0.00 ± 0.02	0.12 ± 0.05	0.23 ± 0.05	0.40 ± 0.08
FeO	20.5	12.61 ± 7.05	15.87 ± 5.15	5.02 ± 1.67	1.42 ± 2.23	8.66 ± 1.08	20.00 ± 1.03	18.36 ± 1.87
MnO	—	0.18 ± 0.11	0.23 ± 0.09	0.06 ± 0.03	0.00 ± 0.02	0.10 ± 0.02	0.28 ± 0.01	0.26 ± 0.02
MgO	10.2	7.96 ± 2.81	8.50 ± 2.07	7.23 ± 2.40	0.61 ± 0.96	8.38 ± 2.49	6.47 ± 0.32	10.21 ± 1.08
CaO	11.0	13.47 ± 2.35	12.53 ± 1.54	15.44 ± 1.43	18.74 ± 1.79	12.58 ± 0.78	12.73 ± 0.29	11.43 ± 1.44
Na_2O	0.28	0.29 ± 0.22	0.31 ± 0.16	0.24 ± 0.17	0.78 ± 0.53	0.37 ± 0.25	0.29 ± 0.06	0.18 ± 0.10
K_2O	0.03	0.06 ± 0.04	0.07 ± 0.11	0.05 ± 0.04	0.06 ± 0.04	0.09 ± 0.02	0.04 ± 0.01	0.04 ± 0.01
P_2O_5	—	0.03 ± 0.03	0.04 ± 0.03	0.02 ± 0.03	0.00 ± 0.01	0.03 ± 0.04	0.03 ± 0.01	0.02 ± 0.01
S	—	0.12 ± 0.13	0.12 ± 0.17	0.04 ± 0.06	0.00 ± 0.02	0.01 ± 0.04	0.19 ± 0.05	0.11 ± 0.08
Total	100.1	100.05	100.13	99.96	100.59	99.15	100.48	99.65

*Bulk soil analyses from Blanchard *et al.* (1978).
†± uncertainties are 1 sigma.

Agglutinates

These compound fragments consist of mineral and lithic clasts bonded by a vesicular, flow-banded glass: the flow banding is defined by schlieren and sub-μm blebs of Fe-metal. Agglutinates display the largest variations among the six samples and are more abundant in the shallower samples.

Plagioclase

These are single mineral fragments of calcic plagioclase.

Pyroxene and Olivine

These are single mineral fragments: core and rim pairs have been analyzed for 112 grains and the results plotted Fig. 4. The average molar FeO/FeO + MgO is .50. About 13% of the pyroxenes and 26% of the olivines are more magnesian than the ranges for these minerals in the low MgO mare basalts as determined by Ryder *et al.* (1977) and indicated on Fig. 4. These magnesian grains may represent either the high MgO basalt series or highland materials. We note that the olivine histogram contains a significant peak at Fo_{75} which corresponds to Ryder *et al.*'s olivine vitrophyre and our high MgO basalt.

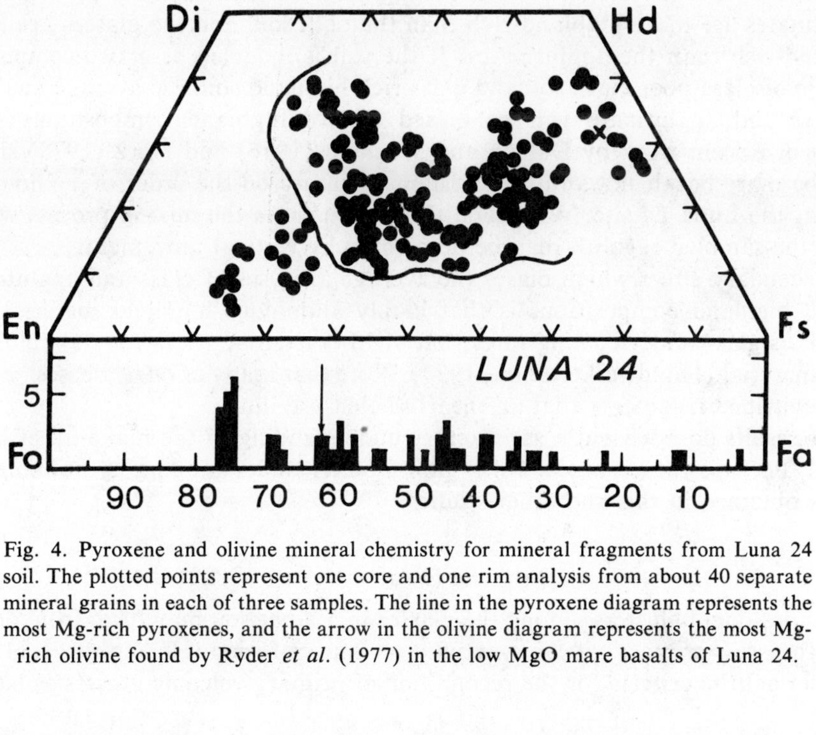

Fig. 4. Pyroxene and olivine mineral chemistry for mineral fragments from Luna 24 soil. The plotted points represent one core and one rim analysis from about 40 separate mineral grains in each of three samples. The line in the pyroxene diagram represents the most Mg-rich pyroxenes, and the arrow in the olivine diagram represents the most Mg-rich olivine found by Ryder *et al.* (1977) in the low MgO mare basalts of Luna 24.

Basalts

These lithic fragments have a "basaltic" texture.

Crystalline Matrix Breccias

These are polycrystalline, holocrystalline fragments with a clast-matrix structure and a range of matrix textures: these fragments are typical of the lunar highlands.

Vitric Matrix Breccias

These polymict fragments contain mineral, lithic, and glass fragments and irregularly shaped, μm-sized pores: these are the "soil breccias" typical of all lunar sites. Unlike the agglutinates, the abundance of these fragments does not vary systematically with depth in the Luna 24 core.

Implications for Luna 24 Regolith Development

Three modal observations help explain the differences among the average soil, the average glass, and the average agglutinate analyses: (1) the glasses, including the agglutinates, do not constitute a major portion of the soil, thus they could be a transported component; (2) the modal abundance of the glasses has a positive correlation with the modal abundance of agglutinates suggesting that the bulk of the glasses are impact-produced and may be considered as a subset of the agglutinates; and (3) the differences among the average analyses is such that the agglutinates are more highland-rich than the bulk soil, and the glasses are more highland-rich than the agglutinates. If the soil is modeled as a two-component mixture of glass-poor mare soil and glass-rich highland soil, the average analyses of glass and agglutinate will be biased toward highland compositions as is observed. Recent work by DeHon and Waskom (1976) and Hörz (1978) shows that the mare basalt flows in the Crisium basin are on the order of 1 kilometer thick at the Luna 24 site. With such thin basalt flows the mixing process which forms the sampled regolith may be dominated by vertical movements.

A secondary effect which biases the average analyses of glass and agglutinate toward highland compositions is that highly aluminous highland material will remain as glass under cooling rates that yield crystalline rocks for the high iron mare material (Handwerker *et al.*, 1977). Note that many of our "glasses" are in fact devitrified fragments that are nearly holocrystalline.

Our results do not yield a satisfactory understanding of the make-up of Luna 24 soil. Perhaps future workers will gain a better understanding by utilizing the results obtained in this and other studies.

SIGNIFICANCE OF THE COMPOSITIONAL CLUSTERS

Compositional clusters among the glasses in a lunar soil may form from one of two processes: primary volcanic activity or impact fusion. Heiken *et al.* (1974) have set out the criteria for the recognition of primary volcanic glasses as having homogeneous compositions, no inclusions, spherical shapes (at least for low viscosity mafic compositions), and low trace siderophile element contents. Simonds *et al.* (1976) have set out the criteria for the recognition of impact-fused glasses as being schlieren and clast-bearing, compositionally heterogeneous on the μm to mm scale, and enriched in trace siderophile elements. However, on the scale of a few grams or centimeters, impact melts are quite homogeneous.

The small compositional scatter among the devitrified glasses and the internal homogeneity and regular shapes of the pale green-glass spheres suggest that these two groups represent primary volcanic liquids. To the contrary, Basu *et al.* (1977a) in a SEM study of these spheres do not find surface features characteristic of lava fire fountaining, and do find iron droplets and trains that they suggest are indicative of an impact-produced origin. The bulk of the green glass fragments, however, display a wider compositional range and are probably impact-produced. Preservation of the compositional cluster probably results from impacts into the local terrane which must have provided a fairly compositionally pure target.

The highland basalt cluster has been observed in samples collected at every landing site (Warner *et al.*, 1972). Its schlieren and other compositional heterogeneities suggest that the Highland Basalt glasses are impact-fused. The ubiquity of the Highland Basalt cluster and its similarity to the highland composition measured from orbit suggest that it is the grand compositional average of the lunar highland crust surrounding each landing site. The small differences in Highland Basalt composition among landing sites are a measure of the differences in the composition of the lunar crust at each site.

Mare Basalts

Two different mare basalt compositional clusters, low MgO and high MgO, have been defined from among the Luna 24 glass spheres and fragments. Both groups contain less TiO_2 (ca. 1.0 wt.%) than has been reported from individual mare basalt rock samples collected at Apollo landing sites, although the Apollo 15 green glasses have even lower TiO_2 contents (ca. 0.4 wt.%, Reid *et al.*, 1973). The Luna 24 basalt clusters are compositionally similar, but not identical, to the "very low TiO_2" basalts discovered at Apollo 17 by Vaniman and Papike (1977a) and Taylor *et al.* (1977). The low MgO (6.5 wt.%) group is compositionally well defined and consists of spherulitic to dendritic-textured devitrified glass fragments. The high MgO (ca. 10 wt.%) group is compositionally more diffuse and consists of pale green glass spheres and fragments. The low MgO group is identified with the bulk of the basaltic lithic fragments discovered in the Luna 24 soil by Vaniman and Papike (1977b), Taylor *et al.* (1977), and Ryder *et al.* (1977). The high MgO group is not correlated with any lithic fragments thus far identified in the Luna 24 soil, but its composition does approach the very MgO-enriched olivine vitrophyre fragments described by Ryder *et al.* (1977). Both mare basalt groups are believed to represent, at least in part, volcanic magmas rather than impact-formed melts (see preceding discussion).

Coish and Taylor (1977) suggested that the two mare basalt clusters are derivable one from the other by near surface crystal fractionation. The experimental petrology of Grove and Vaniman (1977) shows that plagioclase is the liquidus phase for these basalts, but the normative feldspar content of the two basalt types is the same (34.7 vs. 35.2 wt.%) which suggests that fractionation of feldspar is not important. The major normative difference between the two basalts is the olivine content (6.9 wt.% for the low Mg type vs. 15.7 wt.% for the high Mg type) which suggests that perhaps olivine is the critical phase in a fractionation scheme. On Fig. 1 we have sketched olivine fractionation lines using olivine compositions determined by Ryder *et al.* (Fo_{75} for the high MgO group and Fo_{61} for the low MgO group). It is apparent in Fig. 1 that simple olivine fractionation cannot derive one basalt type from the other.

POTASSIUM DEPLETION

Low lithophile element contents (specifically Th) on the moon's eastern limb have been measured by the Apollo orbital gamma-ray experiment (Parker *et al.*, 1977; Bielefeld *et al.*, 1976). This and similar orbital data have led Hubbard and Woloszyn (1977) to suggest that the eastern limb of the moon is compositionally distinct from the moon's central nearside. Note that the Th contents measured for Luna 24 soil (about 0.4 ppm, Blanchard *et al.*, 1977) are well below the Th contents calculated from the orbital data (about 1.8 ppm for Mare Crisium).

Whereas the orbital and bulk soil analyses show that the surface soils are potassium depleted, the Luna 24 glass data demonstrates that this depletion is a characteristic feature of both highland and mare constituents of the soil. The most K-rich cluster (Low-K Fra Mauro cluster in Table 2) which accounts for only 5 of the 226 glasses is similar in composition to many impact-melt rocks of Apollo 16 and 17, except that K_2O in the Luna 24 glasses is depleted by a factor of 3. The three major clusters (Highland Basalt and low and high MgO mare basalt) all contain less than 0.05 wt.% K_2O. It may be argued that the low lithophile trace element abundance in Luna 24 basaltic materials is due to a high degree of partial melting of the source material when the basalts were generated. Such a process would produce Mg-enriched basalts, but not even the high Mg basalts of Luna 24 are particularly Mg-enriched. Rather, Luna 24 basalts are Fe-enriched (and lithophile element poor at the same time) which indicates that they were produced by small amounts of partial melting. Some other explanation must therefore be sought for the low lithophile trace element abundances.

Figure 5, a series of histograms of the K_2O content of glasses in all Luna and Apollo soil samples, demonstrates that the potassium depletion displayed by Luna 24 is shared by Luna 16 and 20 and to a lesser extent by Apollo 11, 16, and 17. Not only are the Luna soils depleted in K_2O, but Fig. 5 suggests that a K-rich constituent is missing—there is essentially no KREEP contained in the Luna soils. This depletion in trace elements in the basaltic constituents of the Luna 24 soil is supported by trace element data of Blanchard *et al.* (1978) and by the Rb-Sr data of Nyquist *et al.* (1977). The lack of a KREEP component is supported by plagioclase mineral analyses as pointed out by Papp *et al.* (1977).

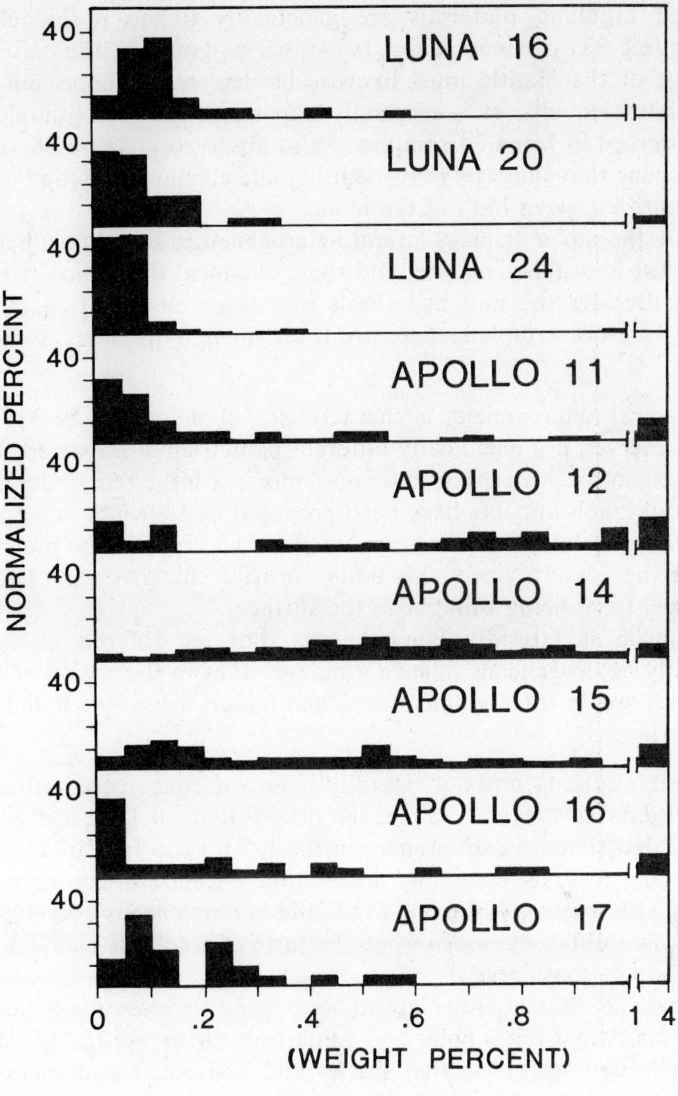

Fig. 5. Histograms of the K_2O content in individual glasses analyzed from the soil at various landing sites. The data are from work performed by the Apollo Soil Survey in the early 1970's. Each histogram is normalized so that the shaded area is equal within each histogram.

Speculations Concerning the Eastern Limb Trace Element Depletion

A key question to consider is the significance of the lithophile trace element depletion in bulk Luna 24 soil analyses and in analyses of individual Luna 24 glass fragments. Because both mare and highland glasses are depleted in potassium and by inference all trace lithophile elements, and because the sources of mare and highland materials are genetically linked in models of lunar petrogenesis (e.g., Taylor and Jakes, 1974), we suggest that the entire crust and at least some of the mantle must likewise be depleted in potassium and other trace lithophile elements. It is especially important that this lithophile element depletion observed in Luna 24 samples is also observed in Luna 16 and Luna 20 samples, because that suggests that the lithophile element depletion is characteristic of the entire eastern limb of the moon.

Apparently the moon displays lateral heterogeneities in the chemical composition of its crust and upper mantle, and these chemical differences have survived for 4.6 A.E. Besides the null hypothesis that trace element-rich materials are present but were not sampled, there are three other hypotheses worth considering:

1. The lateral heterogeneity is characteristic of the moon. It may be formed and preserved if a chemically-different planetesimal impacted the moon at a low enough velocity that it did not "mix" with the remainder of the lunar material (such impacts have been proposed by Safronov, 1969).
2. The moon is laterally homogeneous, but the crust of the moon is thicker under the eastern limb; this added crustal thickness has kept KREEP material from being brought to the surface.
3. The moon is laterally homogeneous, but its differentiation has been laterally heterogeneous in such a manner that in the region of the eastern limb lithophile trace element enriched material has not found its way to the surface.

There are tantalizing hints of laterally heterogeneous accretion on the moon from several data sets. For example, the orbital data of Bielefeld *et al.* (1976) demonstrate that there are differences in the K:Th ratio for different regions of the moon. This ratio is especially interesting because it contrasts a volatile element (K) with a less volatile one (Th), and heterogeneous accretion would be expected to yield different values when the ratio of a volatile element with a less volatile element is considered.

Another data set that suggests lateral heterogeneous accretion is the potassium and Sr-isotopic data from Apollo and Luna soils (Murthy, 1975). Murthy was concerned with demonstrating that the K- and Sr-isotopic contents of each soil are dominated by a Fra Mauro component that was dispersed about the moon. The logical extension of Murthy's arguments is that parts of the moon are lithophile trace element poor and thus the moon's crust is laterally chemically heterogeneous.

We have evaluated the viability of a laterally heterogeneous accretion model by means of a K/Sm vs. Sm plot shown in Fig. 6. This is equivalent to the K:Th ratio in the context of a volatile-less volatile element comparison. In addition to the Luna 24 data on Fig. 6 we have plotted a broad spectrum of lunar samples (including both rocks and soils) from all Apollo missions, a selection of terrestrial basalts, and the average composition of chondritic meteorites. Note that the Luna 24 data have the same K:Sm ratio as is typical for other lunar samples,

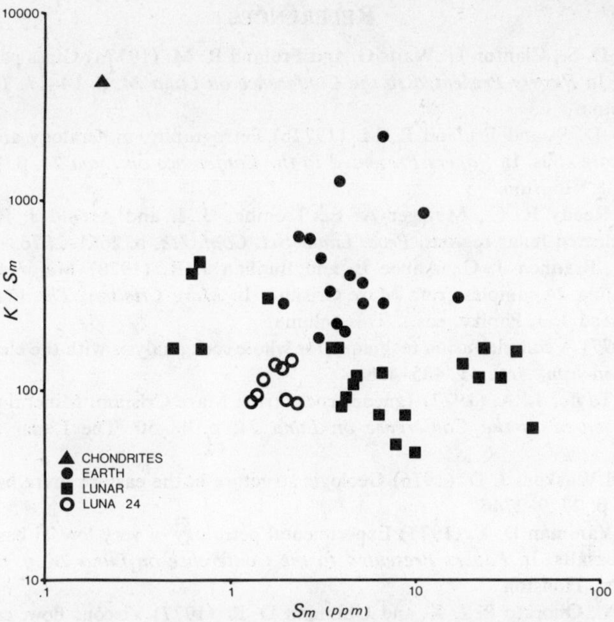

Fig. 6. K:Sm vs. Sm diagram for Luna 24 bulk soil analyses (data from the Luna 24 Conference abstracts), a broad range of lunar soils and rocks from all Apollo landing sites, a selection of terrestrial basalts from Kay, Hubbard and Gast (1969), and the average of chondritic meteorites.

although the absolute amounts are lower than other basalts, but higher than plutonic rocks. This similarity in the K:Sm ratio argues against laterally heterogeneous accretion.

The second hypothesis is based on the inference that KREEP is derived in some manner from a sub-crustal shell of material (see discussions in Warner *et al.*, 1977; McKay and Weill, 1977). It has been suggested that if such a situation exists, then only the largest impacts would have been capable of excavating KREEP. In an area with a thicker crust and none of the largest impacts, a lack of KREEP would be predicted. This is related to the potassium depletion since KREEP is the lunar component that carries most of the potassium and other lithophile trace elements. Such a lack of KREEP would dictate a lack of potassium and other trace lithophile elements in the surface materials.

There is no evidence that the third hypothesis is correct. The eastern limb displays basalts that probably originate at depths similar to the depths of origin of other lunar basalts. At least some of the eastern limb basalts should be enriched in lithophile trace elements.

Acknowledgments—One of the suthors (CHS) was a staff scientist at the Lunar Science Institute which is operated by the Universities Space Research Association under Contract NSR 09-051-001 with the National Aeronautics and Space Administration. This paper constitutes Lunar and Planetary Institute Contribution No. 318.

REFERENCES

Basu A., McKay D. S., Clanton U.,Waits G. and Fruland R. M. (1977a) Glass particles in Luna 24 drill core soils. In *Papers Presented to the Conference on Luna 24*, p. 14–17. The Lunar Science Institute, Houston.

Basu A., McKay D. S. and Fruland R. M. (1977b) Petrography, mineralogy and source rocks of Luna 24 drill core soils. In *Papers Presented to the Conference on Luna 24*, p. 18–21. The Lunar Science Institute, Houston.

Bielefeld M. J., Reedy R. C., Metzger A. E., Trombka J. I. and Arnold J. R. (1976) Surface chemistry of selected lunar regions. *Proc. Lunar Sci. Conf. 7th*, p. 2661–2676.

Blanchard D. P., Brannon J. C., Aaboe E. and Budhan J. R. (1978) Major and trace element chemistry of Luna 24 samples from Mare Crisium. In *Mare Crisium: The View From Luna 24* (R. B. Merrill and J. J. Papike, eds.). This volume.

Brown R. W. (1977) A sample fusion technique for whole rock analysis with the electron microprobe. *Geochim. Cosmochim. Acta* **41**, 435–438.

Coish R. A. and Taylor L. A. (1977) Igneous rocks from Mare Crisium: Mineralogy and petrology. In *Papers Presented to the Conference on Luna 24*, p. 48–50. The Lunar Science Institute, Houston.

DeHon R. A. and Waskom J. D. (1976) Geologic structure in the eastern mare basins. *Proc. Lunar Sci. Conf. 7th*, p. 2729–2746.

Grove T. L. and Vaniman D. T. (1977) Experimental petrology of very low Ti basalts and origin of Luna 24 ferrobasalts. In *Papers Presented to the Conference on Luna 24*, p. 68–71. The Lunar Science Institute, Houston.

Handwerker C. A., Onorato P. I. K. and Uhlmann D. R. (1977) Viscous flow, crystal growth and glass formation of highland and mare basalts from Luna 24. In *Papers Presented to the Conference on Luna 24*, p. 75–77. The Lunar Science Institute, Houston.

Heiken G., McKay D. S. and Brown R. W. (1974) Lunar deposits of possible pyroclastic origin. *Geochim. Cosmochim. Acta* **38**, 1703–1718.

Higuchi H. and Morgan J. W. (1975) Ancient meteoritic component in Apollo 17 boulders. *Proc. Lunar Sci. Conf. 6th*, p. 1625–1651.

Hörz. F. (1978) How thick are lunar mare basalts? (abstract) In *Lunar and Planetary Science IX*, p. 540–542. Lunar and Planetary Institute, Houston.

Hu H.-N. and Taylor L. A. (1977) Luna 24 soils from Mare Crisium: Agglutinate chemistry (abstract). In *Papers Presented to the Conference on Luna 24*, p. 82–84. The Lunar Science Institute, Houston.

Hubbard N. J. and Woloszyn D. (1977) Re-evaluation of the Apollo orbital X-ray fluorescence data. *Proc. Lunar Sci. Conf. 8th*, p. 909–923.

Kay R., Hubbard N. J. and Gast P. W. (1970) Chemical characteristics and origin of oceanic ridge volcanic rocks. *J. Geophys. Res.* **75**, 1585–1613.

McKay G. A. and Weill D. F. (1977) KREEP petrogenesis revisited. *Proc. Lunar Sci. Conf. 7th*, p. 2339–2355.

Murthy V. R. (1975) The source and origin of the exotic components and KREEP-rich material on the moon. *The Moon* **13**, 111–119.

Norman M., Coish R. A. and Taylor L. A. (1977) Glass chemistry and magma evolution at Mare Crisium (abstract). In *Papers Presented to the Conference on Luna 24*, p. 136–138. The Lunar Science Institute, Houston.

Nyquist L. E., Wiesmann H., Bansal B., Wooden J., McKay G. and Hubbard N. (1977) Chemical and Sr-isotopic characteristics of the Luna 24 samples (abstract). In *Papers Presented to the Conference on Luna 24*, p. 139–142. The Lunar Science Institute, Houston.

Papp H., Steele I. M. and Smith J. V. (1977) Luna 24 90–150 micron fraction: Implications for sampling of planetary and asteroidal regoliths (abstract). In *Papers Presented to the Conference on Luna 24*, p. 143–146. The Lunar Science Institute, Houston.

Parker R. E., Haines E. L. and Metzger A. E. (1977) Thorium concentrations in the lunar surface (abstract). In *Lunar Science VIII*, p. 756–758. The Lunar Science Institute, Houston.

Reid A. M., Warner J. L., Ridley W. I. and Brown R. W. (1973) Luna 20 soil: Abundance and composition of phases in the 45–125 micron fraction. *Geochim. Cosmochim. Acta* **37**, 1011–1030.

Ryder G., McSween H. Y. and Marvin U. B. (1977) Basalts from Mare Crisium. *The Moon* **17**, 263–287.

Safronov V. S. (1969) Evolution of the protoplanetary cloud and formation of the Earth and planets. Nauka, Moscow. Translated for NASA, 1972, NASA TTF 667.

Simonds C. H., Warner J. L. and Phinney W. C. (1976) Thermal regimes in cratered terrain with emphasis on the role of impact melt. *Amer. Mineral.* **61**, 569–577.

Taylor G. J., Keil K. and Warner R. D. (1977) Very low-Ti mare basalts. *Geophys. Res. Lett.* **4**, 207–210.

Taylor S. R. and Jakeš P. (1974) The geochemical evolution of the Moon. *Proc. Lunar Sci. Conf. 5th*, p. 1287–1305.

Vaniman D. T. and Papike J. J. (1977a) Very Low Ti (VLT) basalts: A new mare rock type from the Apollo 17 drill core. *Proc Lunar Sci. Conf. 8th*, p. 1443–1471.

Vaniman D. T. and Papike J. J. (1977b) Ferrobasalts from Mare Crisium: Luna 24. *Geophys. Res. Lett.* **4**, 497–500.

Warner J. L., Phinney W. C., Bickel D. E. and Simonds C. H. (1977) Feldspathic granulitic impactites and pre-final bombardment lunar evolution. *Proc. Lunar Sci. Conf. 8th*, p. 2051–2066.

Warner J. L., Reid A. M., Ridley W. I. and Brown R. W. (1972) Major element composition of Luna 20 glasses. *Earth Planet. Sci. Lett.* **17**, 7–12.

Merrill, R. B. and Papike, J. J., editors.
(1978) *Mare Crisium: The View from Luna 24*, Pergamon Press (New York), p. 245–264
Printed in the United States of America

Luna 24: 90–150 micrometer fraction: Implications for remote sampling of regolith

H. A. PAPP, I. M. STEELE and J. V. SMITH

Department of the Geophysical Sciences, University of Chicago, Chicago, Illinois 60637

Abstract—Electron-microprobe analyses were made of all mineral grains and 61 glasses in 12 sections of 150–90 μm fines. Polymineralic grains contain Fe-rich olivine, pyroxene, plagioclase, spinel and ilmenite which match those in "gabbroic" and basaltic fragments from coarser fines described by other workers. Monomineralic grains contain an additional population of Mg-rich olivine (~0.7 <*mg* <0.8) and pyroxene, which are attributed to Mg-rich basalts or gabbros not represented by rock fragments in the Luna 24 core. Minor elements indicate that the host rocks have mare-basalt chemistry. The Mg-rich suite is tentatively assigned to deep basin material (unit 1) identified by Adams *et al.* (1978), and the Fe-rich suite is assigned to later mare fill (unit 2). The units may be related mainly by early cumulation of olivine phenocrysts, thereby explaining the low *mg* of the Luna 24 basaltic and gabbroic rock fragments.

Most of the glasses are related to the mare basalts or local bulk regolith. Non-mare material is represented by glasses of "highland basalt" ("anorthositic gabbro") composition, and by glasses with general resemblance to the Low-K-Fra-Mauro composition. No evidence for high-KREEP and ultrabasic material was found.

The relative value of different types of samples for characterization of a planetary regolith is discussed. From the viewpoint of mineralogy and petrology, a single grab sample at Luna 24 would have been as useful as the core, and a series of grab samples on a long traverse might have been more useful. The fine material (<250 μm) in the Luna 24 core was more informative than the coarser material about the rock types in the Mare Crisium region. Implications for further missions to the moon, asteroids and Mars are briefly discussed.

INTRODUCTION

We deliberately requested samples of the most fine-grained material from the Luna 24 core in order to: (a) determine if the key mineralogical features could be derived from a minimum of material; and (b) obtain pointers to planning automatic collection of samples from planetary and asteroidal regoliths. The data reported briefly by Papp, Steele and Smith (1977) were compared with those reported for several types of core sample at the Conference on Luna 24 and in published papers, and an abstract of the present paper was submitted to the Ninth Lunar and Planetary Science Conference (Smith, Steele and Papp, 1978).

SAMPLE DESCRIPTION AND TECHNIQUE

Twelve polished thin sections were provided of the 90–150 μm fraction from core depths of 77, 109, 149, 174, 182 and 210 cm. Approximately 500 fragments occur in each section, and most are monomineralic grains. Proportions of

Table 1. Samples and modes.

Section	% Pyx	% Plag	% 01	% Other*
**77-51	41	22	12	25
77-52	36	17	9	38
109-46	38	16	8	38
109-47	35	20	11	34
149-55	42	23	10	25
149-56	49	24	15	12
174-44	38	26	8	28
174-45		not counted		
174-46	41	19	12	28
182-43	50	18	12	20
182-44	56	20	12	12
210-49	38	18	15	29
Mean	43	20	11	26

*Others = ilmenite, spinel, glass, silica, agglutinates.
**First numbers are core depth in cm.

pyroxene (43%), plagioclase (20%), and olivine (11%), (Table 1) are similar regardless of depth. The remaining 26% is dominated by agglutinates and glasses.

To obtain a thorough survey of the mineral grains, nearly every one was analyzed for three major cations (pyroxenes, olivine; Ca, Mg, Fe: plagioclase, Na, Ca, K) using a special program (PLOT) for the ARL electron microprobe in which 10-second analyses were made with three fixed spectrometers (15 kV, 0.4 μA beam current). Spectrometer analyses (CHICAGO 1977 system; accuracy 1–5% of major elements; detection level of minor elements ~0.02 wt.% (1σ) were made for: (i) all Mg-rich olivine and pyroxene grains plus selected Fe-rich grains; (ii) 61 selected glasses; and (iii) 75 random plagioclase grains. All minerals in 68 polymineralic grains were analyzed with an energy-dispersive detector (CHICAGO 1977 system). Two analyses were made on most grains, but zoning was usually not indicated. Spectrometer analyses were obtained for 7 spinel, 9 ilmenite and 2 metal grains. Complete lists of analyses can be obtained from J. V. Smith, and selected analyses are given here (Tables 2–6; pyroxene, olivine, plagioclase, opaque minerals, glass).

MINERAL DESCRIPTION

Pyroxene. Figs. 1a-*l* of Papp, Steele and Smith (1977) demonstrate similar Ca-Mg-Fe distributions for monomineralic grains in all twelve thin sections, and two are repeated here as Fig. 1a,b. The *major* concentration range of these grains (Fig. 1d, dashed line) extends from *mg* [= Mg/(Mg + Fe)] ~0.77 to *mg* ~0 with a strong tendency for the wollastonite molecule to lie between 10 and 40%. There is a weak trend towards hedenbergite. Rare analyses in the twelve plots of Figs. 1a-*l* (Papp, Steele and Smith, 1977) show Ca-poor compositions near

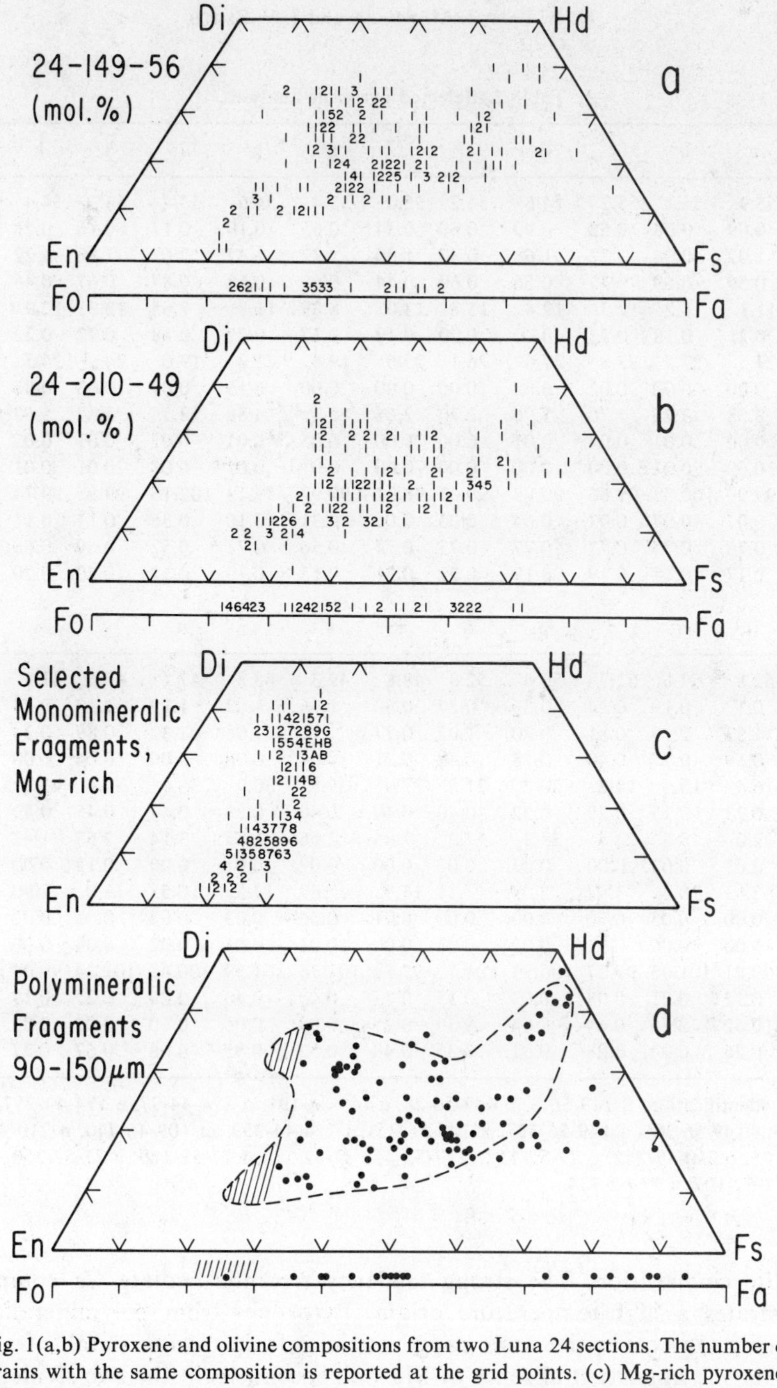

Fig. 1(a,b) Pyroxene and olivine compositions from two Luna 24 sections. The number of grains with the same composition is reported at the grid points. (c) Mg-rich pyroxenes collected from monomineralic fragments in all twelve sections (A = 10, B = 11, etc.) (d) Pyroxene and olivine compositions from polymineralic fragments in all sections. Dashed envelope encompasses most of the analyses for monomineralic pyroxene fragments, while shaded areas indicate pyroxene and olivine compositions absent on polymineralic fragments (except of one olivine).

Table 2. Selected pyroxene analyses.

Label	a	b	c	d	e	f	g	h	i	j	k	l
SiO_2	55.9	56.1	55.2	54.6	54.2	55.0	53.4	54.5	53.4	54.6	54.4	52.5
TiO_2	0.09	0.08	0.62	0.90	0.60	0.34	0.65	0.10	0.17	0.16	0.28	0.42
Al_2O_3	1.02	0.96	1.32	1.68	0.87	1.00	1.77	1.47	2.67	1.88	1.22	1.21
Cr_2O_3	0.59	0.59	0.43	0.56	0.60	0.44	0.63	0.68	0.87	0.67	0.46	0.51
FeO	11.1	11.2	12.3	12.4	13.8	13.0	8.39	11.8	7.65	12.3	12.9	14.7
MnO	0.21	0.18	0.23	0.20	0.20	0.19	0.17	0.23	0.18	0.22	0.22	0.26
MgO	27.5	27.5	27.8	27.6	26.1	25.6	19.6	25.3	19.0	24.6	24.5	23.5
NiO	0.00	0.00	0.02	0.02	0.00	0.00	0.00	0.15	0.09	0.03	0.06	0.02
CaO	3.58	3.75	1.70	1.96	2.70	2.65	15.3	4.88	18.1	5.47	5.67	5.64
Na_2O	0.00	0.01	0.04	0.01	0.00	0.00	0.05	0.01	0.02	0.01	0.02	0.01
K_2O	0.00	0.01	0.00	0.00	0.00	0.00	0.01	0.01	0.01	0.01	0.01	0.00
Sum	99.99	100.38	99.66	99.93	99.07	98.22	99.97	99.13	102.16	99.95	99.74	98.77
Ca	0.07	0.07	0.03	0.04	0.05	0.05	0.31	0.10	0.36	0.11	0.11	0.11
Mg	0.76	0.75	0.77	0.77	0.73	0.74	0.56	0.71	0.52	0.69	0.68	0.67
Fe	0.17	0.17	0.19	0.19	0.22	0.21	0.13	0.19	0.12	0.20	0.20	0.22

	m	n	o	p	q	r	s	t	u	v	w	x
SiO_2	52.8	52.6	52.7	53.6	52.8	48.8	49.3	48.8	47.2	47.6	47.5	48.2
TiO_2	0.17	0.19	0.24	0.36	0.27	0.60	1.16	1.02	1.19	1.13	0.24	0.37
Al_2O_3	1.53	2.19	0.81	0.70	1.02	0.77	1.11	1.08	1.32	0.89	0.24	0.25
Cr_2O_3	0.64	0.77	0.40	0.18	0.48	0.21	0.04	0.06	0.00	0.04	0.00	0.00
FeO	16.4	15.6	18.2	20.2	21.5	27.0	29.5	30.9	36.6	37.4	41.3	46.0
MnO	0.23	0.22	0.29	0.32	0.40	0.40	0.40	0.33	0.47	0.45	0.42	0.63
MgO	22.0	19.6	22.8	23.3	17.2	7.85	2.56	1.85	3.14	2.53	1.42	0.81
NiO	0.01	0.03	0.00	0.04	0.03	0.00	0.02	0.01	0.00	0.15	0.00	0.00
CaO	5.63	10.4	3.92	2.23	7.31	11.9	17.6	17.5	10.5	11.0	7.90	6.39
Na_2O	0.00	0.05	0.00	0.00	0.03	0.01	0.06	0.03	0.04	0.05	0.02	0.01
K_2O	0.00	0.00	0.01	0.00	0.03	0.00	0.01	0.01	0.02	0.00	0.00	0.01
Sum	99.41	101.65	99.37	100.93	101.7	97.54	101.76	101.59	100.48	101.24	99.04	102.67
Ca	0.12	0.21	0.08	0.04	0.15	0.27	0.40	0.40	0.24	0.25	0.19	0.15
Mg	0.62	0.55	0.64	0.64	0.50	0.25	0.08	0.06	0.10	0.08	0.05	0.03
Fe	0.26	0.24	0.28	0.31	0.35	0.48	0.52	0.55	0.66	0.67	0.77	0.93

Grain identification: a 149-56-8, b 149-56-28, c 149-56-191, d 174-44-77, e 174-44-257, f,g 149-56-131, h,i 149-56-300, j 149-56-327, k 149-56-351, l 174-44-259, m 109-46-440, n 210-49-327, o 210-49-105, p 210-49-222, q 77-52-113, r 77-52-4, s 77-52-180, t 77-52-159, u 77-52-350, v 77-52-161, w 77-52-107, x 77-52-734.

ferrosilite composition. The strong tendency for intermediate Ca composition demonstrates a high-temperature origin. Pyroxenes from polymineralic fragments (Fig. 1d) cover the same range as those from monomineralic fragments except for the important absence of Mg-rich specimens in the two shaded areas. Fig. 1c collects the Mg-rich pyroxenes from all twelve sections, and demonstrates: (i) a maximum value of 0.81 for *mg*; and (ii) only a weak separation into Ca-rich and Ca-poor varieties, again indicative of high-temperature crystallization.

Analyses of monomineralic grains in 20–250 μm fractions by Basu *et al.* (1977a), in 150–250 μm fractions by Coish and Taylor (1977), in 150–500 μm fines by Grove and Bence (1977), and in 250–500 μm fines by Taylor *et al.* (1977a) show both the dominant Fe-rich population and the minor Mg-rich population, but Ryder *et al.* (1977) did not find the Mg-rich population in their 250–500 μm monomineralic fragments. Although our polymineralic fragments are too small to yield reliable information on textures, the pyroxenes have compositions which closely match those of basaltic and "gabbroic" fragments of larger grain size reported by Coish and Taylor (1977), Nielsen and Drake (1977), Tarasov *et al.* (1977), and Vaniman and Papike (1977b).

The obvious interpretation of these composition data is that the polymineralic and most of the Fe-rich monomineralic grains derive from high-temperature basaltic and gabbroic rocks at the Luna 24 site, and that there is an additional suite of Mg-rich pyroxenes not represented in polymineralic or rock fragments.

Plots of Ti and Cr vs. Al further characterize the pyroxenes (Fig. 2). Our Fe-rich suite gives a distribution which roughly matches the zoning trends reported by Grove and Bence (1977) for Luna 24 "gabbroic" fragments. Our Mg-rich suite neatly extends the trend for the Mg-rich end of the "gabbroic" suite, and suggests a related origin of the Mg-rich and Fe-rich suites.

Olivine. Compositions of the monomineralic grains are bimodal (Fig. 3) with an Fe-rich suite (mg < ~0.65) which matches olivines from basaltic-gabbroic

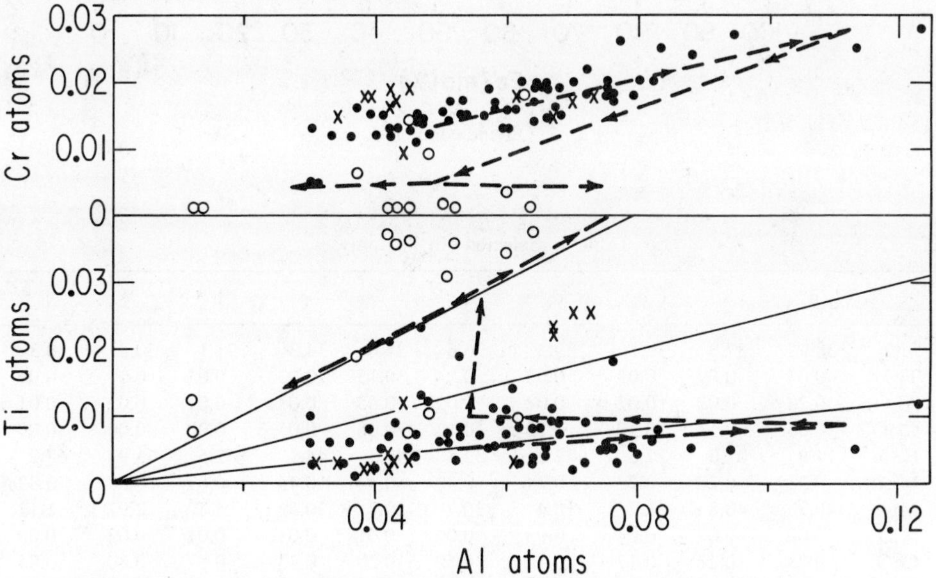

Fig 2. Cr and Ti vs. Al for monomineralic grains of pyroxene. Concentrations referred to six oxygen atoms. The three symbols distinguish pyroxenes with En >70 (cross), <50 (circle), and >50–<70 (dot). Zoning trends (dashed lines) from Grove and Bence (1977, Fig. 2b).

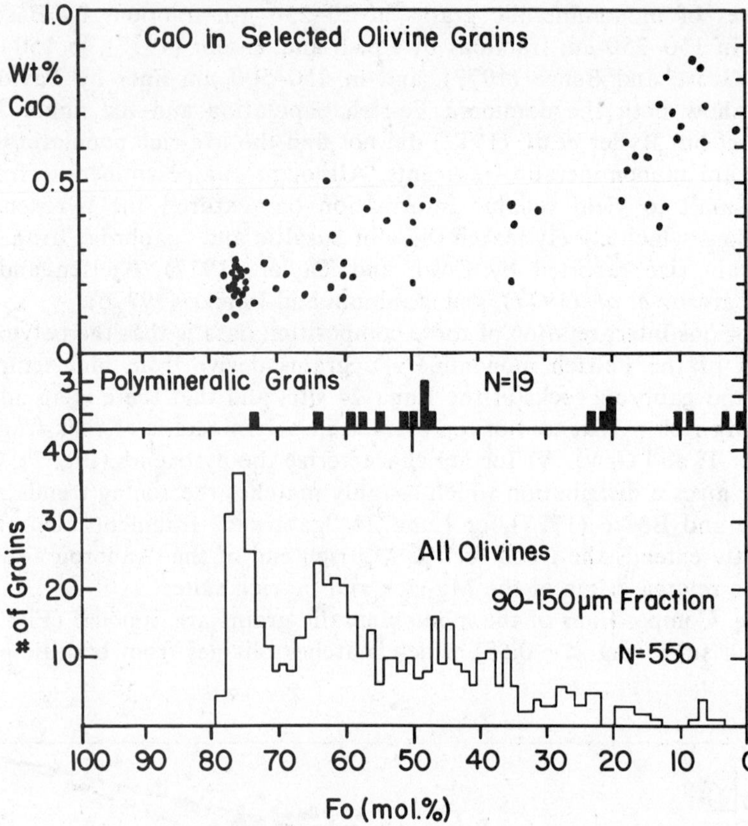

Fig. 3. Olivine compositions.

Table 3. Selected olivine analyses.

	a	b	c	d	e	f	g	h	i	j
SiO_2	38.9	37.8	38.1	39.3	36.6	35.0	34.3	31.1	30.8	30.3
TiO_2	0.03	0.02	0.03	0.03	0.04	0.15	0.03	0.07	0.11	0.09
Al_2O_3	0.24	0.02	0.02	0.02	0.02	0.03	0.01	0.01	0.02	0.01
Cr_2O_3	0.06	0.14	0.10	0.10	0.13	0.08	0.05	0.00	0.00	0.00
FeO	17.9	21.0	22.6	22.2	32.8	43.3	44.1	62.5	63.9	67.2
MnO	0.23	0.22	0.22	0.24	0.40	0.44	0.44	0.67	0.68	0.83
MgO	40.7	40.8	38.2	38.4	30.7	20.7	20.5	5.82	2.97	1.12
NiO	0.00	0.02	0.04	0.03	0.02	0.01	0.01	0.01	0.03	0.00
CaO	0.28	0.22	0.17	0.18	0.50	0.20	0.38	0.57	0.80	0.84
Sum	98.34	100.24	99.48	100.50	101.21	99.91	99.82	100.75	99.31	100.39
mg	0.80	0.78	0.75	0.75	0.63	0.46	0.45	0.14	0.08	0.03

Grain identification: a 149-55-371, b 182-44-201, c 77-52-524, d 77-52-128, e 109-47-457, f 77-51-87, g 77-51-36, h 109-47-300, i 109-47-290, j 109-47-635.

fragments studied by several groups, and a Mg-rich suite (~0.7 <*mg* <0.8). The olivine-bearing polymineralic grains are too few (19) for good statistics, but 18 olivines have *mg* <0.65, while one olivine has *mg* 0.74. All olivines have high CaO positively correlated with Fe (c.f. Tarosov *et al.*, 1977; Basu *et al.*, 1977a), which indicates a high-temperature origin (Smith and Steele, 1976). Minor-element analyses show: (i) Al_2O_3 below 0.06 wt.% except for two values at 0.10; (ii) MnO positively correlated with FeO from MnO 0.2 wt.% at Fo_{75} to 0.8 at Fo_5; (iii) TiO_2 weakly correlated with FeO with 0–0.06 wt.% at Fo_{75} to 0.10–0.14 at Fo_5; and (iv) Cr as Cr_2O_3 negatively correlated with FeO from 0.05–0.18 at Fo_{75} to 0.00 at Fo_5.

Plagioclase. Monomineralic grains (Fig. 4a) have a pronounced maximum at $An_{95\text{-}96}$ and a long tail to An_{76}, which encompasses the trend for plagioclase in gabbroic fragments (Grove and Bence, 1977). Early attempts to use minor elements as indicators of the provenance of plagioclase require further detailed study. The plot of FeO vs. Ab (Fig. 4b) shows few plagioclases with FeO less than 0.4 wt.%, and the plot of MgO vs. FeO (Fig. 4c) shows only three grains with very low MgO and FeO whereas the majority fall in a broad Fe-rich band with a negative correlation between Mg and Fe. All our plagioclases have low K_2O (0.01–0.29 wt.%; Table 4), and have apparently crystallized from K-poor liquids.

The original suggestion (Smith, 1974) that the fields of plagioclase from mare-basalt, KREEP-rich rocks and plagioclase-rich (ANT) rocks were completely separated on an Ab-FeO plot needs detailed revision, though the overall trends remain. The original field of mare-basalt plagioclase was based mainly on Apollo 12 and 15 basalts (cf. Steele and Smith, 1973; Smith, 1974), and data for other mare basalts fall well outside (especially for Apollo 17 basalts; Dymek, Albee and Chodos, 1975), and indeed overlap with the original field for KREEP-rich rocks. Fig. 4b shows the principal composition ranges found by Stöffler and Knöll (1977) from a recent literature review, and demonstrates that most of our Luna 24 plagioclases fall in the mare-basalt field. Fourteen analyses lie just outside the mare-basalt field at $Ab_{4\text{-}6}$,FeO 0.3–0.6 wt.%, and probably should be attributed to the mare-basalt rather than to the ANT field because they lie in the main MgO-FeO trend on Fig. 4c. Probably only three grains belong to the ANT parent. However, we urge caution in the use of minor elements for assignment of plagioclase to host rock until further experimental studies like that of Longhi *et al.* (1976) have been made on the effects of the bulk composition of the host rock, the rate of crystallization, and the extent of sub-solidus processes.

Spinel. Most spinels are Al-chromites with TiO_2 between 1 and 3 wt.%. The Cr_2O_3 values are somewhat lower than for chromites from Apollo 15 and Luna 20 mare basalts, and are closer to values for Apollo 12 chromites (Papp, Steele and Smith, 1977, Fig. 5). One ulvöspinel grain was found. These compositions definitely indicate crystallization in mare basalts, and no evidence was found of spinels characteristic of non-mare rocks (Smith and Steele, 1976). Our data match those of Haggerty (1977) and Tarasov *et al.* (1977).

Ilmenite. All 9 spectrometer analyses showed low MgO (0.2–0.9 wt.%),

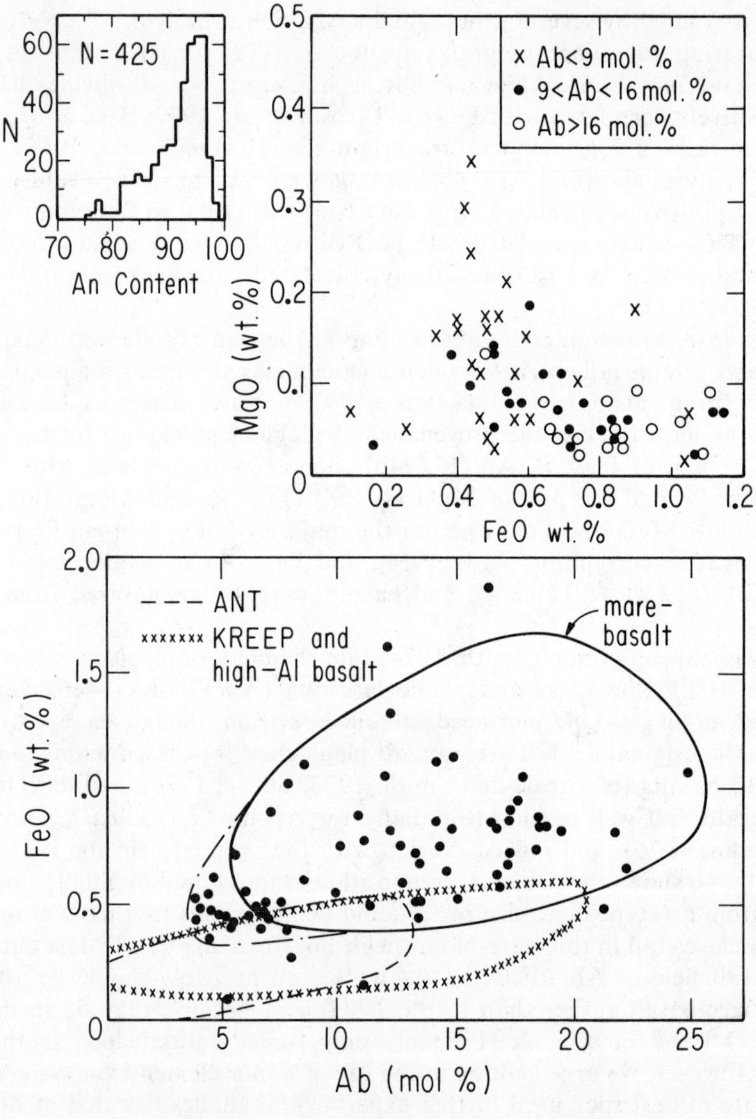

Fig. 4. Plagioclase compositions. (a) Histogram of An content (b) Ab vs. FeO. The reference fields are from Stöffler and Knöll (1977). The dot-dash line is a possible extension of the mare-basalt field to encompass 14 Luna 24 plagioclase grains.

suggestive of late-stage crystallization from an Fe-rich environment (Smith and Steele, 1976). This agrees with the conclusions of Vaniman and Papike (1977b) from textural study of Luna 24 fragments, and of Nehru *et al.* (1975) for Apollo 15 mare basalts. No evidence was found of Mg-rich ilmenite characteristic of some lunar non-mare rocks. Minor element ranges (CaO 0.00–0.14; Cr_2O_3 0.03–

Table 4. Minor elements in plagioclase.

Label	62	230	329	95	93	21	55	465	85	252A	47	24
TiO_2	0.06	0.05	0.04	0.02	0.05	0.03	0.03	0.05	0.03	0.03	0.06	0.05
FeO	0.48	0.82	0.83	0.84	1.05	0.90	0.57	0.64	0.85	0.44	0.51	0.60
MgO	0.13	0.08	0.04	0.04	0.07	0.08	0.05	0.11	0.05	0.10	0.13	0.18
K_2O	0.11	0.06	0.12	0.29	0.15	0.12	0.05	0.05	0.05	0.03	0.09	0.08
An	78.7	80.6	81.3	81.4	82.2	82.7	83.3	85.9	85.9	86.7	86.9	87.4

Label	73	11	124	92	238	96	295	9	296	371*	130*
TiO_2	0.05	0.01	0.06	0.03	0.06	0.02	0.02	0.01	0.01	0.02	0.12
FeO	0.50	0.16	0.77	1.10	0.26	0.40	0.45	0.09	0.46	0.52	1.58
MgO	0.15	0.03	0.08	0.09	0.05	0.17	0.34	0.07	0.06	0.77	1.67
K_2O	0.05	0.04	0.07	0.03	0.12	0.01	0.01	0.16	0.01	0.05	0.01
An	88.7	89.0	89.0	91.5	92.1	93.4	94.8	94.8	95.5	96.2	99.4

All grains are from slide 149-55. Selected to give coverage of An content.
*Note unusually high values of MgO (maskelynite?).

0.22; MnO 0.3–0.5; Al_2O_3 0.00–0.11) match values collected by Smith and Steele (1976).

Troilite. It is very rare, but one interesting occurrence as elongated blebs within Fe-rich olivine ($Fa_{95.5}$) is similar to that reported in Ryder *et al.* (1977).

Metal. Only a few tiny grains were found, and the two analyses showed Ni 7.5 and 9.5 wt.% (Co not determined). A comprehensive study by Friel and Goldstein (1977) also revealed few metal grains.

Silica minerals. Analyses of 8 grains gave K_2O 0.01–0.3 CaO 0.01–0.3 TiO_2 0.1–0.2 Al_2O_3 0.4–0.6 Na_2O 0.02–0.2 wt.%. One grain had a tile texture characteristic of cristobalite. The silica minerals can be assigned to late crystallization of mare basalts.

Table 5. Analyses of spinel-group and ilmenite.

	a	b	c	d	e	f	g	h	i	j	k	l	m
SiO_2	0.07	0.10	0.05	0.14	1.03	0.17	0.07	0.01	0.00	0.09	0.10	0.05	0.04
TiO_2	0.81	1.87	1.26	1.11	2.51	29.6	50.2	50.6	52.2	50.7	51.1	51.7	52.2
Al_2O_3	18.1	18.2	20.7	19.6	18.5	27.3	0.05	0.11	0.00	0.03	0.04	0.01	0.03
Cr_2O_3	48.1	45.6	41.7	40.6	39.5	0.16	0.02	0.04	0.12	0.00	0.16	0.17	0.22
FeO	25.2	28.3	28.5	32.9	35.2	65.0	46.9	46.4	46.3	46.1	45.9	46.1	45.5
MnO	0.24	0.21	0.26	0.25	0.28	0.23	0.48	0.46	0.41	0.37	0.35	0.31	0.41
MgO	6.76	5.54	5.27	3.13	1.92	0.03	0.13	0.17	0.21	0.21	0.40	0.80	0.92
NiO	0.00	0.01	0.06	0.07	0.06	0.00	0.05	0.00	0.00	0.07	0.00	0.00	0.00
CaO	0.00	0.01	0.08	0.03	0.37	0.06	0.14	0.08	0.04	0.02	0.00	0.01	0.00
Sum	99.28	98.84	97.88	97.83	99.37	97.98	98.04	97.87	99.28	97.59	98.05	99.15	99.32

a-f spinel; a 77-51-4, b 77-52-193, c 77-52-147, d 109-51-36, e 109-52-545, f 77-52-148; g-m; ilmenite, g 77-52-58, h 77-52-303, i 77-51-26, j 77-51-35, k 77-52-482, l 77-52-321, m 77-52-271.

Table 6. Analyses of homogeneous glasses.

	a	b	c	d	e	e′	f	g	h
n	1	8	2	2	1	1	1	1	5
SiO_2	45.0	43.6-47.2(45.7)	46.6	46.3	47.7	44.5	50.5	49.3	46.0-47.6(47.0)
TiO_2	1.26	0.16-0.34(0.22)	0.40	0.71	1.21	1.00	1.66	0.40	0.72-1.59(0.91)
Al_2O_3	31.3	26.4-29.4(27.3)	24.0	21.3	18.7	18.0	18.6	15.9	15.4-16.3(15.9)
Cr_2O_3	0.00	0.05-0.16(0.10)	0.14	0.19	0.20	0.14	0.09	0.53	0.14-0.34(0.28)
FeO	2.30	3.7-5.5 (4.33)	6.05	9.80	8.68	13.7	8.63	11.2	12.8-15.7(14.3)
MnO	0.16	0.02-0.09(0.06)	0.07	0.14	0.03	0.18	0.06	0.20	0.16-0.23(0.20)
MgO	0.20	3.5-6.9 (5.69)	7.73	7.56	10.5	10.8	7.01	10.6	6.0-9.3 (8.18)
CaO	18.0	15.5-16.5(16.0)	14.3	13.5	12.2	11.7	13.1	12.7	12.0-12.9(12.4)
Na_2O	0.56	0.07-0.99(0.40)	0.36	0.37	0.64	0.12	0.78	0.35	0.29-0.37(0.34)
K_2O	0.16	0.02-0.09(0.04)	0.07	0.06	0.27	0.03	0.02	0.02	0.04-0.07(0.05)
Sum	98.94	99.84	99.72	99.93	100.13	100.17	100.45	101.20	99.56
Q	2.7	0-1.0 (0.2)	0.7	0	0	0	5.2	0	0-0.7 (0.2)
Or	1.0	0.1-0.5 (0.2)	0.4	0.4	1.6	0.2	0.1	0.1	0.1-0.4 (0.3)
Ab	4.8	0.6-8.3 (3.3)	3.0	3.1	5.4	1.0	6.6	2.9	2.5-3.1 (2.9)
An	83.3	69.0-75.2(72.7)	63.8	56.1	47.2	48.4	46.9	41.2	40.3-43.9(41.8)
Wo	2.9	1.3-3.9 (2.8)	3.1	4.6	5.5	4.0	7.4	8.8	7.3-9.6 (8.4)
En	0.5	2.8-15.7 (10.0)	15.2	15.4	20.5	14.4	17.4	25.6	15.2-21.6(18.4)
Fs	2.5	1.6-8.0 (5.4)	8.7	13.9	10.8	12.7	13.1	19.1	19.5-28.6(22.6)
Fo	0	0-10.0 (2.9)	2.8	2.3	4.0	8.7	0	0.4	0-3.5 (1.4)
Fa	0	0-6.2 (1.8)	1.8	2.5	2.3	8.5	0	0.3	0-4.2 (1.8)
Cm	0	0.1-0.2 (0.2)	0.2	0.3	0.3	0.2	0.1	0.8	0.2-0.5 (0.4)
Il	2.5	0.3-0.6 (0.4)	0.8	1.3	2.3	1.9	3.1	0.8	1.4-3.0 (1.7)

Glass. The normative compositions of 61 homogeneous glasses are strongly dominated by plagioclase, pyroxene and olivine (Table 6), and there is a tendency for a bimodal distribution of normative olivine + pyroxene with one-quarter having 20–35% ol + px and two-thirds having 50–70% (Papp, Steele and Smith, 1977, Fig. 6).

The glasses are the best indication in the Luna 24 sample of the presence of highland rocks on the moon. Empirical sorting on Al_2O_3 and CaO led to the clusters and single analyses listed in order of decreasing Al_2O_3 in Table 6. Our analyses were compared with microprobe analyses of glasses from other lunar missions (e.g., Ridley *et al.*, 1973; Carusi *et al.*, 1974; Glass, 1976; Naney *et al.*, 1976) and from the Luna 24 regolith (Basu *et al.*, 1977b; Grove and Vaniman, 1977; Hu and Taylor, 1977; Norman *et al.*, 1977; Simonds *et al.*, 1977).

Cluster b (8 glasses) broadly matches the compositional range of "highland basalt" or "anorthositic gabbro" reported from all lunar soils. Analyses e,e′,f,g and cluster h (5 glasses) fall in the general range of the Low-K-Fra-Mauro composition (Reid *et al.*, 1977). Cluster j (13 glasses) matches well with brown, green and agglutinitic glasses, as well as bulk regolith (Barsukov, 1977) and regolith fractions (Ma and Schmitt, 1977), meta-basalts and fine-grained basalt

Table 6. (*cont.*)

i 1	j 13	j′ 1	k 1	l 3	m 17	n 2	o 1
43.8	44.2-49.0(46.6)	50.9	46.8	47.5	45.2-48.0(47.2)	47.1	47.2
0.20	0.47-1.39(0.94)	0.73	0.64	0.88	0.71-1.80(1.05)	1.25	2.34
16.0	11.9-15.0(13.8)	13.9	12.8	12.3	9.5-12.2 (10.6)	9.03	6.47
0.14	0.13-0.86(0.37)	0.57	0.13	0.24	0.34-0.59(0.44)	0.21	0.64
13.0	14.6-22.5(17.5)	12.5	11.6	19.2	17.1-22.2(19.2)	22.1	20.2
0.09	0.22-0.31(0.23)	0.22	0.19	0.25	0.19-0.34(0.26)	0.28	0.29
16.6	5.8-11.6 (8.5)	7.2	17.4	6.75	8.63-11.9(10.3)	7.14	10.8
9.06	11.1-15.1(12.2)	14.9	8.48	12.9	8.50-11.6(10.6)	11.8	11.4
0.30	0.15-0.37(0.29)	0.31	0.30	0.26	0.14-0.39(0.25)	0.25	0.14
0.04	0.00-0.17(0.05)	0.05	0.12	0.03	0.02-0.09(0.04)	0.03	0.03
99.23	100.48	101.28	98.46	100.31	99.94	99.19	99.51
0	0-1.5 (0.1)	4.4	0	0	0-1.0 (0.1)	0.6	0
0.2	0.1-1.0 (0.3)	0.3	0.7	0.2	0.1-0.5 (0.2)	0.2	0.2
2.6	1.3-3.1 (2.4)	2.6	2.6	2.2	1.2-3.3 (2.0)	2.1	1.2
42.5	30.3-40.0(35.9)	35.9	33.9	32.1	24.3-32.2(27.8)	23.6	17.0
1.2	7.6-15.8 (10.1)	15.4	3.7	13.1	7.5-12.4 (10.5)	14.7	16.7
12.7	11.5-19.7(16.2)	17.7	26.8	16.2	14.5-25.4(21.5)	16.4	25.4
7.2	13.8-34.8(24.2)	21.6	12.6	32.9	23.7-32.9(28.2)	36.3	31.5
20.3	0-10.7 (3.4)	0	12.1	0.4	0-6.8 (3.0)	1.1	1.1
12.7	0-13.4 (5.0)	0	6.3	0.9	0-11.3 (4.1)	2.2	1.5
0.2	0.2-1.3 (0.6)	0.8	0.2	0.3	0.5-0.9 (0.8)	0.3	0.9
0.4	0.9-2.7 (1.9)	1.4	1.2	1.7	1.3-3.5 (2.1)	2.4	4.5

a 52-436: b range and mean of 8 glasses, 51-23,74,173; 52-127,324,341,428,494; c mean of 2 glasses, 52-312,343: d mean of 2 glasses, 52-255,387: e 51-39: e 52-42: f 52-77: g 51-179: h range and mean of 5 glasses, 52-247,306,316,348,360: i 52-126: j range and mean of 13 glasses, 51-5,37,55,69,73,122,174; 52-106,150,251,344,352,385; j′ 52-277: k 52-547: l mean of 3 near-identical glasses 51-15,76; 52-435: m range and mean of 17 analyses, 51-62,112; 52-21,79,172,181,257,257A,262,267,291,339,452 (2 analyses), 458,459: n mean of 2 similar glasses 52-290,425: o 52-155. The prefixes, 51 and 52, respectively, refer to sections 77-51 and 77-52. n = number of glasses in column.

(McSween *et al.*, 1977; Taylor *et al.*, 1977b), in the Luna 24 samples; these similarities indicate a mutual relationship involving impacts into regolith dominated by basaltic materials, perhaps followed by thermal metamorphism of some impact debris. Cluster m (17 glasses) has a fairly wide composition range (e.g., FeO 17.1–22.2 wt.%), and is probably derived from the dominant basaltic and gabbroic rocks at the Luna 24 site.

Other compositions in Table 6 are mostly represented by only one or two glasses and interpretation is uncertain: briefly, (i) glass a closely matches the feldspar-rich cluster of Simonds *et al.* (1977); (ii) glasses i and k are the most Mg-rich of our glasses; and (iii) glasses n and o resemble Apollo 15 quartz-normative (Reid *et al.*, 1977) and 12015 (Rhodes *et al.*, 1977) basaltic compositions. No K-rich glasses indicative of KREEP rocks were found.

Petrological Implications for the Luna 24 Region

This section utilizes all available data on the Luna 24 core, and the interpretation is set in the context of present knowledge of the moon. It contrasts with the following sections which consider how much information would have been obtained (a) if the Luna 24 sample was the only specimen returned to earth, and (b) if only the 90–150 μm fraction had been collected by a sieve. These later sections are written with the aim of stimulating discussion about the best tactics for remote collection of planetary regolith.

Photogeologic (Butler and Morrison, 1977; Florensky *et al.*, 1977) and telescopic reflectance study of the Mare Crisium region culminated in the following interpretation by Adams *et al.* (1978): (i) the youngest unit (number 3) of the mare fill occupies the central and south-western part of the surface of the basin; (ii) an intermediate unit (number 2) outcrops in an irregular annulus around the northern half of the basin, and as isolated areas in the south, including one area encompassing the Luna 24 landing site; and (iii) the oldest unit (number 1) outcrops along the southeastern rim and also occurs in the ejecta blankets of the craters Picard and Peirce. Unit 2 has a telescopic spectrum indicative of low Ti (<1.5 wt.%) and is probably the source of the basalt and gabbro fragments in the Luna 24 core. Unit 1 has a spectral signature like that at the Luna 16 site. Unit 3 is spectrally similar to the center of Mare Serenitatis. Particularly important to our interpretation of the mineralogy and petrology of the Luna 24 core are: (i) that the orbital X-ray spectrum for the Picard ejecta has higher Mg/Si than adjacent areas; and (ii) the suggestion by Andre *et al.* (1977) and Hubbard and Vilas (1977) that Mare Crisium was initially filled with Mg-rich lavas (unit no. 1) and was later covered by units with lower Mg/Si. Furthermore Andre *et al.* suggested that the finest fraction of the Luna 24 core (<75 μm) is most representative of the surface soil.

From the early petrographic reports of Luna 24 samples by Barsukov *et al.* (1977), Florensky *et al.* (1977) and Tarasov *et al.* (1977) we draw attention to the observation that dolerite and basalt fragments have a greater exposure age (i.e., glass coating, micrometeorite craters) than gabbro fragments. Furthermore the conclusion that Luna 16 and Luna 24 data indicate a province of high-Al mare basalts at the east limb of the near side, fits with the interpretation of orbital X-ray data by Hubbard and Vilas (1977) that Al/Si decreases from east to west for both mare and intermare areas.

We now emphasize the presence of a population of Mg-rich olivines and pyroxenes in the Luna 24 monomineralic grains, and the apparent lack of such Mg-rich minerals in polymineralic and rock fragments. The simplest interpretation is that these Mg-rich minerals and perhaps some of the Fe-rich ones were derived from deeper-seated rocks than those which provided the bulk of the Fe-rich basalt and gabbro fragments, and that the process of transport to the Luna 24 site caused fragmentation of the Mg-rich rocks down to monomineralic grains. In view of the photogeologic and spectral reflectance data, the Mg-rich rocks might derive from the deep-seated unit 1 excavated by craters Picard and Peirce, though not necessarily from the ejecta blankets of these craters. Olivine settling was documented as a cause of differentiation of Apollo 12 basalts (many papers in *Proc. Lunar Sci. Conf. 2nd*), and is a plausible mechanism for differentiation at Mare Crisium.

We see no reason why the Mg-rich and Mg-poor populations of pyroxenes and olivines should not have arisen from a single petrogenetic suite of differentiating mare-basalt liquids. From this viewpoint, the low *mg* of the Luna 24 basalts and gabbros (e.g., Barsukov *et al.*, 1977; Bence and Grove, 1977; Ryder *et al.*, 1977) would be attributed to advanced crystal-liquid differentiation of a parent magma with higher *mg* before separation of olivine. Support for this idea comes from the observation that the maximum *mg* value (~0.78) of the Mg-rich olivine grains is close to that in most other mare basalts (Papike *et al.*, 1976, Fig. 36). Furthermore, the minor-element contents of the Mg-rich and Mg-poor suites of olivines and pyroxenes do not show a hiatus, and are interpretable in terms of differentiation of a single chemical type of basaltic liquid, when account is taken of the complex effects documented for crystal-liquid differentiation in individual lava flows. The relation of Luna 24 basalts to the low-Ti Apollo 17 basalts (Vaniman and Papike, 1977a) must be considered in the context of crystal-liquid differentiation prior to eruption.

Until further details appear of the fragments described by Bence and Grove (1977), it is difficult to discuss whether they arise from a highland source, as proposed by them, or whether they should be re-interpreted as rare rock fragments from Mg-rich mare-basalt rocks. Perhaps the fragment with widely-separated pyroxene compositions was metamorphosed in an ejecta blanket. The other pyroxenes appear to match those in our Mg-rich population, but detailed chemical data are needed for a reliable comparison.

Non-mare (~ highland) fragments are rare in Luna 24 soils (Basu *et al.*, 1977b; Coish and Taylor, 1977; Ryder *et al.*, 1977; Tarasov *et al.*, 1977; Taylor *et al.*, 1977a) as are non-mare monomineralic grains (e.g., feldspar, see earlier section). However, a substantial fraction of glasses (e.g., at least 8 and probably 12 out of 61 glasses in our samples; preceding section) can be assigned to nonmare parents. The high content of MgO (6.8 wt.%) of ilmenite associated with $Mg_{74}Fe_{22}Ca_4$ pyroxene and An_{98} plagioclase in 1 out of 1230 particles studied by Nielsen and Drake (1977) may indicate a high-temperature Mg-rich non-mare parent, based on the distribution of Mg/Fe between ilmenite and pyroxene calibrated experimentally by Bishop (1976).

Lithic fragments with *mg* >0.8 are unrecorded, but a glass with *mg* 0.89 (Bence and Grove, 1977) might indicate the ultrabasic group emphasized by Smith and Steele (1976) as possibly representative of the lunar upper mantle, or lower crust.

KREEP-rich rock fragments are unrecorded, and the Luna 24 mare basalts are low in alkalis even with respect to other mare basalts. Norman *et al.* (1977) recorded "Fra Mauro" glass with 0.35 ± 0.46 wt.% K_2O, and several groups recorded rare glasses and agglutinates with lower but substantial amounts of

K_2O. A glass labeled "high-K mare basalt" with 0.48 ± 0.11 wt.% K_2O was listed by Norman *et al.* (1977). The general lack of K-rich material is consistent with orbital scans of the eastern side of the moon.

A minor component of Low-K-Fra-Mauro basalt is apparently represented in Luna 24 glasses (preceding section).

Dark deposits in Mare Crisium may be represented by olivine vitrophyre (e.g., Ryder *et al.*, 1977) or possibly by Ti-rich agglutinates (e.g., Taylor *et al.*, 1977a).

We note that: (i) two metal fragments with high Co (2.4 and 1.6 wt.%; Friel and Goldstein, 1977) fall in the range of metal from Mg-rich lunar rocks (Smith and Steele, 1976, Fig. 20), which may provide one more tantalizing clue to the lunar interior; and (ii) other metal particles (Friel and Goldstein, 1977; Hutchison and Graham, 1977; this paper) have compositions possibly indicative of a meteoritic origin.

What could be inferred about the moon if the Luna 24 core was the first sample returned to earth?

The answer would depend strongly on the extent of remote studies already performed. Readers will recall the dramatic changes of thinking brought about by: (a) the Turkevich analyses of bulk regolith; and (b) the gamma and X-ray analyses obtained from orbital missions. Coupled with telescopic reflectance studies and various geophysical data, a favorable framework was available for detailed interpretation of the geochemical-mineralogical-petrological measurements. If one assumes that the orbital photogeologic and chemical data were available, as well as the telescopic reflectance spectra, the Luna 24 mineralogy and petrology could lead to powerful conclusions about the moon if the unusual specimens were given similar prominence to the dominant population of mare basalts and gabbros.

Indeed it would be possible to propose major petrologic differentiation of the moon, as was suggested shortly after the Apollo 11 mission by Smith *et al.* (1970) and Wood *et al.* (1970). The absence of microanorthosite in the Luna 24 samples need not have inhibited speculation about plagioclase flotation, since Smith *et al.* (1970) did not initially refer to microanorthosite: furthermore the Apollo 11 microanorthosite did not have cumulate texture, and it was only after several Apollo missions that evidence for cumulate plagioclase appeared. Rare plagioclase-rich fragments and "highland basalt" glasses in the Luna 24 core could have been interpreted as impact debris from the lunar crust (highlands), but the absence of large rocks might have greatly reduced the credibility of the interpretation. The highly-evolved chemistry of the Luna 24 mare basalts and gabbros should have led to similar controversy about the origin of mare volcanism as did the Apollo 11 mare basalts, though the difference in the TiO_2 content would have changed details of the arguments. The general depletion of volatile elements in the moon is just as apparent in the Luna 24 core as in the Apollo 11 samples. That the Luna 24 samples are low in KREEPy elements (including U and Th) would have not been surprising in view of orbital chemical scans, and the high-KREEP region near Oceanus Procellarum would have been targeted for a later lunar mission. Unfortunately Mg-rich (ultrabasic) fragments are conspicuously absent in the Luna 24 core, but this would not have been surprising because of the photo-geologic evidence for absence or near-absence of basin ejecta at the Luna 24 site.

Numerous groups have explicitly or implicitly pointed out the difference between the various fractions of the Luna 24 core, and it suffices here to mention: (a) the greater evidence for non-mare material in glassy than in crystalline material; and (b) the absence of Mg-rich olivines and pyroxenes in polymineralic and rock fragments. There can be no doubt that the fine-grained material (<240 μm) must be examined with very great care, and indeed there is good reason to believe that only part of the useful information had been extracted from glassy and agglutinitic fractions by the time of the Conference on Luna 24. Perhaps these fractions can provide the most detailed information on meteoritic debris and on lunar material transported from great distances.

Certainly there is great value to petrographers and geochemists (especially geochronologists) in having access to large lunar samples (g to kg), but it is worth asking from the viewpoint of *future* remote sampling of planetary regolith how small can a sample be, and how should it be collected. First, we ask some specific questions about the Luna 24 core.

Would a series of grab samples have been more or less useful than the 200cm core?

From the mineralogical-petrological viewpoint, there are no really valuable distinctions between the different sections of the core, and the 2cm layer of coarse gabbroic particles at 170cm depth (Albee *et al.*, 1977) actually provided less *general* information about Mare Crisium than other layers, though the larger sample, of course, allowed careful mineralogic study. Perhaps collection of (say) 6 samples on a traverse of (say) 10km would have been as feasible from the technologic viewpoint as collection of a single 200cm drill core, and might have yielded a greater range of soils especially if the traverse happened to cross rays of ejected material from large craters. Discussion is invited on the desirability of planning a remote mission to the moon in which samples are collected by driving short tubes (~10cm) into the regolith at as many sites as possible (100?) on a long traverse (~100km?). To this end, it is desirable to ask the next question.

Would the fine fraction have yielded enough information about the mineralogy and petrology of the Luna 24 site?

In an ideal world, it is highly desirable to obtain petrographic information from samples at least as large as walnuts, but in a budget-limited world, could the Luna 24 site have been characterized by a thimble-full of fine regolith? Further, should the regolith be sieved to eliminate the larger fragments

(>0.5mm?), thereby yielding a larger number of the smaller fragments? Certainly the finer fractions cannot yield definitive information on textures such as those displayed by several research groups (e.g., Tarasov *et al.*, 1977), and much more reliance must be placed on the mineral chemistry. The 90–150 μm fraction actually provides more information about the Mare Crisium area than an equal weight of >500 μm material. The coarse material lacks the Mg-rich population of olivine and pyroxene, and is very deficient in non-mare glasses and fragments. Given a choice, we would now prefer the finer fraction to the coarse fraction of the regolith, though recognizing that some larger fragments are desirable to give petrographic information. Furthermore, the larger fragments are desirable for consortium studies, especially ones involving chronological and trace-element studies. We conclude with some implications for further exploration of planetary regoliths by transportation of samples back to earth.

Implications for Further Sample Collection

The general framework of knowledge about the moon is now so great that the provenance of a mineral can be confidently established from its chemical signature, especially if further systematic studies are made of the minor and trace elements of minerals selected from well-characterized representative rocks and meteorites. Early studies of minor and trace elements in silicates and oxides have defined the approximate composition ranges as a function of type of host rock, but further studies are needed, especially with the ion microprobe, to expand the data base to more elements, thereby allowing application of n-dimensional statistical techniques (n~12?). The evidences for late volcanism (~ − 3 to − 2Gyr) on the moon, and for geochemical trends with longitude and probably with latitude, provide the basis for requesting further collection missions, but the budget restrictions are highly unfavorable even for relatively cheap missions. With this in mind, it is extremely important to find out how effective are studies of the fine fraction, and how much less than a gram is sufficient to characterize a lunar geologic region.

Now that telescopic reflectance and polarization studies have provided so much valuable and intriguing, but often non-definitive, information about asteroidal surfaces [compare the interpretations of Chapman (1976) and Gaffey and McCord (1978)], it is high time to consider the tactics for direct sampling of asteroidal regolith. It would be presumptuous to discuss here the relative merits of *in situ* bulk chemical analysis (e.g., with a combined alpha-proton-X-ray-fluorescence detection system) and mineralogical-petrological-geochemical study of a sample returned to earth. Let it suffice that study of the tiny aliquots of Luna samples has lead to complete confidence that a grab sample from an asteroid would allow thorough characterization of the surface mineralogy, petrology and geochemistry, and that excellent results could be expected even for a sample weighing less than 1g, so long as it contained a large population of tiny fragments. Of course, much larger samples (kg level) would be useful for many interdisciplinary studies, and the ultimate decisions on sample sizes and numbers must be hammered out by scientists from many disciplines.

Unfortunately, the implications for sampling the Martian surface are not so clear. The sum total of the Viking evidence indicates so much weathering (almost certainly accentuated by photo-chemically produced H_2O_2 and direct photochemical oxidation) that a tiny grab sample of soil may provide little or no evidence about the original igneous rocks. Almost certainly, a major attempt should be made to smash a larger boulder and collect pieces from the inside. If the boulder were composed of just one kind of volcanic rock, the information obtained from the pieces might be restricted (cf. just one piece of lunar mare basalt), and it might be desirable to land on the ejecta blanket of a large crater in order to increase the probability of encountering a boulder of breccia. Another possibility would be to look for a boulder from a landslide from a canyon wall. Obviously there are enormous technical barriers to these possibilities, and it may be necessary to settle for a grab sample. If possible, techniques should be developed for sieving and sorting the regolith in order to look for dense fragments of volcanic rock or breccias whose alteration is only skin-deep.

Conclusion

Detailed mineralogical-petrological-geochemical study of Luna 24 samples provides a good characterization of the Mare Crisium region when: (a) the fine fraction is examined thoroughly, especially for rare fragments; and (b) the data can be interpreted in the context of photo-geologic, telescopic-spectral and orbital-chemical observations. Further study of minor and trace elements of minerals from well-characterized lunar rocks is needed to provide a more reliable reference base for characterization of monomineralic grains. Planning for future missions to the moon, asteroids and Mars should take note of the highly-developed capacity for interpretation to tiny grab samples.

Acknowledgments—We thank the thin-section technicians at NASA-JSC for 12 excellent sections, and O. Draughn and I. Baltuska for technical help at Chicago. The research was supported by NASA 14-001-171 and by Materials Research Laboratory funded by NSF. Helpful reviews by G. A. McKay and G. J. Taylor were greatly appreciated.

References

Adams J. B., Head J. W., III, McCord T. B., Pieters C. and Zisk S. (1978) Mare Crisium: regional stratigraphy and geologic history. *Geophys. Res. Lett.* submitted.

Albee A. L., Chodos A. A. and Dymek R. F. (1977) Petrography of Luna 24 sample 24170 (abstract). In *Papers Presented to the Conference on Luna 24*, p. 5–7. The Lunar Science Institute, Houston.

Andre C. G., Wolfe R. W. and Adler I. (1977) Evidence for a high-magnesium subsurface basalt in Mare Crisium (abstract). In *Papers Presented to the Conference on Luna 24*, p. 8–11. The Lunar Science Institute, Houston.

Barsukov V. L. (1977) Preliminary data for the regolith core brought to earth by the automatic lunar station Luna 24. *Proc. Lunar Sci. Conf. 8th*, p. 3303–3318.

Barsukov V. L. Tarasov L. S., Dmitriev L. V., Kolesov G. M., Shevaleevsky I. D. and Garanin A. V. (1977) The geochemical and petrochemical features of regolith and rocks from Mare Crisium (preliminary data). *Proc. Lunar Sci. Conf. 8th*, p. 3319–3332.

Basu A., McKay D. S. and Fruland R. M. (1977a) Petrography, mineralogy and source rocks of Luna 24 drill core soils (abstract). In *Papers Presented to the Conference on Luna 24*, p. 18–21. The Lunar Science Institute, Houston.

Basu A., McKay D. S., Clanton U., Waits G. and Fruland R. M. (1977b) Glass particles in Luna 24 drill core soils (abstract). In *Papers Presented to the Conference on Luna 24*, p. 14–17. The Lunar Science Institute, Houston.

Bence A. E. and Grove T. L. (1977) The highland component in the Luna 24 core (abstract). In *Papers Presented to the Conference on Luna 24*, p. 22–24. The Lunar Science Institute, Houston.

Bishop F. C. (1976) Ph.D. thesis, University of Chicago.

Butler P., Jr. and Morrison D. A. (1977) Geology of the Luna 24 landing site. *Proc. Lunar Sci. Conf. 8th*, p. 3281–3301.

Carusi A., Cavarretta A., Coradini A., Fanucci O., Funiciello R. and Salomone M. (1974) Lunar rocks and glasses: a univariate and multifactorial analysis of chemical data. *Proc. Lunar Sci. Conf. 5th*, p. 605–620.

Chapman C. A. (1976) Asteroids as meteorite parent bodies: the astronomical perspective. *Geochim. Cosmochim. Acta* **40**, 701-719.

Coish R. A. and Taylor L. A. (1977) Igneous rocks from Mare Crisium: mineralogy and petrology (abstract). In *Papers Presented to the Conference on Luna 24*, p. 48–50. The Lunar Science Institute, Houston.

Dymek R. F., Albee A. L. and Chodos A. A. (1975) Comparative mineralogy and petrology of Apollo 17 mare basalts: Samples 70215, 71055, 74255 and 75055. *Proc. Lunar Sci. Conf. 6th*, p. 49–77.

Florensky C. P., Basilevsky A. T., Ivanov A. V., Pronin A. A. and Rode O. D. (1977) Luna 24: geologic setting of landing site and characteristics of sample core (preliminary data). *Proc. Lunar Sci. Conf. 8th*, p. 3257–3279.

Friel J. J. and Goldstein J. I. (1977) Metal particles in the Luna 24 soil (abstract). In *Papers Presented to the Conference on Luna 24*, p. 57–79. The Lunar Science Institute, Houston.

Gaffey M. J. and McCord T. B. (1978) Asteroid surface materials. Preprint.

Glass B. P. (1976) Major element composition of glasses from Apollo 11, 16 and 17 soil samples. *Proc. Lunar Sci. Conf. 7th*, p. 679–693.

Grove T. L. and Bence A. E. (1977) Petrogenesis of gabbros from Mare Crisium (abstract). In *Papers Presented to the Conference on Luna 24*, p. 64–67. The Lunar Science Institute, Houston.

Grove T. L. and Vaniman D. T. (1977) Experimental petrology of very low Ti basalts and origin of Luna 24 ferrobasalt (abstract). In *Papers Presented to the Conference on Luna 24*, p. 68–71. The Lunar Science Institute, Houston.

Haggerty S. E. (1977) Luna 24: opaque mineral chemistry of gabbroic and basaltic fragments from Mare Crisium. *Geophys. Res. Lett.* **4**, 489-492.

Hu H. and Taylor L. A. (1977) Luna 24 soils from Mare Crisium: agglutinate chemistry (abstract). In *Papers Presented to the Conference on Luna 24*, p. 88–84. The Lunar Science Institute, Houston.

Hubbard N. J. and Vilas F. (1977) From Serenity to Langemak: a regional chemical setting for Crisium (abstract). In *Papers Presented to the Conference on Luna 24*, p. 85–88. The Lunar Science Institute, Houston.

Hutchison R. and Graham A. L. (1977) Basalt and breccias from the Luna 24 core (abstract). In *Papers Presented to the Conference on Luna 24*, p. 89. The Lunar Science Institute, Houston.

Longhi J., Walker D. and Hays J. F. (1976) Fe and Mg in plagioclase *Proc. Lunar Sci. Conf. 7th*, p. 1281–1300.

Ma M.-S. and Schmitt R. A. (1977) Chemical and petrographic studies of 18 Luna 24 lithic fragments (abstract). In *Papers Presented to the Conference on Luna 24*, p. 102–105. The Lunar Science Institute, Houston.

McSween H. Y., Jr., Taylor L. A. and Clark J. C. (1977) Luna 24 metabasalts: possible evidence for assimilation of highlands material? (abstract). In *Papers Presented to the Conference on Luna 24*, p. 118–120. The Lunar Science Institute, Houston.

Naney M. T., Crowl D. M. and Papike J. J. (1976) The Apollo 16 drill core: statistical analysis of glass chemistry and the characterization of a high alumina-silica poor (HASP) glass. *Proc. Lunar Sci. Conf. 7th*, p. 155–185.

Nehru C. E., Prinz M., Dowty E. and Keil K. (1975) Spinel-group minerals and ilmenite in Apollo 15 rake samples. *Amer. Mineral.* **59**, 1220–1235.

Nielsen R. L. and Drake M. J. (1977) Magma types at the Luna 24 site in Mare Crisium (abstract). In *Papers Presented to the Conference on Luna 24*, p. 132–135. The Lunar Science Institute, Houston.

Norman M., Coish R. A. and Taylor L. A. (1977) Glass chemistry and magma evolution at Mare Crisium (abstract). In *Papers Presented to the Conference on Luna 24*, p. 136–138. The Lunar Science Institute, Houston.

Papike J. J., Hodges F. N., Bence A. E., Cameron M. and Rhodes J. M. (1976) Mare basalts: crystal chemistry, mineralogy and petrology. *Rev. Geophys. Space Phys.* **14**, 475–540.

Papp H., Steele I. M. and Smith J. V. (1977) Luna 24 90–150 μm fraction: implications for sampling of planetary and asteroidal regoliths (abstract). In *Papers Presented to the Conference on Luna 24*, p. 143–146. The Lunar Science Institute, Houston.

Reid A. M., Duncan A. R. and Richardson S. H. (1977) In search of LKFM. *Proc. Lunar Sci. Conf. 8th*, p. 2321–2328.

Rhodes J. M., Blanchard D. P., Dungan M. A., Brannon J. C. and Rodgers K. V. (1977) Chemistry of Apollo 12 mare basalts: magma types and fractionation processes. *Proc. Lunar Sci. Conf. 8th*, p. 1305–1338.

Ridley W. I., Reid A. M., Warner J. L., Brown R. W., Gooley R. and Donaldson, C. (1973) Glass compositions in Apollo 16 soils 60501 and 61221. *Proc. Lunar Sci. Conf. 4th*, p. 309–321.

Ryder G., McSween H. Y., Jr. and Marvin U. B. (1977) Basalts from Mare Crisium. *The Moon*, **17**, 263–287.

Simonds C. H., Warner J. L., McGee P. E., Cochran A. and Brown R. W. (1977) Luna 24: Petrochemical potpourri (abstract). In *Papers Presented to the Conference on Luna 24*, p. 170–173. The Lunar Science Institute, Houston.

Smith J. V. (1974) Lunar mineralogy: A heavenly detective story. Presidential Address, Part 1. *Amer. Mineral.* **59**, 231–243.

Smith J. V. and Steele I. M. (1976) Lunar mineralogy: a heavenly detective story. Part II. *Amer. Mineral.* **61**, 1059–1116.

Smith J. V., Steele I. M. and Papp H. (1978) Luna 24: implications for remote sampling of unweathered regolith (abstract). In *Lunar and Planetary Science IX*, p. 1074–1076. Lunar and Planetary Institute, Houston.

Smith J. V., Anderson A. T., Newton R. C., Olsen E. J., Wyllie P. J., Crewe A. V., Isaacson M. S. and Johnson D. (1970) Petrologic history of the moon inferred from petrography, mineralogy, and petrogenesis of Apollo 11 rocks. *Proc. Apollo 11 Lunar Sci. Conf.*, p. 897–925.

Steele I. M. and Smith J. V. (1973) Mineralogy and petrology of some Apollo 16 rocks and fines: General petrologic model of moon. *Proc. Lunar Sci. Conf. 4th*, p. 519–536.

Stöffler D. and Knöll H.-D. (1977) Composition and origin of plagioclase, pyroxene, and olivine clasts of lunar breccias 14006, 14063, 14066, 14311, 14320 and 14321. *Proc. Lunar Sci. Conf. 8th*, p. 1849–1867.

Tarasov L. S., Nazarov M. A., Shevaleevsky I. D., Kudryashova A. F., Gaverdovskaya A. S. and Korina M. I. (1977) Mineralogy and petrography of lunar rocks from Mare Crisium (preliminary data). *Proc. Lunar Sci. Conf. 8th*, p. 3333–3356.

Taylor G. J., Warner R., Wentworth S. and Keil K. (1977a) The Luna 24 regolith: lithologic abundances in the 250–500 μm size fraction, and compositions of agglutinates and nonmare lithic fragments and glasses (absract). In *Papers Presented to the Conference on Luna 24*, p. 186–188. The Lunar Science Institute, Houston.

Taylor G. J., Warner R., Wentworth S., Keil K. and Sayeed U. (1977b) Petrology and chemistry of Luna 24 mare basalts and basaltic glasses (abstract). In *Papers Presented to the Conference on Luna 24*, p. 189–192. The Lunar Science Institute, Houston.

Vaniman D. T. and Papike J. J. (1977a) Very low Ti (VLT) basalts: a new mare rock type from the Apollo 17 drill core. *Proc. Lunar Sci. Conf. 8th*, p. 1443–1481.

Vaniman D. T. and Papike J. J. (1977b) Ferrobasalts from Mare Crisium: Luna 24. *Geophys. Res. Lett.* **4**, 497–500.

Wood J. A., Dickey J. S., Jr., Marvin U. B. and Powell B. N. (1970) Lunar anorthosites and a geophysical model of the moon. *Proc. Apollo 11 Lunar Sci. Conf.*, p. 965–988.

Merrill, R. B. and Papike, J. J., editors.
(1978) *Mare Crisium: The View from Luna 24*, Pergamon Press (New York), p. 265–280
Printed in the United States of America

The Luna 24 sample from Mare Crisium: New structural features in lunar glasses deduced from a study of crystal-field spectra

P. M. Bell, H. K. Mao, and R. M. Hazen

Geophysical Laboratory, Carnegie Institution of Washington, Washington, D. C. 20008

A. L. Mao

George Mason University, Fairfax, Virginia 22030

Abstract—The Luna 24 samples contain colored glass fragments that range in composition from basalt to anorthite. The crystal-field spectra of green glasses in the sample correlate with composition and oxygen fugacity. The green glasses have oxidation states approximately equivalent to those of the same composition synthesized experimentally at 1400°C, $-\log P_{O_2} = 9.1$. The brown glasses have spectra unlike those previously measured for lunar glass. Their spectra have sharp and distinct bands, closely similar to the spectra of augitic pyroxene. Therefore, the high degree of short-range order of these glasses that must exist suggests that they have a structural state intermediate between those of glass and crystal but unlike known glass states or devitrification products. The brown glass spectra do not correlate with composition, oxygen fugacity, or any other experimentally determined properties of glasses. Contributions of brown glass spectra to telescope spectra of Mare Crisium should be taken into account if these glasses are abundant in the Mare Crisium regolith.

Introduction

Silicate glasses sampled from the moon's surface possess a wealth of information. They have been used to identify rock types from which they are thought to be quenched melt portions. Their compositions have been used to model liquid lines of descent from "parental" or "derivative" igneous liquids. The lunar glasses are generally not igneous in origin, however, in the sense that most, if not all, reported samples of lunar glass were produced by shock processes that accompanied meteorite impacts on the moon's surface. The impact processes have, in many instances, introduced complexities of chemical fractionation, elemental sputtering, and other less well understood "shock" effects; thus the lunar glasses may be inadequate for deducing the liquid descent of lunar igneous rocks, but the glasses are suited for interpretation of surface processes that occur in the lunar regolith.

Some of lunar glasses are true "quench products" of impact melting that preserve information on oxidation/reduction conditions in the regolith at the instant of quench. Furthermore, the glasses preserve various structural stages of crystallization to rock textures (such as those of rock 14310) by progressive stages of impact (thermal) metamorphism. In this study absorption spectroscopy is used to assess both oxidation state and degree of order of glasses; some glass types are observed for the first time in Luna 24 glass fragments. The spectral data serve an additional purpose in the interpretation of telescope and remote sensing surveys of the moon's surface, because the absorption bands in lunar glasses contribute significantly to the composite soil or regolith absorption (reflectance) spectra. The new data may have critical bearing on the interpretation of telescope data of Mare Crisium.

THE LUNA 24 SAMPLES

Multiple-grain mounts (several hundred polished grains per mount) were kindly loaned to us by Dr. Larry A. Taylor from his allocation for this study. The data reported in this paper can be considered representative of glasses in the Luna 24 samples listed in Tables 1 and 2.

Glass color and composition

The lunar glasses in general, and the Luna 24 glasses specifically, have been classed by petrographers in a system by color and bulk chemical composition (e.g., Norman *et al.*, 1977; Grove and Vaniman, 1977). In the present study it has been determined that there is always a direct relationship between color and spectral properties, but correlation of color alone with chemical composition is often invalid. Clearly there are causes of color changes and "darkening" other than changes in composition. For example, the brown glasses, "tea-colored" (green-brown) glasses, and green glasses, variously referred to as "ferrobasalt," nondescript, and the Luna 24 green glass or "low-titanium basalt" have essentially the same composition. There is more compositional variation (especially for Mg) within the green glasses than between green and brown glasses. Various other glasses, ranging from light green to colorless (spectrally very light green to ultra light green), have been named "Fra Mauro basalt" and "Highlands basalt," on the basis of chemical composition.

The relation between color and composition of several Luna 24 glasses can be seen in Tables 1 and 2. The brown glasses are relatively uniform in composition, whereas the green glasses have variable magnesium content (Table 1). The light green and colorless glasses have relatively low iron content. The former are relatively high in titanium and aluminum; the latter, very low in titanium, high in calcium, and very high in aluminum. Some glasses appear colored apparently owing to shock or radiation, which is a "darkening" effect not related to chemical composition.

CRYSTAL-FIELD SPECTRA

The crystal-field spectra of lunar glasses have been correlated in detail with experimental data on synthetic glasses relating composition and oxygen fugacity by Bell *et al.* (1976). In Luna 24 glasses, two additional effects that influence the spectra, but are not as yet correlated experimentally, were found. They are

Table 1. Electron microprobe analysis and optical absorption spectral features of glass fragments in multiple grain mount 24077,45.

	Brown glass		Tea-colored glass	Green glass				Light green	Colorless glass	
	d	g	b	e	j	a	h	c	f	i
SiO_2	46.60	46.46	44.71	44.60	45.72	46.67	47.44	50.19	45.26	45.30
TiO_2	1.02	1.00	0.99	1.49	1.00	1.08	0.95	1.96	0.30	0.19
Al_2O_3	12.08	12.16	13.48	10.73	12.29	11.62	11.12	15.98	25.14	27.43
Cr_2O_3	0.29	0.30	0.52	0.47	0.45	0.32	0.30	0.04	0.14	0.14
FeO	20.80	20.82	18.86	19.08	18.82	21.13	19.75	11.87	5.53	5.66
MgO	6.85	6.86	9.61	12.37	10.90	7.65	7.58	6.49	9.92	6.08
CaO	12.38	12.28	11.90	10.22	11.25	11.79	11.50	11.64	14.56	15.85
MnO	0.32	0.33	0.29	0.31	0.45	0.33	0.32	0.16	0.07	0.08
Total	100.33	100.20	100.36	99.27	100.73	100.59	98.96	98.32	100.93	100.73
Slope*	18.7	20.8	18.2	16.0	8.3	14.7	10.0	11.1	—	—
λ_{min}, nm	800	820	660	590	590	590	590	625	—	—
λ_{max}, nm	985	985	1040	1080	1080	1080	1090	1100	1080	1080
α_{max}, cm^{-1}	192	272	170	205	158	178	148	108	37	33

*Slope of absorption edge at 20,000 cm^{-1}, in units of cm^{-1} per 1000 cm^{-1} wave number.

Table 2. Electron microprobe analysis and optical absorption features of glass fragments in multiple grain mounts 24210,48; 24182,48; 24149,49; and 24174,51.

	24210,48				24182,48	24149,49		24174,51	
	Dark brown	Brown glass		Colorless	Colorless	Brown glass	Green glass	Brown glass	Green glass
	c	a	b	d	c	c	e	a	b
SiO_2	46.36	46.69	46.84	44.98	43.88	47.16	47.84	46.11	46.94
TiO_2	0.95	0.96	0.96	0.21	0.00	0.99	1.08	0.15	0.95
Al_2O_3	11.89	11.96	11.84	27.44	34.55	12.00	11.56	29.23	11.83
Cr_2O_3	0.25	0.22	0.25	0.09	0.00	0.28	0.21	0.04	0.21
FeO	19.34	19.46	19.04	3.80	0.46	20.55	20.75	3.21	19.50
MgO	6.69	6.85	6.93	5.67	0.16	7.04	6.50	3.21	6.70
CaO	12.15	12.19	12.32	15.74	19.11	12.81	12.38	16.37	12.08
MnO	0.35	0.33	0.29	0.10	0.01	0.32	0.33	0.09	0.32
Na_2O	0.28	0.28	0.28	0.17	0.48	—	—	0.66	0.24
Total	98.26	98.95	98.74	98.20	98.65	101.15	100.55	99.07	98.77
Slope*	>100	11.5	15.6	0	1.5	1.98	17.7	9.4	12.0
λ_{min}, nm	840	800	850	—	—	830	605	—	590
λ_{max}, nm	970	980	970	—	—	970	1070	—	1040
α_{max}, cm^{-1}	141	87	67	—	—	181	185	—	150

*Slope of absorption edge at 20,000 cm^{-1}, in units of cm^{-1} per 1000 cm^{-1} wave number.

structure (development of short-range order) and degree of devitrification. The four basic effects can be recognized unambiguously from the spectra, and thus this discussion will be subdivided into effects on the spectra caused by composition and oxygen fugacity, short-range order, and devitrification.

The absorption spectra were measured with microparticle techniques developed by Mao and Bell (1973). An important parameter in calculation of the absorption coefficient (α,cm^{-1}) is the grain thickness. The section laboratory at the Johnson Spacecraft Center was unable to state the thickness more accurately than approximately 30 μm because of the irregularities of the present grain mounts (Mr. John Annexstad, pers. comm., 1978). The grain thickness could not be accurately measured by micrometer or by other mechanical device. Therefore, a thickness uncertainty of ± 5 μm and an absorption coefficient uncertainty of $\pm 15\%$ should be applied to all the spectral data in this study.

Spectral Correlation with Oxygen Fugacity and Chemical Composition

The green Luna 24 glasses have normal glass spectra, that is, broad, intense absorption bands caused by octahedrally coordinated Fe^{2+} and a steep apparent absorption edge caused by $Ti^{3+}-Ti^{4+}-Fe^{2+}$ charge-transfer effects (Bell *et al.*, 1976). Representative compositions are given in Tables 1 and 2. Figure 1 shows the spectra of a series of four green glasses from sample 24077,45. These glasses have a range of iron concentration (Table 1) that correlates with experimentally observed wavelengths and intensities of the peak maximum of the octahedral Fe^{2+} crystal-field absorption band centered near 1 μm (Fig. 2). The slope at 20,000 wave numbers also correlates with composition when compared with experimental data on the compositional product, FeO $\times$ TiO_2 (in weight percent) plotted in Fig. 3. Reference to the plot of TiO_2 (weight percent) vs. absorption minimum in Fig. 4 and to the results of Bell *et al.* (1976), who obtained experimental data on spectra and oxygen fugacity of synthetic glasses, indicates that the Luna 24 green glasses are the same as, or slightly more oxidized than, synthetic glasses of the same composition quenched at 1400°C, $-\log P_{O2} = 9.1$.

The correlations of spectral features of the Luna 24 green glasses with experimental data are much the same as those obtained with other lunar glasses. Apparently the configurations of iron and titanium sites in Luna 24 green glass are similar to those in all other green and orange (often referred to as "red") lunar glasses studied.

Spectral Correlation with Short-range Order in Glass

Previously unrecognized in lunar glasses (and in glasses in general) is the development of a highly ordered state, as observed in many of the Luna 24 glasses. These glasses are usually deep brown or have a brown tone, but their spectral patterns have not been reported previously. Figure 5 shows spectra of a brown glass, an intermediate brown-green ("tea-colored") glass, and a normal

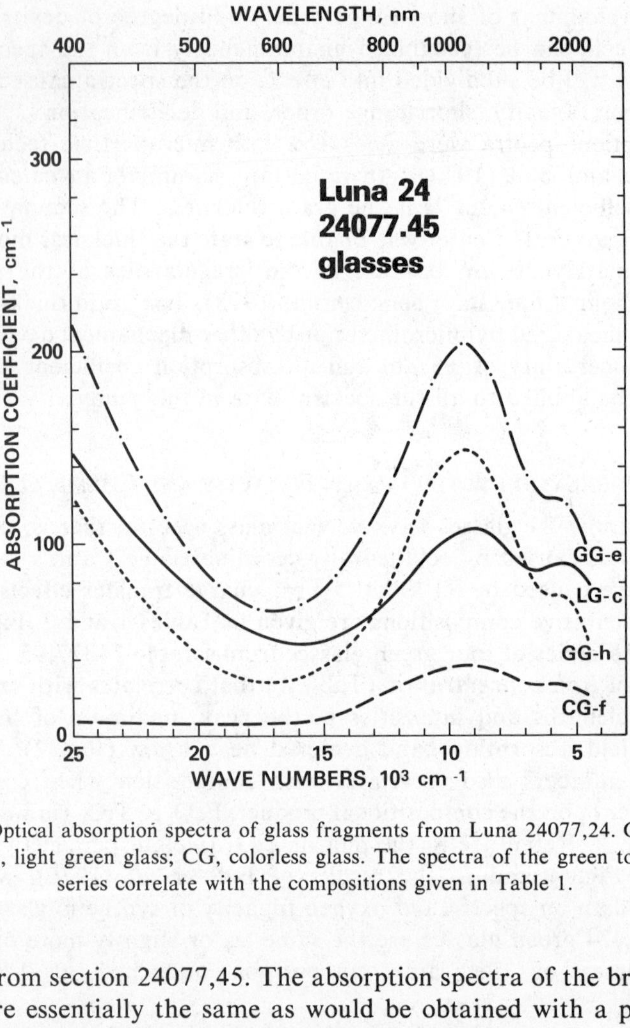

Fig. 1. Optical absorption spectra of glass fragments from Luna 24077,24. GG, green glass; LG, light green glass; CG, colorless glass. The spectra of the green to colorless series correlate with the compositions given in Table 1.

green glass from section 24077,45. The absorption spectra of the brown glass are sharp and are essentially the same as would be obtained with a polycrystalline sample of iron-bearing pyroxene (Bell *et al.*, 1976). These sharp spectral features are unlike the broad absorption bands of green and orange (red) glasses studied in other lunar samples. For example, the band at approximately 1.08 μm in normal glasses is shifted to 0.98 μm in brown glasses, and the broad, low-energy shoulder is observed as a distinct band at approximately 2 μm in the brown glasses. The brown glasses absorb more intensely along the near ultraviolet edge, and in some cases a small peak is observed at 500 nm. The brown glass spectra correlate with the spectra of lunar pyroxene of augite composition.

Figure 6 is a photomicrograph of a common Luna 24 brown glass fragment. These glasses are not devitrified on a microscopic scale (to ×40,000, SEM magnification), but typically they contain a few inclusions of plagioclase crystals (nearly pure anorthite, on the basis of scanning electron microscope nondispersive analysis). The inclusions could have been in the original liquid or could be

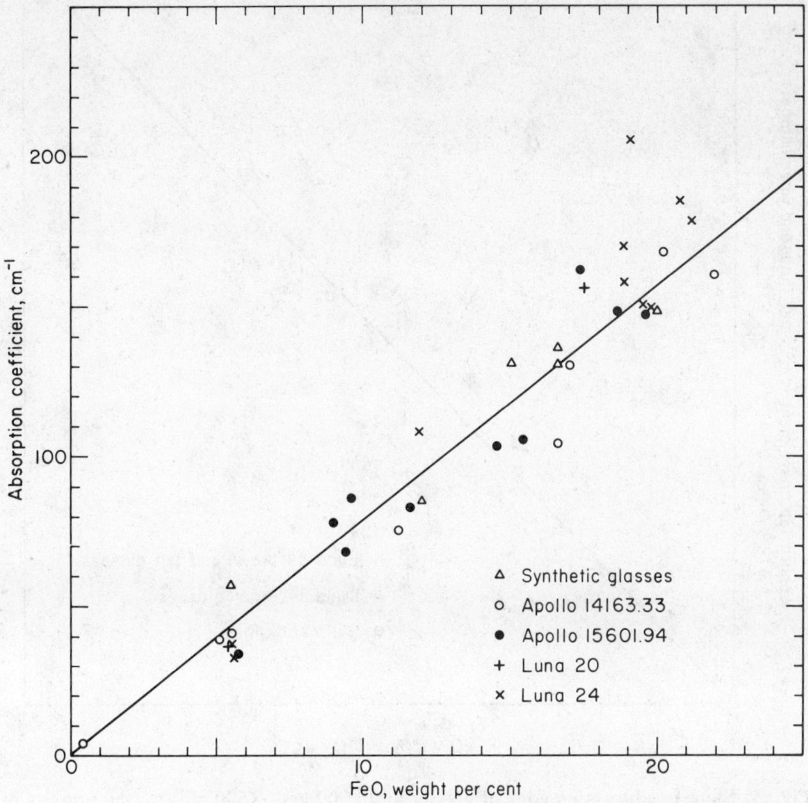

Fig. 2. Intensity of 1 μm crystal field observation band of glasses vs. FeO content (total Fe expressed as FeO). The symbol x represents the present Luna 24 glasses (brown glasses excluded). All other data are from Bell *et al.* (1976).

devitrification products. The grain shown in Fig. 6 shows bleached-appearing zones around the inclusions, but microprobe analysis indicates no compositional gradients. The glass portions of the brown glasses are uniform in composition.

Where extensive submicroscopic devitrification is observed, the intensity of absorption diminishes owing to light diffraction effects. The bulk spectra are significantly different from that of the glass portion. (The devitrification products are pyroxene, olivine, plagioclase, and opaque minerals.) The short-range order of the brown glass is an intermediate, distinct stage that apparently occurred before devitrification.

Figure 7 shows a photomicrograph of a brown glass fragment of uniform composition, most of which contains microcrystalline devitrification products. Only the glassy portions (i.e., left edge of the grain) give a sharp pyroxene-like absorption spectrum (Fig. 8, 24210,48a). The devitrified zones (Fig. 8, 24210,48b) give only weak bands and a strong absorption slope into the near

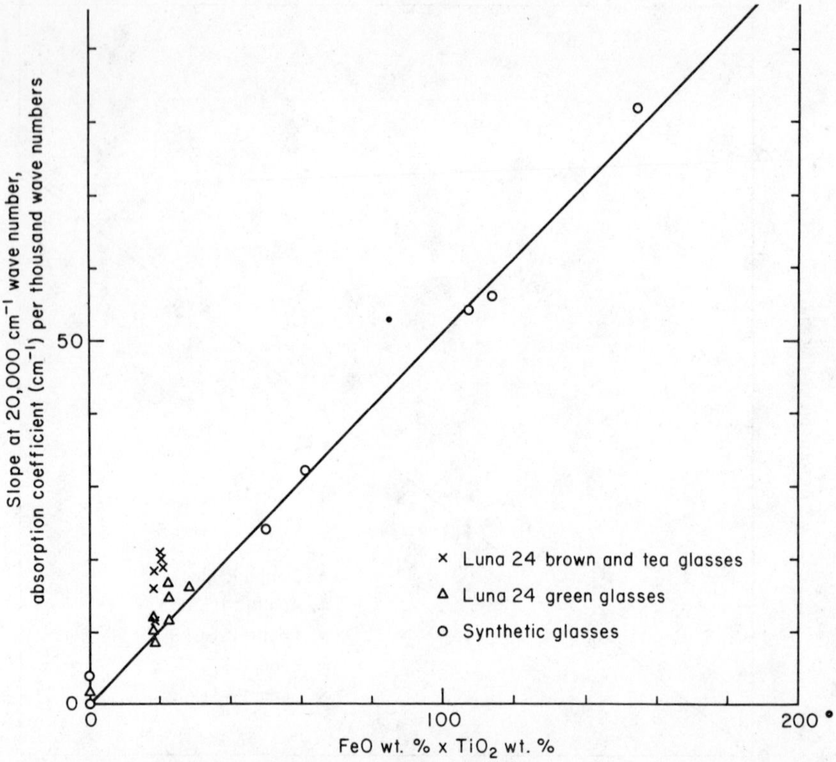

Fig. 3. Slope of absorption edge of glasses at 20,000 cm^{-1} (500 nm) vs. the product of total Fe and total Ti content (expressed as FeO and TiO_2). The data for synthetic glasses are from Bell *et al.* (1976).

ultraviolet region. The most intense and the sharpest absorption spectrum in Fig. 8 (24149,49c) was obtained from the glassy portions of another brown glass of the same composition that had extensive devitrified portions.

One can conclude from these results that spectra or color and composition of the brown glasses do not correlate. This lack of correlation can be seen in Fig. 3, where the brown and "tea-colored" (brownish-green) glasses plot off the composition line. (The Luna 24 glasses are exceptionally low in titanium, and their plot is crowded at one end of the diagram, but the brown glasses plot significantly distant from the composition line.)

Assignment of the absorption bands in the brown glasses is important because, on the basis of the observed spectral features, these glasses have structural states that were unknown previously. Those grains, which have sharp pyroxene-like spectra, represent a structural state observed for the first time in lunar glass. The grains are otherwise indistinguishable from normal lunar glass. The sharp spectral features, however, indicate that most Fe^{2+} cations are in an M2

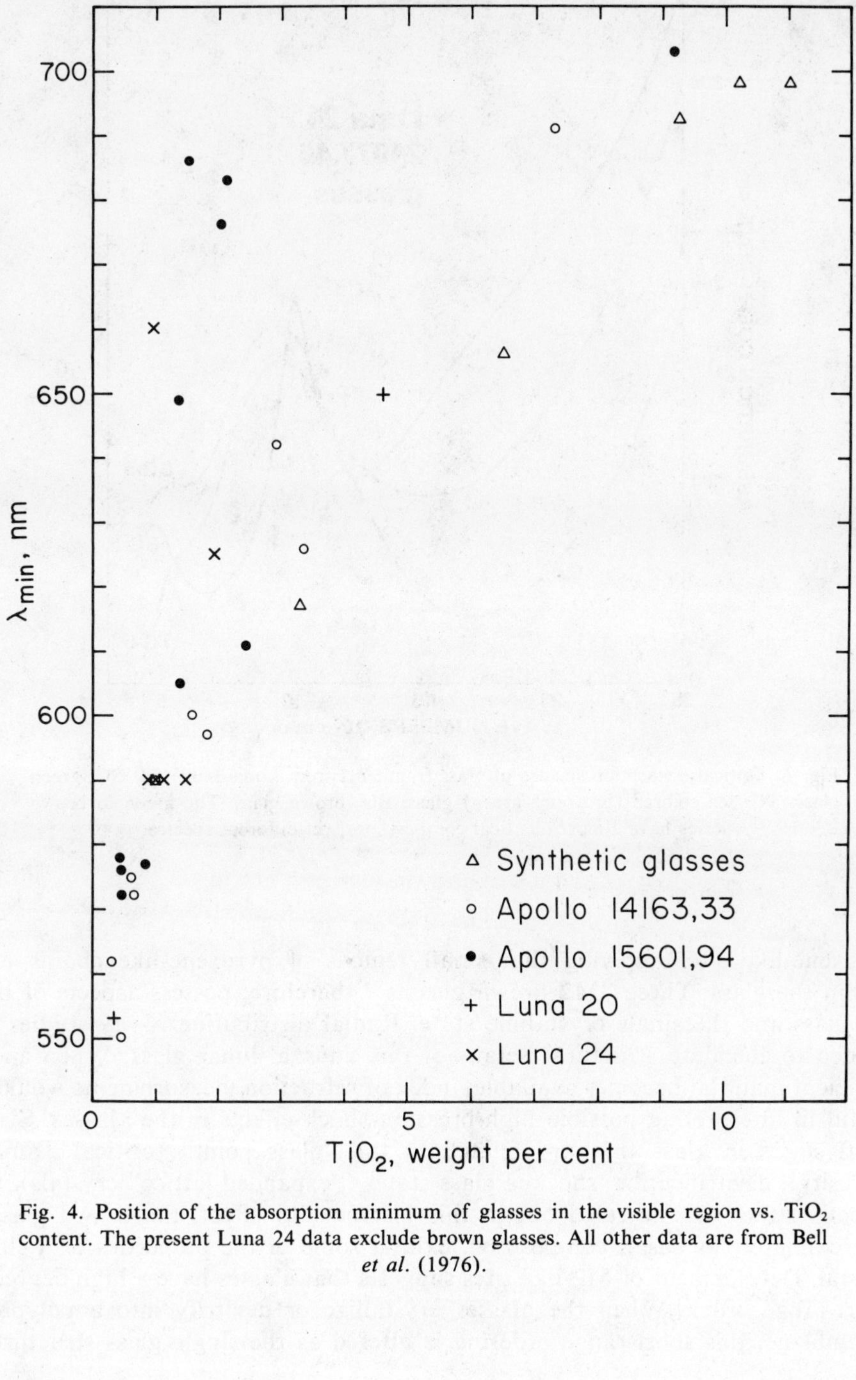

Fig. 4. Position of the absorption minimum of glasses in the visible region vs. TiO_2 content. The present Luna 24 data exclude brown glasses. All other data are from Bell *et al.* (1976).

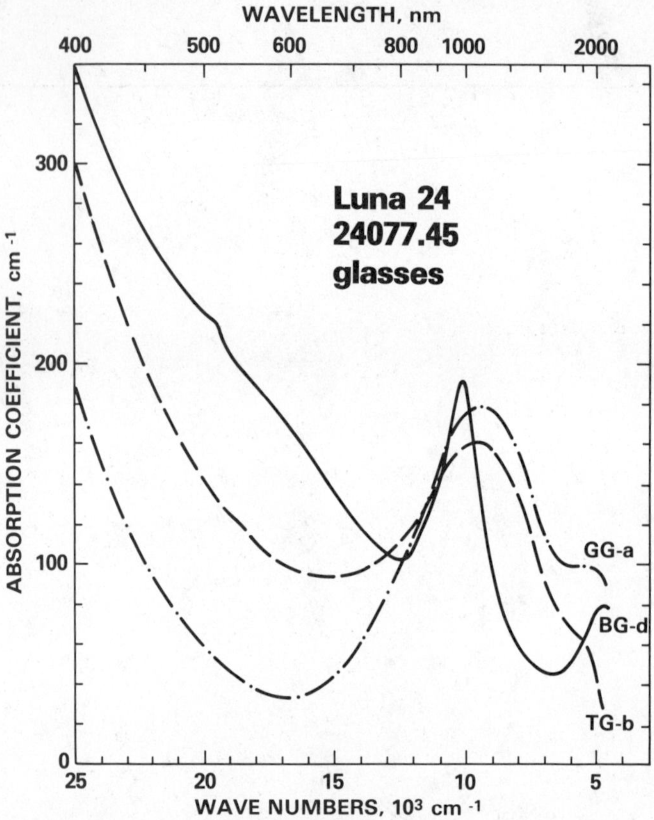

Fig. 5. Optical absorption spectra of glass fragments from Luna 24077,24. GG, green glass; TG, tea-colored (brownish-green) glass; BG, brown glass. The green to brown series have almost identical compositions, yet different spectra.

pyroxene-like site, implying that small regions of pyroxene-like chains exist within the glass. These "M2 brown glasses," therefore, possess aspects of both the glass and the single crystalline state. Radial distribution X-ray studies are needed to elucidate structural details of this unusual lunar glass. When and if sufficient material becomes available, index of refraction measurements would be useful in interpreting possible high-pressure shock effects in the glasses. Structural states in glass that are related to the "glass point" (critical Tamann viscosity), devitrification, shocked glass states ("expanded lattice" crystals), and smectic states can be ruled out in this instance. In effect, the brown glasses, approximately of basalt composition, exhibit some of the properties of a single crystal. Development of M2-like sites suggests that glasses have a high degree of short-range order. When the glasses crystallize or devitrify into a polyphase assemblage, this short-range ordering is altered as the single glass structure is

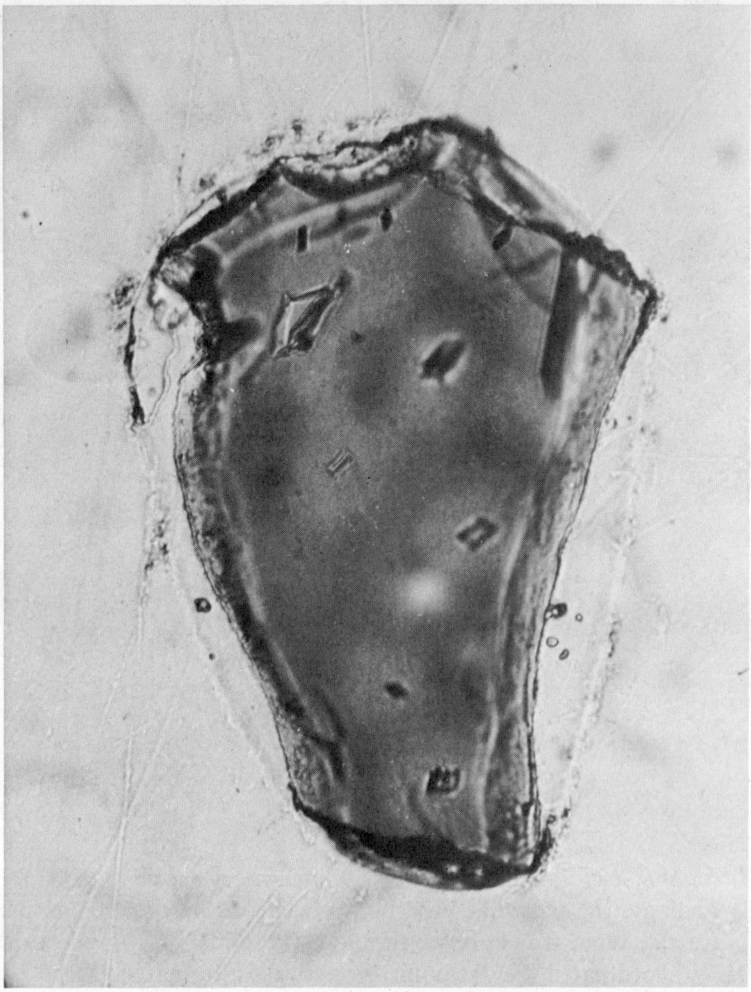

Fig. 6. Photomicrograph of brown glass 24077,45d. The length of the grain is 0.2 mm.

destroyed to reform as olivine, plagioclase, pyroxene, and opaque mineral structures.

The proposed assignment of the absorption bands is much the same as those of pyroxenes (Burns and Vaughan, 1975). The energy-level diagram for d^6 ions (Fe^{2+}, i.e., the Tanabe-Sugano diagrams by Berkes, 1968) indicates that the fundamental $T_{2g} \rightarrow E_g$ electronic transition is multiply degenerate, and both the ground and excited states undergo crystal-field splitting when the d^6 ion is in a pyroxene-like M2 site (Bancroft and Burns, 1967). The sharp absorption bands at wavelengths near 1 and 2 μm correspond to the transitions $^5A_1 - {}^5A_1$ and

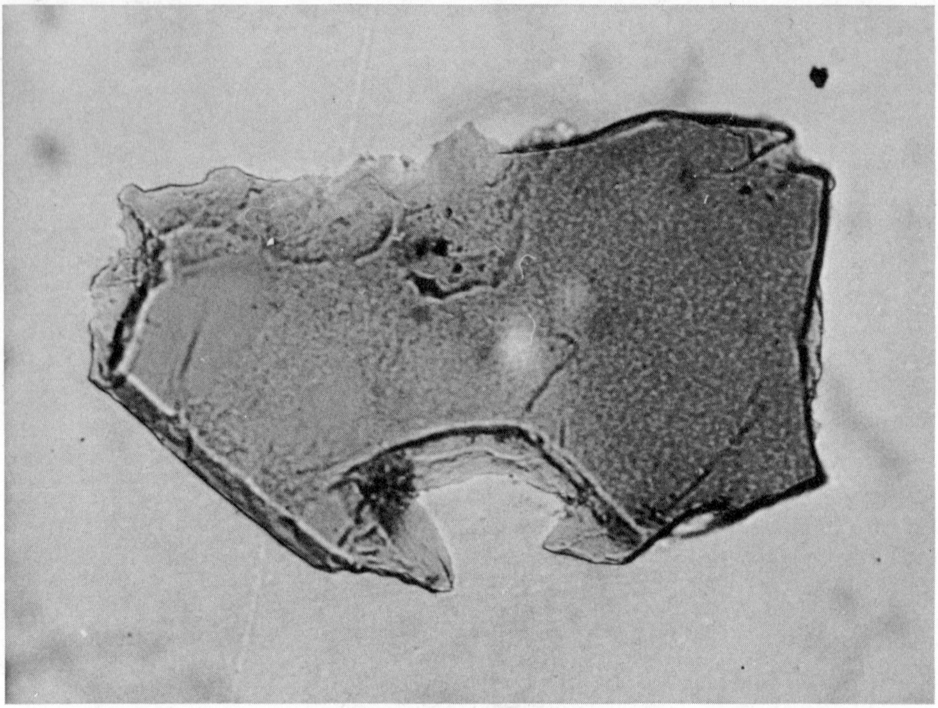

Fig. 7. Photomicrograph of brown glass 24210,48b. The length of the grain is 0.2 mm.

$^5A_1 - {}^5B_1$ for the selection rules of M2 site symmetry (see Fig. 6 of Bell *et al.*, 1975). By analogy, the fact that both bands occur in the brown glasses suggests that they are the same transitions. Although the possibility of the existence of tetrahedrally coordinated Fe^{2+} cannot be entirely eliminated, there is little, if any, evidence in the spectra that could be used to support arguments for its existence.

Figure 8 also shows a relatively sharp but weak absorption feature near 500 nm. This feature corresponds with the weak, sharp band observed in lunar pyroxenes (see, for example, Appollo 14 pigeonite spectra of Bell and Mao in Burns and Vaughan, 1975). These bands are assigned as spin-forbidden bands (change of spin multiplicity) of Fe^{2+} in pyroxenes, and are additional confirmation of the existence of octahedrally coordinated Fe^{2+} ions in the brown glasses.

ANORTHITE GLASS SPECTRA

Glasses of nearly pure anorthite composition comprise a separate set of glasses spectrally. Some of these glasses appear partially or completely brown (Fig. 9), and they should not be confused with the brown glasses already described. The coloring is not due to the presence of iron or other transition elements, but appears to be due to submicroscopic devitrification or perhaps to shock effects or radiation damage. The spectrum of the damaged areas appears as a nearly continuous, sloping absorption background in Fig. 10. Also shown in Fig. 10 is the spectrum of a glass fragment of slightly impure anorthite composition (see Table 2). There is a small feature near 1 μm that could be caused by Fe^{2+}, but otherwise there is simply a sloping absorption continuum. In both types of anorthite glass, the undamaged or clear zones do not show this absorption background. Previous to these observations, the only reported absorption effects of this kind in lunar materials were thought to be caused by metallic coatings on glass or crystals (Bell and Mao, 1977).

Conclusions

The crystal-field spectra of Luna 24 glasses have revealed several effects:

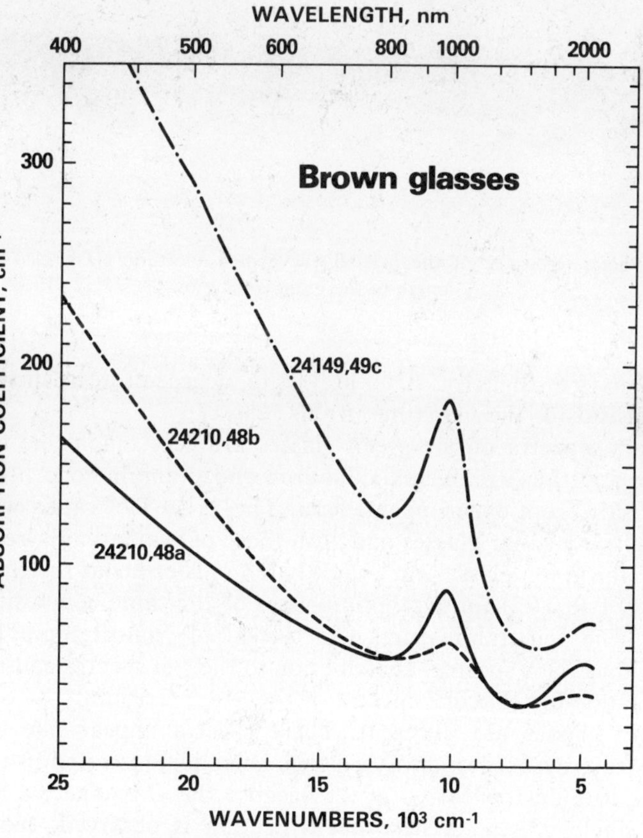

Fig. 8. Optical absorption spectra of three brown glasses with identical chemical compositions.

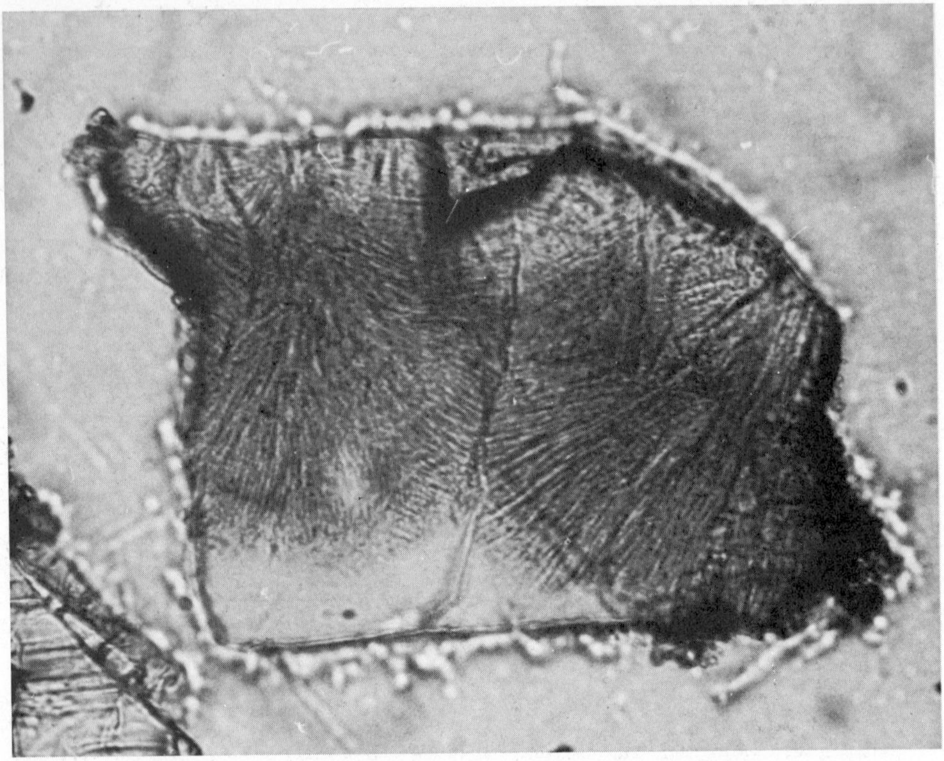

Fig. 9. Photomicrograph of the partially devitrified anorthite glass 24174,51a. The length of the grain is 0.2 mm.

composition, oxygen fugacity, structural state, degree of devitrification. The last two are described for the first time in this report.

In summary, spectra of the green glasses are very much like those of other lunar glasses and thus are normal. The iron and titanium concentrations can be deduced directly from experimental data. The $Ti^{3+}-Ti^{4+}$ ratios were compared with those of other lunar glasses and with those of synthetic glasses. The oxygen fugacities of the green glasses vary slightly but cluster about the equivalent value $-\log P_{O_2} = 9.1$ (±0.05) for synthetic glasses of the same composition quenched at 1400°C. The spectral features are typical of iron-titanium bearing lunar glasses, and thus their spectral contribution to the soil spectra causes no errors in the interpretation of telescope spectra of the Mare Crisium.

The brown glasses are different. These spectra appear like those of iron-titanium bearing pyroxenes, and yet it has been demonstrated in this study that this is not a true devitrification effect because the glass spectra are essentially those of a single crystal. Where devitrification is observed, the product is a polycrystalline basalt. These spectral features vary considerably with the apparent development of order in the glasses, and they do not correlate well with

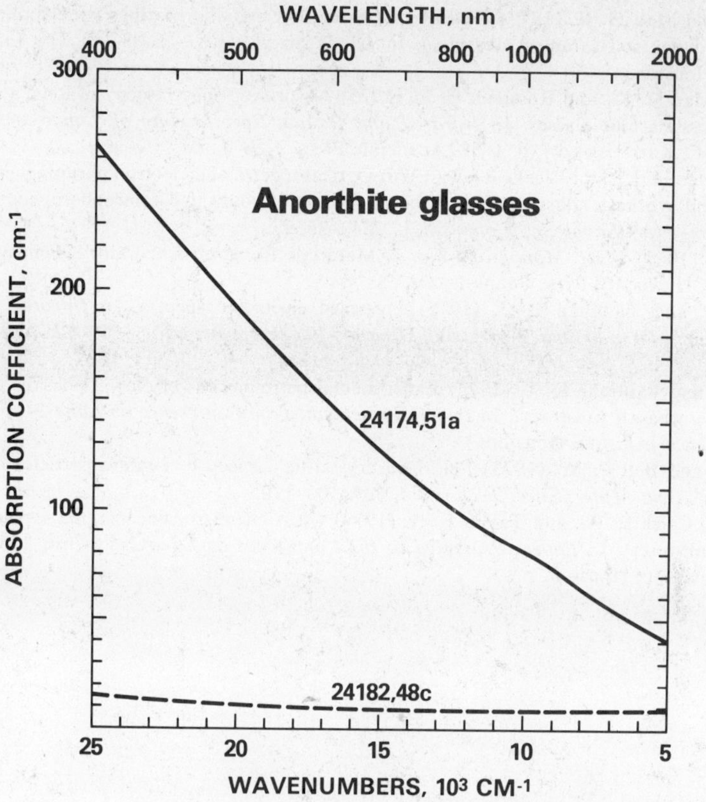

Fig. 10. Optical absorption spectra of the clear anorthite glass 24182,48c and the devitrified anorthite glass 24174,51a.

composition or oxygen fugacity. Fortunately, the spectral features of the brown glasses can be recognized from the present data and should not pose problems of interpretation. Preliminary studies of natural shocked glasses have not revealed these spectral features, but it would now be desirable to examine a range of terrestrial impact basalt glasses.

The brown glasses have a degree of order not previously reported in lunar samples. They are not brown because of composition alone, and thus it would be incorrect to assume that they are a given "basalt" type on the basis of visual observation. It will be important to reassess telescope spectra of Mare Crisium, but the distinct spectral features of the brown glasses should make reassessment possible.

References

Bancroft G. M. and Burns R. G. (1967) Interpretation of the electronic spectra of iron in pyroxenes. *Amer. Mineral.* **52**, 1278–1287.

Bell P. M. and Mao H. K. (1977) Optical spectra of thin metallic coatings with application to the spectra of lunar soil samples (abstract). In *Lunar Science VIII*, p. 88–90. The Lunar Science Institute, Houston.

Bell P. M., Mao H. K., and Rossman G. R. (1975) Absorption spectroscopy of ionic and molecular units in crystals and glasses. In *Infrared and Raman Spectroscopy of Lunar and Terrestrial Minerals* (C. Karr, Jr., ed.), p. 1–38. Academic Press, New York.

Bell P. M., Mao H. K., and Weeks R. A. (1976) Optical spectra and electron paramagnetic resonance of lunar and synthetic glasses: A study of the effects of controlled atmosphere, composition and temperature. *Proc. Lunar Sci. Conf. 7th*, p. 2543–2559.

Berkes J. S. (1968) *MRL Monograph No. 2*. Materials Research Laboratory, Pennsylvania State University, University Park, Pennsylvania.

Burns R. G. and Vaughan D. J. (1975) Polarized electronic spectra. In *Infrared and Raman Spectroscopy of Lunar and Terrestrial Minerals* (C. Karr, Jr., ed.), p. 39–72. Academic Press, New York.

Grove T. L. and Vaniman D. T. (1977) Experimental petrology of very low Ti basalts and origin of Luna 24 ferrobasalt (abstract). In *Papers Presented to the Conference on Luna 24*, p. 68–71. The Lunar Science Institute, Houston.

Mao H. K. and Bell P. M. (1973) Polarized crystal-field spectra of micro-particles of the moon. *Amer. Soc. Test. Mater. Spec. Tech. Publ.* **539**, 100–119.

Norman M., Coish R. A., and Taylor L. A. (1977) Glass chemistry and magma evolution at Mare Crisium (abstract). In *Papers Presented to the Conference on Luna 24*, p. 136–138. The Lunar Science Institute, Houston.

Merrill, R. B. and Papike, J. J., editors.
(1978) *Mare Crisium: The View from Luna 24*, Pergamon Press (New York), p. 281–289
Printed in the United States of America

Glasses in the Luna 24 core and petrogenesis of ferrobasalts

MARC NORMAN, R. A. COISH, and LAWRENCE A. TAYLOR
Department of Geological Sciences, The University of Tennessee, Knoxville, Tennessee 37916

Abstract—Modal abundance and major- and minor-element chemical analysis of homogeneous, non-agglutinitic mare and non-mare glasses from the Luna 24 drill core show that most glasses can be related to known rock types. Mare glasses include: brown glass identical in composition to the fine-grained low-Mg VLT basalt; green glass which might be related to a coarser-grained ferrogabbro; a high-K green glass; and a high-Ti orange glass.

Highland glass compositions include Highland basalt, gabbroic anorthosite, and pure anorthosite (i.e. plagioclase); minor Fra Mauro-type glass may also be present.

It is apparent that fractional crystallization of some primitive basaltic magma occurred at Mare Crisium producing a chemically evolved ferrobasalt and related glasses. An early, high-Mg basin fill, as represented by the olivine vitrophyres, may be the parent magma. Subsequent near-surface fractionation produced a multiply-saturated liquid that finally erupted as the ferrobasalt flows sampled by Luna 24.

INTRODUCTION

The Luna 24 drill core contains 25–45 modal % glassy particles, including agglutinates, quenched impact melts, and volcanic glass beads and fragments (Hu and Taylor, 1977b). A knowledge of the chemistries of these glasses will greatly aid our understanding of the formation of the regolith, as well as the petrology of the igneous rocks involved. Certain of these glasses are volcanic in origin. As such, they represent frozen samples of the endogenous lunar melts and can yield valuable insight into magma evolution at Mare Crisium. This paper presents a modal and chemical characterization of the homogeneous glass beads and fragments, many of which are thought to be of volcanic origin. The distinction between glasses resulting from volcanic versus impact processes is often difficult. Considerable discussion has been generated recently concerning criteria for distinguishing impact glasses from volcanic glasses (e.g. Simonds *et al.*, 1977). Agglutinates are formed by micrometeorite impact on the lunar soil and tend to reflect the bulk composition of that soil (Hu and Taylor, 1977a). They consist of mineral and lithic clasts welded together by glass. This glass usually contains myriads of small native iron blebs and is chemically inhomogeneous within a single fragment.

Butler's (1978) criteria for the recognition of glass droplets formed from an endogenous melt include: chemical homogeneity and regular shape; the absence of schlieren, shocked relict inclusions, and iron metal dust; and the presence of euhedral phenocrysts. We have applied some of these criteria in selecting the glasses discussed in this paper. In addition, we feel that if a glass composition matches that of known volcanic rock fragments at the same site, then it is valid to assume the glass represents a volcanic liquid.

RESULTS

Modal data indicate that the total homogeneous glass population comprises 3–7% of the 90–250 μm portion (Hu and Taylor, 1977b). The glasses were classified initially according to color and texture as brown, green, clear, orange, and devitrified. Absolute abundances of the color groups are shown in Fig. 1.

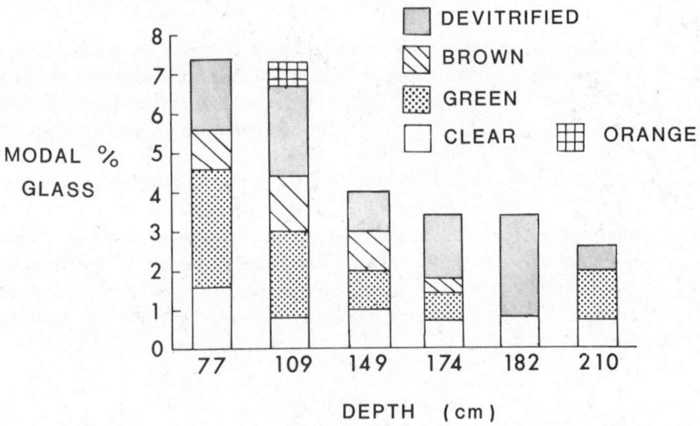

Fig. 1. Variations in modal percent of various glass color groups with depth in the Luna 24 drill core.

Major element chemistry was obtained with an automated MAC 400 S electron microprobe using appropriate natural and synthetic standards. Correction for matrix effects was applied according to procedures described by Bence and Albee (1968) and Albee and Ray (1970). A 40–50 μm diameter beam was employed for the analyses; three to seven analyses were performed on each grain. Mare and non-mare components were subsequently classified chemically on the basis of natural breaks in histograms and binary plots (Glass, 1976) as shown in Figs. 2 and 3. The mare component (<21% Al_2O_3, >14% FeO, CaO/Al_2O_3 ratio >0.80) is reflected in brown, green, and orange glass, while clear glass is attributed to a highland source (>21% Al_2O_3, CaO/Al_2O_3 ≤0.65). Over 60% of the glasses were derived from a mare source. A minor component with intermediate chemistry similar to the Fra Mauro basalt of other lunar soils (<21% Al_2O_3, <14% FeO, CaO/Al_2O_3 0.65–0.80) is also present. Devitrified glasses show textures ranging from spherulitic to myrmekitic and have chemistries which span the entire range of mare and non-mare compositions.

Within this broad classification, subdivisions generally corresponding to known rock types can be made (Reid *et al.*, 1972). A plot of the CIPW normative compositions (Fig. 4) and a variety of binary plots define these subdivisions. The mean of each group is presented in Table 1 along with comparative values from the literature.

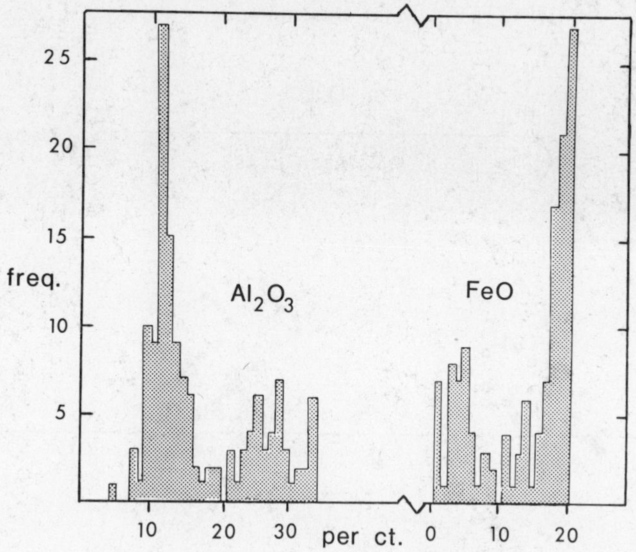

Fig. 2. Histograms of weight percent Al_2O_3 and FeO in the Luna 24 glasses. Note the distinctive bimodal distribution in each plot corresponding to mare and non-mare groups.

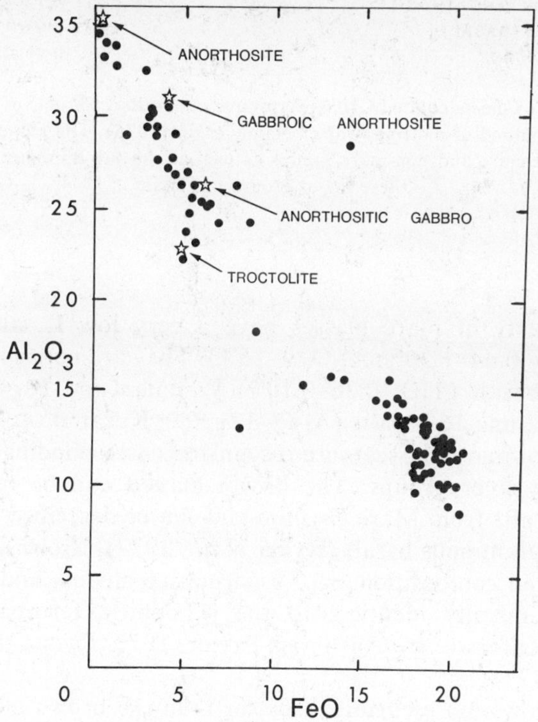

Fig. 3. Variation of Al_2O_3 against FeO in glasses from the Luna 24 drill core. Highland lithologies (Taylor, 1975) are also plotted. Note that all the highland glasses can be related to the highland suite of rocks (anorthosite-gabbroic anorthosite-anorthositic gabbro-troctolite).

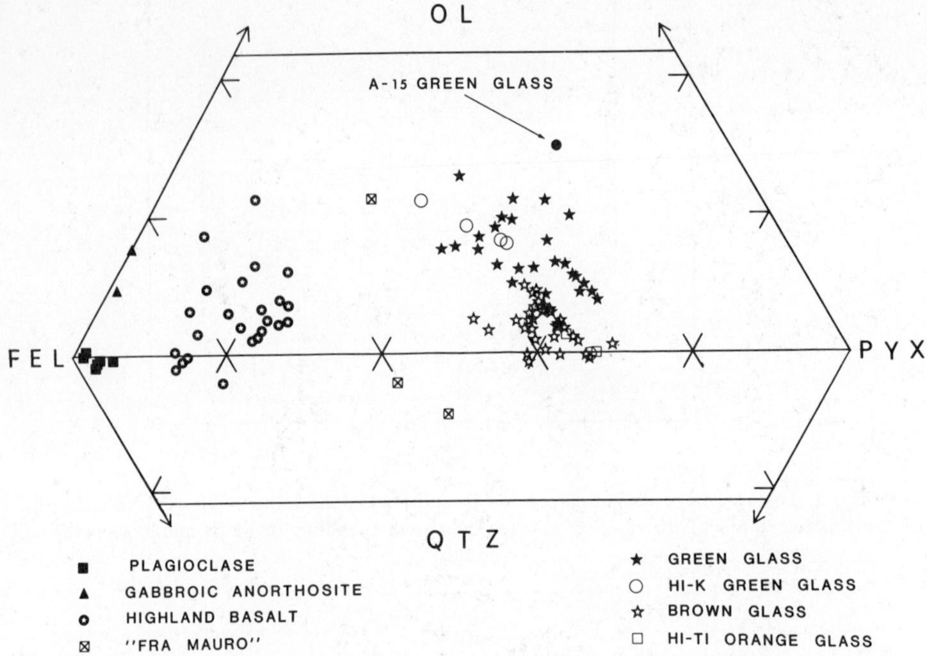

Fig. 4. A plot of the calculated CIPW normative mineralogy for glasses from the Luna 24 soils. The method of plotting is after Naney *et al.* (1976). The plot shows the clear separation into mare and non-mare groups as well as the sub-divisions within each of these major groups of glasses.

Mare glasses

The vast majority of mare glasses have a very low-Ti, aluminous basaltic composition (TiO_2 0.44–1.30%; Al_2O_3 9–16%). They are not as low in TiO_2 as the Apollo 17 VLT basalt (TiO_2 0.36–0.70%; Vaniman and Papike, 1977) nor as aluminous as the Luna 16 basalts (Al_2O_3 12–20%; Kurat *et al.*, 1976). Two main chemical groups of mare glasses were recognized corresponding essentially to the brown and green color groups. The brown glasses can be related directly to existing lunar basalts from Mare Crisium and can be described as equivalent to a low-Ti, low-Mg, aluminous basalt (Ryder *et al.*, 1977). Brown glass is characterized by a restricted composition, <10% normative olivine, and FeO/MgO ratio >2.25. It is chemically identical to the subophitic ferrobasalt reported by numerous investigators (e.g., Coish and Taylor, 1977, Ryder *et al.*, 1977; Grove and Vaniman, 1977).

Green glass shows more chemical scatter than the brown glass; consequently, characterization of its related rock type is difficult. The average green glass, however, has a composition similar to a coarse-grained ferrogabbro thought to exist at the site (Taylor *et al.*, 1977). Chromian spinel is occasionally present as small, equidimensional octahedral crystals which appear deep orange to opaque

Table 1. Luna 24 glass compositions in the 90–250 μm size fractions.

	Brown Glass	Ferrobasalt*	Green Glass	Ferrogabbro#	High-K Mare Glass	High-Ti Mare Glass
SiO_2	46.19 (0.61)	46.40	44.96 (0.92)	47.1	44.90 (1.37)	43.13
TiO_2	0.95 (0.15)	0.97	0.90 (0.20)	0.6	1.42 (0.21)	4.89
Al_2O_3	12.60 (1.19)	12.38	11.52 (1.66)	10.7	12.24 (1.53)	10.58
Cr_2O_3	0.25 (0.09)	0.18	0.43 (0.09)	0.2	0.39 (0.08)	0.38
FeO	19.04 (1.32)	17.68	18.50 (1.23)	20.0	18.61 (1.57)	18.11
MgO	6.85 (0.67)	6.45	10.52 (1.64)	9.0	8.23 (1.03)	9.55
CaO	12.32 (0.58)	13.13	10.81 (0.59)	11.8	11.86 (0.43)	10.20
Na_2O	0.31 (0.09)	0.26	0.27 (0.13)	0.25	0.32 (0.15)	0.13
K_2O	0.01 (0.01)	0.04	0.02 (0.02)	—	0.48 (0.11)	0.04
Total	98.52	97.49	97.93	99.65	98.45	97.01
FeO/MgO	2.78	2.74	1.76	2.22	2.26	1.90
CaO/Al_2O_3	0.98	1.06	0.94	1.10	0.97	0.96
No. Grains	35		23		5	2

*Average of four analyses from Ryder *et al.*, (1977).
#Taylor *et al.* (1977)
Numbers in parentheses are standard deviations.

Table 1. (*Continued*)

	Highland Basalt Glass	Highland Basalt‡	Gabbroic Anortho. Gl.	Plagioclase Glass	Luna 24 "HASP"	"Fra Mauro"
SiO_2	45.18 (1.19)	45.35 (0.91)	42.07	44.32 (1.25)	38.50 (3.05)	48.00 (2.95)
TiO_2	0.34 (0.17)	0.43 (0.22)	0.24	0.16 (0.08)	0.31 (0.10)	1.79 (0.28)
Al_2O_3	26.68 (2.26)	27.63 (1.49)	31.50	33.73 (0.64)	30.43 (2.92)	16.54 (1.64)
Cr_2O_3	0.13 (0.06)	0.07 (0.04)	0.02	0.01 (0.02)	0.08 (0.03)	0.20 (0.16)
FeO	4.98 (1.59)	4.62 (0.97)	3.38	0.55 (0.33)	8.81 (6.39)	11.29 (2.17)
MgO	6.30 (2.55)	6.13 (1.40)	4.45	0.39 (0.41)	6.98 (4.98)	8.10 (2.32)
CaO	15.11 (1.05)	15.64 (0.82)	17.51	18.37 (0.50)	17.11 (1.82)	11.80 (0.55)
Na_2O	0.40 (0.24)	0.42 (0.23)	0.05	0.61 (0.35)	0.01 (0.02)	0.57 (0.08)
K_2O	0.03 (0.03)	0.05 (0.05)	—	0.04 (0.04)	—	0.35 (0.46)
Total	99.15	100.34	99.22	98.18	102.23	98.64
FeO/MgO	0.79	0.75	0.76	1.41	1.26	1.39
CaO/Al_2O_3	0.57	0.57	0.56	0.55	0.56	0.71
No. grains	26		2	6	3	3

‡Reid *et al.* (1973)
Numbers in parentheses are standard deviations.

in transmitted light. Spinel can be seen in both vitrified and devitrified grains, testifying to their common origin. Green glasses are distinguished chemically from their brown counterparts by a higher normative olivine content (10–27%) and lower FeO/MgO (<2.25) as shown in Table 1.

Other mare glasses that may represent possible rock types in or near Mare Crisium include a high-K green glass (K_2O 0.41–0.68%) and two particles containing significantly more TiO_2 (4.69–5.09%) than the other mare glasses. The high-K glass is also somewhat higher in TiO_2 (1.29–1.78%) than the other common mare glasses. It is optically indistinguishable from the common green glass discussed above. The high-Ti fragments are easily recognized by their bright orange color; however, the Apollo 17 orange glass, by comparison, contains almost twice as much TiO_2.

Non-mare glasses

The non-mare component of the glass populations consists of fragments with intermediate composition and beads and fragments having a definite highland affinity. The fragments which are chemically intermediate to the mare and highland components possess a slight quartz-normative component and are low in both Al_2O_3 (<21%) and FeO (<14%). This division is represented by only three devitrified grains, so it is possible that these are the result of impact mixing of mare and highland material.

Highland composition glasses can all be derived from the common ANT suite, which consists of anorthosite, gabbroic anorthosite, anorthositic gabbro (Highland basalt), and troctolite. These rock compositions (after Taylor, 1975) are plotted along with the glasses on Fig. 3. Many of these glasses possess schlieren indicating a probable impact origin. Source material for these glasses is probably the basin rim 40 km from the landing site, or, alternatively, ejecta from large, distant cratering events such as Giordano Bruno (Butler and Morrison, 1977).

Within the highland group of glasses, several clusters corresponding to previously recognized compositions are present (Reid *et al.*, 1972). Highland basalt is represented by clear glass beads and fragments that are remarkably homogeneous both chemically and optically. They are, however, somewhat chemically diverse from grain to grain, ranging from 20% normative olivine to 5% normative quartz, and from 65 to 90% normative feldspar. Gabbroic anorthosite (85% normative feldspar and 15% normative olivine) was identified in minor quantities as was plagioclase glass (≃anorthosite). The plagioclase glass is occasionally associated with extremely shocked pyroxene indicating the glass is probably diaplectic. All highland glasses are characterized by a nearly constant CaO/Al_2O_3 of 0.57, the ratio of calcic plagioclase.

Small amounts of clear glass are too SiO_2 poor (35–41%) to result in a realistic norm analysis. These grains possess high alumina (28–34%), low Na_2O (<0.04%) and no detectable K_2O. Most of the grains have a very low FeO/MgO (<0.33) compared to the average highland component (FeO/MgO ≃0.80). This is consistent with the hypothesis of Naney *et al.* (1976), of differential volatilization of a highland source upon impact. Such a process would surely include a loss of SiO_2, alkalis, and some FeO.

DISCUSSION: ORIGIN OF FERROBASALTS

The brown glass clearly has a composition identical to the Luna 24 ferrobasalts. In this section of the paper, we discuss the origin of the volcanic melts which produced both the basalt and the glass. Because these liquids are multiply-saturated (Grove and Vaniman, 1977), they probably do not represent primary melts; rather, they are derived from a more primitive composition by fractional crystallization. The important problem, then, is to determine a composition of a likely parent magma.

The primitive magma was probably basaltic in composition, enriched in MgO (high MgO/MgO + FeO), and depleted in TiO_2 relative to the ferrobasalt, since olivine is an early crystallizing phase (Grove & Vaniman, 1977). Several investigators have suggested the Apollo 15 green glass as a possible parent due to its primitive mafic composition (Fig. 4) (Taylor *et al.*, 1977; Vaniman & Papike, 1977; Norman *et al.*, 1977). Olivine followed by pyroxene is the crystallization sequence at low pressure (Stolper, 1974). The ferrobasalt, however, appears to be near multiply-saturated in olivine and plagioclase (Grove, 1978). Thus, the Apollo 15 green glass is probably not the parent of the Luna 24 ferrobasalt. This conclusion is supported by rare earth data which indicate that the green glass cannot be fractionated to produce the REE pattern of the ferrobasalt (Ma and Schmitt, 1977). No material matching the composition of the Apollo 15 green glass was found at the Luna 24 site; therefore, the choice of it as a parent is artibrary, at best.

There is considerable evidence to suggest that a Mg-rich basalt existed at Mare Crisium sometime before the emplacement of the ferrobasalts. Andre *et al.* (1977) report a discontinuity about 1400 m beneath the surface of Mare Crisium based on lunar sounder data and attribute this to early, high-MgO mare volcanism. This is supported by remote sensing X-ray fluorescence data on material excavated from the crater Picard, which is believed to have sampled the Mg-rich layers (Schonfeld and Bielefeld, 1977). Stratigraphic studies of the area indicate that this unit is the earliest visible mare unit and that patches of this material occur along the basin rim to the east of the Luna 24 site (Adams *et al.*, 1978).

The apparent existence of primitive mare basalt activity in the Crisium basin indicates that samples of such material should be found in the returned core. Indeed, it appears that there is material in the Luna 24 core which could be interpreted as having been derived from this high-MgO unit. Very Fo-rich olivines and Mg-rich pyroxenes are present in the core as monomineralic fragments. Some of these may be of highlands origin (Bence and Grove, 1977). However, the majority of these particles are probably from a mare basalt rock type (Taylor *et al.*, 1977; Basu *et al.*, 1977b, Papp *et al.*, 1977). Therefore, they may represent a sampling of the early, high-MgO basalt fill, probably derived from the unit near the eastern edge of the basin.

In addition to the monomineralic fragments, primitive, Mg-rich olivine vitrophyres have been reported by several investigators (e.g., Ryder *et al.*, 1977; Taylor *et al.*, 1977; Basu *et al.*, 1977a; Coish and Taylor, 1977; 1978). These rare rock fragments most often consist of euhedral, skeletal to prismatic olivines in a glassy matrix and are somewhat variable in bulk composition. Vitrophryes of this type are important in that they potentially represent the bulk composition of the early mare fill (Andre *et al.*, 1977). They are enriched in MgO and Cr_2O_3 and depleted in Al_2O_3 and TiO_2 relative to the ferrobasalts, and their primitive compositions make them excellent candidates for a possible parent magma. Previously we suggested that the Luna 24 basalts could be produced by fractionating olivine and pyroxene from an olivine vitrophyre composition

(Norman *et al.*, 1977). However, due to phase constraints similar to those discussed for the Apollo 15 green glass, this scheme now appears untenable. Instead, graphical methods indicate that *the ferrobasalt can be derived from an average olivine vitrophyre composition by the simple fractionation of olivine* (Coish and Taylor, 1978).

CONCLUSIONS

1) Brown, green, clear, orange, and devitrified non-agglutinitic glasses are abundant (i.e. 3–7%) in the Luna 24 drill core and represent both mare and non-mare compositions.
2) Most of the glass compositions can be related to known rock types. Brown glass is identical to a low-Mg, low-Ti, aluminous mare basalt. Green glasses are similar but with higher MgO contents. Orange glasses are distinguished by their higher TiO_2 values and have no equivalent rock type in the core. Clear glasses are probably from a highlands source as indicated by their alumina contents and CaO/Al_2O_3 ratios.
3) The multiply-saturated brown glass and related ferrobasalt may have been derived by fractional crystallization of a parent magma with a composition similar to that of some olivine vitrophyres found in the core.

Acknowledgments—We wish to express our appreciation to the Soviet Union for providing the U.S.A. with some of their precious samples and to LSAPT for the efficient manner in which the processing and allocations were conducted. The constructive reviews by Drs. I. Steele and C. Meyers are greatly appreciated. Thanks are also given to the University of Tennessee for providing substantial support for the purchase of the electron microprobe. The research presented in this paper was supported by NASA Grant NGL 43-001-140.

REFERENCES

Adams J. B., Head J. W., III, McCord T. B., Pieters C. and Zisk S. (1978) Mare Crisium: Regional stratigraphy and geologic history. *Geophys. Res. Lett.*, submitted.

Albee A. and Ray L. (1970) Correction factors for electron probe microanalysis of silicates, oxides, carbonates, phosphates, and sulfates. *Anal. Chem.* **42**, 1408–1414.

Andre C. G., Wolfe R. W. and Adler I. (1977) Evidence for a high-magnesium subsurface basalt in Mare Crisium from orbital x-ray fluorescence data (abstract). In *Papers Presented to the Conference on Luna 24*, p. 8–11. The Lunar Science Institute, Houston.

Basu A., McKay D. S., Clanton V., Waits G. and Fruland R. M. (1977a) Glass particles in Luna 24 drill core soils (abstract). In *Papers Presented to the Conference on Luna 24*, p. 14–17. The Lunar Science Institute, Houston.

Basu A., McKay D. S. and Fruland R. M. (1977b) Petrography, mineralogy and source rocks of Luna 24 drill core soils (abstract). In *Papers Presented to the Conference on Luna 24*, p. 18–21. The Lunar Science Institute, Houston.

Bence A. E. and Albee A. (1968) Empirical correction factors for the electron microanalysis of silicates and oxides. *J. Geol.* **76**, 382–403.

Bence A. E. and Grove T. L. (1977) The highland component in the Luna 24 core (abstract). In *Papers Presented to the Conference on Luna 24*, p. 22–24. The Lunar Science Institute, Houston.

Butler P., Jr. (1978) Recognition of lunar glass droplets produced directly from endogenous melts: the evidence from S-Zn coatings (abstract). In *Lunar and Planetary Science IX*, p. 143–145. Lunar and Planetary Institute, Houston.

Butler P., Jr. and Morrison D. A. (1977) Geology of the Luna 24 landing site. *Proc. Lunar Sci. Conf. 8th*, p. 3281–3301.

Coish R. A. and Taylor L. A. (1978) Mineralogy and petrology of basaltic fragments from the Luna 24 drill core. In *Mare Crisium: The View from Luna 24* (R. B. Merrill and J. J. Papike, eds.). This volume.

Coish R. A. and Taylor L. A. (1977) Igneous rocks from Mare Crisium: Mineralogy and petrology (abstract). In *Papers Presented to the Conference on Luna 24*, p. 48–50. The Lunar Science Institute, Houston.

Glass B. P. (1976) Major element composition of glasses from Apollo 11, 16, and 17 soil samples. *Proc. Lunar Sci. Conf. 7th*, p. 679–693.

Grove T. L. (1978) Cooling histories of Luna 24 low-Ti ferrobasalts and ferrogabbros (abstract). In *Lunar and Planetary Science IX*, p. 424–426. Lunar and Planetary Institute, Houston.

Grove T. L. and Vaniman D. T. (1977) Experimental petrology of very low Ti basalts and origin of Luna 24 ferrobasalts (abstract). In *Papers Presented to the Conference on Luna 24*, p. 68–71. The Lunar Science Institute, Houston.

Hu H.-N. and Taylor L. A. (1977a) Lack of chemical fractionation in major and minor elements during agglutinate formation. *Proc. Lunar Sci. Conf. 8th*, p. 3645–3656.

Hu H.-N. and Taylor L. A. (1977b) Luna 24 soils from Mare Crisium: Agglutinate chemistry (abstract). In *Papers Presented to the Conference on Luna 24*, p. 82–84. The Lunar Science Institute, Houston.

Kurat G., Kracher A., Keil K., Warner R. and Prinz M. (1976) Composition and origin of the Luna 16 aluminous mare basalt. *Proc. Lunar Sci. Conf. 7th*, p. 1301–1321.

Ma M.-S. and Schmitt R. A. (1977) Luna 24 soils: A chemical study (abstract). In *Papers Presented to the Conference on Luna 24*, p. 98–101. The Lunar Science Institute, Houston.

Naney M. T., Crowl D. M. and Papike J. J. (1976) The Apollo 16 drill core: Statistical analysis of glass chemistry and the characterization of a high alumina-silica poor (HASP) glass. *Proc. Lunar Sci. Conf. 7th*, p. 155–184.

Norman M., Coish R. A. and Taylor L. A. (1977) Glass chemistry and magma evolution at Mare Crisium (abstract). In *Papers Presented to the Conference on Luna 24*, p. 136–138. The Lunar Science Institute, Houston.

Papp H., Steele I. M. and Smith J. V. (1977) Luna 24 90–150 micron fraction: Implications for sampling of planetary and asteroidal regoliths. In *Papers Presented to the Conference on Luna 24*, p. 143–146. The Lunar Science Institute, Houston.

Reid A. M., Warner J., Ridley W. I., Johnston D. A., Harmon R. S., Jakes P. and Brown R. W. (1972) The major element composition of lunar rocks as inferred from glass compositions in the lunar soils. *Proc. Lunar Sci. Conf. 3rd*, p. 363–378.

Ryder G., McSween H. Y., Jr. and Marvin U. (1977) Basalts from Mare Crisium. *The Moon* **17**, 263-287.

Schonfeld E. and Bielefeld M. J. (1977) Correlation between geochemical "features" and photogeological features in Mare Crisium (abstract). In *Papers Presented to the Conference on Luna 24*, p. 167–169. The Lunar Science Institute, Houston.

Simonds C. H., Warner J. L., McGee P. E., Cochran A. and Brown R. W. (1977) Luna 24: Petrochemical potpourri (abstract). In *Papers Presented to the Conference on Luna 24*, p. 170–173. The Lunar Science Institute, Houston.

Stolper E. M. (1974) Lunar ultramafic glasses. B. A. Thesis, Harvard Univ., 178 pp.

Taylor G. J., Warner R., Wentworth S., Keil K. and Sayeed U. (1977) Petrology and chemistry of Luna 24 mare basalts and basaltic glasses (abstract). In *Papers Presented to the Conference on Luna 24*, p. 189–192. The Lunar Science Institute, Houston.

Taylor S. R. (1975) *Lunar science: A post-Apollo view.* Pergamon, N.Y. 372 pp.

Vaniman D. T. and Papike J. J. (1977) Very low Ti (VLT) basalts: A new mare rock type from the Apollo 17 drill core. *Proc. Lunar Sci. Conf. 8th*, p. 1443–1471.

Merrill, R. B. and Papike, J. J., editors.
(1978) *Mare Crisium: The View from Luna 24*, Pergamon Press (New York), p. 291–302
Printed in the United States of America

Soils from Mare Crisium: Agglutinitic glass chemistry and soil development

HSIEN-NENG HU and LAWRENCE A. TAYLOR

Department of Geological Sciences, The University of Tennessee, Knoxville, Tennessee 37916

Abstract—Detailed chemical and modal analyses were conducted on the agglutinates in 29 polished thin sections of grain mounts from the 90–150 μm, 150–250 μm, and >500 μm fractions of six Luna 24 soil horizons: 24077, 24109, 24149, 24174, 24182, and 24210. Three populations of agglutinitic glass compositions were recognized. In order of decreasing abundance, they are: 1) a high MgO (≃10%), high FeO (≃18%) group representing a coarse-grained basaltic component; 2) a low MgO (≃6%), low FeO (≃8%) group from a highland source; and 3) a low MgO (≃7%), high FeO (≃19%) group probably from the subophitic basalt component. It appears that these 3 soil components each had a previous history prior to mixing at the Luna 24 site, and the *in situ* reworking and mixing has not been sufficient to erase the chemistries, size distributions, and "maturities" which the individual components have imparted to the soil.

The agglutinitic glass compositions display, on the average, an enrichment in plagioclase and a depletion in ferromagnesian elements relative to the bulk soil, in a reverse sense from that predicted by the agglutinate chemical fractionation theory. This is probably due to the significant amount of admixed highland component, with lesser contribution from selective melting of plagioclase.

INTRODUCTION

The drill core returned from the Southeast edge of Mare Crisium by the U.S.S.R. unmanned space probe Luna 24 has provided us with materials from a previously unsampled portion of the Moon. The detailed analyses of these soils can form the basis for advanced understanding of regolith and rock-forming processes. In fact, the chemical processes accompanying soil formation on the lunar surface should be reflected in the chemistry of the impact-produced glasses which bond mineral and lithic fragments together into aggregates called *agglutinates*. The present study was performed in order to chemically characterize the Luna 24 agglutinates and associated glasses and to thereby determine the main source materials for the development of the soil. This detailed examination of the material making up 25% of the soils provides a basis for our understanding the core stratigraphy and evolutionary history of the soils of Mare Crisium.

PROCEDURE

Numerous (i.e. 29) polished-thin sections (PM) of grain mounts from the 90–150 μm (11 PM), 150–250 μm (11 PM), and >500 μm (7 PM) portions of six samples—24077, 24109, 24149, 24174, 24182, and 24210—were examined in this study. Approximately 20–25 agglutinate particles from each sample in the 90–250 μm size range, as well as 9 agglutinates in the >500 μm fractions, were carefully selected according to the criteria of Hu and Taylor (1977a). With each particle, 5 to 10 points of representative agglutinitic glass were analyzed for eight major and minor elements with an automated MAC 400S electron microprobe, using appropriate synthetic and natural standards. All analyses were performed with a defocussed beam of spot size 30–40 μm in diameter. Corrections for matrix effects were applied according to procedures described by Bence and Albee (1968) and Albee and Ray (1970).

Modal analyses of the 90–150 μm and 150–250 μm size fractions were also performed. These size fractions were divided into three general components—agglutinate, breccia, and other (Table 1). The 150–250 μm fractions were further subdivided into seven components (Fig. 1)—plagioclase, olivine & pyroxene, agglutinate & breccia, glass, opaques, coarse-, and fine-grain rock fragments. Phinney *et al.* (1976) in an SEM study developed criteria to distinguish breccia from agglutinate. We used some of these criteria as shown by their reflected light photographs. However, the distinction between these two groups by optical microscope is very subjective and user dependent; hence, we have combined the agglutinate and breccia in Figure 1.

Table 1. Modal analyses of agglutinates in the 90–150 and 150–250 μm fractions of Luna 24.

	24077				24109				24149			
	90–150 μm		150—250 μm		90–150 μm		150–250 μm		90–150 μm		150–250 μm	
	grains	%	grains	%	grains	%	grains	%	grains	%	grains	%
Agglut. + Brec.	181	37.8	238	39.2	185	48.5	193	42.3	247	45.8	109	40.7
Agglutinates	(154)	(32.2)	(195)	(32.1)	(141)	(37.0)	(151)	(33.1)	(110)	(20.4)	(44)	(16.4)
Breccias	(27)	(5.6)	(43)	(7.1)	(44)	(11.5)	(42)	(9.2)	(137)	(25.4)	(65)	(24.3)
Others	297	62.2	369	60.8	196	51.5	263	57.7	293	54.2	159	59.3
Totals	478	100.0	607	100.0	381	100.0	456	100.0	540	100.0	268	100.0

	24174				24182				24210			
	90–150 μm		150–250 μm		90–150 μm		150–250 μm		90–150 μm		150–250 μm	
	grains	%	grains	%	grains	%	grains	%	grains	%	grains	%
Agglut. + Brec.	113	26.0	131	23.1	83	20.2	69	26.1	146	32.9	38	25.0
Agglutinates	(91)	(20.9)	(103)	(17.6)	(55)	(13.4)	(46)	(17.4)	(117)	(26.4)	(28)	(18.4)
Breccias	(22)	(5.1)	(28)	(5.5)	(28)	(6.8)	(23)	(8.7)	(29)	(6.5)	(10)	(6.6)
Others	322	74.0	436	76.9	328	79.8	195	73.9	297	67.1	114	75.0
Totals	435	100.0	567	100.0	411	100.0	264	100.0	443	100.0	152	100.0

RESULTS AND DISCUSSION

The modal analyses of the six soil samples are presented in Figure 1 and Table 1. Figure 2 shows the agglutinate contents of the six samples as determined by the studies of Basu *et al.* (1977), McKay *et al.* (1977), Simonds *et al.* (1977) and the present investigation (Hu and Taylor, 1977b). A decreasing trend in abundance of agglutinate with increasing depth is apparent, although different criteria were obviously used by the different research groups resulting in a spread of values for a given sample. This may involve subtle distinctions between agglutinate and microbreccia. If modal amounts of agglutinates are ever to be used effectively by different investigators, some set of standardized criteria must be developed for recognition of agglutinates.

Using the criteria of McKay *et al.* (1974), the agglutinate contents of the 90–

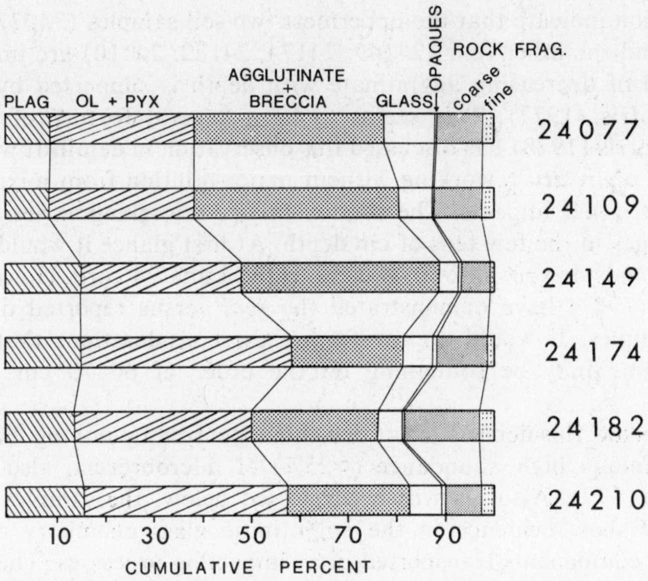

Fig. 1. Modal analysis of the 150–250 μm portion of soils from the Luna 24 core.

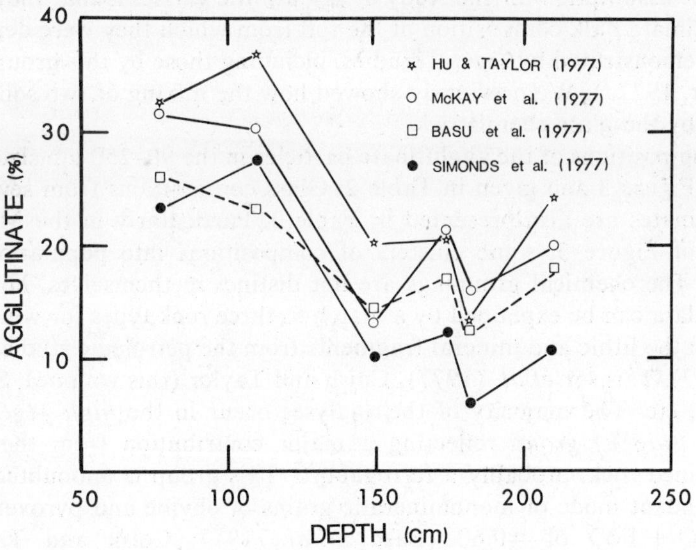

Fig. 2. Agglutinate contents versus postition ("depth") in the Luna 24 core. The data of Basu et al. (1977) are from the 20–250 μm fraction; all other data are from the 90–150 μm fraction.

150 μm portion indicate that the uppermost two soil samples (24077, 24109) are submature and the other four (24149, 24174, 24182, 24210) are immature.

This trend of decreasing agglutinate with depth is supported by the I_s/FeO study of Morris (1977). This seems to be a general feature of soil profiles. Recently, Morris (1978) has discussed this observation in detail. It would seem to be the result of *in situ* reworking without major dilution from mixing produced by somewhat larger impacts. The zone of *in situ* reworking varies from core to core and ranges in the few tens of cm depth. At first glance it would appear that this zone extends down well over 100 cm in the Luna 24 core. However, Bogard and Hirsch (1977) have demonstrated the *real* versus reported depths of the Luna 24 samples. It would appear from these considerations that the *in situ* reworking zone may be something on the order of 60–70 cm for this soil profile.

The core below this depth has not been homogenized by mixing. Sample 24149 contains a unique high abundance ($\simeq$25%) of microbreccia, also reported by Basu *et al.* (1977). Also, as will be discussed below, the samples particularly below 24109 show evidence in the agglutinitic glass chemistry of additional admixed soil components transported in by impacting processes. The variation in agglutinate content, the abundance of 24149 microbreccia, and the mixed-in outside soil components all support the distinct stratigraphy of Barsukov (1977) as defined by lithology.

Glass chemistry

The basic assumption for the study of agglutinitic glasses is that they represent the approximate bulk composition of the soil from which they were derived. This has been demonstrated by several studies, including those by this group (e.g., Hu and Taylor, 1977a). We previously showed how the mixing of two soils could be unraveled by the glass chemistries.

Glass compositions of the agglutinate particles in the 90–250 μm size range are plotted in Figure 3 and given in Table 2. Glass compositions from several >500 μm agglutinates are also presented in Table 3. Particularly in the MgO versus FeO plot of Figure 3, some clusters of compositions into populations can be discerned. The chemical groupings are not distinct in themselves; however, the scatter in data can be explained by a match to three rock types for which there is evidence in the lithic and mineral fragments from the petrologic studies of Ryder *et al.* (1977), Tarasov *et al.* (1977), Coish and Taylor (this volume), Simonds *et al.* (1977), etc. The majority of the analyses occur in the *high MgO ($\simeq$10%), high FeO ($\simeq$18%) group* reflecting a major contribution from the "hidden" coarse-grained rock, probably a ferrogabbro. This group is undoubtedly related to the abundant mode of monomineralic grains of olivine and pyroxene with an MgO/MgO + FeO of $\simeq$0.60 (Basu *et al.*, 1977; Coish and Taylor, this volume).

Another distinct population is the *low MgO ($\simeq$6%), low FeO ($\simeq$8%) group* probably derived from some highland source and carried in as ejecta. This was

Table 2. Chemical compositions of Luna 24 agglutinitic glasses in the 90–250 μm size fractions.

No. of Analyses	24077 102	24109 100	24149 98	24174 105	24182 89	24210 102	Average 596	Average Soil*
FeO	17.16	18.54	15.71	17.64	15.69	16.34	16.86	20.5
MgO	9.14	9.61	8.65	8.91	9.50	8.52	9.05	10.2
TiO_2	0.94	0.96	0.84	0.94	0.93	0.88	0.92	1.1
CaO	11.84	11.49	12.37	12.04	11.86	12.17	11.96	11.0
K_2O	0.05	0.05	0.05	0.05	0.05	0.04	0.05	0.03
Na_2O	0.43	0.42	0.43	0.38	0.49	0.47	0.44	0.28
Al_2O_3	14.47	13.21	16.27	13.79	15.29	15.34	14.73	11.1
SiO_2	44.44	44.34	44.37	44.78	44.78	44.27	44.49	45.4
Totals	98.47	98.62	98.69	98.53	98.59	98.03	98.50	99.6

*Blanchard *et al.* (1977)

Table 3. Chemical compositions of agglutinitic glass in particles larger than 500 microns.

No. of Analyses	24077, 71a 8	24077, 71b 9	24109, 68 6	24109, 70 6	24149, 60 7	24149, 72 8	24174, 69 8	24182, 29a 8	24182, 29b 8	Average 68
FeO	17.88	17.83	18.53	15.27	18.52	20.37	19.31	17.47	19.68	18.21
MgO	10.29	9.99	10.29	10.22	9.82	12.91	11.72	6.51	11.91	10.41
TiO_2	0.99	0.94	0.91	0.76	1.01	1.10	1.06	1.20	1.16	1.01
CaO	11.46	11.50	11.27	11.96	11.37	10.23	10.61	13.05	10.46	11.32
K_2O	0.03	0.03	0.04	0.05	0.03	0.05	0.08	0.04	0.03	0.04
Na_2O	0.39	0.45	0.36	0.46	0.50	0.54	0.44	0.36	0.38	0.43
Al_2O_3	13.64	13.90	13.15	15.42	13.05	9.66	11.28	13.67	11.24	12.78
SiO_2	44.49	44.60	44.07	46.10	44.29	44.25	44.75	46.57	44.54	44.85
Totals	99.17	99.24	98.62	100.24	98.59	99.11	99.25	98.87	99.40	99.05

supported by the finding of some agglutinates in 24182,48 whose glass compositions are similar to the Luna 24 glasses reported by Norman *et al.* (1977a, 1977b) who found that 30% of the nonagglutinitic glasses have highland compositions. McKay *et al.* (1977) found highland lithic fragments as well. In addition, the plagioclase fragments in those agglutinates with highland glass compositions were analyzed (Table 4), and on the basis of their very low FeO and MgO contents, they are considered to be representative of highland feldspar.

There is some indication on Figure 3 of the presence of a *low MgO (≃7%), high FeO (≃19%) group* of compositions. These most likely were derived from the subophitic, VLT, low MgO (6.5%) ferrobasalts which represent the common rock type at the Luna 24 site (e.g., Ryder *et al.*, 1977; Tarakov *et al.*, 1977). These 3 (or more) populations imply that Luna 24 soil was formed by a mixing

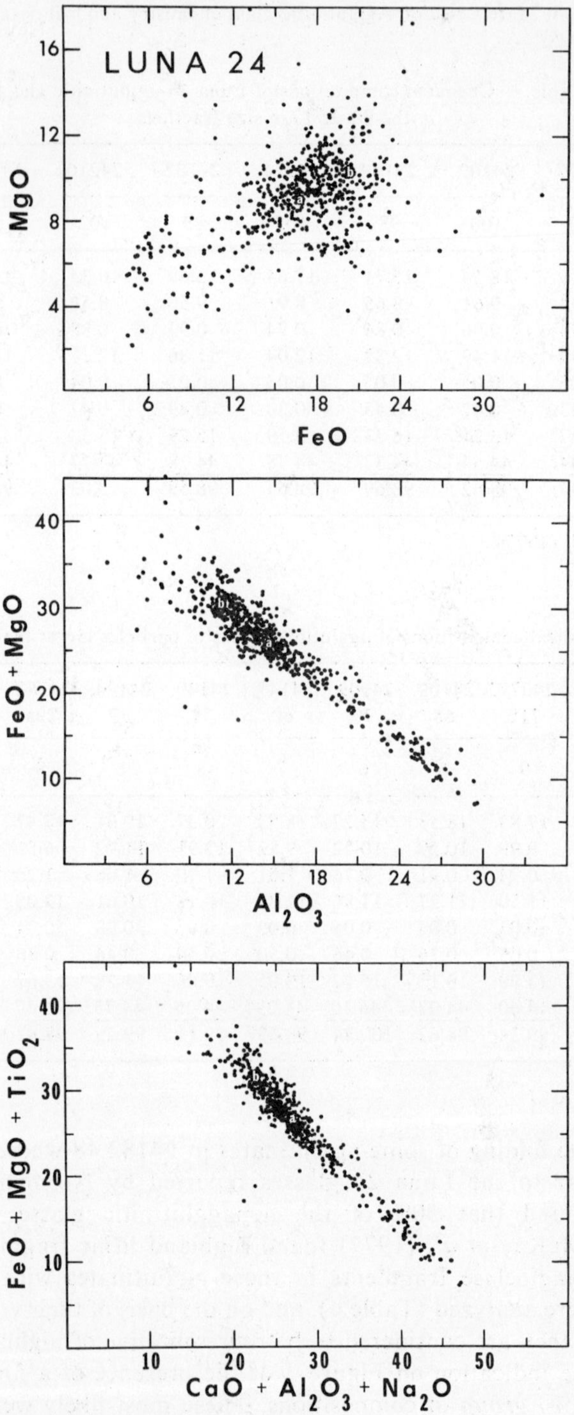

Fig. 3. Agglutinitic glass compositions from the 90–250 μm fraction of six horizons in the Luna 24 core. "a" is the average of all analysis; "b" is the whole soil average from Blanchard *et al.* (1977).

Table 4. Chemical compositions of highland versus mare plagioclase.

	Mare Plagioclase				Highland Plagioclase				Luna 24
	10024*	12021*	15555*	70035*	15415#	15415@	15415@	61241$	24182,48$
FeO	0.55	0.69	0.74	0.55	0.08	0.16	0.09	0.09	0.07
MgO	0.22	0.17	0.20	0.13	0.07	—	—	0.08	0.04
CaO	15.8	17.7	17.8	16.3	19.34	19.66	19.49	19.50	19.89
Na_2O	2.23	1.06	1.06	1.77	0.32	0.22	0.26	0.28	0.39
K_2O	0.28	0.13	0.16	0.29	0.05	—	—	0.01	0.03
TiO_2	0.13	0.09	0.04	0.08	0.04	0.01	0.01	0.02	0.01
Al_2O_3	31.5	33.0	33.6	31.7	36.04	35.77	36.24	35.27	35.34
SiO_2	49.4	47.1	46.8	49.4	43.36	44.19	43.92	43.59	43.20
Total	100.11	99.94	100.40	100.17	99.30	100.01	100.01	98.84	98.97

* = Crawford (1973); # = Stewart *et al.* (1972); @ = Hargraves & Hollister (1972); $ = this study.

process involving at least 3 different source materials. In fact, as shown in Figures 4 and 5, the samples from the lower portions of the core below the *in situ* working zone have not been thoroughly mixed, and thus, the high plagioclase component of the highland material is apparent.

The maturity of these soil samples may not be directly related to exposure age because of the mixing, at the Luna 24 site, of 3 soil components that are themselves soils. Each of the component soils developed its own maturity prior to transportation and mixing at the Luna 24 site. The large amount of *high MgO (≃10%), high FeO (≃18%)* agglutinitic glass and the lack of corresponding lithic fragments would seem to indicate a mature soil component. The *low MgO (≃6%), low FeO (≃8%)* group, representing a highland component, is considered to be mature because of the paucity of >500 μm highland agglutinates or rock fragments and a general increase in highland component in the smaller size fractions of the soils (Blanchard *et al.* 1977). The minor amounts of *low MgO (≃7%), high FeO (≃19%)* glasses are represented by the subophitic basalts; since the ratio of subophitic basalt fragments to corresponding agglutinitic glass is high, this component is considered immature. These conclusions are supported by McKay *et al.* (1977) who concluded that the coincidence between the mean grain size of the fine fractions as well as the agglutinate contents of the soils both indicate that some mature soils were added to the Luna 24 soils. In addition, particle track densities in plagioclase by Blanford and Wood (1977) support this. Thus, it is proposed that *the "maturity" of the Luna 24 soils is actually a combination of the properties of at least 3 different soils.* It is hypothesized that the different ratios of rock fragments to agglutinitic glass analyses in each chemical grouping can be correlated with a different degree of maturity. Therefore, each of the various soils had its own maturity prior to transportation to and mixing with the *in situ* soil.

On Figure 3, the symbol "a" represents the average glass composition of all the

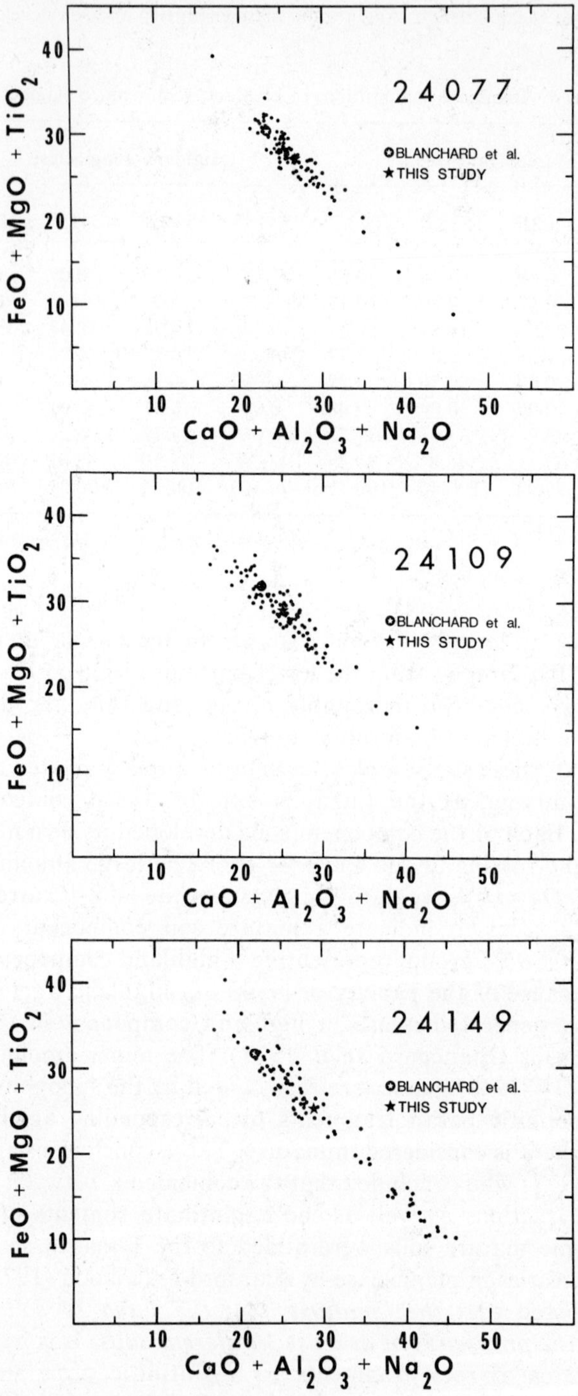

Fig. 4. Agglutinitic glass compositions from the 90–250 μm fraction of Luna 24 soil samples 24077, 24109, and 24149. Note the difference between the average glass composition in each sample versus the respective bulk soil composition for that horizon of Blanchard *et al.* (1977).

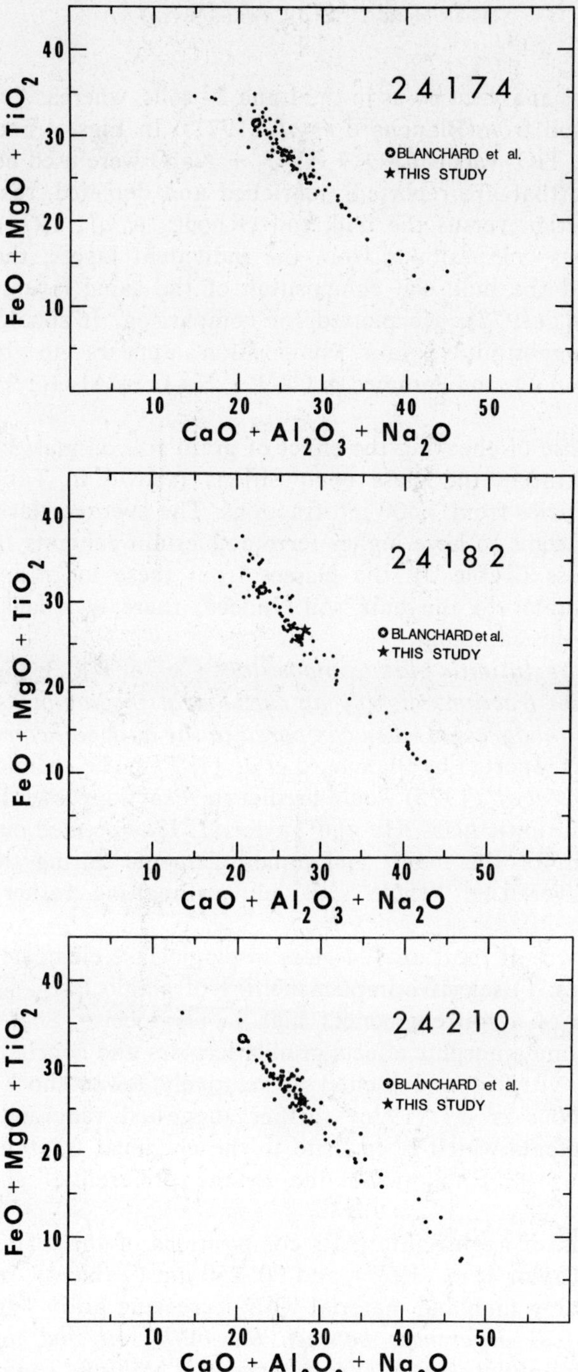

Fig. 5. Agglutinitic glass compositions from the 90–250 μm fraction of Luna 24 soil samples 24174, 24182, and 24210. Note the difference between the average glass composition in each sample versus the respective bulk soil composition for that horizon of Blanchard *et al.* (1977).

agglutinitic glass analyzed by us in the Luna 24 soils, whereas "b" is the average Luna 24 bulk soil from Blanchard *et al.* (1977). In Figs. 4 and 5, the axes of $FeO + MgO + TiO_2$ versus $CaO + Al_2O_3 + Na_2O$ were used because these are the components that are reportedly enriched and depleted, respectively, in an agglutinate fraction versus the bulk soil (Rhodes *et al.*, 1975). These figures present the glass compositions from the individual layers; the average glass composition and the bulk soil composition of the same layer, as reported by Blanchard *et al.* (1977), are plotted for comparison. It should be noted that the average agglutinitic glass composition appears to be depleted in $FeO + MgO + TiO_2$ and enriched in $CaO + Na_2O + Al_2O_3$ relative to the bulk soil.

For the purpose of checking the effect of grain size on glass compositions, we considered separately the glass compositions (shown in Table 3) of the 9 agglutinate particles from >500 μm fractions. The average glass composition of these particles seems to have higher ferromagnesium contents than the average 90–250 μm glass (Table 2); the glasses from these large agglutinates have compositions similar to the bulk soil. Indeed, there is a lack of "highland" agglutinates in this sampling.

The average agglutinitic glass compositions ("a" of Fig. 3; Figs. 4 & 5; Table 2) *of 90–250 μm fractions display an enrichment in plagioclase elements and depletion in ferromagnesian when compared to the bulk soil composition* ("b" of Fig. 3, 4 & 5) as reported by Blanchard *et al.* (1977). The chemical fractionation theory of Rhodes *et al.* (1975) would predict the exact opposite. Based on a study directed at agglutinitic glass, Hu and Taylor (1977a) pointed out a general lack of fractionation for the major and minor elements during the agglutination process and favored a "whole soil" impact melting rather than selective melting.

This enrichment of the Luna 24 glass in plagioclase elements can be effected by two processes: 1) selective impact melting of plagioclase and/or 2) addition from elsewhere of a soil component high in plagioclase. Schaal *et al.* (1976) observed shock metamorphic effects in microcraters and concluded that feldspar was selectively vitrified and melted at relatively lower shock pressures than pyroxene. Gibbons *et al.* (1976) further suggested that any enrichment in plagioclase elements which is opposite to the chemical fractionation theory of Rhodes *et al.* (1975) might be due to the preferential shock melting of plagioclase.

A comparison of agglutinitic glass compositions of the >500 μm (Table 3), 250–500 μm (Taylor *et al.*, 1977), and 90–250 μm (Table 2) fractions reveals a trend of increasing highland material with decreasing grain size. This indicates that *the chemical differences between the bulk soil and agglutinitic glass compositions are mainly due to the presence of a highland component*. This also emphasizes the point we made last year (Hu and Taylor, 1977a) that *only those soils that are entirely of local derivation should be used in any discussion of chemical fractionation during agglutination*. This may in reality severely restrict the number of such qualified soils.

CONCLUSIONS

The formation of Luna 24 soils is dominated by a mixing of 3 or more different source materials. The agglutinitic glass signatures of these components are: 1) high MgO ($\simeq$10%), high FeO ($\simeq$18%) glass from a "hidden" ferrobasalt (Simonds *et al.*, 1977), whose only evidence is in the monomineralic fragments; 2) low MgO ($\simeq$6%), low FeO ($\simeq$8%) glass representing a significant highland component; and 3) low MgO ($\simeq$7%), high FeO ($\simeq$19%) glass representing the subophitic ferrobasalt ($\simeq$6.5% MgO; Ryder *et al.*, 1977).

Each of these components has undergone soil development prior to transportation and mixing at the Luna 24 site. Therefore, the "maturity" of the Luna 24 soils is an integration of the individual maturities of each component soil plus any further modification due to mixing and *in situ* reworking.

The agglutinitic glass, of the 90–250 μm fraction, from each horizon is enriched in plagioclase and depleted in ferromagnesian elements relative to the respective bulk soils. This is in an opposite sense to that predicted by the chemical fractionation theory. This may be due to 1) selective melting of plagioclase before olivine/pyroxene (Schaal *et al.*, 1976) and/or 2) addition of a highland component. The latter seems to be the major factor based on several lines of evidence (e.g. the presence of "highland" plagioclase).

However, no matter how confident we may be about our conclusions, we are reminded of the sage advice of Nagle and Walton (1977), "interpretive conclusions based on study of particles of one lithology may not directly apply to the soil as a whole."

Acknowledgments—We wish to express our appreciation to the Soviet Union for providing the U.S.A. with some of their precious samples and to LSAPT for the efficient manner in which the processing and allocations were conducted. The constructive review of the manuscript by Drs. Stu Nagle and Arden L. Albee are greatly appreciated. Thanks are also given to the University of Tennessee for providing substantial support for the purchase of the electron microprobe. The research presented in this paper was supported by NASA grant NGL 43-001-140.

REFERENCES

Albee A. and Ray L. (1970) Correction factors for electron probe microanalysis of silicates, oxides, carbonates, phosphates, and sulfates. *Anal. Chem.* **42**, 1408–1414.

Basu A., McKay D. D., and Fruland R. M. (1977) Petrography, mineralogy and source rocks of Luna 24 drill core soils (abstract). In *Papers Presented to the Conference on Luna 24*, p. 18–21. The Lunar Science Institute, Houston.

Barsukov V. L. (1977) Preliminary data for the regolith core brought to earth by the automatic lunar station Luna 24. *Proc. Lunar Sci. Conf. 8th*, p. 3303–3318.

Bence A. E. and Albee A. (1968) Empirical correction factors for the electron microanalysis of silicates and oxides. *J. Geol.* **76**, 382–403.

Blanchard D. P., Haskin L. A., Brannon J. C., and Aaboe E. (1977) Chemistry of soils and particles from Luna 24 (abstract). In *Papers Presented to the Conference on Luna 24*, p. 37–40. The Lunar Science Institute, Houston.

Blanford G. E. and Wood G. C. (1977) Particle track densities in the Luna 24 core (abstract). In *Papers Presented to the Conference on Luna 24*, p. 41–43. The Lunar Science Institute, Houston.

Bogard D. D. and Hirsch W. C. (1977) Noble gases in Luna 24 core soils (abstract). In *Papers Presented to the Conference on Luna 24,* p. 44–47. The Lunar Science Institute, Houston.

Coish R. A. and Taylor L. A. (1978) Mineralogy and petrology of basaltic fragments from the Luna 24 drill core. In *Mare Crisium: The View from Luna 24* (R. B. Merrill and J. J. Papike, eds.). This volume.

Crawford M. L. (1973) Crystallization of plagioclase in mare basalts. *Proc. Lunar Sci. Conf. 4th,* p. 705–717.

Gibbons R. V., Horz F., and Schaal R. B. (1976) The chemistry of some individual lunar soil agglutinates. *Proc. Lunar Sci. Conf. 7th,* p. 405–421.

Hargraves R. B. and Hollister L. S. (1972) Mineralogic and petrologic study of lunar anorthosite slide 15415,18. *Science* **175,** 430–432.

Hu H.-N. and Taylor L. A. (1977a) Lack of chemical fractionation in major and minor elements during agglutinate formation. *Proc. Lunar Sci. Conf. 8th,* p. 3645–3656.

Hu H.-N. and Taylor L. A. (1977b) Luna 24 soils from Mare Crisium: Agglutinate chemistry (abstract). In *Papers Presented to the Conference on Luna 24,* p. 82–84. The Lunar Science Institute, Houston.

McKay D. S., Basu A., Waits G., Clanton U., Fuhrman R., and Fruland R. (1977) Grain size and evolution of Luna 24 soils (abstract). In *Papers Presented to the Conference on Luna 24,* p. 114–117. The Lunar Science Institute, Houston.

McKay D. S., Fruland R. M. and Heiken G. H. (1974) Grain size and the evolution of lunar soils. *Proc. Lunar Sci. Conf. 5th,* p. 887–906.

Morris R. V. (1977) FMR and magnetic properties of Luna 24 soils and >1 mm soil particles (abstract). In *Papers Presented to the Conference on Luna 24,* p. 121–124. The Lunar Science Institute, Houston.

Morris R. V. (1978) In situ reworking (gardening) of the lunar surface: Evidence from the Apollo cores (abstract). In *Lunar and Planetary Science IX,* p. 757–759. Lunar Planetary Institute, Houston.

Nagle J. S. and Walton W. J. A. (1977) *Luna 24 catalog and preliminary description.* NASA—JSC, 85 pp.

Norman M., Coish R. A., and Taylor L. A. (1977a) Luna 24 core samples from Mare Crisium: Glass chemistry and magma evolution (abstract). *EOS (Trans. Amer. Geophys. Union)* **59,** 223.

Norman M., Coish R. A. and Taylor L. A. (1977b) Glass chemistry and magma evolution at Mare Crisium (abstract). In *Papers Presented to the Conference on Luna 24,* p. 136–138. The Lunar Science Institute, Houston.

Phinney W. C., McKay D. S., Simonds C. H., and Warner J. L. (1976) Lithification of vitric- and clastic-matrix breccia: SEM petrography. *Proc. Lunar Sci. Conf. 7th,* p. 2469–2492.

Rhodes J. M., Adams J. B., Blanchard D. P., Charette M. P., Rodgers K. V., Jacobs J. W., Brannon J. C., and Haskin L. A. (1975) Chemistry of agglutinate fractions in lunar soils. *Proc. Lunar Sci. Conf. 6th,* p. 2291–2307.

Ryder G., McSween H. Y., and Marvin V. B. (1977) Basalts from Mare Crisium, *The Moon* **17,** 263-287.

Schall R. B., Horz F., and Gibbons R. V. (1976) Shock metamorphic effects in lunar microcraters. *Proc. Lunar Sci. Conf. 7th,* p. 1039–1054.

Simonds C, H., Warner J. L., McGee P. E., Cochran A., and Brown, R. W. (1977) Petrochemical potpourri (abstract). In *Papers Presented to the Conference on Luna 24,* p. 170–173. The Lunar Science Institute, Houston.

Stewart D. B., Ross M., Morgan B. A. Appleman D. E., Huebner J. S., and Commeau R. R. (1972) Mineralogy and petrology of lunar anorthosite 15415 (abstract). In *Lunar Science III,* p. 726–728. The Lunar Science Institute, Houston.

Tarasov L. S., Nazarov M. A., Shevaleevsky I. D., Kudryashova A. F., Gaverdovskaya A. S. and Korina M. I. (1977) Mineralogy and petrography of lunar rocks from Mare Crisium (Preliminary data). *Proc. Lunar Sci. Conf. 8th,* p. 3333–3356.

Taylor G., Warner R., Wentworth S., Keil K., and Sayeed, U. (1977) Petrology and chemistry of Luna 24 mare basalts and basaltic glasses (abstract). In *Papers Presented to the Conference on Luna 24,* p. 189–192. The Lunar Science Institute, Houston.

Merrill, R. B. and Papike, J. J., editors.
(1978) *Mare Crisium: The View from Luna 24*, Pergamon Press (New York), p. 303–320
Printed in the United States of America

Luna 24 lithologies: Petrochemical relationships among lithic fragments, mineral fragments, and glasses

G. JEFFREY TAYLOR, RICHARD D. WARNER, SUSAN WENTWORTH, KLAUS KEIL and USMAN SAYEED*

Department of Geology and Institute of Meteoritics, University of New Mexico, Albuquerque, New Mexico 87131

Abstract—We used petrographic and electron microprobe techniques to study 13 thin sections containing 700 fragments from the 250–500 μm size fraction of 6 Luna 24 soils, and two sections from gabbro layer 24170. We identified the following lithologies: (1) Ferrobasalts with ophitic to subophitic textures. Their bulk compositions show that they are Fe-rich, Mg-poor VLT mare basalts; (2) Metabasalts, which have microgranular textures and homogeneous mineral compositions. Their bulk compositions suggest that they are metamorphosed ferrobasalts; (3) Ferrogabbros, which appear to be coarse-grained versions of the ferrobasalts. This is supported by average mineral compositions, pyroxene compositional trends, and an estimated bulk composition; (4) Mg-rich gabbros, which are mainly represented among monomineralic fragments in the soils. They also appear to be VLT mare basalts; (5) Olivine-rich fragments, with textures that suggest they formed by impact melting; and (6) Nonmare fragments, most of which have been shock-melted.

INTRODUCTION

The Luna 24 regolith contains lithic fragments of mare basalt and gabbro that are characterized by very low TiO_2 contents (~1 wt.%) and high Fe/Mg (Barsukov *et al.*, 1977). Their compositions resemble VLT (very low-Ti) mare basalts, which were first described as lithic fragments and glasses in Apollo 17 soils and breccias (Vaniman and Papike, 1977a; Taylor *et al.*, 1977a). The Luna 24 samples thus provide an opportunity to study additional samples of this new type of mare basalt in considerable detail.

We received 15 thin sections of seven soil samples from the Luna 24 core. Two sections were from gabbro layer 24170; the remainder were from the 250–500 μm fraction from six bulk soils. We classified the fragments on these sections and then determined the chemical compositions of lithic and monomineralic fragments by electron microprobe analysis. The results are reported here, with emphasis on the compositions of the major lithologies in the Luna 24 soil. We discuss the petrogenesis of Luna 24 and other VLT mare basalts in a companion paper (Taylor *et al.*, 1978).

---

*Permanent address: Florida International University
Tamiami Trail
Miami, Florida 33199

EXPERIMENTAL TECHNIQUES

We have previously described our experimental techniques and microprobe correction procedures (e.g., Dowty *et al.*, 1973). The one difference in our normal procedures involved broad-beam microprobe analyses of lithic fragments. We corrected the analyses reported here by the normative correction procedure described by Bower *et al.* (1977). The procedure involved computing a norm from the uncorrected data, then applying Bence-Albee corrections to the oxides in each normative mineral. A new norm was then calculated from the corrected data, and the procedure was repeated.

RESULTS AND DISCUSSION

Lithologic abundances

We classified the 700 fragments on the 13 sections of the six soils other than 24170 into the following categories: agglutinates, soil breccias, aphanitic fragments, glasses, mare basalt lithic fragments (subdivided into gabbro, fine-grained basalt, and metabasalt), monomineralic fragments, and other lithic fragments (Table 1).

The number of fragments in any given sample is too small for rigorous statistical treatment, but several observations are pertinent. First, the abundance of nonmare lithic fragments is low. The soil appears to be derived predominantly

Table 1. Relative proportions (%) of lithic fragments, mineral fragments and glasses in the 250–500 μm size range in six Luna 24 soils.

	24077	24109	24149	24182	24174	24210
No. of particles	109	176	75	22	273	45
Agglutinates	23.9	16.5	1.3	—	7.7	6.7
Soil Breccias	10.1	21.0	48.0	9.1	12.8	22.2
Aphanitic frags.	2.8	2.3	4.0	—	3.7	4.4
Glasses						
colored	4.6	1.1	—	—	—	2.2
colorless	—	0.6	—	—	0.7	—
Mare Basalt lithic frags.						
gabbros	7.3	8.0	5.3	18.2	12.4	—
fine-grained	9.2	8.5	8.0	9.1	7.0	8.9
metabasalts	2.8	—	2.7	13.6	4.0	—
Nonmare lithic frags.	0.9	0.6	—	4.5	1.1	6.7
Other lithic frags.	2.8	1.7	—	—	1.1	8.9
Mineral Fragments						
plagioclase	8.3	9.1	6.7	13.6	11.4	15.6
pyroxene	24.8	25.6	21.3	22.7	30.0	17.8
olivine	2.8	4.5	2.7	9.1	7.7	6.7
other	—	0.6	—	—	0.4	—
Total	100.3	100.1	100.0	99.9	100.0	100.1

from mare lithologies with a very small contribution from nonmare sources. Soil compositions confirm this observation (Blanchard *et al.*, 1977; Ma and Schmitt, 1977). Second, the dominant mare lithology is gabbroic, as represented by identifiable coarse-grained lithic fragments and by monomineralic fragments. Compositions of monomineralic fragments (see below) suggest that almost all of them are derived from mare basalts or related rocks. Third, in all but one case (24077), the abundance of vitric soil breccias exceeds that of agglutinate. (We distinguished agglutinates from vitric soil breccias by the presence of round vesicles in agglutinates.) Fourth, our results agree with those reported by Ryder *et al.* (1977) on the same size fraction. Fifth, our results are also consistent with those reported by Basu *et al.* (1977a), Nielsen and Drake (1977), and Papp *et al.* (1977) on smaller size fractions, although a few differences stand out. Specifically, the <250 μm fraction contains a higher abundance of agglutinates and soil breccias, more agglutinates relative to soil breccias, and fewer gabbroic lithic fragments. Nevertheless, there is general agreement, so it should be safe for us to infer bulk soil properties from the lithologic abundances present in the 250–500 μm fraction. We return to this topic in a later section.

Basalt and metabasalt lithic fragments

We observed two textural types of fine-grained basaltic lithic fragments in Luna 24 soils, in agreement with Ryder *et al.* (1978) and Vaniman and Papike (1977b). Most basaltic lithic fragments have ophitic to subophitic textures (Fig. 1a), with crystal sizes typically smaller than 100 μm. They contain pyroxene (~60%), plagioclase (~35%), olivine (~5%), and lesser amounts of chromite, ulvöspinel, ilmenite, and silica. In contrast, a few lithic fragments are finer-grained (crystals usually smaller than 25 μm) and have granulitic textures (Fig. 1b). They contain the same minerals in roughly the same proportions, but usually contain ulvöspinel as their only opaque oxide. Their textures, as well as their mineral compositions (see below), led Ryder *et al.* (1978) to call them *metabasalts*.

We determined the bulk compositions of 36 basalt and 7 metabasalt lithic fragments. The analyses scatter quite widely (Figs. 2&3). Since the fragments are at most 500 μm across, we attribute most of the scatter to sampling errors. This seems to be confirmed by the data in Figs. 2 and 3, where the analyses string out along lines that extend from the average compositions of the fragments to the average compositions of their two major phases, pyroxene and plagioclase. (We do not rule out, however, the possibility that some of the variation seen in Figs. 2 and 3 records differentiation of Luna 24 ferrobasalt flows.) Obviously, an analysis of an individual metabasalt and, particularly, a basalt lithic fragment is meaningless. Consequently, we list only the average compositions of each type in Table 2. For comparison, we also list in Table 2 the average of Simonds *et al.*'s (1977) "low-Mg glass cluster" and Norman *et al.*'s (1977) "brown glasses".

Three points emerge from a study of these data. First, it appears that the basalt and metabasalt lithic fragments have essentially identical compositions,

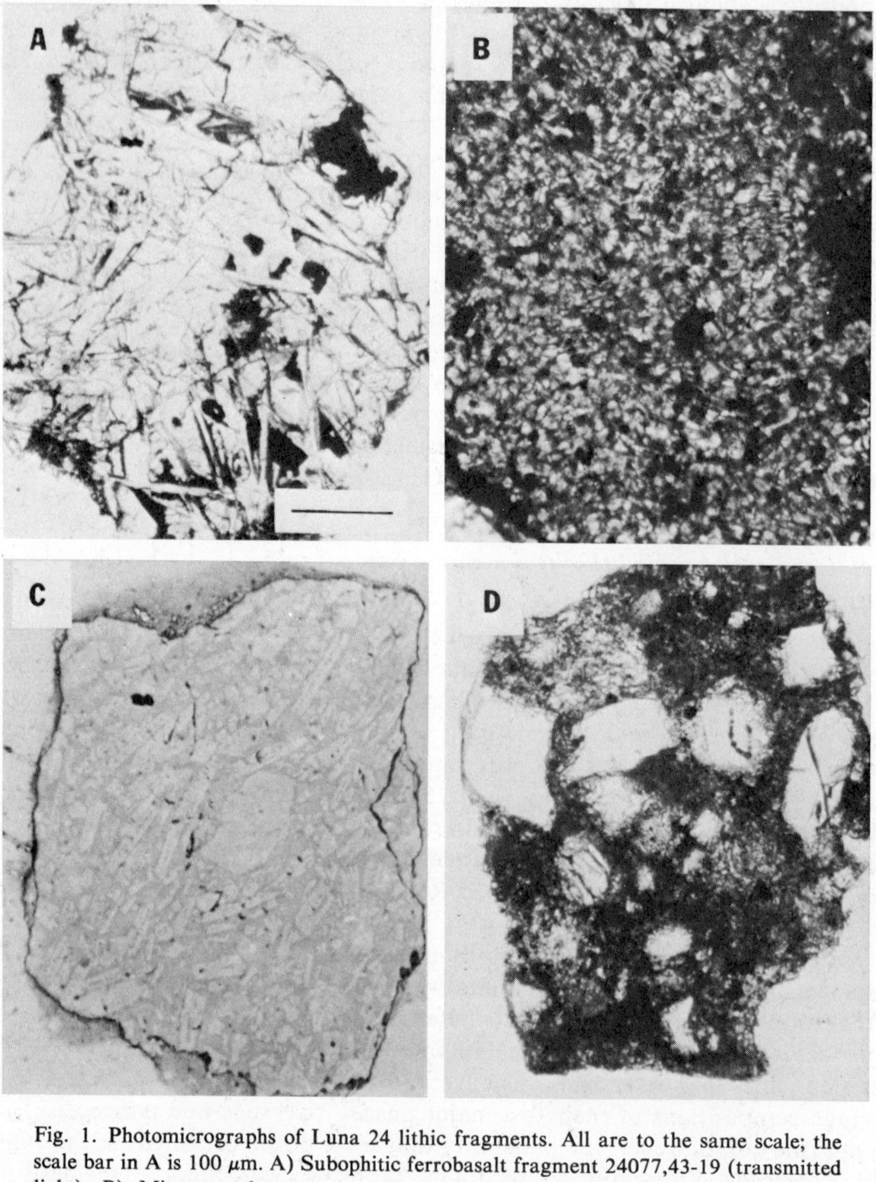

Fig. 1. Photomicrographs of Luna 24 lithic fragments. All are to the same scale; the scale bar in A is 100 μm. A) Subophitic ferrobasalt fragment 24077,43-19 (transmitted light). B) Microgranular texture typical of metabasalt lithic fragments; fragment 24182,51-11, transmitted light. C) Olivine vitrophyre lithic fragment 24174,59-12 (reflected light) consisting of skeletal olivine crystals in a matrix of pale-green glass. Note the relic olivine clast in the center. D) Olivine-rich lithic fragment 24149,42-18 (transmitted light), showing olivine and pyroxene clasts in a fine-grained, igneous-textured matrix.

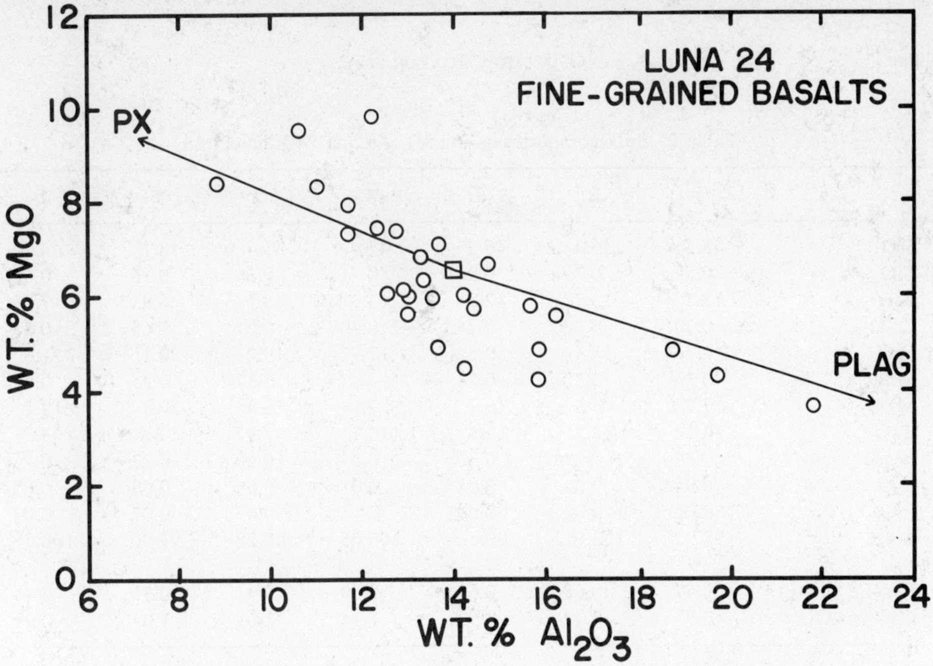

Fig. 2. MgO vs Al_2O_3 from broad-beam microprobe analyses of Luna 24 ferrobasalt lithic fragments. Square in center is the average composition. The tendency for the analyses to disperse along a line pointing to average pyroxene and plagioclase compositions indicates that the scatter is due to sampling errors.

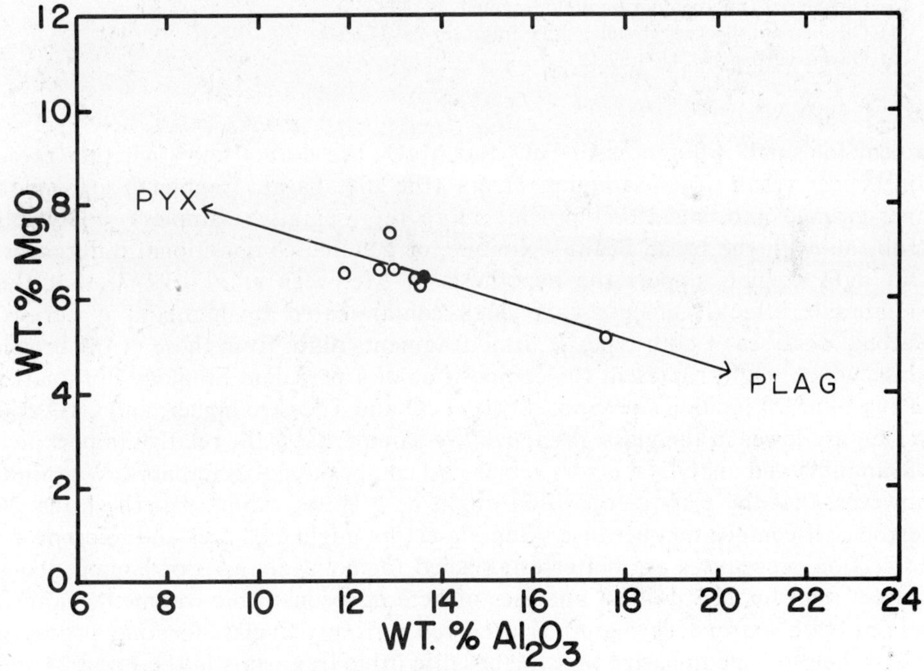

Fig. 3. MgO vs Al_2O_3 for metabasalt lithic fragments. Note that the data scatter along lines that extend from the average lithic fragment composition towards the average pyroxene and plagioclase compositions, suggesting that the scatter is due to sampling errors.

Table 2. Bulk compositions (wt.%) of Luna 24 lithologies.

	1	2	3	4	5	6	7
SiO_2	46.5	46.4	46.2	47.9	45.4	43.1	44.2
TiO_2	0.67	0.79	0.96	0.71	0.66	0.15	0.62
Al_2O_3	13.3	13.7	12.8	12.3	8.9	6.9	7.7
Cr_2O_3	0.30	0.20	0.24	0.20	0.21	0.65	0.53
FeO	17.2	18.5	19.5	17.9	19.2	20.1	21.6
MnO	0.29	0.30	0.28	—	0.37	0.25	0.43
MgO	7.2	6.5	6.7	7.1	15.4	20.8	18.2
CaO	13.1	13.3	12.5	13.7	8.9	6.9	7.0
Na_2O	0.30	0.28	0.30	0.23	0.20	0.12	0.15
K_2O	0.04	0.04	0.02	0.01	0.05	0.03	0.05
P_2O_5	0.02	<0.02	0.03	—	<0.02	<0.02	0.03
Total	98.92	100.01	99.52	100.05	99.29	99.00	100.51
Fe/(Fe + Mg)	0 57	0.61	0.62	0.59	0.41	0.35	0.40
CaO/Al_2O_3	0.98	0.97	0.98	1.14	1.0	1.0	0.91

(1) Average of 36 broad-beam microprobe analyses of basaltic lithic fragments.
(2) Average of 7 broad-beam analyses of metabasalt lithic fragments.
(3) Glass analyses: average of Simonds *et al.*'s (1977) low-Mg cluster and Norman *et al.*'s (1977) brown glass.
(4) Bulk composition of gabbroic lithology in 24170; estimated from modal data and average mineral compositions (see text).
(5) Olivine-rich, microporphritic lithic fragment 24149,42-18.
(6) Olivine-rich, shocked, granular lithic fragment 24174,65-7.
(7) Olivine vitrophyre 24174,59-12.

except for small differences in FeO and MgO. We do not know whether these differences result from sampling errors (the metabasalt fragments are much finer-grained and, therefore, provide more representative samples), analytical problems with the broad-beam technique, or genuine compositional differences. Our data do not support the hypothesis of McSween *et al.* (1977) that the metabasalt lithic fragments have been contaminated by highland materials. Second, analyses of both types of lithic fragments differ from those of the brown glass, which might represent the composition of a prevalent lithology in the area of the Luna 24 landing site. Specifically, FeO and TiO_2 are higher, and CaO and Al_2O_3 are lower in the glass. As above, we cannot assess the relative importance of sampling and analytical errors versus real compositional differences. We think, however, that the glass composition might be a better estimate of the Luna 24 ferrobasalt composition because sampling errors might be lower and microprobe correction procedures are better established for point analyses of homogeneous glasses than for broad beam analyses of heterogeneous lithic fragments (unless mixing with extraneous materials occurred during the glass-forming process). Nevertheless, we emphasize that the basaltic lithic fragments in the Luna 24 soil might indeed be different in composition from the brown glass. Third, in spite of uncertainties in the precise nature of the Luna 24 basalt lithic fragments, it is

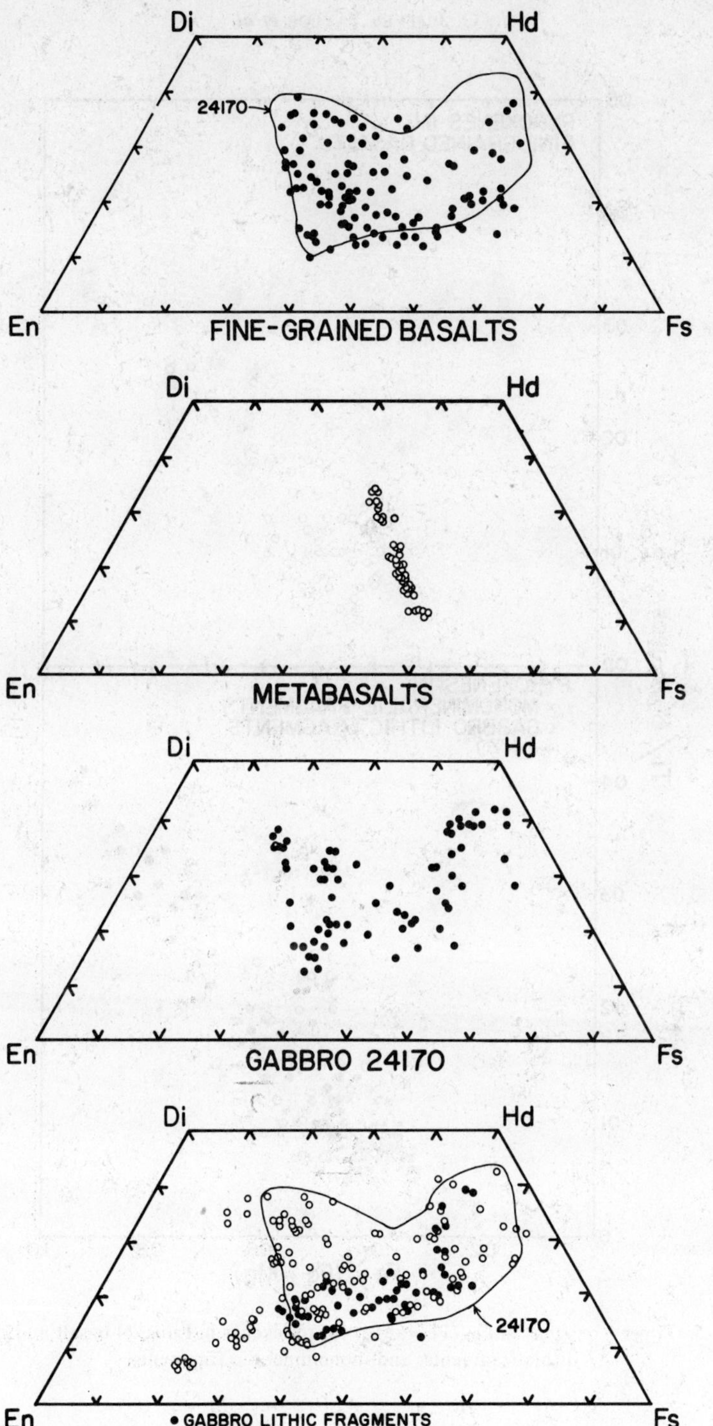

Fig. 4. Pyroxene compositions (mole %) in terms of ideal end members for the indicated Luna 24 lithic and monmineralic fragments.

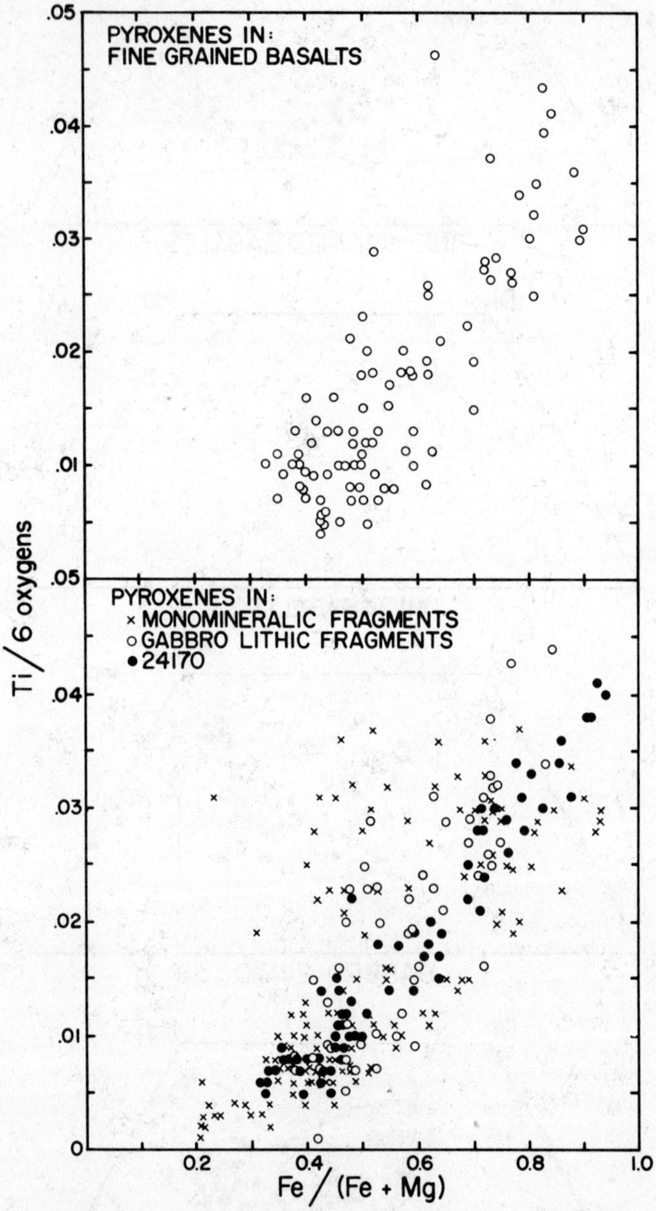

Fig. 5. Ti per 6 oxygens vs Fe/(Fe + Mg) for pyroxenes in Luna 24 basalt and gabbro lithic fragments, and monomineralic fragments.

clear that they are distinct from other mare basalts, particularly in TiO_2 content and Fe/(Fe + Mg) ratios. They resemble Apollo 17 VLT mare basalt lithic fragments, though Luna 24 basalt lithic fragments are somewhat higher in TiO_2

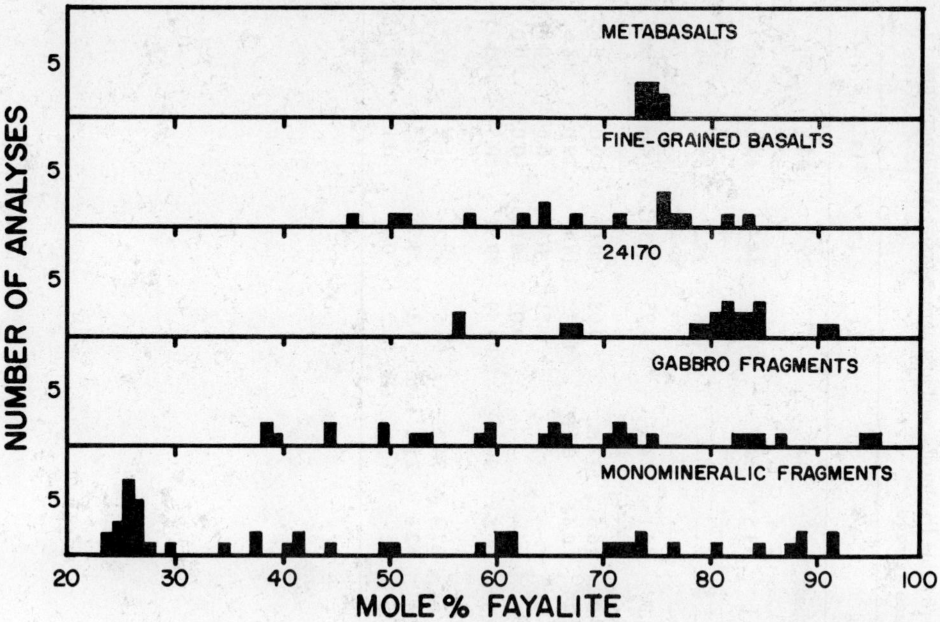

Fig. 6. Histograms of olivine compositions in Luna 24 lithic and monomineralic fragments.

and much lower in MgO. We discuss differences among VLT mare basaltic lithic fragments and glasses elsewhere in this volume (Taylor *et al.*, 1978). To emphasize their high Fe/Mg ratios, we will refer to the Luna 24 basalts as ferrobasalts (Vaniman and Papike, 1977b). The term "Luna 24 fine-grained basalt" is synonymous with "ferrobasalt".

We present mineral compositional data for ferrobasalt and metabasalt lithic fragments in Figs. 4–6; Table 3 gives average mineral compositions. (These averages were obtained by measuring the centers and edges of numerous crystals in each fragment studied, rather than in a point-counting fashion and so do not necessarily represent the true average composition. We doubt that they are significantly different from a true average, however). Pyroxene and olivine compositions clearly distinguish basalt from metabasalt lithic fragments. The basalt fragments have compositionally zoned minerals that range widely in composition. In contrast, metabasalt fragments have more uniform, and, therefore, equilibrated mineral compositions (Fig. 4&6). Since the average metabasalt and basalt pyroxene and olivine, as well as the bulk lithic fragment compositions, are similar, we conclude that metabasalts are simply metamorphosed equivalents of Luna 24 ferrobasalts. D. T. Vaniman (oral presentation at the Luna 24 Conference) showed quantitatively that the slight differences in mineral compositions are due to predictable reactions that took place during the metamorphism of Luna 24 ferrobasalt to metabasalt.

Table 3. Average Mineral compositions (wt.%) in basalt, metabasalt, and gabbro lithic fragments, and among monomineralic fragments. N is the number of analyses in each average.

	1	2	3	4	5		6	7	8	9	10		11	12	13	14
N	104	42	72	182	8		12	8	20	44	34		30	29	20	30
SiO_2	49.4	49.4	50.8	50.4	54.6		33.3	32.5	31.6	33.5	38.2		44.5	44.4	45.1	46.0
TiO_2	0.58	0.65	0.58	0.54	0.12		—	—	—	—	—		—	—	—	—
Al_2O_3	1.65	0.99	1.26	1.33	1.04		—	—	—	—	—		34.6	35.2	35.2	34.6
Cr_2O_3	0.41	0.24	0.33	0.37	0.57		—	—	—	—	—		—	—	—	—
FeO	26.0	27.8	24.2	23.3	13.7		49.3	56.8	58.3	49.1	23.3		0.95	0.91	0.52	0.61
MnO	0.37	0.33	0.46	0.38	0.36		—	—	—	—	—		—	—	—	—
MgO	11.6	10.4	11.1	12.8	26.4		16.6	10.9	9.0	16.5	37.6		0.08	<0.05	<0.05	0.05
CaO	10.6	10.0	12.5	11.2	3.7		0.44	0.29	0.49	0.44	0.21		18.7	18.6	19.0	18.6
Na_2O	—	—	—	—	—		—	—	—	—	—		0.71	0.79	0.69	0.86
K_2O	—	—	—	—	—		—	—	—	—	—		0.05	0.06	0.03	0.03
Total	100.61	99.81	101.23	100.32	100.49		99.64	100.49	99.39	99.54	99.31		99.59	99.96	100.54	100.75
En	34.3	31.3	33.0	37.8	71.8	Fo	37.5	25.5	21.6	37.4	74.2	An	96.5	96.1	96.2	95.2
Fs	43.2	47.0	40.3	38.5	20.9	Fa	62.5	74.5	78.4	62.6	25.8	Ab	3.3	3.7	3.7	4.7
Wo	22.4	21.7	22.4	23.7	7.2							Or	0.2	0.2	0.1	0.1

(1) Pyroxenes in ferrobasalt lithic fragments.
(2) Pyroxenes in metabasalt lithic fragments.
(3) Pyroxenes in 24170,7 (gabbro).
(4) Monomineralic pyroxene fragments, except those with En >70.
(5) Monomineralic pyroxene fragments with En >70.
(6) Olivines in ferrobasalt lithic fragments.
(7) Olivines in metabasalt lithic fragments.
(8) Olivines in 24170,7 (gabbro).
(9) Monomineralic olivine fragments, except those with Fo >70.
(10) Monomineralic olivine fragments with Fo >70.
(11) Plagioclases in ferrobasalt lithic fragments.
(12) Plagioclases in metabasalt lithic fragments.
(13) Plagioclase in 24170,7 (gabbro).
(14) Monomineralic plagioclase fragments.

Minor element compositional trends among pyroxenes in Luna 24 ferrobasalts, as illustrated by Ti vs Fe/(Fe + Mg) (Fig. 5), are typical for those that crystallize from low-Ti magmas. Except for the first pyroxenes being richer in Fe, the trend of increasing Ti with increasing Fe/(Fe + Mg) is identical to that observed in Apollo 17 VLT mare basalt lithic fragments (Vaniman and Papike, 1977a; Taylor *et al.*, 1977a). Plagioclase is not notably zoned in either basalt or metabasalt lithic fragments and is relatively high in FeO, as it is in all mare basalts. Like that of Apollo 17 VLT mare basalt lithic fragments, plagioclase of Luna 24 ferrobasalt and metabasalt lithic fragments is very low in alkalis (Ab <6, Or <0.2).

Gabbroic and monomineralic fragments

Sample 24170 is from a light-colored layer thought to be a disaggregated rock broken as the Luna 24 core drilled through the regolith (Florensky *et al.*, 1977). The rock is clearly coarse-grained, with many mineral grains over a millimeter long. We originally called it a coarse-grained basalt (Taylor *et al.*, 1977b) but most investigators called it a gabbro (Florensky *et al.*, 1977; Tarasov *et al.*, 1977; Bence *et al.*, 1977; Ryder *et al.*, 1977). Albee *et al.* (1977) give a thorough description of the sample and point out that the rock appears to have a subophitic texture. It consists of pyroxene, plagioclase, and olivine, with minor amounts of opaques and mesostasis. The mesostasis contains small crystals (~5 μm across) of fayalite, silica, and Fe-rich pyroxene (Ryder *et al.*, 1977).

Gabbro 24170 is not the only sample of coarse-grained, or gabbroic rock. All six of the soils (250–500 μm) we studied contain gabbroic lithic fragments, which consist of a few large (usually >250 μm) crystals. There are also monomineralic grains usually >250 μm in size, most of which appear to be fragments of gabbroic rocks. We studied these fragments with two goals in mind: First, we wanted to determine if the monomineralic and gabbroic lithic fragments were of the same lithology as gabbro 24170 and, if not, to what lithologies they might be related. Second, we wanted to establish the bulk composition of the gabbroic rock represented by 24170. We could not obtain a meaningful bulk composition from modal and average mineral analyses of our thin section of 24170 because it consisted of a group of hand-picked grains, which is clearly a biased sample. Our discussion in this section centers on these two objectives.

Pyroxene compositions in 24170, both in terms of ranges and average compositions, are very similar to those in the fine-grained ferrobasalt lithic fragments (Figs. 4&5). This suggests that gabbro 24170 has a VLT mare basaltic composition. Furthermore, although pyroxenes in gabbroic lithic fragments are generally similar to those in 24170, the Ca-rich part of the 24170 trend (Fig. 4) seems to be absent, and some of the analyses deviate from the rather narrow field on the Ti vs Fe/(Fe + Mg) diagram (Fig. 5). Thus, we conclude that some of these gabbroic fragments are related to a second gabbro type different from 24170. Also, olivine compositions (Fig. 6) of the gabbro lithic fragments extend to more magnesian compositions than those in 24170. This also suggests the existence of a second gabbro type. Many monomineralic fragments have compositions indistinguishable from those in 24170. However, a significant number of the pyroxene and over half the olivine analyses are clearly more magnesian. We conclude again that these magnesian olivines and pyroxenes probably do not come from the same lithology sampled by 24170, but are related to a second gabbro type. Thus, we list the average compositions of these olivines and pyroxenes separately from the monomineralic fragments judged to be related to 24170 (Table 3). We have, however, calculated the average compositions of all monomineralic pyroxene, olivine, and plagioclase (Table 4) for use in a later section to estimate the bulk composition of the Luna 24 soil. (Some of the magnesian pyroxenes and olivines might be from nonmare sources. However, monomineralic plagioclase compositions are essentially all from mare-related rocks (Papp *et al.*, 1977). Therefore, we assume that all Mg-rich pyroxenes are from mare lithologies).

We estimated the bulk composition of gabbro 24170 as follows: We first assumed that there were only two gabbro compositions represented in the Luna 24 soils. One has Mg-rich olivines and pyroxenes, the other has mineral compositions like those in 24170. The calculation of the 24170 composition would be straightforward if we knew its mode. We estimated the mode from the relative abundances of monomineralic fragments, only counting pyroxenes and olivines with compositions in the range observed for these minerals in 24170, and from the mode of all the gabbro lithic fragments (Table 5). The weighted average of these two modes (Table 5) may provide an estimate of the mode of the Fe-rich gabbroic lithology. Combining this mode with the average mineral compositions in 24170 (Tables 3, 4) we estimate the bulk chemical composition of the Fe-rich gabbro 24170 (Table 2). Our estimated composition is similar to that of the Luna 24 ferrobasalt lithic fragments, although noticeably higher in CaO/Al_2O_3.

The data on monomineralic fragments, gabbroic lithic fragments, and 24170 lead us to the following two conclusions: First, gabbro 24170 and associated fragments represent a coarse-grained, Fe-rich, VLT mare basaltic lithology at the Luna 24 site. This conclusion is in accord with that of Bence *et al.* (1977), who estimated the bulk composition of 24170 from pyroxene compositions. Second, at least one other gabbroic lithology is present at the Luna 24 site. Compositions of monomineralic fragments suggest that this rock type (or types) is considerably higher in Mg than 24170.

Olivine-rich lithic fragments

Three of the lithic fragments in the samples we studied have distinctive compositions, characterized by high normative olivine contents and high bulk MgO (Table 2). The fragments are rich in MgO and FeO, low in TiO_2, alkalis, and P_2O_5. Their Cr_2O_3 contents (0.2 to 0.6 wt.%) clearly indicate that they are related to mare materials. This mafic, low-TiO_2 composition led T. L. Grove and co-workers (oral presentation at the Luna 24 Conference) to single out these olivine-rich fragments as possible parent magmas to the Luna 24 ferrobasalts.

Table 4. Average compositions (wt.%) and abundances of soil components in six Luna 24 soils, and the estimated soil composition. N is the number of particles.

	1	2	3	4	5	6	7	8	9	10	11	12	13
N	39	183	71	1	1	3	211	57	64	19	8	657	—
SiO_2	35.6	50.5	46.0	100.0	—	49.3	44.5	46.5	46.9	46.4	44.5	46.3	45.4
TiO_2	—	0.52	—	—	52.5	0.43	0.92	0.67	0.64	0.79	0.29	0.67	1.1
Al_2O_3	—	1.32	34.6	—	—	0.30	14.7	13.3	19.5	13.7	25.0	12.6	11.1
Cr_2O_3	—	0.38	—	—	0.20	<0.03	0.32	0.30	0.39	0.20	0.11	0.28	0.52
FeO	37.6	22.9	0.61	—	46.8	42.1	16.9	17.2	11.9	18.5	6.6	17.6	20.5
MgO	25.9	13.4	0.05	—	0.6	1.4	9.1	7.2	6.5	6.5	7.0	9.8	10.2
CaO	0.34	10.9	18.6	—	—	6.2	12.0	13.1	13.9	13.3	14.8	12.0	11.0
Na_2O	—	—	0.86	—	—	<0.03	0.44	0.30	0.47	0.28	0.43	0.32	0.28
K_2O	—	—	0.03	—	—	0.06	0.05	0.04	0.02	0.04	0.07	0.03	0.03
Total	99.44	99.54	100.75	100.0	100.1	100.47	98.92	98.92	100.21	99.71	98.8	99.60	100.1

(1) Monomineralic olivine fragments.
(2) Monomineralic pyroxene fragments.
(3) Monomineralic plagioclase fragments.
(4) Monomineralic silica fragment (composition assumed).
(5) Monomineralic ilmenite fragment.
(6) Mesostasis from gabbroic rocks.
(7) Agglutinates and soil breccias (composition of agglutinates from Hu and Taylor (1977), except for Cr, which is from Simonds *et al.* (1977).
(8) Ferrobasalt lithic fragments.
(9) Gabbro lithic fragments (composition from mode and average monomineralic fragment compositions).
(10) Metabasalt lithic fragments.
(11) Nonmare lithic fragments.
(12) Soil composition (weighted average of columns 1–11).
(13) Average of six soil analyses (Blanchard *et al.*, 1977).

Table 5. Estimated average modal composition (vol.%) of gabbro in Luna 24 soils.

	A	B	C
pyroxene	65.0	37.9 } 41.1	59.8
olivine	5.4	3.2 }	5.0
plagioclase	27.6	54.9	32.8
opaques	0.4	1.9	0.7
silica	0.4	1.9	0.7
mesostasis	1.2	0.1	1.0
	100.0	99.9	100.0

(A) Computed from abundances of 257 monomineralic fragments, excluding these whose compositions lie outside of the range established for minerals in section 24170,7.

(B) Mode of 61 gabbro lithic fragments. Pyroxene and olivine in same ratio as column A.

(C) Weighted average of columns A and B.

Textures among the three fragments differ considerably. Sample 24174,59-12 is vitrophyric with aligned, skeletal olivines in a groundmass of pale-green glass (Fig. 1c). A probable remnant clast, not a phenocryst, occurs in the center of the thin section. This clast is much richer in FeO (Fo_{58}) than either the skeletal olivines (Fo_{81-64}) or the calculated liquidus olivine (Fo_{83}). The presence of this relic clast supports the idea expressed by A. Basu and co-workers (oral presentation at the Luna 24 Conference) that the olivine vitrophyres are impact melts. This is also supported by the analysis of olivine vitrophyre 24149,45-13 which contains 1.4 wt.% TiO_2 (Ryder *et al.*, 1977), perhaps indicating an admixture of mare basalt with higher TiO_2 contents than in VLT mare basalt.

Sample 24149,42-18 also appears to be an impact melt. It is superficially microporphyritic, with olivine and pyroxene crystals surrounded by a finer-grained groundmass of igneous texture (Fig. 1d). However, the olivine and pyroxene crystals are not euhedral, as would be expected of phenocrysts, but are angular, suggesting that they are relic fragments. Our third sample, 24174,65-7, is highly shocked and appears to have a microgranular texture.

Mineral compositions of the three olivine-rich lithic fragments appear in Fig. 7. Note that although the olivine compositions span a range from Fo_{82-41}, they tend to cluster between Fo_{60-70}. Thus, they do not represent the olivines in the large cluster at Fo_{75} among monomineralic olivine fragments (Fig. 6). Pyroxenes are less abundant than olivines. They are mainly pigeonitic and have Fe/(Fe + Mg) that ranges from 0.71 to 0.37: These compositions approximate those of the Mg-rich monomineralic pyroxene fragments that fall outside of the 24170 compositional range (Fig. 4).

In summary, the olivine-rich lithic fragments have a variety of textures that suggest that they are impact melts. Petrologists who advocate that these fragments have the compositions of a possible parent magma for the Luna 24 ferrobasalts must keep this property in mind. The olivine-rich fragments are

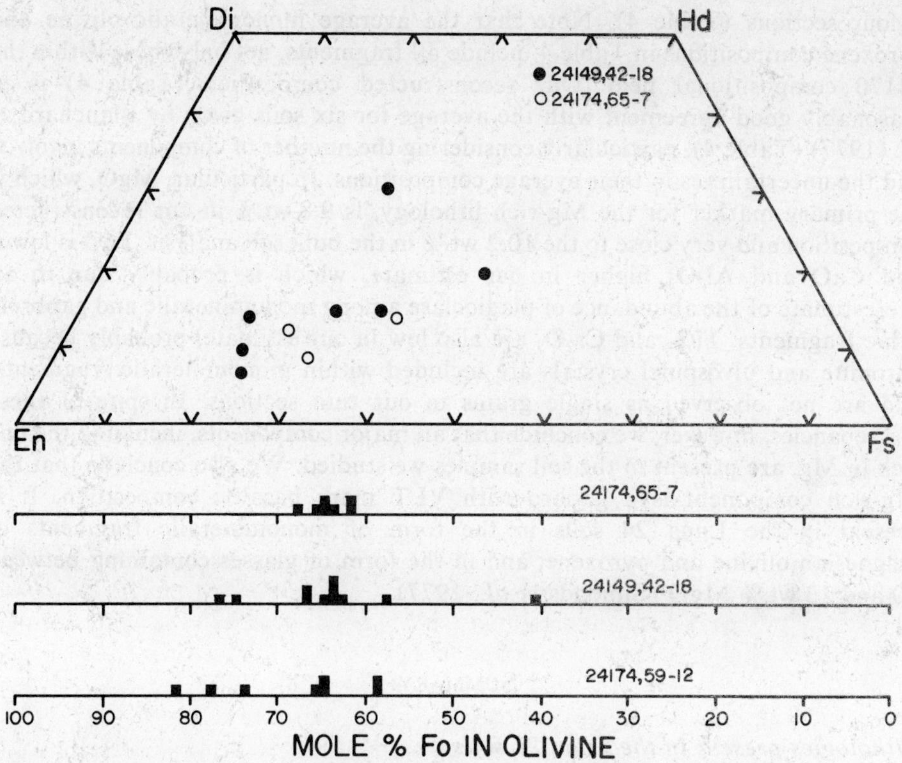

Fig. 7. Pyroxene and olivine compositions in olivine-rich lithic fragments from Luna 24.

clearly related to the mare basalts, but they may also have been modified by mixing during impact melting, which may have altered their original compositions. Thus, they may not represent the parent material of the Luna 24 ferrobasalts.

Bulk soil composition

A great deal of discussion at the Conference on Luna 24 centered around the nature of a supposed Mg-rich component present in the soil. Its presence was inferred from bulk soil compositions, which indicate that the soils contain ~10 wt.% MgO (Blanchard *et al.*, 1977; Ma and Schmitt, 1977), yet the most abundant lithic fragments appear to contain only 6–7 wt.% MgO (Table 2). Before we can attempt to determine the nature of the Mg-rich lithology, we must decide whether the samples we studied are representative of the average Luna 24 soil and, therefore, contain this Mg-rich component. Consequently, we first reconstructed the bulk chemical composition of the soil from the average compositions of the identifiable components and from their relative abundances in our sections (Table 4). Note that the average monomineralic olivine and pyroxene compositions in Table 4 include all fragments, not only those within the 24170 compositional field. Our reconstructed composition (Table 4) is in reasonably good agreement with the average for six soils given by Blanchard *et al.* (1977) (Table 4), particularly considering the number of components involved and the uncertainties in their average compositions. In particular, MgO, which is the primary marker for the Mg-rich lithology, is 9.8 wt.% in our reconstructed composition and very close to the 10.2 wt.% in the bulk soil analysis. FeO is lower and CaO and Al_2O_3 higher in our estimate, which is probably due to an overestimate of the abundance of plagioclase among monomineralic and gabbroic lithic fragments. TiO_2 and Cr_2O_3 are also low in our estimate, probably because chromite and ulvöspinel crystals are included within monomineralic fragments, and are not observed as single grains in our thin sections. In spite of these discrepancies, however, we conclude that all major components, including the one rich in Mg, are present in the soil samples we studied. We also conclude that the Mg-rich component is a gabbro with VLT mare basaltic composition. It is present in the Luna 24 soils in the form of monomineralic fragments of magnesian olivine and pyroxene, and in the form of glasses containing between 10 and 13 wt.% MgO (Simonds *et al.*, 1977).

SUMMARY:

Lithologies present in the Luna 24 soils

The data presented above and the inferences made from them allow us to conclude that the following lithologies are present in the Luna 24 soils.

Ferrobasalts: These are fine-grained, ophitic to subophitic basalts. Their bulk compositions (Table 2) show that they are low-Mg, high-Fe, VLT mare basalts.

Metabasalts: These are microgranular and have equilibrated mineral compositions. Their bulk compositions suggest that they are metamorphosed ferrobasalts.

Ferrogabbro: This lithology is best represented by sample 24170. It appears to be a coarse-grained version of Luna 24 ferrobasalt, as is indicated by similarities in average mineral and bulk lithic fragment compositions between 24170 and the ferrobasalts.

Other gabbros: The compositions of the monomineralic olivine and pyroxene fragments suggests that many of them were derived from gabbroic rocks richer in Mg than typical Luna 24 ferrogabbro such as 24170. The hypothetical Mg-rich soil component is clearly one of these gabbros. The group of olivine fragments with compositions of $\sim Fo_{75}$ and the low-Ca pyroxene fragments with compositions of $\sim En_{70}$ might represent a cumulate rock formed during the fractional crystallization of the Mg-rich gabbro. In fact, the Mg-rich gabbro might represent the parent magma for the Luna 24 ferrobasalts and ferrogabbros.

Olivine-rich samples: These are usually vitrophyric, but occasionally are microporphyritic or even granular. Their textures suggest that many of them are impact melts. Therefore, their compositions may have been altered by impact mixing which makes it very difficult to define precisely their original compositions.

Nonmare rocks: The nonmare contribution to the Luna 24 soils is very small (Table 1). Both ANT and KREEP are present.

Petrogenesis of Luna 24 VLT mare basalts

We discuss the chemical relationships among all VLT mare basalts and their petrogenesis in a separate paper in this volume (Taylor *et al.*, 1978). However, we conclude here by emphasizing three points: First, we do not know the precise compositions of Luna 24 ferrobasalts, thus making high -T,P experiments and theoretical calculations on their petrogenesis somewhat doubtful. Second, some proposed parent magmas, such as the olivine vitrophyres, appear to be impact melts, whereas others, such as the Mg-rich gabbro, are ill-defined compositionally. In either case, the composition of the parent magma is not known with certainty. Third, in spite of these reservations, it seems likely that the cluster of Mg-rich olivines (Fig. 6) among the monomineralic fragments represents crystals that fractionated from the parent magmas that ultimately formed the Luna 24 ferrobasalt and ferrogabbros.

Acknowledgments—We thank George Conrad for doing the numerous broad-beam analyses. This work is supported by NASA grant NGL 32-004-063.

REFERENCES

Albee A. L., Chodos A. A. and Dymek R. F. (1977) Petrography of Luna 24 sample 24170 (abstract). In *Papers Presented to the Conference on Luna 24*, p. 5–7. The Lunar Science Institute, Houston.

Barsukov V. L., Tarasov L. S., Dmitriev L. V., Kolesov G. M., Shevaleevsky I. D. and Garanin A. V. (1977) The geochemical and petrochemical features of regolith and rocks from Mare Crisium (preliminary data). *Proc. Lunar Sci. Conf. 8th*, p. 3319–3332.

Basu A., McKay D. E. and Fruland R. M. (1977a) Petrography, mineralogy, and source rocks of Luna 24 drill core soils (abstract). In *Papers Presented to the Conference on Luna 24*, p. 18–21. The Lunar Science Institute, Houston.

Bence A. E., Grove T. L. and Scambos T. (1977) Gabbro from Mare Crisium: An analysis of the Luna 24 soil. *Geophys. Res. Lett.* **4**, 493–496.

Blanchard D. P., Haskin L. A., Brannon J. C. and Aaboe E. (1977) Chemistry of soils and particles from Luna 24 (abstract). In *Papers Presented to the Conference on Luna 24*, p. 37–40. The Lunar Science Institute, Houston.

Bower J. F., Wood J. A., Richardson S. M., McSween H. Y. and Ryder G. (1977) Rock compositions by defocussed beam analysis (abstract). In *Abstracts of VIII International Conf. of X-ray Optics and Microanalysis* (Boston, 1977). In press.

Dowty E., Prinz M., and Keil K. (1973) Composition, mineralogy, and petrology of 28 mare basalts from Apollo 15 rake samples. *Proc. Lunar Sci. Conf. 4th*, p. 423–444.

Florensky C. P., Basilevsky A. T., Ivanov A. V., Pronin A. A. and Rode O. D. (1977) Luna 24 Geologic setting of landing site and characteristics of sample core (preliminary data). *Proc. Lunar Sci. Conf. 8th*, p. 3257–3279.

Hu H. -N. and Taylor L. A. (1977) Luna 24 soils from Mare Crisium: Agglutinate chemistry (abstract). In *Papers Presented to the Conference on Luna 24*, p. 82–84. The Lunar Science Institute, Houston.

Ma M.-S. and Schmitt R. A. (1977) Luna 24 soils: A chemical study (abstract). In *Papers Presented to the Conference on Luna 24*, p. 98–101. The Lunar Science Institute, Houston.

McSween H. Y., Jr., Taylor L. A. and Clark J. C., (1977) Luna 24 metabasalts: Possible evidence for assimilation of highlands materials? (abstract). In *Papers Presented to the Conference on Luna 24*, p. 118–120. The Lunar Science Institute, Houston.

Nielsen R. L. and Drake M. J. (1977) Magma types at the Luna 24 site in Mare Crisium (abstract). In *Papers Presented to the Conference on Luna 24*, p. 132–135. The Lunar Science Institute, Houston.

Norman M., Coish R. A. and Taylor L. A. (1977) Glass chemistry and magma evolution at Mare Crisium (abstract) In *Papers Presented to the Conference on Luna 24*, p. 136–138. The Lunar Science Institute, Houston.

Papp H., Steele I. M. and Smith J. V. (1977) Luna 24 90–150 micron fraction: Implications for sampling of planetary and asteroidal regoliths (abstract). In *Papers Presented to the Conference on Luna 24*, p. 143–146. The Lunar Science Institute, Houston.

Ryder G., McSween H. Y., Jr. and Marvin U. B. (1977) Basalts from Mare Crisium. *The Moon* **17**, 263-287.

Simonds C. H., Warner J. L., McGee P. E., Cochran A. and Brown R. W. (1977) Luna 24: Petrochemical potpourri (abstract). In *Papers Presented to the Conference on Luna 24*, p. 170–173. The Lunar Science Institute, Houston.

Tarasov L. S., Nazarov M. A., Shevaleevsky I. D., Kudryashova A. F., Gaverdovskaya A. S. and Korina M. I. (1977) Mineralogy and petrography of lunar rocks from Mare Crisium (preliminary data). *Proc. Lunar Sci. Conf. 8th*, p. 3333–3356.

Taylor G. J., Keil K. and Warner R. D. (1977a) Very low-Ti mare basalts. *Geophys. Res. Lett.* **4**, 207–210.

Taylor G. J., Warner R. D., Wentworth S., Keil K. and Sayeed U. (1977b) Petrology and chemistry of Luna 24 mare basalts and basaltic glasses (abstract). In *Papers Presented to the Conference on Luna 24*, p. 189–192. The Lunar Science Institute, Houston.

Taylor G. J., Warner R. D. and Keil K. (1978) VLT Mare Basalts: Impact mixing, parent magma types, and petrogenesis. *Mare Crisium: The View From Luna 24* (R. B. Merrill and J. J. Papike, eds.) This volume.

Vaniman D. T. and Papike J. J. (1977a) Very low Ti (VLT) basalts: A new mare rock type from the Apollo 17 drill core. *Proc. Lunar Sci. Conf. 8th*, p. 1443–1471.

Vaniman D. T. and Papike J. J. (1977b) Ferrobasalts from Mare Crisium: Luna 24. *Geophys. Res. Lett.* **4**, 497–500.

Merrill, R. B. and Papike, J. J., editors.
(1978) *Mare Crisium: The View from Luna 24*, Pergamon Press (New York), p. 321–337
Printed in the United States of America

Origin and modal petrography of Luna 24 soils

ABHIJIT BASU[1,2], DAVID S. MCKAY[3], RUTH M. FRULAND[3]

[1]The Lunar Science Institute, Houston, Texas 77058
[2]Department of Geology, Indiana University, Bloomington, Indiana 47401
[3]NASA-Johnson Space Center, Houston, Texas 77058

Abstract—Modal petrography of 7383 photographically documented particles in polished grain mounts of four sieved size fractions from each of six Luna 24 soils reveals a uniquely high abundance (~52%) of monomineralic phases. The soils are obviously immature with only ~19% agglutinates. Igneous textured crystalline lithic fragments, if identifiable, are almost exclusively mare basalt clasts. Highland materials make up probably <5% of the soil. Such a dominance of mare lithology is also seen in low grade vitric breccias as well as in the high grade recrystallized breccias and lithics.

Electronprobe analyses of the monomineralic silicate phases in the size range of 150–250 μm confirm that most of the particles are derived from local mare basalts and gabbros found in the coarser sizes (>250 μm). In addition, the existence of a magnesian olivine rich mare cumulate is inferred from olivine and pyroxene chemistry. This cumulate must have contributed about 10% of this regolith which explains why the bulk soil composition is more magnesian and chromian than the average composition of the mare basalt fragments sampled. Although the preponderance of mare basalts in this soil is overwhelming in all other particle types, the homogeneous glass particles, as revealed by cluster analysis of their compositions, show an important (~42%) highland affinity.

INTRODUCTION

We have performed detailed petrographic modal analyses of polished grain mounts of four sieve fractions in the size range of 20 μm to 250 μm for each of the following samples from Mare Crisium regolith: 24077,9; 24109,13; 24149,15; 24174,10; 24182,15; 24210,9. We have examined optically 7383 photographically documented soil particles. Additionally, we have made about 900 electronprobe microanalyses of the monomineralic and homogeneous glass particles. As the returned Luna 24 sample contains no large rock fragments, it is particularly important to infer the nature of the parent or source rocks from the core soil samples. Based on the regolith particle types, their relative abundances, and their chemical compositions, we have attempted to deduce the nature of source rocks for this regolith. We also attempt to estimate the relative contributions of the major source rocks.

EXPERIMENTAL PROCEDURE

Polished grain mounts (thin sections) of aliquots of four grain size fractions viz. 20–45 μm, 45–90 μm, 90–150 μm, and 150–250 μm of the six allocated soils were photographed and numbered according to the standard system (e.g., Taylor *et al.*, 1972). Modal analyses of these polished grain mounts were carried out with a Zeiss ultraphot optical microscope. Approximately three hundred particles were counted from each thin section. Each particle was examined with a high power (×40) objective in both transmitted and reflected light. We believe that both transmitted and reflected light optics at high magnification are essential for lunar soil petrography, especially for identification of agglutinates, internal textures of breccias and other composite particles.

Plagioclase was analyzed with the ETEC Autoprobe at Indiana University. A focused electron beam of 1 μm-2 μm size was used at 15 kV and approximately 0.22 mA. Counting time was 10 seconds for each of the 9 elements (Na, Mg, Al, Si, K, Ca, Ti, Mn, Fe) analyzed. All other analyses were performed with the ARL electron probe at the Johnson Space Center with a focused beam at 20 kV and approximately 0.4 mA. Counting time was 20 seconds for each of the 12 elements (Na, Mg, Al, Si, P, K, Ca, Ti, Cr, Mn, Fe, Ni) analyzed. Correction factors of Bence and Albee (1968) and Albee and Ray (1970) were used for data reduction of all analyses. All of our analyses were made on 150–250 μm sized particles.

CLASSIFICATION AND DESCRIPTION OF PARTICLE TYPES

A brief description of each of the particle types is given below. For the purpose of this paper, "particles" will denote each discrete entity in a soil, and "grain" will denote the phases that make up a particle. Thus, a few grains of plagioclase and pyroxene occur in a basalt particle of the soil. Similarly, the soil is sieved into 4 size classes of particles; the grain size of the phases in a mare basalt particle (150–250 μm) could be ~40 μm on an average.

Monomineralic Particles. Any particle that consists principally (>90%) of only one phase belongs to this broad group. The only exception is for anorthositic fragments which are polycrystalline and are considered to be crystalline lithic fragments. Almost all of the other monomineralic fragments are also monocrystalline. Monomineralic fragments were divided into the three principal silicates, olivine, pyroxene, and plagioclase; opaques and silica phases (cristobalite etc.) were also recognized as separate classes. The opaques were further subdivided into oxides (ilmenite, chromite, etc.), sulfides (troilite), and Fe-metal.

Crystalline Lithics. All igneous textured lithic particles as well as those conforming to the descriptions of the ANT suite of rocks (e.g., Prinz and Keil, 1977), unless brecciated or recrystallized, belong to this group. The ANT particles were subdivided into two groups, anorthositic (plagioclase >67%) and gabbroic (plagioclase <67%). Mare basalt particles were recognized on the basis of: (1) the dominance of Fe-rich pyroxene; and (2) the general paucity of mesostasis. Higher mesostasis constitutes a larger part of non-mare basalts and has a distinctive texture (Basu and Bower, 1976, 1977; Irving, 1975). Mare basalt particles were subdivided into two groups based on the presence or absence of olivine as a major phase. The non-mare basalts are either feldspathic (>67% plagioclase) and with rare or no silica, K-feldspar, phosphate, and mesostasis, or, they are KREEPy with larger areas of high-Si high-K mesostasis (>10%) having phosphate crystals. A large number of lithic fragments are polymineralic, but their textures cannot be determined at the size ranges of the soil particles. This is commonly due to the presence of large grains in these particles. Such particles are considered to be 'indeterminate' in nature, and are further subdivided on the basis of the dominant phase. It is probable that most of these particles are mare basalts.

Breccias and Recrystallized Lithics. These particles were classified into two broad groups based on the glass content of the groundmass. Low grade breccias contain >5% glass while those with <5% glass are classified as high grade. Thus, all the low grade breccias are vitric and are either 'dark matrix' or 'light matrix' (*cf.*, Basu and Bower, 1977). All soil breccias belong to this group. High grade breccias and recrystallized lithics are poikilitic (including micropoikilitic textures), or melt-matrix (usually showing a criss-cross pattern of plagioclase laths in the matrix), or other textures. The meta-basalts (e.g., McSween *et al.*, 1977) were classified in this third category of high grade breccia and recrystallized lithics. An unusual type of melt matrix breccia was found in these soils (Basu *et al.*, 1977b). This breccia type is vuggy (up to ~20% vugs) and has a matrix which exhibits quench-crystalline texture. Very fine Fe-metal and troilite in the form of globules and crystals are dispersed in the matrix. Pyroxene and plagioclase in an approximate ratio of 60:40 make up the groundmass. Clasts are mostly monomineralic olivine, pyroxene and plagioclase. Lithic clasts are extremely rare. A few olivine and pyroxene clasts were analyzed which show that they are magnesian ($Mg/Mg + Fe > 0.5$). The pyroxene is not Ca-poor. We infer a mare source for these particles which could be the disintegration product of a gas rich melt sheet generated by a medium size impact on a well irradiated mare regolith.

Agglutinates. Petrographic descriptions by Heiken (1975) and Basu and Bower (1977) were used to identify the agglutinates which are characterized by the inhomogeneous nature of their glassy groundmass.

Glass. Homogeneous glass, usually in the form of droplets or fragments thereof, belong to this class. Glass particles have been classified on the basis of their color in thin sections.

RESULTS

In order to best represent the composition of the analyzed soils, we have combined our modal analysis data of the size fractions into a single set for each of the samples. We have taken the sieve data from McKay *et al.* (1978) and have weighted the individual size-class modal data accordingly (Table 1). Thus, our average modal analysis of 7383 particles represents about 80% by weight of the <250 μm size fraction of the Luna 24 soil. The most remarkable feature about the modal data is the very high abundance of monomineralic particles (~52%). No other set of soils in any of the other sample return missions contains such a high proportion of monomineralic particles. In this respect alone the Luna 24 regolith is unique.

DISCUSSION

Our petrographic survey of Luna 24 soils has been restricted to the sub-250 μm

Table 1. Weighted modal analysis of 20 μm–250 μm grain size fractions of Luna 24 drill core soils.

Sample No.	−077,9	−109,13	−149,15	−174,10	−182,15	−210,9	Average
wt.% of <250 μm fraction	73.52%	80.22%	82.95%	79.50%	77.01%	84.47%	79.61 ± 3.98
Monomineralic fragments	47.53	48.42	51.22	56.34	58.24	50.04	51.97 ± 4.36
Plagioclase	9.78	9.95	9.95	12.05	15.82	12.71	11.71 ± 2.36
Pyroxene	32.49	33.64	36.14	38.87	36.20	31.87	34.87 ± 2.67
Olivine	4.44	3.81	3.42	3.92	5.03	4.49	4.19 ± 0.58
Opaques	0.68	0.82	1.00	1.19	0.69	0.45	0.81 ± 0.26
Ilmenite, Chromite, etc.	0.00	0.50	0.40	0.79	0.06	0.07	0.30 ± 0.31
Troilite	0.05	0.17	0.08	0.10	0.06	0.00	0.08 ± 0.06
Fe metal	0.63	0.17	0.80	0.38	0.55	0.38	0.49 ± 0.22
SiO_2 phases	0.14	0.21	0.45	0.40	0.56	0.52	0.38 ± 0.17
Xlline Lithics	10.99	11.52	11.43	14.68	13.57	13.24	12.57 ± 1.47
ANT	0.45	0.36	0.47	1.14	0.53	0.94	0.65 ± 0.31
Anorthositic	0.31	0.36	0.16	0.73	0.40	0.63	0.43 ± 0.21
Gabbroic	0.14	0.00	0.31	0.42	0.14	0.31	0.22 ± 0.15
Mare basalts	5.52	5.78	5.30	6.04	5.71	4.76	5.52 ± 0.45
Olivine-bearing	1.56	1.57	1.82	1.21	1.45	1.10	1.45 ± 0.26
Olivine-free	3.97	4.25	3.48	4.83	4.26	3.66	4.08 ± 0.48
Non-mare basalts	0.0	0.06	0.00	0.0	0.00	0.13	0.03 ± 0.05
Feldspathic	0.0	0.06	0.00	0.0	0.00	0.13	0.03 ± 0.05
KREEPy	0.0	0.00	0.00	0.0	0.00	0.00	0.0 ± —
Indeterminate	5.02	5.31	5.64	7.50	7.31	7.41	6.37 ± 1.16
Plagioclase rich	0.53	1.75	1.27	2.59	2.39	1.82	1.73 ± 0.75
Pyroxene rich	3.52	2.78	3.81	4.42	4.10	4.91	3.92 ± 0.74
Olivine rich	0.24	0.52	0.34	0.24	0.64	0.53	0.43 ± 0.16
Opaque rich	0.10	0.09	0.24	0.28	0.17	0.08	0.16 ± 0.08
SiO_2 rich	0.04	0.17	0.00	0.00	0.06	0.09	0.06 ± 0.06
Breccia & Rexllzd Lithics	10.32	9.94	17.43	9.72	9.19	13.32	11.65 ± 3.18
Low grade, vitric	8.43	7.03	15.19	8.26	6.64	11.33	9.48 ± 3.25
Dark matrix	7.86	6.83	14.80	8.10	6.09	10.57	9.04 ± 3.21
Light matrix	0.57	0.20	0.40	0.15	0.60	0.63	0.43 ± 0.21
High grade, rexllzed	1.89	2.89	2.24	2.13	2.53	2.14	2.30 ± 0.35
Poikilitic	0.20	0.62	0.42	0.63	0.29	0.08	0.37 ± 0.22
Melt-matrix	0.97	1.00	0.84	0.28	1.04	0.97	0.85 ± 0.29
Other	0.72	1.27	0.96	0.57	1.23	0.73	0.91 ± 0.29
Agglutinates	26.05	23.68	14.64	15.64	12.70	18.02	18.71 ± 5.18
Glass	5.03	6.27	5.66	3.26	6.05	6.06	5.39 ± 1.13
Black, brown, etc.	2.90	2.76	2.20	1.56	3.00	1.93	2.39 ± 0.58
Green, yellow, etc.	1.18	1.12	1.41	0.55	0.56	1.40	1.04 ± 0.39
Colorless, grey, etc.	0.97	2.39	2.05	1.13	2.55	2.77	1.98 ± 0.76
Miscellaneous	0.00	0.09	0.24	0.14	0.00	0.00	0.08 ± 0.10
Total %	99.92	99.92	100.62	99.78	99.75	100.68	100.37
Total # of grains	1229	1219	1245	1233	1236	1221	7383

size fraction. Detailed petrography and chemistry of the lithic fragments in large sizes (>250 μm) have been carried out by other groups (Albee *et al.*, 1977; Bence and Grove, 1977; Bence *et al.*, 1977; Ma *et al.*, 1977; Marvin *et al.*, 1977; McSween *et al.*, 1977; Ryder *et al.*, 1977a, 1977b; Taylor *et al.*, 1977a,b; Vaniman and Papike, 1977a,b). In addition, lithic fragments from the sub-250 μm fraction have also been studied by Coish and Taylor (1977), Nielson and Drake (1977) and Papp *et al.* (1977). The principal rock types reported by these other workers are: Ferrobasalts and ferrogabbros of various grain sizes (hereafter called local mare basalts), VLT basalts (which are a part of the local mare basalts), one fragment of a low Ti (2.3%) mare basalt with 0.55% Na_2O, melt rocks of highland composition, and metabasalts. In this discussion we shall examine our data in the light of the above rock types.

Crystalline Lithics. More than 90% of the identifiable rock fragments that we have examined are mare basalts. If the same proportion is also extended to the 'indeterminate' lithic fragments, as has been done by McKay *et al.* (1977), then about 11% of the Luna 24 soil consists of crystalline mare basalts. It is not possible to assign these particles to the various basalt categories of Luna 24 on the basis of optical petrography alone without fairly extensive electronprobe analyses of the individual phases, especially the pyroxene. We have not attempted these analyses because there are about 750 such particles in our thin sections requiring a total of at least 5000 analyses within a short time frame. However, the fact remains that we, as well as others, are unable to *quantitatively* classify the mare basalt fragments. It is, therefore, possible that mare basalt types not yet identified in the larger size fractions are represented in the <250 μm fraction.

Optical identification of ANT suite of rocks and non-mare basalts is always difficult in the particle size range that we have studied. However, these types do not make up more than 4% of the lithic fragments.

Breccias. The soil breccias (vitric, dark matrix) are immature in internal texture consisting mainly of detrital grains from local mare basalts. Nearly 10% of the soil is made up of these particles. The abundance of high grade breccias is low (2.3%) in these soils. More than 80% of these breccias are metabasalts and/or vuggy melt matrix breccias related to mare sources. Thus, the breccia population indicates a very low highland component, perhaps not more than 0.5% in the total soil. We have not seen any K-feldspar or phosphate-mineral in these breccias. KREEP, if present at all in this soil, is probably not in the breccias.

Glass. Homogeneous glass droplets in lunar soils are either impact derived or were formed in a volcanic fire fountain (Heiken *et al.*, 1974; Heiken and McKay, 1977). Examination of the surfaces of 200 glass droplets with a high resolution scanning electron microscope (Basu *et al.*, 1977a) failed to reveal any feature similar to those on Apollo 15 and Apollo 17 spherules which Heiken and McKay (1977) ascribe to fire fountaining. We have performed a limited number (about 180) of 12 element electron probe analyses on Luna 24 homogeneous glass particles. We ran the data through a cluster analysis program (Davis, 1973). Our data show that there are three principal clusters which conform to anorthosite (~15%), anorthositic gabbro (~27%), and mare basalts (~58%). Two subsidiary classes can be recognized in the mare basalt (Basu *et al.*, 1977a). We found no statistically significant KREEP cluster; only one KREEPy glass (LKFM with high alumina) particle occurs in the population we have analyzed. Simonds *et al.* (1977) carrying out a similar study, in the 90–150 μm sized fraction and with a different cluster analysis program, identified all the five components mentioned above. While we infer about 42% highland component with no KREEP in our data set, Simonds *et al.* (1977) estimate about 55% highland component including a distinct KREEP cluster (~3%). In part, because Simonds *et al.* (1977) included devitrified nearly holocrystalline "glass" in their data base and also used a very different clustering technique, our results are not identical to theirs, although the two results are similar. Norman *et al.* (1977) also note a significant highland basalt component in their glass analyses. The high (~42%) abundance of the highland component in the glass means that nearly 2.5% of the bulk soil is highland glass. This could be contrasted to the very low highland breccia component (~0.5%) in these soils. Perhaps glass can survive longer distances of transport than highland breccias which presumably are reduced to the individual constituents, and thus lose their identity. But glass once formed remains as glass despite all mechanical fragmentation.

Monomineralic Particles. More than half of the <250 μm Luna 24 soils are composed of monomineralic particles. Monomineralic particles are also the most abundant in the size range of 250 μm–500 μm (Taylor *et al.*, 1977a; and our unpublished modal data). They are also common (~22%) in the 0.5 to 1.00 mm size range (Nagle and Walton, 1977). However, of the twelve >1.0 mm particles only one is a monomineralic but polycrystalline (anorthosite?) particle (Nagle and Walton, 1977). The high abundance of monomineralic particles in the soil suggests that the coarse grained parent rocks of many of these particles are not present in the bulk samples. Finer grained equivalents of all of these coarse grained parent rocks may be present but that has not been shown. It may be that at least some of the coarse grained parent rocks have no fine grained equivalent.

Pyroxene is by far the most abundant monomineralic particle in Luna 24 soils making up about 35% of the total. Chemical zoning in pyroxene is the best indicator of its origin (Bence and Papike, 1972). Chemical zoning of pyroxene in lithic fragments has been used extensively to characterize Luna 24 rock types (e.g., Vaniman and Papike, 1977a; Taylor *et al.*, 1977b; Bence and Grove, 1977). In monomineralic fragments, however, chemical zoning trends have to be interpreted with caution because it cannot be known how a grain has broken off, or if it has gone through an agglutination cycle (Basu and Bower, 1976). Therefore, unless a large number of grains are analyzed, the data may not be representative. Furthermore, data of pyroxene from various parent rocks may have overlapping fields so it may not be possible to assign individual pyroxene crystals to specific rock types.

We have plotted 265 analyses of monomineralic pyroxene in the four common display modes (Fig. 1a-d). Most of the plots fall within the fields of pyroxene from the ferrobasalts and ferrogabbros of Luna 24 (Albee *et al.*, 1977; Nielson and Drake, 1977; Ryder *et al.*, 1977a, 1977b; Taylor *et al.*, 1977a; Vaniman and Papike 1977a,b). This is true not only for the major element variation (Fig. 1a,b) but also for the trends shown by the minor elements (Fig. 1c,d). Note that the arrows in these figures represent the approximate iron enrichment (Fe/Fe + Mg) trend. Most monomineralic pyroxene appears to have come from the local basalts represented in the lithic fragments. Surprisingly, pyroxferroite compositions are totally lacking in our analyses. However, some analyses plot outside the fields of local basalts. For example, there is a small cluster of low-Ca Mg-rich pyroxene and two high-Ca Mg-rich pyroxenes (Fig. 1a) which are obviously derived from other sources. These pyroxene crystals also contain very low Ti and plot almost at the base of the Al^{VI}-Ti-Cr triangle (Fig. 1c). These fragments individually show very little zoning, suggesting slow equilibrated growth. It is possible that all of these are of non-mare origin and represent a highland component (Bence and Grove, 1977). However, typical plots of highland pyroxene usually show equally weighted clusters of both bronzites plus pigeonites and diopsidic augites (Irving, 1975). In our data, collected entirely from the 150–250 μm fraction, there is a paucity of diopsidic augites and of the low-Ca clusters; only one is a bronzite (Wo <5) and the rest are pigeonites. We believe that it is equally possible that some of these "non-mare" pyroxene crystals come from a cumulate (therefore slowly cooled) unit of the local mare basalts. The very low abundance of highland lithic fragments supports the interpretation that some of these pyroxene crystals do not come from highland sources. We conclude that only a part of this exotic component is highland derived and the rest are from a mare cumulate, not represented in the larger crystalline lithics because of its coarse grain size. Coish and Taylor (1977) report a cumulate textured particle which may well be a fragment from the proposed cumulate body.

We estimate that about 85%–90% of the pyroxene particles are derived from local mare units, about 5%–8% are from the proposed cumulate and probably less than 3% are from highland rock types. Additional evidence for the proposed cumulate body comes from the compositional variation of monomineralic olivine particles discussed below.

Olivine shows three distinct populations in terms of its Fe/Fe + Mg ratios which vary from about Fo_{78} to $Fo_{0.5}$ (Fig. 2a; 181 analyses). The highest mode occurs in the range of Fo_{73-78}, followed by a subsidiary mode at about Fo_{65}, and a smaller mode at Fo_{45-50}. The olivine crystals, particularly the Mg-rich varieties, are virtually unzoned indicating that they probably had slow growth, and if they are magmatic, cooling must have been fairly slow during olivine crystallization. The three populations, seen in the frequency bar diagrams (Fig. 2a) can also be seen in a plot of CaO% vs. Fo content (Fig. 2b). In general, CaO increases with increasing Fe. The clusters at Fo_{75} and Fo_{65} have a mode of 0.18% CaO (varying from 0.04% to 0.29%) and appear to belong to a CaO-Fe enrichment trend different from that shown by the more Fe-rich olivine.

Olivine in the analyzed local basalts and gabbros are almost never more magnesian than Fo_{65}, and more frequently than not they are much more Fe-rich (Bence *et al.*, 1977; Nielson and Drake, 1977; Ryder *et al.*, 1977a, 1977b; Taylor *et al.*, 1977a; Albee *et al.*, 1977). Bence *et al.* (1977) report a high-Mg gabbro (24109,51, #34) with an Fo_{68} olivine and high-Mg pigeonite and augite. They also report Mg-olivine, from Luna 24 soils, up to Fo_{81}—presumably monomineralic fragments—which they suspect to be of highland origin. Coish and Taylor (1977) report an Fo_{75} olivine from a cumulate textured lithic fragment (150–250 μm) which they interpret to be of mare origin.

We interpret our olivine data (Fig. 2b) to indicate at least three, if not four, source rocks. All of the monomineralic olivine particles which are more Fe rich

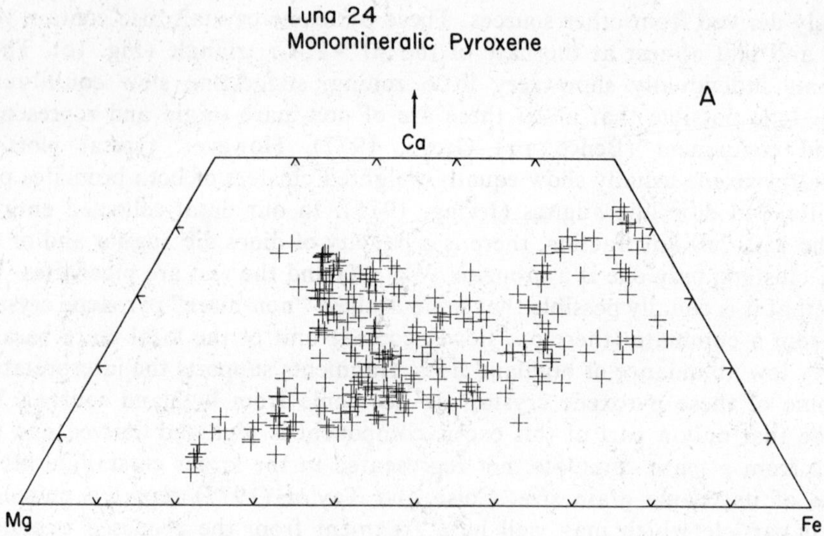

Fig. 1a. Trends of major element zoning in monomineralic pyroxene particles.

than Fo_{65} and have >0.22% CaO are from the local mare basalts. The olivine cluster at ~Fo_{70}-Fo_{65} and with CaO between approximately 0.1% to 0.25% are from a high-Mg gabbro similar (but not identical) to 24109,51 #34 (Bence *et al.*, 1977). The compositions of these olivine particles are somewhat similar to those from the olivine vitrophyre in the Luna 24 site (Ryder *et al.*, 1977a; Coish and Taylor, 1977; Basu *et al.*, 1977b). However, the olivine grains in vitrophyre are usually much smaller, skeletal, zoned, and with a slightly higher Fo content. If the olivine vitrophyres are impact melts (Basu *et al.*, 1978), then this high-Mg gabbro body could have been the target. The third and the most abundant cluster of olivine is also the most Mg-rich (Fo_{74-78}) and the CaO% varies from 0.04 to 0.29. We believe that all the olivine particles in this cluster containing <0.15% CaO probably have a highland origin. But those with >0.15% CaO and making up the bulk of the cluster, in our opinion, are mare related and are most probably derived from a mafic cumulate of a mare basalt suite. We suggest that this cumulate body was layered. The lower layer may have consisted of mostly olivine (~Fo_{77}) and the upper layer may have been Mg-rich noritic olivine (Fo_{65}) gabbro.

We estimate that 50% of the monomineralic olivine was contributed by the local ferrobasalts and ferrograbbros while ~25%–30% came from the olivine-rich part of the proposed cumulate body and ~20% came from the high-Mg gabbro part of the cumulate body. Perhaps <5% of the monomineralic olivine particles are highland derived. We believe that lithic fragments of the proposed cumulate body have not been found because its texture is too coarse grained. Judging from the size of the monomineralic olivine and pyroxene crystals and their abundances, we infer that the grain size of the cumulate body is likely to be on the order of 1–2

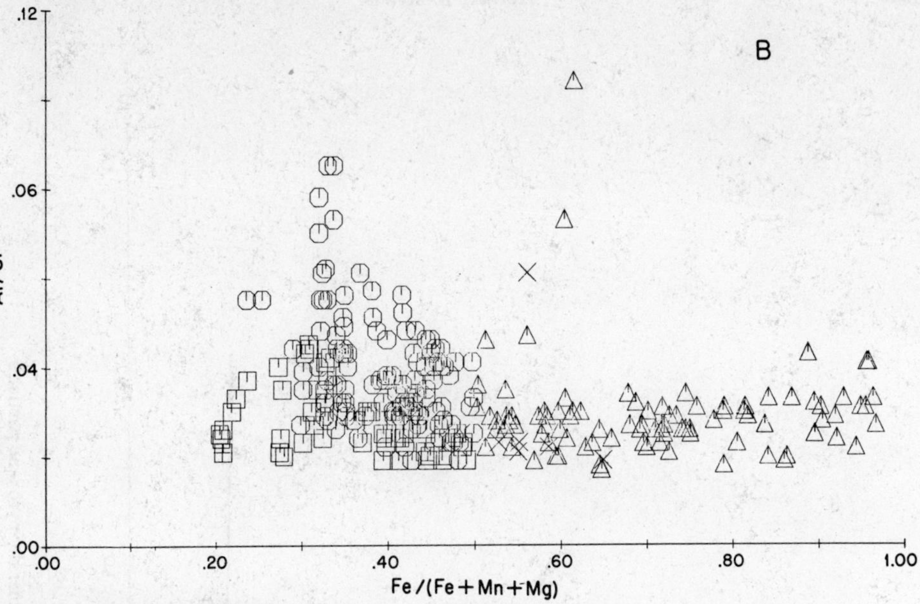

Fig. 1b. Plots of Al/Si vs. Fe/Fe + Mg of monomineralic pyroxene particles.

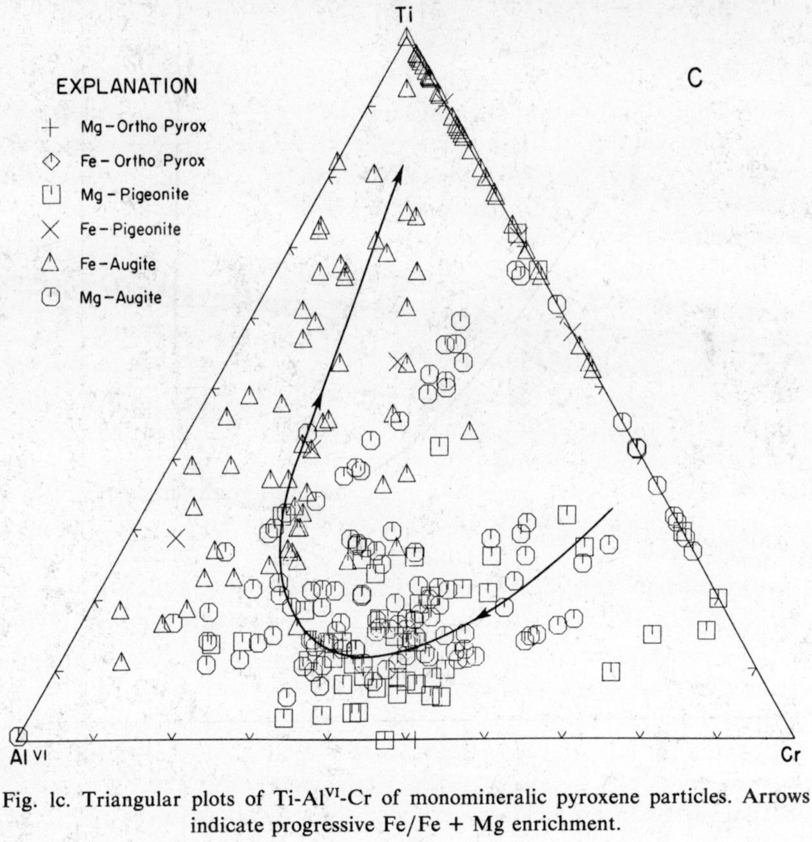

Fig. 1c. Triangular plots of Ti-Al^{VI}-Cr of monomineralic pyroxene particles. Arrows indicate progressive Fe/Fe + Mg enrichment.

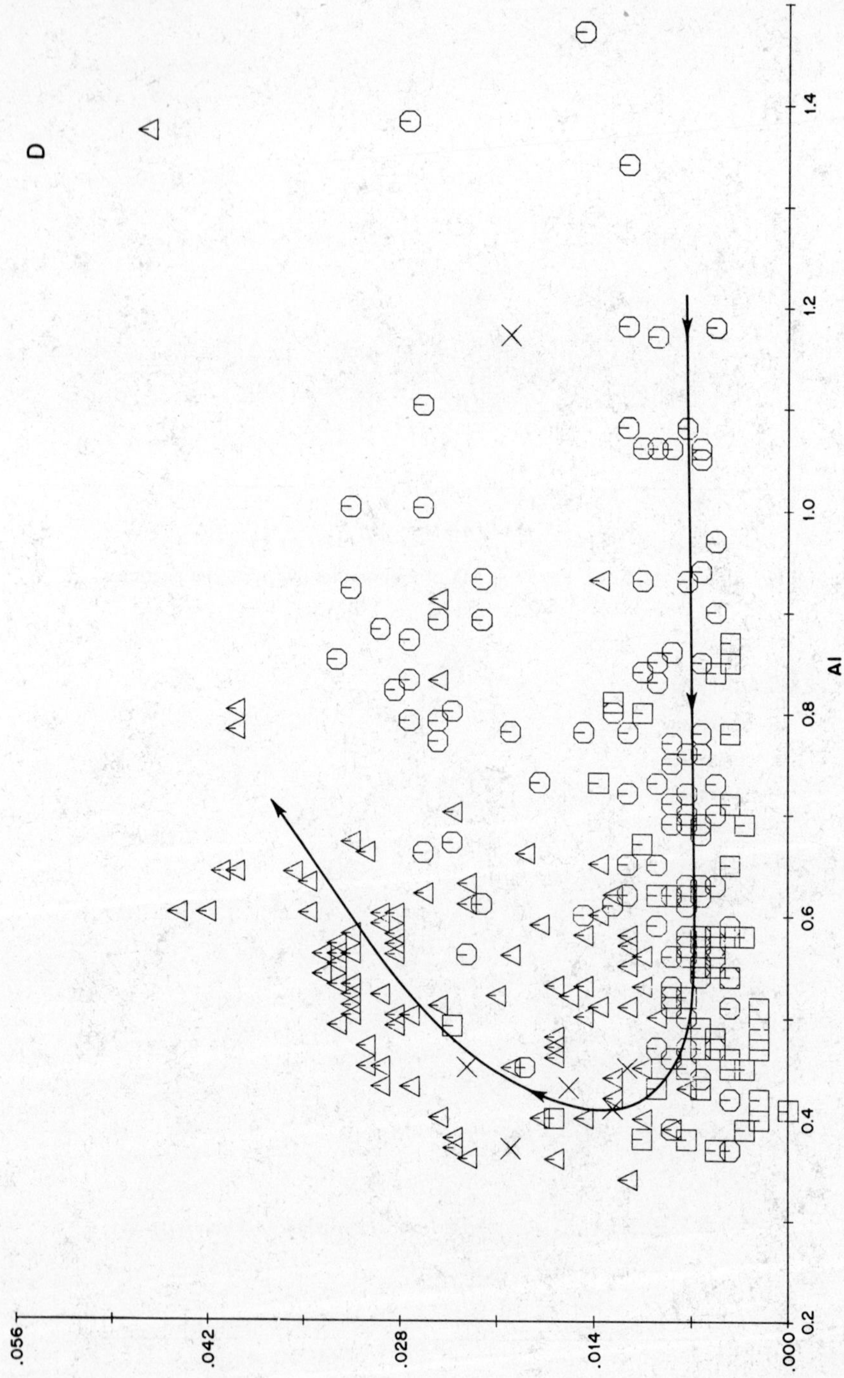

Fig. 1d. Plots of Ti vs. Al (cations per 6 oxygens) of monomineralic pyroxene particles. Arrows indicate progressive Fe/Fe + Mg enrichment.

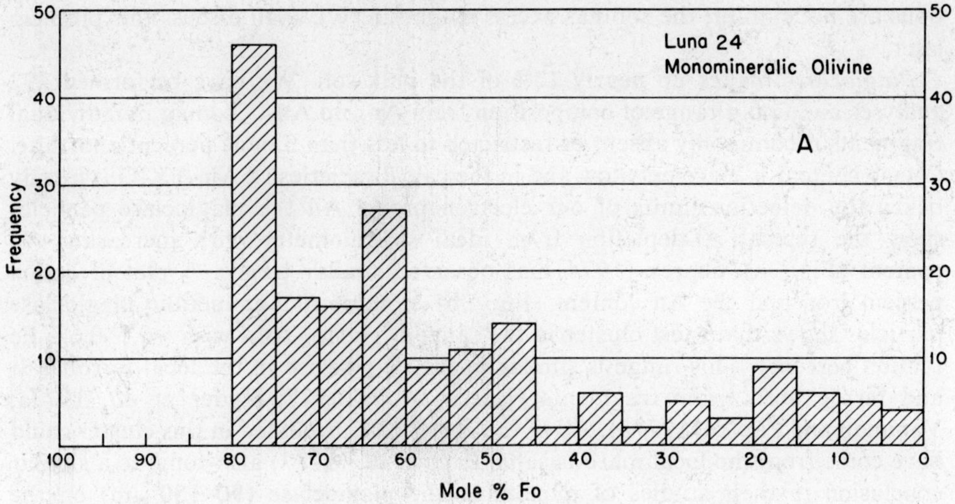

Fig. 2a. Frequency distribution of monomineralic olivine compositions.

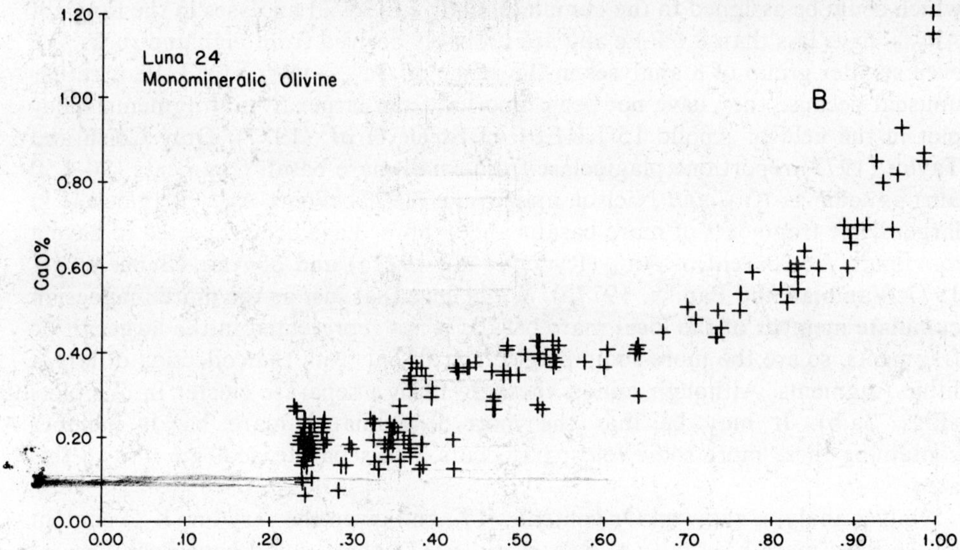

Fig. 2b. Plot of CaO percent vs. Fo content in monomineralic olivine particles.

mm. Unless the bulk composition of this cumulate body is similar to the bulk composition of the local basalts there is no reason to expect that the bulk composition of the soil would be similar to the bulk composition of local basalts found as lithic fragments. In fact, as discussed by Morris (1977), the compositions are not similar; the soil has excess magnesia. We shall discuss this problem later.

Plagioclase makes up nearly 12% of the bulk soil. We have performed 292 analyses and find a range of composition from An_{98} to An_{75}. Zoning in individual fragments is commonly absent or restricted to less than 5 mole percent anorthite. Potash content is extremely low and in the calcic varieties ($>An_{85}$) K_2O is mostly below the detection limits of our electron probe. All the plagioclase particles show the typical Al-depletion from ideal stoichiometry with increasing Ab content (Fig. 3a). Storey (1974) has observed similar trends. A plot of cation percent iron and the An content (Fig. 3b) of these monomineralic plagioclase particles shows that most cluster in the range of $\sim An_{83}$-An_{97} with ~0.1–0.6% Fe (cation percent). This range is similar to the plagioclase in the local ferrobasalt and ferrogabbro lithic fragments (Bence *et al.*, 1977; Ryder *et al.*, 1977a; Vaniman and Papike, 1977a). All of the particles represented in this cluster could have come from the local mare basalts. Papp *et al.* (1977) also come to a similar conclusion in their studies of monomineralic plagioclase (90–150 μm) on the basis of a similar plot. However, it is also possible that some of these plagioclases came from the high-Mg gabbro part of the proposed mare related cumulate body. At this time we are unable to differentiate any subpopulation of the plagioclase which could be assigned to the cumulate. Only 11 (3.8%) analyses in the range of An_{93-97} have less than 0.1% Fe and are probably derived from highland rocks. An even smaller group of 8 analyses in the range of An_{75-80} with >0.3% Fe is rather unusual because they have not been reported from larger lithic fragments. They plot in the field of Apollo 15 KREEP of Steele *et al.* (1977). Only Coish and Taylor (1977) report one plagioclase from small mare basalt fragments (90–150 μm) as sodic as An_{78} and Nielson and Drake (1977) report An_{75}. Plagioclase in larger lithic fragments of mare basalts and gabbros have been reported to have a maximum Ab content of An_{80} (Ryder *et al.*, 1977a) and of An_{88} (Bence *et al.*, 1977; Vaniman and Papike, 1977b). We believe that just as the more magnesian cumulate member of the local mare basalts is not represented in the larger lithic fragments, so are the more sodic plagioclase absent from the collection of larger lithic fragments. Although minor, these do form a separate cluster in our plots (Figs. 3a,b). It may be that the more fractionated mare basalt member containing these more sodic feldspars occurs as a separate geologic unit in this area.

Only 3 analyses show an Or content >1%, and given the very low K_2O content of the local mare basalts (e.g., Ma *et al.*, 1977), these could have been derived from another minor source of high-K mare basalt or from a minor KREEP component. Such a source did not contribute more than a fraction of a percent of the regolithic material at this site.

Cristobalite (or some other SiO_2 phase) is present in a very minor quantity (0.38%) in this soil. Some of the grains are large and present in the 150–250 μm fraction. Although Barsukov *et al.* (1977) and Tarasov *et al.* (1977) report cristobalite gabbros, the cristobalite grain reported by Albee *et al.* (1977) in a gabbro is only ~50 μm in size. Large cristobalite grains are known only from

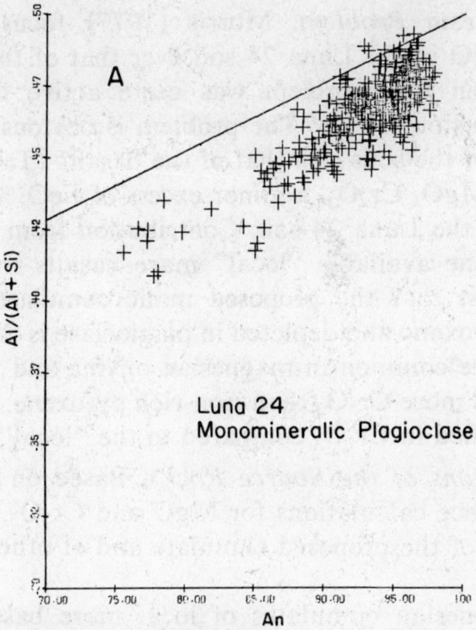

Fig. 3a. Plot of Al/Sl + Si vs. An content of monomineralic plagioclase particles. Solid line denotes the loci of ideal stoichiometry.

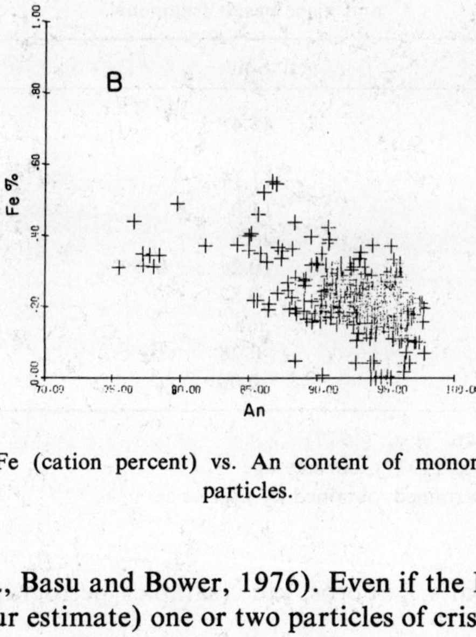

Fig. 3b. Plot of Fe (cation percent) vs. An content of monomineralic plagioclase particles.

KREEP basalts (e.g., Basu and Bower, 1976). Even if the KREEP component of this soil is <0.5% (our estimate) one or two particles of cristobalite may still have survived attrition. Alternatively, and more likely, these grains are from a coarse cristobalite gabbro which has not been sampled in the larger particle size fractions.

The Excess Magnesia Problem. Morris (1977) focused attention on the problem of excess MgO in the Luna 24 soil over that of the local mare basalts, although discussion on this problem was quite active during the Luna 24 Conference (LSI, Houston, 1977). The problem is obvious if one compares the average composition of the soils and that of the basalts (Table 2). It is seen that there is an *excess* of MgO, Cr_2O_3, a minor excess of FeO, and a slight *depletion* of CaO and Al_2O_3 in the Luna 24 soil. Contribution from a source more mafic and less felsic than the available "local" mare basalts must account for this difference. We suggest that the proposed mafic cumulate, rich in magnesian olivine, magnesian pyroxene and depleted in plagioclase is this source. Aluminian chromite inclusions are common in magnesian olivine and pyroxene; magnesian pyroxene also contains more Cr_2O_3 than iron-rich pyroxene. The cumulate is thus also much more enriched in Cr_2O_3 compared to the "local" mare basalts.

Relative Contributions of the Source Rocks. Based on modal analyses data and simple mass balance calculations for MgO and Cr_2O_3, we can estimate the relative contributions of the proposed cumulate and of other source rocks to the Luna 24 soils.

1. Early high magnesian cumulates of local mare basalt magma (no confirmed sample analyzed). 10%

Table 2. Average compositions of Luna 24 soils and mare basalt fragments.

	Soil[a]	Basalt[b]
SiO_2	45.4	45.7[c]
TiO_2	1.1	1.2
Al_2O_3	11.1	13.0
CaO	11.0	12.2
FeO	20.5	19.5
MgO	10.2	7.6
Cr_2O_3	0.52	0.23
MnO	n.d.	0.24
Na_2O	0.28	0.31
K_2O	0.03	n.d.

[a]Blanchard *et al.* (1977)
[b]Ma *et al.* (1977); our average
[c]not determined; obtained by difference

2. Local mare basalts, gabbros, etc. (samples petrographically studied and analyzed). 85%
3. Residual fractionation products of local mare basalt magma containing sodic plagioclase (An_{75}) (no confirmed sample analyzed). 1%
4. Undifferentiated highland rocks, mostly glass and ANT. 3%
5. KREEP 0.5%

It is likely that all the mare related source rocks identified above (#1–3) are genetically related to each other. We have a strong support for this conclusion from the opaque mineral studies of Haggerty (1977) who writes "The Luna 24 samples have provided what may be the first suite of lunar samples for which there is reasonably good evidence that an intrusive rock is genetically and chemically linked to an extrusive suite."

CONCLUSIONS

The Luna 24 regolith contains a very high proportion (~52%) of monomineralic fragments and is, therefore, unique among the lunar regoliths sampled so far. The chemistry of the monomineralic fragments shows that the large (>250 μm) mare basalt fragments in Luna 24 soils do not represent all the local rock types. We infer the existence of a magnesian-olivine-rich mare related cumulate body which has not been detected by the petrological studies of the >250 μm fraction of the soil. We infer that 85% of this regolith is derived from local mare basalts, gabbros, etc. and about 10% is derived from the cumulate. Highland sources did not contribute more than 3% of this regolith.

Acknowledgments—Indiana Memorial Foundation and NSF grant #GA36186 to C. Klein supported the part of this research carried out at Indiana University. Part of this research was conducted by one of us (AB) as a Visiting Scientist at the Lunar and Planetary Institute which is operated by the Universities Space Research Association under contract No. NSR-09-051-001 with the National Aeronautics and Space Administration. This paper constitutes the Lunar and Planetary Institute Contribution No. 320.

One of us (AB) is grateful to T. R. McGetchin and the staff of the LPI for many personal favors, and to J. Mitra of N.Y.U. for unlimited encouragement.

REFERENCES

Albee A. L. and Ray L. (1970) Correction factors for electron probe microanalysis of silicates, oxides, carbonates, phosphates and sulfates. *Anal. Chem.* **42**, 1408–1414.

Albee A. L., Chodos A. A. and Dymek R. F. (1977) Petrography of Luna 24 sample 24170. In *Papers Presented to the Conference on Luna 24*, p. 5–7. The Lunar Science Institute, Houston.

Barsukov V. L., Tarasov L. S., Dmtriev L. V., Kolesov G. M., Shevaleevsky I. D. and Garanin A. V. (1977) The geochemical and petrochemical features of regolith and rocks from Mare Crisium (preliminary data). *Proc. Lunar Sci. Conf. 8th*, p. 3319–3332.

Basu Abhijit and Bower Janice F. (1976) Petrography of KREEP basalt fragments from Apollo 15 soils. *Proc. Lunar Sci. Conf. 7th*, p. 659–678.

Basu Abhijit and Bower Janice F. (1977) Provenance of Apollo 15 deep drill core sediments. *Proc. Lunar Sci. Conf. 8th*, p. 2841–2867.

Basu Abhijit, McKay D. S., Clanton U. S., Waits G. and Fruland R. M. (1977a) Glass particles in Luna 24 drill core soils. In *Papers Presented to the Conference on Luna 24*, p. 14–17. The Lunar Science Institute, Houston.

Basu Abhijit, McKay D. S. and Fruland R. M. (1977b) Petrography, mineralogy and source rocks of Luna 24 drill core soils (abstract). In *Papers Presented to the Conference on Luna 24*, p. 18–21. The Lunar Science Institute, Houston.

Basu Abhijit, McKay D. S. and Fruland R. M. (1978) Impact melt origin of Luna 24 olivine vitrophyres (abstract). In *Lunar and Planetary Science IX*, p. 50–52. Lunar and Planetary Institute, Houston.

Bence A. E. and Albee A. L. (1968) Empirical correction factors for the electron microanalysis of silicates and oxides. *J. Geol.* **76**, 382–403.

Bence A. E. and Papike J. J. (1972) Pyroxenes as recorders of lunar basalt petrogenesis: Chemical trends due to crystal-liquid interaction. *Proc. Lunar Sci. Conf. 3rd*, p. 431–469.

Bence A. E. and Grove T. L. (1977) The highland component in the Luna 24 core (abstract). In *Papers Presented to the Conference on Luna 24*, p. 22–24. The Lunar Science Institute, Houston.

Bence A. E., Grove T. L. and Scambos T. (1977) Gabbros from Mare Crisium: An analysis of the Luna 24 soil. *Geophys. Res. Lett.* **4**, 493–496.

Blanchard D. P., Haskin L. A., Brannon J. C. and Aaboe E. (1977) Chemistry of soils and particles from Luna 24 (abstract). In *Papers Presented to the Conference on Luna 24*, p. 37–40. The Lunar Science Institute, Houston.

Coish R. A. and Taylor L. A. (1977) Igneous rocks from Mare Crisium: Mineralogy and petrology (abstract). In *Papers Presented to the Conference on Luna 24*, p. 48–50. The Lunar Science Institute, Houston.

Davis J. C. (1973) *Statistics and Data Analysis in Geology.* John Wiley & Sons, N.Y. 550 pp.

Haggerty S. E. (1977) Luna 24: opaque mineral chemistry of gabbroic and basaltic fragments from Mare Crisium. *Geophys. Res. Lett.* **4**, 489–492.

Heiken G. (1975) Petrology of lunar soils. *Rev. Geophys. Space Phys.* **13**, 567–587.

Heiken G. H. and McKay D. S. (1977) A model for eruption behavior of a volcanic vent in eastern Mare Serenitatis. *Proc. Lunar Sci. Conf. 8th*, p. 3243–3255.

Heiken G. H., McKay D. S. and Brown R. W. (1974) Lunar deposits of possible pyroclastic origin. *Geochim. Cosmochim. Acta* **38**, 1703–1718.

Irving A. J. (1975) Chemical, mineralogical, and textural systematics of non-mare melt rocks: Implications for lunar impact and volcanic processes. *Proc. Lunar Sci. Conf. 6th*, p. 363–394.

Ma M. S., Schmitt R. A., Taylor G. J., Warner R. D., Lange D. E. and Keil K. (1977) Chemical and petrographic studies of 18 Luna 24 lithic fragments (abstract). In *Papers Presented to the Conference on Luna 24*, p. 102–105. The Lunar Science Institute, Houston.

Marvin U. B., Ryder G. and McSween H. Y. Jr. (1977) Some exotic particles in the Luna 24 core sample (abstract). In *Papers Presented to the Conference on Luna 24*, p. 106–109. The Lunar Science Institute, Houston.

McKay D. S., Basu A. and Waits G. (1978) Grain size and the evolution of Luna 24 soils. In *Mare Crisium: The View from Luna 24* (R. B. Merrill and J. J. Papike, eds.). This volume.

McKay D. S., Basu A., Waits G., Clanton U., Fuhrman R. and Fruland R. M. (1977) Grain size and evolution of Luna 24 soils (abstract). In *Papers Presented to the Conference on Luna 24*, p. 114–117. The Lunar Science Institute, Houston.

McSween H. Y. Jr., Taylor L. A. and Clark J. C. (1977) Luna 24 metabasalt: Possible evidence for assimilation of highland materials. In *Papers Presented to the Conference on Luna 24*, p. 118–120. The Lunar Science Institute, Houston.

Morris R. V. (1977) Is the Luna 24 soil really from the Luna 24 site? Oral presentation at the Conference on Luna 24.

Nagle J. S. and Walton W. J. A. (1977) Luna 24 (samples presented to the U.S.A., 1977): Catalog and preliminary description. NASA-JSC, Houston. 85 pp.

Nielson R. L. and Drake M. J. (1977) Magma types at the Luna 24 site in Mare Crisium (abstract). In *Papers Presented to the Conference on Luna 24*, p. 132–135. The Lunar Science Institute, Houston.

Norman M., Coish R. A. and Taylor L. A. (1977) Glass chemistry and magma evolution at Mare Crisium (abstract). In *Papers Presented to the Conference on Luna 24*, p. 136–138. The Lunar Science Institute, Houston.

Papers Presented to the Conference on Luna 24 (1977) The Lunar Science Institute, Houston. 213 pp.

Papp H., Steel I. M. and Smith J. V. (1977) Luna 24 90–150 micron fraction: Implications for sampling of planetary and asteroidal regoliths (abstract). In *Papers Presented to the Conference on Luna 24*, p. 143–146. The Lunar Science Institute, Houston.

Prinz M. and Keil K. (1977) Mineralogy, petrology and chemistry of ANT-suite rocks from the lunar highlands. *Phys. Chem. Earth* **10**, 215–237.

Ryder G., McSween H. Y. Jr. and Marvin U. B. (1977a) Basalts from Mare Crisium. *The Moon* **17**, 263–287.

Ryder G., McSween H. Y. Jr. and Marvin U. B. (1977b) Luna 24 VLT basalt: Character and origin (abstract). In *Papers Presented to the Conference on Luna 24*, p. 160–163. The Lunar Science Institute, Houston.

Simonds C. H., Warner J. L., McGee P. E., Cochran A. and Brown R. W. (1977) Luna 24: Petrochemical potpourri (abstract). In *Papers Presented to the Conference on Luna 24*, p. 170–173. The Lunar Science Institute, Houston.

Steele I, Irving A. J. and Smith J. V. (1977) Apollo 15 breccia rake samples—mineralogy of lithic and mineral clasts. *Proc. Lunar Sci. Conf. 8th*, p. 1925–1941.

Storey W. C. (1974) Anomalous lunar plagioclase composition and the composition of the coexisting melt. *Nature* **251**, 403–405.

Tarasov L. S., Nazarov M. A., Shevaleevsky I. D., Kudryashova A. F., Gaverdovskaya A. S. and Karina M. I. (1977) Mineralogy and petrography of lunar rocks from Mare Crisium (preliminary data). *Proc. Lunar Sci. Conf. 8th*, p. 3333–3356.

Taylor G. J., Marvin U. B., Reid J. B. and Wood J. A. (1972) Norite fragments in the Apollo 14 and 12 soils and the origin of Oceanus Procellarum. *Proc. Lunar Sci. Conf. 3rd*, p. 995–1014.

Taylor G. J., Warner R., Wentworth S. and Keil K. (1977a) The Luna 24 regolith: Lithologic abundances in the 250–500 μm size fraction, and compositions of agglutinates and non-mare lithic fragments and glasses (abstract). In *Papers Presented to the Conference on Luna 24*, p. 186–188. The Lunar Science Institute, Houston.

Taylor G. J., Warner R., Wentworth S., Keil K. and Sayeed U. (1977b) Petrology and chemistry of Luna 24 mare basalts and basaltic glasses (abstract). In *Papers Presented to the Conference on Luna 24*, p. 189–192. The Lunar Science Institute, Houston.

Vaniman D. T. and Papike J. J. (1977a) Ferrobasalts from Mare Crisium: Luna 24. *Geophys. Res. Lett.* **4**, 497–500.

Vaniman D. T. and Papike J. J. (1977b) Luna 24 ferrobasalt: Petrology and comparisons with other mare basalt types (abstract). In *Papers Presented to the Conference on Luna 24*, p. 196–199. The Lunar Science Institute, Houston.

Merrill, R. B. and Papike, J. J., editors.
(1978) *Mare Crisium: The View from Luna 24*, Pergamon Press (New York), p. 339–355
Printed in the United States of America

On the origin of Luna 24 basalts and soils

GRAHAM RYDER[1] and URSULA B. MARVIN

Center for Astrophysics, Smithsonian Astrophysical
Observatory and Harvard College Observatory,
60 Garden Street, Cambridge, Massachusetts 02138

[1]Now at Department of Geology, University of California, Davis, Davis, California 95616

Abstract—The scatter in the various analyses of fine-grained Luna 24 VLT basalt is unusual in trend and is most likely to result from sampling errors associated with the approximate methods (DBAs and modal reconstruction) involved. The small variation in MgO contents of fine-grained basalts and the tight clustering of the compositions of brown glasses (possible pyroclastic equivalents of Luna 24 VLT basalt) suggest that a single homogeneous magma is represented. The homogeneity and the phase relationships of the fine-grained basalts suggest that the Luna 24 VLT basalt is not related to any other VLT basalt by fractional crystallization. Even at high pressures, a parent liquid would have to be even more aluminous than the high alumina Luna 24 VLT basalt, and this seems unlikely. A direct partial melting origin is most likely, and the source may have contained plagioclase, even though it lay below the lunar crust. Limited differentiation of Luna 24 VLT basalt within surface flows seems possible.

The high-Mg component in the Luna 24 regolith is enigmatic, but olivine vitrophyres and their degraded products (monomineralic olivines and so on) are the most likely candidates. The high-Mg component is probably more degraded than the Luna 24 VLT basalt, and is distributed throughout the drill core. These characteristics suggest that the high-Mg component is an older, Mg-rich mare basalt that is local to the Luna 24 site, and was not delivered to it by impact events such as Picard or Fahrenheit.

INTRODUCTION

Petrographic studies of the Luna 24 drill core have revealed that most of the basaltic fragments and at least a large proportion of the coarser mineral fragments represent a distinctive mare basalt type (Tarasov *et al.*, 1977; Ryder *et al.*, 1977a, 1977b; Taylor *et al.*, 1977b; Vaniman and Papike, 1977a; Grove and Bence, 1977). This basalt, which has a variable grain size, is characterized by very low titanium, low magnesium, low alkalis, low chromium, and high aluminum contents (Table 1, Fig. 1). Herein we refer to this basalt type as Luna 24 VLT basalt. Metabasalt fragments, impact melt fragments, and brown glasses in the core all have compositions similar to the Luna 24 VLT basalt. Apart from this basalt and its degraded products, the soil includes agglutinates, vitric soil breccias, vesicular and homogeneous glasses, and rare highlands lithologies.

A modally minor particle type observed in the core is an olivine vitrophyric basalt (e.g. Ryder *et al.*, 1977b; Coish and Taylor, 1977). We believe that the most magnesian of the olivine vitrophyres are fragments of a bona fide basalt type from Mare Crisium: these magnesian olivine vitrophyres do not have clastic

TABLE 1. Microprobe analyses of materials in the Luna 24 regolith.

	1	2	3	4	5
SiO_2	46.5	45.0	44.5	49.9	38.1
TiO_2	1.00	1.44	0.70	1.07	10.2
Al_2O_3	12.5	6.2	9.6	13.1	6.3
Cr_2O_3	0.2	0.56	0.63	0.43	0.63
FeO	18.0	19.0	19.1	17.0	20.7
MnO	0.25	0.25	0.32	0.29	0.26
MgO	6.5	16.9	15.5	5.2	14.0
CaO	13.5	7.9	8.9	13.0	7.4
Na_2O	0.35	0.02	0.24	0.32	0.35
K_2O	<0.05	0.07	0.00	0.00	0.08
P_2O_5	<0.04	0.02	0.02	0.05	0.02
Σ	98.89	97.34	99.51	100.37	98.04
#pts	—	19	5	5	3
beam size (μm)	—	100	100	—	100

1. Average Luna 24 VLT basalt, (best-estimate from probe data).
2. Olivine vitrophyre 24149, 45-13*
3. Olivine vitrophyre 24174, 63-20
4. Glass in olivine vitrophyre 24174, 63-20 (focussed beam)
5. Red-brown spherule in 24149, 47.

*24149, 45 is the NASA thin section number, -13 is identification of particle on SAO mapping photographs

textures, and their compositions are not represented among the numerous glass analyses of Simonds *et al.* (1977). The olivine vitrophyric basalt type is very low in titanium, but is distinct from the Luna 24 VLT basalt in having a much higher magnesium and a lower aluminum content (Ryder *et al.*, 1977a and this paper, Marvin *et al.*, 1977).

The Luna 24 regolith is chemically homogeneous from layer to layer (Blanchard *et al.*, 1977; Ma and Schmitt, 1977). However, while the Luna 24 VLT basalt dominates the basaltic fragment population, the bulk composition of the Luna 24 soil is much more magnesian and chromian, and less aluminous, than Luna 24 VLT. Therefore, the soil cannot be derived from simple comminution of Luna 24 VLT basalt flows; at least one higher magnesium component is present. One candidate for this component is, of course, the olivine vitrophyre (e.g. Nyquist *et al.*, 1977) although olivine-rich cumulates have been suggested (e.g. McKay *et al.*, 1977; Basu *et al.*, 1977b).

This paper addresses itself to the interpretation of material from the drill core. In particular we are interested in the origin and derivation of the Luna 24 VLT basalts, and the possibility of their being related to other lunar basalt compositions by fractional crystallization. We are also interested in the soil composition—what is the physical manifestation of the high-magnesium component, and what is its origin?

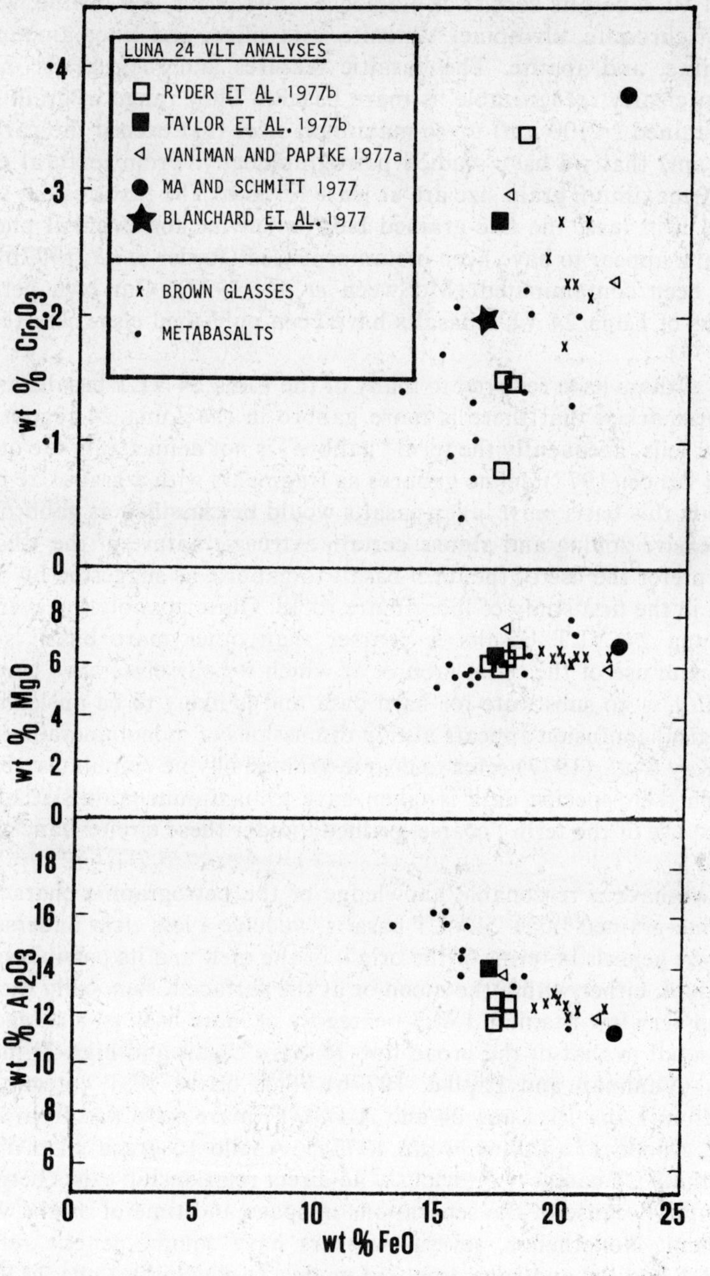

Fig. 1. Selected oxide variation diagrams for Luna 24 VLT basalts and related materials. Glass and metabasalt data from various sources.

Luna 24 VLT basalts

Luna 24 VLT basalts consist of plagioclase, pyroxene, and olivine, with minor amounts of chromite, ulvöspinel, ilmenite, and silica, and trace amounts of Fe-metal, troilite, and apatite. The basaltic textures and their mafic mineralogy make them clearly recognizable as mare basalts. They range in grain size from very fine-grained (<100 μm) up to maximum sizes that exceed the particle sizes (250–500 μm) that we have studied petrographically. From textural considerations, these maximum grain size are at least 1.5 mm. The textures are subophitic to ophitic; in at least the fine-grained basalts, olivine forms small phenocrysts. Some basalts appear to have been metamorphosed (Ryder *et al.*, 1977b) but may also have been contaminated (McSween *et al.*, 1977). Our own petrographic descriptions of Luna 24 VLT basalts have been published elsewhere (e.g. Ryder *et al.*, 1977b).

Several authors have referred to many of the Luna 24 VLT basalts as gabbros, and have remarked that there is more gabbro in the Luna 24 regolith than in other mare soils. Frequently the term "gabbro" is not defined. On the other hand, Grove and Bence (1977) define gabbros as fragments with a grain size more than 0.1 mm—on this basis *most* lunar basalts would be classified as gabbros. In view of the extensive zoning and almost certain extrusive nature of the VLT basalts, we would prefer the use of the term basalt to gabbro, as suggested by Schmitt *et al.* (1970) in the first study of lunar mare rocks. Quite possibly the average grain size of Luna 24 VLT basalts is coarser than other mare basalt suites, but indiscriminate use of the term gabbro, of which we ourselves have been to some extent guilty, is no substitute for hard data and is likely to be misleading. Such terminological confusion appears also in discussions of monomineralic fragments. Thus McKay *et al.* (1977) refer to coarse-grained olivine cumulates; the samples from which their specific data is taken have a maximum grain size of 250 μm. Clearly the use of the term "coarse-grained" under these circumstances is at best misleading.

While we have a reasonable knowledge of the petrographic characters of at least the fine-grained Luna 24 VLT basalts, we have a less clear understanding of their specific genesis in terms of the origin of the melt and its possible subsequent fractionations, either within the moon or at the surface. Chemically they fall into a sweeping very low titanium (VLT) category of mare basalts, a category with a standing equal to that of the broad low-Ti mare basalt and high-Ti mare basalt groupings (Vaniman and Papike, 1977b). This broad VLT category includes material from Luna 16, Luna 24 and Apollo 17 mare soils, and from Boulder 1, Station 2, Apollo 17 (Taylor *et al.*, 1977a). Apollo 15 green glass, if volcanic, falls into the VLT category. Obviously no direct relationship exists between all of these basalts, because of the separations in space and time of origin which exist among them. Nonetheless, several workers have sought genetic relationships among VLT basalts, and have proposed models in which the Luna 24 VLTs are a product of fractional crystallization of some other VLT basalt *composition* (e.g. Taylor *et al.*, 1977c; Vaniman and Papike, 1977c; Grove, pers. comm.). We have alternatively rejected such models and proposed that the Luna 24 VLT basalt group was derived from the lunar interior independently of other lunar mare basalts.

The relevant information lies fundamentally in the chemical composition of Luna 24 VLT basalts and other VLT basalts (including trace elements), and in the petrographic characteristics of the basalts. With these data as a basis, experimental data and least-squares calculations can potentially provide insights unobtainable from sample inspection alone. Unfortunately all the samples of VLT basalts are small soil particles or equally small clasts in Boulder 1, Station 2, Apollo 17. Therefore refined data, other than petrographic, is not available for most of the VLT fragments. Nonetheless, at least one statement may confidently be made concerning the origin of Luna 24 VLT basalts: they are not derived from the fractional crystallization of a high-Ti or a low-Ti basalt parent. In such a genesis, a Ti-rich phase must have crystallized from the parent, and the Luna 24 VLT would have a Ti-rich phase on its liquidus. Petrographically it can be seen that the Ti-rich phases in the Luna 24 VLT basalts occur very late in the crystallization sequence, and even at high pressure would not be on the liquidus. Thus any possible parent for the Luna 24 VLT must itself have been a VLT basalt.

Relations of Luna 24 VLTs to other VLTs

In general, schemes which attempt to relate Luna 24 VLT basalts to Apollo 17 VLTs or Apollo 15 green glass (or both) by fractional crystallization (unless perhaps at high pressure) have been abandoned. Grove and Vaniman (1977) have shown experimentally that Apollo 15 green glass is not a suitable parent because Luna 24 VLTs lie close to an olivine-plagioclase cotectic, and crystallization of Apollo 15 green glass is not towards this cotectic. Apollo 17 VLTs similarly do not fractionate towards Luna 24 VLT basalt compositions. Grove (pers. comm.) now proposes that even Apollo 17 VLT compositions cannot be produced by fractional crystallization of Apollo 15 green glass, and Taylor *et al.* (1978) suggest the possibility that the Apollo 17 VLT basalts themselves may be contaminated with plagioclase. Boulder 1, Station 2, Apollo 17 and Luna 16 VLT basalts are more alkaline than Luna 24 VLTs, and thus could not represent parental magmas. The Boulder 1 VLTs are at least 600 m.y. older than the Luna 24 VLTs, and a link would have to involve the derivation of two similar liquids 600 m.y. apart in time. While this is common on the earth, we know of no documented instance of such similarity of magma composition separated by such a time scale in lunar suites, and we reject it as a genetic link here also (indeed even magmas produced approximately contemporaneously in a given lunar region seem to be chemically quite diverse and unrelated; e.g. Apollo 12, Rhodes *et al.* (1977).

However, a relationship between Luna 24 VLT basalt and *some* VLT basalt composition by fractional crystallization is not ruled out merely by rejecting specific examples. Grove (pers. comm.) believes that the Luna 24 VLT basalt may have a parent among the green glasses found at the Luna 24 site itself. We point out that if we search long enough, we could find *some* composition that would satisfy, in terms of fractional crystallization, the characteristics required of a basalt parent: whether that is the way in which the basalt actually formed is a different matter. Coish and Taylor (1977) have proposed that the Luna 24 olivine vitrophyric basalt compositions may be parental to Luna 24 VLT basalts. A realistic rejection or acceptance of a relation of Luna 24 VLT basalt to these or any other VLT composition requires a closer inspection of the data, especially as compared to what we might expect of a derivative liquid.

Chemistry of Luna 24 VLT basalt

An important constraint on the origin and derivation of Luna 24 VLT basalt is its chemistry, and the variation in chemistry between samples. These chemical characteristics are important in assessing how homogeneous the magma was, whether or not surface differentiation occurred, and whether we are in fact dealing with a single magma. While we are less interested, in the present discussion, in surface fractionation, we would like to be able to distinguish between surface fractionation effects on a homogeneous magma on the one hand and extrusions of variable magmas related by fractional crystallization at shallow depth on the other.

Because of the nature of the sample, a regolith core, the chemical data *presently* available are not comparable in quality to that from analyzed Apollo rock suites. Most of the data are from modal reconstructions or from defocussed-beam microprobe analyses (DBAs) of particles less than 500 μm in diameter. One fragment described as a fine-grained basalt was analyzed by Ma *et al.* (1977) using Instrumental Neutron Activation techniques. Some trace element data is available (e.g. Ma *et al.*, 1977; Laul, 1977; etc.) for particles and soils, but need not be specifically considered in the present discussion.

(i) *Fine-grained basalts*

It is most instructive to consider primarily the finer-grained basalts, as these are likely to represent liquid compositions, whereas the coarser-grained material *could* include cumulates (by finer-grained we mean in general as compared with the fragment sizes we have studied, and thus basalts with grain-sizes less than about 200 μm are considered fine-grained). An estimate of average fine-grained Luna 24 VLT basalt is included in Table 1. Some of the data for material described as "fine-grained" are abstracted in Figs. 1 and 2. Clearly the data show some scatter, and we suggested earlier (Ryder *et al.*, 1977b) that the total real range in composition of the Luna 24 VLT basalts may be similar to that of other specific mare basalt subgroups. However, the nature of the scatter for fine-grained basalts (in Fig. 1) is different from that of other analyzed mare basalt suites, and we now conclude differently. The FeO content of Luna 24 VLT basalt *apparently* varies from ~16.5% to ~21.5% (Fig. 1), a much larger range than other conventionally-analyzed mare basalt suites. The Apollo 17 high-Ti basalts

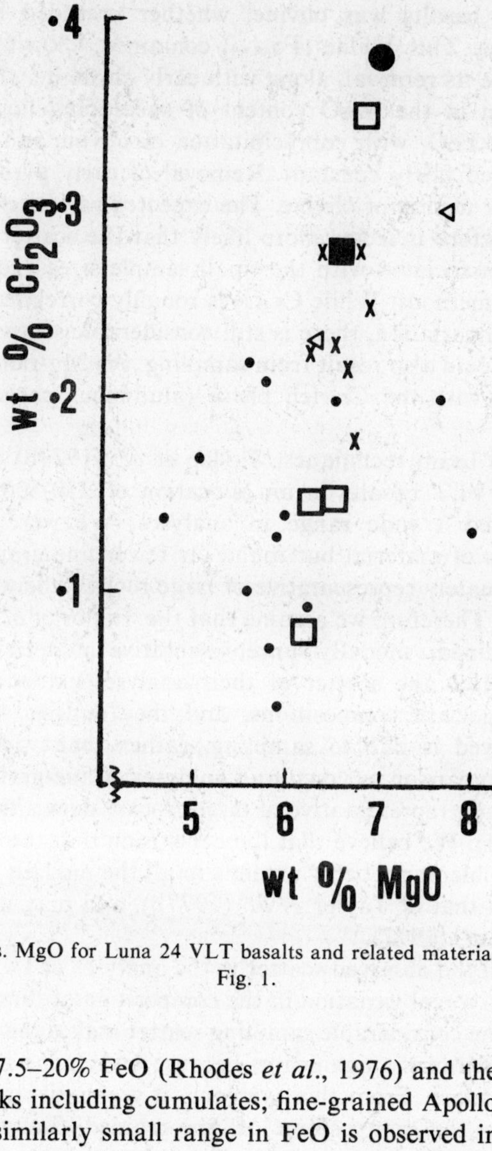

Fig. 2. Cr_2O_3 vs. MgO for Luna 24 VLT basalts and related materials. Symbols as in Fig. 1.

are restricted to 17.5–20% FeO (Rhodes *et al.*, 1976) and these samples include coarse-grained rocks including cumulates; fine-grained Apollo 17 basalts show a smaller range. A similarly small range in FeO is observed in Apollo 12 olivine and pigeonite basalt suites (Rhodes *et al.*, 1977). Conversely, the MgO content of Apollo mare basalt suites varies considerably (much more than FeO) but only varies from ~5.5 to ~7.5% in the Luna 24 VLT basalt suite. This difference in variation trends compared to other mare basalt suites leads us to suspect the data.

This suspicion is strengthened by the fact that it is not easy to envisage a lunar fractionation process which could result in considerable variation in FeO but little in MgO. Petrographic (e.g. Ryder *et al.*, 1977b) and experimental studies (Grove and Vaniman, 1977) indicate that the first mafic mineral to crystallize in the Luna 24 VLT basalts was olivine, whether preceded by plagioclase or preceding plagioclase. This olivine (Fo_{60-65}) contained ~30 wt.% MgO and ~30 wt.% FeO. Therefore its removal, along with early chromite, should have caused substantial depletion in the MgO content of succeeding liquids, but a much smaller depletion in FeO; with coprecipitation of olivine and plagioclase, FeO should have remained fairly constant. Removal of early pyroxene should have had an effect similar to that of olivine. The expected variation is the opposite of that observed. Therefore it seems more likely that the scatter in FeO is due to sampling problems associated with the small sample sizes and analytical errors associated with the methods. While Cr_2O_3 is roughly correlated with MgO (Fig. 2), suggesting a real variation, there is still considerable scatter in the data, and such a correlation would also result from sampling: the Mg-rich phase (olivine) is spatially associated with the Cr-rich phase (aluminous chromite) in Luna 24 VLT basalts.

Using defocussed beam techniques, Taylor *et al.* (1977b) analyzed 30 fragments of Luna 24 VLT basalts in an allocation of 250-500 μm particle thin sections, and showed a wide range in analyses. We were allocated similar quantities and types of material but found far fewer fine-grained basalts which we considered adequately representative of large rocks to be worth analysis with a defocussed beam. Therefore we assume that the Taylor *et al.* (1977b) analyses include coarser-grained, modally unrepresentative materials. Taylor *et al.* (1977b) point out that the scatter of their analyses extends towards average pyroxene and plagioclase compositions, and they suggest that the range in compositions observed is due to sampling, rather than variations in magma compositions. In comparison our own four analyses of fine-grained basalts, which we believe are modally representative of larger rocks, show a tight cluster (Ryder *et al.*, 1977b, Fig. 4). We believe that the conclusion that the scatter observed is due to sampling problems probably extends to all the modally reconstructed and DBA data, not just that of Taylor *et al.* (1977b), and may include the particle analyzed by Ma *et al.* (1977).

We conclude that the observed scatter in the analyses of Luna 24 VLT basalts cannot be related to a real variation in the composition of these basalts. The fact that we do have some considerable sampling scatter makes the fact that MgO has very little scatter extremely significant, because even some of this very limited scatter must be random. From this evidence, it seems likely that there is very little variability in the composition of fine-grained (liquid) Luna 24 VLT basalts.

(ii) *Brown glasses*

The regolith sampled at the Luna 24 site contains a group of brown glasses whose compositions form a distinct cluster on oxide varation diagrams. This cluster corresponds closely to the average of the Luna 24 VLT basalt analyses (Fig. 1). Wood (1975) showed that most lunar glasses are not compositionally similar to the local basalts. Instead, most of the glasses in the regolith represent local bulk soil compositions, with other small clusters perhaps representing exotic material or "pyroclastic" material, e.g. Apollo 17 orange glass and Apollo 15 green glass. While Wood (1975) drew his conclusion from the Apollo 11 and Apollo 17 sites, other data e.g. von Englehardt *et al.* (1971) demonstrate that it is true of all lunar sites. The Luna 24 brown glasses are *not* Luna 24 soil compositions, such a distinction applying to some of the greenish glasses which are far more common in the core. The chemical similarity of brown glasses and the Luna 24 VLT basalt is unlikely to be coincidental, and the glasses are probably either "pyroclastic" equivalents or they were formed by impacts into a soil which was made up entirely of Luna 24 VLT basalt. If the latter genesis is true, then that soil no longer exists at the Luna 24 site, but has been considerably contaminated. Several "impact melt" particles containing clastic fragments are observed in the core, and their compositions also correspond closely to those of the Luna 24 VLT basalts and brown glasses (Ryder *et al.*, 1977b). This supports an impact origin for the brown glasses, but by no means rules out a volcanic origin. Regardless of their precise mode of origin, the tight clustering of compositions of the brown glasses supports the concept of a homogeneous Luna 24 VLT basalt type at the site.

(iii) *Metabasalts*

Particles from all levels of the core include fragments with compositions similar to Luna 24 VLT basalts, but with granular textures and equilibrated mineral compositions (Ryder *et al.*, 1977b; Taylor *et al.*, 1977b). We suggested that these were fragments of metamorphosed flow margins, and as such could represent bulk liquid compositions. Metabasalt analyses show more scatter than basalt analyses (Fig. 1), and the pyroxene composition varies slightly from fragment to fragment, indicating, if the genetic model is correct, that there was variation in the Luna 24 liquid compositions. However, it has been proposed that these metabasalts were contaminated with highlands material (or at least plagioclase-rich materials) prior to metamorphism (McSween *et al.*, 1977). All the variation in metabasalt compositions cannot be ascribed to contamination. For instance, TiO_2 contents as low as 0.3% indicate ~60% contamination, a proportion not reflected by the major elements, and some of this variation may be due to sampling problems. In any case, the variation in metabasalt compositions cannot be taken as evidence of variation in Luna 24 VLT basalt compositions. It is possible that the metabasalts do in fact represent primary magma compositions from which other compositions could have been derived by plagioclase fractionation under near-surface conditions. This seems unlikely in view of the restricted compositions of brown glasses, which would either have to be unrelated to Luna 24 VLT basalts, or not include original surface liquid compositions.

(iv) *Conclusion based on chemistry*

The evidence from fine-grained basalts and brown glasses suggests that the

Luna 24 VLT basalt was a single homogeneous magma with a bulk chemistry distinct from any other known mare basalt type. [The data are inadequate to test the hypothesis of the existence of more than one magma, similar in major elements but different in trace elements, such as has been proposed from mineral analyses by Nielsen and Drake (1977), or which may be indicated by future trace element data.] This homogeneity is an important factor in assessing the origin of the basalt. Grove (pers. comm.) has suggested that the basalts are daughter products of fractional crystallization of a Luna 24 green glass composition. *If* this is the case, it is extremely difficult to understand why the Luna 24 VLT composition should be so restricted, and why there is no regular variation of compositions between brown glass (i.e., Luna 24 VLT) and green glass (see Simonds *et al.*, 1977 and Basu *et al.*, 1977a). Further, why should the parent coincidentally have a composition similar to the bulk soil compositions found at the Luna 24 site? It is more likely that these green glasses are impact melts of soil compositions. Some of the same reasoning applies to the possible derivation of Luna 24 VLT from olivine vitrophyre compositions (Coish and Taylor, 1977): no gradation of compositions exists. If such a fractionation sequence did exist, it would be the most extremely differentiated single sequence known within the mare basalt group, spanning the range from the most Mg-rich to the most Mg-poor. Improbably, the Luna 24 core would contain only the parent and its most differentiated product, and not representatives of the intermediate liquids. This leads us to conclude that the Luna 24 VLT basalt, with its homogeneous and distinct chemistry, is not related by fractional crystallization to any other VLT basalt, except perhaps at extreme depths where we have no possible way (at present) of distinguishing partial melting and fractional crystallization effects. The mineralogy of the basalts themselves also supports this conclusion, as outlined in the following section.

Crystallization sequence of Luna 24 VLT basalt

Derivation of Luna 24 basalts from an olivine vitrophyric magma requires separation of pyroxene as well as olivine (e.g. Coish and Taylor, 1977). Thus the Luna 24 VLT should have pyroxene as a liquidus phase. More generally, for shallow-level fractionation the Luna 24 VLT basalt must be multiply saturated, since olivine separation alone will not produce Luna 24 VLT compositions from green glass or vitrophyric compositions (e.g. Simonds *et al.*, 1977). The crystallization sequence and possibility of multiple saturation of Luna 24 VLT basalts are points of controversy. Grove and Vaniman (1977), using a brown glass composition, showed that plagioclase was the low-pressure liquidus phase, followed by olivine and then pyroxene. This order changes according to the rate of cooling such that with faster cooling plagioclase is suppressed and olivine becomes the liquidus phase. Our petrographic studies indicated that olivine was the liquidus phase, followed by plagioclase and then by pyroxene, at least for fine-grained basalts (Ryder *et al.*, 1977a, 1977b). This is in agreement with Grove and Vaniman (1977) for faster cooling rates, but the "olivine-first" observation *might* result from a real magma more mafic than that used by Grove and Vaniman (1977). In such a case, plagioclase may never have been a liquidus phase at shallow levels, even with slow cooling. There are several conclusions to be drawn from the experimental and petrographic data. First, pyroxene never preceded plagioclase at low pressure, and thus all least-squares calculations (or equivalents) which require pyroxene removal but not plagioclase removal from a potential parent to produce Luna 24 VLT basalt compositions are not consistent with the inferred order of crystallization. [Schemes involving olivine (± chromite) fractionation without pyroxene could be acceptable (G. J. Taylor, pers. comm.)]. For the composition used by Grove and Vaniman (1977), pyroxene is not an equilibrium liquidus phase until a pressure of over 5 kb is reached (Grove, pers. comm.), when it is a liquidus phase with plagioclase, and olivine is no longer stable. Second, since plagioclase crystallization precedes that of pyroxene in the Luna 24 VLT basalt compositions, plagioclase must precede pyroxene in the olivine vitrophyre parent postulated. However, the olivine vitrophyres appear to crystallize pyroxene in the fine-grained material between the olivine phenocrysts rather than plagioclase. Third, since the plagioclase is an early crystallizing phase in the Luna 24 VLT basalts, and is almost certainly the first at slow cooling conditions at a shallow depth in the crust (even if the magma is mafic enough to crystallize olivine first at the surface), then the postulated parent must be even higher in alumina than the Luna 24 VLT basalt, which is itself a high-alumina basalt. Clearly neither olivine vitrophyres nor any other known mare basalt compositions are suitable parents given this constraint.

Origin of Luna 24 VLT basalt liquids

In summary, Luna 24 VLT basalt liquids, as expressed by fine-grained fragments, appear to have had a very restricted range of compositions. This is indicated by the small range in MgO contents, and the homogeneity suggests that the basalts are not related by fractional crystallization to other VLT basalt compositions, because we would expect to see evidence of intermediate compositions. Furthermore, the crystallization sequence established petrographically and experimentally shows that fractional crystallization models must involve plagioclase, and therefore a high-alumina parent. However, pyroxene cannot crystallize without plagioclase. These factors effectively eliminate known magma types, including the olivine vitrophyres or materials similar to them, from consideration as parents. In total, this reduces the hypothesis of the derivation of Luna 24 VLT basalt by fractional crystallization to an exercise involving unknown and unusual parent magmas at depth within the moon. Such a hypothetical parent must be high in alumina, and very low in alkalis. At the same time it must be very low in titanium, and have a high Fe/Mg ratio. We see no reason to postulate anything other than a direct partial melting origin for the Luna 24 VLT basalt composition. It is possible that the source consisted of plagioclase and pyroxene, and lay just below the lunar crust [based on the position of the experimentally determined multiple-saturation point (Grove, pers. comm.)]. A plagioclase-bearing source is compatible with Rb/Sr considerations (Nyquist *et al.*, 1977). Obviously, confirmation of such a source would emphasize the internal heterogeneity of the moon, and its probable zoned pattern of chemistry and mineralogy.

Surface differentiation of Luna 24 VLT basalts

Coarse-grained basalts in the core may not represent liquid compositions, but potentially provide information concerning surface fractionation. Unfortunately, the coarser basalts have grain-sizes exceeding the fragment sizes we have studied, and thus bulk compositional estimates by DBA are impossible. Modal reconstructions are unreliable because most of the soils are contaminated with a high-Mg component which may influence the estimated composition. Sample 24170, which seems to be a single crushed coarse basalt, has a mode which suggests a composition similar to Luna 24 VLTs. It is mineralogically slightly different from VLTs: it has little chromite, no ulvöspinel, and its olivine does not include such Mg-rich compositions as the finer-grained basalts. Nonetheless, other coarse monomineralic fragments include olivine as Mg-rich as those in fine-grained basalts. Some particles with a medium grain size (in our samples) consist predominantly of olivine and plagioclase; if these particles represent larger volumes of rock then they must be cumulates, and some fractionation is suggested. On the limited evidence available, it seems quite probable that some differentiation took place at the surface, within flows, but the amount of such differentiation, the thickness of the flows in which such differentiation may have taken place, and whether we have samples from one or more than one flow, are topics beyond the scope of this paper.

The high-Mg component in the Luna 24 soil

In our initial study of the Luna 24 regolith, we concluded that it was dominated by the Luna 24 VLT basalt and its degraded products (Ryder *et al.*, 1977b). Bulk analyses of the soil show that there is a significant amount of a high-Mg component in the soils in addition to the Luna 24 VLT basalt (Blanchard *et al.*, 1977; Ma and Schmitt, 1977). That this component is not immediately obvious petrographically suggests that it may be more degraded than Luna 24 VLT basalt. This would be compatible with suggestions that the soil is a mixture of two components with different maturities and/or grain size (Nagle, 1977; McKay *et al.*, 1977). Presumably the Luna 24 VLT would represent the immature component and the high-Mg component the mature component. Of course, the reality might be much more complex than this.

The soil differs from Luna 24 VLT basalt not only in being richer in MgO (~10% soil cf. ~6.5% VLT) but also in Cr_2O_3 (~.4–.5% cf. ~.2%) and in being poorer in Al_2O_3 and CaO. The soil and Luna 24 VLT basalt have similar SiO_2, FeO and TiO_2 contents (the latter perhaps slightly higher in the soil) and similar low trace element abundances (Blanchard *et al.*, 1977; Ma and Schmitt, 1977).

The KREEP content of the soil is almost negligible. The high-Mg component is clearly of mare derivation, being low in Al. Highlands dunite is excluded as a source candidate by the low Fe of the dunites, and it is difficult to envisage a highlands dunite, a rare rock type, being intimately mixed with a mare soil without the addition of other highlands materials.

Several of the particle types in the regolith are richer in Mg than the Luna 24 VLT basalts. Predominant among these are the agglutinate fragments and the green glasses, but their compositions are very similar to the bulk soil; they probably represent impact meltings of the bulk soil. By themselves they are inadequate to mix with the Luna 24 VLT basalts to produce bulk soil compositions. In our 250-500 μm size fraction, such particles comprise only about ⅓ of the total. One high-Mg component is represented by a red-brown spherule which we found in our 250-500 μm allocation (Marvin *et al.*, 1977). An analysis is given in Table 1. The spherule is chemically similar to orange-red glasses from the Apollo 11 soils, but we do not know if it is local to Mare Crisium or was transported from elsewhere. In any case it is both inadequate in abundance and in composition (low SiO_2, high TiO_2) to be the high-Mg component in the soil.

Two particle varieties in the soil do have compositions appropriate for the high-Mg component. The most obvious are the olivine vitrophyre fragments, which have been recognized in the Luna 24 soils by several groups. We concur with Basu *et al.* (1978) that many of these are of impact melt origin, but this conclusion does not necessarily apply to the most Mg-rich ones. We believe that the latter may represent bona fide volcanics, and that in any case they indicate the existence of an Mg-rich mare basalt type in the Luna 24 region, even if it was not vitrophyric. Analyses of the two olivine vitrophyres we have found are given in Table 1. They differ slightly from the most magnesian olivine vitrophyres observed by others, and are not identical. The bulk composition of the olivine vitrophyres is not precisely known, and we have no data on trace elements. However, they have the correct major element chemical characteristics for the high-Mg component: high Mg, Fe, Cr, and low Ti, Al. Their abundance seems to be inadequate to provide the high-Mg component—we found only two olivine vitrophyres out of our total allocation of about 500 randomly chosen particles.

A second high-Mg component of the soils consists of mafic monomineralic particles whose compositions do not fall within the fields defined for Luna 24 VLT basalt particles. These consist predominantly of olivine, but also include low-Ca pyroxenes. These Mg-rich mafic minerals seem to be most common in the <250 μm (Basu *et al.*, 1977b; Papp *et al.*, 1977) but also occur in the 250–500 μm fraction (Taylor *et al.*, 1977b; Grove and Bence, 1977). It seems quite possible that at least some of the Mg-rich olivines are derived from the olivine vitrophyres—the compositional mode of such particles is similar to the compositions of phenocrysts in olivine vitrophyres i.e., $Fo_{70\text{-}75}$ (Basu *et al.*, 1977b; Papp *et al.* 1977). The magnesian pyroxenes are less easy to account for, since pyroxene is only a minor phase in the vitrophyres, but obviously they could occur in coarser-grained chemical equivalents of vitrophyres. The Mg-rich monomineralic particles seem to be sufficiently abundant, at least in the finer fraction of the soils, to be the high-Mg component; Papp *et al.* (1977) calculated a bulk soil composition with 14% MgO based on petrographic and microprobe analyses of the 90–150 μm fraction. It seems quite possible that the mafic minerals are the high-Mg component, and that many of these were ultimately derived from olivine-vitrophyres and their coarser-grained equivalents.

This model for mixing Luna 24 VLT basalt and olivine vitrophyric compositions is not entirely satisfactory. As noted above, the mafic monomineralic particles appear to be more common among the finer grain sizes (although admittedly the data are poor), which would imply that the high-Mg component is more common in the finer grain sizes. However, a comparison of the <125 μm fraction analyzed by Ma and Schmitt (1977) with the bulk soil analyses of Blanchard *et al.* (1977) reveals no tendency for Mg to be enriched in the finer fraction. Further, analyses of separate size fractions by Blanchard *et al.* (1977) indicate that Cr is *lower* in the finer size fractions, and this is supported by the data for the <125 μm fraction reported by Ma and Schmitt (1977). It follows that if the finer size fractions are enriched in the high-Mg component, a *separate* component is carrying the chromium; however, the olivine vitrophyre is high in both Mg and Cr. Clearly the high-Mg component remains enigmatic. It seems that the olivine vitrophyres (or coarser-grained chemical equivalents) are the most likely candidates for the component, but if so they occur mainly in a degraded (though not fine-grained) state. Better data on the percentages of high-Mg mafic phases (unrelated to the Luna 24 VLT basalt) are required for all size fractions before this hypothesis can be firmly supported.

Some authors have proposed that a coarse-grained olivine cumulate contributes to the regolith (e.g. Basu *et al.*, 1977b). We point out that coarse-grained in this case refers to >250 μm, and this can easily be accommodated by coarse-grained equivalents of the olivine vitrophyres. By analogy with other lunar suites, (e.g. Apollo 12 olivine basalts), it is likely that coarser-grained equivalents would be olivine porphyritic, and would include cumulates. Thus materials produced from them would include coarse-grained olivines. There are also geological problems involved in adding only cumulate material to a soil during impacting—in particular, what happened to the fractionated, derivative material overlying the cumulate? Such material would have been more readily excavated by a cratering event and should be apparent in the Luna 24 core. We conclude that it is insufficient to postulate only a Mg-rich *cumulate*: the whole flow from which the component was added must have been Mg-rich compared to the Luna 24 VLT basalt. Further, if the Mg-rich material was added from the Fahrenheit event, then the added material at such a distance should have been of *shallow* origin, not deep, if our understanding of cratering is correct.

A significant constraint on the origin of the Luna 24 regolith is the fact that the bulk composition at each level is similar (with the probable exception of lighologically distinct layers such as the one represented by 24170 in the U. S. allocation). The high-Mg component, whether related to olivine vitrophyres or not, must be distributed throughout the core, as are the Luna 24 VLT basalts, brown glasses, metabasalts, and so on. Because the inconspicuous nature of the high-Mg component suggests that it is more degraded than the Luna 24 VLT basalt component, and because it is well-mixed into the regolith, we believe that it was not delivered to the site in an impact such as the Picard event (Andre *et al.*, 1977) or the Fahrenheit event (Basu *et al.*, 1977b) which would have delivered fresh material. The high-Mg component is probably older than the Luna 24 VLT basalt, and is local to the Luna 24 site. The heterogeneous nature of the region (Pieters *et al.*, 1977) makes further conclusions about the precise distribution of Luna 24 VLT basalt and "high-Mg mare basalt" around the Luna 24 site speculative.

Conclusions

Luna 24 VLT basalt, and its possible pyroclastic equivalent brown glass, are compositionally homogeneous. This fact, associated with the phase relationships, makes it unlikely that the Luna 24 VLT basalt is related to any other VLT basalt by fractional crystallization. Even a fractional crystallization origin at depth requires a parent which is even more aluminous. We conclude that the Luna 24 VLT basalt was derived from the lunar interior independently of any other lunar basalts, possibly from a source containing plagioclase.

The high-Mg component in the soil is enigmatic because it is not obvious petrographically. It is most likely to be derived from material similar to the imprecisely known olivine vitrophyre and its degraded products, or from some other high-Mg, very low Ti basalt. Its distribution throughout the core suggests that it is local to the Luna 24 site, and its assumed degraded nature suggests that it is older than the Luna 24 VLT basalt.

Acknowledgments—This research was partially supported by NASA grant NGL-09-015-150 (J. A. Wood, Principal Investigator). Reviews and suggestions by M. Prinz, R. F. Dymek, G. J. Taylor, A. Basu, and K. Molylewski improved the manuscript.

References

Andre C. G., Wolfe R. W. and Adler I. (1977) Evidence for a high-magnesium subsurface basalt in Mare Crisium (abstract). In *Papers Presented to the Conference on Luna 24*, p. 8–11. The Lunar Science Institute, Houston.

Basu A., McKay D. S., Clanton U., Waits G. and Fruland R. M. (1977a) Glass particles in Luna 24 drill core soils (abstract). In *Papers Presented to the Conference on Luna 24*, p. 14–17. The Lunar Science Institute, Houston.

Basu A., McKay D. S. and Fruland R. M. (1977b) Petrography, mineralogy and source rocks of Luna 24 drill core soils (abstract). In *Papers Presented to the Conference on Luna 24*, p. 18–21. The Lunar Science Institute, Houston.

Basu A., McKay D. S. and Fruland R. M. (1978) Impact melt origin of Luna 24 olivine vitrophyres (abstract). In *Lunar and Planetary Science IX*, p. 50–52. Lunar and Planetary Institute, Houston.

Blanchard D. P., Haskin L. A., Brannon J. C. and Aaboe E. (1977) Chemistry of soils and particles from Luna 24 (abstract). In *Papers Presented to the Conference on Luna 24*, p. 37–40. The Lunar Science Institute, Houston.

Coish R. A. and Taylor L. A. (1977) Ingeous rocks from Mare Crisium: Mineralogy and petrology (abstract). In *Papers Presented to the Conference on Luna 24*, p. 48–50. The Lunar Science Institute, Houston.

von Engelhardt W., Arndt J., Muller W. F. and Stöffler D. (1971) Shock metamorphism and origin of regolith and breccias at the Apollo 11 and Apollo 12 landing sites. *Proc. Lunar Sci. Conf. 2nd*, p. 833–854.

Grove T. L. and Bence A. E. (1977) Petrogenesis of gabbros from Mare Crisium (abstract). In *Papers Presented to the Conference on Luna 24*, p. 64–67. The Lunar Science Institute, Houston.

Grove T. L. and Vaniman D. T. (1977) Experimental petrology of very low Ti basalts and origin of Luna 24 ferrobasalt (abstract). In *Papers Presented to the Conference on Luna 24*, p. 68–71. The Lunar Science Institute, Houston.

Laul J. C. (1977) Chemical composition of Luna 24 melt rocks (24077,4; 24077,62; 24174,7; 24182,12; 24210,50) and gabbro (24182,8) (abstract). In *Papers Presented to the Conference on Luna 24*, p. 94–97. The Lunar Science Institute, Houston.

Ma M.-S. and Schmitt R. A. (1977) Luna 24 soils: a chemical study (abstract). In *Papers Presented to the Conference on Luna 24*, p. 102–105. The Lunar Science Institute, Houston.

Ma M.-S., Schmitt R. A., Taylor G. J., Warner R. D., Lange D. E. and Keil K. (1977) Chemical and petrographic studies of 18 Luna 24 lithic fragments (abstract). In *Papers Presented to the Conference on Luna 24*, p. 102–105. The Lunar Science Institute, Houston.

Marvin U. B., Ryder G. and McSween H. Y., Jr. (1977) Some exotic particles in the Luna 24 core sample (abstract). In *Papers Presented to the Conference on Luna 24*, p. 106–109. The Lunar Science Institute, Houston.

McKay D. S., Basu A., Waits G., Clanton U., Fuhrman R. and Fruland R. (1977) Grain size and evolution of Luna 24 soils (abstract). In *Papers Presented to the Conference on Luna 24*, p. 114–117. The Lunar Science Institute, Houston.

McSween H. Y. Jr., Taylor L. A. and Clark J. C. (1977) Luna 24 metabasalts: possible evidence for assimilation of highlands material? (abstract). In *Papers Presented to the Conference on Luna 24*, p. 118–120. The Lunar Science Institute, Houston.

Nagle J. S. (1977) Luna 24, clast populations (abstract). In *Papers Presented to the Conference on Luna 24*, p. 129–131. The Lunar Science Institute, Houston.

Nielsen R. L. and Drake M. J. (1977) Magma types at the Luna 24 site in Mare Crisium (abstract). In *Papers Presented to the Conference on Luna 24*, p. 132–135. The Lunar Science Institute, Houston.

Nyquist L., Weismann H., Bansal B., Wooden J., McKay G. and Hubbard N. J. (1977) Chemical and Rb-isotopic characteristics of the Luna 24 samples (abstract). In *Papers Presented to the Conference on Luna 24*, p. 139–142. The Lunar Science Institute, Houston.

Papp H., Steele I. M. and Smith J. V. (1977) Luna 24 90–150 micron fraction: implications for sampling of planetary and asteroidal regoliths (abstract). In *Papers Presented to the Conference on Luna 24*, p. 143–146. The Lunar Science Institute, Houston.

Pieters C., McCord T. B. and Adams J. B. (1977) Regional basalt types in the Luna 24 landing area as derived from remote observations. *Geophys. Res. Lett.* **3**, 697–700.

Rhodes J. M., Hubbard N. J., Wiesmann H., Rodgers K. V., Brannon J. C. and Bansal B. M. (1976) Chemistry, classification and petrogenesis of Apollo 17 mare basalts. *Proc. Lunar Sci. Conf. 7th*, p. 1467–1489.

Rhodes J. M., Blanchard D. P., Dungan M. A., Brannon J. C. and Rodgers K. V. (1977) Chemistry of Apollo 12 mare basalts: magma types and fractionation processes. *Proc. Lunar Sci. Conf. 8th*, p. 1305–1338.

Ryder G., McSween H. Y. Jr. and Marvin U. B. (1977a) Luna 24 VLT basalt: character and origin (abstract). In *Papers Presented to the Conference on Luna 24*, p. 160–163. The Lunar Science Institute, Houston.

Ryder G., McSween H. Y. Jr. and Marvin U. B. (1977b) Basalts from Mare Crisium. *The Moon* **17**, 263–287.

Tarasov L. S., Nazarov M. A., Schevaleevsky I. D., Kudryashova A. F., Gaverdovskaya A. S. and Korina M. I. (1977) Mineralogy and petrography of lunar rocks from Mare Crisium (preliminary data). *Proc. Lunar Sci. Conf. 8th*, p. 3333–3356.

Schmitt H. H., Lofgren G., Swann G. A. and Simmons G. (1970) The Apollo 11 samples: introduction. *Proc. Apollo 11 Lunar Sci. Conf.* , p. 1–54.

Simonds C. H., Warner J. L., McGee P. E., Cochran A. and Brown R. W. (1977) Luna 24: Petrochemical potpourri (abstract). In *Papers Presented to the Conference on Luna 24*, p. 170–183. The Lunar Science Institute, Houston.

Taylor G. J., Warner R. D. and Keil K. (1977a) Chemical distinctions among very low-Ti mare basalts (abstract). In *Papers Presented to the Conference on Luna 24*, p. 183–185. The Lunar Science Institute, Houston.

Taylor G. J., Warner R. D., Wentworth S., Keil K. and Sayeed U. (1977b) Petrology and chemistry of Luna 24 mare basalts and basaltic glasses (abstract). In *Papers Presented to the Conference on Luna 24*, p. 189–192. The Lunar Science Institute, Houston.

Taylor G. J., Keil K. and Warner R. D. (1977c) Very low-Ti mare basalts. *Geophys. Res. Lett.* **4**, 207–210.

Taylor G. J., Warner R. D. and Keil K. (1978) Chemical trends among VLT mare basalts and glasses: clues to the nature of their parent magmas (abstract). In *Lunar and Planetary Science IX*, p. 1143–1145. Lunar and Planetary Institute, Houston.

Vaniman D. T. and Papike J. J. (1977a) Luna 24 ferrobasalt: petrology and comparisons with other mare basalt types (abstract). In *Papers Presented to the Conference on Luna 24*, p. 196–199. The Lunar Science Institute, Houston.

Vaniman D. T. and Papike J. J. (1977b) Very low Ti (VLT) basalts: a new mare rock type from the Apollo 17 drill core. *Proc. Lunar Sci. Conf. 8th*, p. 1443–1471.

Vaniman D. T. and Papike J. J. (1977c) Ferrobasalts from Mare Crisium: Luna 24. *Geophys. Res. Lett.* **4**, 497–500.

Wood J. A. (1975) Glass compositions as a clue to unsampled mare basalt lithologies (abstract). Papers Presented to the Conference on *Origins of Mare Basalts and Their Implications for Lunar Evolution*, p. 194–198. The Lunar Science Institute, Houston.

Merrill, R. B. and Papike, J. J., editors.
(1978) *Mare Crisium: The View from Luna 24*, Pergamon Press (New York), p. 357–370
Printed in the United States of America

VLT mare basalts: Impact mixing, parent magma types, and petrogenesis

G. JEFFREY TAYLOR, RICHARD D. WARNER and KLAUS KEIL

Department of Geology and Institute of Meteoritics, Universtiy of New Mexico, Albuquerque, New Mexico 87131

Abstract—We have examined available major-element compositions of VLT mare basalt lithic fragments and glasses. Many of the lithic fragments contain relic mineral and sometimes lithic clasts, which suggests that many samples of VLT mare basalt glasses and lithic fragments were formed by impact melting. Therefore, some may have compositions that represent mixtures of VLT mare basalts and other types of rocks, not authentic igneous rock compositions. In spite of this problem, we have identified a number of possible parent magma compositions. These hypothetical magmas vary in CaO/Al_2O_3(0.9-1.1), FeO (17 to 24 wt.%), Fe/(Fe + Mg) (0.34-0.46), and K_2O (0-0.2 wt.%). Each may have differentiated to produce a wide variety of derivative magmas. Calculations involving olivine fractionation from the proposed parent magmas indicate that Luna 24 ferrobasalts might have formed from a magma whose composition is similar to that of a group of high-MgO (~13–14 wt.%) Apollo 17 VLT mare basalt glasses.

INTRODUCTION

Very low-Ti (VLT) mare basalts have distinctly lower TiO_2 contents (<1 wt.%) than other mare basalts. They were first recognized in the Apollo 17 deep drill core (Vaniman and Papike, 1977; Taylor *et al.*, 1977a), but have since been identified in other Apollo 17 samples: they occur as clasts in soil breccias 78568 and 78547 (Warner *et al.*, 1978a) and Boulder 1, Station 2 breccia 72235 (Ryder *et al.*, 1975); and green glass vitrophyre 78526 has the bulk composition of a VLT mare basalt (Warner *et al.*, 1978b). Furthermore, as the data in this volume demonstrate, VLT mare basalts dominate the Luna 24 soils.

The Apollo 15 green glass might be related chemically to VLT mare basalts, though it contains more MgO. Similar glass compositions have been reported from all missions. These analyses represent another body of data that we can use when trying to unravel the chemical relationships among this interesting suite of mare basaltic materials.

In this paper, we summarize the compositional trends observed for VLT mare basalt lithic fragments and glasses. Our two main goals are to identify the major parental magma types and to evaluate the role that low-P fractionation may have played in their evolution. One of the most significant problems we face in trying to achieve these goals is that almost all of the available bulk chemical analyses are of glasses and small (~1 mm) lithic fragments. Some of these fragments and glasses, perhaps most, might have formed by impact melting and, therefore, could represent mixtures, not authentic igneous rock compositions. Indeed, the evidence favors an impact origin for VLT mare basalt rake sample 78526 (Warner *et al.*, 1978b). Consequently, we have a third goal: to evaluate the extent to which impact mixing has played a role in the evolution of the materials we identify as VLT mare basalts.

TEXTURES OF LITHIC FRAGMENTS

VLT mare basalt lithic fragments display a variety of textures, including vitrophyric (glassy), microporphyritic, ophitic to subophitic, and granular. Some have unambiguous igneous textures. For example, fragment 70007,333-L1, from the Apollo 17 drill core, has euhedral phenocrysts of pigeonite and olivine set in a fine-grained, variolitic groundmass (Fig. 1A). Fragment 70009,295-L1, also from the Apollo 17 drill core, has a subophitic texture (Fig. 1B). Neither of these samples shows any textural evidence for origin by impact melting.

In contrast to these apparent volcanic rocks, many VLT lithic fragments have textures that suggest they formed by impact melting. Vitrophyric fragment 70007,321-G1, for example, has skeletal olivine and pyroxene crystals in a pale-green glass matrix. The olivines and pyroxenes clearly crystallized from a melt, but the fragment is probably not volcanic. It contains a large, blocky pigeonite crystal (Fig. 1C) that has rather irregular edges. We suggest that this crystal is a *relic*, not a phenocryst; therefore, the fragment formed by impact melting. Rake sample 78526 is very similar compositionally and texturally to vitrophyre 70007,321-G1 and contains convincing evidence for an impact origin (Warner *et al.*, 1978b). Another example of a possible impact melt is sample 72235,59-3, a lithic fragment in Boulder 1, from Apollo 17 Station 2. This sample has a microporphyritic texture with olivine and lesser amounts of pigeonite set in a finer-grained matrix. However, the large crystals (particularly olivine) are not generally euhedral as would be expected of phenocrysts, but instead appear to be fragments of larger crystals (Fig. 1 D). Thus, we interpret these crystals as remnant clasts.

The textural evidence for an impact origin for some of these lithic fragments, while not conclusive, emphasizes the need for caution when interpreting the bulk compositions of VLT basalt lithic fragments. [In many cases, such as green glass vitrophyre 78526 (Warner *et al.*, 1978b), the textural evidence is supported by chemical data]. We do not argue that all VLT basalt lithic fragments are impact melts or, even if they were, that all have been contaminated with other types of lunar rocks. We simply point out that some VLT basalt lithic fragments may have compositions that do not reflect the composition of a genuine lunar magma.

BULK COMPOSITIONS OF LITHIC FRAGMENTS

We have previously listed all available bulk analyses of VLT lithic fragments (Taylor *et al.* 1977b). The data define five main groups. The first group occurs in Apollo 17 soils and breccias. The fragments have either vitrophyric or micropor-

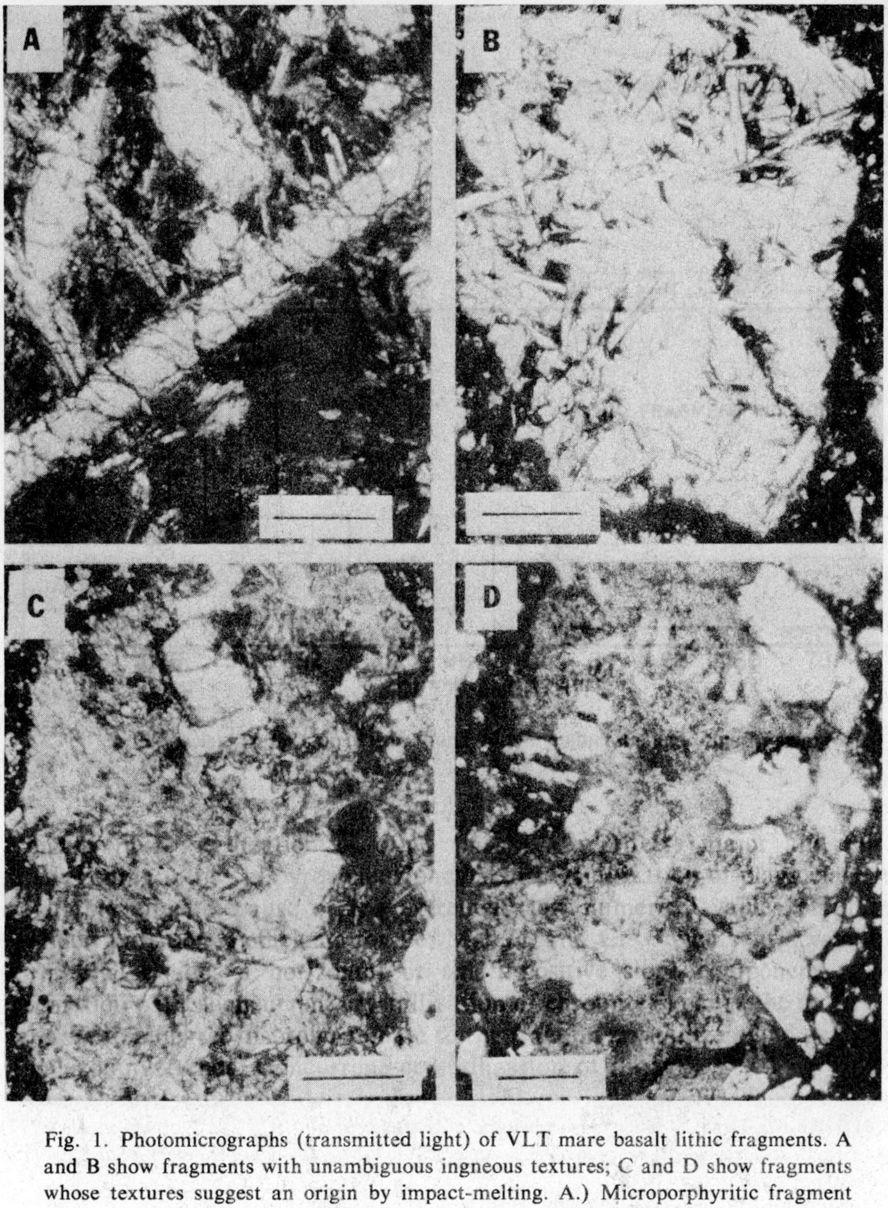

Fig. 1. Photomicrographs (transmitted light) of VLT mare basalt lithic fragments. A and B show fragments with unambiguous ingneous textures; C and D show fragments whose textures suggest an origin by impact-melting. A.) Microporphyritic fragment 70007,333-L1; scale bar is 200 μm. B.) Subophitic fragment 70009,295-L1; scale bar is 200μm. C.) Glassy fragment 70007,321-G1; scale bar is 400 μm. Note large pigeonite clast in top center. The groundmass is glassy and contains numerous quench crystals of olivine and pyroxene. D.) Microporphyritic fragment 72235,59–3; scale bar is 200 μm. Many of the large crystals look like clasts, rather than phenocrysts.

phyritic textures. Bulk compositions (Table 1, columns 1 and 2) of each textural type are essentially identical, though the vitrophyres are slightly higher in FeO and MgO and lower in Al_2O_3 and CaO. They are characterized by MgO contents of ~11 wt.%, atomic Fe/(Fe + Mg) (FFM) of ~0.48, weight ratio CaO/Al_2O_3 of ~0.9, and very low alkali contents (Na_2O <0.2 wt.%; K_2O <0.05 wt.%). A group of Luna 24 green glasses (Norman *et al.*, 1977; Simonds *et al.*, 1977) have very similar compositions (Table 1, column 3).

The second group consists mainly of Luna 24 ferrobasalts which occur as fine-grained, ophitic to subophitic lithic fragments in the Luna 24 soils. Their bulk compositions (Taylor *et al.*, 1978) are similar to those of Luna 24 brown glasses (Simonds *et al.*, 1977; Norman *et al.*, 1977). Since point analyses of glasses are more accurate than broad-beam analyses of heterogeneous lithic fragments, we have used the glass analyses to define the average composition of the Luna 24 ferrobasalts (Table 1, column 4). The average composition is characterized by FeO of 19.5 wt.%, low MgO of 6.7 wt.%, FFM of 0.62, CaO/Al_2O_3 of 0.98, and

Table 1. VLT mare basalt compositions (wt.%).

	1	2	3	4	5	6	7	8	9	10
SiO_2	48.0	49.2	45.0	46.2	48.4	47.6	46.6	47.5	45.1	46.5
TiO_2	0.68	0.41	0.93	0.96	0.73	0.66	0.54	0.35	0.60	0.67
Al_2O_3	11.1	11.5	12.1	12.8	12.7	15.2	11.3	12.7	9.6	9.9
Cr_2O_3	0.71	0.51	0.42	0.24	0.27	0.58	0.60	0.32	0.42	0.45
FeO	18.0	17.4	18.4	19.5	18.0	13.8	16.5	15.8	19.3	18.8
MnO	0.27	0.30	0.26	0.28	0.34	0.24	0.28	0.27	0.21	0.18
MgO	11.5	10.5	10.4	6.7	7.0	9.4	13.0	10.2	14.9	14.3
CaO	10.1	10.6	11.1	12.5	12.9	12.9	10.1	11.9	9.4	9.7
Na_2O	0.14	0.13	0.22	0.30	0.16	0.19	0.18	0.60	0.26	0.24
K_2O	0.02	0.03	0.03	0.02	0.03	0.04	0.18	0.20	0.04	0.02
P_2O_5	0.02	<0.02	0.02	0.03	<0.02	0.03	<0.02	0.05	0.03	0.02
Total	100.54	100.58	98.88	99.52	100.53	100.64	99.28	99.89	99.86	100.78
Fe/(Fe + Mg)	0.47	0.48	0.50	0.62	0.59	0.45	0.42	0.47	0.42	0.42
CaO/Al_2O_3	0.91	0.92	0.92	0.98	1.02	0.85	0.89	0.94	0.98	0.98

(1) Average of 7 Apollo 17 vitrophyric VLT mare basaltic fragments from Taylor *et al.* (1977b) and Vaniman and Papike (1977).

(2) Average of 5 Apollo 17 microporphyritic VLT lithic fragments from Taylor *et al.* (1977b) and Vaniman and Papike (1977).

(3) High-Mg Luna 24 green glass (average of Simonds *et al.*, 1977 and Norman *et al.*, 1977).

(4) Luna 24 ferrobasalt; average of glass analyses (low-Mg cluster) of Simonds *et al.* (1977) and (brown glass) of Norman (1977).

(5) Apollo 17 lithic fragment 70009, 295-L1.

(6) Average of granular lithology in 78526 (Warner *et al.*, 1978b) and Apollo 17 lithic fragment 70004, 485-L1.

(7) Average of lithic fragments 72235, 59–3 and 72235,59–4 from Boulder 1, Station 2, Apollo 17.

(8) Luna 16 lithic fragment 313–15 (Green *et al.*, 1972).

(9) Apollo 17 homogeneous glass 78546,3–34.

(10) Apollo 17 homogeneous glass 78546,3–38.

like most Apollo 17 VLT mare basalts, a low alkali content. One of the lithic fragments from the Apollo 17 drill core, subophitic fragment 70009,295-Ll, has a very similar composition (Table 1, column 5), except for lower FeO (18 wt.%).

The third group consists of two Apollo 17 lithic fragments. Both have granular textures. One occurs as a clast in rake sample 78526 (Warner *et al.*, 1978b), the other as a lithic fragment in drill core sample 70004,485. The average of broad beam analyses of the two fragments appears in Table 1, column 6. They contain more Al_2O_3 (~15 wt.%) and less FeO (~14 wt.%) than other VLT mare basalt fragments and have the lowest CaO/Al_2O_3 ratios (0.85). Like other VLT mare basalts, they have very low contents of alkalis (~0.2 wt.% Na_2O, <0.05 wt.% K_2O).

The fourth group of VLT mare basalt fragments was originally described by Ryder *et al.* (1975). The fragments occur in Apollo 17 breccia 72235; Ryder *et al.* (1975) called them olivine-normative pigeonite basalts. Ryder and Taylor (1976) pointed out that these basaltic fragments had more affinities to mare basalts than to nonmare rocks, even though they were discovered in a highland breccia from South Massif. Their compositions are broadly similar to VLT mare basalts (Table 1, column 7). They are characterized by FeO contents of 16 wt.%, MgO of 13 wt.%, FFM of 0.42, and relatively low CaO/Al_2O_3, ~0.89. Their most distinctive feature is their high K_2O contents (~0.2 wt.%), coupled with low Na_2O (0.2 wt.%). In spite of higher K_2O than most other VLT mare basalt compositions, the Boulder 1, Station 2 fragments are not, as Vaniman and Papike (1977) state, KREEP-rich:P_2O_5 contents are less than 0.02 wt.%. Although all VLT mare basalts require more study, the Boulder 1, Station 2 group deserves special attention because the fragments occur in larger clasts in the boulder; these large clasts have been dated as 4.01 b.y. old (Leich *et al.*, 1975). Thus, the VLT lithic fragments within these clasts are at least that old and, hence, are among the oldest mare basalts. This, in turn, makes it even more crucial to determine whether they are impact melts or true volcanic rocks.

The last group actually consists of only one small lithic fragment from Luna 16. We are not confident that its analysis is representative of the larger rock it came from and list it here only for completeness and because it has a distinctive composition (Table 1, column 8). It contains ~16 wt.% FeO, 10 wt.% MgO, FFM of 0.47, and CaO/Al_2O_3 of 0.94. Its unique feature is a high alkali content: Na_2O is 0.60 (double any other VLT mare basalt composition) and K_2O is 0.20 wt.%.

Glasses: Clues to the Nature of VLT Parent Magmas

Many more analyses have been reported for glasses of VLT mare basalt composition than for lithic fragments. However, only Apollo 15 green glass contains evidence of being volcanic and, therefore, representing the composition of a magma. Some, conceivably even all, of the rest might be impact melts. Some of the impact-melted samples might have the composition of a VLT basalt lava, but others might have the composition of a mixture of VLT basalts with other lunar rocks such as high-Ti mare basalts or ANT rocks. Furthermore, some might be impact melts of cumulate rocks. In spite of these uncertainties, we can gain useful information from a study of the available data.

We plotted analyses of glasses and lithic fragments in Figs. 2–5. To qualify as a VLT basalt composition, an analysis had to have $CaO/Al_2O_3 > 0.75$ and $TiO_2 < 1.1$ wt.%. Note the wide range in composition that the whole set of data exhibits. MgO ranges from ~22 to 6 wt.%. Also note the general trend of

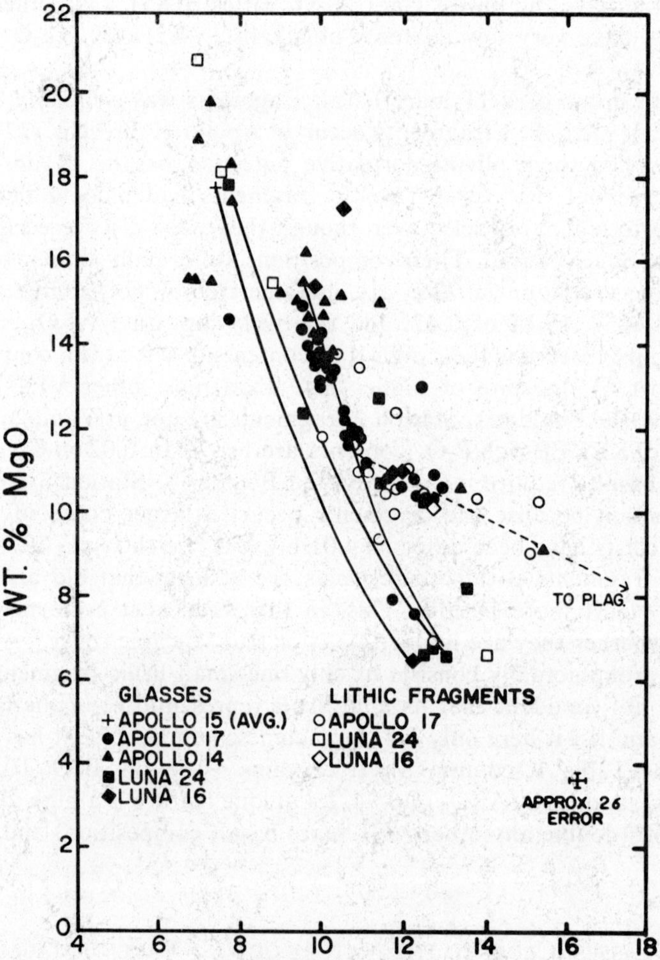

Fig. 2. MgO vs. Al_2O_3 for VLT basalt lithic fragments and glasses. Most data are from our laboratory; others are from Vaniman and Papike (1977) and the references in Table 2. Solid lines represent trends produced by fractional crystallization of olivine and begin at an arbitrarily chosen parent. Dashed line illustrates mixing with plagioclase-rich rocks. Error bars refer to analytical uncertainty.

decreasing MgO with increasing Al_2O_3, which suggests that fractional crystallization has played a role in causing the wide range in composition observed. The way the data cluster along a line from the olivine corner in Fig. 5 supports the idea. But other processes may have also operated to produce the ranges in CaO/Al_2O_3 (Fig. 4) and FeO (Fig. 3). To understand the evolution of the VLT mare basalt suite, we must determine what those processes are.

The first step is to identify potential parent magmas. The most promising candidates are those analyses with >15 wt.% MgO. Table 2 lists the analyses, arranged in order of decreasing CaO/Al_2O_3. These materials exhibit a substantial range in composition. Significant variations occur in CaO/Al_2O_3, FFM, bulk MgO, FeO, and K_2O. It seems clear that these hypothetical parent magmas could fractionate to produce a similarly wide range of derivative lavas. We discuss the possible relationships between the potential parent compositions listed in Table 1 and the less magnesian VLT basaltic compositions in the next section.

We suggested (Taylor *et al.*, 1977a) that Luna 24 VLT mare basalts might be derived from Apollo 17 VLT basalts (specifically one with the composition of 78526) by fractionation of olivine, pigeonite, and chromite. Detailed modeling of that fractionation trend (Taylor *et al.*, 1977b), coupled with the knowledge that 78526 is an impact melt, leads us to retract this idea. However, some Apollo 17 glasses have more appropriate compositions. We have listed two of these in Table 1 (columns 9 and 10). They differ from Apollo 17 lithic fragments in having higher MgO and higher CaO/Al_2O_3. In the next section, we test the idea that these glasses have the appropriate compositions to be parental magmas to the Luna 24 ferrobasalts.

Role of Olivine Fractionation

Every analysis plotted in Fig. 5 falls in the olivine primary crystallization field. The shift in the positions of the olivine-plagioclase cotectic and the pyroxene-olivine-plagioclase peritectic point with an increase in FFM might place the Luna 24 ferrobasalts slightly out of the olivine field (Grove and Vaniman, 1977), but certainly most points fall in it. This, plus the tendency for the analyses to trend directly away from the olivine corner, suggest that olivine fractionation played a significant role in the evolution of the VLT mare basalt suite.

We have modeled the fractional crystallization of olivine by using the procedure described by Drake (1976). We used Roeder and Emslie's (1970) value of 0.3 for K_D (olivine-liquid) and assumed that the crystallized olivine was totally removed from the melt. The calculations were carried out as follows: The liquidus olivine composition was calculated from Roeder and Emslie's K_D, then 2% of this was removed from the melt, resulting in a new melt composition. The calculation was then repeated. The liquidus olivine composition typically decreased by ~0.5 to 1.5 mole % Fo in each step.

Typical fractionation trends for selected hypothetical parent magmas appear as solid lines in Figs. 2–4. The results show that olivine fractionation does not

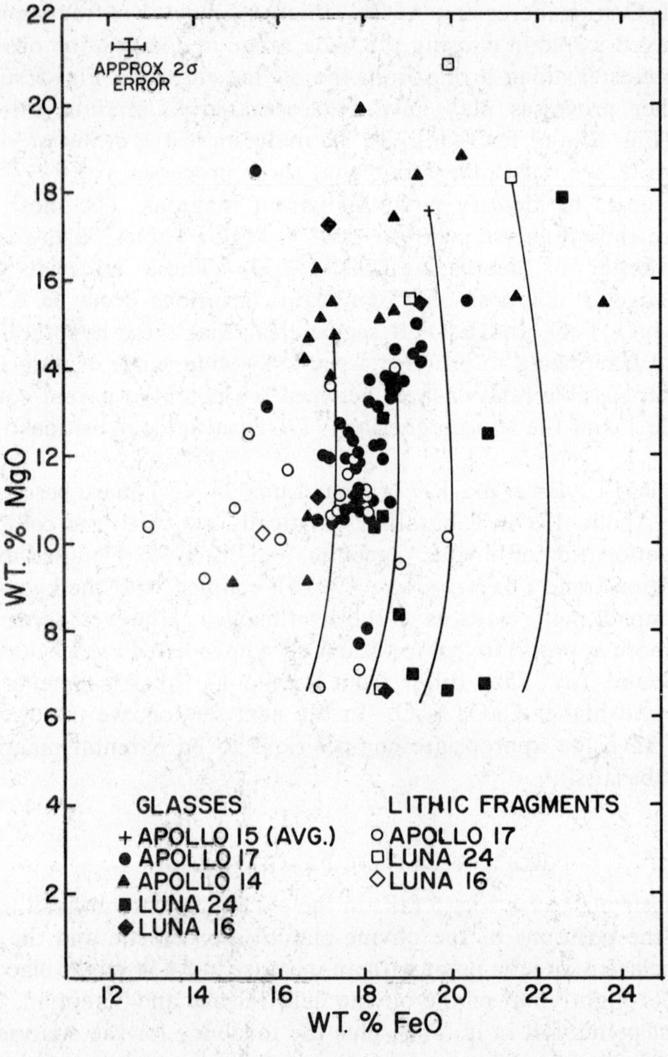

Fig. 3. MgO vs. FeO for VLT basalt lithic fragments and glasses. Sources of data given in Fig. 2. Solid lines show trends produced by olivine fractionation. Note that FeO content changes only slightly during differentiation.

change the CaO/Al_2O_3 ratios (Fig. 4) of the residual magmas (which is not surprising since olivine does not incorporate significant quantities of either element), nor does it change the FeO content appreciably. Consequently, some VLT basalt compositions cannot be related to one another by olivine fractionation. For example, Apollo 15 green glass cannot fractionate to produce the Luna 24 ferrobasalts, because CaO/Al_2O_3 is ~1.1 in Apollo 15 green glass (Table 2) but only 1.0 in Luna 24 ferrobasalts (Table 1). T. L. Grove and co-workers (oral

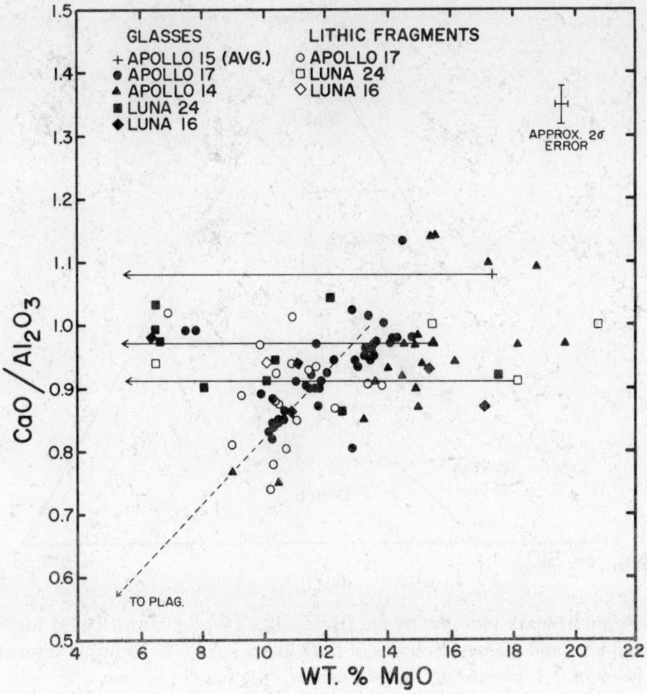

Fig. 4. Weight ratio CaO/Al_2O_3 vs. MgO for VLT mare basalt lithic fragments and glasses. Sources of data listed in Fig. 2. Solid lines show trends produced by olivine fractionation; note that CaO/Al_2O_3 does not change. Dashed line shows mixing with plagioclase.

presentation at the Conference on Luna 24) made the same observation, based on the compositions of experimental liquids. Trace element data (Ma *et al.*, 1978) support this conclusion. Actually, Apollo 15 green glass has such high CaO/Al_2O_3 that it cannot have fractionated to produce any of the *observed* VLT mare basalt compositions.

Another important, though disappointing, observation is that most of the analyses do not fall on (or even parallel to) the calculated fractionation trends. Several processes may have caused this lack of correlation. Possibly olivine was not the only fractionating phase. This explanation seems unlikely except for small amounts of chromite fractionation (see below), since most VLT basalt compositions fall in the central region of the olivine primary crystallization field in Fig. 5. Perhaps we are dealing with analyses of many individual flows with different parent magmas. We cannot rule out this possibility, but unfortunately, we cannot test it with existing data. Or possibly the hypothetical parent magmas have no relationship with the observed basaltic compositions. After all, why should a glass fragment found at Apollo 14 represent the composition of a magma that evolved to produce the Apollo 17 VLT lithic fragments? Another idea is that most of these analyses represent mixtures of VLT basalts (which may have followed the

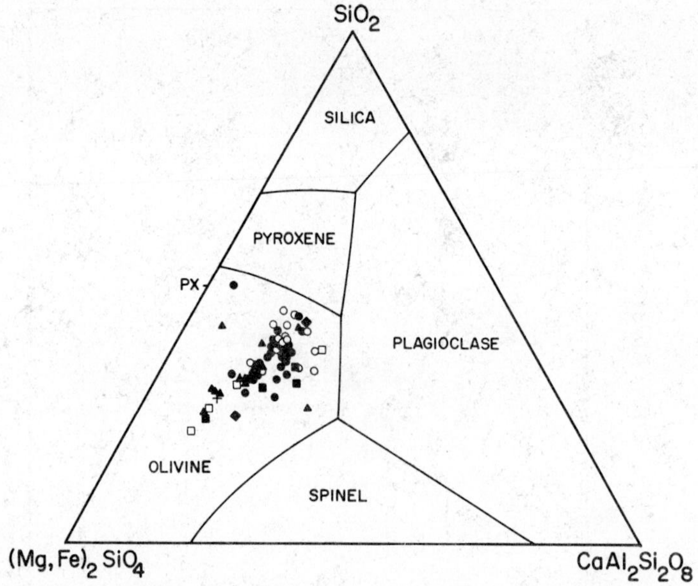

Fig. 5. Pseudoternary plot olivine-anorthite-silica (Walker *et al.* 1973) for VLT basalt lithic fragments and glasses. Sources of data listed in Fig. 2. Multiple saturation curves are for FFM of 0.3; increasing Fe/Mg causes shift toward olivine-silica join. Note that the analyses fall in the olivine primary crystallization field and that they trend away from olivine corner. This suggests that olivine fractionation has played a role in producing VLT mare basalts. (Some analyses, such as Luna 24 ferrobasalts, might lie outside of olivine field due to shift of cotectic lines with increasing Fe/Mg, but most probably plot in it.)

fractionation trends drawn in Figs. 2–4) and varying amounts of other rocks. We return to this topic in the next section.

We used the results of our calculations to reevaluate the hypothesis that Luna 24 VLT ferrobasalts formed by fractionation of Apollo 17 VLT basalts. It seems clear that except for ferrobasalt 70009,295-L1, Apollo 17 lithic fragments (Table 1, columns 1–3) cannot possibly represent a suitable parent magma because their FeO contents and CaO/Al_2O_3 ratios are too low. However, the more magnesian Apollo 17 VLT basalt glass fragments (Table 1, columns 9 and 10) are good candidates. Table 3 compares calculated compositions of Luna 24 ferrobasalt and Apollo 17 lithic fragment 70009,295-L1 to their observed compositions. The agreement is within analytical uncertainty, except for Cr_2O_3 in both compositions and Na_2O in 70009,295-L1. The discrepancy in Cr_2O_3 suggests that a small amount (~1% of the fractionating assemblage) of chromite might also have fractionated along with the olivine. Although the agreement is encouraging, we note that we have no guarantee that the proposed parent magmas really represent magma compositions. Furthermore, major elements alone are not sufficient to establish a magma parent-daughter relationship; trace element concentrations in

Table 2. Glass (and one vitrophyre) compositions (wt.%) that might represent parental magmas of VLT mare baslats.

	1	2	3	4	5	6	7	8	9	10	11	12	13	14
SiO_2	45.7	46.4	45.4	47.7	43.5	45.6	45.6	46.4	45.2	44.8	43.6	44.2	44.9	43.8
TiO_2	0.88	0.41	0.42	0.55	0.55	0.58	0.55	0.48	0.93	0.7	0.88	0.62	0.57	0.60
Al_2O_3	6.9	7.0	7.7	9.4	7.3	7.8	9.6	9.6	9.0	9.8	7.8	7.7	10.5	10.5
Cr_2O_3	0.08	0.39	0.43	0.42	—	—	0.43	0.45	—	0.6	0.32	0.53	0.28	—
FeO	23.7	20.4	19.6	18.5	18.1	19.4	17.4	17.0	18.8	19.5	22.8	21.6	17.4	17.1
MnO	—	—	—	—	—	—	—	—	—	0.3	0.18	0.43	—	—
MgO	15.4	18.8	17.5	15.0	19.9	18.3	15.5	16.2	15.2	15.4	17.7	18.2	17.2	15.1
CaO	7.8	7.7	8.3	9.2	7.1	7.6	9.3	9.0	8.5	9.1	7.2	7.0	9.1	9.1
Na_2O	0.23	0.19	0.12	0.23	0.22	0.14	0.29	0.25	0.24	0.0	0.38	0.15	0.26	0.28
K_2O	0.06	0.06	0.01	0.08	0.16	0.00	0.10	0.17	0.02	0.0	0.05	0.05	0.04	0.21
P_2O_5	—	—	—	—	—	—	—	—	—	0.0	0.04	0.03	—	—
Total	100.75	101.35	99.48	101.08	96.83	99.42	98.77	99.55	97.89	100.2	100.95	100.51	100.25	99.69
CaO/Al_2O_3	1.14	1.09	1.08	0.98	0.97	0.97	0.97	0.94	0.94	0.93	0.92	0.91	0.87	0.87
Fe/(Fe + Mg)	0.46	0.38	0.39	0.41	0.34	0.37	0.39	0.37	0.41	0.42	0.42	0.40	0.36	0.39
Ref.	A	A	B	A	C	C	A	A	C	D	E	E	F	C

(1) 14156, 2D008; (2) 14156, 2E054; (3) Apollo 15 Green Glass; (4) 14156, 2E008; (5) 14259, 108A2451GV; (6) 14259, 102A1711GH; (7) 14156, 2C002; (8) 14156, 2AO58; (9) 14259, 102AO671GH; (10) Luna 16, 308–75; (11) 24077,43–34; (12) 24174,59–12; (13) Luna 16, A3A1AG130A; (14) 14259,103B1031 GD.

(A) A.M. Reid *et al.* (1971); (B) A.M. Reid *et al.* (1972); (C) Brown *et al.* (1971a); (D) J.B. Reid *et al.* (1972); (E) Taylor *et al.* (1977b); (F) Brown, *et al.* (1971b).

Table 3. Comparisons of compositions (wt.%) of VLT mare basalts with those calculated by removing olivine from proposed parent magmas.

	A		B		C	
	Obs.	Calc.	Obs.	Calc.	Obs.	Clac.
SiO_2	48.4	47.5	46.2	48.6	46.6	47.2
TiO_2	0.73	0.88	0.96	0.82	0.54	0.64
Al_2O_3	12.7	13.0	12.8	13.2	11.3	11.2
Cr_2O_3	0.27	0.59	0.24	0.57	0.60	—
FeO	18.0	18.3	19.5	18.8	16.5	17.2
MnO	0.34	—	0.28	—	0.28	—
MgO	7.0	7.0	6.7	6.8	13.0	13.3
CaO	12.9	12.8	12.5	12.9	10.1	9.7
Na_2O	0.16	0.32	0.30	0.36	0.18	0.30
K_2O	0.03	0.03	0.02	0.06	0.18	0.22
P_2O_5	<0.02	0.02	0.03	0.04	<0.02	—
Total	100.53		99.52		99.28	
Fe/(Fe + Mg)	0.59	0.59	0.62	0.61	0.42	0.42
CaO/Al_2O_3	1.02	0.98	0.98	0.98	0.89	0.87

(A) Obs.: Lithic fragment 70009, 295-Ll; Calc.: removal of 24% olivine (Fo_{82} to Fo_{71}) from composition of glass 78546,3–38 (Table 1, column 10)

(B) Obs.: Luna 24 ferrobasalt (Table 1, Column 6); Calc.: removal of 27% olivine (Fo_{82} to Fo_{71}) from composition of glass 78546,3–34 (Table 1, Column 9).

(C) Obs.: Average Boulder 1, Station 2, Apollo 17 VLT lithic fragment (Table 1, Column 8); Calc.: removal of 6% olivine (Fo_{84} to Fo_{83}) from composition of glass 14259, 103B1031GD (Table 2, Column 14).

both proposed parent and derivative magmas must be determined and then quantitatively compared.

Grove and Vaniman (oral presentation at the Conference on Luna 24) suggested that Luna 24 olivine vitrophyres and related glasses represent appropriate parent magma compositions for Luna 24 basalts. One of our calculated fractionation trends in Figs. 2–4 begins at an olivine vitrophyre that we analyzed in our Luna 24 soils (Taylor *et al.*, 1978). The trends are reasonable, though CaO/Al_2O_3 is much too low in the vitrophyre (Fig. 4). More importantly, these small fragments appear to have formed by impact melting (Basu *et al.*, 1978; Taylor *et al.*, 1978). They have remnant clasts strewn about in a glassy groundmass. Basu *et al.* (1978) even describe a clast of *silica* immersed in an olivine vitrophyre. This clast is obviously not a phenocryst. Since the Luna 24 olivine vitrophyres were formed by impact-melting, their compositions may not represent the composition of a true parent magma.

We also tested whether the supposed parent magmas that contain higher K_2O than usual (Table 2) might represent the parental magmas for the Apollo 17 (Boulder 1, Station 2) VLT mare basalts. The results (Table 3) suggest that the hypothesis is reasonable. Once again, however, we wonder: Are the Apollo 17 (Boulder 1, Station 2) VLT basalts impact-produced mixtures? Could their supposed parent magma be a mixture?

ROLE OF IMPACT MIXING

As mentioned above, most of the analyses in Figs. 2–4 do not conform to olivine-fractionation trends. The data in Figs. 2 and 4 indicate that the major deviations from fractionation trends occur along mixing lines with plagioclase. (In each figure, the same set of analyses follow the mixing lines). This would appear to be strong evidence for mixing, probably by impact, of VLT mare basalts with plagioclase-rich rocks. Since K_2O and P_2O_5 contents are low in almost all VLT glasses and lithic fragments, ANT-suite rocks, rather than KREEP, would be the best candidates for the contaminants.

We examined quantitatively the notion that compositions that lie along the plagioclase lines in Figs. 2 and 4 are mixtures by making least-squares mixing calculations. Specifically, we tested whether the granular VLT mare basalt fragments (Table 1, column 6), which lie furthest along the plagioclase mixing lines, might be mixtures of Apollo 17 VLT mare basalts (Table 1, columns 1–3) and an aluminous highland lithology, such as highland basalt, anorthositic gabbro, or anorthositic norite. All calculations resulted in poor fits (e.g., the calculated and observed compositions differed by >1 wt.% in MgO, CaO, and Al_2O_3). We tried improving the results by adding additional plagioclase to the mix, but the calculated compositions still differed significantly from the observed composition. It appears that VLT lithic fragments, even if they are impact melts, have not been contaminated with nonmare rocks. On the other hand, rake sample 78526 is clearly a mixture of at least three lithologies (Warner *et al.*, 1978b): granular, aluminous VLT mare basalt, porphyritic VLT mare basalt similar to most Apollo 17 VLT basalts, and low-Ti (1.6 to 2.2 wt.% TiO_2) mare basalt. This suggests that the compositions of VLT basaltic glasses and lithic fragments may have been affected by impact mixing of a variety of mare basaltic lithologies.

One might also argue that the trends along plagioclase mixing lines indicate that assimilation reactions took place in the feldspathic crust as the basalts traveled from their source regions in the moon to the surface. Existing data do not permit us to rule out this possibility. We note, however, that assimilation with outer crustal material does not seem to have played a role in the evolution of low-Ti and high-Ti mare basalts. In any case, the mere fact that we cannot rule out assimilation as a mechanism shows how essential it is that we consider the roles impact melting and mixing have played in the evolution of VLT mare basalts, determine the number of VLT parent and derivative magmas, and establish where VLT mare basalts have erupted on the lunar surface.

Origins and Source Regions

In spite of our frequent warnings about uncertainties in the compositions of VLT basalt types and their parent magmas, it seems clear that an extensive suite of VLT basalts exists on the lunar surface. Accepting conventional petrologic thought, it also seems reasonable that these basalts formed by partial melting in the lunar interior. The work reported here does not bear directly on the problem of the nature of the source areas for VLT mare basalts, and a discussion of the problem is beyond the scope of this paper. However, our work does point out two important factors. First, it is quite possible that fractionation of primary VLT magmas was accomplished almost entirely by the removal of olivine alone. The REE patterns of VLT basalts (assuming no impact-mixing), therefore, would be essentially unchanged from their parent magmas, except for a general increase in concentrations. Consequently, the REE pattern of the source is easier to deduce from VLT mare basalts than from other mare basalts. Second, we have identified potential parent magmas. High-T, P experiments on these compositions might yield useful information about the lunar interior. Grove and Lindsley's (1978) experiments on Apollo 15 green glass composition have yielded very useful results, but Apollo 15 green glass, although the best-studied VLT basalt material, may, ironically, not be the most suitable parent material for the observed VLT basalt compositions (Fig. 4).

Acknowledgment—This work was supported by NASA Grant No. NGL 32-004-063.

References

Basu A., McKay D. S., and Fruland R. M. (1978) Impact melt origin of Luna 24 olivine vitrophyres (abstract). In *Lunar and Planetary Science IX*, p. 50–52. The Lunar and Planetary Institute, Houston.

Brown R. W., Harmon R. S., Jakeš P., Reid A. M., Ridley W. I., and Warner J. L. (1971a) Microprobe analyses of glasses and minerals from Luna-16 soil. NASA TM X-58082. Johnson Space Center, Houston.

Brown R. W., Reid A. M., Ridley W. I., Warner J. L., Jakeš P., Butler P., Williams R. J., and Anderson D. H., (1971b) Microprobe analyses of glasses and minerals from Apollo 14 sample 14259. NASA TM X-58080. Johnson Space Center, Houston.

Drake M. J. (1976) Evolution of major mineral compositions and trace element abundances during fractional crystallization of a model lunar composition. *Geochim. Cosmochim. Acta* **40**, 401–412.

Green J. A., Kurat G., Keil K., and Prinz M. (1972) Electron microprobe analyses of lithic fragments, glasses, chondrules, and minerals from Luna 16 fines. Institute of Meteoritics Spec. Publ. No. 5, Univ. New Mexico, Albuquerque.

Grove T. L. and Lindsley D. H. (1978) Compositional variation and origin of lunar ultramafic green glasses (abstract). In *Lunar and Planetary Science IX*, p. 430–432. The Lunar and Planetary Institute, Houston.

Grove T. L. and Vaniman D. T. (1977) Experimental petrology of very low Ti basalts and origin of Luna 24 ferrobasalt (abstract). In *Papers Presented to the Conference on Luna 24*, p. 68–71. The Lunar Science Institute, Houston.

Leich D. A., Kahl S. B., Kirschhaum A. R., Niemeyer S., and Phinney D. (1975) Rare gas constraints on the history of Boulder 1, Station 2, Apollo 17. *The Moon* **14**, 407–444.

Ma M.-S., Schmitt R. A., Taylor G. J., Warner R. D., Lange D. E., and Keil K. (1978) Chemistry and petrology of Luna 24 lithic fragments and <250 μm soils: Constraints on the origin of VLT mare basalts. In *Mare Crisium: The View from Luna 24* (R. B. Merrill and J. J. Papike, eds.). This volume.

Norman M., Coish R. A., and Taylor L. A. (1977) Glass chemistry and magma evolution at Mare Crisium (abstract). In *Papers Presented to the Conference on Luna 24*, p. 136–138. The Lunar Science Institute, Houston.

Reid A. M., Ridley W. I., Jakeš P., and Warner J. L. (1971) Microprobe analyses of glasses from Apollo 14 sample 14156. NASA TM X-58081. Johnson Space Center, Houston.

Reid A. M., Warner J. L., Ridley W. I., and Brown R. W. (1972) Major element composition of glasses in three Apollo 15 soils. *Meteoritics* **7**, 395–415.

Reid J. B. Jr., Taylor G. J., Marvin U. B., and Wood J. A. (1972) Luna 16: Relative proportions and petrologic significance of particles in the soil from mare Fecunditatis. *Earth Planet. Sci. Lett.* **13**, 286–298.

Roeder P. L. and Emslie R. F. (1970) Olivine-liquid equilibrium. *Contrib. Mineral. Petrol.* **29**, 275–289.

Ryder G., Stoeser D. B., Marvin U. B., Bower J. F. and Wood J. A. (1975) Boulder 1, Station 2, Apollo 17: Petrology and petrogenesis. *The Moon* **14**, 327–357.

Ryder G. and Taylor G. J. (1976) Did mare-type volcanism commence early in lunar history? *Proc. Lunar Sci. Conf. 7th*, p. 1741–1755.

Simonds C. H., Warner J. L., McGee P. E., Cochran A., and Brown R. W. (1977) Luna 24: Petrochemical potpourri (abstract). In *Papers Presented to the Conference on Luna 24*, p. 170–173. The Lunar Science Institute, Houston.

Taylor G. J., Keil K., and Warner R. D. (1977a) Very low-Ti mare basalts. *Geophys. Res. Lett.* **4**, 207–210.

Taylor G. J., Warner R. D., and Keil K. (1977b) Chemical distinctions among very low-Ti mare basalts (abstract). In *Papers Presented to the Conference on Luna 24*, p. 183–185. The Lunar Science Institute, Houston.

Taylor G. J., Warner R. D., Wentworth S., Keil K. and Sayeed U. (1978) Luna 24 lithologies: Petrochemical relationships among lithic fragments, mineral framents, and glasses. In *Mare Crisium: The View from Luna 24* (R. B. Merrill and J. J. Papike, eds.). This volume.

Vaniman D. T. and Papike J. J. (1977) Very low Ti (VLT) basalts: A new mare rock type from the Apollo 17 drill core. *Proc. Lunar Sci. Conf. 8th*, p. 1443–1471.

Walker D., Grove T. L., Longhi J., Stolper E. M., and Hays J. F. (1973) Origin of lunar feldspathic rocks. *Earth Planet. Sci. Lett.* **20**, 325–336.

Warner R. D., Keil K., Nehru C. E., and Taylor G. J. (1978a) *Catalogue of Apollo 17 Rake Samples from Stations 1A, 2 7, and 8*. Institute of Meteoritics Spec. Publ. No. 18, Univ. New Mexico, Albuquerque. 88 pp.

Warner R. D., Taylor G. J., Keil K., Planner H. N., Nehru C. E., Ma M.-S., and Schmitt R. A. (1978b) Green glass Vitrophyre 78526: An impact melt of very low-Ti mare basalt composition. Submitted to *Proc. Lunar Planet Sci. Conf. 9th*.

Merrill, R. B. and Papike, J. J., editors.
(1978) *Mare Crisium: The View from Luna 24*, Pergamon Press (New York), p. 371–401
Printed in the United States of America

Luna 24 ferrobasalts and the mare basalt suite: Comparative chemistry, mineralogy, and petrology

J. J. PAPIKE and D. T. VANIMAN

Planetary Regolith Studies, Department of Earth and Space Sciences
State University of New York, Stony Brook, New York 11794

Abstract—Although the lunar mare basalts constitute less than 1% of the lunar crust, they are abundant at the surface (17% of lunar surface area), and they are important petrologic indicators of compositions and conditions in zones of melting at depth. Although sample studies have revealed rare mare-type compositions >4 B.y. old, and cratering studies indicate mare flows as young as 2.5 B.y., the ages of sampled mare flows range from 3.15 to 3.96 B.y. Recent studies have greatly expanded our knowledge of the mare basalts that occur within this time span. Since 1976 there has been a revision of the Apollo 12 low-Ti mare basalt suite and the discovery of a new very low-Ti ("VLT": <1% TiO_2) basalt suite at Apollo 17 and in the new Soviet samples from Mare Crisium (Luna 24). Current studies suggest that the VLT basalts may be in some way related to the enigmatic "green glasses" which are found in the soils from every lunar landing site. These new discoveries indicate that mare basalts can no longer be classified into the two simple groups of older, high-Ti basalts and younger, low-Ti basalts. Telescopic studies of spectral reflectance and crater systematics also show that basalts of varying Ti content were extruded throughout the history of mare volcanism. Within each of the sub-groups of mare basalts, there is a range of grain sizes and textures that suggests variable cooling histories. Near-surface fractionation histories are recorded which principally involve olivine in the low-Ti and VLT basalts and olivine and Fe-Ti oxides in the high-Ti basalts. The important minerals in mare basalts are silicates (pyroxene, feldspar, olivine) and oxides (spinel, ilmenite, armalcolite). There is still much contention over the source regions for mare basalts. Three general models of mare basalt origin include (1) remelting of cumulates which originated in early lunar differentiation, (2) melting of primitive, undifferentiated lunar material, and (3) a hybrid model that includes both primitive and differentiated sources. With the discovery of VLT basalts and possible new evidence for the petrologic significance of green glasses, our understanding of primitive or cumulate mare source regions may improve in the near future.

INTRODUCTION

Mare basalts were derived by partial melting at lunar mantle depths of ~150–500 km. Thus, they hold important clues to the nature of the lunar interior and its thermal history. According to Head (1976) these rocks comprise only 1% of the lunar crust and 17% of the lunar surface area. Papike *et al.* (1976) reviewed the crystal chemistry, mineralogy and petrology of these rocks and Head (1976) reviewed the broader aspects of lunar volcanism based largely on photogeologic and remote-sensing results. One might assume that since these reviews were written such a short time ago, there is little new to be said. This is not the case.

Several important post-1976 studies have changed our thinking markedly. Rhodes *et al.* (1977) and Dungan and Brown (1977) re-examined a large suite of Apollo 12 mare basalts and have given us some important new insights into these rocks. Vaniman and Papike (1977a) discovered an important new mare basalt type characterized by very low TiO_2 contents (VLT). This mare type which was discovered in the Apollo 17 core materials is described in more detail by G. J. Taylor *et al.* (1977a) and Vaniman and Papike (1977b). Coincident with the discovery of Apollo 17 VLT, the initial results of the Luna 24 mission were presented at the 8th Lunar Science Conference. The Luna 24 mare basalts are also low in TiO_2, but have higher FeO and are more fractionated. Thus, instead of our broad two-fold division of mare basalts into high Ti and low Ti groups, we must now add a third, very low Ti or VLT.

The U. S. Geological Survey (Flagstaff) has a computerized system to synthesize and display on a common lunar grid such diverse data as: x-ray spectra, x-ray fluorescence spectra, geological maps, gravity maps, magnetometer surveys, electron reflection measurements, visible and near-infrared spectras, reflectance maps, mare relative age maps based on crater morphology, albedo maps, and laser altimetry profiles. This ambitious task has led to some remarkable new insights and some aspects of these results will be discussed below. Papers based on this synthesis appeared in the 8th Proceedings of the Lunar Science Conference and include: Arnold *et al.* (1977); Bielefeld *et al.* (1977); Johnson *et al.* (1977a,b); Russell *et al.* (1977); Eliason and Soderblom (1977); Metzger *et al.* (1977); and Soderblom *et al.* (1977). With the addition of new ground truth from a Conference on Luna 24 (held at the Lunar Science Institute, Houston, Texas, December 1–3, 1977), it is clear that our ideas and models concerning mare basalt petrogenesis must be updated.

Regional Setting and Temporal Relations

Figure 1 summarizes the lunar sampling locations and the distribution of three spectrally distinct mare lithologies. The diagram is modified from Head (1976) who, in turn, modified it after Strom (1972) and Whitaker (1972). The map represents combined ultraviolet negatives and infrared positives of the full moon. It is thought that the color differences represent, among other things, the Ti contents of the mare units. Blue might represent high TiO_2 basalts (8–14 wt.% TiO_2), orange low TiO_2 basalts (1.5–5 wt.% TiO_2) and red very low TiO_2 (VLT) basalts ($<$1.5 wt.% TiO_2). This is obviously an over-simplification and much more detailed insights are gained by looking at all the remote sensing data on a common grid as stated above. Nevertheless this whole-moon map is a useful backdrop for discussions of regional setting. Apollo 11 is located on the edge of Mare Tranquillitatis in the blue mare unit, and it appears that these samples are quite representative of the surface flows over much of this basin. Apollo 12 is located in the orange unit which, although not completely mapped, probably also extends up to the Apollo 15 site. Apollo 17 is also located in the blue unit, and the Apollo 17 samples which are high in TiO_2 are similar to Apollo 11. It is clear that the Apollo 17 basalts are not representative of all of the Serenitatis Basin. At least two additional units occur within this Basin (Bryan and Adams, 1973; Boyce and Dial, 1973). It is possible that the orange annulus is comprised of yet

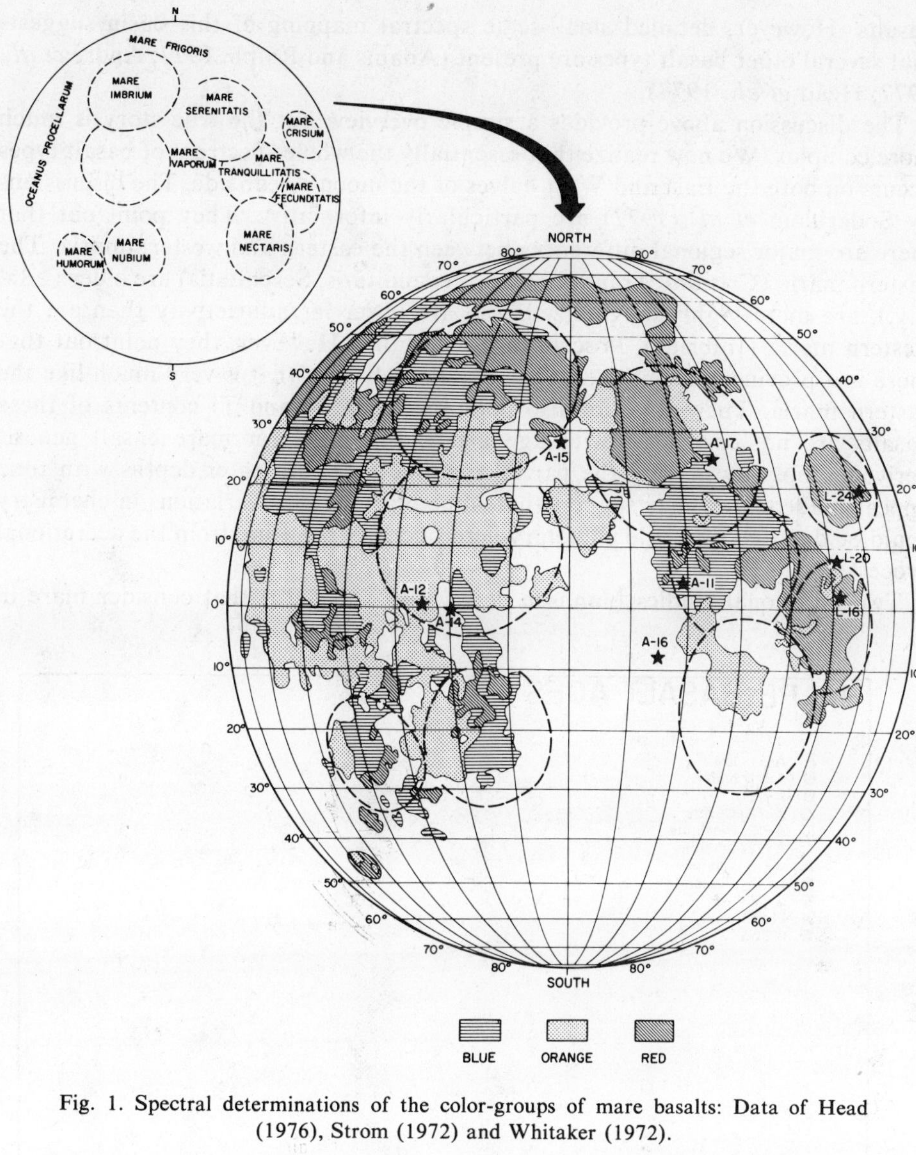

Fig. 1. Spectral determinations of the color-groups of mare basalts: Data of Head (1976), Strom (1972) and Whitaker (1972).

unsampled mare basalts while the red represents the VLT basalts that were discovered as a minor component in the Apollo 17 soil (Vaniman and Papike, 1977a). Luna 16 is located in a red region of Mare Fecunditatis but the basalts are not VLT as would be predicted by our simplistic approach. However, VLT basalts are found in the Luna 16 collection (Green *et al.*, 1972) and perhaps both low-Ti and VLT basalts occur in close proximity in this basin. Luna 24 is located in a red region of Mare Crisium, and, as would be predicted, is comprised of VLT basalts. However, detailed small-scale spectral mapping of this basin suggests that several other basalt types are present (Adams and Ralph, 1977; Andre *et al.*, 1977; Head *et al.*, 1977).

The discussion above provides a simple overview but the true story is much more complex. We now realize that essentially the whole spectrum of basalt types occurs on both the East and West halves of the moon's near side. The discussions by Soderblom *et al.* (1977) are particularly informative. They point out that there are major regional differences between the eastern and western maria. The eastern maria (Crisium, Tranquillitatis, Fecunditatis, Serenitatis) are older (≥3.4 B.y.), are more highly magnetized, and are lower in radioactivity than are the western maria (Imbrium, Procellarum, Nubium). However, they point out that there is a province in southwest Oceanus Procellarum that is very much like the eastern maria. These authors also point out that Fe and Ti contents of these basalts are not correlated with age and thus models for mare basalt genesis derived simply by moving the partial melting zone to greater depths with time appear inadequate. They suggest that some of the lateral variations in chemistry could be due to large-scale planetary heterogeneity resulting from the accretional process.

Two additional studies (Johnson *et al.*, 1977a, 1977b) that consider mare in

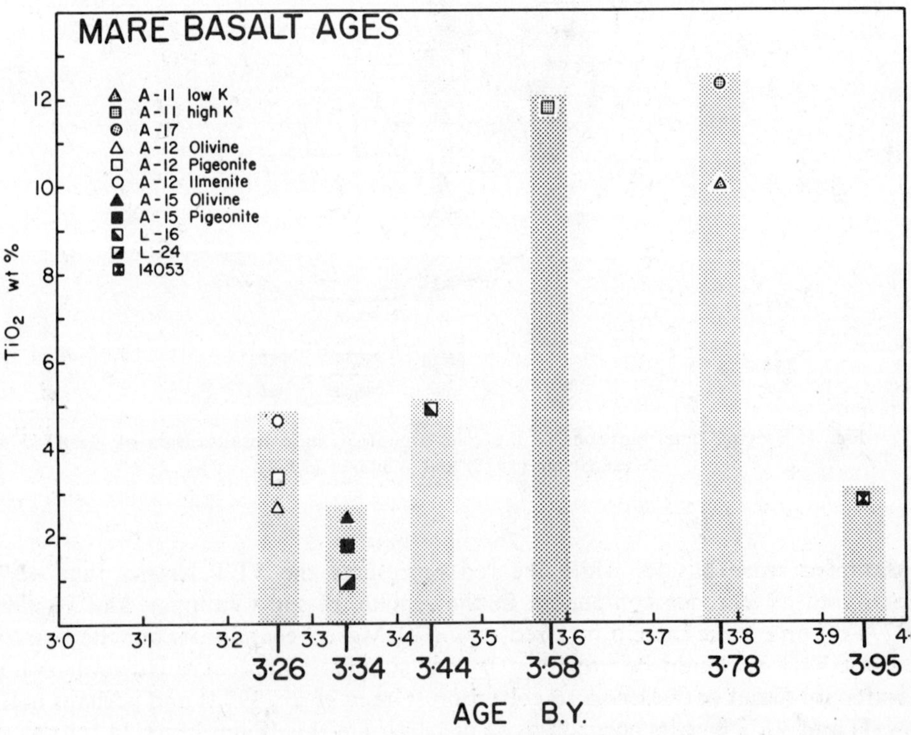

Fig. 2. Radiometric ages of mare basalts compared with TiO_2 content.

the lunar northern hemisphere confirm the wide distribution of VLT basalts which apparently match the low- and high-Ti basalts in areal distribution.

Figure 2 summarizes the ages determined on lunar samples. Although the samples that have been collected perhaps do not represent all of the major mare units, several important things are clear: (1) the samples represent an interval of volcanism between from 3.2 to 4.0 m.y.; (2) there is no simple correlation of TiO_2 and age; (3) samples from eastern maria have old and young ages, e.g., Apollo 11 and 17 compared to Luna 16 and 24; (4) a large span of volcanism (~200 m.y.) is represented in samples from the Apollo 11 site; (5) very old mare samples exist in the Apollo 14 collection and probably represent the earliest extensive mare flooding. However, it is apparent that we have not sampled the entire temporal span of mare volcanism.

The data of Boyce (1976) based on crater degradation studies strongly indicate that basalts <3.0 B.y. old are present on the lunar surface. It is quite likely that such basalts exist in the lunar soil samples but have not yet been recognized.

Major-Element Chemistry

The major-element chemistry of mare basalts was reviewed in detail by Papike *et al.* (1976). Therefore, this paper will repeat only the highlights of that discussion and incorporate the new results on VLT basalts. As part of the NASA-funded project "Basaltic Volcanism in the Terrestrial Planets" sponsored by the Lunar and Planetary Institute, Houston, a suite of mare basalts was selected that is large enough to include all of the mare types that were sampled and small enough to be incorporated into the total planetary basalt suite. The major-element analyses of these basalts are presented in Table 1 and the data are plotted in Figs. 3–9. More detailed variation diagrams may be found in Papike *et al.* (1976) where fractionation trends are discussed in some detail. Figure 4 is a variation diagram of TiO_2 wt.% versus Mg/Mg + Fe (atomic) = Mg′. TiO_2 is the most useful single chemical discriminant for separating the mare basalt groups. The high-Ti basalts include A-11 low-K, A-11 high-K and A-17; low-Ti basalts include A-12 olivine, A-12 pigeonite, A-12 ilmenite, A-15 olivine and A-15 pigeonite; very low-Ti basalts include A-17 VLT and Luna 24 VLT. In addition, there are feldspathic basalts (e.g., Luna 16 and 14053) which based on TiO_2 content are in the low-Ti group. Sample 14053 (251 grams) is a large breccia clast and is the largest mare basalt fragment returned from the Apollo 14 mission. Sample 14063 deserves special mention. Basalt fragments in this Apollo 14 breccia were discussed by Ridley (1975). One important feature of this basalt type is that it has a TiO_2 content right in the middle of "the TiO_2 gap" between the high- and low-Ti basalts. These samples warrant intensive additional study and every attempt should be made to obtain an age. Two important mare glasses are also plotted on these diagrams, the Apollo 15 emerald green glass and Apollo 17 orange glass.

The high-Ti basalts show a trend of TiO_2 decrease with fractionation. This reflects the subtraction of an Fe-Ti oxide as a liquidus or near-liquidus phase. An important implication of this observation is that if high-Ti lavas erupt from a fissure and flow long distances, fractionation can result in significant Ti gradients along flow lines. Spectral mapping could interpret this titanium difference as reflecting more than one flow unit when, in fact, it may be only one continuous flow undergoing fractionation. Combined TiO_2 and crater degradation maps might resolve this ambiguity. The low-Ti and VLT basalts behave in just the opposite way with respect to titanium. Here fractionation causes an increase in TiO_2.

The VLT suite displays a range of Mg′ values from those characteristic of Apollo 15 green glass to the fractionated Luna 24 ferrobasalts. Several authors

Table 1. Major-element compositions of mare basalts.

	Apollo 11 Low K			Apollo 11 High K		Apollo 12 Olivine			
Sample No.	(1) 10003	(2) 10020	(3) 10047	(4) 10049	(5) 10072	(6) 12002	(7) 12009	(8) 12035	(9) 12040
SiO_2	39.76	39.79	42.16	41.00	40.26	43.56	45.03	43.17	43.88
TiO_2	10.50	10.62	9.43	11.30	12.07	2.6	2.90	2.28	2.45
Al_2O_3	10.43	9.93	9.89	9.5	7.93	7.87	8.59	8.03	7.27
FeO	19.80	19.21	19.11	18.7	19.58	21.66	21.03	22.20	21.09
MnO	0.30	0.26	0.28	0.25	0.23	0.28	0.28	0.29	0.27
MgO	6.69	7.93	5.67	7.03	7.8	14.88	11.55	15.49	16.45
CaO	11.13	11.32	12.15	11.0	10.41	8.26	9.42	8.08	8.01
Na_2O	0.40	0.39	0.45	0.51	0.53	0.23	0.23	0.21	0.17
K_2O	0.06	0.04	0.11	0.36	0.28	0.05	0.06	0.05	0.05
Cr_2O_3	(0.25)*	0.39	0.18	0.32	0.36	0.96	0.55	0.49	0.63
Σ	99.32	99.88	99.43	99.97	99.45	100.35	99.64	100.29	100.27
Mg/Mg + Fe	0.38	0.42	0.35	0.40	0.42	0.55	0.50	0.55	0.58
CaO/Al_2O_3	1.07	1.14	1.29	1.16	1.31	1.05	1.10	1.01	1.10

Table 1. (*cont.*)

Apollo 12 Pigeonite				Apollo 12 Ilmenite				Apollo 15 Olivine		
(10) 12011	(11) 12021	(12) 12052	(13) 12064	(14) 12005	(15) 12008	(16) 12022	(17) 12051	(18) 15016	(19) 15545	(20) 15555
46.63	46.68	46.40	46.30	41.56	42.75	42.77	45.31	44.08	45.21	44.57
3.29	3.53	3.28	3.99	2.76	4.45	4.85	4.68	2.28	2.41	2.10
9.77	10.78	10.16	10.73	5.30	7.98	9.08	9.95	8.38	8.59	8.69
19.53	19.31	20.15	19.89	22.27	21.94	21.75	20.22	22.74	22.15	22.53
0.29	0.26	0.27	0.27	0.30	0.30	0.25	0.28	0.32	0.30	0.29
8.26	7.39	8.22	6.49	19.97	12.33	11.01	7.01	11.30	10.28	11.36
10.63	11.38	10.80	11.77	6.31	8.97	9.47	11.39	9.27	9.82	9.40
0.25	0.31	0.27	0.28	0.16	0.25	0.38	0.29	0.27	0.31	0.27
0.06	0.07	0.07	0.07	0.04	0.05	0.07	0.06	0.04	0.04	0.09
0.59	0.40	0.52	0.37	0.75	0.61	0.56	0.31	0.85	0.68	0.61
99.30	100.11	100.14	100.16	99.42	99.63	100.19	99.50	99.53	99.79	99.91
0.43	0.40	0.42	0.37	0.62	0.50	0.47	0.38	0.47	0.45	0.47
1.09	1.06	1.06	1.10	1.19	1.12	1.04	1.14	1.11	1.14	1.08

Table 1. (*cont.*)

	Apollo 15 Pigeonite				Apollo 17 Group A		Apollo 17 Group B		Apollo 17 Group C		
	(21)	(22)	(23)	(24)	(25)	(26)	(27)	(28)	(29)	(30)	(31)
Sample No.	15058	15499	15597	A-15 Emerald Green Glass	75055	76136	70215	70275	74245	74255	74275
SiO_2	48.14	47.81	47.98	44.14	40.60	38.6	37.79	39.37	38.59	38.18	38.43
TiO_2	1.69	1.78	1.80	0.37	10.79	12.64	12.97	11.90	11.92	12.47	12.70
Al_2O_3	8.89	9.11	9.44	7.81	9.67	8.65	8.85	10.23	8.72	8.69	8.72
FeO	19.86	20.19	20.23	21.05	18.01	19.12	19.66	18.61	18.06	18.05	18.14
MnO	0.27	0.28	0.30	—	0.29	0.28	0.27	0.28	0.27	0.28	0.26
MgO	9.28	9.33	8.74	16.72	7.05	8.61	8.44	6.09	9.65	10.61	10.36
CaO	10.27	10.34	10.43	8.41	12.35	10.53	10.74	11.65	10.59	10.27	10.32
Na_2O	0.28	0.32	0.32	0.13	0.43	0.38	0.36	0.38	0.36	0.37	0.35
K_2O	0.03	0.05	0.06	0.03	0.08	0.06	0.05	0.06	0.06	0.09	0.07
Cr_2O_3	0.66	0.57	0.48	0.33	0.27	0.44	0.41	0.26	0.54	0.60	0.65
Σ	99.37	99.78	99.78	98.99	99.54	99.31	99.54	98.83	98.76	99.61	100.00
Mg/Mg + Fe	0.45	0.45	0.44	0.59	0.41	0.45	0.43	0.37	0.49	0.51	0.50
CaO/Al_2O_3	1.16	1.14	1.10	1.08	1.28	1.22	1.21	1.14	1.21	1.18	1.18

Table 1. (*cont.*)

Apollo 17 Group U		A-17 Orange Glass	Luna 16 B1	Luna 24		Apollo 17 VLT			Other Feldspathic	
(32)	(33)	(34)	(35)	(36)	(37)	(38)	(39)	(40)	(41)	(42)
70017	70035	74220		24182, 25-100	24109, 52-18	70008, 356-P	70007, 328-V	78526,1	14053	14063
38.54	37.84	38.57	43.80	46.82	45.2	48.1	48.8	46.7	46.4	45.9
12.99	12.97	8.81	4.90	1.00	0.89	0.36	0.69	0.92	2.64	7.3
8.65	8.85	6.32	13.65	12.58	13.8	11.2	10.0	10.0	13.6	14.4
18.25	18.46	22.04	19.35	20.46	20.5	18.2	17.9	18.6	16.8	10.7
0.25	0.28	—	0.20	0.24	0.27	0.26	0.30	0.24	0.26	—
9.98	9.89	14.44	7.05	6.65	6.35	11.0	11.8	12.2	8.48	6.7
10.28	10.07	7.68	10.40	12.49	12.7	10.2	9.4	10.0	11.2	11.3
0.39	0.35	0.36	0.33	0.10	0.24	0.15	0.06	0.12	—	1.0
0.05	0.06	0.09	0.15	0.02	0.01	0.01	0.02	<0.01	0.10	0.1
0.50	0.61	0.75	0.28	0.16	0.19	0.60	0.69	0.74	—	0.2
99.88	99.38	99.06	100.11	100.52	100.2	100.1	99.7	99.52	99.48	97.60
0.49	0.49	0.54	0.39	0.37	0.36	0.52	0.54	0.54	0.48	0.53
1.19	1.14	1.22	0.76	0.99	0.92	0.90	0.93	1.00	0.82	0.78

have suggested this might represent a fractionation sequence (G. J. Taylor *et al.*, 1977a; Vaniman and Papike, 1977b,c). However, Grove and Vaniman (1978) show that Luna 24 ferrobasalts cannot be obtained by low-pressure fractionation of an A-15 green glass parent.

The $Al_2O_3 \times Mg'$ diagram (Fig. 5) shows that the VLT basalts are separated in Al_2O_3 content from the low- and high-Ti groups, but the latter groups have strong overlap in this chemical variable. We also see an Al_2O_3 increase with Mg' decrease. Good group separation is observed on a $Na_2O \times Mg'$ diagram (Fig.

6): i.e., high-Ti basalts have higher Na_2O contents than low-Ti basalts. A Na_2O increase with fractionation is also observed. The behavior of K_2O is more complicated than Na_2O (Fig. 7). Here the high-Ti and low-Ti groups show significant overlap with the exception of the A-11 high-K basalts. However, the K_2O content of the VLT basalts is significantly lower than for the other two groups. On the CaO × Mg′ diagram (Fig. 8) no clear group separation is apparent, and the trend is mainly one of increasing CaO with decreasing Mg′. Likewise, the groups overlap in the Cr_2O_3 × Mg′ diagram (Fig. 9); however, here Cr_2O_3 decreases with decreasing Mg′.

An important diagram that helps us select relatively unfractionated basalts is a histogram of Mg′ (Fig. 10). Papike *et al.* (1976) pointed out that on such a diagram both high- and low-Ti basalts display an Mg′ ≈0.5 for the least fractionated members; basalts from these groups with Mg′ >0.5 are thought to be cumulates. We thus speculated that the Mg′ value of 0.5 had special lunar significance. In fact, it must, but perhaps only for the high- and low-Ti suites. The A-17 VLT basalts show Mg′ >0.6 and a VLT vitrophyre from Luna 24 has Mg′ ≈0.6 (G. J. Taylor *et al.*, 1977b). The VLT basalts thus have Mg′ values approaching A-15 green glass. This could imply that either the VLT basalts resulted from partial melting of cumulates much deeper in the lunar mantle or that they were derived from below the cumulate zones from primitive lunar material.

An additional useful chemical variation diagram is a three-axis plot of Mg′ × (CaO/Al_2O_3 wt. ratio) × TiO_2 wt.% (Fig. 11). This diagram not only effectively separates the mare basalt groups from each other but also separates the mare basalts from the highland melt rocks. The plotted data include the average compositions for the high-Ti and low-Ti basalt groups from Papike *et al.* (1976); A-17 VLT and A-17 yellow/green glass cluster from Vaniman and

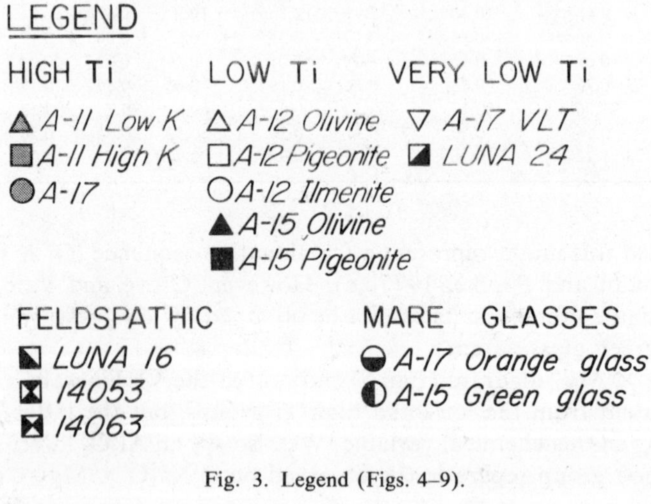

Fig. 3. Legend (Figs. 4–9).

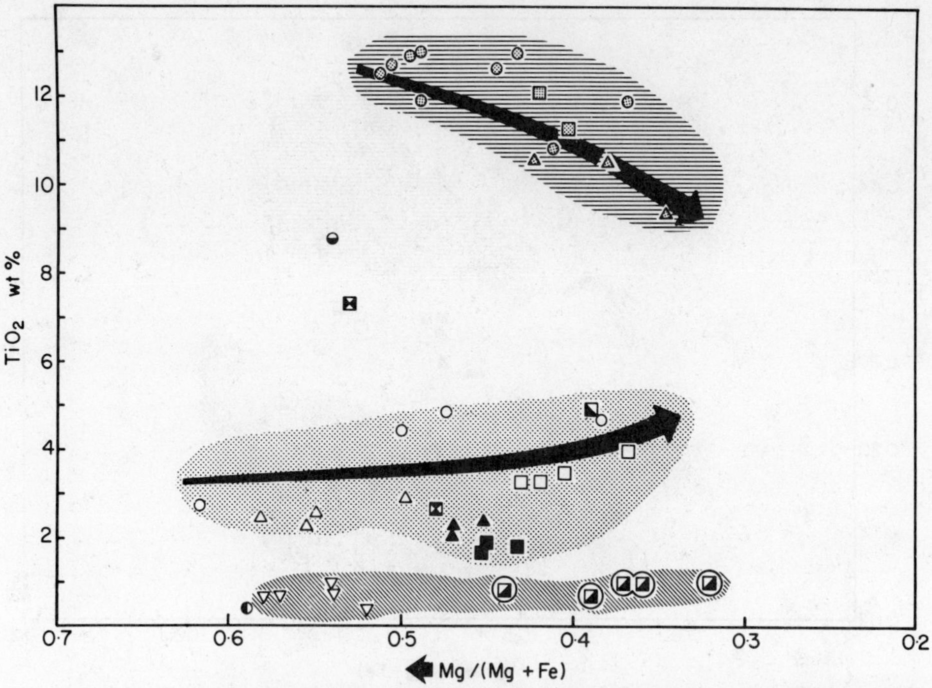

Fig. 4. TiO_2 (wt.%) versus atom ratio Mg/(Mg + Fe) in mare basalt major-element chemistry. Legend in Fig. 3.

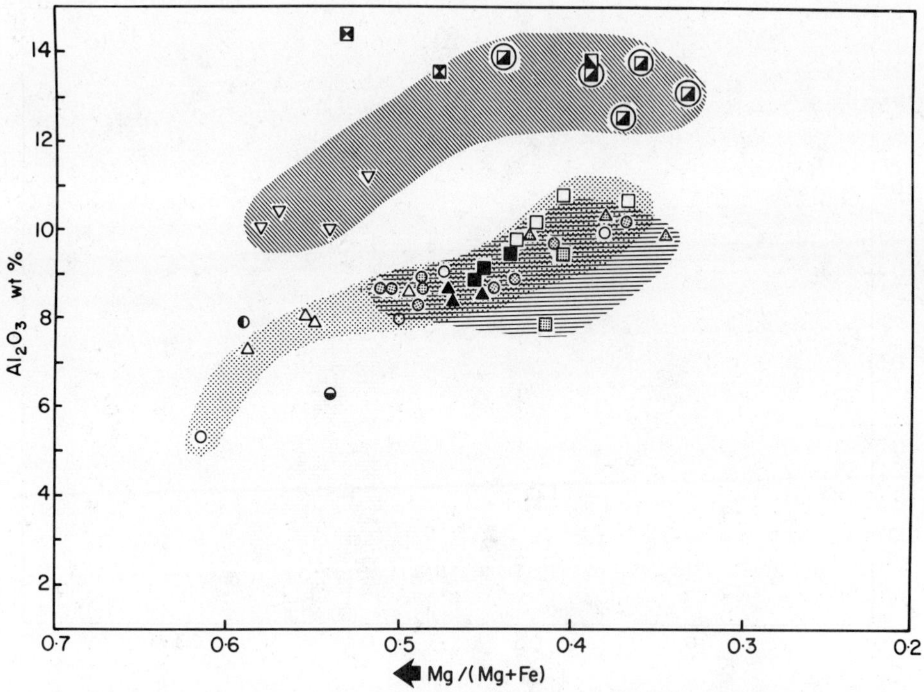

Fig. 5. Al_2O_3 (wt.%) versus atom ratio Mg/(Mg + Fe) in mare basalt major-element chemistry. Legend in Fig. 3.

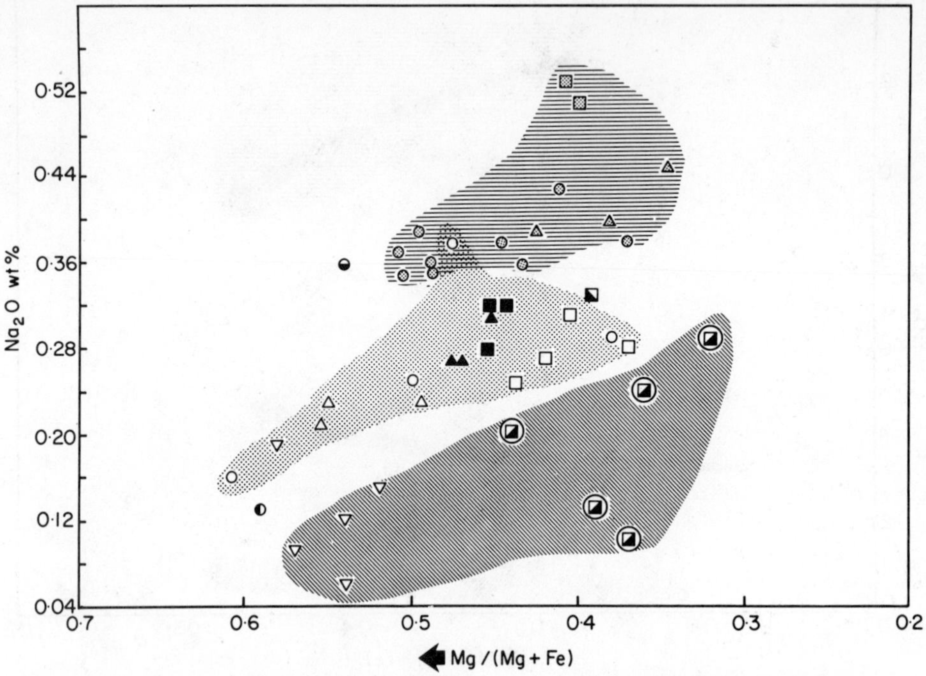

Fig. 6. Na_2O (wt.%) versus atom ratio Mg/(Mg + Fe) in mare basalt major-element chemistry. Legend in Fig. 3.

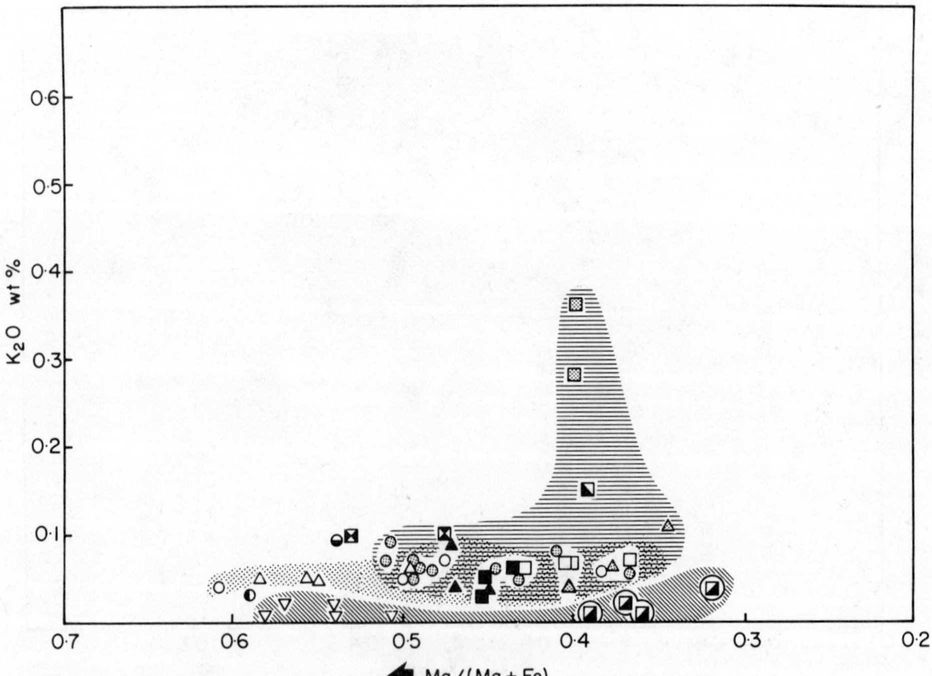

Fig. 7. K_2O (wt.%) versus atom ratio Mg/(Mg + Fe) in mare basalt major-element chemistry. Legend in Fig. 3.

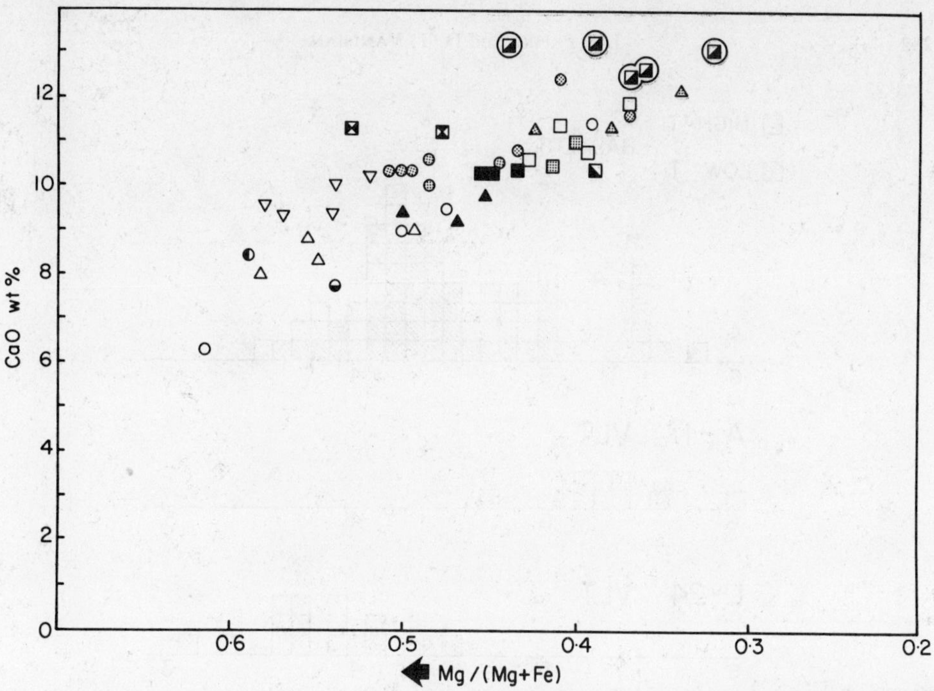

Fig. 8. CaO (wt.%) versus atom ratio Mg/(Mg + Fe) in mare basalt major-element chemistry. Legend in Fig. 3.

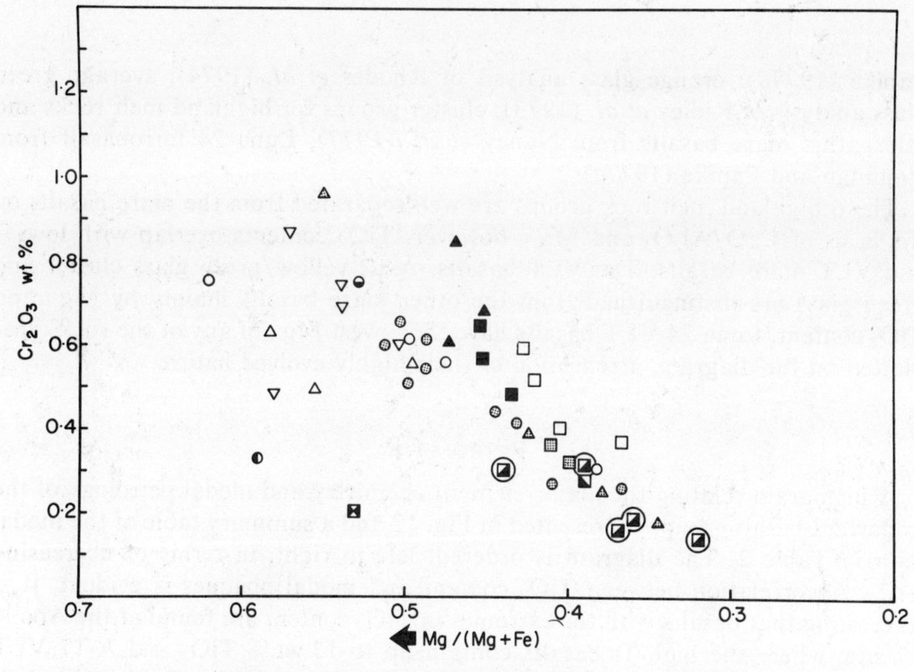

Fig. 9. Cr_2O_3 (wt.%) versus atom ratio Mg/(Mg + Fe) in mare basalt major-element chemistry. Legend in Fig. 3.

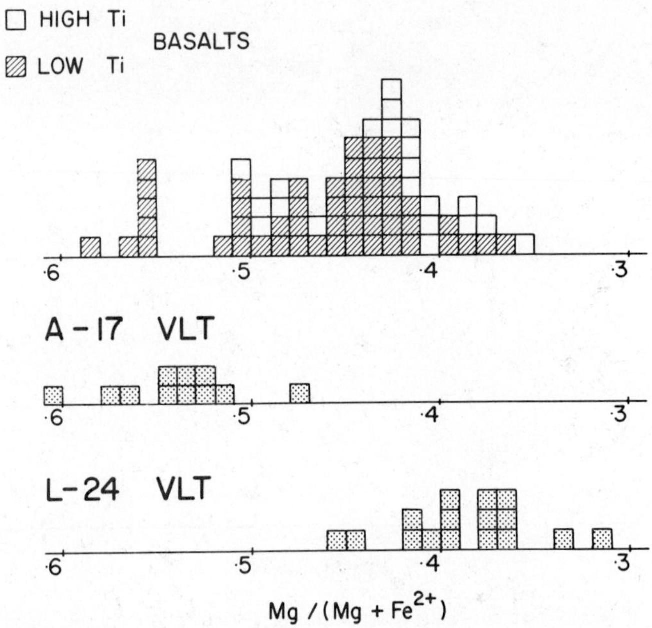

Fig. 10. Comparative histograms of atom ratio Mg/(Mg + Fe) in High-Ti, Low-Ti and VLT mare basalts.

Papike (1977d); orange glass analysis of Rhodes *et al.* (1974); average green glass analysis of Ridley *et al.* (1973); cluster groups for highland melt rocks and feldspathic mare basalts from Naney *et al.* (1977); Luna 24 ferrobasalt from Vaniman and Papike (1977c).

The 6 highland melt rock groups are well separated from the mare basalts on the basis of CaO/Al_2O_3 and Mg′; however, TiO_2 contents overlap with low-Ti and VLT mare basalts. The VLT basalts, A-17 yellow/green glass cluster and green glass are distinguished from the other mare basalts mainly by Mg′ and TiO_2 content. Luna 24 VLT basalts have the lowest Mg′ of any of the rock types plotted on the diagram, a reflection of their highly evolved nature.

PETROLOGY

A histogram relating the major-element chemistry and modal petrology of the important basalt groups is presented in Fig. 12 and a summary table of the modal data in Table 2. The diagram is ordered, left to right, in terms of decreasing TiO_2. A correlation between TiO_2 content and modal opaques is evident. It is interesting that basalts with the extremes in TiO_2 content are found at the Apollo 17 site, where the high-Ti basalts contain up to 13 wt.% TiO_2 and A-17 VLT ~0.6 wt.% TiO_2. The interpretation of this observation was discussed above. It is apparent that in the Serenitatis Basin the early flooding (~3.8 B.y.) involved

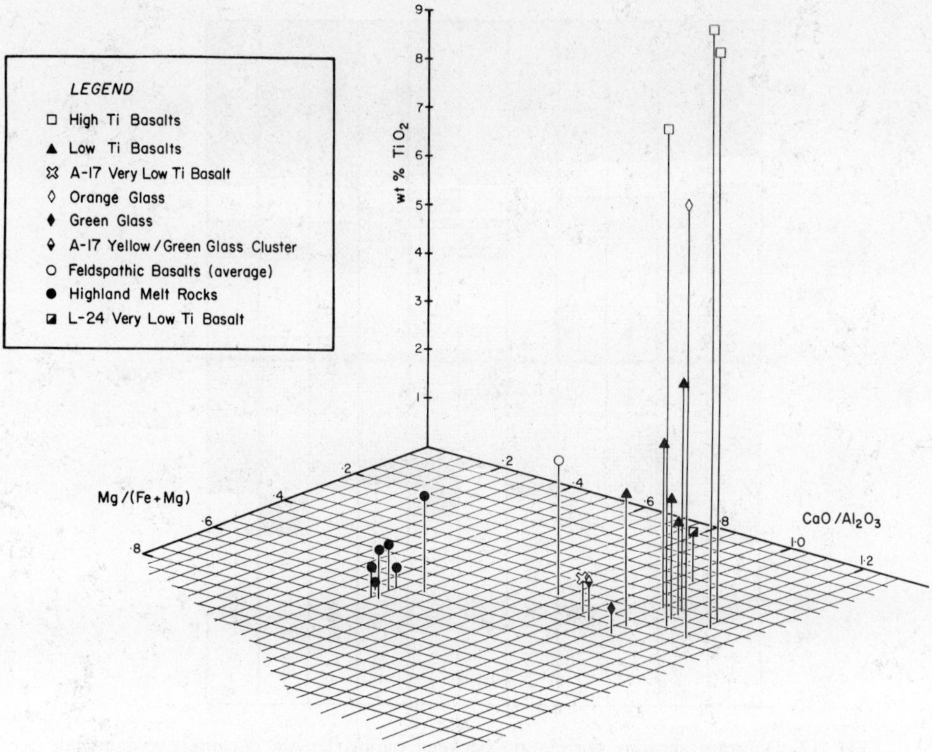

Fig. 11. Atom ratio Mg/(Mg + Fe), weight % ratio CaO/Al_2O_3, and weight % TiO_2 in mare basalts and highland melt rocks. Averages of mare basalt types are compared to the averages of six highland melt rock groups plus the average feldspathic basalt.

basalts highly enriched in Ti, whereas the younger central basin fill is at the lower extreme in Ti content. The diagram also points out that Luna 24 ferrobasalts are enriched in Al_2O_3, CaO and FeO and depleted in MgO relative to Apollo 17 VLT. These chemical differences result in higher modal feldspar in Luna 24 ferrobasalts than in Apollo 17 VLT. The higher modal olivine for Luna 24 basalts is less easily explained. The reason for higher modal olivine in Luna 24 than in Apollo 17 VLT probably results from the lower SiO_2 content of Luna 24 basalts and the different late-stage cooling histories of these two basalt groups. In A-17 VLT basalts, pyroxenes crystallized metastably in the "forbidden region" (Lindsley and Munoz, 1969) and did not form the stable assemblage hedenbergite-SiO_2-olivine. Thus, some of the "olivine content" in Apollo 17 VLT basalts is still tied up in the pyroxenes. Another factor (Grove, 1978, pers. comm.) is that at the higher (Fe/Fe + Mg) of Luna 24 ferrobasalts the pyroxene and olivine no longer are in a reaction relation; therefore, more olivine crystallizes.

The remaining discussion in this section will concern the petrology of some of

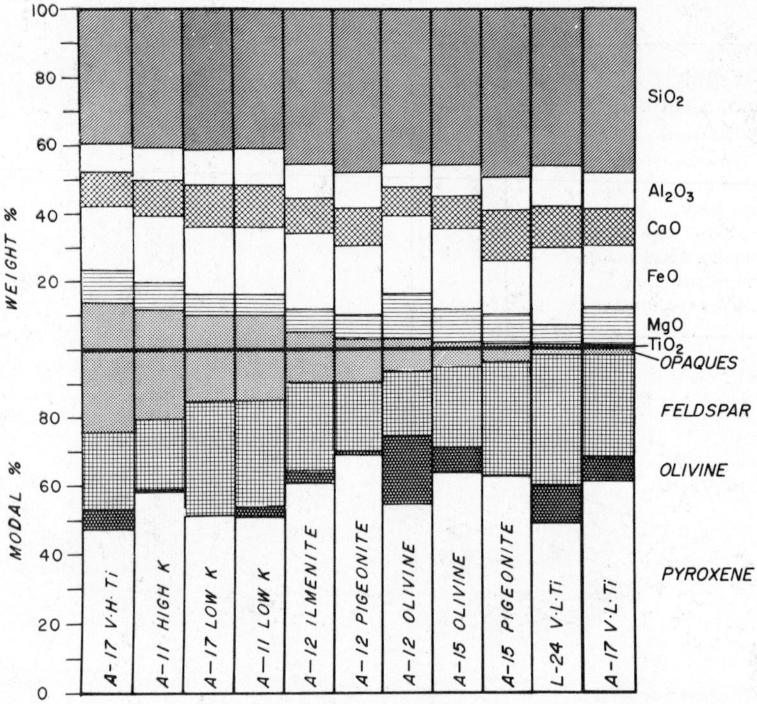

Fig. 12. Diagram showing correlation between major-element chemistry and modal mineralogy for mare basalts. Histogram is ordered from left to right by decreasing TiO_2 (or opaque oxide) content.

Table 2. Summary modal data* for the mare basalts.

	Opaques	Pyroxene	Feldspar	Olivine
A-17 very high Ti	24.4	47.7	23.4	4.6
A-11 high K	20.6	57.5	21.7	0.1
A-17 low K	15.1	51.6	33.3	—
A-11 low K	14.6	50.9	32.2	2.3
A-12 ilmenite	9.3	61.1	25.9	3.6
A-12 pigeonite	9.1	68.4	21.1	1.4
A-12 olivine	7.1	53.5	19.2	20.2
L-16	7.1	51.5	41.2	0.1
A-15 olivine	5.5	63.3	24.1	7.0
A-15 pigeonite	3.7	62.5	33.8	—
14053	3.2	53.8	43.0	—
L-24 ferrobasalt	1.8	48.6	39.1	10.4
L-24 {Recrystallized Ferrobasalt	1.4	60.2	34.2	4.2
A-17 "VLT"	1.0	61.7	31.9	5.4

*Normalized to 100% for the four phases considered. Ordered from top to bottom in terms of decreasing modal content of opaques.

the basalt groups. However, since this topic was reviewed extensively by Papike *et al.* (1976), the discussion will emphasize only the new results on Apollo 12 mare basalts, Apollo 17 VLT basalts, and Luna 24 VLT basalts.

Apollo 17 high-Ti basalts.

An extensive chemical study by Rhodes *et al.* (1976) throws additional light on this suite of rocks. An early two-fold classification of the Apollo 17 basalts is an oversimplification. The additional data of Rhodes *et al.* (1976) demonstrate that the hiatus between the two proposed basalt types is occupied by additional analyzed samples, resulting in a broad compositional continuum between the most mafic olivine-normative basalts and the chemically and mineralogically fractionated quartz-normative type. Their study defines three self-consistent basalt types on the basis of fine-grained, rapidly chilled samples. The chemical variation within each group is attributed to moderate amounts of near surface crystal fractionation (from 4 to 22%) dominated by the removal of olivine, armalcolite/ilmenite, and chrome-spinel. Rhodes *et al.* (1976) suggest further that the three groups of basalts and their respective parental magmas cannot be explained by near-surface crystal fractionation processes. The differences are attributed to partial melting of a heterogeneous source region.

Apollo 12 basalts.

The mare basalts returned by the Apollo 12 mission consist of four chemically and texturally distinct rock groups: (1) olivine basalts, (2) pigeonite basalts, (3) ilmenite basalts, and (4) feldspathic basalts.

The Apollo 12 olivine basalts are a group of fine-to-medium-grained olivine-rich basalts ranging from olivine vitrophyre to microgabbro. The chemical, modal, and textural trends indicate that many of the coarser-grained samples are olivine cumulates. There is a good correlation between chemistry and grain size, with the coarser-grained rocks being distinctly more magnesian. Experimental studies (Green *et al.*, 1971; Newton *et al.*, 1971) indicate that compositions similar to those of coarse-grained samples should crystallize olivine considerably more magnesian than that found in the samples. Other studies of finer-grained samples 12002 (Walker *et al.*, 1976) and 12009 (Green *et al.*, 1971; Donaldson *et al.*, 1975) indicate that they may be representative of liquids.

The Apollo 12 pigeonite basalts are a group of slightly vesicular to highly porphyritic samples characterized by large, zoned phenocrysts of pyroxene. Major-element data indicate that the Apollo 12 pigeonite basalts could have been derived from the Apollo 12 olivine basalts through olivine fractionation.

The discussion of the Apollo 12 suite by Papike *et al.* (1976) was limited by the rather small number of samples that had been studied up to that time. This situation was rectified in a recent study by Rhodes *et al.* (1977), where they obtained additional major and trace-element chemical data on a large suite of Apollo 12 mare basalts. Their new data confirm earlier classifications of the

Apollo 12 basalts into four compositional groups as mentioned above. The new data also confirm the fact that the olivine and pigeonite basalts are comagmatic and are related by olivine fractionation. Rhodes *et al.* (1977) demonstrate that the ilmenite basalts are a major basalt type, indigenous to the Apollo 12 site and with a compositional range equivalent to that of the olivine-pigeonite basalts. Rhodes *et al.* (1977) suggest further that the flows at the Apollo 12 site are thicker than the majority of lunar flows, and that the younger ilmenite basalts overlie the olivine-pigeonite basalts. Chemical variation within these basalt groups is attributed to extensive, olivine-dominated, near-surface crystal fractionation (about 30%), from parental magmas with compositions equivalent to the vitrophyres in each group. Dungan and Brown (1977) carried out a detailed petrologic study of the Apollo 12 ilmenite basalts. They point out that the chemical variations in the ilmenite suite can be reconciled with differentiation within a single cooling unit. The suite includes olivine porphyritic basalts, medium-grained basalts with evolved compositions and coarse-grained olivine cumulates. The chemical variation can be accounted for by fractionation of olivine equivalent to the composition of phenocrysts in vitrophyre 12008 which might represent the parental magma for the ilmenite basalt suite.

Apollo 17 VLT basalts.

The Apollo 17 drill core contains a minor amount of phaneritic fragments, vitrophyres and glass beads of a new very low-Ti (VLT) mare basalt (Vaniman and Papike, 1977a,b; G. J. Taylor *et al.*, 1977a). The basalt is characterized by TiO_2 ~0.5 wt.%, Mg′ ~0.52, CaO/Al_2O_3 (wt. ratio) ~0.9 and low alkali content (Na_2O ~0.1 wt.%, K_2O ~0.0 wt.%). Phaneritic VLT basalts from the Apollo 17 drill core are medium-grained (0.2–0.4 mm) subophitic rocks. Subhedral pyroxene rims are intergrown with euhedral laths of later-crystallized plagioclase. The earliest phases to crystallize are chromian spinel and olivine. Euhedral spinels up to 0.05 mm are characteristic of VLT basalt. Zoned and rounded olivine grains have resorbed boundaries against pyroxene. Intergrain (late-stage) areas contain glass and anhedral crystals of CaFe pyroxene, ulvöspinel, ilmenite, silica, Fe-sulfide and metallic iron. Vitrophyres contain chromian spinel and olivine ± pyroxene in a matrix of pale yellow/green to clear glass; olivine and pyroxene form skeletal crystals with grain sizes ranging from 0.001 mm to 0.1 mm. Spinel is ubiquitous and occurs as small (0.005 mm) equant octahedral or cruciform crystals in the vitrophyres. Yellow/green glass spheres (0.02–0.2 mm diameter) with VLT compositions occur in the A-17 soil and form a tight compositional cluster (Vaniman and Papike, 1977d).

Olivine (Fo_{74-75}) and chromian spinel are liquidus or near-liquidus phases; all vitrophyres have at least these two minerals. Pigeonite coexists with olivine in many vitrophyres, but in phaneritic rocks the olivine is zoned to Fo_{55} and the rims are rounded and resorbed against adjacent pyroxene. No plagioclase was observed in the vitrophyres. Phaneritic VLT basalts have plagioclase laths which are intergrown only with the pyroxene rims. These observations suggest that the onset of plagioclase crystallization occurred late in the pyroxene crystallization history. Oxide crystallization had a two-stage history: (1) iron-enrichment of chromian spinel, with a slight increase in Ti, followed by cessation of spinel crystallization and (2) a late-stage crystallization of ulvöspinel with Fe-sulfide and Fe metal.

It has been suggested that Apollo 17 VLT basalts can be related to Apollo 15 green glass compositions (G. J. Taylor *et al.*, 1977a; Vaniman and Papike, 1977b). Grove and Vaniman (1978) conducted low-pressure equilibrium experiments in iron capsules on synthetic analogs of Apollo 15 green glass and Luna 24 brown glass to determine whether crystal fractionation of the Apollo 15 mafic green glasses could generate Apollo 17 VLT and/or Luna 24 VLT. Their results indicate that Apollo 17 VLT's can be derived by low-pressure crystal-liquid fractionation from a green glass type that is similar, but not identical to Apollo 15 emerald green glass. Because of these relationships Vaniman and Papike (1977b) speculated that the age of Apollo 17 VLT's might be similar to green glass and the low-Ti basalts (A-12 and A-15) and predicted an age of 3.3 B.y. A fragment of Apollo 17 VLT has not yet been dated, but ages on the similar Luna 24 VLT's appear to confirm this prediction (Papanastassiou *et al.*, 1977; Schaeffer *et al.*, 1978).

Luna 24 VLT basalts.

Several recent abstracts have been published on the Luna 24 ferrobasalts in conjunction with the Luna 24 Conference in Houston, Texas. Most of the discussion reported here is from Vaniman and Papike (1977c).

Soils from the Soviet Luna 24 mission to Mare Crisium contain fragments of ferrobasalt and ferrogabbro. The ferrobasalt is very low in Ti content (~1.0 wt.% TiO_2), similar to Apollo 17 VLT. They are also characterized by Mg′ ~0.3–0.4, Al_2O_3 ~13%, CaO ~13% and Cr_2O_3 ~0.2–0.3%. Fragments of ferrogabbro are dominant within the soil, with lesser amounts of ferrobasalt (Barsukov *et al.*, 1977).

Luna 24 VLT ferrobasalts are ophitic, with pyroxene and olivine phenocrysts (up to 0.4 mm), small euhedral chrome spinels, and subhedral ulvöspinels up to 0.5 mm in diameter. Residual patches of ulvöspinel, ilmenite, silica, Fe-pyroxene and K-enriched glass (0.5% K_2O) occur in intergrain regions.

The first pyroxenes to crystallize range from Ca-rich to Ca-poor varieties. Zoning trends of Fe-enrichment pass through a low calcium point $Ca_{25}Mg_{25}Fe_{50}$, and then either rise in calcium toward hedenbergite or zone directly to a final Fe-pyroxene $Wo_{20}En_5Fs_{75}$. This residual pyroxene composition is identical to that which occurs in the Apollo 17 VLT basalt. However, the initial pyroxenes in Apollo 17 VLT are more magnesian, consistent with the more magnesian nature of those rocks. Olivine is more abundant in Luna 24 ferrobasalt than in Apollo 17 VLT and Luna 24 olivines are zoned to much more Fe-rich compositions ($Fo_{58}-Fo_5$) than the olivine of Apollo 17 VLT basalt ($Fo_{76}-Fo_{53}$). Feldspar in Luna 24 ferrobasalts is ophitically intergrown with both olivine and pyroxene. It is zoned from $An_{90}Ab_4$ to $An_{86}Ab_{14}$ and Or content is low, but increases from ~0.0% in calcic plagioclase to ~1.0% in the mare sodic varieties. Haggerty (1977) shows that the first opaque to crystallize in Luna 24 ferrobasalt is chromian spinel. With subsequent crystallization the spinel becomes enriched in Fe and Ti and ulvöspinel + ilmenite appear as residual phases. This spinel crystallization trend is approximately the same as that observed for Apollo 17 VLT (Vaniman and Papike, 1977b). Differences between these two basalts in Mg/Fe and Cr are reflected in minor differences between their spinel crystallization trends. The Luna 24 ferrobasalt spinels have initial compositions more Fe-rich than spinels from Apollo 17 VLT.

Based on the results of the Apollo 17 VLT and Luna 24 VLT investigations, Vaniman and Papike (1977c) speculated on the vertical plumbing beneath the Serenitatis and Crisium Basins. They assumed that green glass-type liquids rose from mantle depths under both basins. It is now clear based on the results of recent experimental investigations (Grove and Vaniman, 1978) that this model requires several different parental liquids related to, but not identical to, Apollo 15 emerald green glass. These liquids, which could have been derived at ~200–400 km (Taylor and Jakeš, 1974) moved to the surface under the Serenitatis and Crisium Basins at ~3.3 B.y. Green glass was extruded by fire fountaining (Chou *et al.*, 1975; Meyer *et al.*, 1975), while parental liquids for Apollo 17 VLT and Luna 24 VLT were trapped in holding chambers under the Serenitatis and Crisium Basins, respectively. At shallow levels fractionation of olivine/pyroxene and chrome spinel took place. Apollo 17 VLT liquids were tapped earlier in their fractionation path than the Luna 24 VLT's which are highly evolved.

Apollo 14 mare basalt fragments from breccia 14063.

A group of mare basalts found in Apollo 14 breccia 14063 (Ridley, 1975) are aluminous, but have higher TiO_2, Na_2O, Mg/Fe and lower FeO than other members of this group. The basalts have a uniform texture defined by lathy to acicular plagioclase, lathy and anhedral ilmenite, intergranular pyroxene, and rare pale-brown mesostasis glass (Ridley, 1975). The texture suggests rapid cooling from a melt in which plagioclase and ilmenite nucleated prior to pyroxene. As mentioned above these basalts are particularly interesting because they have TiO_2 contents of ~7% which is right in the middle of the gap between the low and high-Ti basalts.

Apollo 15 green glass.

Ridley *et al.* (1973) discuss the chemistry of the Apollo 15 green glasses. They point out that green glass spheres are a distinctive component in several Apollo 15 soils. They analyzed 28 spheres or fragments from spheres in the <1 mm soil fraction and found each grain devoid of inclusions and homogeneous. Chou *et al.* (1975), by using leaching and etching experiments, demonstrated that Zn, Ge, Cd, In and Au are concentrated in surficial deposits on the Apollo 15 green-glass spherules. They speculate that the surface deposits appear to be condensates from the magmatic gas associated with the eruptive process by which the spheres were extruded from the lunar interior. Meyer *et al.* (1975) obtained ion microprobe mass analyses of the surface film on the green glass spheres and showed that the film is composed of Zn, Ga, Pb, Cu, Tl, S, F, Cl and other elements. They also conclude that these elements condensed as a sublimate on the outside surface of these glasses in lava fountains.

Stolper (1974) and Wood and Ryder (1977) point out that in fact there is a range of green glass compositions with a systematic trend. Stolper *et al.* (1974) described a trend of olivine control that would require an olivine composition of Fo_{70}, a composition which is considerably at odds with the low-pressure liquidus olivine ($\sim Fo_{82}$). However, Grove and Lindsley (1978), based on reversed high-pressure experiments and a re-evaluation of the compositional variations within the Apollo 15 green glass data, conclude that the chemical variation cannot be explained simply by olivine of Fo_{70} composition. They observe that the chemical variations involve Al_2O_3, CaO, MgO and FeO and follow a trend of silica depletion or enrichment. They suggest that variable degrees of batch partial melting of a two-pyroxene + olivine ± garnet source at depths >400 km may give rise to the compositional variations.

Apollo 17 orange glass.

A deposit of orange glass was discovered by the Apollo 17 astronauts at Shorty Crater. Initial speculation was that this might be a young fumerole deposit enriched in volatiles and perhaps even water. Detailed studies of this glass require a different scenario. The soil consists almost entirely of glass spheres (0.1–0.2 mm) which are either orange in color or black because of devitrification resulting in crystallization of Fe-Ti oxides. The high TiO_2 content (~9 wt.%) imparts the orange color. The high-Ti content, close association with the high-Ti basalts at the Apollo 17 site, and old age led many investigators to consider the orange glass and Apollo 17 basalts as a closely related group. There is some uncertainty about the true age of Apollo 17 orange glass and at least three ages are reported: 3.54 B.y. (Huneke *et al.*, 1973), 3.71 B.y. (Husain and Schaeffer, 1973), and 3.66 B.y. (Turner, 1977). In any case the age is apparently somewhere in the interval between the 3.58 B.y. age of Apollo 11 high-K basalts and 3.78 age of Apollo 11 low-K and Apollo 17 mare basalts (see Fig. 2). Meyer *et al.* (1975) conducted an ion probe study of the surface coated volatiles and as discussed above for the green glass concluded that the glasses must have formed during a fire-fountaining event. Hubbard and Rhodes (1976) questioned the classification of Apollo 17 orange glass with Apollo 17 mare basalts. By examining the major and trace-element characteristics of orange glass and mare basalts, they conclude that orange glass is most closely associated with the Apollo 12 ilmenite basalts. The chemical variation diagrams presented in Figs. 4–9 show that orange glass has a higher Mg′ (~0.54) than low- and high-Ti mare basalts (excluding the rocks thought to be cumulates) but lower than Apollo 15 emerald green glass

(Mg′ ~0.58). When compared to the mare basalt groups, the orange glass is seen to have TiO_2 ~9 wt.%, right in the middle of the "TiO_2 gap" between high- and low-Ti basalts. It is also lower in Al_2O_3 than most mare basalts, but with Na_2O and K_2O comparable to high-Ti basalts, and CaO and "Cr_2O_3" contents falling on a trend common to all of the mare basalts. Green *et al.* (1975) conducted high-pressure experiments on orange glass compositions. They conclude that the liquids could have been derived by partial melting from olivine + orthopyroxene at ≥400 km depths.

MINERALOGY

The crystal chemistry and mineralogy of the mare basalt phases pyroxene, feldspar, olivine, spinel, ilmenite and armalcolite have been discussed extensively by Papike *et al.* (1976) and many high quality microprobe analyses are tabulated in that paper. Thus, our discussion presented here will be brief and will concentrate on major-element variations.

Pyroxene.

The Apollo 17 VLT pyroxene trends illustrated (Fig. 13) correspond to a "slower" cooling trend in the phaneritic fragments and a "faster" cooling trend in

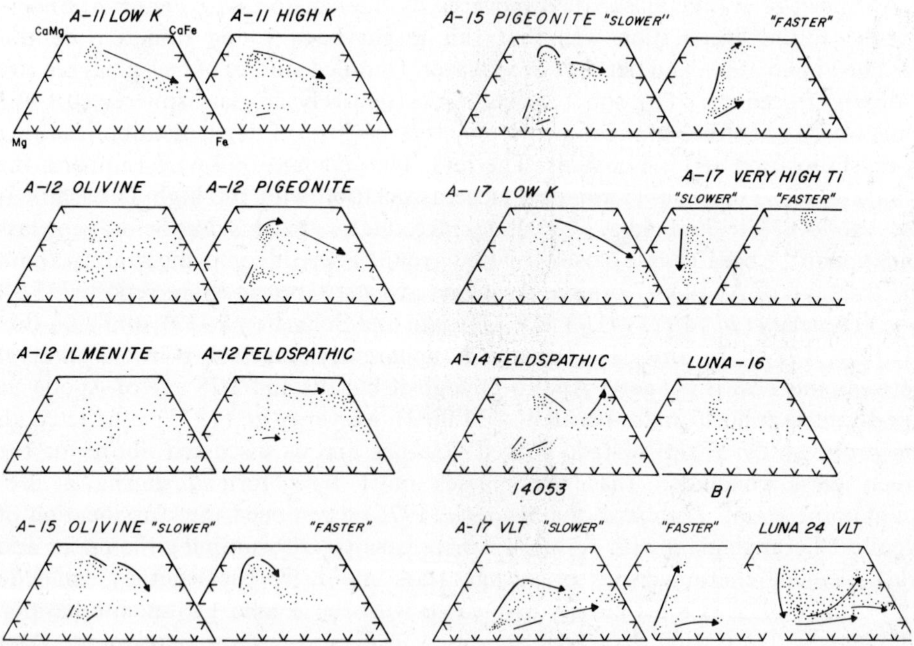

Fig. 13. Ca-Fe-Mg variation in pyroxenes of mare basalts. Diagrams are schematic, with arrows indicating certain crystallization trends.

the A-17 VLT vitrophyres. The data and discussion are from Vaniman and Papike (1977b). The first pyroxene to crystallize in both the phaneritic and vitrophytic A-17 VLT's is a pigeonite $Ca_6Mg_{68}Fe_{26}$. Pyroxenes in the vitrophyres ("faster") develop strong sectoral differences in Ca abundances, but these sectoral differences are suppressed in phaneritic rocks ("slower"). The Ca-enrichment trend in phaneritic VLT shows an inflection when plagioclase begins to crystallize and remove Ca from the melt. In contrast, the Ca-enrichment trend is not interrupted in VLT vitrophyres, where plagioclase nucleation is suppressed. The first pyroxenes formed in Luna 24 ferrobasalts range from a dominant Ca-rich pyroxene at $Ca_{37}Mg_{41}Fe_{22}$ to a Ca-poor pyroxene at $Ca_{12}Mg_{52}Fe_{36}$. Zoning trends pass through a low-Ca point $Ca_{25}Mg_{25}Fe_{50}$ and then either rise toward hedenbergite or zone directly across the "forbidden" region to $\sim Ca_{20}Mg_5Fe_{75}$. The Luna 24 VLT pyroxene grains start crystallizing at more Fe-rich compositions than the A-17 VLT, a reflection of the highly fractionated nature of the Luna 24 ferrobasalt.

Olivine.

The modal abundance of mare basalt olivine ranges from 0 to 36% (Papike *et al.*, 1976) and thus plays a variable role in these rocks. However, when it is present it is quite commonly an early-crystallizing phase that is a chief contributor to the fractionation process. The Fe-Mg variations of olivines from the major basalt groups are illustrated in Fig. 14. The most magnesian olivines reported are Fo_{75} to Fo_{80} and are commonly zoned to Fo_{30}. Relatively few analyses have been reported in the range $Fo_{30}-Fo_{10}$ until recently, when it was found that Luna 24 basalts show a range of olivine compositions from Fo_{60} to Fo_0. The Fe-rich early olivine in Luna 24 ferrobasalts is consistent with their highly evolved nature. For mare basalts in general the very Fe-rich olivines ($Fo_{10}-Fo_0$) are part of the stable three-phase assemblage (hedenbergite-fayalite-silica) that crystallizes from relatively slowly cooled late-stage iron-enriched portions of the basalt melts. Mare basalts that cooled rapidly during the late stages of crystallization contain Fe-rich pyroxene (in some cases zoning to pyroxferroite) which crystallizes metastably relative to the stable three-phase assemblage. The most magnesian olivines in the mare basalt suite ($Fo_{75}-Fo_{80}$) should be in equilibrium with a melt with $Mg' \sim 0.47$–0.57 (K_D estimates range from 0.26 for high-Ti basalts to 0.33 for low-Ti basalts; Longhi *et al.*, 1975). This is consistent with our estimates for the most primitive melts for the high- and low-Ti mare basalt suites but is too low for some of the VLT vitrophyres ($Mg' \sim 0.6$), where an olivine of $\sim Fo_{83}$ would be expected, but the olivine grains are too small to probe. The plotted olivine Fo contents in Fig. 14 are from the phaneritic lithic fragments that have lower Mg' values.

Feldspar.

The Na-Ca (Ab-An) variations of feldspar grains from the major mare basalt

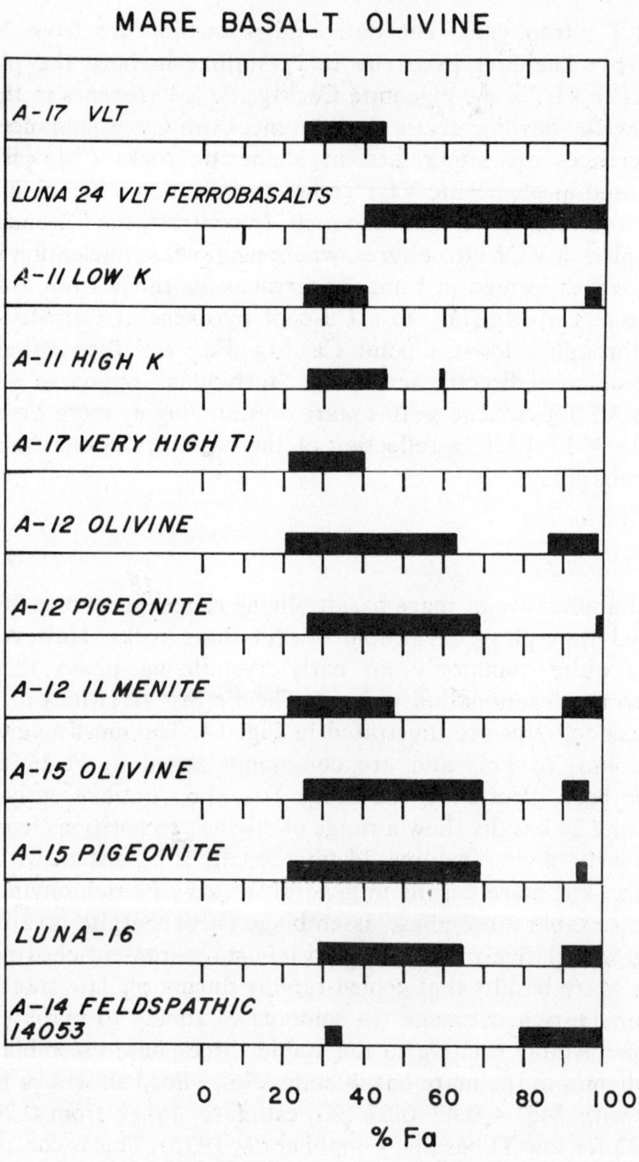

Fig. 14. Fayalite content of mare basalt olivines.

groups are illustrated in Fig. 15. A striking feature of the feldspar Ab-An variations ($An_{73}-An_{98}$) is the limited solid solution toward the albite ($NaAlSi_3O_8$) end-member compared to terrestrial basalts (e.g., $An_{40}-An_{80}$, mid-ocean ridge basalts "MORB"). Low Ab and Or ($KAlSi_3O_8$) contents of the mare feldspar grains are a logical consequence of the low alkali contents of these basalts (e.g., Na_2O for mare basalts ~0.21–0.55 wt.%; Na_2O for MORB ~1.6–

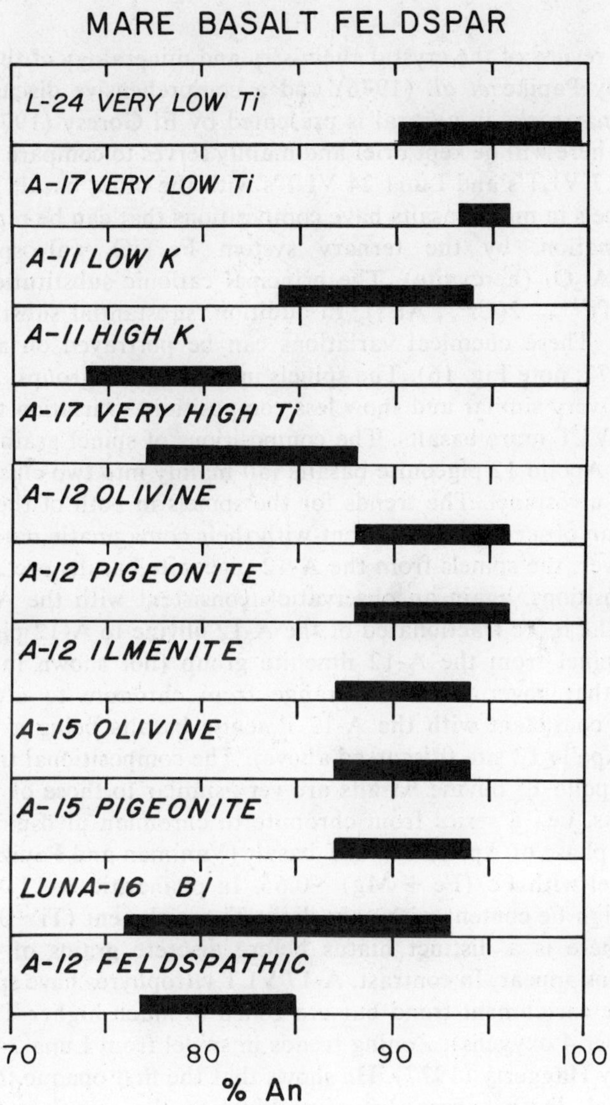

Fig. 15. Anorthite content of mare basalt feldspars.

2.7 wt.%). Despite the low albite content of mare feldspar, there is some correlation of the Ab content of the feldspar and the Na_2O content of the mare basalt (note Fig. 6). For example, the high-Ti basalts and Luna 16 basalts which have relatively high Na_2O contents have the most Ab-enriched feldspar; the low-Ti basalts which have intermediate Na_2O contents have feldspars with intermediate Ab contents, and the VLT basalts (especially A-17 VLT) which have the lowest Na_2O contents have the least Ab-enriched feldspars.

Spinel.

A complete review of the crystal chemistry and mineralogy of the Fe-Ti oxides is presented by Papike *et al.* (1976) and a comprehensive discussion of oxide minerals in lunar rocks in general is presented by El Goresy (1976). Therefore, our discussion here will be kept brief and mainly serves to compare the new spinel data from A-17 VLT's and Luna 24 VLT's with the other basalt groups.

Typical spinels in mare basalts have compositions that can be represented, as a first approximation, by the ternary system Fe_2TiO_4 (ulvöspinel)-$FeCr_2O_4$ (chromite)-$FeAl_2O_4$ (hercynite). The principal cationic substitutions are of the type, $Fe^{2+} + Ti^{4+} \rightleftarrows 2(Cr^{3+}, Al^{3+})$. In addition, substantial substitution of Mg for Fe occurs. These chemical variations can be portrayed on a spinel prism (Haggerty, 1972: note Fig. 16). The spinels in the high-Ti groups, Apollo 11 and Apollo 17, are very similar and show less compositional variation than spinels in the low-Ti or VLT mare basalts. The compositions of spinel grains from Apollo 12 olivine and Apollo 12 pigeonite basalts fall mainly into two clusters, chromite and chromian ulvöspinel. The trends for the spinels in both of these groups are quite similar, an observation consistent with their comagmatic nature (discussed above). However, the spinels from the A-12 pigeonite basalts are zoned to more Fe-rich compositions, again an observation consistent with the A-12 pigeonite basalts being the more fractionated of the A-12 olivine to A-12 pigeonite series. In contrast, spinel from the A-12 ilmenite group (not shown in Fig. 16) has compositions that cover the entire range from chromite to ulvöspinel. This observation is consistent with the A-12 ilmenite basalts being a discrete mare group at the Apollo 12 site (discussed above). The compositional trends of spinel grains from Apollo 15 olivine basalts are very similar to those of the Apollo 12 ilmenite basalts, i.e., a series from chromite to chromian ulvöspinel. The near-liquidus oxide phase of Apollo 17 VLT basalt (Vaniman and Papike, 1977b) is a chromian spinel with $Fe/(Fe + Mg) \sim 0.65$. In phaneritic A-17 VLT's, spinels are zoned to high Fe content with only slight Ti enrichment (Ti ~0.03 atoms per 4 oxygens); there is a distinct hiatus before discrete grains of late anhedral ulvöspinel grains appear. In contrast, A-17 VLT vitrophyres have spinels with the same initial Fe-enrichment trend but are zoned to much higher Ti contents (Ti ~0.45 atoms per 4 oxygens). Zoning trends in spinel from Luna 24 VLT basalts are reported by Haggerty (1977). He shows that the first opaque to crystallize is chromian spinel. With subsequent crystallization the spinel becomes Fe, Ti-enriched and zones to ulvöspinel. The spinel zoning trend is strikingly similar to that of the A-17 VLT's. Differences between these two basalts in bulk Mg/Fe and Cr content are reflected in minor differences between their spinel crystallization trends. The Luna 24 spinels have initial compositions more Mg-poor than A-17 VLT spinels ($Mg/Mg + Fe = 0.33$ vs. 0.40) and late-stage Ti-enriched spinels have lower Cr/Al (<3) than spinel grains in A-17 VLT (Cr/Al >3).

PETROGENESIS

Experimental studies on mare basalts were reviewed by Kesson and Lindsley

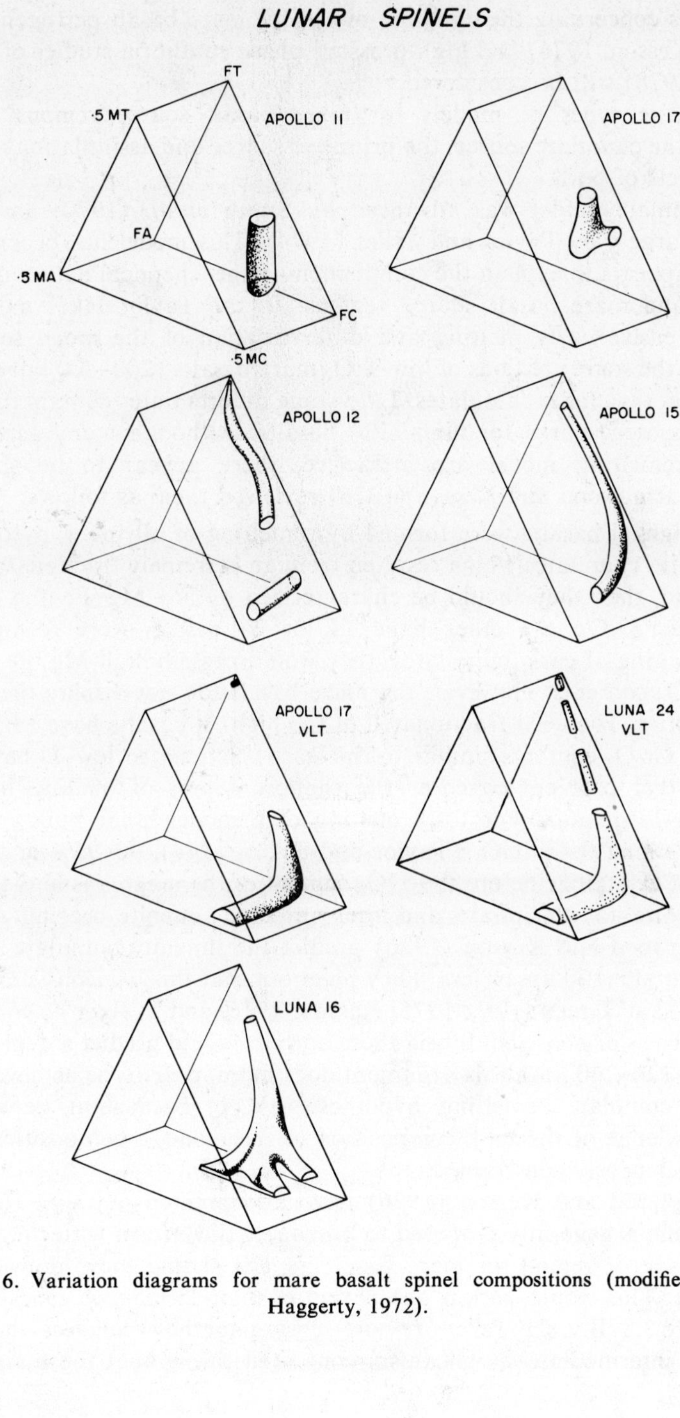

Fig. 16. Variation diagrams for mare basalt spinel compositions (modified after Haggerty, 1972).

(1976) and a brief review of mare basalt petrogenesis was presented by Papike *et al.* (1976). Aspects of those discussions will be included here. In addition, post-1976 results concerning the dynamic model for mare basalt petrogenesis (Ringwood and Kesson, 1976) and high-pressure phase equilibria studies of Grove and Lindsley (1978) will be considered.

Three basic types of models for mare basalt source regions have been presented: the cumulate source, the primitive source and assimilation models that utilize aspects of both.

The cumulate model was advanced by Smith *et al.* (1970) and has been developed largely by Taylor and Jakeš (1974). This model has become popular because it appears to explain the complementary geochemical nature of the lunar highlands and mare basalt source regions. In the Taylor-Jakeš model, which invokes extensive early melting and differentiation of the moon to depths of ~1000 km, the source regions of low-TiO_2 mare basalts (200–400 km depths) are composed of ultramafic cumulates. Late-stage differentiates contain ilmenite and are the proposed source for high-TiO_2 basalts. Although many aspects of the cumulate remelting model are attractive, there appear to be some major difficulties. Ringwood and Kesson (1976) reviewed them as follows:

(1) If high-Ti basalts were formed by remelting of olivine + pyroxene + ilmenite cumulates which resulted from an extremely fractionated residual liquid, then they should be characterized by low Mg/Fe and low Cr_2O_3 contents. On the other hand, if low-Ti basalts were formed by the remelting of early cumulates, they would possess high Mg/Fe ratios and Cr_2O_3 contents. However, the mare basalts do not display these characteristics. The least fractionated of the high-Ti basalts have Mg/Fe ratios and Cr_2O_3 contents similar to the least fractionated low-Ti basalts.

(2) Another problem concerns the shallow depths of melting implied for high-Ti magma. For total melting of a model lunar bulk composition containing 0.4% TiO_2 (Taylor and Jakeš, 1974), 90–95% of the system must crystallize before the TiO_2 content of the magma is high enough for ilmenite to precipitate and thus produce ilmenite-bearing cumulates. Ringwood and Kesson (1976) predict the ilmenite cumulate horizon to occur at ~100 km or less. They point out that this is inconsistent with the results of Green *et al.*, 1975, Kesson, 1975, and Walker *et al.*, 1975. On the basis of phase equilibria these authors would predict a depth of origin of ~120–240 km. This argument does not appear to be as devastating to the cumulate remelting hypothesis as (1) because it depends on a knowledge of the bulk composition of the moon, a composition which is rather poorly constrained.

(3) Ringwood and Kesson (1976) have also pointed out that the thermal regime is generally proposed to propagate downward with time, and thus the source region for mare basalt genesis should move downward with time. This would permit one period of high-Ti magma genesis at about ~3.6–3.8 B.y. However, remote-sensing methods suggest that high-Ti and intermediate-Ti volcanism continued throughout the main phase of

low-Ti basalt activity (Boyce *et al.*, 1974). This is another severe problem with the cumulate remelting model.

The second model, the primitive source model, proposes that the mare source regions were composed of primitive material of lunar bulk composition at the time of magma genesis (Ringwood, 1970; Ringwood and Essene, 1970; Green *et al.*, 1971). The range of basalt chemistries is attributed to different degrees of partial melting of this homogeneous source. Some problems confronting this model are:

(1) It is unable to explain the geochemical characteristics of high- and low-Ti basalts, e.g., their REE patterns, Eu anomalies, and TiO_2 contents (Ringwood and Kesson, 1976).
(2) It cannot account for the early differentiation event which is recorded in mare basalt Sm-Nd (Lugmair *et al.*, 1975) and U-Pb isotopic systematics (Tera and Wasserburg, 1975).

Ringwood and Kesson (1976) state that both of the above models are inadequate but some aspect of each must be saved. They consider that the Sm-Nd and U-Pb requirements for an early differentiation at ~4.4 B.y. are compelling, and thus they agree that cumulate formation must have occurred. They have tried to reconcile this differentiation with the evidence for an undifferentiated (chondritic) mare component. This reconcilation leads to the third class of models involving hybridization or assimilation.

Ringwood and Kesson (1976) point out that two groups, Anderson (1971) and Hubbard and Minear (1975), were on the right track, but did not come up with adequate models. Hubbard and Minear (1975) suggested that magmas from the primitive interior selectively assimilated parts of the 4.6–4.4 B.y. late-stage residuum, thereby giving rise to mare basalt hybrid liquids. Ringwood and Kesson (1976) point out that such a process should produce basalt liquids with extremely high Fe/Mg ratios; this is not observed.

Based on the above observations, Ringwood and Kesson (1976) present their latest version of a "dynamic assimilation" model. They propose that only the outer few hundred kilometers of the moon were involved in the 4.6–4.4 B.y. lunar differentiation event. The Fe, Ti and incompatible element-rich late-stage liquids thus formed were segregated in "pods" (of dimensions 5–20 km) beneath the crust and above early cumulates and refractory residua. This gravitationally unstable situation resulted in the sinking of ilmenite-pyroxenite pods through the underlying cumulate and residua sequence. Meanwhile temperatures in the primordial interior were beginning to rise due to radiogenic heating, and the sinking pods started to melt. The Fe and Ti-rich liquids thus produced reacted and reequilibrated with a large surrounding giving rise to high-Ti mare basalts with high Mg/Fe and Cr_2O_3 as well as high TiO_2 contents, and still having the isotopic imprint of the ilmenite pyroxenites. Ringwood and Kesson (1976) further speculate that by ~3.3 B.y. continued radioactive heating resulted in partial melting of a zone near the base of the differentiated sequence. Smaller and less efficiently differentiated ilmenite-pyroxenite pods sank much more slowly through the differentiated sequence. The infall of these relatively cool pods into the partially molten zone allegedly produced low-Ti hybrid magmas. These low-Ti hybrid liquids would also carry the imprint of the 4.6–4.4 B.y. differentiation and the ilmenite pyroxenite cumulates. Ringwood and Kesson (1976) suggest that high-Al_2O_3 mare basalts were produced at the crust-mantle interface, when parental magmas from the deep interior assimilated limited amounts of plagioclase-rich crustal material.

Although many aspects of the recent Ringwood/Kesson model may be inaccurate, these authors have cast serious doubt on whether *any* of the basalt liquids sampled on the lunar surface are capable of giving us direct constraints on the source region mineralogy using the high-pressure phase equilibrium approach. This approach requires direct partial melts uncontaminated by additions or subtractions enroute to the lunar surface after the last equilibration.

Based on the above discussions, the newly discovered Apollo 17 VLT and Luna 24 VLT take on additional importance. Grove and Vaniman (1978) present arguments suggesting that VLT's can be related to members of the green glass family by fractionation. Green glass itself could be the only relatively pristine primitive liquid derived either from melting deep cumulates or the primitive undifferentiated moon. Thus, the recent experiments of Grove and Lindsley (1978) take on special importance. Based on new high-pressure experiments on green glass compositions and a re-evaluation of the chemical variations, they conclude that the green glass family originated by variable degrees of batch partial melting of a two-pyroxene + olivine ± garnet source region at depths in excess of 400 km. Thus, if the cumulate zone is relatively shallow (i.e., <400 km deep), the green glasses carry clues about the primitive lunar material. If, on the other hand, the cumulate zones extend to much greater depths, green glass carries clues as to the nature of these deep cumulate horizons. In any case, the green-glass liquids thus derived may be capable, through fractionation both at depth and near surface, of generating the newly discovered VLT suite.

CONCLUDING STATEMENT

Major advances in our understanding of mare basalt petrogenesis are still being made. However, our efforts have been and will be compromised by the fact that we have been working with a very limited number of samples. This situation must be rectified by (1) continuing to quantify and correlate remote sensing results; (2) continuing to look for and find small mare basalt fragments in regolith samples and breccias that are not represented in the large rock sample suite; (3) continuing to carry out unmanned sampling of strategic locations on the lunar surface (e.g., far lunar west side and backside).

Acknowledgments—This research was supported by NASA Grant #NSG-9044, for which we are extremely grateful.

REFERENCES

Adams J. B. and Ralph R. L. (1977) Spectral reflectance of Luna 24 soils (abstract). In *Papers Presented to the Conference on Luna 24*, p. 1–4. The Lunar Science Institute, Houston.

Anderson A. T. Jr. (1971) Exotic armalcolite and the origin of Apollo 11 ilmenite basalts. *Geochim. Cosmochim. Acta* **35**, 969–973.

Andre C. G., Wolfe R. W. and Adler I. (1977) Evidence for a high magnesium subsurface basalt in Mare Crisium from orbital x-ray fluorescence data (abstract). In *Papers Presented to the Conference on Luna 24*, p. 8–11. The Lunar Science Institute, Houston.

Aronld J. R., Metzger A. E. and Reedy R. C. (1977) Computer-generated maps of lunar composition from gamma-ray data. *Proc. Lunar Sci. Conf. 8th*, p. 945–948.

Barsukov V. L., Tarasov L. S., Dmitriev L. V., Kolesov G. M., Shevaleevsky I. D. and Garanin A. V. (1977) The geochemical and petrochemical features of regolith and rocks from Mare Crisium (preliminary data). *Proc. Lunar Sci. Conf. 8th*, p. 3319–3332.

Bielefeld M. J., Andre C. G., Eliason E. M., Clark P. E., Adler I and Trombka J. I. (1977) Imaging of lunar surface chemistry from orbital x-ray data. *Proc. Lunar Sci. Conf. 8th*, p. 901–908.

Boyce J. M. (1976) Ages of flow units in the lunar near-side maria based on Lunar Orbiter IV photographs. *Proc. Lunar Sci. Conf. 7th*, p. 2717–2728.

Boyce J. M. and Dial A. L. (1973) Relative ages of some near-side mare units based on Apollo 17 metric photographs. *Apollo 17 Prelim. Sci. Rep.* NASA SP-330, p. 29-26 to 29-28.

Boyce J. M., Dial A. L. and Soderblom L. A. (1974) Ages of the lunar near-side eight plains and maria. *Proc. Lunar Sci. Conf. 5th*, p. 11–23.

Bryan W. B. and Adams M. L. (1973) Some volcanic and structural features of Mare Serenitatis. *Apollo 17 Prelim. Sci. Rep.* NASA SP-330, p. 30-9 to 30-12.

Chou C. L., Boynton W. V., Sundberg L. L. and Wasson J. T. (1975) Volatiles on the surface of Apollo 15 green glass and trace element distributions among Apollo 15 soils. *Proc. Lunar Sci. Conf. 6th*, p. 1701–1727.

Donaldson C. H., Usselman T. M., Williams R. J. and Lofgren G. E. (1975) Experimental modeling of the cooling history of Apollo 12 olivine basalts. *Proc. Lunar Sci. Conf. 6th*, p. 843-869.

Dungan M. A. and Brown R. W. (1977) The petrology of the Apollo 12 ilmenite basalt suite. *Proc. Lunar Sci. Conf. 8th*, p. 1339-1381.

El Goresy A. (1976) Oxide minerals in lunar rocks. *Mineral. Soc. Amer. Short Course Notes*, p. EG-1 to EG-46.

Green D. H., Ringwood A. E., Hibberson W. O. and Ware N. G. (1975) Experimental petrology of Apollo 17 mare basalts. *Proc. Lunar Sci. Conf. 6th*, p. 871–893.

Green D. H., Ware N. G., Hibberson W. O. and Major A. (1971) Experimental petrology of Apollo 12 basalts, I. Sample 12009. *Earth Planet. Sci. Lett.* **13**, 85–96.

Green J. A., Kurat G., Keil K. and Prinz M. (1972) Electron microprobe analyses of lithic fragments, glasses, chondrites and minerals from Luna 16 fines. *U.N.M. Institute of Meteorites, Spec. Publ. 5*, p. 1–16.

Grove T. L. and Lindsley D. H. (1978) Compositional variation and origin of lunar ultramafic green glasses (abstract). In *Lunar and Planetary Science IX*, p. 430–432. Lunar and Planetary Institute, Houston.

Grove T. L. and Vaniman D. T. (1978) Experimental petrology of very low Ti basalts and origin of Luna 24 ferrobasalt (abstract). In *Papers Presented to the Conference on Luna 24*, p. 68–71. The Lunar Science Institute, Houston.

Haggerty S. E. (1972) Luna 16: An opaque mineral study and a systematic examination of compositional variations of spinels from Mare Fecunditatis. *Earth Planet. Sci. Lett.* **13**, 328–352.

Haggerty S. E. (1977) Luna 24: Opaque mineral chemistry of gabbroic and basaltic fragments from Mare Crisium. *Geophys. Res. Lett.* **4**, 489–492.

Head J. W. (1976) Lunar volcanism in space and time. *Rev. Geophys. Space Phys.* **14**, 265–300.

Head J. W., Zisk S., Adams J., McCord T. and Pieters C. (1977) Geologic setting and topography of the Crisium Basin and the Luna 24 landing region (abstract). In *Papers Presented to the Conference on Luna 24*, p. 78–81. The Lunar Science Institute, Houston.

Hubbard N. J. and Minear J. W. (1975) A chemical and physical model for the genesis of lunar rocks: Part II. Mare basalts (abstract). In *Lunar Science VI*, p. 405–407. The Lunar Science Institute, Houston.

Hubbard N. J. and Rhodes J. M. (1976) The chemical classification of Apollo 17 orange glass (abstract). In *Lunar Science VII*, p. 390–392. The Lunar Science Institute, Houston.

Huneke J. C., Jessberger E. K., Podosek F. A. and Wasserburg G. J. (1973) $^{40}Ar/^{39}Ar$ measurements in Apollo 16 and 17 samples and the chronology of metamorphic and volcanic activity in the Taurus-Littrow region. *Proc. Lunar Sci. Conf. 4th*, p. 1725–1756.

Husain L. and Schaeffer O. A. (1973) Lunar volcanism: Age of the glass in the Apollo 17 orange soil. *Science* **180**, 1358–1360.

Johnson T. V., Mosher J. A. and Matson D. L. (1977a) Lunar spectral units: A northern hemispheric mosaic. *Proc. Lunar Sci. Conf. 8th*, p. 1013-1028.

Johnson T. V., Saunders R. S., Matson D. L. and Mosher J. A. (1977b) A TiO_2 abundance map for the northern maria. *Proc. Lunar Sci. Conf. 8th*, p. 1029–1036.

Kesson S. E. (1975) Mare basalts: Melting experiments and petrogenetic interpretations. *Proc. Lunar Sci. Conf. 6th*, p. 921–944.

Kesson S. E. and Lindsley D. H. (1976) Mare basalt petrogenesis—A review of experimental studies. *Rev. Geophys. Space Phys.* **14**, 361–374.

Lindsley D. H. and Munoz J. L. (1969) Subsolidus relations along the join hedenbergite-ferrosilite. *Amer. J. Sci.* **267A**, 295–324.

Longhi J., Walker D. and Hays J. F. (1975) Fe-Mg distribution between olivine and lunar basaltic liquids (abstract). *EOS (Trans. Amer. Geophys. Union)* **56:6**, 471.

Lugmair G. W., Scheinin N. B. and Marti K. (1975) Sm-Nd age of Apollo 17 basalt 75075: Two-state igneous processes in mare basalt genesis. In *Lunar Science VI*, p. 807–809. The Lunar Science Institute, Houston.

Metzger A. E., Lin R. P. and Russell C. T. (1977) On a correlation between surface remanent magnetism and chemistry for the lunar frontside and limbs. *Proc. Lunar Sci. Conf. 8th*, p. 1187–1190.

Meyer C. Jr., McKay D. S., Anderson D. H. and Butler P. (1975) The source of sublimates on the Apollo 15 green and Apollo 17 orange glass samples. *Proc. Lunar Sci. Conf. 6th*, p. 1673–1699.

Naney M. T., Papike J. J. and Vaniman D. T. (1977) Lunar highland melt rocks: Major element chemistry and comparisons with lunar highland glass groups (abstract). In *Lunar Science VIII*, p. 717–719. The Lunar Science Institute, Houston.

Newton R. C., Anderson A. T. and Smith J. V. (1971) Accumulation of olivine in rock 12040 and other basaltic fragments in the light of analysis and synthesis. *Proc. Lunar Sci. Conf. 2nd*, p. 575–582.

Papanastassiou D. A., McCulloch M. T., DePaolo D. J., Huneke J. C. and Wasserburg G. J. (1977) Ages and evolution of Luna 24 samples (abstract). Unpublished abstract.

Papike J. J., Hodges F. N., Bence A. E., Cameron M. and Rhodes J. M. (1976) Mare basalts: Crystal chemistry, mineralogy and petrology. *Rev. Geophys. Space Phys.* **14**, 475–540.

Rhodes J. M., Blanchard D. P., Dungan M. A., Brannon J. C. and Rodgers K. V. (1977) Chemistry of Apollo 12 mare basalts: Magma types and fractionation. *Proc. Lunar Sci. Conf. 8th*, p. 1305–1338.

Rhodes J. M., Hubbard N. J., Wiesmann H., Rodgers K. V., Brannon F. C. and Bansal B. M. (1976) Chemistry, classification and petrogenesis of Apollo 17 mare basalts. *Proc. Lunar Sci. Conf. 7th*, p. 1467–1489.

Rhodes J. M., Rodgers K. V., Shih C., Bansal B. M., Nyquist L. E., Wiesmann H. and Hubbard N. J. (1974) The relationships between geology and soil chemistry of the Apollo 17 landing site. *Proc. Lunar Sci. Conf. 5th*, p. 1097–1117.

Ridley W. I. (1975) On high-alumina mare basalts. *Proc. Lunar Sci. Conf. 6th*, p. 131–145.

Ridley W. I., Reid A. M., Warner J. L. and Brown R. W. (1973) Apollo 15 green glass. *Phys. Earth Planet. Inter.* **7**, 133–136.

Ringwood A. E. (1970) Petrogenesis of Apollo 11 basalts and implications for lunar origin. *J.*

Geophys. Res. **75**, 6453–6479.

Ringwood A. E. and Essene E. (1970) Petrogenesis of Apollo 11 basalts: Internal consitution of the moon. *Proc. Apollo 11 Lunar Sci. Conf.*, p. 769–799.

Ringwood A. E. and Kesson S. E. (1976) A dynamic model for mare basalt petrogenesis. *Proc. Lunar Sci. Conf. 7th*, p. 1697–1722.

Russell C. T., Weiss H., Coleman P. J., Soderblom L. A., Stuart-Alexander D. E. and Wilhelms D. E. (1977) Geologic-magnetic correlations on the moon: Apollo subsatellite results. *Proc. Lunar Sci. Conf. 8th*, p. 1171–1185.

Schaeffer O. A., Bence A. E., Eichorn G., Papike J. J. and Vaniman D. T. (1978) ^{39}Ar-^{40}Ar and petrologic study of Mare Crisium: Age and petrology of Luna samples 24077,13 and 24077,63 (abstract). In *Lunar and Planetary Science IX*, p. 1007–1009. Lunar and Planetary Institute, Houston.

Smith J. V., Anderson A. T., Newton R. C., Olson E. J., Wyllie P. J., Crewe A. V., Isaacson M. S. and Johnson D. (1970) Petrologic history of the moon inferred from petrography, mineralogy, and petrogenesis of Apollo 11 rocks. *Proc. Apollo 11 Lunar Sci. Conf.*, p. 897–925.

Soderblom L. A., Arnold J. A., Boyce J. M. and Lin R. P. (1977) Regional variations in the lunar maria: Age, remanent magnetism and chemistry. *Proc. Lunar Sci. Conf. 8th*, p. 1191–1199.

Stolper E. M. (1974) Lunar ultramafic glasses. B. A. Thesis, Harvard University, Cambridge, Massachusetts. 178 pp.

Stolper E. M., Walker D., Longhi J. and Hays J. F. (1974) Compositional variation in lunar ultramafic glasses (abstract). In *Lunar Science V*, p. 749–751. The Lunar Science Institute, Houston.

Strom R. G. (1972) Lunar mare ridges, rings, and volcanic ring complexes. In *The Moon* International Astronomical Union Symposium 47 (S. Runcorn and H. Urey, eds.), p. 187–215. Reidel, Dordrecht, Holland.

Taylor G. J., Keil K. and Warner R. D. (1977a) Very low-Ti basalt. *Geophys. Res. Lett.* **4**, 207–210.

Taylor G. J., Warner R., Wentworth S., Keil K. and Sayeed U. (1977b) Petrology and chemistry of Luna 24 mare basalts and basaltic glasses (abstract). In *Papers Presented to the Conference on Luna 24*, p. 189–192. The Lunar Science Institute, Houston.

Taylor S. R. and Jakeš P. (1974) The geochemical evolution of the moon. *Proc. Lunar Sci. Conf. 5th*, p. 1287–1305.

Tera F. and Wasserburg G. J. (1975) The evolution and history of mare basalts as inferred from U-Th-Pb systematics (abstract). In *Lunar Science VI*, p. 807–809. The Lunar Science Institute, Houston.

Turner G. (1977) Potassium-Argon chronology of the Moon. *Phys. Chem. Earth* **10**, 145–195.

Vaniman D. T. and Papike J. J. (1977a) The Apollo 17 drill core: Chemistry and stratigraphy of monomineralic fragments and the discovery of a new very low Ti (VLT) mare basalt (abstract). In *Lunar Science VIII*, p. 952–954. The Lunar Science Institute, Houston.

Vaniman D. T. and Papike J. J. (1977b) Very low Ti (VLT) basalts; A new mare rock type from the Apollo 17 drill core. *Proc. Lunar Sci. Conf. 8th*, p. 1443–1471.

Vaniman D. T. and Papike J. J. (1977c) Ferrobasalts from Mare Crisium: Luna 24. *Geophys. Res. Lett.* **4**, 497–500.

Vaniman D. T. and Papike J. J. (1977d) The Apollo 17 drill core: Modal petrology and glass chemistry (Sections 70007, 70008, 70009). *Proc. Lunar Sci. Conf. 8th*, p. 3161–3193.

Walker D., Kirkpatrick R. J., Longhi J. and Hays J. F. (1976) Crystallization history of lunar picritic basalt sample 12002: Phase equilibria and cooling-rate studies. *Bull. Geol. Soc. Amer.* **87**, 646–656.

Walker D., Longhi J., Stolper E. M., Grove T. L. and Hays J. F. (1975) Origin of titaniferous lunar basalts. *Geochim. Cosmochim. Acta* **39**, 1219–1235.

Whitaker A. E. (1972) Mare Imbrium lava flows and the relationships to color boundaries. *Apollo 15 Prelim. Sci. Rep.* NASA SP-289, p. 25-83 to 25-84.

Wood J. A. and Ryder G. (1977) The Apollo 15 green clods and the green glass enigma. In *Lunar Science VIII*, p. 1026–1028. The Lunar Science Institute, Houston.

Merrill, R. B. and Papike, J. J., editors.
(1978) *Mare Crisium: The View from Luna 24*, Pergamon Press (New York, p. 403–417
Printed in the United States of America

Mineralogy and petrology of basaltic fragments from the Luna 24 drill core

R. A. Coish and Lawrence A. Taylor

Department of Geological Sciences, The University of Tennessee, Knoxville, Tennessee 37916

Abstract—The Luna 24 drill core contains 60% rock and mineral fragments. Rock fragments include subophitic ferrobasalt, metabasalt, late-stage rocks, olivine vitrophyres and non-mare lithic material. Monomineralic grains, which are much more abundant than rock fragments, include pyroxene, plagioclase, olivine, ilmenite, chromite-ulvöspinel, troilite and native Fe.

Subophitic rock fragments are very low-Ti, low MgO, aluminous basalts (VLT). Early pyroxene in the subophitic basalts has a high Fe/Fe + Mg ratio with later-formed pyroxene exhibiting a trend towards ferrohedenbergite and pyroxferroite. Olivine ranges in composition from Fo_{59} to Fo_5. Metabasalts are recrystallized rocks with equilibrated mineral compositions. Late-stage fragments contain very Fe-rich olivine and pyroxene in association with cristobalite. Olivine vitrophyres consist of skeletal, feathery olivine phenocrysts in a glassy matrix, and are characterized by primitive bulk compositions (Fe/Fe + Mg = 0.42). A non-mare lithic fragment comprises high-Mg olivine (Fo_{75}) and pyroxene (En_{75}), and An-rich plagioclase. Some monomineralic olivine and pyroxene grains have higher MgO contents than those in the fine-grained ferrobasaltic fragments. These high-Mg minerals may be from a mare cumulate.

The fine-grained subophitic basalt is probably a derivative liquid from a VLT parent. The parent composition may be similar to some olivine vitrophyres. Crystal fractionation of olivine from this parent can give rise to a liquid with the composition of the subophitic basalt.

Introduction

In August, 1976, an unmanned Soviet sampling device recovered 2 meters of soil from the southern part of Mare Crisium. Samples from six levels (77, 109, 149, 174, 182, and 210 cm.) of the drill core, totalling approximately 3 gms, were transferred to the U.S.A. for subsequent distribution and study.

The mineral and rock fragments in 22 polished thin sections of grain mounts from the 90-250 μm size fraction from all six levels of the Luna 24 soil, and selected fragments from the >500 μm fraction have been examined petrographically and analyzed with the electron microprobe. We have identified five components among the rock fragments: 1) subophitic basalts; 2) metabasalts; 3) late-stage fragments; 4) olivine vitrophyres; and 5) non-mare lithic fragments of possible cumulate origin. The mineral fragments consist of pyroxene, plagioclase, olivine, chromite-ulvöspinel, native Fe, troilite, and ilmenite.

In this paper, the petrology of the rock fragments and monomineralic grains is described and a petrogenetic scheme is presented for the derivation of Mare Crisium basalts.

ANALYTICAL METHODS

Mineral analyses were performed on an automated MAC 400 S electron microprobe, using appropriate synthetic and natural standards. Whole rock analyses of the fine-grained rock fragments were obtained by defocussed beam analysis (DBA) using the procedure of Albee *et al.* (1977).

ROCK FRAGMENTS

Subophitic basalts

Fine-grained subophitic basalt is the most abundant type of rock fragment in the Luna 24 drill core. This rock type also has been referred to as ferrobasalt, fine-grained basalt and VLT mare basalt (Ryder *et al.*, 1977; Taylor *et al.*, 1977a; Vaniman and Papike, 1977b). Mineralogically, this rock consists of olivine phenocrysts and pyroxene microphenocrysts in a subophitic matrix of plagioclase and pyroxene. Chromite-ulvöspinel, troilite and native Fe are common accessory opaques. The low abundance of ilmenite is notable. K-rich mesostasis was not observed in the Luna 24 subophitic basalts. An average mode is 49% pyroxene, 39% plagioclase, 10% olivine, and 2% opaques (Vaniman and Papike, 1977b).

Chemically, the fragments of subophitic basalt are very low titanium, low magnesium, aluminous basalts (Table 1; Fig. 1; Ryder *et al.*, 1977; Vaniman and Papike, 1977c) and, in these respects, are significantly different from most other

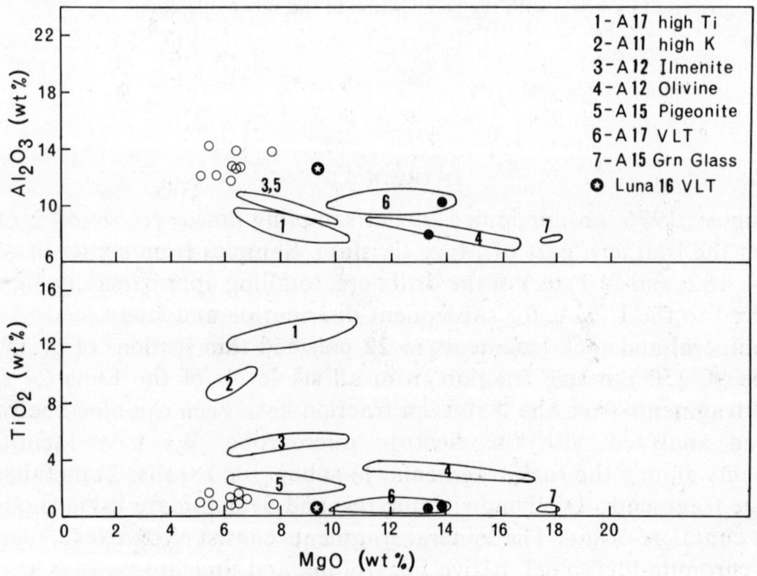

Fig. 1. TiO_2 and Al_2O_3 variations against MgO for subophitic basalt fragments (o) and olivine vitrophyres (●) in the Luna 24 soil compared to other mare basalts. Data from Ryder *et al.*, (1977), Taylor *et al.* (1977b) Kurat *et al.* (1976), and this study.

Table 1. Whole rock analyses of Luna 24 rock fragments.

	1	2	3	4	5	6	7	8
SiO_2	44.64	45.60	45.90	46.50	46.56	45.67	45.08	45.57
TiO_2	1.62	0.75	0.94	0.38	0.55	0.38	0.45	0.66
Al_2O_3	14.34	12.17	14.40	17.96	10.24	13.44	8.01	10.40
Cr_2O_3	0.31	0.12	0.31	0.14	0.35	0.14	0.54	0.64
FeO	20.18	20.77	16.92	13.09	18.46	18.40	19.00	17.64
MgO	5.40	5.59	5.65	4.66	9.04	5.97	13.36	13.83
CaO	11.27	12.04	12.99	15.36	11.08	13.21	9.65	9.70
Na_2O	0.41	0.36	0.43	0.41	0.14	0.31	0.07	0.24
Total	98.17	97.40	97.79	98.70	96.70	97.78	96.16	98.68
CIPW Normative Minerals								
Qtz	—	—	—	—	—	—	—	—
Ab	3.53	3.12	3.73	3.52	1.23	2.69	0.62	2.06
An	37.96	32.40	38.27	47.84	28.30	36.14	22.39	27.65
Di	16.24	24.75	23.61	25.04	24.09	26.44	22.72	17.43
Hy	34.17	33.99	31.08	22.42	43.60	28.09	36.36	31.77
Fo	1.19	1.04	0.31	0.08	0.44	1.65	7.66	9.51
Fa	3.32	3.04	0.71	0.17	0.71	4.04	8.54	9.35
Ilm	3.13	1.46	1.83	0.73	1.08	0.74	0.89	1.27
Chr	0.47	0.18	0.47	0.21	0.54	0.21	0.83	0.96
Fe/Fe + Mg	0.68	0.68	0.63	0.61	0.53	0.63	0.44	0.42

1—subophitic basalt, 24077,67; 2—subophitic basalt, 24174,74; 3—metabasalt, 24109,71; 4—metabasalt, 24210,44; 5—metabasalt, 24109,40; 6—metabasalt, 24109,40; 7—olivine vitrophyre, 24109,40; 8—olivine vitrophyre, 24109,41.

mare basalts. Their titanium contents are similar to Apollo 17 very low titanium (VLT) basalts (Vaniman and Papike, 1977a), but they have lower MgO and higher Al_2O_3 contents. Some Luna 16 VLT basalts (Kurat *et al.*, 1976) have similar Al_2O_3 and TiO_2 values but are richer in MgO. The low MgO (and consequently high Fe/Fe + Mg ratio) of the Luna 24 subophitic basalts suggests that they crystallized from a chemically evolved liquid.

The compositions of the ferromagnesian minerals in the subophitic basalt also reflect the evolved nature of the liquid. The earliest pyroxenes (most Mg-rich) have Fe/Fe + Mg ratios >.35 (Table 2, Fig. 2) whereas the most Mg-rich olivine is Fo_{59} (Table 3; Fig. 2). The pyroxenes trend towards ferrohedenbergite and pyroxferroite, and the olivines range to compositions as Fe-rich as Fo_5 (Fig. 2). Ti and Al in pyroxenes are low in comparison to other lunar pyroxenes (Ryder *et al.*, 1977). In addition, Ti and Ti/Al both increase and Cr_2O_3 decreases with increasing Fe/Fe + Mg ratio in the pyroxene (Fig. 3). The Ti/Al ratio increases from less than ¼ to ½. The minor element variations in the pyroxenes indicate that a major Ti-bearing phase did not crystallize early and that plagioclase co-crystallized with pyroxene at some stage in the cooling history.

Plagioclase compositions range from An_{96} to An_{83} with an average of around

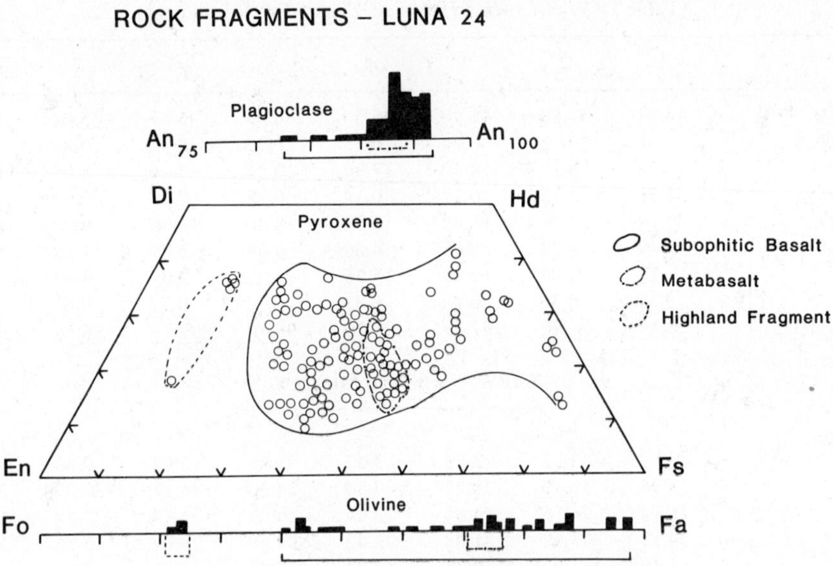

Fig. 2. Compositions of major minerals in rock fragments from the Luna 24 drill core. Note the wide range of mineral compositions in subophitic basalts but the restricted range for metabasalts and highland rock fragment.

An_{93} (Fig. 2). Fe and Mg contents are normal for plagioclase from mare basalt (Table 3).

Spinels in Luna 24 subophitic basalts are members of the chromite-ulvöspinel-hercynite solid solution series (Fig. 4; Haggerty, 1977). Their textures are reminiscent of Apollo 12 and 15 spinels in as much as these are Ti-chromite rimmed by Cr-ulvöspinel. The spinel chemistry is somewhat unusual in having high Al_2O_3 (up to 14.2 wt.%) and low MgO (0.6-1.6 wt.%).

Metabasalts

A second major lithic type in the Luna 24 soil is metabasalt (Ryder *et al.*, 1977; McSween *et al.*, 1977). Metabasalt fragments are readily distinguishable from the subophitic basalts because of their fine-grained, equigranular texture. Many plagioclase and pyroxene grains exhibit irregular, embayed boundaries. Modal data for metabasalts show that they contain less olivine and silica, but more pyroxene than subophitic ferrobasalts (Vaniman and Papike, 1977b). The greater abundance of pyroxene in the metabasalts has been explained by the reaction of olivine with free silica to form pyroxene.

Pyroxene, olivine and plagioclase in metabasalts have very restricted compositions (Fig. 2; McSween *et al.*, 1977). Pyroxene falls in an intermediate compositional group in the pyroxene quadrilateral (between En_{40}-En_{30}); olivines are Fo_{20}-Fo_{30}; and plagioclase compositions are between An_{91} and An_{94}. The

Table 2. Selected analyses of pyroxene from subophitic basalts in the Luna 24 drill core.

	1	2	3	4	5	6	7	8	9	10	11	12	13	14	15
SiO_2	49.55	46.17	49.20	51.60	49.98	47.84	51.99	52.07	49.04	52.17	51.07	51.18	49.50	51.77	49.40
TiO_2	0.30	1.27	0.76	0.22	0.24	0.55	0.16	0.32	0.95	0.53	0.21	0.23	0.52	0.23	0.91
Al_2O_3	0.76	1.38	1.43	1.35	0.74	1.25	0.83	1.30	1.27	2.30	0.91	1.31	0.97	0.99	1.23
Cr_2O_3	0.29	0.18	0.32	0.69	0.34	0.39	0.38	0.54	0.20	0.92	0.32	0.60	0.26	0.48	0.30
FeO	30.28	29.02	24.92	17.02	26.42	29.73	20.87	21.65	30.33	23.73	23.64	15.61	31.70	22.05	26.31
MgO	12.43	4.46	8.74	17.16	15.97	10.41	19.47	14.54	7.48	11.14	17.58	16.14	11.05	18.21	10.48
CaO	5.86	16.31	14.79	12.05	5.46	7.67	6.63	10.21	10.98	12.46	5.25	14.34	6.66	6.87	11.62
Total	99.46	98.79	100.14	100.09	99.15	97.84	100.33	100.62	100.25	100.32	98.98	99.40	100.66	100.60	100.26
							Cations per 6 Oxygen								
Si	1.963	1.906	1.939	1.941	1.948	1.942	1.953	1.974	1.956	1.969	1.963	1.940	1.953	1.951	1.938
Ti	.008	.039	.021	.005	.007	.017	.004	.009	.028	.014	.005	.006	.015	.005	.026
Al	.035	.066	.065	.060	.033	.060	.036	.058	.060	.102	.041	.058	.045	.043	.056
Cr	.008	.004	.009	.020	.010	.012	.011	.015	.005	.026	.009	.018	.007	.014	.009
Fe	1.003	1.002	.819	.535	.861	1.009	.654	.687	1.011	.454	.759	.495	1.045	.694	.864
Mg	.733	.273	.513	.961	.927	.629	1.090	.821	.443	.780	1.007	.911	.649	1.023	.612
Ca	.248	.721	.623	.485	.228	.333	.267	.414	.469	.602	.215	.582	.281	.276	.488

Depth (cm) and section number of samples: 1—24174,74; 2—24174,74; 3—24174,74; 4—24077,67; 5—24077,67; 6—24077,67; 7—24077,68; 8—24077,68; 9—24077,68; 10—24077,45; 11—24077,45; 12—24109,40; 13—24109,40; 14—24149,48; 15—24210,44.

Table 3. Selected analyses of olivine and plagioclase from Luna 24 subophitic basalt fragments.

	Olivine						Plagioclase						
	1	2	3	4	5	6	7	8	9	10	11	12	13
SiO_2	35.41	35.72	31.66	34.31	35.44	30.52	44.69	44.53	44.33	44.29	45.73	43.80	44.03
Al_2O_3	—	—	—	—	—	—	34.91	34.04	35.20	36.07	32.61	35.17	34.57
FeO	40.80	37.37	58.97	42.75	37.97	60.77	0.46	0.43	0.51	0.60	0.85	0.70	1.35
MgO	23.83	27.09	9.25	22.71	26.13	7.70	0.29	0.25	0.29	0.09	0.24	0.14	0.29
CaO	0.55	0.58	—	—	—	—	19.68	19.64	18.39	18.65	18.60	19.20	19.09
Na_2O	—	—	—	—	—	—	0.50	0.45	0.63	0.68	1.04	0.55	0.61
K_2O	—	—	—	—	—	—	0.01	0.00	0.01	0.04	0.01	0.05	0.01
Total	100.59	100.76	99.88	99.77	99.54	98.99	100.54	99.34	99.36	100.42	99.08	99.62	99.95
	Cations per 4 Oxgyen						Cations per 8 Oxygen						
Si	.999	.987	1.000	.992	1.001	.991	2.060	2.076	2.061	2.038	2.137	2.040	2.050
Al	—	—	—	—	—	—	1.896	1.871	1.928	1.957	1.796	1.940	1.897
Fe	.963	.865	1.561	1.034	.896	1.642	.018	.016	.019	.023	.033	.033	.052
Mg	1.002	1.118	.436	.979	1.100	.373	.019	.016	.019	.005	.016	.005	.019
Ca	.016	.017	—	—	—	—	.971	.981	.915	.920	.931	.949	.953
Na	—	—	—	—	—	—	.044	.040	.056	.059	.093	.081	.054
K	—	—	—	—	—	—	.000	.000	.000	.001	.000	.001	.000

1—24109,40; 2—24109,40; 3—24077,68; 4—24077,68; 5—24077,68; 6—24174,74; 7—24077,45; 8—24077,45; 9—24109,40; 10—24109,40; 11—24149,48; 12—24210,44; 13—24182,47.

equilibrated nature of the mineral compositions supports the contention that reheating and recrystallization has occurred in the metabasalts (Ryder *et al.*, 1977; McSween *et al.*, 1977).

Are metabasalts simply recrystallized ferrobasalts or are they derived from a different rock type? Vaniman and Papike (1977b) favour the former interpreta-

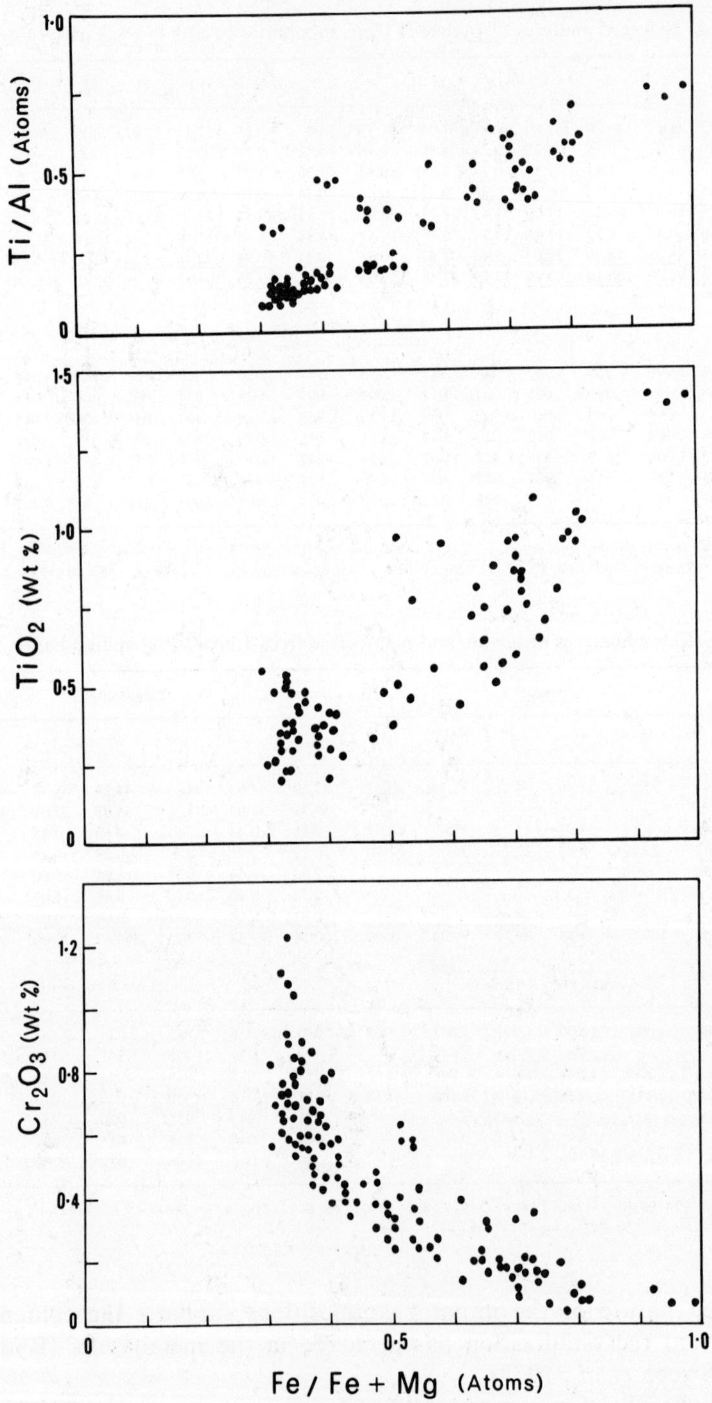

Fig. 3. Variation of Fe/Fe + Mg ratio against Ti/Al, TiO_2, and Cr_2O_3 in pyroxenes from the Luna 24 drill core.

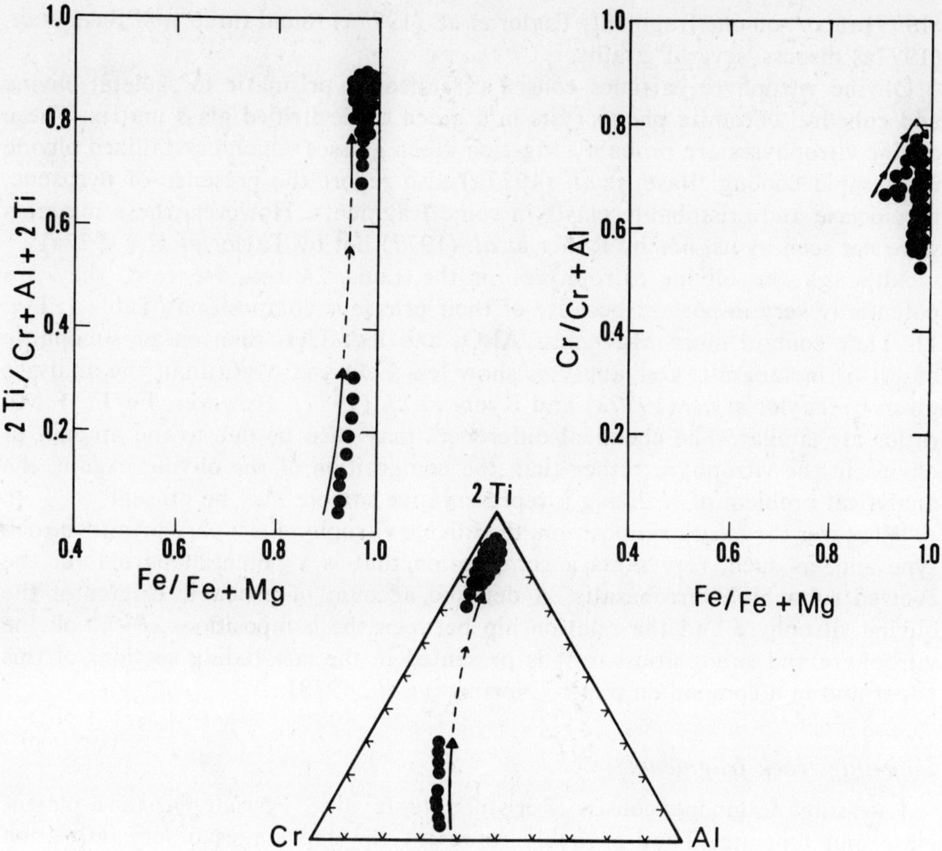

Fig. 4. Ti, Cr and Al variations, and Fe/Fe + Mg versus 2Ti/Cr + Al + 2Ti, Cr/Cr + Al in spinel from Luna 24 subophitic basalts.

tion because the bulk rock compositions (Vaniman, oral comm., Luna 24 Conf.) of the subophitic ferrobasalts and metabasalts are identical. On the other hand, McSween *et al.* (1977) show that some whole rock compositions of metabasalts have significantly more CaO and Al_2O_3 than subophitic basalts. This was interpreted as an original difference in composition between the ferrobasalts and parents of the metabasalts. The metabasalt parent was believed to have been contaminated by highland material.

It is possible that the discrepancy in interpretations is due to analytical problems. With such small rock fragments, it is extremely difficult to know if the specimen is representative of the rock type.

Olivine Vitrophyre

Olivine vitrophyre fragments make up an exceedingly small portion of the Luna 24 soil. We recognized only two fragments in 22 thin sections; Ryder *et al.*

(1977) report on one fragment; Taylor *et al.* (1977a) found three, and Basu *et al.* (1977a) discuss 'several' grains.

Olivine vitrophyre particles consist of euhedral, prismatic to skeletal olivine and euhedral chromite phenocrysts in a green or devitrified glass matrix. These olivine vitrophyres are probably Mg-rich green glasses which crystallized olivine upon rapid cooling. Basu *et al.* (1977a) also report the presence of pyroxene, plagioclase and cristobalite clasts in some fragments. However, these minerals were not seen by us, nor by Ryder *et al.* (1977) not by Taylor *et al.* (1977a).

Although the olivine vitrophyres in the Luna 24 soil are rare, they are potentially very important because of their primitive composition (Table 1; Fig. 1). They contain more MgO, less Al_2O_3 and less TiO_2 than either subophitic basalt or metabasalt. Our analyses show less FeO and MgO than the analyses given by Taylor *et al.* (1977a) and Ryder *et al.* (1977). However, Fe/Fe + Mg ratios are similar. The chemical differences may then be due to the amount of olivine in the vitrophyre rather than the composition of the olivine. Again, the analytical problem of realizing a representative sample may be present.

Whatever the exact composition, the olivine vitrophyre is a very primitive rock type and, as such, represents a composition that is a potential parent for the evolved subophitic ferrobasalts. A detailed account of probable origins of the olivine vitrophyre and the relationship between the compositions of the olivine vitrophyre and subophitic basalt is presented in the concluding sections of this paper and in a companion paper (Norman *et al.*, 1978).

Late-stage rock fragments

Late-stage fragments consist of cristobalite, fayalite, Fe-rich pyroxene, plagioclase and ilmenite. They probably represent the end stages of crystallization and/or fractionation of a basalt magma. In the 150-250 μm grain size fraction of the Luna 24 soil, all five minerals are rarely found in a single fragment. Usually, cristobalite (the distinguishing mineral of this rock type) occurs with only one or two other minerals. The silica mineral is present as either sub-rounded, fractured blotches or as euhedral crystals. Grain sizes range up to 300 μm.

The iron-rich chemistry of olivine ($Fo_{5\text{-}15}$) and pyroxene ($Wo_{20\text{-}26}Fs_{45\text{-}55}En_{35\text{-}19}$) clearly indicates the late stage nature of these fragments (Table 4).

Iron-rich minerals such as occur in these late-stage rocks have usually been found as fine-grained mesostasis in lunar basalts. In those cases, the development of these phases is attributed to *in-situ* crystallization of residual liquids in a basalt magma. The Luna 24 late-stage fragments are coarse grained relative to previously described mesostasis areas. Perhaps, then, the late stage rocks are a separate rock type formed by extreme amounts of fractionation at Mare Crisium.

Non-mare lithic fragments

Highlands rock fragments are exceedingly rare in the Luna 24 drill core (Bence and Grove, 1977; Coish and Taylor, 1977). We recognized only one such

Table 4. Pyroxene, olivine and plagioclase analyses from late stage fragments in the Luna 24 drill core.

	Pyroxene				Olivine					Plagioclase				
	1	2	3	4	5	6	7	8	9	10	11	12	13	14
SiO_2	49.34	48.91	47.91	49.17	30.89	31.08	31.16	29.25	29.54	43.37	44.47	44.27	44.45	43.80
TiO_2	0.58	0.75	1.07	0.70	0.11	0.07	0.11	0.19	0.16	—	—	—	—	—
Al_2O_3	1.48	1.46	1.17	1.31	—	—	—	—	—	35.29	35.21	34.75	34.76	34.60
Cr_2O_3	0.46	0.41	0.18	0.46	0.14	0.06	0.06	0.10	0.07	—	—	—	—	—
FeO	26.48	26.48	31.79	29.15	62.63	63.43	58.77	68.78	67.55	0.64	0.63	0.43	0.41	0.41
MgO	11.49	10.05	11.71	9.05	5.85	4.90	10.45	1.79	1.94	0.19	0.20	0.16	0.20	0.17
CaO	8.93	10.52	11.71	9.05	0.59	0.69	0.46	0.86	0.68	19.44	19.31	19.06	19.02	19.14
Na_2O	—	—	—	—	—	—	—	—	—	0.43	0.43	0.55	0.84	0.50
K_2O	—	—	—	—	—	—	—	—	—	0.01	0.02	0.00	0.01	0.00
Total	98.77	98.57	99.92	99.88	100.21	100.25	101.01	100.97	99.94	99.38	100.28	99.23	99.68	98.61
	Cations per 6 oxygens				Cations per 4 oxygens					Cations per 8 oxygens				
Si	1.953	1.950	1.943	1.951	0.995	1.005	0.980	0.973	0.987	2.026	2.053	2.064	2.064	2.056
Ti	.016	.022	.032	.020	.001	.001	.001	.004	.004	—	—	—	—	—
Al	.068	.067	.055	.060	.003	.004	.003	.004	.002	1.942	1.917	1.908	1.902	1.914
Cr	.014	.013	.004	.014	.003	.001	.001	—	.002	—	—	—	—	—
Fe	.876	.882	1.078	.966	1.689	1.715	1.521	1.915	1.890	.024	.023	.016	.015	.015
Mg	.677	.596	.367	.593	.281	.235	.481	.089	.096	.012	.013	.011	.012	.011
Ca	.378	.449	.508	.384	.020	.023	.014	.030	.023	.972	.956	.952	.946	.962
Na	—	—	—	—	—	—	—	—	—	.038	.038	.049	.075	.045
K	—	—	—	—	—	—	—	—	—	.000	.001	.000	.000	000

1—6, 10, 11 from sample 24182,47; 7,12 from 24174,52; 8,9,13,14 from 24174,51.

fragment. It consists of sub-rounded olivine (50%) and pyroxene grains in interstitial plagioclase. Native iron is also present. The unusual proportion of olivine and the truncation of plagioclase grains imply that this fragment is not a representative sample of the rock from which it was derived. Therefore, no bulk chemical analysis was attempted.

The chemistry of the ferromagnesian minerals in this rock fragment is significantly different from that of minerals in other Luna 24 rocks (Fig. 2; Table 5). The olivine grains have *uniform* compositions near Fo_{76}, both pyroxene types are Mg-rich, and the plagioclase is An-rich with very low Fe and Mg contents. This rock fragment is considered to be of highlands origin because of the above mineral-chemical characteristics. It is also tentatively suggested that the rock is a cumulate (albeit a fine-grained cumulate) because of the texture of sub-rounded olivine grains poikilitically enclosed by plagioclase, and also because olivine compositions do not vary from grain to grain.

Monomineralic Fragments

Mineral fragments account for up to 50% of the Luna 24 drill core. In order of abundance, pyroxene, plagioclase, olivine, ilmenite, chromite-ulvöspinel and native iron occur. The predominance of mineral grains greater than 150 μm suggests that the major rock type at the Luna 24 site was a relatively coarse-grained basalt.

Table 5. Analyses of pyroxene, olivine and plagioclase in highland rock fragment (cumulate?) from 24077,45.

	Pyroxene					Olivine					Plagioclase			
SiO_2	52.94	53.27	52.27	53.09	53.18	39.40	38.65	37.09	37.39	37.87	44.73	45.23	44.99	44.51
TiO_2	0.87	0.99	0.88	0.85	1.35	0.11	0.06	—	—	—	—	0.05	0.03	0.00
Al_2O_3	1.64	1.85	1.75	1.86	2.16	0.11	0.16	—	—	—	34.73	34.88	34.28	34.06
Cr_2O_3	0.40	0.69	1.19	0.72	0.61	0.07	0.02	—	—	—	—	—	—	—
FeO	8.57	7.66	7.97	7.68	7.39	22.15	23.07	23.46	23.03	21.99	0.03	0.00	0.16	0.14
MgO	25.41	17.59	19.14	17.94	17.38	38.65	38.69	39.69	39.59	40.10	0.48	0.13	0.11	0.16
CaO	11.05	18.20	16.48	17.32	18.54	0.18	0.18	—	—	—	19.08	18.95	18.54	18.65
Na_2O	—	—	—	—	—	—	—	—	—	—	0.48	0.51	0.70	0.65
Total	100.88	100.24	99.68	99.47	100.61	100.66	100.83	100.24	100.01	99.96	99.53	99.23	98.12	97.57
	Cations per 6 oxygens					Cations per 4 oxygens					Cations per 8 oxygens			
Si	1.896	1.945	1.921	1.949	1.936	1.013	.998	.970	.977	.984	2.080	2.093	2.108	2.100
Ti	.023	.026	.023	.022	.036	.002	.001	—	—	—	—	—	—	—
Al	.068	.079	.075	.080	.092	.003	.004	—	—	—	1.873	1.903	1.892	1.893
Cr	.015	.019	.034	.020	.017	.001	.001	—	—	—	—	—	—	—
Fe	.256	.233	.245	.235	.224	.476	.498	.512	.503	.477	.003	.000	.005	.005
Mg	1.355	.957	1.048	.981	.942	1.481	1.489	1.547	1.542	1.554	.009	.006	.006	.007
Ca	.423	.711	.648	.681	.723	.004	.004	—	—	—	.904	.940	.929	.944
Na	—	—	—	—	—	—	—	—	—	—	.104	.048	.065	.060
Fe/Fe + Mg	0.16	0.20	0.19	0.19	0.19	0.24	0.25	0.25	0.25	0.23	0.03	—	0.45	0.38

Monomineralic olivine and pyroxene fragments have a much larger compositional range than minerals in subophitic basalts (Fig. 5, *cf.* Fig. 2). Monomineralic fragments include very Mg-rich olivine and pyroxene grains that do not occur in the fine-grained subophitic basalts (Fig. 5; Basu *et al.*, 1977b). The occurrence of these Mg-rich phases suggests that there is a coarse-grained rock type at the Luna 24 site which is more primitive than the subophitic basalts. This high-Mg, coarse-grained rock probably accounts for the fact that the MgO content of the soil (10 wt.%, Blanchard *et al.*, 1977) is higher than that of the subophitic basalts (6 wt.%). Compositions of monomineralic pyroxene grains range to ferrohedenbergite. Minor element trends are the same for pyroxene in monomineralic and rock fragments. Compositions of monomineralic plagioclase and those in subophitic basalts are similar.

DISCUSSION

Investigations of the Luna 24 drill core have revealed much new information on the nature and diversity of very low-Ti mare basalts. At Mare Crisium, there is a range of VLT basalt compositions with the subophitic basalt fragments probably having more evolved chemistries than the rock types which were comminuted to give the monomineralic fragments. In fact, three mare rock types have been identified in the Luna 24 drill core: (1) fine-grained subophitic basalt (also referred to as ferrobasalt with MgO = 6.0 wt.%); (2) coarse-grained ferrogabbro (MgO = 10 wt.%, Taylor *et al.*, 1977a) represented in monomineralic fragments, and (3) coarse-grained rock with MgO >10 wt.% evidenced by

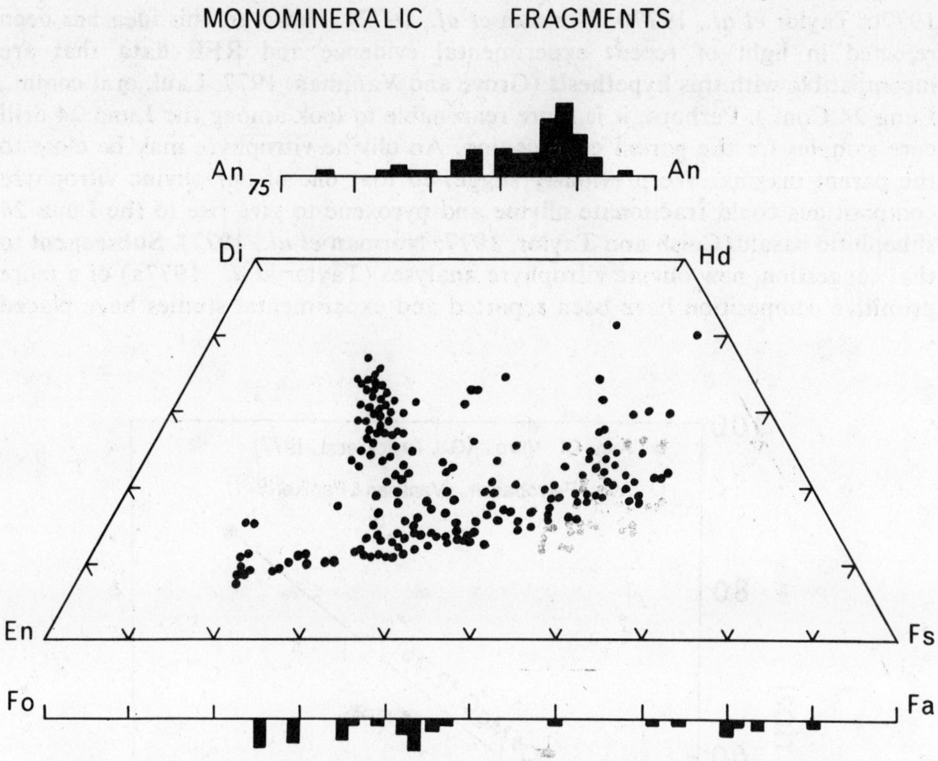

Fig. 5. Compositions of monomineralic plagioclase, pyroxene and olivine in the Luna 24 drill core. Note the occurrence of Mg-rich pyroxenes and olivines which are not present in the subophitic ferrobasalts (Fig. 2).

the presence of *very-high* Mg olivines and pyroxenes. Possible origins of these mare rock types are discussed in the following section.

Origin of Luna 24 fine-grained subophitic basalts

Subophitic basalt can be either direct partial melts of a lunar mantle source or derivatives of a more primitive magma. Based on our terrestrial bias for primary magmas, the high Fe/Fe + Mg ratios and the multiply-saturated nature (Grove and Vaniman, 1977) of the Luna 24 subophitic basalts would seem to rule out these rocks being direct melts of an ultramafic mantle. It is more likely that they were derived from a more primitive magma by fractional crystallization. The important question then is: what is the parent magma of these subophitic basalts?

The parent magma will be richer in MgO (low Fe/Fe + Mg) and impoverished in TiO_2 compared to the subophitic basalt. A previous candidate for the parent magma was an Apollo 15 green glass composition (Vaniman and Papike,

1977c; Taylor *et al.*, 1977b; Norman *et al.*, 1977). However, this idea has been rejected in light of recent experimental evidence and REE data that are incompatible with this hypothesis (Grove and Vaniman, 1977; Laul, oral comm., Luna 24 Conf.). Perhaps, it is more reasonable to look among the Luna 24 drill core samples for the parent composition. An olivine vitrophyre may be close to the parent magma. We previously suggested that one of our olivine vitrophyre compositions could fractionate olivine and pyroxene to give rise to the Luna 24 subophitic basalt (Coish and Taylor, 1977; Norman *et al.*, 1977). Subsequent to that suggestion, new olivine vitrophyre analyses (Taylor *et al.*, 1977a) of a more primitive composition have been reported and experimental studies have placed

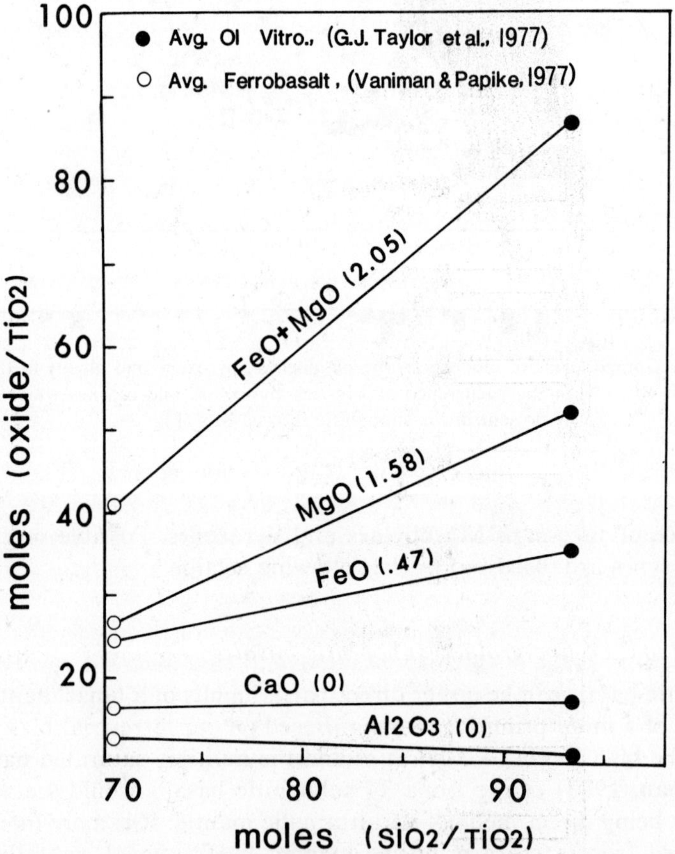

Fig. 6. Moles SiO_2/TiO_2 versus moles FeO + MgO/TiO_2, FeO/TiO_2, MgO/TiO_2, CaO/TiO_2, and Al_2O_3/TiO_2 for average olivine vitrophyre and fine-grained subophitic ferrobasalt. Method of plotting is after Pearce (1968). Numbers in parentheses are slopes of the respective lines. The slope (2.05) of the FeO + MgO/TiO_2 line is consistent with olivine fractionation from the olivine vitrophyre to derive the subophitic basalt.

constraints on the minerals which can fractionate to give Luna 24 ferrobasalts (Grove and Vaniman, 1977).

Taking these new developments into account, it can be shown that simple olivine fractionation of an average olivine vitrophyre composition will yield a Luna 24 ferrobasalt (Fig. 6). Figure 6 is plotted by the method of Pearce (1968, 1970). By this method, all oxides (in moles) are divided by moles of an oxide which is not removed by the fractionation process. Therefore, that oxide changes in weight percent only because of removal of other material. By ratioing other oxides to this "constant" oxide, true (absolute) variations are seen among the various elements. TiO_2 was chosen as the constant oxide in the Luna 24 magma because it is unlikely that Ti was removed by fractionation. Fe-Ti oxides are not early phases to separate in low-Ti basalts and pyroxene fractionation is not compatible with experimental studies (Grove and Vaniman, 1977). The slope of $FeO + MgO/TiO_2:SiO_2/TiO_2$ line between olivine vitrophyre and ferrobasalt is approximately 2—this variation can be explained by fractionation of olivine alone (Fe + Mg:Si in olivine is 2). If pyroxene were involved as well, this slope would be less than 2 since Mg + Fe:Si in pyroxene is <1. The near horizontal lines for CaO and Al_2O_3 indicate no plagioclase fractionation. Since olivine alone can account for the variation, then the slope of the MgO and FeO lines should give an average olivine composition, which in this case is Fo_{76}. This value lies between the most Mg-rich olivine in the olivine vitrophyre (Fo_{83}) and that in the ferrobasalt (Fo_{60-64}). The total amount of fractionation is greater than 30% as ascertained from the change in TiO_2 content between olivine vitrophyre (0.64 wt.%) and ferrobasalt (0.86 wt.%).

It is cautioned that these two end-member compositions alone are not ideal for application of this method; it is much more revealing to have a series of intermediate compositions. Towards this end, we attempted to integrate some of the glass bead chemistries into the fractionation scheme. The brown glass compositions cluster around the ferrobasalt average and thus are not intermediate fractionation products. Some green glass analyses plot between the ferrobasalt and olivine vitrophyre. However, there is a large amount of scatter in these data.

Because of some variations in analyses, we can conclude only that a composition *close to* that of the olivine vitrophyres could have been the parent magma for the subophitic (ferro) basalts. It is likely that such primitive volcanic liquids did exist at Mare Crisium since high-Mg olivines (Fo_{75-80}, Basu *et al.*, 1977b) occur as monomineralic fragments. These olivines must have precipitated from a very Mg-rich volcanic liquid.

Significance of monomineralic compositions

The monomineralic fragments may represent two rock types, both of which are different in composition from the fine-grained subophitic basalt fragments. Taylor *et al.* (1977a) estimated the composition of a coarse-grained rock type at the Luna 24 site. They obtained an average mode of the coarse mineral fragments (using modes of both microgabbro 24170 and monomineralic fragments) and combined this mode with average mineral compositions to arrive at a bulk composition. Olivine with Fo>70 and pyroxene with En>70 were excluded from the calculations. The resulting rock composition has an MgO content of around 10 wt.%, which corresponds to the high-Mg basaltic glass cluster of Simonds *et al.* (1977). The second rock type contributing to the monomineralic fragments may consist principally of high-Mg olivine (Fo>70) with minor high-Mg pyroxene and perhaps even some plagioclase. This rock would have a MgO content much greater than 10 wt.%.

The origin of both these postulated rock types still lies in the realm of speculation. The ferrogabbro (MgO = 10%) may simply be a less differentiated equivalent of the subophitic basalt, i.e. it may lie along the fractionation path between the olivine vitrophyre and ferrobasalt. The very high-Mg olivines may be from a cumulate which formed as the parent magma-olivine vitrophyre-fractionated. The high-Mg pyroxenes may be intercumulus material of this cumulate.

CONCLUSIONS

1. The Luna 24 ferrobasalt is a derivative magma from a composition similar to that of the olivine vitrophyre. Approximately thirty percent olivine fractionation from an olivine vitrophyre magma can give rise to a ferrobasalt composition.
2. The high-Mg monomineralic olivine fragments not represented in the fine-grained basalt fragments may be from a cumulate. The accumulation may have taken place in small, shallow-level magma chambers.
3. Metabasalts are recrystallized basalt flows which may have been contaminated by plagioclase grains.
4. The highlands component in the Luna 24 drill core is represented by a small number of lithic fragments with very Mg-rich olivine and pyroxene, and low-Fe plagioclase.

Acknowledgments—We wish to thank Drs. H. Y. McSween, Jr., R. P. Gamble, T. L. Grove, D. T. Vaniman and G. J. Taylor, and Mr. Marc Norman for helpful discussions. Mr. J. C. Clark provided assistance with the microprobe analyses. The final version of the manuscript was greatly improved by the suggestions of the reviewers, Drs. R. F. Dymek and J. L. Warner. We are very grateful to the U.S.S.R. for kindly donating the samples from Mare Crisium. We also thank the Univeristy of Tennessee for substantial support for the purchase of the electron microprobe. This research was supported by NASA Grant NGL 43-001-140.

REFERENCES

Albee A. L., Quick J. E. and Chodos A. A. (1977) Source and magnitude of errors in "broad-beam analysis" (DBA) with the electron probe (abstract). In *Lunar Science VIII*, p. 7–9. The Lunar Science Institute, Houston.

Basu A., McKay D. S., Clanton U., Waits G. and Fruland R. M. (1977a) Glass particles in Luna 24 drill core soils. In *Papers Presented to the Conference on Luna 24*, p. 14–17. The Lunar Science Institute, Houston.

Basu A., McKay D. S. and Fruland R. M. (1977b) Petrography, minerology and source rocks of Luna 24 drill core soils. In *Papers Presented to the Conference on Luna 24*, p. 18–21. The Lunar Science Institute, Houston.

Bence A. E. and Grove T. L. (1977) The highland component in the Luna 24 core. In *Papers Presented to the Conference on Luna 24*, p. 22–24. The Lunar Science Institute, Houston.

Blanchard D. P., Haskin L. A., Brannon J. C. and Aaboe E. (1977) Chemistry of soils and particles from Luna 24. In *Papers Presented to the Conference on Luna 24*, p. 37–40. The Lunar Science Institute, Houston.

Coish R. A. and Taylor L. A. (1977) Igneous rocks from Mare Crisium: mineralogy and petrology. In *Papers Presented to the Conference of Luna 24*, p. 48–50. The Lunar Science Institute, Houston.

Coish R. A., Hu H.-N. and Taylor L. A. (1978) Luna 24 core samples from Mare Crisium: Mineralogy and petrology (abstract). *EOS (Trans. Amer. Geophys. Union)* **59**, 223.

Grove T. L. and Vaniman D. T. (1977) Experimental petrology of very low Ti basalts and origin of Luna 24 ferrobasalt. In *Papers Presented to the Conference on Luna 24*, p. 68–71. The Lunar Science Institute, Houston.

Haggerty S. E. (1977) Luna 24: systematics in spinel mineral chemistry. In *Papers Presented to the Conference on Luna 24*, p. 72–74. The Lunar Science Institute, Houston.

Kurat G., Kracher A., Keil K., Warner R. and Prinz M. (1976) Composition and origin of Luna 16 aluminous mare basalts. *Proc. Lunar Sci. Conf. 7th*, p. 1301–1321.

McSween H. Y., Jr., Taylor L. A. and Clark J. C. (1977) Luna 24 metabasalts: possible evidence for assimilation of highlands material? In *Papers Presented to the Conference on Luna 24*, p. 118–120. The Lunar Science Institute, Houston.

Norman M., Coish R. A. and Taylor L. A. (1977) Glass chemistry and magma evolution at Mare Crisium. In *Papers Presented to the Conference on Luna 24*, p. 136–138. The Lunar Science Institute, Houston.

Norman M., Coish R. A. and Taylor L. A. (1978) Glasses in the Luna 24 core and petrogenesis of ferrobasalts. In *Mare Crisium: The View from Luna 24* (R. B. Merrill and J. J. Papike, eds.) This volume.

Pearce T. H. (1968) A contribution to the theory of variation diagrams. *Contrib. Mineral. Petrol.* **19**, 142–157.

Pearce T. H. (1970) Chemical variations in the Palisades Sill. *J. Petrol.* **11**, 15–32.

Ryder G., McSween H. Y., Jr. and Marvin U. B. (1977) Basalts from Mare Crisium. *The Moon* **17**, 263-287.

Simonds C. H., Warner J. L., McGee P. E., Cochran A. and Brown R. W. (1977) Luna 24: petrochemical potpourri. In *Papers Presented to the Conference on Luna 24*, p. 170–173. The Lunar Science Institute, Houston.

Taylor G. J., Warner R., Wentworth S., Keil K. and Sayeed U. (1977a) Petrology and chemistry of Luna 24 mare basalts and basaltic glasses. In *Papers Presented to the Conference on Luna 24*, p. 189–192. The Lunar Science Institute, Houston.

Taylor G. J., Keil K. and Warner R. D. (1977b) Very low-Ti mare basalts. *Geophys. Res. Lett.* **4**, 207–210.

Vaniman D. T. and Papike J. J. (1977a) Very low Ti (VLT) basalts: a new mare rock type from Apollo 17 drill core. *Proc. Lunar Sci. Conf. 8th*, p. 1443–1471.

Vaniman D. T. and Papike J. J. (1977b) Luna 24 ferrobasalt: petrology and comparison with other mare basalt types. In *Papers Presented to the Conference on Luna 24*, p. 196–199. The Lunar Science Institute, Houston.

Vaniman D. T. and Papike J. J. (1977c) Ferrobasalts from Mare Crisium: Luna 24. *Geophys. Res. Lett.* In press.

Merrill, R. B. and Papike, J. J., editors.
(1978) *Mare Crisium: The View from Luna 24*, Pergamon Press (New York), p. 419–428
Printed in the United States of America

The case for at least three mare basalt magmas at the Luna 24 landing site

ROGER L. NIELSEN* and MICHAEL J. DRAKE**

Lunar and Planetary Laboratory, University of Arizona, Tucson, Arizona 85721

* Also Department of Geosciences

** Also Department of Planetary Sciences

Abstract—We report a petrographic and electron microprobe study of 1230 particles contained on six polished thin sections of samples from the Luna 24 drill-core. The particle census is consistent with the core remaining relatively undisturbed, except for gardening, since emplacement. Igneous lithic fragments and monomineralic particles are overwhelmingly of mare origin. ANT samples are rare, and KREEP is represented by a single feldspar crystal. Major and minor element chemical compositions of pyroxenes are utilized to distinguish a minimum of three distinct mare basalt magmas. One of these magmas is represented primarily by Mg-rich monomineralic particles of olivine and pyroxene. The remaining two magma types, collectively called "Luna 24 VLT basalts" by other workers, are distinguished by the Ti/(Ti+Cr) ratio at constant Fe/(Fe+Mg) ratio in pyroxenes from polymineralic igneous lithic fragments. Evidence for these distinct magmas is also present in "whole rock" chemical analyses for V and Cr. Breccias contain mineral fragments from all three proposed mare basalt magmas, and minor quantities of ANT fragments. Breccias appear to be primarily locally derived, and their formation postdates mare basalt emplacement.

INTRODUCTION

In August, 1976, the Soviet automatic probe Luna 24 landed in southeastern Mare Crisium, and subsequently returned to Earth a core of approximately 160 cm length. We were allocated six polished thin sections containing a total of 1230 particles in the 150 μm–250 μm size range from the material made available to the United States. Thin sections studied are, in order of increasing depth, 24077,47; 24109,42; 24149,51; 24174,53; 24182,49; 24210,45. In this paper we report a petrographic and microprobe study of these 1230 particles. Specifically, we shall report the particle census, discuss mineral chemistry, and present evidence for at least three mare basalt magma types in the Luna 24 drill-core. Preliminary reports on this work were presented by Nielsen and Drake (1977, 1978).

PARTICLE CENSUS

A census of particle types present in the Luna 24 core is given in Table 1. Six distinct types of particles are recognized:

(1) Igneous rocks—lithic fragments with basaltic or gabbroic textures, and mean crystal dimensions of greater than 30 μm.

Table 1. Distribution of fragment types in thin sections from Luna 24 (in % unless otherwise noted).

	Glass	Breccias	Aggluti-nates	Igneous Rocks	Single Crystals	Melt Rocks	Total Number of Fragments
24077,47	12	16	21	5	37	8	271
24109,42	7	16	18	10	41	8	256
24149,51	5	14	13	6	57	4	157
24174,53	3	3	6	19	60	9	232
24182,49	3	9	6	10	66	6	155
24210,45	6	18	12	13	40	11	159
Total # of Fragments	81	155	165	132	599	98	1230

(2) Melt rocks—lithic fragments with mean crystal dimensions of less than 30 μm which appear to have crystallized from quenched impact melts.
(3) Crystal fragments—monomineralic particles.
(4) Glass fragments—particles of homogeneous glass.
(5) Breccias—monomict and polymict.
(6) Agglutinates—fragments of breccias and minerals cemented primarily by inhomogeneous glass.

Approximately 11% of all particles were igneous lithic fragments, and to these we shall devote most attention. Of the mineral fragments, 58% were pyroxene, 20% were plagioclase, 20% were olivine, and 2% were opaque phases. No other mineral types were represented.

As seen in Table 1 distinct differences are observed in particle-type distribution in the core. With the exception of the deepest sample, 24210,45, there is a monotonic decrease with increasing depth of agglutinates, breccias, and glass fragments, and a monotonic increase with depth of single crystals and igneous rocks. The abundance of melt rocks remains approximately constant throughout the core. These monotonic trends are consistent with the core remaining relatively undisturbed since emplacement except for gardening, a conclusion which is in agreement with the results of other workers (e.g., Morris, 1977).

PETROLOGY OF IGNEOUS FRAGMENTS

The major element compositions of pyroxenes from Luna 24 polymineralic fragments are shown in Fig. 1. Most pyroxenes are of mare basalt origin, and occupy the irregularly shaped envelope. Open circles in Fig. 1 denote pyroxenes from finer grained basaltic rocks, while closed circles denote pyroxenes from coarser grained gabbroic rocks. The ranges of composition overlap but gabbroic pyroxenes exhibit a greater range of Fe-enrichment.

Two pyroxenes, designated by open squares, are significantly more Mg-rich and Ca-poor. Their compositions are similar to Luna 20 ANT rock pyroxenes.

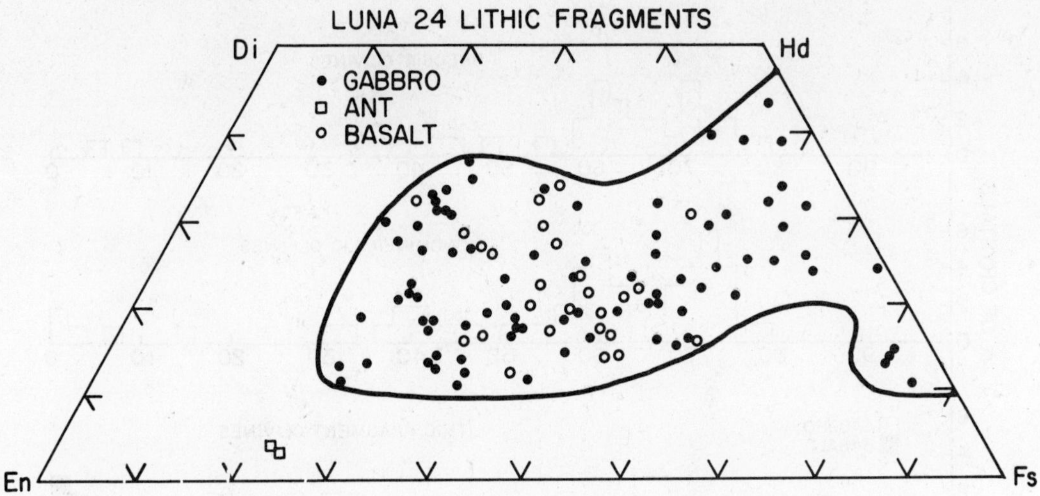

Fig. 1. Pyroxene compositions from Luna 24 igneous lithic fragments.

Both fragments are characterized by gabbroic texture, but the small size of the fragments precludes further textural analysis. An ANT origin for these pyroxenes is also supported by their association with anorthitic plagioclase (An_{98-95}) containing less than 0.22 wt.% FeO. Associated with one of the ANT pyroxenes is a unique (amongst the 1230 particles studies) magnesian ilmenite (6.82 wt.% MgO).

Olivine compositions in polymineralic lithic fragments are shown in Fig. 2. The distribution of compositions becomes bimodal when bins of 5 mole % forsterite are used, with peaks at approximately Fo_{50} and Fo_{20}. The more Fe-rich compositions appear to dominate the distribution, but two Mg-rich olivines are represented in the igneous lithic fragments. These Mg-rich olivines are both associated with minerals of mare composition. In one case, the Mg-rich olivine is associated with plagioclase (An_{90}, 0.85 wt.% FeO). The other Mg-rich olivine is associated with pyroxene (0.8 wt.% TiO_2, 1.1. wt.% Cr_2O_3). This fragment is a special case and will be discussed later.

Plagioclase compositions are illustrated in Fig. 3. The distinction between plagioclase of ANT, KREEP and mare basalt origins was made on the basis of Fe and K content. Crystals with less than 0.2 wt.% FeO were designated ANT, and the one KREEP plagioclase was categorized on the basis of its high K content (1.2 wt.% K_2O). In plagioclase feldspar of mare basalt origin potassium concentrations never exceeded 0.5 wt.% K_2O, magnesium concentrations were less than 0.5 wt.% MgO, and the iron concentration varied between 0.4 and 2.1 wt.% FeO.

Three major opaque oxide phases are present. Aluminian chromites containing 12–16 wt.% Al_2O_3 are found as euhedral crystals in association with the more Mg-rich mare basalt pyroxenes, but are not observed in association with ilmenite

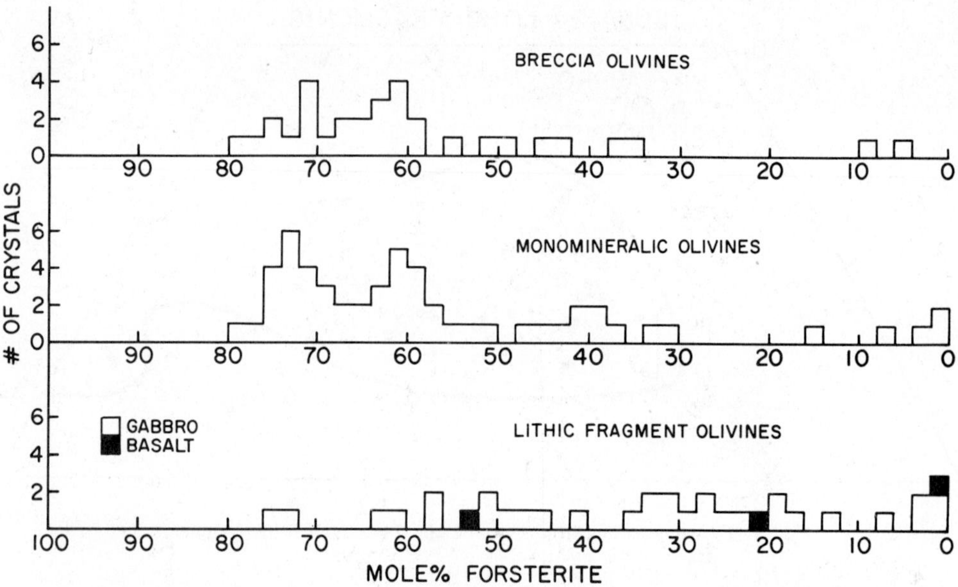

Fig. 2. Olivine compositions from Luna 24 lithic, monomineralic and breccia fragments.

or ulvöspinel. Ilmenites, usually containing less than 1 wt.% MgO are associated with the more Fe-rich mare basalt pyroxenes, and often occur in association with troilite, cristobalite, (K, Si)-rich glass, fayalite, and pyroxferroite. Chromian ulvöspinels, containing up to 8 wt.% Cr_2O_3 are often associated with ilmenite but tend to occur with mare basalt pyroxenes which are intermediate in composition relative to the most Mg- and Fe-rich pyroxenes. (The complete data base is available from the second author upon request.)

DISTINCT MAGMAS RECOGNIZED IN POLYMINERALIC IGNEOUS LITHIC FRAGMENTS

There is evidence in the major and minor element chemical compositions of mare basalt pyroxenes for at least two distinct very-low titanium (VLT) basalts at the Luna 24 site. This conclusion is based upon the plot of Ti/(Ti+Cr) versus Fe/(Fe+Mg) in the pyroxenes (Fig. 4). The rationale for the plot is that magmas with different initial compositions will crystallize pyroxenes which evolve along distinct fractionation paths from lower left to upper right. These trends are a consequence of the numerical values of pyroxene/melt partition coefficients for Fe, Mg, Ti, and Cr. Above 1000°C in a pyroxene-saturated basalt, pyroxene/melt partition coefficients for Ti and Fe are always less than unity, while partition coefficients for Cr and Mg are always greater than unity (Grove and Bence, 1977).

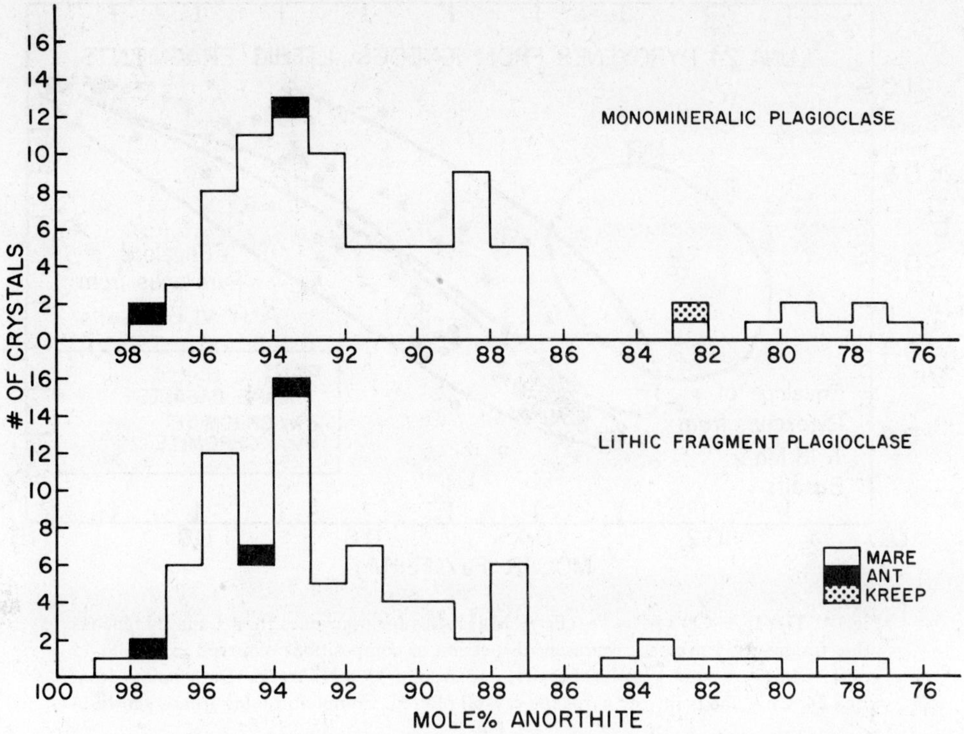

Fig. 3. Plagioclase compositions from Luna 24 monomineralic and lithic fragments.

Concurrent crystallization of other phases also affects major and minor element zoning trends in pyroxene. Concurrent crystallization of plagioclase has a negligible effect because of the extremely low numerical values of plagioclase/melt partition coefficients for Fe, Mg, Ti, and Cr. Concurrent crystallization of olivine adds a horizontal or small positive slope component to the fractionation path. Concurrent crystallization of chromite depletes the magma in Cr and Fe to a degree disproportionate to the fraction crystallized, and generates vertical zoning trends. Concurrent crystallization of ilmenite generates zoning trends moving from upper left to lower right in a plot of Ti/(Ti+Cr) versus Fe/(Fe+Mg), i.e., approximately opposite to the zoning trend generated by pyroxene crystallization.

The envelopes shown on Fig. 4 represent the range of major and minor element compositions in pyroxenes from Apollo 15 low-Ti basalts and ANT rocks (e.g. Lindstrom *et al.*, 1977), and Apollo 17 very low-Ti (VLT) basalts (Vaniman and Papike, 1977). The purpose of the Apollo 15 low-Ti and Apollo 17 VLT envelopes is to demonstrate the dispersion present in other mare magma types. These envelopes represent the *minimum* dispersion for a single magma type since there is evidence that there are at least two magma types present in the Apollo 15

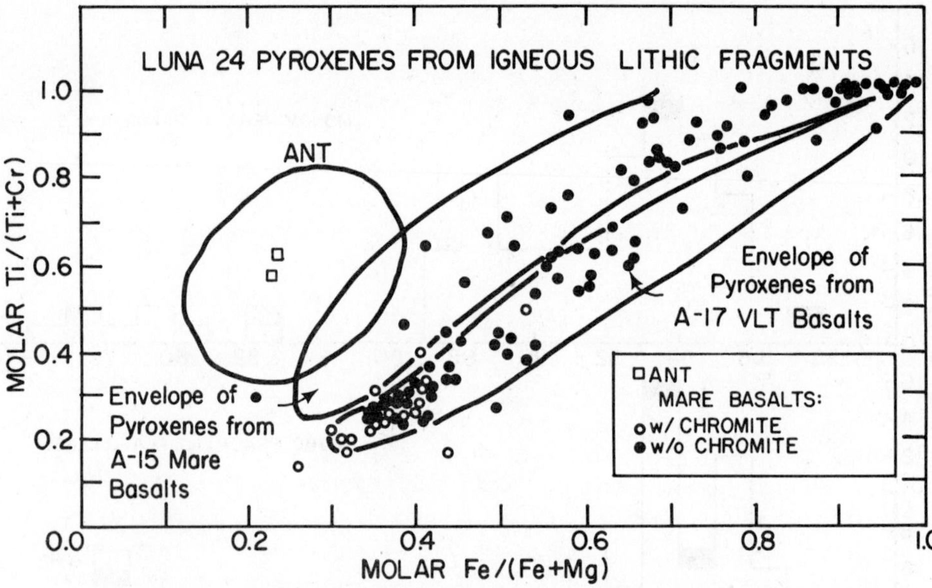

Fig. 4. Ti/(Ti + Cr) versus Fe/(Fe + Mg) ratios for pyroxenes from Luna 24 igneous lithic fragments. Envelopes represent dispersion of compositions observed in Apollo 15 low-Ti basalts, Apollo 17 VLT basalts, and Apollo 15 ANT rocks. Data points from Luna 24. Only one point per pyroxene crystal plotted. Complete zoning trends shown in Fig. 5.

low-Ti basalts, and it is unknown whether there is more than one magma type represented in the Apollo 17 VLT basalts. Pyroxenes from Luna 24 exhibit a dispersion in the Ti/(Ti+Cr) ratio for a constant Fe/(Fe+Mg) ratio of approximately twice that characteristic of other mare basalt groups.

The dispersion in the Ti/(Ti+Cr) ratios for Luna 24 basalts is not due to a systematic increase or decrease in Ti or Cr. Single unzoned pyroxenes show a range of Ti and Cr compositions with the *same* Ti/(Ti+Cr) ratio. We argue therefore, that there are at least two magma types represented in the Luna 24 polymineralic lithic fragments, one with a similar fractionation trend to the Apollo 15 low-Ti basalts, the other with a similar fractionation trend to the Apollo 17 VLT basalts.

COULD THE PYROXENE ZONING TRENDS BE ARTIFACTS?

There are several possible reasons why the scatter of data points in Fig. 4 could be real and the conclusion that at least two magmas were sampled false. The first, and most obvious, would be if one envelope was occupied by Ca-rich clinopyroxenes and the other by Ca-poor clinopyroxenes. This is not the case, however. In all cases, coexisting low-Ca pyroxene and high-Ca pyroxene plot within a single envelope.

Could the dispersion of the Ti/(Ti+Cr) ratio at constant Fe/(Fe+Mg) ratio be greater for Luna 24 VLT basalts than for Apollo 15 low-Ti basalts or Apollo 17 VLT basalts? We believe the answer is *no.* In Fig. 5 we plot zoning trends in individual pyroxenes from lithic and monomineralic fragments. Zoning trends in individual pyroxene crystals do not extend from one envelope to the other but, with one exception, are subparallel to the direction of elongation of the envelopes. In lithic fragments containing two-or-more pyroxene crystals, all pyroxenes fall in the same envelope or are subparallel to the envelopes and lie between them. In contrast, breccias frequently contain pyroxenes from both envelopes (Fig. 6). Thus we argue that there are at least two VLT basalt magmas represented among the Luna 24 lithic fragments, and that breccias contain pyroxenes from both types.

Could crystallization of chromite, from magmas which initially crystallize pyroxenes which plot in the Apollo 17 VLT basalt field, generate magmas which crystallize pyroxenes which plot in the Apollo 15 mare basalt field? Again, we believe the answer is *no.* Petrographic observations suggest that chromites crystallized concomitantly with pyroxenes for at least part of the crystallization sequence. Thus we would expect to see pyroxene zoning trends which are essentially vertical in Fig. 5 and which cross from one envelope to another. With one possible exception, such trends are not observed. Thus we conclude that the pyroxene zoning trends are not artifacts and our identification of at least two mare basalts magma types among the Luna 24 lithic fragments is supported by the data.

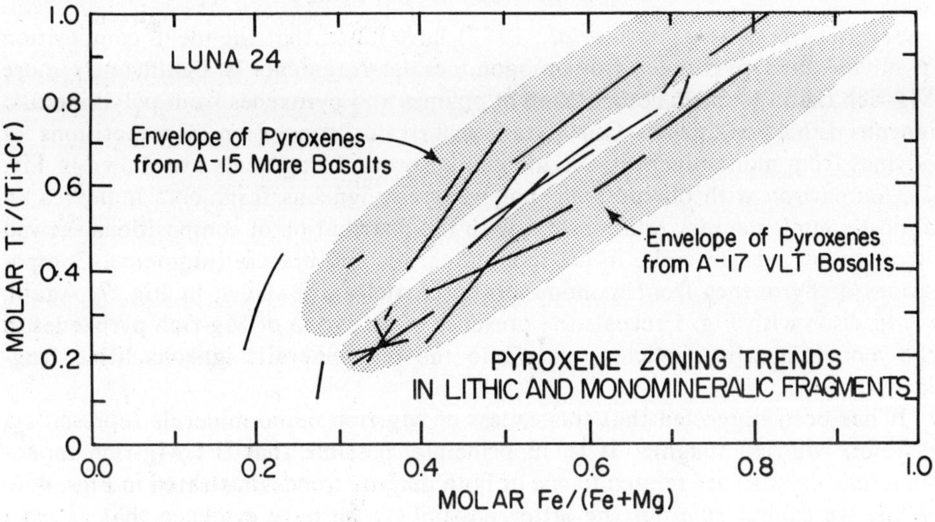

Fig. 5. Ti/(Ti + Cr) versus Fe/(Fe + Mg) zoning trends in pyroxenes from Luna 24 igneous fragments. Envelopes as in Fig. 4.

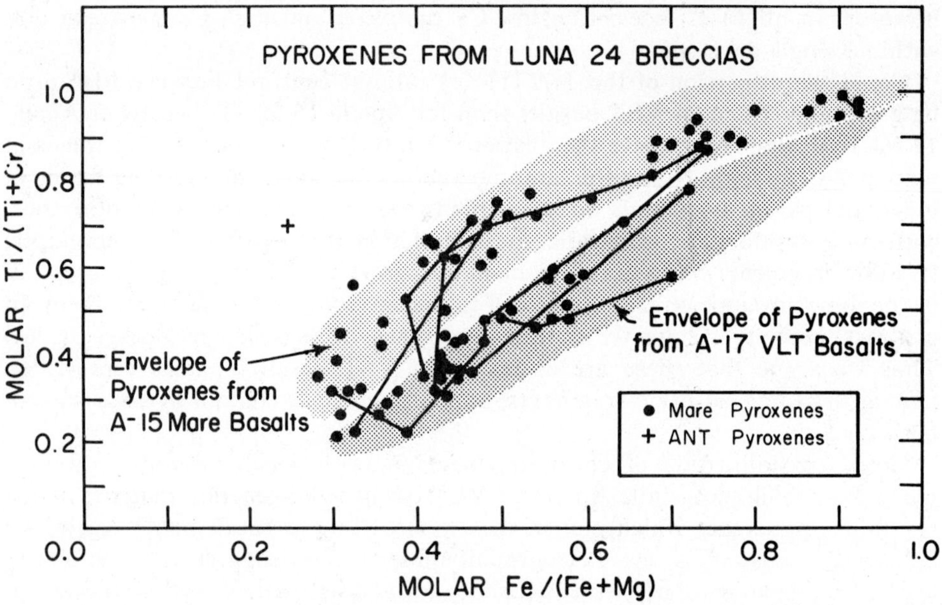

Fig. 6. Ti/(Ti + Cr) versus Fe/(Fe + Mg) ratios for pyroxenes from Luna 24 breccias. Lines connect individual pyroxene crystals from the same breccia, but do not represent zoning trends. Envelopes as in Fig. 4.

Is a Third, Mg-Rich Mare Basalt Magma Represented in the Luna 24 Drill Core?

Several groups (e.g. Basu *et al.*, 1977) have noted that the mean composition of olivines and pyroxenes from monomineralic fragments is significantly more Mg-rich than the mean composition of olivines and pyroxenes from polymineralic igneous lithic fragments. Our data support this conclusion. Compositions of olivines from monomineralic fragments and breccia fragments are shown in Fig. 2. Comparison with olivine from polymineralic igneous fragments indicates an almost complementary relationship, with the distribution of compositions skewed strongly towards forsterite in the monomineralic and breccia fragments. Compositions of pyroxenes from monomineralic particles are shown in Fig. 7. Again, comparison with Fig. 1 reveals the presence of an excess of Mg-rich pyroxenes in the monomineralic particles relative to the polymineralic igneous lithic fragments.

It has been suggested that this excess of Mg-rich mafic minerals represents a distinct, Mg-rich magma. It is, in principle, possible that the Mg-rich monomineralic crystals are related to one or both magma trends illustrated in Figs. 4–6. While we cannot rule out the latter possibility, we have evidence that at least some of the Mg-rich pyroxenes have crystallized from a distinct magma. The most Mg-rich zoning trend in Fig. 5 is for pyroxene from a coarse grained polymineralic fragment. This fragment appears to be related to the Mg-rich monomineralic particles in that it also contains Mg-rich olivine (Fo_{75}). Note that the pyroxene zoning trend does not lead into the upper envelope, but is consistent with crystallization from a magma with a higher Ti/(Ti+Cr) ratio. Other polymineralic basalts from which the Mg-rich olivines and pyroxenes crystallized may be present at the Luna 24 site. If they are present, they are presumably too coarse grained to be sampled as polymineralic fragments by the Luna 24 drill-core. The excess of Mg-rich olivines and pyroxenes among monomineralic particles is presumably derived from these basalts. Thus we conclude that the Mg-rich olivines and pyroxenes are samples of a third, more Mg- and, possibly Ti-rich magma.

Breccias

As noted above and shown in Fig. 6, breccias in the Luna 24 drill-core contain pyroxenes from both zoning trends, and they contain Mg-rich pyroxenes and olivines (Fig. 2). Thus breccias appear to contain mineral fragments from all three proposed magma groups. In addition, the breccias contain fragments of ANT rocks. This mineral and lithic fragment association indicates that the breccias are locally-derived, and that breccia formation postdates not only emplacement of the ANT crust, but also mare basalt flooding.

Conclusions

We have presented evidence for the presence of *at least three mare basalt magmas at the Luna 24 site.* One of these magma types is represented primarily by the Mg-rich monomineralic fragments. The other two groups, collectively called "Luna 24 VLT basalts" by other workers, are represented primarily

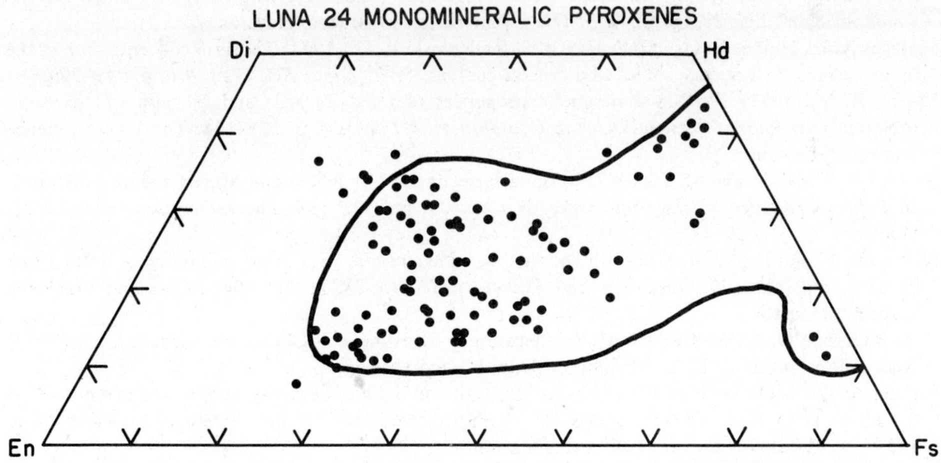

Fig. 7. Pyroxene compositions from Luna 24 monomineralic fragments.

among the polymineralic igneous lithic fragments and are distinguished by the minor element chemistry of their pyroxenes (Figs. 4–6). Additional evidence for the presence of at least two magmas within the VLT polymineralic lithic fragments is found in the V and Cr abundances reported by Ma *et al.*, (1977). Two distinct ages of 3.65±0.12 AE (Ar^{40}–Ar^{39}) and 3.30±0.05 AE (Sm-Nd) have been reported for Luna 24 samples by Stettler and Albarede (1977) and Papanastassiou *et al.*, (1977) respectively. It would be premature, however, to conclude that these distinct, unconfirmed ages also support the presence of several magmas. We note that more than three magma types may be present in the Luna 24 drill-core, but that our data lack the resolution necessary for further subdivision. Finally, the low abundance of ANT rocks (see Fig. 4) and virtual absence of KREEP rocks (one K-rich plagioclase in 1230 particles) is consistent with orbital X-ray and γ-ray maps (Andre *et al.*, 1977; Metzger *et al.*, 1977).

Acknowledgments—This work was supported by NASA grant NGR 03-002-388. The University of Arizona scanning electron microprobe facility was developed with the aid of a grant from the National Science Foundation (DES74-23721).

REFERENCES

Andre C. G., Wolfe R. W., Alder I., Clark P. E. Weidner J. R. and Philpotts J. A. (1977) Chemical character of the partially flooded Smythii Basin based on Al/Si orbital X-ray data. *Proc. Lunar Sci. Conf. 8th*, p. 925–931.

Basu A., McKay D. S. and Fruland R. M. (1977) Petrography, mineralogy and source rocks of Luna 24 drill core soils (abstract). In *Papers Presented to the Conference on Luna 24*, p. 18–21. The Lunar Science Institute, Houston.

Grove T. L. and Bence A. E. (1977) Experimental study of pyroxene-liquid interactions in quartz normative basalt 15597. *Proc. Lunar Sci. Conf. 8th*, p. 1549–1579.

Lindstrom M. M., Nielsen R. L. and Drake M. L. (1977) Petrology and geochemistry of lithic fragments separated from the Apollo 15 deep-drill core. *Proc. Lunar Sci. Conf. 8th*, p. 2869–2888.

Ma M. S., Schmitt R. A., Taylor G. J., Warner R. D., Lange E. E. and Keil K. (1977) Chemical and petrographic studies of 18 Luna 24 lithic fragments (abstract). In *Papers Presented to the Conference on Luna 24*, p. 102–105. The Lunar Science Institute, Houston.

Metzger A. E., Haines E. L., Parker R. E. and Radocinski R. G. (1977) Thorium concentration in the lunar surface. I: Regional values and crustal content. *Proc. Lunar Sci. Conf. 8th*, p. 949–999.

Morris R. V. (1977) FMR and magnetic properties of Luna 24 soils and >1 mm soil particles (abstract). In *Papers Presented to the Conference on Luna 24*, p. 121–124. The Lunar Science Institute, Houston.

Neilsen R. L. and Drake M. J. (1977) Magma types at the Luna 24 site in Mare Crisium (abstract). In *Papers Presented to the Conference on Luna 24*, pp. 132–135. The Lunar Science Institute, Houston.

Nielsen R. L. and Drake M. J. (1978) The case for at least three VLT mare magma types in the Luna 24 drill core (abstract). In *Lunar and Planetary Science IX*, p. 802–804. Lunar and Planetary Institute, Houston.

Papanastassiou D. A., McCulloch M. T., Depaolo D. J., Huneke J. C. and Wasserburg G. J. (1977) Ages and evolution of Luna 24 samples. Unpublished abstract.

Stettler A. and Albarede F. (1977) Ar^{39}–Ar^{40} pattern and light noble gas systematics of two mm-sized fragments from Mare Crisium (abstract). In *Papers Presented to the Conference on Luna 24*, p. 175–178. The Lunar Science Institute, Houston.

Vaniman D. T. and Papike J. J. (1977) Very low-Ti (VLT) basalts: a new mare rock type from the Apollo 17 drill core. *Proc. Lunar Sci. Conf. 8th*, p. 1443–1471.

Merrill, R. B. and Papike, J. J., editors.
(1978) *Mare Crisium: The View from Luna 24*, Pergamon Press (New York), p. 429–444
Printed in the United States of America

The Luna 24 highland component

A. E. BENCE and TIMOTHY L. GROVE

Department of Earth and Space Sciences, State Univeristy of New York,
Stony Brook, New York 11794

Abstract—Highland lithologies represent a very small, but significant, component of the Luna 24 drill core. They consist dominantly of melt rocks and glasses of probable impact origin, which are distinguished from their mare counterparts by their CaO/Al_2O_3 (weight) ratios of ≤ 0.8 and $Fe/(Fe + Mg)$ (atomic) ratios ≤ 0.4. Compositions equivalent to highland basalt dominate the melt rock and glass fraction. Minor lithologies include possible plutonic lithic and monomineralic fragments. The high proportion of highland basalt relative to the other highland lithologic types, the absence of KREEP basalts, and the low KREEP content of the highland melt rocks and glasses in the Luna 24 soil are also characteristics of the Luna 20 site located in the highland terrain south of Mare Crisium.

INTRODUCTION

The Luna 24 spacecraft returned a 160 cm core of the lunar regolith from southwest Mare Crisium (Florensky *et al.*, 1977). The landing site is located approximately 40 km from highlands terrain, which rises 3–5 km above the mare surface (Butler and Morrison, 1977). Nominal sampling depth by the coring device, which drilled at an angle of 30° to the vertical, was 200 cm from the surface (Basrukov, 1977). Preliminary analysis of the core (Barsukov, 1977; Barsukov *et al.*, 1977; Tarasov *et al.*, 1977) revealed that it is comprised largely of components of mare origin. In these early reports no mention was made of a highland component in the core. A detailed analysis of the 0.09 to 0.50 mm fraction from seven levels of the Luna 24 core revealed that this fraction is dominated by mare components consisting of ferrogabbro lithic and monomineralic fragments, ferrobasalt, mare glass and devitrified glass, and agglutinates of local provenance (Bence *et al.*, 1977). Highland components, though only sparsely represented, provide information on the chemistry of the highlands in the vicinity of Mare Crisium. The highland samples may provide "ground truth" for the remote sensing experiments (see, for example, Adams and Ralph, 1977; Pieters and McCord, 1977) and provide a comparison for other highland sites such as the nearby Luna 20 site.

Through the FOCUS consortium, we obtained eighteen thin sections of 0.09 to 0.50 mm fragments from 7 levels in the core containing a total of 1908 fragments. On the basis of its dominant lithologic character, each fragment was given a petrographic designation according to the classification given in Table 1. Subsequently, a modal analysis of each section was obtained (Table 1). The mare

Table 1. Particle abundances (%).

		Mare									
					Monomineralic						
Sample No.	Σ fragments	Glass	Basalts	Gabbro	Plag	Cpx	Ol	Crist	Opaques	Σ Monomineralic	Σ Mare
24077,41 (0.25-0.50 mm)	37	13.5	10.8	10.8	10.8	10.8	5.4	0	0	27.0	62.2
24109,50 (0.25-0.50 mm)	48	4.2	6.3	16.7	6.3	16.7	2.1	0	0	25.0	52.1
24109,51 (0.25-0.50 mm)	55	5.5	9.1	14.5	12.7	9.1	1.8	0	0	23.6	52.7
24109,52 (0.25-0.50 mm)	57	7.0	14.0	10.5	10.5	7.0	3.5	0	0	21.1	52.6
24109,54 (0.25-0.50 mm)	18	0	11.1	5.6	0	27.8	0	0	0	27.8	44.4
24149,40 (0.25-0.50 mm)	29	0	13.8	10.3	6.9	17.2	10.3	0	0	34.5	58.6
24149,41 (0.25-0.50 mm)	41	1.2	6.1	14.6	7.3	19.5	4.9	0	0	31.7	53.7
24149,50 (0.15-0.25 mm)	138	0.2	7.2	0.2	14.9	37.7	10.9	0	0	63.4	75.0
24170,4 (0.25-0.50 mm)	12	0	0	25.0	8.3	66.7	0	0	0	75.0	100.0
24170,8 (0.25-0.50 mm)	259	0	0	12.7	29.4	45.6	8.3	2.6	0.5	0	0
24174,55 (0.25-0.50 mm)	46	4.3	9.8	20.7	10.9	32.6	2.2	0	0	45.6	80.4
24174,56 (0.25-0.50 mm)	52	2.9	9.6	17.3	14.4	35.6	1.9	0	0	51.9	81.7
24174,57 (0.25-0.50 mm)	71	3.5	5.6	8.5	16.9	29.6	7.0	0	0	0	0
24174,49 (0.09-0.15 mm)	469	4.5	6.6	0.9	18.3	40.7	12.2	0.9	0.6	0	0
24174,50 (0.09-0.15 mm)	401	7.2	3.5	2.2	18.4	37.9	13.2	1.0	1.0	0	0
24182,25 (0.15-0.25 mm)	110	9.1	2.7	5.5	14.5	43.6	9.1	0.9	3.6	71.8	89.0
24182,50 (0.25-0.50 mm)	24	8.3	0	16.7	8.3	29.1	10.4	0	0	47.9	72.9
24210,40 (0.25-0.50 mm)	41	4.9	15.8	7.3	12.1	22.0	4.9	0	0	39.0	67.1
Total:	1908										

component at the Luna 24 site is characterized by high alumina and low titania, and breakdown of the melt rocks into mare and highland groups by petrographic techniques is difficult. Thus, it was necessary to develop chemical criteria to distinguish highland from mare fragments. A survey of the compositions of the melt rock, glass, lithic, and monomineralic fragments provided the necessary

Table 1. (*Cont.*)

Highlands										
		Monomineralic								
Glass	Melt Rocks	Opx	Plag	Oliv	Cpx	Σ Monomin.	ANT	Σ Highlands	Agglutinates & Soil Breccias	Others
0	2.7	0	0	0	0	0	0	2.7	35.1	0
0	2.1	0	0	0	0	0	0	2.1	45.8	0
0	7.3	0	0	0	0	0	0	7.3	38.2	1.8
0	8.8	0	0	0	0	0	1.8	10.5	36.8	0
0	5.6	5.6	0	0	0	5.6	0	11.1	44.4	0
0	10.3	0	0	0	0	0	0	10.3	31.0	0
0	0	0	0	0	0	0	0	0	46.3	0
0.1	0.2	0	0	0	0	0	0.2	3.6	21.4	0
0	0	0	0	0	0	0	0	0	0	0
0	0	0	0	0	0	0	0	0	8.7	<0.1
0	0	0	0	0	0	0	0	0	19.6	2.2
0	0	0	1.9	1.9	0	0	0	3.8	14.4	0
1.4	0	0	0	0	0	0	0	1.4	27.5	0
0.9	0.6	0.1	0	0	0.9	1.1	0.1	4.1	12.2	0
1.5	0.5	0	0	0	1.2	1.2	0.5	3.7	11.7	0
4.5	1.8	0	0	0	0	0	0	6.4	4.5	0
0	8.3	0	0	0	0	0	0	8.3	18.8	0
2.4	2.4	0	0	0	0	0	0	0	23.2	4.9

data. From these data, the following distinguishing criteria were developed:

(1) Highland melt rocks were distinguished by textural criteria, bulk Fe/(Fe + Mg) (atomic) ≤ 0.4, and CaO/Al_2O_3 (weight) <0.8.

(2) Glasses were distinguished on compositional grounds similar to those applied to the melt rocks.

(3) The distinguishing criterion for single plagioclase grains was the combination of very low Fe ($\leq$0.03/formula unit) and negligible Mg.
(4) Olivine of probable highland origin was distinguished by Fe/(Fe + Mg) $\leq$0.20 combined with textural criteria.
(5) All single grains of orthopyroxene were assigned a highland origin.

The bulk soil at the Mare Crisium site has a distinct mare character (CaO/Al_2O_3 ~0.99 and Fe/(Fe + Mg) ~0.53); however, the average soil is more magnesian than the mare ferrobasalt and ferrogabbro fragments within the core. This is due to a significant magnesian olivine (~Fo_{75}) fraction (Basu *et al.*, 1977; Papp *et al.*, 1977) and a less abundant Mg-rich mare green glass fraction which are from a mare basalt type(s) not yet identified in the mare lithic fragments. The abundance of highland lithic fragments in the soil is extremely low, and we suspect that the abundance of highland monomineralic fragments is correspondingly low.

Analytical Procedures

Glass and mineral analyses were obtained in the Stony Brook electron microprobe laboratory using data reduction procedures of Bence and Albee (1968) with a modified a-matrix. Melt rock analyses were obtained using broadbeam data and a reduction program modified from Albee *et al.* (1977), that attempts to correct for sample crystallinity. It appears that alumina is systematically high (by several percent of the amount present) in the analyses of melt rocks whose grain sizes exceed ~5μm.

Petrography and Chemistry of the Highland Component

In our fraction, highland lithologies are represented by melt rocks, glasses, annealed breccias, single mineral grains and relatively coarse-grained lithic fragments of possible plutonic origin. The overall abundance of highland fragments in our sample is less than about 2%, but varies from 0% in sample 24170 to ~10% in sample 24109 (Table 1).

Other workers (Basu *et al.*, 1977; Papp *et al.*, 1977; Ryder *et al.*, 1977; Simonds *et al.*, 1977) have reported modal data for their Luna 24 sample allocations. Unfortunately, either they do not single out possible highland components or, where they do, their classifications are not accompanied with an explanation of the criteria used. Taylor *et al.* (1977) measured the relative proportions of non-mare lithic fragments in their Luna 24 samples. However, they did not provide a breakdown of the glasses or monomineralic fragments. Consequently, it is not possible to compare our results with those of these workers.

Lithic and Single Grain Fragments

Lithic fragments are exceedingly rare but two of possible plutonic origin were noted in our sample. They include a "norite" and an "anorthosite."

The "anorthosite," 24109,52,22 (Fig. 1a) consists of a large (~0.4 mm) shocked plagioclase (An_{94}) fragment containing several small olivine (Fo_{80}) droplets. The shape of the olivine grains and their very low CaO contents (Table 2) is consistent with annealing. The Mg-rich composition of the olivine and the minor element (Fe and Mg) content of the plagioclase are consistent with a highlands origin for this fragment.

Longhi *et al.* (1976) show that the plagioclase-liquid distribution coefficients for FeO and MgO are lower for plutonic than for basaltic environments and that the Fe-Mg exchange $K_D = \dfrac{FeO^P\ MgO^L}{Mg^PO\ FeO^L}$ is higher in the plutonic case. This suggests that for a particular magma the amount of MgO and FeO included in plagioclase should be less, and the Fe/(Fe + Mg) greater in the plutonic case. Steele and Smith (1973) illustrate the effects in a plot similar to Fig. 2 where they distinguish a "non-basalt" field. We were able to distinguish plagioclase with very low FeO and MgO below detectability limits, and the plagioclase in our "anorthosite" has this chemistry. Thus, we suggest that these plagioclases constitute plutonic highland plagioclase. In cases where textural evidence points to a highland origin (e.g., plagioclase clasts in a highland melt rock, Fig. 1b), the compositional information was also consistent.

The "norite" fragment, 24210,40,40 contains orthopyroxene ($Wo_{1.9}En_{68.2}$), augite ($Wo_{45.9}En_{42.7}$), and plagioclase ($An_{95.6}$) (Table 2 and Fig. 3). The partitioning of FeO and MgO between coexisting pyroxenes defines a $K_D=[X^{FeO}_{cpx}/X^{MgO}_{cpx})\ (X^{MgO}_{opx}/X^{FeO}_{opx})] = 0.603$. This is consistent with an annealing temperature of less than 810°C (Lindsley *et al.*, 1974).

Several olivine (Fo_{80-84}) and plagioclase (An_{94-97}) single grains of probable highland origin were observed in agglutinate fragments. They are restricted exclusively to the two uppermost samples in the core which were made available to us (24077 and 24109). The olivine was inferred to be of highland origin on the basis of its high forsterite and low CaO contents. The criteria described above were used to distinguish plagioclase of probable highland origin from mare plagioclase.

One pyroxene grain of probable highland origin was recognized. This grain, which occurs in sample 24174,49 contains both orthopyroxene ($Wo_{1.7}En_{64.2}Fs_{34.1}$) and augite ($Wo_{43.6}En_{43.7}Fs_{12.7}$) (Fig. 3a, Table 2). The K_D (Lindsley *et al.*, 1974) of $\simeq 0.55$ indicates an equilibration temperature of <810°C. Both phases contain low concentrations of Ti, Cr, and Al. The abundances and proportions of these elements provide a means of distinguishing these highland pyroxenes from mare pyroxenes. Pyroxenes from this grain and the noritic fragment described above are plotted on Fig. 3, where they are distinguishable from the mare pyroxenes largely on the basis of the low-Ca content of the orthopyroxene.

Melt Rocks and Glasses

Very fine-grained melt rocks and glasses dominate the highland population in our sample. They occur as discrete fragments or as clasts within agglutinate

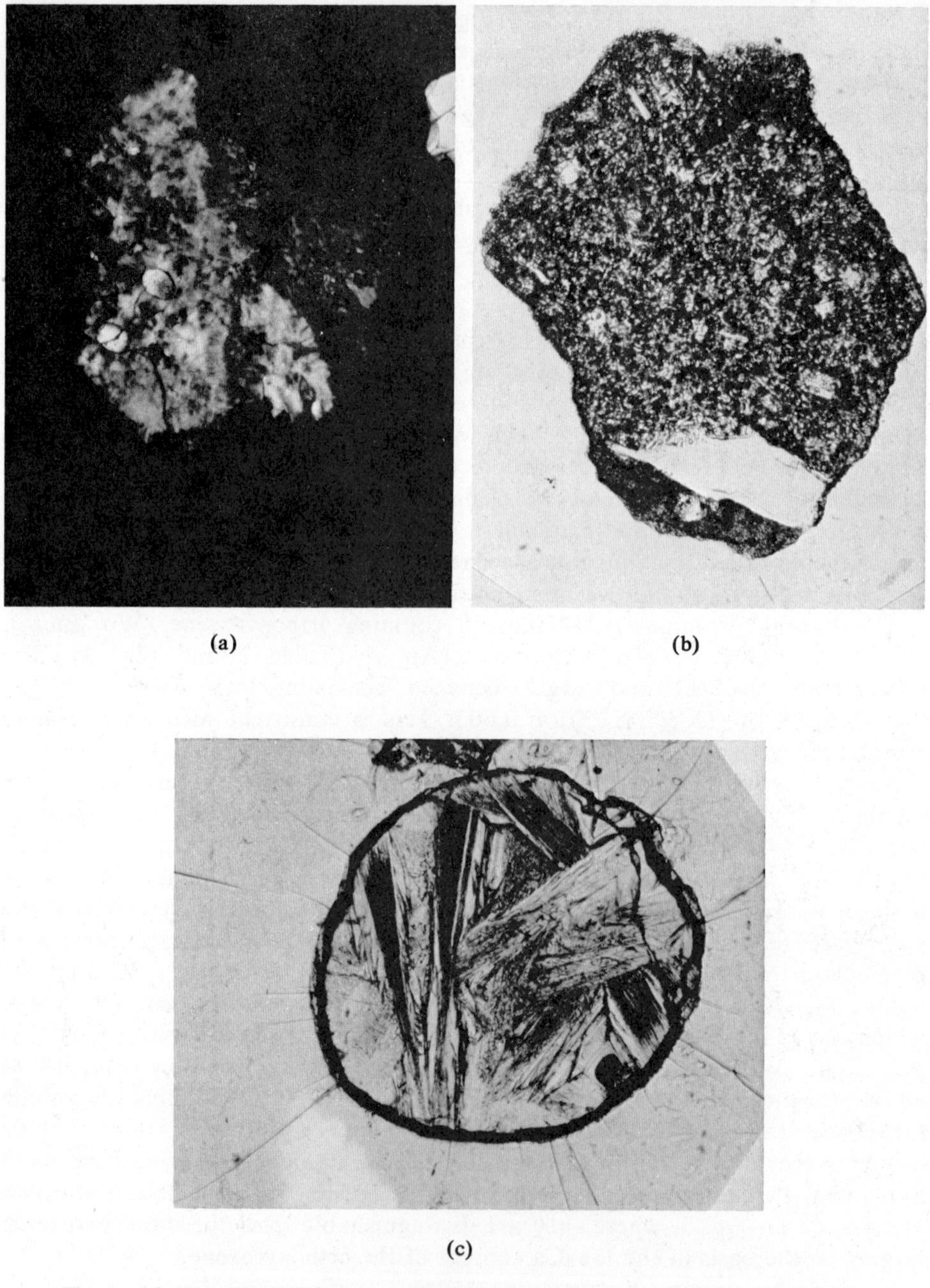

(a) (b) (c)

Fig. 1. (a) "Anorthosite" fragment 24109,52,22 containing round olivines. Field 0.5 × 0.7 mm. X-nicols. (b) Gabbroic anorthosite melt rock 24109,52,43 with plagioclase clast. Field 0.5 × 0.7 mm. (c) Sphere of spinel normative glass fragment 24174,57,47. Field 0.5 × 0.7 mm.

Table 2. Selected Mineral Analyses.

	1	2	3	4	5	6	7
SiO_2	38.8	44.5	53.7	52.5	44.4	52.4	52.6
Al_2O_3	0.00	35.7	1.03	2.08	33.8	1.20	0.76
TiO_2	0.00	—	0.22	0.69	—	0.60	0.29
FeO	19.1	0.00	19.2	7.02	1.69	7.77	21.1
MnO	0.12	—	0.31	0.24	—	0.09	0.30
MgO	41.4	0.02	25.0	15.2	0.37	15.2	23.2
CaO	0.17	18.8	0.96	22.7	18.77	21.1	0.82
Na_2O	—	0.47	—	—	0.48	0.02	0.00
K_2O	—	0.09	—	—	0.00	—	—
Cr_2O_3	0.00	—	0.25	0.40	—	0.26	0.15
Σ	99.6	99.5	100.8	100.8	99.5	98.6	99.2
Si	0.997	2.060	1.959	1.929	2.076	1.967	1.967
Al^{IV}	0.000	1.949	0.041	0.071	1.863	0.033	0.033
Al^{VI}			0.003	0.019		0.020	0.001
Ti	0.000	—	0.006	0.019	—	0.017	0.008
Fe	0.411	0.000	0.587	0.216	0.066	0.244	0.658
Mn	0.003	—	0.010	0.008	—	0.003	0.010
Mg	1.588	0.001	1.359	0.832	0.026	0.848	1.293
Ca	0.005	0.931	0.038	0.895	0.940	0.847	0.033
Na	—	0.042	—	0.000	0.043	0.001	0.000
K	—	0.005	—	—	0.000	—	—
Cr	0.000	—	0.007	0.012	—	0.008	0.005
	Fo 79.32	Ab 4.33	Wo 1.88	45.88	Ab 4.40	Wo 43.62	1.66
	Fay 20.68	An 95.13	En 68.19	42.66	An 95.60	En 43.67	64.85
		Or 0.54	Fs 29.93	11.46	Or 0.00	Fs 12.71	33.50

1. Anorthosite fragment 24109,52,22. Olivine 0 ≡ 4.
2. Anorthosite fragment 24109,52,22. Plagioclase 0 ≡ 8.
3. Norite fragment 24210,40,40. Orthopyroxene 0 ≡ 6.
4. Norite fragment 24210,40,40. Augite 0 ≡ 6.
5. Norite fragment 24210,40,40. Plagioclase 0 ≡ 8.
6. Pyroxene fragment 24174,49,439. Augite 0 ≡ 6.
7. Pyroxene fragment 24174,49,439. Orthopyroxene 0 ≡ 6.

fragments. Only the discrete melt rock and glass fragments have been considered in our modal abundance measurements. To distinguish these fragments from the more abundant aluminous and low TiO_2 mare ferrobasalts, high-Mg mare glasses, and locally generated agglutinate glass we have arbitrarily selected Fe/(Fe + Mg) (atomic) ≤ 0.40 and CaO/Al_2O_3 (weight) ≤ 0.8. Figure 4 indicates that discrimination on the basis of these two values is justified. A number of our analyses do, however, plot in the region between the mare and highland fields (Figs. 4 and 5) and may represent mixtures of highland and mare components.

The melt rocks are fine to very fine-grained, clast-laden, and have textures ranging from vitrophyric to subophitic. The majority are intersertal. The glasses are colorless to pale green and are present as spherules (Fig. 1c), spherule

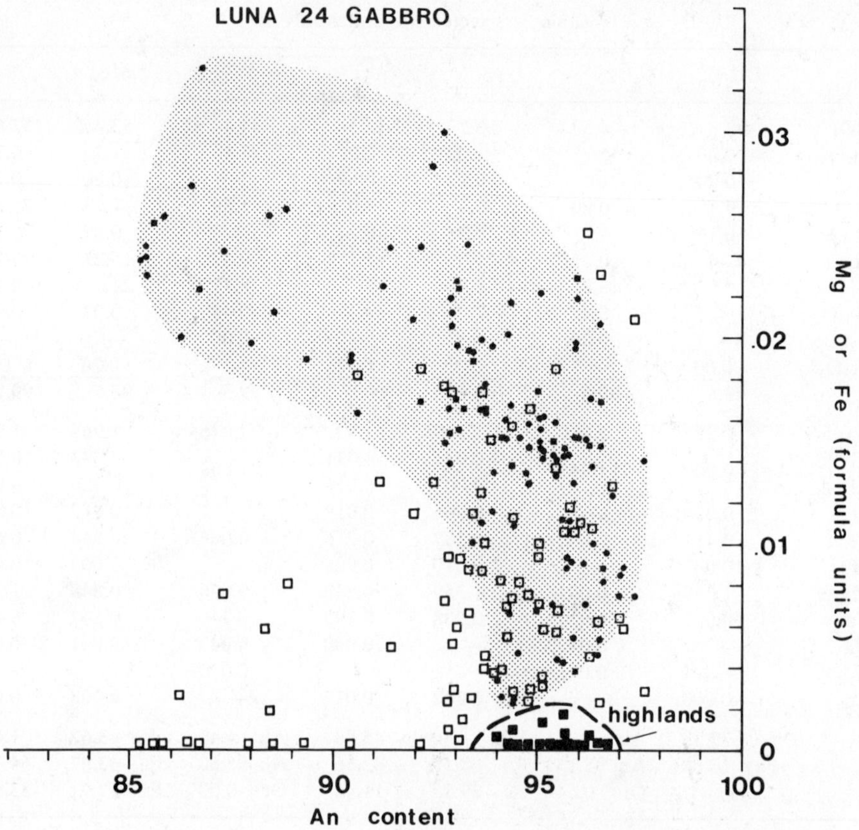

Fig. 2. Luna 24 plagioclase minor element content. Mare plagioclase: Fe solid circle, Mg open squares. Highland plagioclase: Fe solid squares, Mg below detection (0.005 formula units).

fragments, or shards. The compositions of these samples span the range of several of the compositional groups recognized among the melt rocks and glasses from other highland localities. With the exception of the low silica glasses and a high Mg-Al glass of uncertain origin, the Luna 24 compositions plot in the plagioclase field of the olivine-anorthite-silica pseudoquaternary (Fig. 6). Each data point on Figs. 4, 5, and 6 represents the average of 4–8 analyses of a single fragment. The averages reported in Table 3 were calculated from these individual fragment compositions. The very small numbers of analyses preclude further statistical treatment of the data.

Highland basalt compositions (Reid, 1974) dominate the highland melt rock and glass component (Table 3). The average of 16 fragments is virtually identical to the average of 67 Luna 24 points reported by Simonds *et al.* (1977) and is close to the Luna 20 anorthositic norite average (63 lithic fragments) reported by

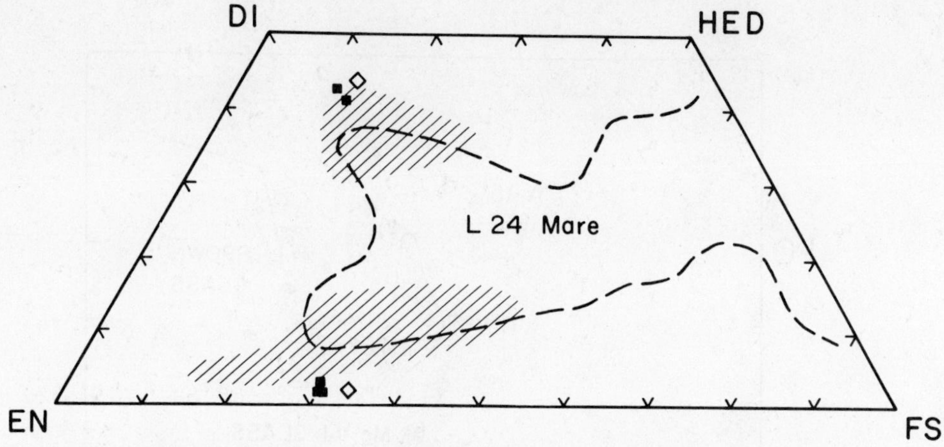

Fig. 3. Luna 24 pyroxene major element chemistry. Diamond; exsolved pyroxene in fragment 24174,49,439. Squares: norite fragment 24210,40,40. Hachured field; Luna 20 highland pyroxenes.

Prinz *et al.* (1973) and Luna 20 and Apollo 16 highland basalt glass averages (Reid *et al.*, 1973; Ridley *et al.*, 1973). The major differences among these average highland basalt compositions are higher TiO_2 in the Luna 20 and Apollo 16 samples and higher average FeO and MgO contents in the Luna 20 highland basalts.

The highland basalt compositional type at all sites has a relatively restricted range of CaO/Al_2O_3 (weight) (0.5–0.65) which reflects the dominance of the plagioclase component. The variability in the Fe/(Fe + Mg) ratio and TiO_2 and K_2O abundances within and between fragments at each site suggests that this composition, which is very close in composition to estimates of the average highland crust (Taylor and Jakeš, 1974), is a simple mechanical mixture of varying proportions of KREEP and low Fe/(Fe + Mg) components with a dominant anorthositic component. It is highly unlikely that any of our samples are representative of endogenous, >4.0 AE, aluminous basalts intruded into the lunar highland crust. The preponderance of highland basalt compositions in the Luna 24 highland fraction suggests that this is the major compositional type in the highlands surrounding Mare Crisium.

Gabbroic anorthosite and anorthosite compositions which occur as glasses or devitrified glasses are very rare in the Luna 24 highland component. Three fragments of gabbroic anorthosite composition noted in our sample have the average composition and compositional range given in Table 3. Except for lower TiO_2 and higher Fe/(Fe + Mg) ratio, this composition is similar to the Luna 20 average noritic and troctolitic anorthosite group of the ANT suite (Prinz *et al.*, 1973). The Luna 24 average composition is plotted on the olivine-anorthite-

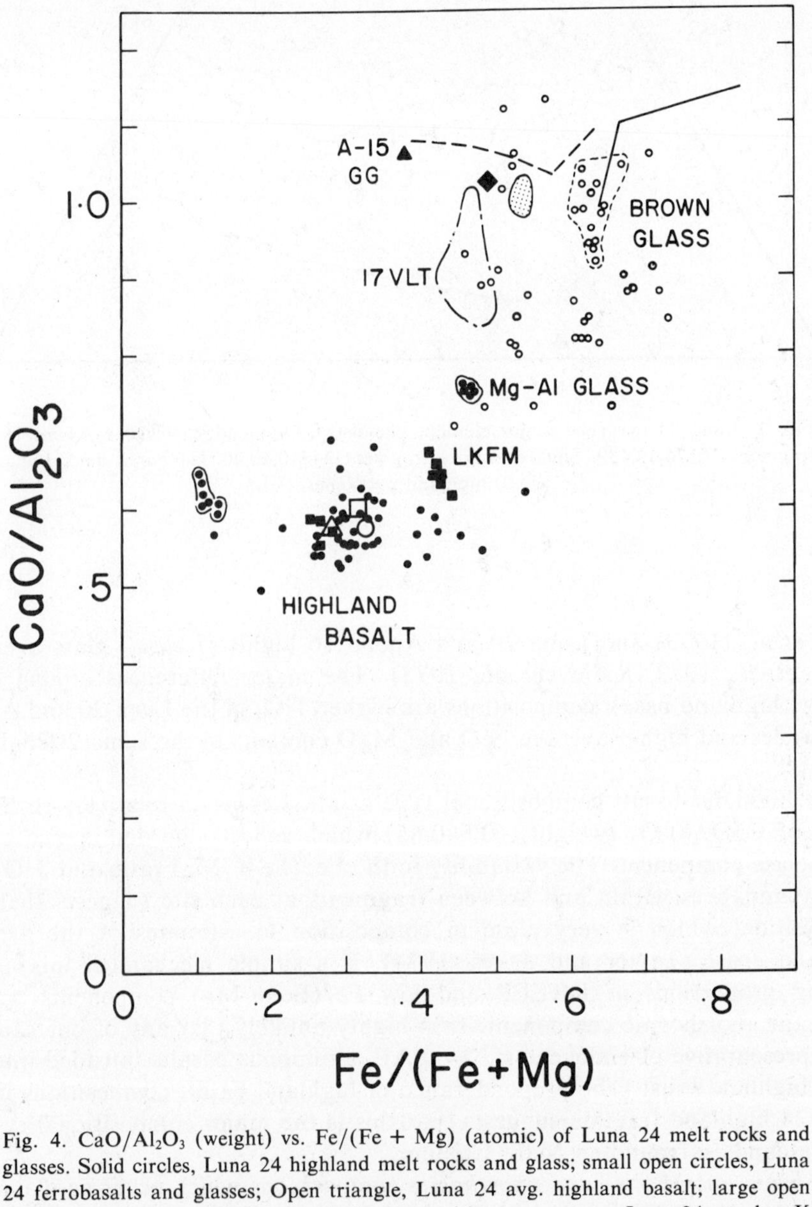

Fig. 4. CaO/Al_2O_3 (weight) vs. Fe/(Fe + Mg) (atomic) of Luna 24 melt rocks and glasses. Solid circles, Luna 24 highland melt rocks and glass; small open circles, Luna 24 ferrobasalts and glasses; Open triangle, Luna 24 avg. highland basalt; large open circle Luna 24 avg. gabbroic anorthosite composition; open square Luna 24 avg. low-K Fra Mauro basalt composition; solid diamond, Mg-mare basalt; solid triangle, Apollo 15 green glass avg; dotted field, Luna 24 soils (Blanchard *et al.*, 1977); dashed line, A-15 green glass liquid line of descent; solid line Luna 24 brown glass liquid line of descent (Grove and Vaniman, 1978).

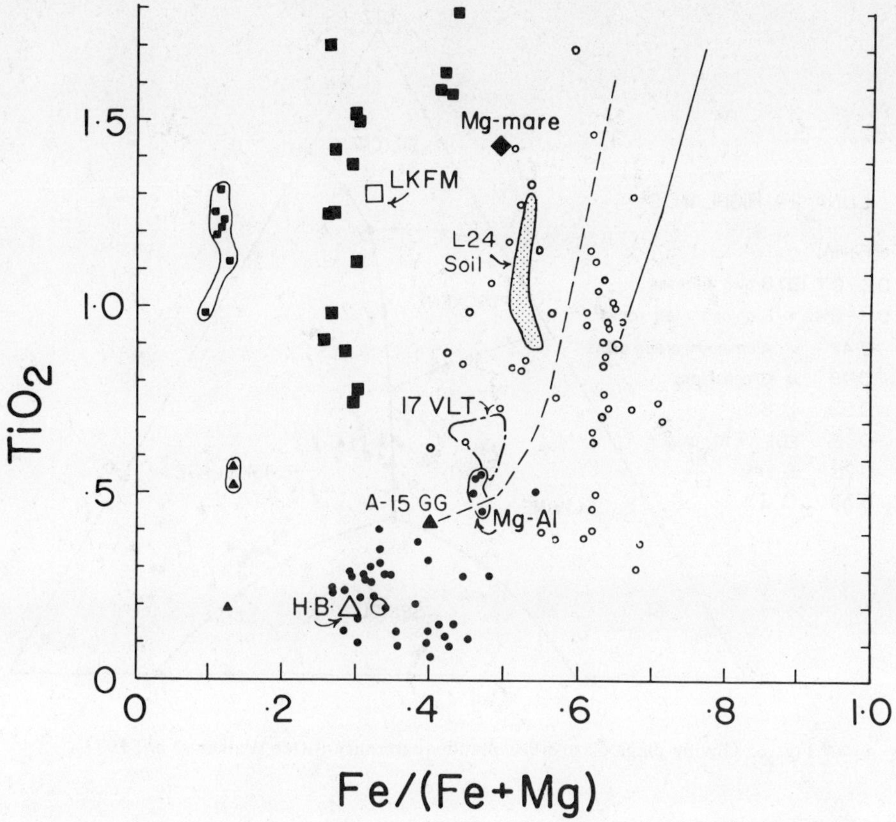

Fig. 5. TiO_2 vs. Fe/(Fe + Mg) of Luna 24 melt rocks and glasses. Symbols as for Fig. 4.

quartz pseudoquaternary (Fig. 6). One anorthositic glass fragment of possible highland provenance occurs in our sample.

Low-K Fra Mauro glasses and melt rocks form the second most abundant highland compositional group in our sample. No medium or high-K Fra Mauro compositions were found. The average composition of four low-K Fra Mauro fragments (Table 3) is similar to the low-K Fra Mauro cluster obtained for five Luna 24 fragments by Simonds *et al.* (1977), although a higher CaO/Al_2O_3 (weight) ratio indicates a somewhat larger clinopyroxene component in their samples. Reid *et al.* (1973) also reported that low-K Fra Mauro basalt glass is the second most abundant component in the Luna 20 soil. The Apollo 16 glass study (soil samples 60501 and 61221) of Ridley *et al.* (1973) indicated that this component is also subordinate to both plagioclase glass and highland basalt at that site. This comparison suggests that the Luna 24 highland component is more comparable in lithologic proportions to the nearby Luna 20 site than to the more

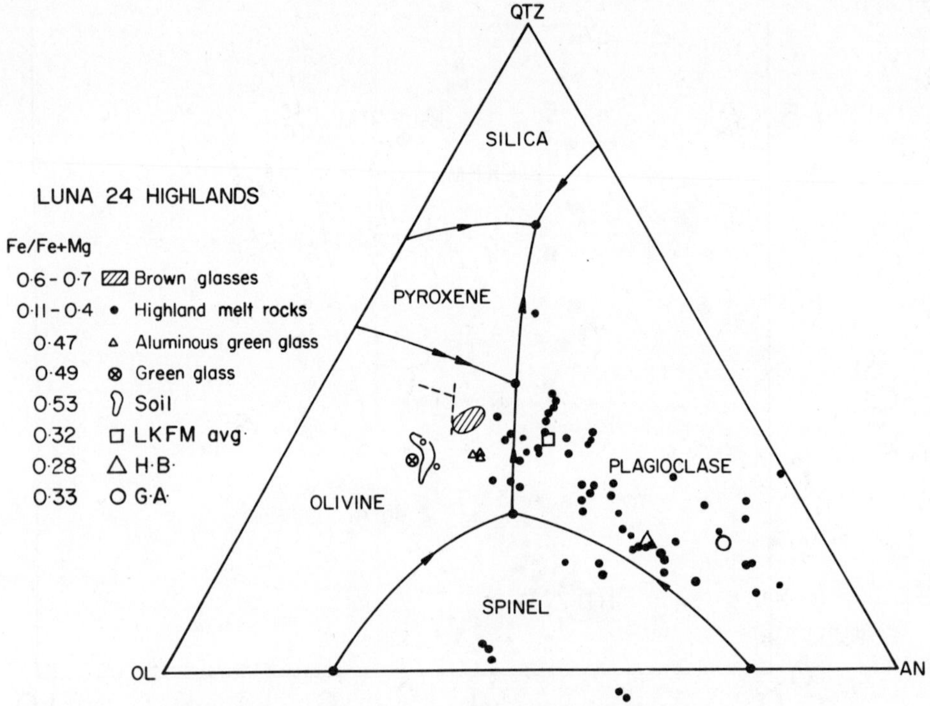

Fig. 6. Olivine-plagioclase-silica pseudoquaternary (after Walker *et al.*, 1973).

distant Apollo 16 site at Descartes. The lack of a significant KREEP component in both the Luna 20 and Luna 24 samples is consistent with this conclusion.

One very fine-grained vitrophyric fragment, 24077,41,7 (Table 3) is siliceous but K_2O-poor. Its composition plots well above point B along the pyroxene-plagioclase cotectic of the pseudoquaternary (Fig. 6). It has an evolved Fe/(Fe + Mg) ratio (0.43), high TiO_2, and moderate (for the Luna 24 core) K_2O. With the exception of the low K_2O and a slightly higher silica content, it is comparable to the Fra Mauro basalt glasses from the Luna 20 core (Reid *et al.*, 1973). Apollo 16 KREEP glass [Ridley *et al.* (1973) contains a factor of 4 higher K_2O and less silica, but is in all other respects quite similar]. This composition could be a fractionated melt formed by impacts into a KREEP deficient highland crust.

A *spinel normative glass* sphere, sphere fragment, and melt rock characterized by low Fe/(Fe + Mg) ratios (0.25-0.11) occur in our sample. Their compositions (Table 3) project into the spinel field of the pseudoquaternary (Fig. 6). The two glasses, which project close to the olivine-anorthite join (one is silica deficient), are very homogeneous. A third, melt rock fragment, is relatively inhomogeneous. Its bulk composition projects into the spinel field near the spinel-plagioclase

Table 3. Highland glasses and melt rocks.

	1	2	3	4	5	6	7	8	9	10
SiO_2	45.2 (44.3 -45.8)	45.1 (41.2 -47.4)	47.7 (47.1 -48.6)	46.1	53.6	42.2	41.6	45.5	46.1	44.6
Al_2O_3	30.2 (29.2 -30.9)	26.6 (24.7 -28.4)	20.6 (19.1 -22.8)	19.7	16.7	20.5	26.6	24.2	15.2	11.0
TiO_2	0.19 (0.13- 0.23)	0.20 (0.14- 0.30)	1.31 (1.06- 1.64)	1.42	2.07	1.27	0.55	0.19	0.51	1.43
FeO	3.07 (2.59- 3.62)	4.77 (4.09- 7.37)	7.73 (6.29- 9.32)	8.07	9.40	4.01	3.24	3.37	15.5	20.3
MnO	0.05 (0.01- 0.09)	0.06 (0.01- 0.10)	0.10 (0.08- 0.12)	0.07	0.18	0.05	0.07	0.08	0.19	0.35
MgO	3.45 (2.36- 5.55)	6.81 (5.12- 9.84)	9.40 (7.15-11.0)	12.3	7.04	18.3	12.2	13.4	9.69	10.9
CaO	17.2 (16.9 -17.6)	15.2 (13.8 -16.7)	12.1 (11.3 -12.8)	11.5	10.6	13.0	16.0	13.7	11.4	11.2
Na_2O	0.32 (0.19- 0.56)	0.24 (0.02- 0.45)	0.70 (0.44- 0.92)	0.37	0.05	0.00	0.46	0.02	0.04	0.23
K_2O	0.04 (0.03- 0.06)	0.09 (0.02- 0.19)	0.16 (0.11- 0.22)	0.10	0.21	0.01	0.06	0.05	0.02	0.09
Cr_2O_3	0.07 (0.04- 0.11)	0.14 (0.03- 0.22)	0.16 (0.11- 0.19)	0.15	0.16	0.17	0.10	0.21	0.39	0.53
	99.8	99.2	100.0	99.9	99.9	99.5	100.9	100.6	99.0	100.6
Fe/(Fe + Mg) (At.)	0.33	0.28	0.32	0.27	0.43	0.11	0.13	0.12	0.47	0.49
CaO/Al_2O_3(wt.)	0.57	0.57	0.59	0.58	0.63	0.63	0.60	0.57	0.75	1.02

1. Gabbroic anorthosite glasses and melt rocks. Average of 3 fragments. Total range in parentheses.
2. Highland basalt glasses and melt rocks. Average of 16 fragments.
3. Low-K Fra Mauro glasses and melt rocks. Average of 4 fragments.
4. Fragment 24109,52,43.
5. Fragment 24077,41,7.
6. Spinel normative glass 24182,50,1.
7. Spinel normative glass 24077,41,17.
8. Spinel normative melt rock 24174,57,70.
9. High alumina glass shard 24109,51,24.
10. Mg-mare glass 24077,41,35.

cotectic (Fig. 6). These compositions have comparable CaO/Al_2O_3 (wt.) ratios and Fe/(Fe + Mg) ratios (Table 3). However, they have quite different concentrations of TiO_2. One of the fragments (24182,50,1) which has an Fe/(Fe + Mg) (atomic) ratio of 0.11 and CaO/Al_2O_3 (wt.) = 0.65 also contains less than 25 ppm Ni. We believe that these silica deficient samples are impact melts of more siliceous, and possibly more iron-rich, highland compositions that underwent vapor fractionation during the impact event. The titania differences reflect initial differences in the target material and the relatively low volatility of TiO_2. The low soda contents of two of the samples are consistent with this origin.

Alternatively, at least two of these samples could be impact-melted troctolites having variable TiO_2 contents.

Mg-Al glasses form a unique glass compositional group (represented by four fragments of virtually identical composition) which has characteristics that are transitional between the highland and mare glass components. These green glasses are aluminous, both Fe- and Mg-rich, but have an intermediate Fe/(Fe + Mg) ratio (0.53). A representative analysis of one of the fragments (24109,51,24) is reported in column 9 of Table 3. This composition, which has a CaO/Al_2O_3 (weight) ratio of 0.75 (Fig. 4) and TiO_2 ~0.5 (Fig. 5), cannot be generated from either the Apollo 15 green glass or Luna 24 brown glass compositions through low-pressure fractionation (Figs. 4 and 5). It could, however, be a consequence of the mixing of highland and mare components during impact melting. Two alternate possibilities must also be considered: (I) This composition represents a new and unique mare magma type; or (II) it represents a new, highland glass component. It is not possible to choose between these or other possible models at this time.

CONCLUSIONS

The following conclusions arise from this study. The highland component is volumetrically a very minor constituent of the Luna 24 core. This fraction consists largely of melt rocks and glasses as well as rare lithic and single grain fragments. The melt rocks and glasses contain anorthosite, anorthositic gabbro, highland basalt, low-K Fra Mauro, and spinel normative compositions. Fragments with highland basalt composition dominate. The KREEP content of the melt rocks and glasses is exceedingly low, and in this respect the Luna 24 highland component is comparable to the Luna 20 soil sample recovered from the highlands south of Mare Crisium, but differs from that sampled by Apollos 15, 16 and 17. These differences confirm the heterogeneous distribution of KREEP in the lunar highlands.

Acknowledgments—We wish to acknowledge the generosity of the U.S.S.R. Academy of Sciences in making the Luna 24 sample available to U. S. scientists. This research was funded by NASA Grant #NGL-33-015-130.

References

Adams J. B. and Ralph R. L. (1977) Spectral reflectance of Luna 24 soils (abstract). In *Papers Presented to the Conference on Luna 24*, p. 1–4. The Lunar Science Institute, Houston.

Albee A. L., Quick J. E. and Chodos A. A. (1977) Source and magnitude of errors in "broad-beam analysis" (DBA) with the electron probe (abstract). In *Lunar Science VIII*, p. 7–9. The Lunar Science Institute, Houston.

Barsukov V. L. (1977) Preliminary data for the regolith core brought to earth by the automatic lunar station Luna 24. *Proc. Lunar Sci. Conf. 8th*, p. 3303–3318.

Barsukov V. L., Tarasov L. S., Dmitriev L. V., Kolesov G. M., Shevaleevsky I. D. and Garanin A. V. (1977) The geochemical and petrochemical features of regolith and rocks from Mare Crisium (preliminary data). *Proc. Lunar Sci. Conf. 8th*, p. 3319–3332.

Basu A., McKay D. S. and Fruland R. (1977) Petrography, mineralogy and source rocks of Luna 24 drill core soils (abstract). In *Papers Presented to the Conference on Luna 24*, p. 18–21. The Lunar Science Institute, Houston.

Bence A. E. and Albee A. L. (1968) Empirical correction factors for electron microanalysis of silicates and oxides. *J. Geol.* **76**, 382–403.

Bence A. E., Grove T. L. and Scambos T. (1977) Gabbros from Mare Crisium: An analysis of the Luna 24 soil. *Geophys. Res. Lett.* **4**, 493–496.

Blanchard D. P., Haskin L. A., Brannon J. C. and Aaboe E. (1977) Chemistry of soils and particles from Luna 24 (abstract). In *Papers Presented to the Conference on Luna 24*, p. 37–40. The Lunar Science Institute, Houston.

Butler P., Jr. and Morrison D. A. (1977) Geology of the Luna 24 landing site. *Proc. Lunar Sci. Conf. 8th*, p. 3281–3301.

Florensky C. P., Basilevsky A. T., Ivanov A. V., Pronin A. A. and Rode O. D. (1977) Luna 24: Geologic setting of landing site and characteristics of sample core (preliminary data). *Proc. Lunar Sci. Conf. 8th*, p. 3257–3279.

Grove T. L. and Vaniman D. T. (1978) Experimental petrology of very low Ti (VLT) basalts. In *Mare Crisium: The View from Luna 24* (R. B. Merrill and J. J. Papike, eds.). This volume.

Lindsley D. H., King H. E., Jr. and Turnock A. C. (1974) Compositions of synthetic augite and hypersthene coexisting at 810°C: Application to pyroxenes from lunar highlands rocks. *Geophys. Res. Lett.* **1**, 134–136.

Longhi J., Walker D. and Hays J. F. (1976) Fe and Mg in plagioclase. *Proc. Lunar Sci. Conf. 7th*, p. 1281–1300.

Papp H., Steele I. M. and Smith J. V. (1977) Luna 24 90–150 micron fraction: Implications for sampling of planetary and asteroidal regoliths (abstract). In *Papers Presented to the Conference on Luna 24*, p. 143–146. The Lunar Science Institute, Houston.

Pieters C. and McCord T. B. (1977) Mare Crisium: A second look at regional spectral properties (abstract). In *Papers Presented to the Conference on Luna 24*, p. 147. The Lunar Science Institute, Houston.

Prinz M., Dowty E., Keil K. and Bunch T. E. (1973) Mineralogy, petrology and chemistry of lithic fragments from Luna 20 fines: Origin of the cumulate ANT suite and its relationship to high-alumina and mare basalts. *Geochim. Cosmochim. Acta* **37**, 979–1006.

Reid A. M. (1974) Rock types present in lunar highland soils. *The Moon* **9**, 141–146.

Reid A. M., Warner J. L., Ridley W. I. and Brown R. W. (1973) Luna 20 soil: abundance and composition of phases in the 45–125 micron fraction. *Geochim. Cosmochim. Acta* **37**, 1011–1030.

Ridley W. I., Reid A. M., Warner J. L., Brown R. W., Gooley R. and Donaldson C. (1973) Glass compositions in Apollo 16 soils 60501 and 61221. *Proc. Lunar Sci. Conf. 4th*, p. 309–321.

Ryder G., McSween H. Y. and Marvin U. B. (1977) Luna 24 VLT basalt: Character and origin (abstract). In *Papers Presented to the Conference on Luna 24*, p. 160–163. The Lunar Science Institute, Houston.

Simonds C. H., Warner J. L., McGee P. E., Cochran A. and Brown R. W. (1977) Luna 24: Petrochemical potpourri (abstract). In *Papers Presented to the Conference on Luna 24*, p. 170–173. The Lunar Science Institute, Houston.

Steele I. M. and Smith J. V. (1973) Mineralogy and petrology of some Apollo 16 rocks and fines: General petrologic model of the moon. *Proc. Lunar Sci. Conf. 4th*, p. 519–536.

Tarasov L. S., Nazarov M. A., Shevaleevsky I. D., Kudryashova A. F., Gaverdovskaya A. S. and Korina M. I. (1977) Mineralogy and petrography of lunar rocks from Mare Crisium (preliminary data). *Proc. Lunar Sci. Conf. 8th*, p. 3333–3356.

Taylor G. J., Warner R. D. and Keil K. (1977) Chemical distinctions among very low-Ti mare basalts (abstract). In *Papers Presented to the Conference on Luna 24*, p. 183–185. The Lunar Science Institute, Houston.

Taylor S. R. and Jakeš P. (1974) The geochemical evolution of the moon. *Proc. Lunar Sci. Conf. 5th*, p. 1287–1305.

Walker D., Longhi J., Grove T. L., Stolper E. and Hays J. F. (1973) Experimental petrology and origin of rocks from the Descartes Highlands. *Proc. Lunar Sci. Conf. 4th*, p. 1013–1032.

Merrill, R. B. and Papike, J. J., editors.
(1978) *Mare Crisium: The View from Luna 24*, Pergamon Press (New York), p. 445–471
Printed in the United States of America

Experimental petrology of very low Ti (VLT) basalts

TIMOTHY L. GROVE and D. T. VANIMAN
Department of Earth and Space Sciences,
State University of New York, Stony Brook, New York 11794

Abstract—The results of iron capsule experiments on synthetic analogs of Luna 24 ferrobasalt and an Apollo 15 green glass allow tests of two hypotheses for the generation of Luna 24 ferrobasalt: (1) Luna 24 ferrobasalt may be derived by near-surface fractional crystallization of a Luna 24 green glass magma. The Luna 24 green glass magma is compositionally distinct from green glasses found at other mare and highland sites and is represented at Mare Crisium as rare glasses and vitrophyre fragments. At least two Mg-rich mare basalt types occur at the Luna 24 site. One of these Mg-rich basalts is compositionally intermediate between Luna 24 ferrobasalt and the Luna 24 green glass parent. Ferrobasalt, Mg-rich basalt and Luna 24 green glass may be samples of a single Luna 24 green glass flow (or flows) that underwent differentiation by olivine settling at (or near) the lunar surface. An ideal fractional crystallization calculation shows that removal of ~48% olivine from a Luna 24 green glass yields Luna 24 ferrobasalt. (2) Fractional crystallization of a liquid similar to Apollo 15 green glass in a subsurface magma chamber might yield Luna 24 ferrobasalt. High-pressure synthesis experiments indicate that Luna 24 ferrobasalt magma could be in equilibrium with a plagioclase and high-Ca clinopyroxene residuum at depths of ~100 km.

The low-pressure phase equilibria are used to explore the relations between very low Ti (VLT) basalt types. The Apollo 17 VLT basalts cannot be related to the Apollo 15 green glass or Luna 24 green glass by near-surface fractional crystallization. The explanation for the differences between these basalt types may lie in the processes of melt generation at depth in the lunar interior.

INTRODUCTION

The Soviet Luna 24 mission to Mare Crisium returned samples of very low Ti (VLT) ferrobasalts and ferrogabbros; mare basalts characterized by low TiO_2 content (~1.0 wt.%), high Fe/(Fe + Mg) ~(0.6 to 0.7 molar) and a high CaO/Al_2O_3 wt. ratio (~1.02). We have carried out equilibrium experiments in iron capsules sealed in silica glass tubes and synthesis experiments in iron capsules at high pressures to determine the processes that generated the Luna 24 ferrobasalts. Luna 24 ferrobasalt is the second crystalline representative of VLT basalt identified in our returned lunar samples. The first member of the VLT suite was recognized in Apollo 17 soils (Vaniman and Papike, 1977a; G. J. Taylor *et al.*, 1977a) and other candidates have been identified at Apollo 15 (Steele *et al.*, 1977). These basalts may be related to low-Ti mafic glasses that have been identified at every lunar sample locality and that are especially abundant at the Apollo 15 and Apollo 14 sites. In order to delineate the low-pressure phase relations pertinent to the crystallization of VLT basalts and glasses, iron capsule, sealed silica tube experiments were performed on a synthetic analog of average Apollo 15 green glass (Warner *et al.*, 1972; Ridley *et al.*, 1973). The results of these green glass and Luna 24 ferrobasalt experiments constrain the low-pressure fractionation processes that control liquid compositions, and indicate that Luna 24 ferrobasalts cannot be generated by near-surface fractionational crystallization of an Apollo 15 green glass magma. However, the experiments allow us to propose that Luna 24 ferrobasalt can be derived by olvine fractionation of a Luna 24 green glass. Apollo 17 VLT basalts are not related to the Apollo 15 green glass or the Luna 24 green glass parents by any low-pressure fractionation process, but may have been derived from a different magma represented at Mare Serenitatis as a green glass chemistry (Vaniman and Papike, 1977a).

EXPERIMENTS

Iron Capsule-Sealed Silica Tube Experiments

Experiments were performed using synthetic starting materials prepared after the method of Lindsley *et al.* (1974). An oxide mixture of appropriate bulk composition was fused at temperatures 50°C above the equilibrium liquidus and quenched to a glass. Subsequent experiments were performed using the glass or crystallized glass as starting material. The synthetic composition chosen was similar to the average of ten brown glass compositions (Table 6, col. 5) and is nearly identical to bulk compositions estimated by other workers using larger crystalline ferrobasalt fragments (Table 6, cols. 6–8). The Luna 24 ferrobasalt composition also falls within the range of estimates derived from pyroxene minor element chemistry and modal recombination of ferrobasalt lithic fragments (Table 6, cols. 1–4).

Equilibrium experiments were performed in high purity (Johnson Matthey grade 1) iron capsules sealed in evacuated silica tubes as previously described by Grove and Bence (1977). Temperature was monitored by Pt-10Rh thermocouples calibrated against the melting points of gold and lithium metasilicate on the IPTS (1968) temperature scale (Biggar, 1972). The results of experiments on the Luna 24 and Apollo 15 compositions are tabulated in Tables 1 and 3 and the compositions of liquids and coexisting phases are given in Tables 4 and 5. The iron capsule technique may not precisely duplicate the crystallization conditions of these VLT compositions, but it is felt that the technique was satisfactory for the following reasons: (1) The bulk composition of the charge remained constant with negligible Fe gain or loss in all experiments. This was verified in the partially crystalline experiments by modally recombining the compositions of the silicate and liquid phases to estimate bulk composition and by the preservation of the bulk liquid composition in runs held at above liquidus temperatures (Table 4). (2) Silicate phases similar in composition to those found in Luna 24 ferrobasalt were produced in the Luna 24 experiments.

All experiments contained Fe blebs in the silicate and liquid portions of the charge. It should be noted that the Luna 24 ferrobasalts were not Fe-saturated during the early stages of crystallization (Friel and Goldstein, 1977) as evidenced by the absence of Fe blebs in early formed silicates. Thus, f_{O_2} in the experiments was somewhat lower than that imposed during crystallization under lunar conditions (Kesson and Lindsley, 1976). Essentially all alkalis were lost from the silicate glass starting material and spinel is absent in the experiments, but present in Luna 24 ferrobasalt as an early phase in the paragenetic sequence (Haggerty, 1977). The lack of spinel in the experiments may result from an error in estimating the Cr_2O_3 content of Luna 24 ferrobasalt (see Table 6) or from the lower f_{O_2} imposed by the iron capsule technique.

High-Pressure Synthesis Experiments—Luna 24 Ferrobasalt

Synthesis experiments in the pressure range 5 to 15 kbar were performed in a solid-media, piston-cylinder apparatus (Boyd and England, 1960). The piston-out technique with a 2 kb overshoot was employed (see Kesson, 1975). Temperature control was achieved using $W_{25}Re_{75}W_3Re_{97}$ thermocou-

Table 1. Results of equilibrium experiments on a synthetic analog of Luna 24 ferrobasalt performed in iron capsules held in sealed silica tubes.

Run #	T°C	Duration (Hours)	Starting Material	Run Products
7	1192	27.6	glass	glass
37	1189	2.9	#8*	glass
8	1182	25.5	glass	plag + glass
3	1179	172.6	glass	plag + glass
43	1163	7.5	#21⁺	plag + glass
6	1155	113.4	glass	plag + ol + glass
41	1151	8.2	#21⁺	plag + ol + glass
4	1143	148.5	glass	plag + ol + pyx + glass
5	1129	425.7	glass	plag + ol + pyx + glass
9	1099	236.0	glass	plag + ol + pyx + glass
10	1071	361.0	glass	plag + ol + pyx + glass

*A portion of experiment #8 was used as starting material for this run.

⁺Run #21 was a linear cooling rate experiment carried out at 1.6°C/hr from 1202 to 1049°C. The run product consisting of plag + ol + pyx + liq was used as starting material.

glass = quenched liquid, plag = plagioclase, ol = olivine, pyx = clinopyroxene of variable composition.

Table 2. Results of high-pressure synthesis experiments on the Luna 24 ferrobasalt composition. Starting material is synthetic glass.

Pressure (kb)	T°C	Duration (Hours)	Run Products
5	1213	2.4	glass
	1203	2.6	plag + pyx + glass
	1193	3.7	plag + pyx $Wo_{28}En_{44}$ + glass
	1173	6.0	plag + pyx $Wo_{26}En_{40}$ + glass
7.5	1236	3.0	pyx + glass
	1219	3.7	plag + pyx + glass
10	1301	2.0	glass
	1255	2.1	pyx + glass
	1236	1.5	plag + pyx $Wo_{35}En_{35}$ + glass
	1220	7.3	plag + pyx $Wo_{37}En_{35}$ + glass
15	1339	1.4	pyx + glass
	1310	1.5	pyx + glass
	1290	1.8	plag + pyx + glass

See Table 1 for abbreviations.

ples with the sample container located in the hot spot and a temperature correction applied for the position of the thermocouple. No pressure correction to thermocouple emf was applied. The results of the experiments are summarized in Table 2 and illustrated in Fig. 3. The purpose of the high-pressure experiments was to evaluate the possibility of generating the Luna 24 ferrobasalt magma at depth by partial melting of a lunar mantle source or as a sample of residual liquid tapped from a fractionating magma chamber.

Table 3. Results of equilibrium experiments on a synthetic analog of Apollo 15 green glass performed in iron capsules held in sealed silica tubes.

Run #	T°C	Duration (Hours)	Starting Material	Run Products
2	1408	1.2	glass	glass[a]
4	1392	1.5	glass	glass + ol
6	1359	3.1	glass	glass + ol
5	1339	3.5	glass	glass + ol
7	1302	3.9	glass	glass + ol
8	1273	10.8	glass	glass + ol
10	1248	24.4	glass	glass + ol
3	1221	41.1	glass	glass + ol
20	1198/1210	65.8/26.1	glass*	glass + ol
15	1183/1202	25.3/5.6	glass*	glass + ol + pyx
14	1192	27.6	glass	glass + ol + pyx
9	1179	172.6	glass	glass + ol + pyx
16	1163	7.5	#11[+]	glass + ol + pyx
13	1155	113.4	glass	glass + ol + pyx + plag
11	1143	148.5	glass	glass + ol + pyx + plag

[a]See Table 1 for abbreviations.

*In runs #15 and 20 the charge was held at an initial low temperature to nucleate ol + pyx, then temperature was raised in an attempt to dissolve the pyroxene.

[+]A portion of the run product from experiment #11 was used as starting material.

Analytical Procedures

Compositions of liquids and coexisting minerals in the experimental charges were obtained on an automated 4-spectrometer ARL-EMX SM microprobe using real time data reduction. The matrix correction procedures of Bence and Albee (1968) and modifications of Albee and Ray (1970) were used.

DISCUSSION

Crystallization of Luna 24 Ferrobasalt

Several groups have studied the ferrobasalts and ferrogabbros and each has deduced a unique scheme for the crystallization sequence of these samples. Ryder *et al.* (1977) suggest the paragenetic sequence olivine and chromite followed by plagioclase, and finally by pyroxene based on textures preserved in the ferrobasalt lithic fragments. Bence *et al.* (1977) suggest that the liquid that crystallized to form the Luna 24 ferrogabbro was nearly multiply-saturated with olivine, two pyroxenes and spinel followed rapidly by plagioclase. Tarasov *et al.* (1977) propose the paragenetic sequence: plagioclase, followed by augite, and finally olivine. The phase appearances (Fig. 1) deduced from the experiments provide an explanation for this variety of textural interpretations. Plagioclase is the liquidus phase and is joined by olivine after a modest temperature interval (~25°C) and 5% plagioclase crystallization. Dynamic crystallization experiments on Luna 24 ferrobasalt are olivine + plagioclase saturated on the liquidus at cooling rates similar to those experienced by the Luna 24 ferrobasalt (Grove, 1978a). Pyroxene joins plagioclase and olivine after another 15°C (at 20% crystallinity). Thus, Luna 24 ferrobasalts were multiply saturated with olivine and plagioclase at lunar surface conditions. One might expect that the textures produced by crystallization of a multiply-saturated liquid would be difficult to interpret. No phenocryst phase is produced because all phases appear nearly simultaneously and a graphic or ophitic texture is generated (Grove, 1978a; Walker *et al.*, 1978). Within a single sample, different crystallization sequences might be deduced from areas of differing texture.

If the bulk composition of Luna 24 basalts represents a liquid composition, the phase chemistries in the experiments should be nearly identical to those in the ferrobasalts. Walker *et al.* (1977) point out that the application of equilibrium experiments is restricted to the interpretation of slowly cooled (<0.1°C/hr)

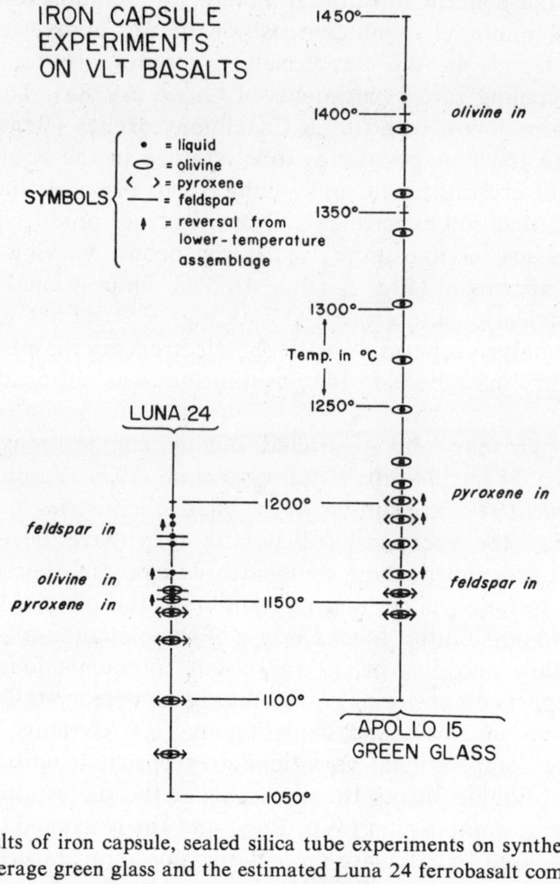

Fig. 1. Results of iron capsule, sealed silica tube experiments on synthetic analogs of Apollo 15 average green glass and the estimated Luna 24 ferrobasalt compositions. See Tables 1 and 3 for experimental conditions and Tables 4 and 5 for analyses of liquids and coexisting phases.

basalts, and even then the effects of kinetic processes must be considered. Thus, any interpretation of phase chemistries and crystallization sequence must be substantiated by dynamic crystallization experiments, which are in progress (Grove, 1978a) and indicate that the Luna 24 ferrobasalts cooled from liquids of their own bulk composition at cooling rates on the order of 0.5 to 1.7°C/hr.

The experimental olivine-liquid $K_D^{ol\text{-}liq}$ (= $X_{FeO}^{ol}X_{MgO}^{liq}/X_{FeO}^{liq}X_{MgO}^{ol}$) is 0.34 to 0.35, in close agreement with that determined by Longhi *et al.* (1977) for low Ti basalts in the temperature range 1150–1050°C. Olivines of Fo_{60} (Table 4) are the first to appear in the experiments. The Fe/(Fe + Mg) range of Luna 24 ferrobasalts is (0.60 to 0.65) and the reported olivine core compositions vary between Fo_{56} and Fo_{64}, depending on the investigator. This range of Fo contents may reflect a variation in Fe/(Fe + Mg) of the ferrobasalt suite, or as Walker *et al.* (1977) and L. A. Taylor *et al.* (1977a, 1977b) have pointed out, subsolidus Fe-Mg intercrystalline diffusion. Nevertheless, the experiments produce olivines that are similar to those preserved in the ferrobasalt, and it is felt that the composition does not contain substantial amounts of cumulus olivine.

The major and minor element composition of the pyroxene cores and the observed zoning trends in the ferrobasalt pyroxenes are reproduced in the controlled linear cooling rate experiments of Grove (1978a). The slowly cooled ferrogabbro contains low Ca and high Ca clinopyroxenes (Bence *et al.*, 1977) that are similar to the first pyroxenes to crystallize in the equilibrium experiments, but the later crystallization and zoning trends resemble those produced in the dynamic crystallization experiments. The pyroxenes produced in the equilibrium experiments are compositionally heterogeneous, varying greatly in Ca content for each experiment (Fig. 2, Table 4). The compositional variability may result from the nucleation and growth of both high Ca and low Ca domains that are too small to analyze separately with the electron microprobe. On the other hand, the variability may indicate that equilibrium was not attained during the experiment.

The variable pyroxene and restricted olivine compositions found in the recrystallized Luna 24 ferrobasalts (McSween *et al.*, 1977; Vaniman and Papike, 1977b; Ryder *et al.*, 1977) are similar to the phase chemistries in the equilibrium experiments. Thus, the recrystallized basalts may have been fragments of ferrobasalt that were annealed at temperatures close to the solidus for time periods sufficient to homogenize and re-equilibrate the phases. McSween *et al.* (1977) explore two possibilities for the origin of these metabasalts: (1) as contact metamorphosed flow margins; or, (2) as basalt fragments foundered in lunar lava. They also report compositional variability in some recrystallized basalts that involves enrichment in Al_2O_3 and depletion in TiO_2 contents with respect to ferrobasalt. These compositional variations are consistent with the removal of late-stage residual liquids during the annealing of the recrystallized basalt. The least calcic pyroxene composition ($Wo_{15}En_{32}$) and the restricted olivine composition (Fo_{26}) can be used to calculate an equilibration temperature. The temperature-dependent $K_D^{xtal\text{-}liq} = X_{FeO}^{xtal}X_{MgO}^{liq}/X_{FeO}^{liq}X_{MgO}^{xtal}$ for low-Ca clinopyroxene (from Grove, 1978b and the Luna 24 experiments) and the temperature independent

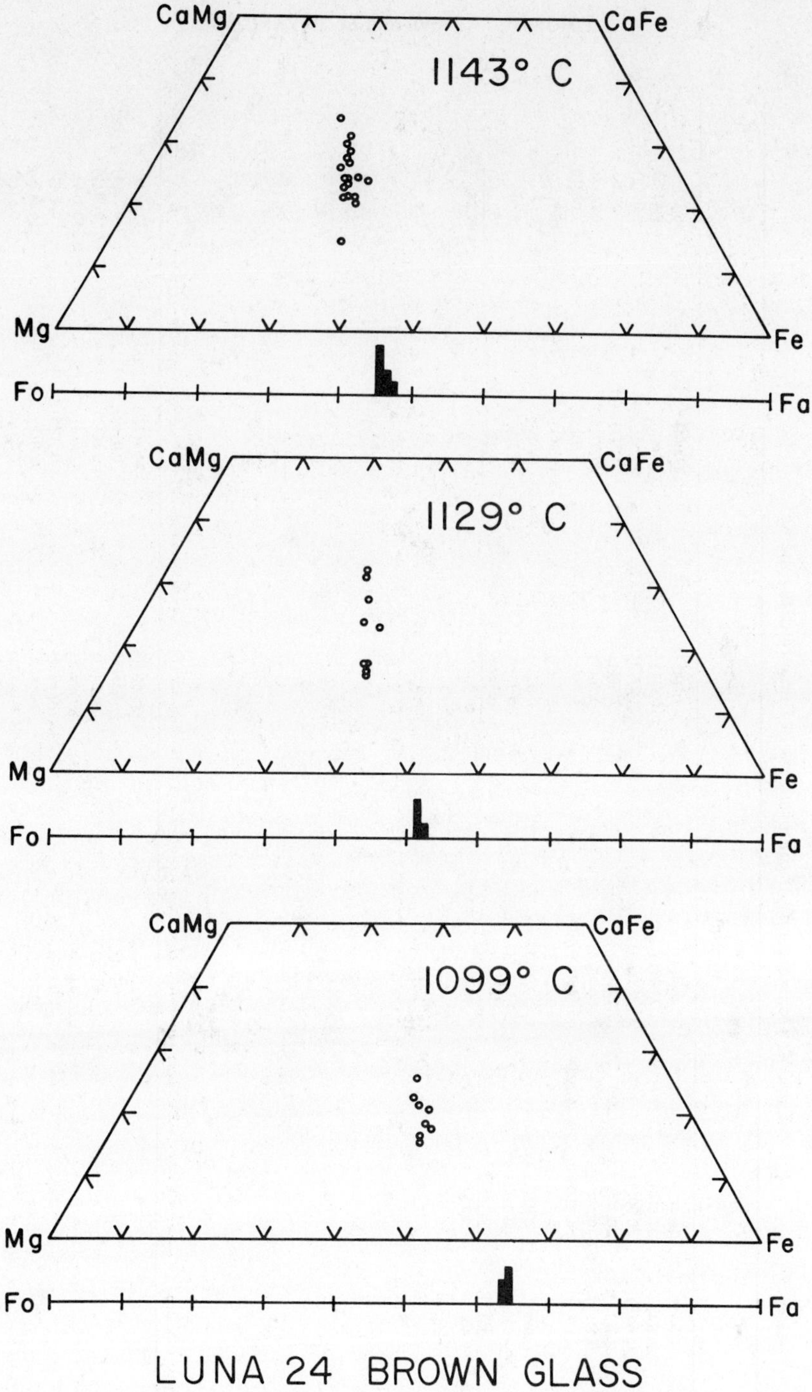

Fig. 2. Experimentally produced pyroxenes and olivines in Luna 24 basalt experiments. The compositional variation in the experimental pyroxenes is similar to that observed in recrystallized ferrobasalts. Representative analyses of the extremes in chemical composition are found in Table 4.

Table 4. Compositions of run products in Luna 24 experiments.

Run #	7	8	3	6	4	5	9
# of analyses	8	7	2	5	4	5	5
Liquid SiO_2	46.4(2)*	46.4(2)	46.4	46.4(4)	47.2(4)	47.0(4)	46.2(1)
Al_2O_3	12.4(4)	12.3(4)	12.6	11.2(1)	10.9(1)	10.6(1)	9.42(5)
TiO_2	0.88(5)	0.90(3)	0.93	1.00(4)	1.24(3)	1.63(7)	2.09(2)
FeO	20.3(2)	20.2(1)	20.1	21.1(2)	22.6(1)	24.5(2)	27.3(2)
MnO	0.22(3)	0.22(2)	0.31	0.27(5)	0.36(4)	0.37(3)	0.26(2)
MgO	6.17(5)	6.21(8)	6.22	5.91(5)	5.15(4)	4.50(6)	3.06(10)
CaO	12.8(3)	12.8(3)	12.4	12.3(1)	12.10(6)	11.8(1)	10.9(5)
Na_2O	0.02(1)	0.02(1)	0.02	0.07(3)	0.07(1)	0.01	0.0
K_2O	0.02(1)	0.01(1)	0.0	0.03(1)	0.0	0.0	0.01
Cr_2O_3	0.19(3)	0.19(2)	0.28	0.25(1)	0.28(2)	0.23(2)	0.12(2)
Σ	99.40	99.25	99.26	98.53	99.90	100.64	99.36
Plagioclase							
SiO_2		+	+	44.0	+	44.3	+
Al_2O_3				34.0		35.3	
FeO				1.05		1.54	
MgO				0.27		0.27	
CaO				19.6		19.3	
Na_2O				0.22		0.10	
K_2O				0.01		0.01	
Σ				99.15		100.82	
An content				98	98	99	
Olivine							
# of analyses				6	4	4	3
SiO_2				35.3(4)	35.6(4)	34.4(2)	33.4(3)
Al_2O_3				0.05(7)	0.35(30)	0.07(9)	0.22(1)
TiO_2				0.07(1)	0.04(3)	0.10(2)	0.11(1)
FeO				35.6(4)	38.4(2)	42.7(4)	50.1(4)
MnO				0.40(3)	0.33(4)	0.50(5)	0.43(1)
MgO				28.9(2)	25.6(5)	22.8(2)	16.3(2)
CaO				0.55(3)	0.68(3)	0.79(4)	0.81(6)
Cr_2O_3				0.32(2)	0.24(2)	0.29(1)	0.18(3)
Σ				101.19	101.24	101.65	101.55
Fo content				59	53–54	48–49	36–37

	High Ca	Low Ca	High Ca	Low Ca	High Ca	Low Ca
Pyroxene[a]						
SiO_2	51.5	51.7	51.1	51.0	50.7	50.2
Al_2O_3	1.68	3.17	1.62	1.53	1.31	1.38
TiO_2	0.29	0.31	0.42	0.36	0.38	0.41
FeO	15.5	18.6	17.5	21.7	23.2	26.3
MnO	0.37	0.29	0.36	0.39	0.29	0.26
MgO	15.4	16.6	14.4	16.4	12.3	13.6
CaO	14.5	10.4	15.3	7.35	12.3	7.21
Na_2O	0.0	0.0	0.0	0.0	0.0	0.0
Cr_2O_3	0.78	0.63	0.69	0.48	0.41	0.39
Σ	100.02	101.70	101.39	99.21	100.89	99.75
	Σ/6O	Σ/6O	Σ/6O	Σ/6O	Σ/6O	Σ/6O
SiO_2	1.941	1.916	1.925	1.953	1.952	1.953
Al_2O_3	0.075	0.138	0.072	0.069	0.059	0.063
TiO_2	0.008	0.009	0.012	0.010	0.011	0.012
FeO	0.489	0.577	0.550	0.696	0.745	0.856
MnO	0.012	0.009	0.011	0.013	0.009	0.009
MgO	0.869	0.916	0.809	0.938	0.705	0.791
CaO	0.585	0.414	0.617	0.301	0.507	0.301
Na_2O	0.000	0.000	0.000	0.000	0.000	0.000
Cr_2O_3	0.023	0.018	0.021	0.015	0.012	0.012
Σ	4.001	3.997	4.017	3.995	4.001	3.997
Wo	30.1	21.7	31.2	15.5	25.9	15.4
En	44.7	48.0	40.9	48.5	36.0	40.6
Fs	25.2	30.3	27.9	36.0	38.0	44.0

[a]Clinopyroxene composition varies substantially in the experiments. The Ca-rich and Ca-poor end members are tabulated.

*Parenthesized units represent one standard deviation of replicate analyses in terms of least units cited. Therefore 46.4(2) should read 46.4 ± 0.2. See Table 1 for run conditions.

+Phase was present in experiment but not analyzed.

K_D for olivine allow the calculation of an olivine-pigeonite thermometer. This thermometer gives a temperature estimate of 1035°C, which is in the range of the solidii for equilibrium (<1070°C) and cooling rate (<990°C) experiments. Equation (8) of Jaeger (1957) gives an expression for contact temperature. When applied using the highest reasonable initial magma temperatures for the Luna 24 basalt flow, the contact temperature estimate is in the range of 700 to 800°C. Our temperature estimate is much higher than that expected for contact metamorphism. We favor the process of foundering of ferrobasalt fragments in a lava flow, based on the compositional variation which is consistent with partial fusion and on the temperature estimate.

Crystallization of Apollo 15 Green Glass

The results of equilibrium experiments on the Apollo 15 green glass composition are summarized in Fig. 1. Olivine is the liquidus phase (~1400°C) and the $K_D^{ol\text{-}liq}$ for all experiments in the range 1390° to 1140°C is 0.33, identical to that of Longhi *et al.* (1977) for iron capsule experiments on low-Ti lunar basalt compositions. A low-Ca pyroxene (Wo_8En_{68}) crystallizes at 1206°C and plagioclase joins the paragenetic sequence at 1160°C. Residual liquid compositions and phase chemistries are summarized in Table 5. The glass starting material (Table 5, #2) is identical to the average green glass of Ridley *et al.* (1973). These experiments delineate the low-pressure cotectics in the low-Ti basalt system and aid in the identification of parental liquids for Luna 24 and Apollo 17 VLT basalts.

Origin of Luna 24 Basalts

We have argued that the Luna 24 ferrobasalt was multiply saturated with olivine and plagioclase under lunar surface conditions. Therefore, this basalt composition may represent a liquid derived by near-surface fractionation processes. Other possibilities might be considered; for example, the generation of the Luna 24 ferrobasalt composition by partial melting of the lunar interior or by tapping of liquid generated during fractional crystallization in a magma chamber at depth. The high-pressure synthesis experiments (Fig. 3) provide some limited information concerning these hypotheses. Plagioclase is the liquidus phase to 5 kbar. At 5 kbar (corresponding to a depth of ~100 km), the liquid is multiply saturated with $Wo_{30}En_{40}$ clinopyroxene and plagioclase. With increasing pressure pyroxene becomes the liquidus phase. Thus, we might propose that this Fe-rich low-Ti liquid is derived by a partial melting process at 100 km depth in which plagioclase and pyroxene are left in the residuum. Alternatively, the ferrobasalt might be a sample of a residual liquid that was derived from a more olivine normative low-Ti magma that underwent fractional crystallization in a chamber at ~100 km depth. The latter possibility deserves further consideration.

As previously noted (Grove and Vaniman, 1977) low-pressure fractionation of Apollo 15 green glass will not produce a residual liquid that is similar to Luna 24 ferrobasalt. Under near-surface conditions the parent magma must have a crystallization sequence, olivine, followed by olivine + plagioclase. This is not the case for Apollo 15 green glass, which fractionates olivine, followed by olivine + pyroxene and reaches olivine + plagioclase + pyroxene multiple saturation at a bulk composition that is unlike the Luna 24 ferrobasalt composition. With increasing pressure the olivine-pyroxene cotectic shifts toward more olivine normative compositions (Stolper, 1974; Walker *et al.*, 1975), and at approximately 5 kbar (or depths of ~100 km) the olivine-pyroxene cotectic projects into the Luna 24 ferrobasalt composition. Thus, a possible high-pressure fractionation scheme involving an Apollo 15 green glass parent would be olivine removal, followed by olivine-pyroxene fractionation and saturation with olivine, pyroxene and plagioclase. Further fractionation involving reaction of olivine and liquid occurred and the resulting liquid sampled from 100 km depth as Luna 24 ferrobasalt was pyroxene-plagioclase cosaturated.

Thus, we cannot rule out a process of high-pressure fractionation of an Apollo 15 green glass parent. However, other more olivine-normative basalts are present in the Luna 24 sample, basalts from which Luna 24 ferrobasalts may be derived by near-surface fractional crystallization involving olivine subtraction. In the following section we will explore an origin for the Luna 24 ferrobasalt composition by near-surface fractionation processes.

Relation of Ferrobasalt to Other Mare Crisium Basalts

The bulk soil composition (Blanchard *et al.*, 1977) indicates that the Luna 24 sample is predominantly mare-derived material. This low-Ti mare soil contains lithic and monomineralic fragments that may represent a degraded sample of a low-Ti basalt flow (or flows). Identification of the basalt types in the Luna 24 soil is difficult at best, since we can never be certain with such small sample sizes whether fine-grained igneous rocks represent liquid compositions or a rapidly chilled impact-generated melt of a mixture of rock types. Unfortunately, the only material available from the Mare Crisium site is submillimeter-size soil fragments and millimeter-size "rocks." Therefore, in the following discussion the reader must realize that there is some amount of uncertainty as to whether or not a particular sample is truly representative of an igneous rock.

Several magnesian glass and lithic fragments that we will call Luna 24 green glass (Table 6) have been identified in the soil. G. J. Taylor *et al.* (1977b) describe a vitrophyre (24174,59-12), an olivine-rich microporphyritic fragment (24149,42-18), a granular lithic fragment (24174,65-7) and a glass (24077,43-42) that are strikingly similar in compositon to ultramafic green glasses. The most magnesian olivine in vitrophyre 24174,59-12 is Fo_{81} (G. J. Taylor, pers. comm.), close to the equilibrium liquidus olivine for the bulk composition of the fragment (Fe/Fe + Mg = 0.40 for the fragment and the liquidus olivine should be Fo_{82}, using $K_D^{ol\text{-}liq} = 0.33$). G. J. Taylor (pers. comm.) also finds a clast of Fo_{58} olivine in the vitrophyre. The clast could be interpreted in two ways: (1) the vitrophyre is a shock-melted cumulate and the Fo_{58} olivine represents an

Table 5. Compositions of run products in Apollo 15 green glass experiments.

Run #	2	4	6	5	7	8	10	3	14	9	13
# of analyses	13	4	8	6	9	11	9	8	6	5	6
Liquid SiO_2	45.9(4)*	45.9(1)	46.2(3)	46.8(2)	47.0(4)	47.9(3)	48.3(3)	49.4(3)	49.7(2)	48.4(2)	47.9(3)
Al_2O_3	7.94(43)	7.65(45)	8.50(25)	8.47(23)	9.70(7)	9.89(36)	10.6(2)	11.1(2)	11.7(2)	12.8(2)	11.4(1)
TiO_2	0.43(3)	0.43(3)	0.47(4)	0.48(4)	0.51(3)	0.52(3)	0.55(4)	0.61(5)	0.62(4)	0.70(5)	1.53(3)
FeO	19.3(3)	19.1(2)	19.5(2)	19.2(1)	19.1(3)	18.6(2)	18.1(1)	16.9(2)	16.6(2)	16.4(1)	19.4(3)
MnO	0.22(3)	0.22(3)	0.22(3)	0.21(3)	0.23(2)	0.22(3)	0.23(3)	0.21(3)	0.18(2)	0.27(6)	0.24(3)
MgO	17.4(3)	16.9(1)	15.0(2)	14.1(1)	12.3(2)	11.0(2)	9.82(2)	8.92(14)	8.05(7)	7.42(30)	6.39(9)
CaO	8.43(25)	8.51(8)	9.06(51)	9.26(13)	9.94(37)	10.8(4)	11.2(4)	11.9(5)	12.6(3)	12.8(2)	12.4(2)
Na_2O	0.0	0.0	0.02	0.0	0.02	0.01	0.02	0.0	0.01(1)	0	0.08(3)
K_2O	0.01	0.02	0.01	0.01	0.01	0.01	0.03	0.02	0.02(1)	—	0.05(2)
Cr_2O_3	0.36(4)	0.41(3)	0.41(3)	0.42	0.44(2)	0.43(4)	0.40(2)	0.43(3)	0.39(2)	0.43(4)	0.31(2)
Σ	99.99	99.14	99.40	98.95	99.25	99.38	99.25	99.49	99.87	99.22	99.70
Olivine											
# of analyses		4	5	5	5	7	5	4	3	4	3
SiO_2		39.2(3)	38.3(3)	38.7(2)	37.7(3)	37.6(3)	37.5(2)	37.2(1)	37.7(5)	36.7(2)	36.4(1)
Al_2O_3		0.20(1)	0.13(2)	0.14(5)	0.14(5)	0.14(1)	0.14(3)	0.15(2)	0.12(1)	0.13(2)	0.05(4)
TiO_2		0.0	0.0	0.0	0.01(1)	0.01	0.01	0.0	0.02	0.03(1)	0.04(3)
FeO		16.6(1)	18.3(3)	19.0(1)	20.7(3)	22.2(2)	23.3(1)	23.8(2)	25.1(4)	26.6(3)	31.7(2)
MnO		0.16(3)	0.14(3)	0.19(1)	0.20(3)	0.21(2)	0.21(4)	0.25(2)	0.27(1)	0.27(2)	0.31(3)
MgO		44.7(3)	43.7(4)	42.5(3)	41.1(2)	39.3(3)	38.5(6)	38.5(3)	36.9(5)	36.3(4)	31.9(1)
CaO		0.36(2)	0.31(3)	0.36(3)	0.35(2)	0.39(5)	0.39(2)	0.46(9)	0.46(1)	0.42(2)	0.53(6)
Cr_2O_3		0.29(1)	0.26(2)	0.30(3)	0.33(4)	0.34(4)	0.36(3)	0.42(2)	0.45(3)	0.41(4)	0.31(9)
Σ		101.51	101.14	101.19	100.53	100.79	100.41	100.78	101.02	100.86	101.24
Fo content		83	81	80	78	76	75	74	72	70–71	64

# of analyses		4		2		4	
Pyroxene			Σ/6 O		Σ/6 O		Σ/6 O
SiO_2		54.0(2)	1.952	52.2(2)	1.943	52.6(1)	1.941
Al_2O_3		1.49(38)	0.063	1.51(9)	0.066	1.60(21)	0.070
TiO_2		0.11(1)	0.003	0.14(1)	0.004	0.23(9)	0.007
FeO		15.2(2)	0.458	15.2(2)	0.473	16.7(9)	0.516
MnO		0.23(1)	0.007	0.23(3)	0.008	0.25(7)	0.008
MgO		24.8(1)	1.336	23.5(1)	1.301	20.5(6)	1.127
CaO		4.13(22)	0.160	4.65(6)	0.185	8.01(90)	0.317
Na_2O		0.0	0.000	0.0	0.000	0.0	0.000
Cr_2O_3		0.80(3)	0.023	0.93(4)	0.028	0.76(7)	0.022
Σ		100.76	4.002	98.36	4.008	100.65	4.008
	Wo	8.2		9.4		16.2	
	En	68.4		66.4		57.5	
	Fs	23.4		24.1		26.3	
						Plagioclase	
						An_{99}	

*Parenthesized units represent one standard deviation of the replicate analyses in terms of least units cited. Therefore, 45.9(4) should read 45.9 ± 0.4. See Table 3 for run conditions.

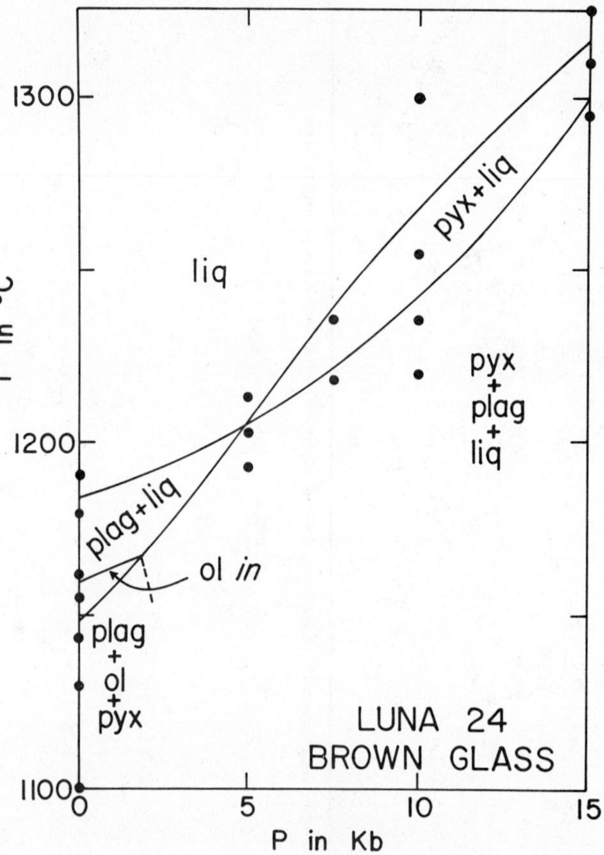

Fig. 3. Summary of iron capsule synthesis experiments on the Luna 24 ferrobasalt at high pressures. Run conditions are summarized in Table 3. Olivine is not a liquidus phase at any pressure. The olivine disappearance at 2 kbar is estimated from the low-Ti basalt data of Walker *et al.* (1976a).

unmelted relict, or (2) the Fe-rich olivine was a mineral fragment incorporated during the extrusion of the liquid across the lunar surface. Sample 24174,65-7 contains olivine in the compositional range Fo_{67} to Fo_{61}, considerably more iron-rich than the calculated liquidus olivine for this bulk composition (Fe/Fe + Mg = 0.35, liquidus olivine = Fo_{85}). Thus, this fragment is interpreted as a cumulate.

Tarasov *et al.* (1977) identified a vitrophyric basalt (1516, Table 6, col. 15) that consists of olivine phenocrysts and laths of plagioclase in glassy groundmass. This basalt's bulk chemistry is similar to the bulk soil composition (Table 6, col. 16; Blanchard *et al.*, 1977) and to a glass chemistry identified by several groups (Table 6, col. 9; Simonds *et al.*, 1977; Norman *et al.*, 1977; G. J. Taylor *et al.*, 1977b) and designated Mg-mare glass by Simonds *et al.* Tarasov *et al.* (1977)

Table 6. Comparison between estimated compositions of the Luna 24 ferrobasalt, with additional data of Luna 24 Mg-rich mare compositions and bulk soil.

	1	2	3	4	5	6	7	8	9	10	11	12	13	14	15	16
SiO_2	47.9	46.9	46.7	45.7	47.1	46.5	—	45.3	45.6	43.6	46.2	43.6	44.2	43.1	44.8	45.4
Al_2O_3	9.0	11.3	12.9	13.8	12.6	11.8	11.6	12.4	12.2	7.8	12.3	7.8	7.7	6.9	11.1	11.1
TiO_2	0.70	0.70	0.83	0.86	0.88	1.1	1.3	1.2	1.09	0.88	1.39	0.55	0.62	0.15	0.82	1.1
FeO	21.9	21.6	21.0	19.0	20.2	22.9	22.4	20.3	20.3	22.8	21.6	21.1	21.6	20.1	21.0	20.5
MnO	0.26	0.26	0.29	0.28	0.25	0.27	0.26	0.27	0.19	0.18	0.29	0.16	0.43	0.25	0.29	—
MgO	5.7	5.6	6.90	7.06	6.31	5.8	7.0	7.5	10.2	17.7	5.7	18.6	18.2	20.8	10.4	10.2
CaO	14.4	13.3	11.5	13.0	12.8	11.5	12.3	12.2	11.3	7.2	11.4	7.9	7.0	6.9	9.9	11.0
Cr_2O_3	0.24	0.24	0.26	0.25	0.23	0.30	0.38	0.23	0.37	0.32	0.51	0.14	0.53	0.65	0.41	0.52
Na_2O	—	—	0.27	0.22	0.20	0.26	0.29	0.37	0.23	0.38	0.59	0.12	0.15	0.12	0.32	0.28
K_2O	—	—	0.01	0.01	0.02	0.02	—	0.03	0.04	0.05	0.07	0.01	0.05	0.03	0.18	0.03
Σ	100.1	99.9	100.7	100.2	100.6	100.45	—	99.80	101.52	100.91	100.05	99.98	100.48	99.00	99.22	100.13
CaO/Al_2O_3	1.60	1.18	0.89	0.94	1.02	0.97	1.06	0.98	0.92	0.92	0.92	1.01	0.91	1.00	0.89	0.99
Mg/(Fe + Mg)	0.32	0.32	0.37	0.40	0.36	0.31	0.36	0.40	0.47	0.58	0.32	0.62	0.60	0.65	0.47	0.47

Columns (1–8) = Ferrobasalt compositions. Columns (1,2): Calculated analyses using maximum and minimum CaO/Al_2O_3, from K_D data of Grove and Bence (1977) applied to pyroxenes of the Luna 24 ferrogabbro; Column (3): Average of two modal recombinations of recrystallized (metamorphosed) ferrobasalts, 24109,51 frag. 11 and 24109,51 frag. 35; Column (4)(Average of two modal recombinations of ophitic ferrobasalts, 24109,52 frag. 18 and 24077,41 frag. 21; Column (5): Average analysis of ten brown glass fragments; Column (6): INAA analysis of ophitic ferrobasalt 24174,7: J. C. Laul (1977) analyst; Column (7): INAA analysis of ophitic ferrobasalt 24109,78: M.-S. Ma *et al.* (1977) analyst; Column (8): Electron beam analysis of rock powder 24143,4–006.2; Tarasov *et al.* (1977) analyst. Columns (9–16) = Luna 24 Mg-rich mare basalts and bulk soil. Column (9): Mg-rich Luna 24 mare glass 24109,51–39; Column (10): Mg-rich Luna 24 mare glass 24077, 43–34: G. J. Taylor *et al.*, 1977b; Column (11): Mare glass 24077,43–34 of Taylor *et al.*, previous column, with calculated fractionation of 46 wt.% olivine; Column (12): Average brown glass, cited in column 5, with 48 wt.% olivine addition; Column (13): Mg-rich Luna 24 olivine vitrophyre 24174,59–12: G. J. Taylor *et al.*, 1977b; Column (14): Mg-rich Luna 24 olivine-rich granular rock 24174,65–7: G. J. Taylor *et al.*, 1977b; Column (15): Mg-rich Luna 24 vitrophyre 1516: Tarasov *et al.*, 1977; Column (16): Average bulk soil of Blanchard *et al.*, 1977.

provide only a bulk analysis for basalt (1516), but we can calculate the liquidus olivine. Fe/(Fe + Mg) for basalt 1516 is 0.53 and in the range 0.50 to 0.55 for the glass and soil chemistries, so the liquidus olivine for the basalt (assuming $K_D^{ol\text{-}liq} = 0.33$) would be Fo_{73}, or in the range Fo_{71} to Fo_{75} for the other chemistries. This Mg-rich mare basalt is also represented in the monomineralic fragments in the soil. Papp *et al.* (1977); Bence and Grove (1977); Nielsen and Drake (1977); Basu *et al.* (1977) and G. J. Taylor (pers. comm.) recognized a magnesian olivine component (Fo_{70} to Fo_{75}) that suggests a more magnesian basalt related to the Luna 24 ferrobasalt by olivine addition.

The workers cited above also recognized magnesian low-Ca clinopyroxenes (Wo_7En_{70} to $Wo_{12}En_{62}$) as rare monomineralic and lithic fragments with olivine. Estimates of the magma that generated these pyroxenes were made using the pyroxene-liquid partition coefficients of Grove and Bence (1977). The results indicate that this other basalt magma becomes olivine-pyroxene saturated before reaching plagioclase saturation and has a TiO_2 content that may be higher (estimates vary from 0.8 to 2.5 wt.% TiO_2 depending on the pyroxene analysis chosen). As suggested by Nielsen and Drake (1977, 1978) at least two Mg-rich basalt types are represented in the soil and some melt breccias (e.g., 24077,4 analyzed by J. C. Laul, 1977) may be impact mixtures of both basalts. This second basalt type is not considered in the remainder of the discussion and must represent another magnesian VLT or low-Ti mare basalt, that cannot be related to Luna 24 ferrobasalt by near-surface fractionation.

Relationships between Basalt Compositions at Mare Crisium

The experiments on Luna 24 ferrobasalt and Apollo 15 green glass provide the necessary information to test whether or not the Luna 24 green glass, the Mg-rich basalt and the Luna 24 ferrobasalt composition are related by near-surface fractionation processes. We will use subprojections in the system Ol-An-Cpx-Qtz (see Appendix), TiO_2 content and CaO/Al_2O_3 vs. Fe/(Fe + Mg) variations to explore the relationships. The experimentally determined liquid lines of descent for the Luna 24 and Apollo 15 green glass experiments are shown in the subprojections Ol-An-Qtz, Ol-Cpx-Qtz and Ol-An-Cpx (Fig. 4). The Luna 24 green glass + lithic fragments, the Mg-mare basalt, the bulk soils and the ferrobasalt are also plotted in the three subprojections. The compositions are scattered, but lie in the vicinity of an olivine addition or subtraction line. We suggest that these three rock types are related by near-surface fractionation processes involving olivine settling in a basalt flow or a shallow magma chamber. The subprojections allow this relationship, but do not provide complete information. Therefore, calculations based on experimentally derived constraints are desirable. The necessary experimental data are: a $K_D^{ol\text{-}liq}$ appropriate for olivine removal from the Luna 24 green glass magma, and the characteristics of residual liquid compositions that are olivine + plagioclase saturated. With this information we can quantitatively calculate an ideal fractional crystallization path for the Luna 24 green glass by removing a small amount of olivine of the appropriate composition (2% by wt.), renormalizing the liquid to 100%, recalculating the appropriate olivine composition, removing 2% of this olivine, etc. Combined with this iterative olivine fractionation process, we test for saturation with plagioclase. The Apollo 15 green glass, Luna 24 ferrobasalt and 15597 (Grove and Bence, 1977) experiments are plagioclase saturated when their liquids contain molar olivine/anorthite ratios of ~1.48 (using the projection scheme described in the Appendix). Thus, the calculation must include plagioclase removal when this ratio is reached in the fractionation calculation. The results are indicated by the dash-dot line for green glass 24077,43-34. If this glass is representative of the initial magma composition at Luna 24, then it would undergo about 46 wt.% Fo_{81} to Fo_{59} olivine fractionation and become plagioclase saturated. Both the projections and the calculated residual liquid composition (Table 6) show that this green glass will generate a composition that is similar, but not identical to Luna 24 ferrobasalt.

The reverse of the ideal fractional cyrstallization calculation was performed by starting with the Luna 24 ferrobasalt bulk composition, and assuming that it preserved the magma composition at which plagioclase saturation occurred. Olivine (2 wt.%) of the appropriate composition is added, a new liquid composition is calculated, and 2% olivine of the appropriate composition is added, etc. After the addition of ~48 wt.% olivine of Fo_{60} to Fo_{82} (Table 6) this calculation generates a Luna 24 green glass composition that is similar but not identical to the green glass and lithic fragments. Therefore, it is possible to generate ferrobasalt-like magmas from Luna 24 green glass-like starting materials and to carry out the reverse process. The CaO/Al_2O_3 and TiO_2 vs Fe/(Fe + Mg) plots (Fig. 5) show that the variations produced in Fe/(Fe + Mg) in both calculations span the appropriate range. A green glass composition with Fo_{82} liquidus olivine is generated by the olivine addition calculation and an iron-rich olivine (Fo_{59}) is produced when the residual liquid becomes plagioclase saturated in the olivine subtraction calculation. The failure of either calculation to provide a direct relation between a specific green glass parent and the fractionated ferrobasalt magma is probably related to our inability to identify a representative Luna 24 green glass parent material. The compositional spread of the glass and lithic fragment analyses (Figs. 4 and 5) may result from analytical biases between different microprobe facilities or from mixing of rock types at Mare Crisium during the glass-forming events. Our calculations do not account for the removal (or addition) of a small amount of Cr-spinel, which is found in lithic fragments (Haggerty, 1977). The presence of spinel is suggested by the Cr_2O_3 content of the soil, and it is probably a near-liquidus phase in the Mg-rich mare basalt and Luna 24 green glass.

Therefore, we suggest that a Luna 24 green glass magma underwent fractional crystallization involving the subtraction of olivine + minor Cr-rich spinel to produce the Mare Crisium Mg-rich basalt and the Luna 24 ferrobasalt. Approximately 28 wt.% olivine fractionation (Fo_{82} to Fo_{74}) would produce the Mg-rich mare component and an additional 20 wt.% (Fo_{74} to Fo_{60}) olivine removal generates the Luna 24 ferrobasalt. We do not feel that we have

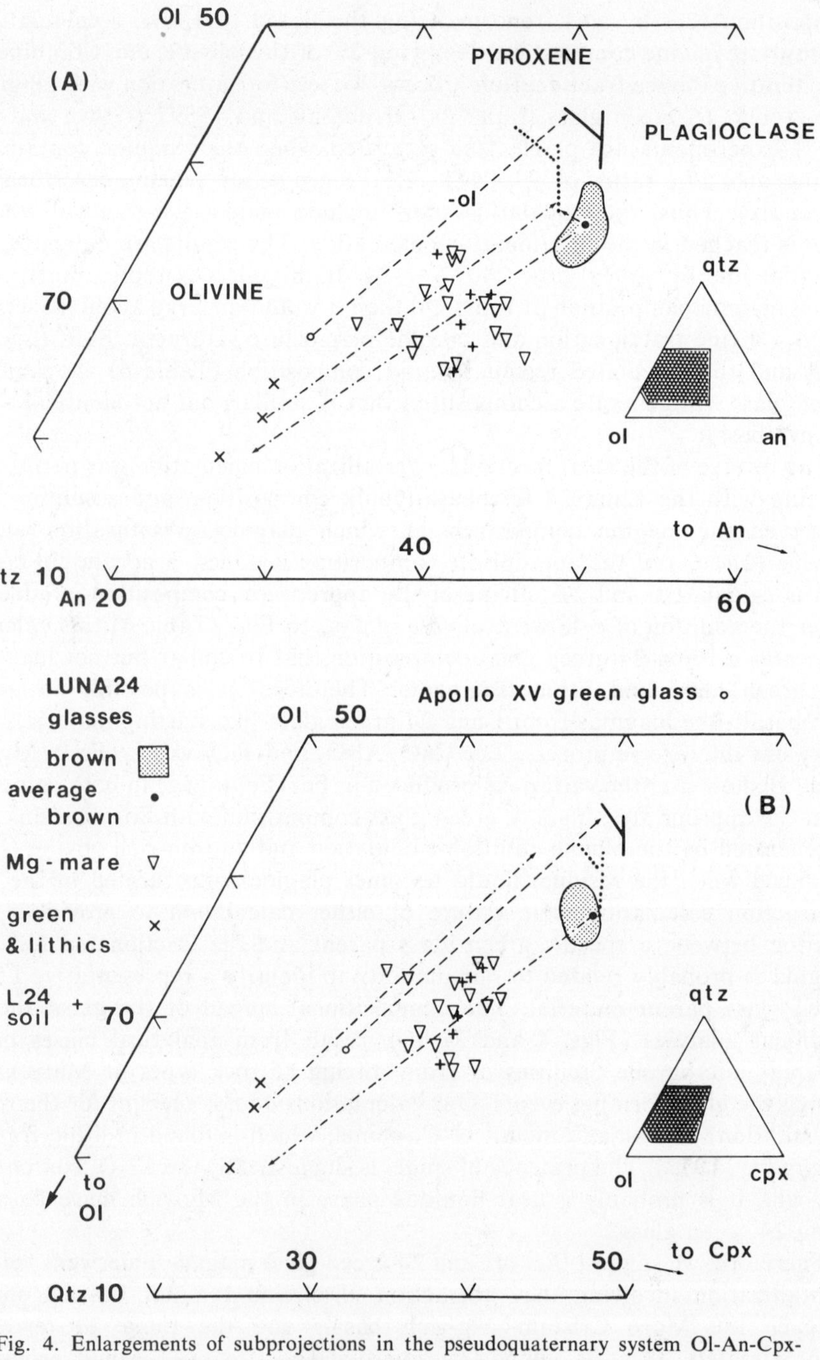

Fig. 4. Enlargements of subprojections in the pseudoquaternary system Ol-An-Cpx-Qtz. These projections along with diagrams that illustrate Fe/(Fe + Mg) variation are used to constrain low-pressure phase relations in VLT basalts. The projection scheme is described in the Appendix. (a) Portion of the subprojection Ol-An-Qtz. The primary

phase volumes for anorthite, olivine and pyroxene are outlined by the experimentally determined cotectics. The dashed lines are the olivine subtraction and plagioclase subtraction lines for the Apollo 15 and Luna 24 experiments, respectively. Solid cotectics are for the Apollo 15 experiments and dotted cotectics are determined from the Luna 24 experiments. These cotectics are not extended beyond the experimentally determined limits, because they are only applicable to the bulk composition upon which the experiment was performed. (b) Portion of the subprojection Ol-Cpx-Qtz. (c) Portion of the subprojection Ol-An-Cpx. Mg-rich mare glasses are from G. J. Taylor *et al.* (1977b), Norman *et al.* (1977), Simonds *et al.* (1977) and our laboratory. Soil analyses are from Blanchard *et al.* (1977). Luna 24 green glasses are from G. J. Taylor *et al.* (1977b).

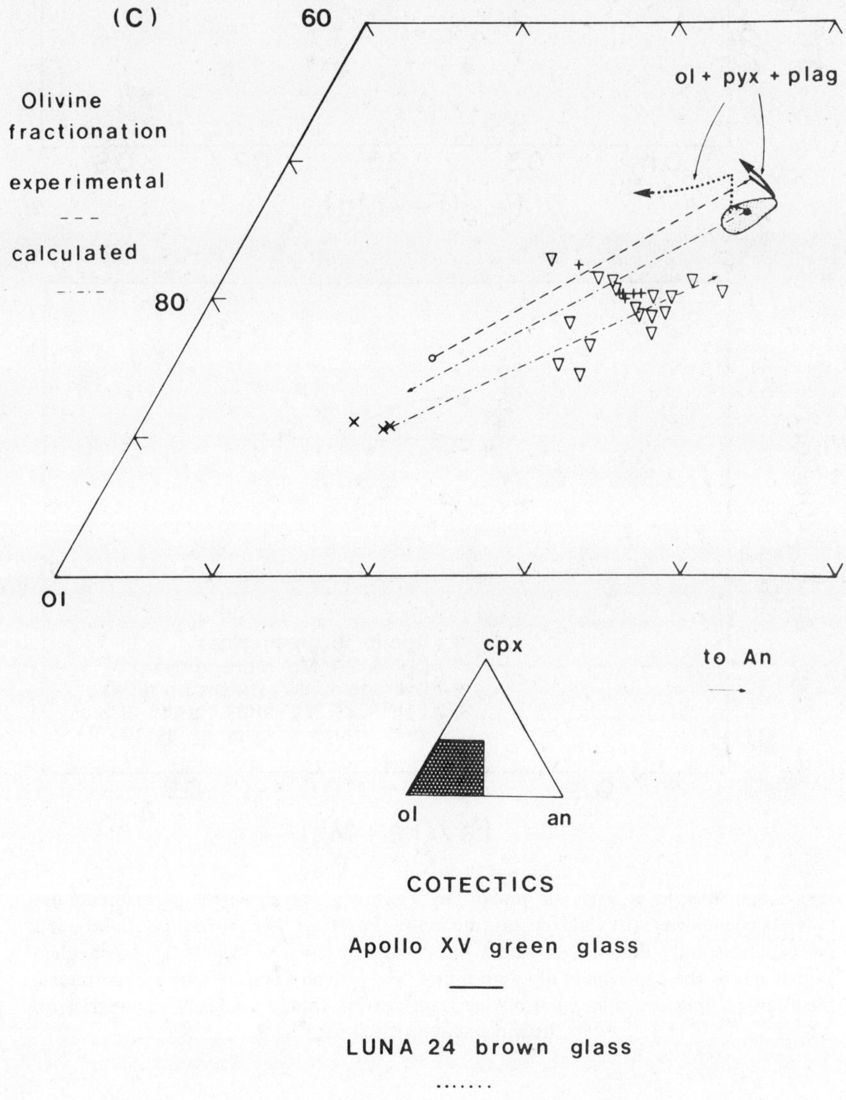

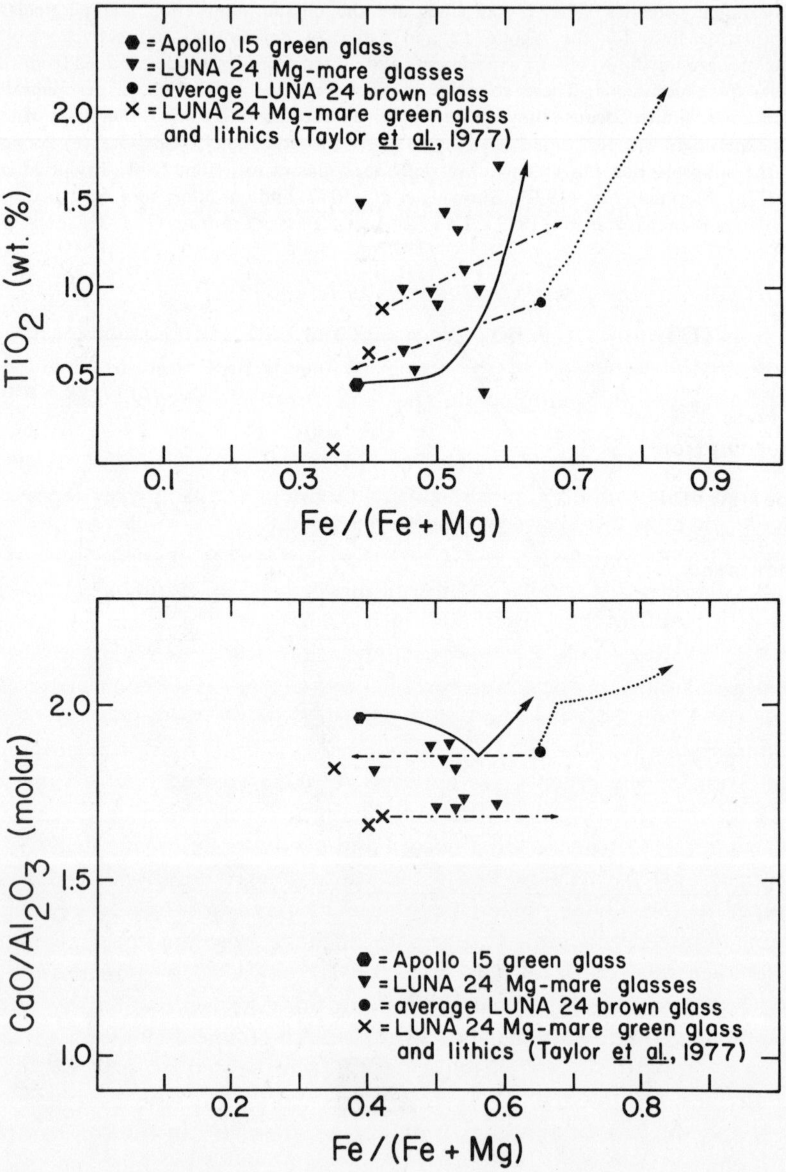

Fig. 5. (a) Weight % TiO_2 vs. molar Fe/(Fe + Mg) variation for experiments and Luna 24 chemistries. (b) CaO/Al_2O_3 (molar) vs. Fe/(Fe + Mg) variations. Solid line is the experimentally determined liquid line-of-descent for the Apollo 15 green glass. Dotted line is the experimentally determined line for the Luna 24 basalt experiments. Dash-dotted lines are calculated olivine fractionation trends. Luna 24 chemistries are from the same sources cited for Fig. 4.

positively identified the parental magma composition, because the compositions of the three basalt types may not be truly representative. One would like a larger data base, but we feel that our hypothesis has some merits: (1) The ferrobasalt is an abundant well-characterized rock type and perhaps the best estimate of the Luna 24 parent magma can be gained by working backwards from this material; (2) The Mg-rich mare member of the suite is a little more difficult to pick out, but it appears to be a well-represented rock type; and (3) The Luna 24 green glasses and lithic fragments provide the most Mg-rich parental candidates.

Reconstruction of the Mare Crisium Green Glass Flow

Our consideration of the rock types and the bulk soil chemistry allows us to speculate on the petrology of the Luna 24 basalt flow that was broken up by meteorite impacts and sampled by the drill core. We suggest that at least six mare rock types are represented in the soil: (1) Luna 24 ferrobasalt; (2) ferrogabbro; (3) Mg-vitrophyre (1516); (4) a coarse-grained Mg-mare basalt (represented by the monomineralic Fo_{70-75} olivine; (5) vitrophyres representing a chill margin which preserved the green glass composition; and (6) another flow represented by the magnesian, low-Ca pyroxene. Further, it is possible that first 5 rock types were produced by a differentiation process involving chiefly olivine settling in a flow (or flows) with the bulk composition of Luna 24 green glass. Our model of the Luna 24 flow assumes that the green glass magma was extruded as a liquid on the lunar surface, and subsequently cooled and differentiated at the Luna 24 site. The calculations of olivine settling in an Apollo 12 picrite magma by Walker *et al.* (1976b) serve as a basis for our discussion of the differentiation of the green glass magma. A reconstructed flow is illustrated in Fig. 6. Chill margins preserve the bulk magma composition as Luna 24 green glass vitrophyres. Olivine settling occurs and produces an olivine depleted region in the upper part of the flow and a cumulate zone. The abundant ferrobasalts crystallized in the olivine depleted region. The ferrogabbros also crystallized in the olivine depleted zone, but at a greater distance from the upper contact of the flow. Between the chill margin and the ferrobasalts, a moderately olivine-depleted liquid would crystallize. This liquid may be represented by vitrophyre 1516. Sandwiched between the zone of maximum olivine depletion and the basal olivine cumulates, the Mg-rich mare basalt (represented by the monomineralic Fo_{70-75} olivine) crystallized. Only this upper portion of the flow needs to be sampled, and the basal cumulates may not be abundant in the soil sample. The soil bulk chemistry is consistent with a sample of only the upper portion of the green glass flow, because the bulk soil is intermediate between the green glass parent and the differentiated ferrobasalt. If a complete flow had been sampled, the soil composition should represent the magma composition.

A similar reconstruction could be made assuming that the green glasses represent impact remelted samples of a basal cumulate, vitrophyre 1516 a sample of chill margin that preserved the initial magma composition, and the soil a representative sample of the lava flow. Again, the ferrobasalts and ferrogabbros

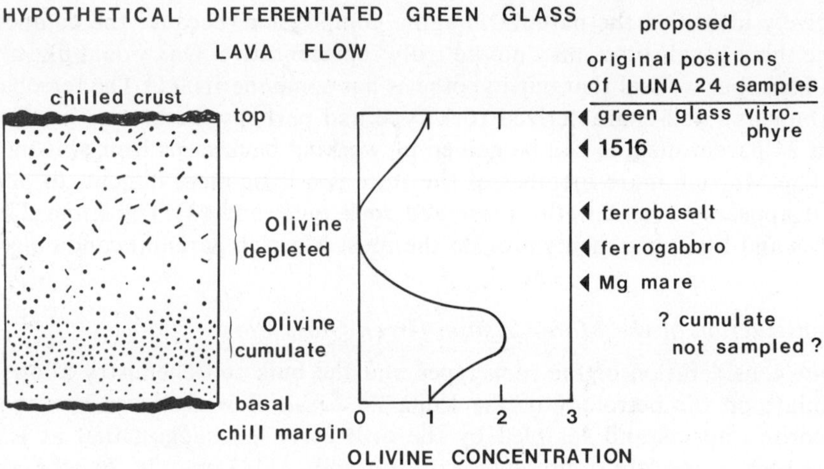

Fig. 6. Schematic diagram to illustrate the differentiation of Luna 24 green glass magma (see Walker *et al.*, 1976b).

would represent olivine depleted basalts from the upper interior portion of the flow. There is no compelling reason to choose between the two models. Both models may be extremely over-simplified; several flows may be represented in the soil or a sample of predominantly olivine-depleted ferrobasalt may be mixed with minor amounts of basal olivine cumulate to form the soil. We prefer the choice of Luna 24 green glass as the parent magma composition because abundant glasses of similar chemical composition have been found at other mare sites. One might expect that such a widespread magma type would appear somewhere on the moon as basalt flows, and we propose that at Mare Crisium we have sampled a Luna 24 green glass magma that crystallized and differentiated at the lunar surface. Ideas for the origin of green glasses never fail to stir up a lively discussion, but there is compelling evidence to accept the glass chemistry as representing a basaltic liquid. The presence of volcanic sublimates on the glass spheres (Butler and Meyer, 1976; Meyer *et al.*, 1975) is probably the strongest argument for the origin of the glass spheres by volcanic fire fountaining. A magma composition as olivine normative as green glass would only be preserved as vitrophyres and rapidly chilled glasses, and would undergo substantial differentiation by olivine settling in a lava flow. Therefore, the identification of the parent magma of a green glass flow would not be an easy task, since most of a lava flow (of substantial thickness) would consist of cumulates or olivine depleted basalts.

Relationships between Green Glass Types

Grove and Vaniman (1977) concluded that Apollo 15 green glass and Apollo 17 VLT basalts were related by near-surface fractionation processes involving olivine fractionation. We were guilty of not considering all subprojections in the projection scheme that we employed. Apollo 17 VLT basalts and Apollo 15 average green glass cannot be related to one another by processes of near-surface fractional crystallization. Apollo 17 VLT basalts were derived from magmas with substantially lower CaO/Al_2O_3 ratios and higher TiO_2 and Cr_2O_3 contents. If the Luna 24, Apollo 17 VLT and Apollo 15 green glasses are truly igneous and represent melts produced in the lunar interior, the compositional differences between the three magmas (Fig. 7) require three distinct source regions and/or different degrees of partial melting in each source.

This is shown in the three subprojections by differences in anorthite/clinopyroxene ratio as well as in olvine content for each glass type (Fig. 7). It has been suggested that the Apollo 15 green glass chemistry was derived by partial melting of lunar material at depth (Walker *et al.*, 1975; Green and Ringwood, 1973). We concur with an origin by partial melting, but we are not prepared to take sides on the nature of the source material or the degree of partial melting. Where abundant green glasses have been collected (e.g., Mare Imbrium) they show a large compositional variation (Hlava *et al.*, 1973; Stolper, 1974; Stolper *et al.*, 1974) These compositional differences cannot be produced by olivine fractionation at depth or near the lunar surface. Thus, the magmas are not related by a simple mechanism. Grove and Lindsley (1978) have explored the compositional variations in Apollo 15 green glass and concluded that a continuous melting mechanism of a lunar mantle source at depths of 400 km or greater may give rise to the compositional variation in Apollo 15 glass chemistry. If the three magmas shown in Fig. 7 originated by melting of lunar mantle material, the compositional variations within each glass group may provide a clue to the depth at which each melt was generated. Experiments to test this possibility are in progress in our laboratory.

Conclusions

We conclude that: (1) Near-surface olivine fractionation (48 wt.%, Fo_{82} to Fo_{60}) of a Luna 24 green glass parent can generate the Luna 24 ferrobasalt. (2) The Luna 24 soil may be representative of basalts that underwent less olivine fractionation (20–26 wt.%, Fo_{82} to Fo_{74}). (3) Luna 24 green glass and lithic fragments of vitrophyres may be samples of the parent magma that differentiated at or near the Mare Crisium site to produce the Luna 24 ferrobasalts. Calculation of simple olivine fractionation fails to relate ferrobasalt to the Luna 24 green glass. The failure of this exercise results from a lack of good compositional control on the green glass parent, and from neglecting to account for a small quantity of Mg-chromite that must have been involved in the fractionation. In spite of our inability to identify the material parental to Luna 24 ferrobasalt, we have quantitatively outlined the near-surface processes that produce this fractionated liquid composition and propose a model for generating this basalt in a flow or sill.

We suggest that the enigmatic lunar green glasses represent basalts that were

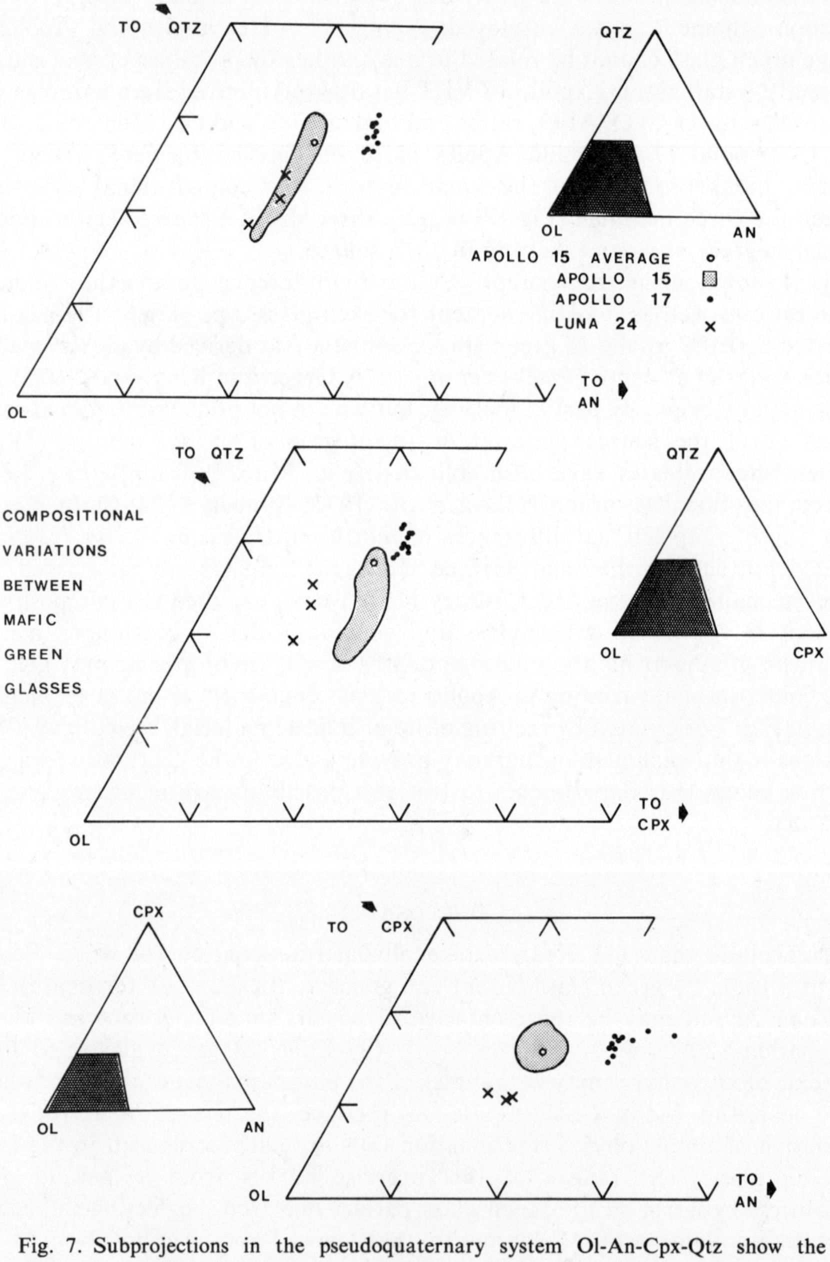

Fig. 7. Subprojections in the pseudoquaternary system Ol-An-Cpx-Qtz show the chemical variations between ultramafic green glasses from Mare Crisium, Mare Imbrium and Mare Serenitatis. The Apollo 15 green glass analyses are from Hlava *et al.* (1973) and the average green glass is that of Warner *et al.* (1972) (also used for the experiments).

generated in the lunar interior. These basalts were erupted as volcanic fire fountains and as lava flows. Samples from Luna 24, Apollo 17 (Vaniman and Papike, 1977a) and possibly Apollo 15 (Steele *et al.*, 1977) contain basalts that are near-surface differentiates of green glass magmas.

Acknowledgments—This work was supported by NASA grant NGL-33-015-130 (TLG) and NSG-9044 (DTV). We thank A. E. Bence and J. J. Papike for assistance in probe analysis of the Luna 24 samples. G. J. Taylor generously provided analyses of Luna 24 fragments. Walter Holzwarth maintained the electron microprobe facility.

References

Albee A. L. and Ray L. (1970) Correction factors for electron probe microanalysis of silicates, oxides, carbonates, phosphates and sulfates. *Anal. Chem.* **42**, 1408–1414.

Basu A., McKay D. S., Clanton U., Waits G. and Fruland R. M. (1977) Glass particles in the Luna 24 drill core soils (abstract). In *Papers Presented to the Conference on Luna 24*, p. 14–17. The Lunar Science Institute, Houston.

Bence A. E. and Albee A. L. (1968) Empirical correction factors for electron microanalysis of silicates and oxides *J. Geol.* **76**, 382–403.

Bence A. E. and Grove T. L. (1977) The highland component in the Luna 24 core (abstract). In *Papers Presented to the Conference on Luna 24*, p. 22–24. The Lunar Science Institute, Houston.

Bence A. E., Grove T. L. and Scambos T. (1977) Gabbros from Mare Crisium: An analysis of the Luna 24 soil. *Geophys. Res. Lett.* **4**, 493–496.

Biggar G. M. (1972) Diopside, lithium metasilicate, and the 1968 temperature scale. *Mineral. Mag.* **38**, 768–770.

Blanchard D. P., Haskin L. A., Brannon J. C. and Aaboe E. (1977) Chemistry of soils and particles from Luna 24 (abstract). In *Papers Presented to the Conference on Luna 24*, p. 37–40. The Lunar Science Institute, Houston.

Boyd F. R. and England J. L. (1960) Apparatus for phase equilibrium measurements at pressures up to 50 kilobars and temperatures up to 1750°C. *J. Geophys. Res.* **65**, 741–748.

Butler P., Jr. and Meyer C., Jr. (1976) Sulfur prevails in coatings on glass droplets: Apollo 15 green and brown glasses and Apollo 17 orange and black (devitrified) glasses. *Proc. Lunar Sci. Conf. 7th*, p. 1561–1581.

Friel J. J. and Goldstein J. I. (1977) Metallic phases in the Luna 24 soil samples. *Geophys. Res. Lett.* **4**, 481–483.

Green D. H. and Ringwood A. E. (1973) Significance of a primitive lunar basaltic composition present in Apollo 15 soils and breccias. *Earth Planet Sci. Lett.* **19**, 1–8.

Grove T. L. (1978a) Cooling histories of Luna 24 low-Ti ferrobasalts and ferrogabbros (abstract). In *Lunar and Planetary Science IX*, p. 424–426. Lunar and Planetary Institute, Houston.

Grove T. L. (1978b) Experimentally determined FeO-MgO-CaO partitioning between pyroxene and liquid in lunar basalts (absract). *EOS (Trans. Amer. Geophys. Union)* **59**, 401.

Grove T. L. and Bence A. E. (1977) Experimental study of pyroxene-liquid interaction in quartz-normative basalt 15597. *Proc. Lunar Sci. Conf. 8th*, p. 1549–1579.

Grove T. L. and Vaniman D. T. (1977) Experimental petrology of very low Ti basalts and origin of Luna 24 ferrobasalt (abstract). In *Papers Presented to the Conference on Luna 24*, p. 68–71. The Lunar Science Institute, Houston.

Grove T. L. and Lindsley D. H. (1978) Compositional variation and origin of lunar ultramafic green glasses (abstract). In *Lunar and Planetary Science IX*, p. 430–432. Lunar and Planetary Institute, Houston.

Haggerty S. E. (1977) Luna 24: Opaque mineral chemistry of gabbroic and basaltic fragments from Mare Crisium. *Geophys. Res. Lett.* **4**, 489–492.

Hlava P., Green J., Prinz M., Nehru C., Keil K., Dowty E. and Bunch T. (1973) Apollo 15 rake sample microbreccias and non-mare rocks: bulk rock, mineral and glass electron probe analyses. Inst. Meteoritics Spec. Publ. 11, Univ. of New Mexico, Albuquerque.

Jaeger J. C. (1957) The temperature in the neighborhood of a cooling intrusive sheet. *Amer. J. Sci.* **255**, 306–318.

Kesson S. E. (1975) Mare basalts: Melting experiments and petrogenetic interpretations. *Proc. Lunar Sci. Conf. 6th*, p. 921–944.

Kesson S. E. and Lindsley D. H. (1976) Mare basalts: A review of experimental studies. *Rev. Geophys. Space Phys.* **14**, 361–373.

Laul J. C. (1977) Chemical composition of Luna 24 melt rocks (24077,4; 24077,62; 24174,7; 24182,12; 24210,50) and gabbro (24182,8) (abstract). In *Papers Presented to the Conference on Luna 24*, p. 94–97. The Lunar Science Institute, Houston.

Lindsley D. H., Kesson S. E., Hartzman M. J. and Cushman M. K. (1974). The stability of armalcolite: Experimental studies in the system MgO-Fe-Ti-O. *Proc. Lunar Sci. Conf. 5th*, p. 521–534.

Longhi J., Walker D. and Hays J. F. (1977) The distribution of Fe and Mg between olivine and lunar basaltic liquids. Submitted to *Geochim. Cosmochim. Acta.*

Ma M.-S., Schmitt R. A., Taylor G. J., Warner R. D., Lange D. E. and Keil K. (1977) Chemical and petrographic studies of 18 Luna 24 lithic fragments (abstract). In *Papers Presented to the Conference on Luna 24*, p. 102–105. The Lunar Science Institute, Houston.

Meyer C., Jr., McKay D. S., Anderson D. H. and Butler P., Jr. (1975) The source of sublimates on the Apollo 15 green and Apollo 17 orange glass samples. *Proc. Lunar Sci. Conf. 6th*, p. 1673–1699.

McSween H. Y., Jr., Taylor L. A. and Clark J. C. (1977) Luna 24 metabasalts: Possible evidence for assimilation of highland materials? (abstract). In *Papers Presented to the Conference on Luna 24*, p. 118–120. The Lunar Science Institute, Houston.

Nielsen R. L. and Drake M. J. (1977) Magma types at the Luna 24 site in Mare Crisium (abstract). In *Papers Presented to the Conference on Luna 24*, p. 132–135. The Lunar Science Institute, Houston.

Nielsen R. L. and Drake M. J. (1978) The case for at least three VLT magma types in the Luna 24 drill core (abstract). In *Lunar and Planetary Science IX*, p. 802–804. Lunar and Planetary Institute, Houston.

Norman M., Coish R. A. and Taylor L. A. (1977) Glass chemistry and magma evolution at Mare Crisium (abstract). In *Papers Presented to the Conference on Luna 24*, p. 136–138. The Lunar Science Institute, Houston.

Papp H., Steele I. M. and Smith J. V. (1977) Luna 24 90–150 micron fraction: Implications for sampling of planetary and asteroidal regoliths (abstract). In *Papers Presented to the Conference on Luna 24*, p. 143–146. The Lunar Science Institute, Houston.

Ridley W. I., Reid A. M., Warner J. L. and Brown R. W. (1973) Apollo 15 green glass. *Phys. Earth Planet. Inter.* **7**, 133–136.

Ryder G., McSween H. Y., Jr., and Marvin U. B. (1977) Basalts from Mare Crisium. *The Moon* **17**, 263–287.

Simonds C. H., Warner J. L., McGee P. E., Cochran A. and Brown R. W. (1977) Luna 24: Petrochemical potpourri (abstract). In *Papers Presented to the Conference on Luna 24*, p. 170–173. The Lunar Science Institute, Houston.

Steele I. M., Irving A. J. and Smith J. V. (1977) Apollo 15 breccia rake samples-mineralogy of lithic and mineral clasts. *Proc. Lunar Sci. Conf. 8th*, p. 1925–1941.

Stolper E. (1974) Lunar ultramafic glasses. A. B. thesis, Harvard Univ., 178 pp.

Stolper E., Walker D., Longhi J. and Hays J. F. (1974) Compositional variation in lunar ultramafic glasses (abstract). In *Lunar Science V*, p. 749–751. The Lunar Science Institute, Houston.

Tarasov L. S., Nazarov M. A., Shevaleevsky I. D., Kudryashova A. F., Gaverdovskaya A. S. and Korina M. I. (1977) Mineralogy and petrography of lunar rocks from Mare Crisium (preliminary data). *Proc. Lunar Sci. Conf. 8th*, p. 3333–3356.

Taylor G. J., Keil K. and Warner D. (1977a) Very low-Ti mare basalts. *Geophys. Res. Lett.* **4**, 207–210.

Taylor G. J., Warner R., Wentworth S., Keil K. and Sayeed U. (1977b) Petrology and chemistry of Luna 24 mare basalt and basaltic glasses (abstract). In *Papers Presented to the Conference on Luna 24*, p. 189–192. The Lunar Science Institute, Houston.

Taylor L. A., Onorato P. I. K. and Uhlmann D. R. (1977a) Cooling rate estimations based on kinetic modelling of Fe-Mg diffusion in olivine. *Proc. Lunar Sci. Conf. 8th*, p. 1581–1592.

Taylor L. A., Onorato P. I. K., Uhlmann D. R. and Coish R. A. (1977b) Cooling rates of Luna 24 sub-ophitic basalts (abstract). In *Papers Presented to the Conference on Luna 24*, p. 193–195. The Lunar Science Institute, Houston.

Vaniman D. T. and Papike J. J. (1977a) Very low Ti (VLT) basalts: A new mare rock type from the Apollo 17 drill cores. *Proc. Lunar Sci. Conf. 8th*, p. 1443–1471.

Vaniman D. T. and Papike J. J. (1977b) Luna 24 ferrobasalt: Petrology and comparison with other mare basalt types (abstract). In *Papers Presented to the Conference on Luna 24*, p. 196–199. The Lunar Science Institute, Houston.

Walker D., Longhi J. and Hays J. F. (1975) Differentiation of a very thick magma body and implications for the source regions of mare basalts. *Proc. Lunar Sci. Conf. 6th*, p. 1103–1120.

Walker D., Longhi J. and Hays J. F. (1976a) Heterogeneity in titaniferous lunar basalts. *Earth Planet. Sci. Lett.* **30**, 27–36.

Walker D., Longhi J., Kirkpatrick R. J. and Hays J. F. (1976b) Differentiation of an Apollo 12 picrite magma. *Proc. Lunar Conf. 7th*, p. 1365–1389.

Walker D., Longhi J., Lasaga A. C., Stolper E. M., Grove T. L. and Hays J. F. (1977) Slowly cooled microgabbros 15555 and 15065. *Proc. Lunar Sci. Conf. 8th*, p. 1521–1547.

Walker D., Powell M. A. and Hays J. F. (1978) Dynamic crystallization of a eucritic basalt (abstract). In *Lunar and Planetary Science IX*, p. 1196–1198. Lunar and Planetary Institute, Houston.

Warner J., Ridley W. I., Reid A. M. and Brown R. W. (1972) Apollo 15 glasses and the distribution of non-mare crustal rock types. In *The Apollo 15 Lunar Samples* (J. W. Chamberlain and C. Watkins, eds.), p. 179–181. The Lunar Science Institute, Houston.

APPENDIX—PROJECTION SCHEME FOR LOW-TI BASALTS

The projection scheme used for the VLT basalts is appropriate for liquids with low TiO_2, Cr_2O_3 and alkalis. An oxide analysis is recalculated to molar units and then recast into the quaternary system olivine, clinopyroxene, anorthite and silica, where

$$\begin{aligned} Ol &= (Mg,Fe)_2SiO_4 \\ Cpx &= Ca(Mg,Fe)Si_2O_6 \\ An &= CaAl_2Si_2O_8 \\ Qtz &= SiO_2. \end{aligned}$$

The expressions that transform molar components to mineral components are given below normalized to 100%.

$$Ol = 100X(FeO + MgO - CaO + Al_2O_3)/2(SiO_2 - CaO)$$
$$Cpx = 100X(CaO - Al_2O_3)/(SiO_2 - CaO)$$
$$An = 100X\ Al_2O_3/(SiO_2 - CaO)$$
$$Qtz = 100\times[SiO_2 - FeO/2 - MgO/2 - 3/2(CaO) - 1/2(Al_2O_3)]/(SiO_2 - CaO)$$

From this pseudoquaternary system the four subprojections can be used to view compositional variation in low-Ti liquids, to estimate liquidus phases and to deduce crystallization sequences. There are several limitations to this projection scheme. Fe-Mg variation is suppressed, so additional projections involving Fe/(Fe + Mg) are required for testing relationships between basalt liquids. Compositional variation in low-Ti basalts is complex and does not lend itself to projection into two-dimensional space. Therefore, these projections must be used with caution and should not be applied to other basalt systems. The problems that can arise by not considering all possible compositional variations are well illustrated by the low Ti basalts. For example, in the Ol-An-Qtz subprojection (Grove and Vaniman, 1977) Apollo 17 VLT basalts appeared to be related to Apollo 15 green glass. However, consideration of the Ol-An-Cpx subprojection showed that the two compositions were distinct and could not be related by olivine fractionation.

Merrill, R. B. and Papike, J. J., editors.
(1978) *Mare Crisium: The View from Luna 24*, Pergamon Press (New York), p. 473–481
Printed in the United States of America

Subophitic basalts from Mare Crisium: cooling rates

LAWRENCE A. TAYLOR

The Department of Geological Sciences, The University of Tennessee, Knoxville, Tennessee 37916

P. I. K. ONORATO and D. R. UHLMANN

Department of Material Sciences and Engineering, Massachusetts Institute of Technology, Cambridge, Massachusetts 02139

R. A. COISH

The Department of Geological Sciences, The University of Tennessee, Knoxville, Tennessee 37916

Abstract—Subophitic basalt is the most common rock type at Mare Crisium. The cooling rate of a sample of this rock was determined by: (1) an olivine cooling speedometer and (2) Zr partitioning between ilmenite and ulvöspinel. The kinetic modeling of the olivine cooling speedometer starts with a calculation of the compositional profile of olivine (the "as-solidified profile") and proceeds to re-equilibration by diffusion as a function of cooling rate. The estimated cooling rate for subophitic basalts from the Luna 24 site is in the range of 2°C/day (~0.1°C/hr) which is reasonably well corroborated by dynamic crystallization studies of Grove (1978).

INTRODUCTION

Subophitic basalt is the most common basalt rock type at Mare Crisium and was the subject of considerable interest at the Luna 24 Science Conference. However, the presentations were mainly about the Mineralogy/Petrology and Geochemistry, and little attention was paid to the cooling rate of this rock, a parameter that, to a large extent, controls these features. Therefore, a study was undertaken to characterize the cooling rate of subophitic basalt as based on: 1) a refinement of the olivine diffusion technique (Taylor *et al.*, 1977) and 2) the kinetics of the re-equilibration of Zr partitioning between ilmenite and ulvöspinel (Taylor *et al.*, 1975).

PROCEDURE

A subophitic basalt fragment (24077,68) approximately 1 × 1.2 mm was used for this study. Compositional data on olivine and coexisting ilmenite and ulvöspinel were obtained with an automated MAC 400S electron microprobe, using appropriate synthetic and natural standards. Corrections for matrix effects were applied according to procedures described by Bence and Albee (1968) and Albee and Ray (1970). Care must be taken with olivine traverses to use crystals which display compositional flattening over the "core", thereby suggesting that the section cuts the olivine near the center of the grain.

SUBOPHITIC BASALTS

Cooling rates were determined on a basalt chip, 24077,68, in the grain size fraction >500 μm. The fragment is a medium-grained subophitic basalt. It consists of phenocrysts of olivine (0.6 by 0.4 mm) and microphenocrysts of olivine and pyroxene set in a matrix of subophitically intergrown plagioclase laths (0.15 mm long) and pyroxene. Minor amounts of chromite-ulvöspinel and ilmenite also occur. The olivine phenocrysts have core compositions of Fo_{56-60} whereas smaller (and presumably later) grains have Fo_{20}–Fo_{25} contents. Pyroxenes show a wide range of compositions–$Wo_{13}En_{51}Fs_{35}$ to $Wo_{24}En_{23}Fs_{53}$. The compositions of the early-formed olivines indicate that this rock fragment crystallized from a liquid with a high Fe/Fe + Mg ratio. This high fractionation ratio is typical of all Luna 24 subophitic ferrobasalt fragments. The details of this rock type are presented by Coish and Taylor (this volume).

OLIVINE COOLING SPEEDOMETER

Recently, Taylor *et al.* (1977) reported on a method for estimating the cooling rate from a compositional profile in olivine. A major limitation of that kinetic model was a lack of data on the "as-solidified profile" of olivine due to primary crystallization. A step function Fo concentration profile based on mass balance considerations was assumed, and this resulted in minimum cooling rate estimates. Our present analysis takes specific account of the partial equilibrium which occurs during solidification and is based on an initial Fo profile across the olivine grain. That is, the "as-solidified profile" is now calculated as the first-step improvement of the overall kinetic model of the cooling process.

Walker *et al.* (1977) presented a different approach to calculating cooling rates from compositional profiles in olivines. These workers neglected the compositional dependence of the diffusion coefficient and assumed an initial compositional profile in the grain where the concentration varies as exp $(-Br^3)$. Like the original analysis of Taylor *et al.* (1977), homogenization of the initial profile was assumed to begin at some closure temperature. The assumed initial profile of Walker *et al.* (1977) is physically plausible, but the neglect of the compositional dependence of the diffusion coefficient—which was required to obtain their analytical solution—introduces an uncertainty into the calculated cooling rates. A detailed comparison of the predictions of the models of Walker *et al.* (1977), Taylor *et al.* (1977), the present model, and a fourth model involving diffusion-controlled growth of the olivine crystals (Onorato *et al.*, 1978) will be presented in the last-mentioned paper.

The cooling rate has been estimated from measurements of compositional profiles in olivines. The analysis considers diffusional equilibration in the olivine grains, and takes specific account of the partial equilibration which takes place during solidification. In estimating the initial forsterite profile of an olivine grain, the following assumptions were made: (1) Growth starts at some temperature slightly below the liquidus, in this case at 1150°C; (2) an olivine grain grows in a small spherical volume (radius = 570 μm) of well-stirred melt; (3) the composition of the olivine solidifying at any time is in equilbrium with the melt; (4) intracrystalline diffusion may take place during solidification; (5) growth continues until one of the components of olivine in the small volume surrounding the grain are depleted to essentially zero. The assumed size of the small volume is that necessary to form a sphere of the observed size. Due to the finite steps taken in growth of the sphere and to the form of the partition relation, pure fayalite is not formed when the MgO in the liquid is depleted. Therefore, the calculated composition at the rim of the grain upon solidification does not realistically describe that in the sample. When the liquid is depleted in either MgO or FeO the growth of the grain may be limited by the diffusion of these components in the liquid to the olivine-melt interface.

In order to calculate the initial olivine profile, the melt composition for a typical subophitic basalt is estimated at 18.5 wt.% FeO and 6.0 wt.% MgO (Ryder *et al.*, 1977). This corresponds to a melt composition of about 11.5 mole% FeO and 7.5 mole% MgO.

Roeder and Emslie (1970) determined that the K_D for olivine-liquid equilibria was about 0.30 for terrestrial melts and was independent of temperature. Longhi (*pers. comm.*, 1978) and Bianco and Taylor (1977) found that a K_D of $\simeq$0.33 is more appropriate for lunar systems. Therefore, the crystal-liquid partitioning is:

$$\frac{C^{ol}_{FeO}}{C^{liq}_{FeO}} \cdot \frac{C^{liq}_{MgO}}{C^{ol}_{MgO}} = 0.33 \qquad (1)$$

This results in an initial olivine core of 66.4% forsterite, is not a function of temperature, but depends primarily on the composition of the liquid at the interface. Olivine starts to crystallize at about 1150°C in the Luna 24 ferrobasalt system (Grove and Vaniman, 1977). The growth rate of the olivine is a function of temperature and is estimated from growth measurements by Bianco and Taylor (1977), in a high-iron composition similar to lunar sample 15555. In the Luna 24 composition, the growth rate is probably different than that in 15555, but this is the only available data and should give a reasonable estimate of the growth rate of olivine in the Luna 24 melt. The growth rate of olivine in the 15555 composition is approximately 4×10^{-8} cm sec^{-1} at 1150°C, and is assumed to be isotropic. The growth rate decreases uniformly to 2×10^{-8} cm sec^{-1} at 1125°C. At this temperature, growth stops because of the depletion of MgO in the liquid.

The diffusion coefficient has been shown by the detailed investigation of Buening and Buseck (1973) to be a function of: (1) Temperature; (2) Composition; (3) Oxygen partial pressure; and (4) Crystallographic direction. All of these factors were taken into account. The measured profiles are approximately in the *c* crystallographic direction. The oxygen fugacity was assumed to be about 0.5 log atm. below the iron/wustite buffer curve:

$$\log P_{O_2} = 0.15\ T\ (°K) - 34.6 \qquad (2)$$

The diffusion equation in spherical coordinates

$$\frac{\delta C}{\delta t} = \frac{1}{r^2}\frac{\delta}{\delta r}\{D(C,T,P_{O_2})r^2\frac{\delta C}{\delta r}\} \tag{3}$$

was replaced by its backward difference analog, and the resulting set of simultaneous equations was solved numerically. During growth, the number of spatial increments was increased at every step in time, with the size of the additional increment equal to the change in radius during that time period. Due to the temperature dependence of the growth rate, the spatial increments are in general of different sizes. Therefore, the finite difference analog of the differential equation was derived from the Taylor series appropriate when the spatial increment Δr is not uniform.

The first derivative of the concentration with respect to distance at a point i and at an unknown time n + 1 is assumed to be the average of the derivatives at points $i + \frac{1}{2}\Delta r_i$ and $i - \frac{1}{2}\Delta r_{i-1}$. That is:

$$\left.\frac{\delta C}{\delta r}\right|_i = \frac{1}{2}\left(\frac{C_{i+1,n+1} - C_{i,n+1}}{\Delta r_i} + \frac{C_{i,n+1} - C_{i-1,n+1}}{\Delta r_{i-1}}\right) \tag{4}$$

where $\Delta r_i = r_i + 1 - r_i$ and $\Delta r_{i-1} = r_i - r_{i-1}$.
The second derivative was derived from the Taylor series expansion about points at $i + \Delta r_i$ and $i - \Delta r_{i-1}$. The difference expression for the second derivative is:

$$\left.\frac{d^2C}{\delta r^2\, m}\right|_i = 2\,\frac{(C_{i+1} + C_{i-1} - 2C_i - \frac{\delta C}{\delta r_i}(\Delta r_i - \Delta r_{i-1}))}{\Delta r_i^2 + \Delta r_{i-1}^2} \tag{5}$$

Equations (3), (4), and (5) are then combined to form the difference equation:

$$\frac{-C_{i,n}}{\Delta t} = C_{i-1,n+1}\left\{\frac{D}{\Delta r_i^2 + \Delta r_{i-1}^2}\left(1 + \frac{\Delta r_i}{\Delta r_{i-1}}\right) - \frac{(A + \frac{2D}{r_i})}{2\Delta r_{i-1}}\right\}$$

$$+ C_{i,n+1}\left\{-\frac{1}{\Delta t} + \frac{D}{\Delta r_i^2 + \Delta r_{i-1}^2}\left(-2 - \frac{\Delta r_{i-1}}{\Delta r_i} - \frac{\Delta r_i}{\Delta r_{i-1}}\right) + A + \frac{2D}{r_i}\frac{1}{2\Delta r_{i-1}} - \frac{1}{2\Delta r_i}\right\}$$

$$+ C_{i+1,n+1}\left\{\frac{D}{\Delta r_i^2 + \Delta r_{i-1}^2}\left(1 + \frac{\Delta r_{i-1}}{\Delta r_i}\right) + \frac{A + 2D/r_i}{2\Delta r_i}\right\} \tag{6}$$

where $A = \frac{\delta D}{\delta C}\frac{\delta C}{\delta r}$. Because of the nonlinearity of this equation it was necessary to iterate to approach the correct solution.

During solidification, the boundary conditions are:

$$\frac{\delta C}{\delta r}(O,t) = 0; \; C(R,t) = C_{eq} \quad (7)$$

where C_{eq} is that given by the partition relation. The liquid composition is changed to account for the spherical shell of volume dv that has solidified during a period dt as well as any diffusion at the boundary (Brody and Fleming, 1966):

$$dC_{MgO}^{liq} = -dv\, C_{eq}/V_{liq} - \frac{SD(R,t)}{V_{liq}} \cdot \left.\frac{\delta C}{\delta r}\right|_{r=R} dt \quad (8)$$

$$dC_{FeO}^{liq} = -dv\, C_{eq}/V_{liq} - \frac{SD(R,t)}{V_{liq}} \cdot \left.\frac{\delta C}{\delta r}\right|_{r=R} dt \quad (8a)$$

where V_{liq} is the volume of liquid remaining and S is the surface area of the grain.

The generation of the initial olivine profile can be summarized: the olivine is allowed to grow a small extent, its composition dictated by Equation (1). The growth is stopped, a new composition calculated for the liquid, and growth reinstated with a slightly less Fo-rich composition of the olivine, again dictated by the exchange equation, reiteration, etc.

Figure 1 depicts the "equilibrium" profile that would result if there were no diffusion operative during solidification. According to this model, the equilibrium and "as-solidified" profiles are not necessarily identical. The difference between these profiles is a measure of the extent of diffusion during solidification. This is

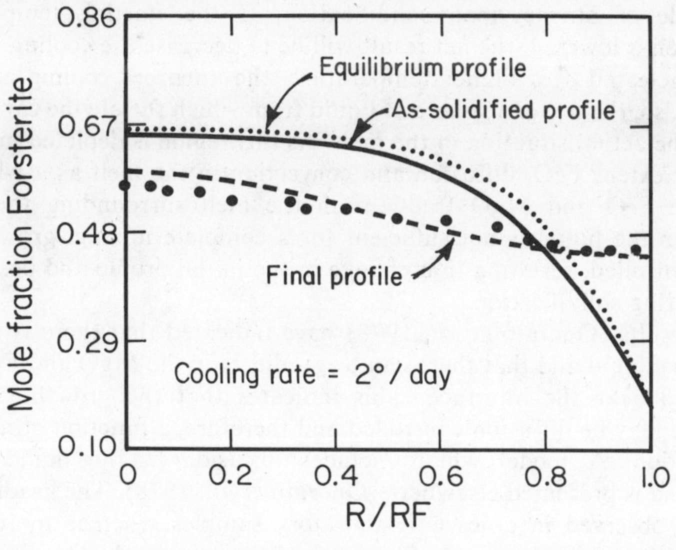

Fig. 1. Olivine compositional profiles used in calculating the cooling rate of subophitic basalt 24077, 68. The black dots represent EMP analyses. The computer fit (solid line) to the real compositional profile is near perfect.

significant (Fig. 1) for a growth rate of 4×10^{-8} cm sec^{-1} starting at 1150°C, but it is negligible for a growth rate $\geq 10^{-6}$ cm sec^{-1}.

After solidification of the olivine is *complete*, the olivine is assumed to be an isolated system—i.e., no flux exists across the crystal-liquid interface or between grains. Hence, the outer boundary condition is changed to:

$$\frac{\delta C}{\delta r}(R,t) = 0 \tag{9}$$

The analysis proceeds as in the previous modeling (Taylor *et al.*, 1977), and cooling is continued until the Fo profile no longer changes significantly with time, in this case about 900°C.

It was found that *cooling from 1150°C*, of a melt with 18.5 wt.% FeO and 6.0 wt.% MgO, at a *linear rate of 2°C/day* results in Fo profile which closely matches the measured profile in the *c* direction (Fig.1).

There are several sources of error which may affect the final forsterite profile and hence the cooling rate. The diffusion coefficient is greatest in the *c* direction, so any deviation from this direction in the measured sample would tend to lower the predicted cooling rate. No olivine growth-rate data exist for a low-Ti basalt melt such as in the Luna 24 system in question. Use of the 15555 data for this problem is speculative at best. If the growth rate is slower than the values used here, there would be more diffusion during solidification but less diffusion after solidification is complete. Whether the apparent cooling rate is increased or decreased depends on the temperature at which the olivine grain stops growing.

The greatest uncertainty in the model probably lies in the determination of the initial profile of olivine upon solidification. If the starting temperature of crystallization is lowered, the net result will be to decrease the cooling rate. If the olivine is nucleated at a higher temperature, the apparent cooling rate will be increased. Also, the use of a sphere of liquid from which the olivine can grow does not reflect the actual situation in the melt. As this region is depleted in MgO and to a smaller extent FeO, diffusion and convection in the melt as a whole would increase the FeO and MgO fractions in the melt surrounding the grain. If convection in the liquid is not sufficient for a complete mixing, growth may be diffusion-controlled, resulting in a change in the initial profile and possibly more diffusion during solidification.

Recent results (Onorato *et al.*, 1978) have indicated that there is incomplete mixing in the liquid and that there can be gradients in the MgO and FeO content in the liquid near the interface. This indicates that the growth rate at any temperature may be diffusion controlled and therefore, a function of time as well as composition. A model which includes diffusion in the liquid has been developed and is presented elsewhere (Onorato *et al.*, 1978). The gradients in the liquid were observed in quiescent laboratory samples whereas more complete mixing may take place during the formation of the lunar rock. The 570 μm radius sphere from which the olivine grows is somewhat smaller than the 690μm radius sphere which would correspond to half the inter-olivine center distance if the rock is 10% olivine and the grains are 320 μm in radius. This indicates that growth may be limited by transport in the melt in addition to the availability of FeO + MgO in the melt as a whole.

Zr Cooling Speedometer

The partitioning of Zr between coexisting ilmenite and ulvöspinel is a function of temperature. The kinetic model of the re-equilibration of this partitioning has been used to estimate absolute cooling rates of many lunar rocks (Taylor *et al.*, 1975). In as much as the subophitic basalt fragment in 24077,68 contains intergrowths of these phases, an attempt was made to apply this speedometer.

The Luna 24 subophitic basalts, in general, contain only sparse assemblages of coexisting ilmenite and ulvöspinel. However, where intergrown the ilmenite is usually <15 μm wide and the ulvöspinel is considerably larger (e.g., 100–150 μm). This does present some problems for the kinetic model since the ilmenite is not large enough to be considered an "infinite" volume. In addition, the amounts of Zr in both phases are low (i.e., usually <0.20% in ilmenite) such that the Zr compositions as determined by EMP have relatively large precision errors associated with them. These are limitations to the application of this speedometer and sources of error.

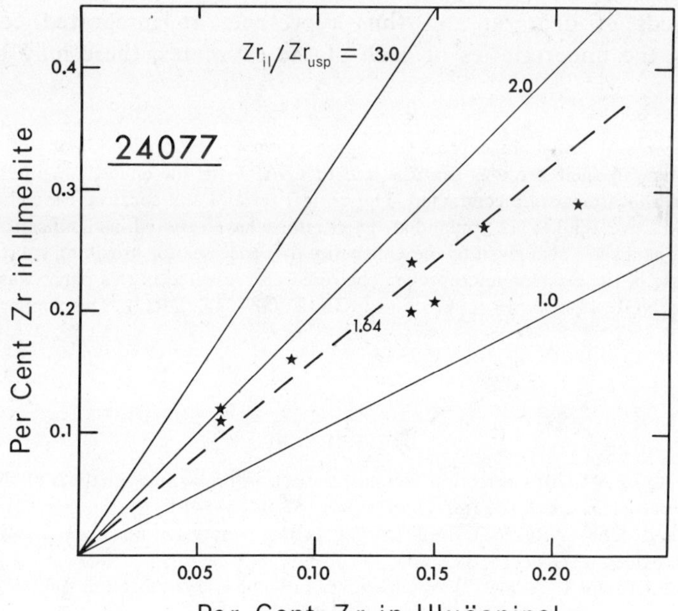

Fig. 2. The Zr contents of coexisting ilmenite and ulvöspinel in subophitic basalt 24077,68.

The Zr contents of coexisting ilmenite and ulvöspinel were determined for sample 24077,68 as shown in Fig. 2. The mean of the K_D values is 1.64 with a standard deviation of 0.23. Using the kinetic model of Taylor *et al.* (1975), this results in a characteristic temperature of 887°C. In turn, this temperature indicates a linear cooling rate of 2.2°C/day. However, the limitations discussed above translate into an uncertainty in the cooling rate estimate of at least a factor of 3–5.

CONCLUSION

The cooling rates of two fragments of Luna 24 subophitic basalt were determined by both the olivine cooling speedometer and the Zr partitioning model. The olivines indicate a cooling rate of 2°C/day, and the Zr model estimate is also 2°C/day. Therefore, the cooling rates of Luna 24 subophitic basalts are estimated to be on the order of 2°C/day. This estimate should be good to better than order-of-magnitude accuracy.

The cooling rate of Luna 24 ferrobasalt has been estimated by Grove (1978), based on pyroxene chemistry as determined by dynamic crystallization experiments, to be between 0.5° and 1.75°C/hr. This compares well with, but is somewhat higher than, our estimate presented above of about 0.1°C/hr. In reality, the pyroxene chemistry reflects a cooling rate over a relatively small temperature interval far removed from the liquidus when the cooling rate was perhaps rapid. The olivine records crystallization and diffusive re-equilibration over hundreds of degrees, and thus represents an integrated cooling rate. Considering the uncertainties in each of the estimates, these may be identical values.

Acknowledgments—We wish to express our appreciation to the Soviet Union for providing the U.S.A. with some of their precious samples and to LSAPT for the efficient manner in which the processing and allocations were conducted. The critical reviews and constructive comments by Drs. Tim Grove and Dave Walker are applauded. Their efforts have resulted in significant improvements in the paper. Thanks are also given to the University of Tennessee for providing substantial support for the purchase of the electron microprobe. The research presented in this paper was supported by NASA Grants NGL 43-001-140 (LAT) and NGR 22-009-646 (DRU). This support is gratefully acknowledged.

REFERENCES

Albee A. and Ray L. (1970) Correction factors for electron probe microanalysis of silicates, oxides, carbonates, phosphates, and sulfates. *Anal. Chem.* **42**, 1408–1414.

Bence A. E. and Albee A. (1968) Empirical correction factors for the electron microanalysis of silicates and oxides. *J. Geol.* **76**, 382–403.

Bianco A. S. and Taylor L. A. (1977) Applications of dynamic crystallization studies: Lunar olivine-normative basalts *Proc. Lunar Sci. Conf. 9th*, p. 1593–1610.

Brody H. D. and Fleming M. C. (1966) *Trans. Met. Soc. AIME* **236**, 615–624.

Buening D. K. and Buseck P. R. (1973) Fe-Mg lattice diffusion in olivine. *J. Geophys. Res.* **78**, 6852–6862.

Coish R. A. and Taylor L. A. (1978) Mineralogy and petrology of basaltic fragments from the Luna 24 drill core. *Mare Crisium: The View From Luna 24* (R. B. Merrill and J. J. Papike, eds.) This volume.

Grove T. L. and Vaniman D. T. (1977) Experimental petrology of very low Ti basalts and origin of Luna 24 ferrobasalt (abstract). In *Papers Presented to the Conference on Luna 24*, p. 68–71. The Lunar Science Institute, Houston.

Grove T. L. (1978) Cooling histories of Luna 24 low-Ti ferrobasalts and ferrogabbros (abstract). In *Lunar and Planetary Science IX*, p. 424–426. Lunar and Planetary Institute, Houston.

Onorato P. I. K., Uhlmann D. R., Taylor L. A. and Coish R. A. (1978) An olivine cooling speedometer. In *Proc. Lunar and Planet. Sci. Conf. 9th.* This volume.

Roeder P. L. and Emslie R. F. (1970) Olivine-liquid equilibrium. *Contrib. Mineral. Petrol.* **24**, 275–289.

Ryder G., McSween H. Y., Jr. and Marvin U. (1977) Basalts from Mare Crisium. *The Moon* **17,** 263-287.

Taylor L. A., Onorato P. I. K. and Uhlmann D. R. (1977) Cooling rate estimations based on kinetic modelling of Fe-Mg diffusion in olivine. *Proc. Lunar Sci. Conf. 8th*, p. 1581–1592.

Taylor L. A., Uhlmann D. R., Hooper R. W. and Misra K. C. (1975) Aboslute cooling rates of lunar rocks: Theory and application. *Proc. Lunar Sci. Conf. 6th*, p. 181–191.

Walker D., Longhi J., Lasaga A. C., Stolper E. M., Grove T. L. and Hays, J. F. (1977) Slowly cooled microgabbros 15555 and 15065. *Proc. Lunar Sci. Conf. 8th*, p. 1521–1547.

Merrill, R. B. and Papike, J. J., editors.
(1978) *Mare Crisium: The View from Luna 24*, Pergamon Press (New York), p. 483–493
Printed in the United States of America

Viscous flow, crystal growth, and glass formation of highland and mare basalts from Luna 24

C. A. HANDWERKER, P. I. K. ONORATO, and D. R. UHLMANN
Department of Materials Science and Engineering, Massachusetts Institute of Technology
Cambridge, Mass. 02139

Abstract—A kinetic treatment of glass formation has been applied to two Luna 24 compositions. The compositions examined here are average compositions of clusters of 12-element glass analyses. These compositions are a low Ti, low Mg mare basalt composition and the highland basalt composition determined by Simonds *et al.* (1977). The crystal growth rates have been determined over the temperature range of 850° to 1140°C for the mare basalt composition and for 850° to 1350°C for the highland basalt composition. The viscous flow behavior of both compositions was determined at both high and low temperatures. From these measurements of crystal growth rates and viscous flow behavior, the critical conditions for glass formation of these compositions have been calculated. The results, notably the minimum critical cooling rates, have been combined with heat flow analyses to estimate the dimensions of the largest body which can be formed with the degree of crystallinity seen in the Luna 24 samples.

The data on growth rate and viscosity have been combined to construct time-temperature-transformation (TTT) curves and CT curves for both constant-rate and logarithmic cooling corresponding to various degrees of crystallinity, and in particular to the degrees of crystallinity observed in the two Luna 24 compositions (greater than 50 volume percent crystallized for the low Mg, low Ti mare basalt composition and less than 1 volume percent for the highland basalt composition). The critical cooling rate for the mare basalt composition with a volume fraction crystallized of 0.50 is estimated as 5.4°C/sec; and for the highland basalt composition with a volume fraction crystallized of 10^{-2} as 1.2°C/sec.

Heat flow analyses have also been carried out, and have been used to estimate the dimensions of the largest possible sphere of each composition which can be obtained with the observed degree of crystallinity. For the mare basalt composition, the maximum size body for volume fraction crystallized of 0.50 is estimated to be 1 cm. For the highland basalt composition the maximum size body for a volume fraction crystallized of 10^{-2} is estimated to be 2 cm.

I. INTRODUCTION

A kinetic theory of glass formation applicable to both inorganic and organic systems has been developed using the formal theory of transformation kinetics [Uhlmann (1972)]. This approach to glass formation has been successfully applied to a number of glass-containing samples returned from the lunar surface. These include lunar soils, small homogeneous glass spheres, glassy intrusions, and breccias of various sizes, geometries, and degrees of crystallinity [e.g., Scherer *et al.* (1972), Uhlmann *et al.* (1976), Handwerker *et al.* (1977)].

The description of the crystallization kinetics is based on the construction of continuous cooling-transformation (CT) curves corresponding to the degree of crystallinity observed in the lunar samples. The CT curves are calculated for different cooling histories of interest—e.g., constant-rate cooling and logarithmic cooling—from time-temperature-transformation (TTT) curves for the specified degree of crystallinity. The TTT curves are constructed using the Avrami equation with measured values of the crystal growth rate over a wide range of temperature together with values of the nucleation frequency calculated from the standard treatment of nucleation kinetics and measured values of the liquid viscocity as functions of temperature.

The use of the standard treatment of nucleation kinetics and the magnitude of the nucleation barrier assumed for lunar compositions has received strong support from measurements of the nucleation kinetics of lunar composition 70019 [Klein *et al.* (1977)]. From the constructed TTT and CT curves, it is possible to estimate the rate at which the sample cooled.

The critical conditions derived from this analysis for the formation of lunar glasses and bodies having various degrees of partial crystallinity have been coupled with heat flow calculations for various sample geometries to elucidate the thermal histories of the samples.

This treatment is now applied to the two Luna 24 compositions. These compositions are average compositions of clusters of 12-element glass analyses identified by Warner *et al.* (1978). The glass fragments, whose sizes range from 90 to 250 μm, were analyzed using an electron microprobe with a 15 μm beam spot. The compositions investigated are shown in Table 1. Similar compositions were identified by Taylor *et al.* (1978) for particles in the 90–250 μm range. The mare basalt compositional cluster has a low Mg (~6.5 wt.%) and low Ti content (~1.0 wt.%) for lunar mare compositions. As reported by Simonds, the sample group is compositionally well defined and shows a dendritic to spherulitic devitrification microstructure. Some largely glassy particles were noted within this sample cluster, which consists mostly of completely crystalline and largely crystalline fragments. The largely glassy particles were generally spherical in shape, whereas most of the highly crystalline particles were angular.

The highland basalt composition corresponds to a highly vitric highland basalt cluster which consists of colorless glass fragments. The size range studied was

TABLE 1. Luna 24 Compositions Investigated (wt.%)

Mare Basalt Composition		Highland Basalt Composition
46.26	SiO_2	44.97
0.98	TiO_2	0.30
12.98	Al_2O_3	26.74
20.00	FeO	4.89
6.47	MgO	6.86
12.73	CaO	15.64
0.29	Na_2O	0.23
0.04	K_2O	0.05

similar to that of the low Mg mare basalt cluster (90–250 μm). Some glassy spherical particles are found in this sample.

The highland basalt composition is similar to the overall highland composition measured from orbit. The composition is also quite close to that of lunar soil Sample 68502 [Uhlmann *et al.* (1974)].

While it has been suggested that the mare basalt composition clusters were formed by volcanic action and the highland basalt samples were impact modified, the evidence for a definite formation mechanism for individual particles seems inconclusive for all samples, since sample size precludes the spectrum of analyses necessary to differentiate impact from volcanic materials.

II. Experimental Procedure

The materials studied in this investigation were prepared from reagent grade chemicals. The melting procedure, sample preparation method and measuring techniques for the viscous flow study have been described previously [Cukierman *et al.* (1972)]. The melting was carried out in multi-kilogram batches in a large-volume, Mo-wound furnace. This promotes homogeneity in composition and avoids many problems associated with the volatility of constituents in the batch. Viscosities in the molten range were determined using a rotating-cylinder viscosimeter, while viscosities greater than about 10^8 poise were determined using a bending-beam viscosimeter. Both viscosimeters were provided with an atmosphere at 5% H_2-forming gas; both were calibrated using NBS reference materials.

The measurements of crystal growth rates were carried out on small glass specimens cut from the material used in the viscous flow study. The oxygen fugacity was controlled to the conditions suggested by Sato *et al.* (1973). A solid ceramic electrolyte cell was used to monitor the oxygen fugacity continuously in the furnace [Williams (1972)].

The liquidus temperatures were determined by differential thermal analysis and by visually observing the melting of crystallized bodies. The liquidus temperatures determined by both techniques agreed within 5°C.

III. Viscous Flow and Crystallization Behavior

The crystal growth rates of the two composition as a function of temperature are shown in Fig. 1. The growth rates have been determined over the temperature range of 850° to 1140°C for the mare basalt composition (Composition 24-1) and from 850° to 1350°C for the highland basalt composition (Composition 24-2). At all temperatures, the crystallization of both compositions proceeded from the external surfaces of the samples, which had been dusted with seed crystals from the same composition prior to the crystallization heat treatment. In all cases, the crystal layer thicknesses increased linearly with time.

The maximum growth rate of the highland basalt composition is about 9×10^{-2} cm min^{-1}. It occurs at about 1150°C, an undercooling of about 215°C below the liquidus temperature which was determined as 1365°C ± 10°C. The maximum growth rate of the mare basalt is about 2×10^{-2} cm min^{-1}. It occurs at about 962°C, an undercooling of about 223°C below the liquidus which was determined as 1185°C ± 10° C.

The shapes of the curves shown in Fig. 1 are characteristic of glassforming materials, with the growth rate exhibiting distinct maxima as a function of

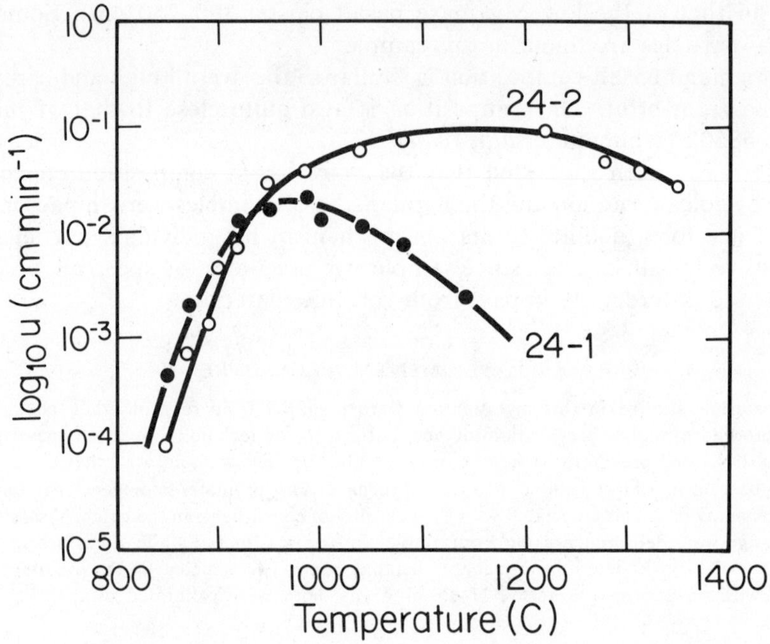

Fig. 1. Crystal growth rate, u, as a function of a temperature for the mare basalt (24-1) and highland basalt (24-2) composition.

temperature. These maxima reflect a competition between the driving force for crystallization, which increases with increasing undercooling, and the molecular mobility, which decreases with increasing undercooling.

The variations of viscosity with temperature for both compositions are shown in Figs. 2 and 3. The dashed interpolations of each curve represent regions of temperature for which reliable flow data could not be obtained because of crystallization. Data in the high-temperature range below the liquidus were obtained by superheating the specimens about 100°C above the respective liquidus temperatures for about 1 hour prior to cooling to eliminate retained crystal embryos.

The viscous flow behavior of the mare basalt composition (24-1) is compared in Fig. 2 with that of the glassy intrusion of lunar sample 15286 which has a similar SiO_2 + Al_2O_3 content [Uhlmann *et al.* (1976)]. The viscosity vs. temperature relation for the highland basalt composition (24-2) is compared in Fig. 3 with that of lunar sample 68502, which is compositionally similar [Uhlmann *et al.* (1974)]. The viscosity behavior of the highland basalt composition is almost identical to that of composition 68502.

The glass transition temperature ($\eta = 10^{13}$ poise) of the mare basalt is about 650°C and that of the highland basalt is about 705°C. The apparent activation energies for flow in the high temperature region are about 45 kcal mole^{-1} for the mare basalt composition (24-1) and about 55 kcal mole^{-1} for the highland basalt composition (24-2); the corresponding values in the low temperature region as the glass transition is approached are about 120 kcal mole^{-1} for 24-1 and 170 kcal mole^{-1} for 24-2.

The variations of viscosity with temperature for the two Luna 24 compositions are within the range of viscosities observed previously for other lunar compositions. In particular, they are similar to the viscosity temperature relations of the materials with which they are compared in Figs. 2 and 3.

The mare basalt composition has a moderate $SiO_2 + Al_2O_3$ content of about 59.2 wt.%, while the highland basalt composition has a high $SiO_2 + Al_2O_3$ content of 71.7 wt.%. Both show strong temperature dependences of their viscous flow behavior; both have fairly high viscosities at their respective liquidus temperatures and show rapid increases in viscosity with falling temperature below the liquidus. Based on their viscous flow and crystallization behavior, it is expected that both compositions would be reasonably good glass formers. This expectation is considered further in the following section.

IV. Cooling Rates and Thermal Histories

The data on growth rates and viscosity have been combined to construct time-temperature-transformation or TTT curves corresponding to the degrees of crystallinity observed in the two Luna 24 sample clusters. Such curves are constructed using the Avrami equations which describe the volume fraction crystallized as a function of temperature and time. For a given volume fraction crystallized (V_c/V) spherical nuclei and crystallites, and a nucleation frequency (I_v) and growth rate (u) independent of time, the time required for a given V_c/V to form can be calculated using the relation:

$$\frac{V_c}{V} \approx I_v u^3 t^4 \tag{1}$$

The assumptions that I_v and u are independent of time are justified from experiment. From crystal growth rate measurements on these two lunar compositions and other similar silicate compositions, it is found that crystal growth rates at all temperatures are independent of time. This crystallization behavior is consistent with the suggestion that the crystal length is large in comparison with the size of the diffusion field and that steady-state conditions of growth exist—i.e., that the size of diffusion field does not change with time.

For the small volume fractions crystallized in these highly viscous glass forming systems, it is reasonable to assume that I_v is not a function of time. Since the total number of nucleation events is small and the crystal growth rates are slow, newly-formed nuclei do not impinge on previously nucleated regions; and

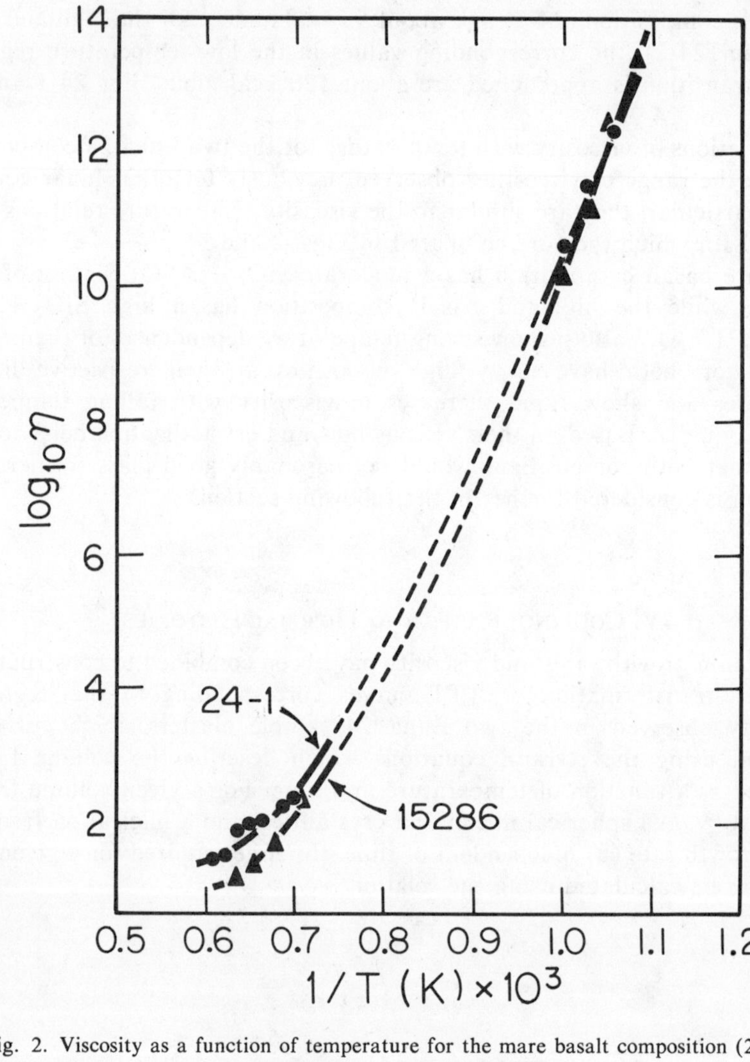

Fig. 2. Viscosity as a function of temperature for the mare basalt composition (24-1) and lunar sample 15286.

for such small clusters of atoms, the nucleation process is unlikely to exhibit a time dependence characteristic of large-scale diffusion control.

The observed volume fraction crystallized in the mare basalt cluster is greater than 50%, whereas the degree of crystallinity of the highland basalt is less than 1%. For both Luna 24 compositions, the nucleation frequencies required in the kinetic treatment have been calculated from the measured viscosities using the classical theory of homogeneous nucleation. In these calculations, the nucleation barrier was taken as 55 kT at a relative undercooling, ΔT_r, of 0.2. Here $\Delta T_r = \Delta T/T_E$, where ΔT is the undercooling and T_E is the liquidus temperature.

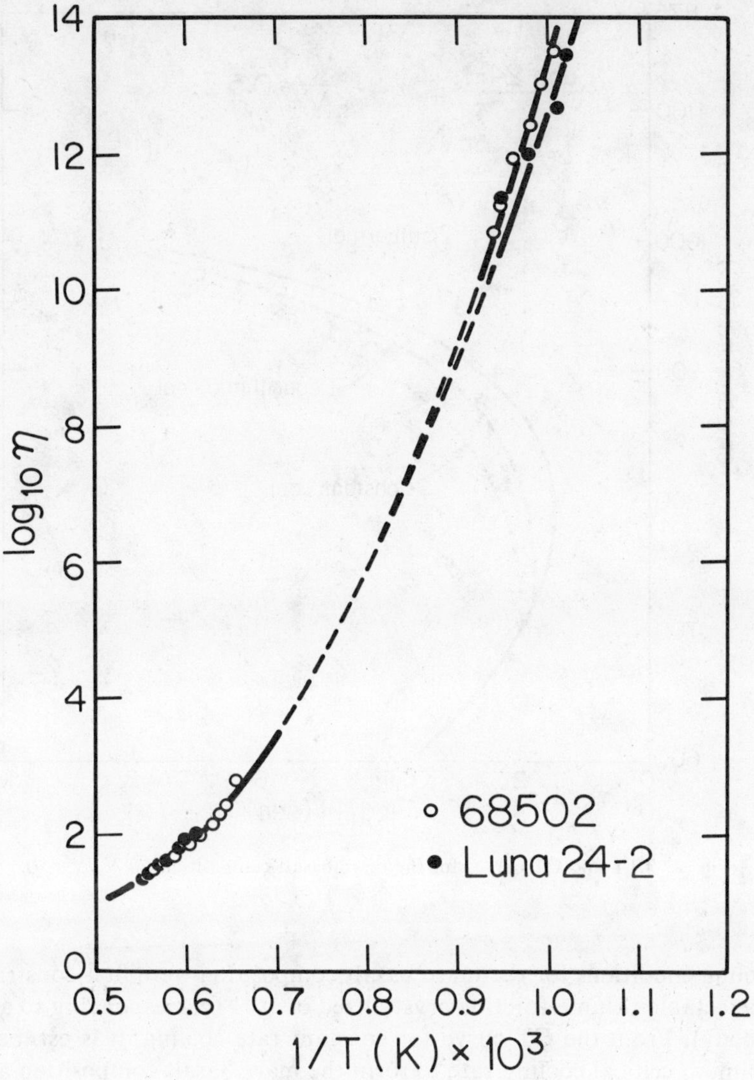

Fig. 3. Viscosity as a function of temperature for the highland basalt composition (24-2) and lunar sample 68502.

The use of classical nucleation theory and the range of assumed nucleation barriers are in agreement with the results obtained by Klein and Uhlmann (1974) on lunar composition 70019.

Continuous cooling (CT) curves were constructed using the approach of Grange and Keifer (1941). This approach was elaborated to include a variety of cooling histories and was applied to the problem of glass formation by Onorato and Uhlmann (1976). Continuous cooling curves for logarithmic and constant-

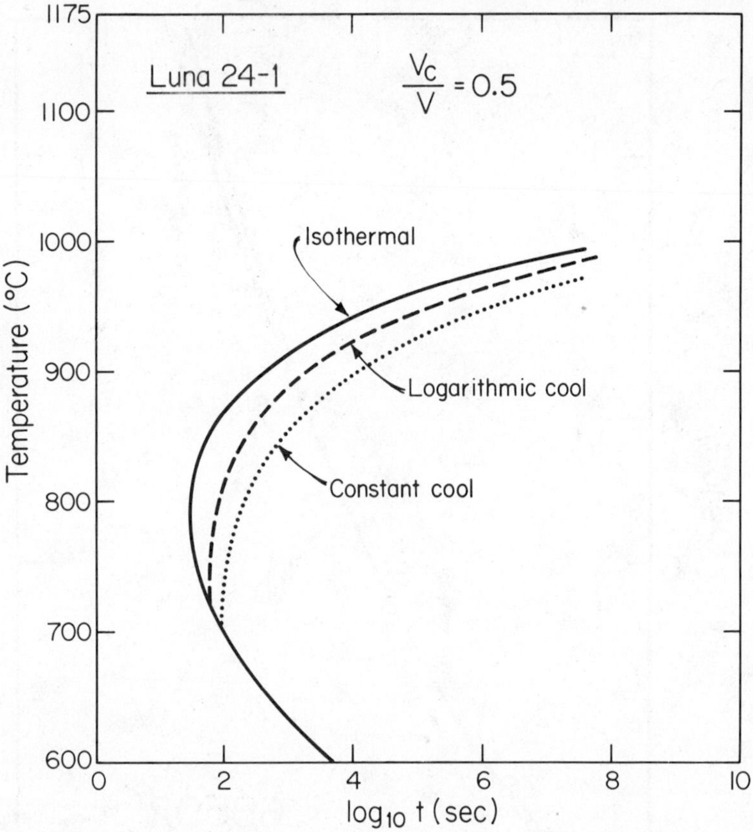

Fig. 4. TTT and CT curves for the mare basalt composition for $V_c/V = 0.5$

rate cooling conditions for the mare basalt composition were first constructed for a just-detectable volume fraction crystallized of 10^{-6} (corresponding to a sensibly glassy body). From the CT curve for constant rate cooling, it is estimated that the minimum critical cooling rate to form the mare basalt composition as a glass is 146°C sec^{-1}. TTT and CT curves have also been constructed for a 50% volume fraction crystallized, which is the observed degree of crystallinity in the angular mare basalt particles. These curves are shown in Fig. 4. The estimated critical cooling rate in this case is about 5.4°C sec.$^{-1}$.

The TTT and CT curves for the highland basalt composition have also been constructed for a volume fraction crystallized of 10^{-6}. From the CT analysis, the critical cooling rate to form this material as a glass is estimated as about 158°C sec^{-1}. For a volume fraction crystallized of 10^{-2}, which is the upper limit of the degree of crystallinity observed in the samples, the CT analysis predicts a critical cooling rate of about 16°C sec^{-1}. The TTT and CT curves for $V_c/V = 10^{-2}$ are shown in Fig. 5

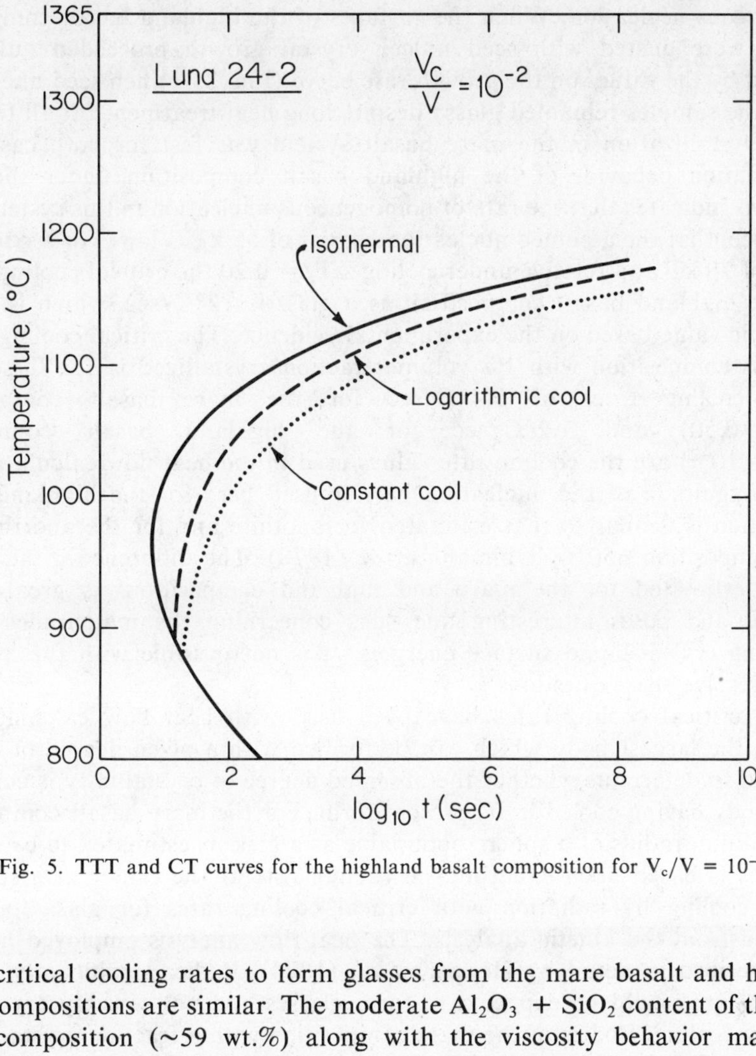

Fig. 5. TTT and CT curves for the highland basalt composition for $V_c/V = 10^{-2}$

The critical cooling rates to form glasses from the mare basalt and highland basalt compositions are similar. The moderate Al_2O_3 + SiO_2 content of the mare basalt composition (~59 wt.%) along with the viscosity behavior make this composition a moderately good glass former. The high Al_2O_3 + SiO_2 content of the highland basalt composition (~72 wt.%) and its viscosity-temperature behavior suggest that this composition is a good glass former. Moreover, the liquidus temperature which is inversely related to glass-forming ability is high for this composition (1365°C ± 10°C). This suggests that the highland basalt composition is only a moderately good glass former, as is the mare basalt composition.

Melting and quenching experiments have shown that the glass forming abilities of the two compositions are not similar. The mare basalt composition is much more difficult to form as a glass than the highland basalt composition. The source of the discrepancy between calculated and observed cooling rates required to form these compositions as glasses is found in the assumed barrier for homogeneous necleation. When the surfaces of the highland basalt composition samples were dusted with seed nulcei, crystal growth proceeded quickly as indicated by the values on the growth rate curve (Fig. 2). When seed nuclei were absent, the samples remained glassy despite long heat treatments at all temperatures. Crystallization in the mare basalt system was fast for both cases. The crystallization behavior of the highland basalt composition under these two conditions indicates that the rate of homogeneous nucleation in this system is low and, hence, that the assumed nucleation barrier of 55 kT is low. For a nucleation barrier of 70 kT at a relative undercooling $\triangle T_r = 0.20$ the critical cooling rate to form the highland basalt composition as a glass is 12°C/sec. which is a more reasonable value based on the experimental evidence. The critical cooling rate to form this composition with 1% volume fraction crystallized is 1.2°C/sec. The critical cooling rates of 5.4°C/sec for the mare basalt compositions ($V_c/V = 0.50$) and 1.2°C/sec for the highland basalt composition ($V_c/V = 10^{-2}$) are the cooling rate values used in the heat flow calculations.

The magnitude of the nucleation barrier used here for the highland basalt composition is similar to that estimated for anothite and for the anorthite-rich lunar composition 65016 (Uhlmann *et al.*, 1978). The difference in nucleation barriers estimated for the mare and highland compositions is greater than expected, and poses interesting questions concerning the magnitudes of the underlying crystal-liquid surface energies. It is not possible with the available data to resolve these questions.

These critical cooling rates have been used with heat flow calculations to estimate the largest body which can be formed with a given degree of crystallinity and to determine whether the observed degree of crystallinity is consistent with a body having cooled in its present form. For the mare basalt composition, the maximum radius of a sphere obtainable as a glass is estimated to be about 2 mm. This is based upon equating the cooling rate at the center of a spherical particle cooling by radiation with critical cooling rates for glass formation estimated from the kinetic analysis. The heat flow analysis employed here has been described in detail by Hopper *et al.* (1974). Bodies significantly smaller than this size would be expected to form glasses and not sensibly crystallized bodies on cooling. Bodies of the mare composition with a fraction crystallized of 50% are expected to be formed when the size of the cooling particle is about 1 cm. For the highland basalt composition, the maximum size sphere obtainable as a glass is estimated to be about 5 mm in radius. When a 1% volume fraction crystallized is assumed, the maximum size is 2 cm.

These results indicate that bodies of the highland basalt composition of the size of the Luna 24 samples studied, that is 90–250 μm, are expected to be formed as glasses; and it is expected that larger bodies could also be formed as glasses. In the case of the mare basalt composition, bodies of the size of the Luna 24 samples are also expected to cool as glasses. While some glassy particles are seen, many of the particles have sizeable degrees of crystallinity (50% or more). The present work indicates that these particles did not cool in their present form but must be fragments of larger bodies—and their angular appearance is consistent with this suggestion.

Application of the analysis suggests that the size of the body which would cool with the observed fraction crystallized is about 1 cm; and it is suggested that Luna 24 mare basalt samples studied by Warner, Taylor, and their coworkers originally cooled in bodies of about this size.

Acknowledgments—Financial support for the present work was provided by the National Aeronautics and Space Administration. This support is gratefully acknowledged.

REFERENCES

Cukierman M., Tutts P. M. and Uhlmann D. R. (1972) Viscous flow behavior of Lunar Compositions 14259 and 14310. *Proc. Lunar Sci. Conf. 3rd,* p. 2619–2625.

Grange R. A. and Kiefer J. M. (1941) Transformation of austenite on continuous cooling and its relation to transformation at constant temperature. *Trans. ASM* **29**, 85.

Handwerker C. A., Klein L. C., Onorato P. I. K., and Uhlmann D. R. (1977) Matrix glass vs. intruded glass in Lunar breccia 15286. *Proc. Lunar Sci. Conf. 8th,* p. 2581–2592.

Hopper R. W., Onorato P. and Uhlmann D. R. (1974) Thermal histories and crystal distributions in partly devitrified lunar glasses cooled by radiation. *Proc. Lunar Sci. Conf. 5th,* p. 2257–2253.

Klein L. C. and Uhlmann D. R. (1976) The kinetics of lunar glass formation, revisited. *Proc. Lunar Sci. Conf. 7th,* p. 1113–1121.

Onorato P. I. K. and Uhlmann D. R. (1976) Nucleating heterogeneities and glass formation. *J. Non-Cryst. Solids* **22**, 367–378.

Sato M., Hickling N. L., and McLane J. E. (1973) Oxygen fugacity values of Apollo 12, 14, and 15 lunar samples and reduced state of lunar magmas. *Proc. Lunar Sci. Conf. 4th,* p. 1061–1079.

Scherer G., Hopper R. W., and Uhlmann D. R. (1972) Crystallization behavior and glass formation of selected lunar compositions. *Proc. Lunar Sci. Conf. 3rd.* p. 2627–2637.

Taylor L. A., Norman M. and Coish R. A. (1978) Glass chemistry and magma evolution at Mare Crisium. In *Mare Crisium: The View from Luna 24* (R. B. Merrill and J. J. Papike, eds.). This volume.

Uhlmann D. R. (1972) A kinetic treatment of glass formation. *J. Non-Cryst. Solids* **7**, 337–348.

Uhlmann D. R., Klein L. C., Kritchevsky G., and Hopper R. W. (1974) The formation of lunar glasses. *Proc. Lunar Sci. Conf. 5th,* p. 2317–2331.

Uhlmann D. R. and Klein L. C. (1976) Crystallization kinetics, viscous flow, and thermal histories of lunar breccias 15286 and 15498. *Proc. Lunar Sci. Conf. 7th,* p. 2529–2541.

Uhlmann D. R., Handwerker C. A., Goncz D. and Onorato P. I. K. (1978) The formation kinetics of lunar glasses. *Proc. Lunar Planet. Sci. Conf. 9th.* In press.

Warner J. L., Simonds C. H., McGee P. E., and Cochran A. (1978) Luna 24: Lateral heterogeneity in the lunar crust. In *Mare Crisium: The View from Luna 24* (R. B. Merrill and J. J. Papike, eds.) This volume.

Williams R. J. (1972) A solid ceramic electrolyte system for measuring redox conditions in high temperature gas mixing systems. NASA TMX-58105.

Merrill, R. B. and Papike, J. J., editors.
(1978) *Mare Crisium: The View from Luna 24*, Pergamon Press (New York), p. 495–522
Printed in the United States of America

Melt inclusions in Luna 24 soil fragments

EDWIN ROEDDER

959 National Center, U.S. Geol. Survey, Reston, Virginia 22092

P. W. WEIBLEN

Dept. of Geology and Geophysics, Univ. of Minnesota, Minneapolis, Minnesota 55455

Abstract—Optical examination of 28 slides (polished grain mounts) of Luna 24 soil fragments (0.09–0.50 mm size range) revealed melt inclusions in grains of olivine, plagioclase, pyroxene, spinel, and ilmenite, as well as interstitial inclusions. Electron microprobe analyses (63) are given of the inclusions and their daughter, host, and associated minerals. In contrast with the Apollo samples, the Luna 24 samples contain sulfide melt inclusions indicating that saturation with respect to an iron sulfide melt took place throughout much of the crystallization history, even while olivine was crystallizing. The Luna 24 silicate-melt inclusions have recorded a more extensive differentiation toward higher iron magmas than have the Apollo inclusions, but they have recorded some inexplicably low aluminum values; silicate immiscibility eventually took place and formed "high-Fe," and "high-silica," high-K_2O granitic melts, all generally higher in Fe than their Apollo counterparts. Two uniquely high-SiO_2 melt inclusions in olivine (87.2 and 93.8% SiO_2) were found; they may have formed by metastable subliquidus immiscibility in these unusually Fe-rich melts. Swarms of microinclusions in plagioclase appear to be of olivine composition and may represent exsolution products.

INTRODUCTION

The Soviet Luna 24 spacecraft returned a 160-cm core sample from the regolith in southwest Mare Crisium. As a part of the FOCUS (Friends of Crisium Unmanned Sampling) consortium*, we searched 28 polished slides (multiple-grain mounts) optically at 200× for melt inclusions, representing the trapping of former liquids. The individual slides were made up of sized grains in various ranges from 90 to 500 μm and consist of material from each of the seven sampling locations along the core provided to NASA by the U.S.S.R. The NASA slide numbers were <u>24077,41</u>; 24077,45; 24077,53; <u>24109,50</u>; 24109,51; 24109,52; 24109,54; <u>24109,57</u>; 24149,40; 24149,41; <u>24149,49</u>; <u>24149,50</u>; 24149,57; <u>24170,4</u>; 24170,8; 24174,47; 24174,49; <u>24174,50</u>; 24174,51; 24174,55; <u>24174,56</u>; 24174,57; 24182,20; 24182,25; <u>24182,46</u>; 24182,48; <u>24182,50</u>; and 24210,48. In each number, the third through fifth digits give the centimeter depth in the core barrel as recovered. The specific grains studied are identified in this report by a FOCUS consortium-assigned grain number following the slide number. Most of our studies were on the eight slides

*Consortium members are A. E. Bence, J. Friel, J. Goldstein, T. Grove, S. Haggerty, J. Head, J. J. Papike, E. Roedder, D. T. Vaniman, and P. W. Weiblen.

underlined above; the other 20 were examined only optically, but showed inclusions that appeared similar to those that were analyzed.

Melt inclusions were present in many grains of olivine, plagioclase, pyroxene, spinel, and ilmenite, but relatively few were found that were both at the slide surface and large enough for analysis. These inclusions were further studied by scanning electron microscopy with energy dispersive x-ray analysis (referred to here simply as SEM), and electron microprobe with wave length-dispersive x-ray analysis. Except as stated, the analytical data reported here were all obtained by the electron microprobe and were reduced using methods described earlier (Roedder and Weiblen, 1973).

The major goal of the examination of the abundance, significance, and nature of these melt inclusions was to determine what information they might yield on the petrology of the lavas that were disaggregated to yield these soil fragments. The mineral grains studied were either in lithic fragments of ferrogabbro (Bence *et al.*, 1977) and ferrobasalt (Vaniman and Papike, 1977), or in single crystal fragments that were presumably derived mainly from these two rock types. On the basis of their grain size, we presume that the bulk of the single-crystal fragments studied came from the disaggregation of ferrograbbro (grain size >1 mm; Bence *et al.*, 1977) rather than ferrobasalt (grain size 0.1–0.2 mm); minor-element analyses may permit more specific assignments of origin to individual grains.

DATA

Early-stage inclusions in olivine

General occurrence Most of the olivine grains containing inclusions in these samples are individual crystal clasts in the 0.125–0.250-mm range and are probably from the coarser ferrogabbro at the site, but similar inclusions were found in the olivine of ferrobasalt fragments. Euhedral opaque chromium-rich spinel crystals ($\leq 20\ \mu m$) occur in a few olivine grains (e.g., 24109,50-41). Silicate-melt inclusions (mostly 3–50 μm diameter) were found in many grains, but only five were amenable to analysis. Most silicate-melt inclusions in olivine in these samples, although probably originally trapped as a single homogeneous melt phase, now are multiphase. Thus all inclusions larger than a few micrometers now have a "vapor bubble" (this phase is probably close to a vacuum), which formed by differential shrinkage and which generally occupies ~15 volume percent. Those inclusions in which the silicate melt has largely crystallized to various phases in addition to crystallization of olivine on the walls now appear to have a somewhat larger vapor bubble, but the volume percent vapor in such inclusions can only be crudely estimated.

Even some very small silicate-melt inclusions in olivine contain an opaque globule, which has been found to be iron sulfide in those rare inclusions where it is at the surface (Fig. 1). Similar to the Apollo samples (Roedder and Weiblen, 1971), this sulfide generally constitutes ~0.5 vol. % of the inclusions, corresponding to ~2000 ppm S, but a few inclusions in olivine in the Luna 24 samples contain much more sulfide, as much as ~10 vol. % (Fig. 2), and some in olivine and pyroxferroite described below under the heading "Miscellaneous inclusions," are all sulfide.

Inclusions containing crystalline silicate phases In addition to vapor and sulfide phases, about half the silicate-melt inclusions in these olivines have crystallized some silicate phases. This is true of inclusions in olivine from both ferrogabbro and ferrobasalt; we have not recognized any differences in the inclusions from these two rock types in these samples, but this is perhaps expectable, as the differences in grain size and hence probable cooling rates for the two types are not large. The presence of such crystalline inclusions is actually useful in the recognition of olivine, as even small grains may contain rather large inclusions (Fig. 3).

Analyses of three inclusions and their optically unzoned hosts, using an appropriately sized defocused beam where necessary, are given in Table 1 (Anals. 1–8). Many contain lath-shaped, relatively low-Mg, high-Ca crystals; these are presumably pyroxene, but are too small for accurate analysis (Anal. 7). They are embedded in a more siliceous, presumably glassy matrix (Figs. 4–6). An electron-microprobe traverse (Fig. 10), made by counting at 1-μm intervals across such a crystalline inclusion in grain 24109,57-3 (Figs. 4,5) shows some rather high SiO_2 values (>70%) in the glass areas; other inclusions show small areas that may be an actual silica phase (Fig. 7). The inclusions shown in Figs. 4 and 5 each contain a small sulfide globule and a roughly tabular, transparent yellowish-brown daughter crystal, presumably a spinel, but both spinel and sulfide are sub-surface and therefore not analyzed. Both spinel (?) crystals are elongated parallel to (100) of the enclosing olivine, just as are the ilmenite daughter crystals in the inclusions in the higher-Ti Apollo mare basalt olivines (Roedder and Weiblen, 1970,1971), suggesting an epitaxial relationship. The only previously recorded example of this presumed spinel-olivine epitaxy in melt inclusions was reported by Roedder (1965, last entry in Table 3d) in inclusions in a picrite basalt from the Oregon Coast Range, but Kohlstedt and Vander Sande (1975) found another spinel phase, magnetite, among the oriented precipitates in oxidized olivine crystals from Alaska. Several other inclusions in the Luna 24 sample show very thin opaque plates arranged parallel to (100) of the enclosing olivine; these are presumably ilmenite, as in Apollo 11, but none are at the surface and hence available for analysis.

Inclusions without crystalline silicate phases The other half of the melt inclusions in these olivines consists of colorless homogeneous glass plus a shrinkage bubble of ~15 vol. %; most also contain an opaque globule of iron

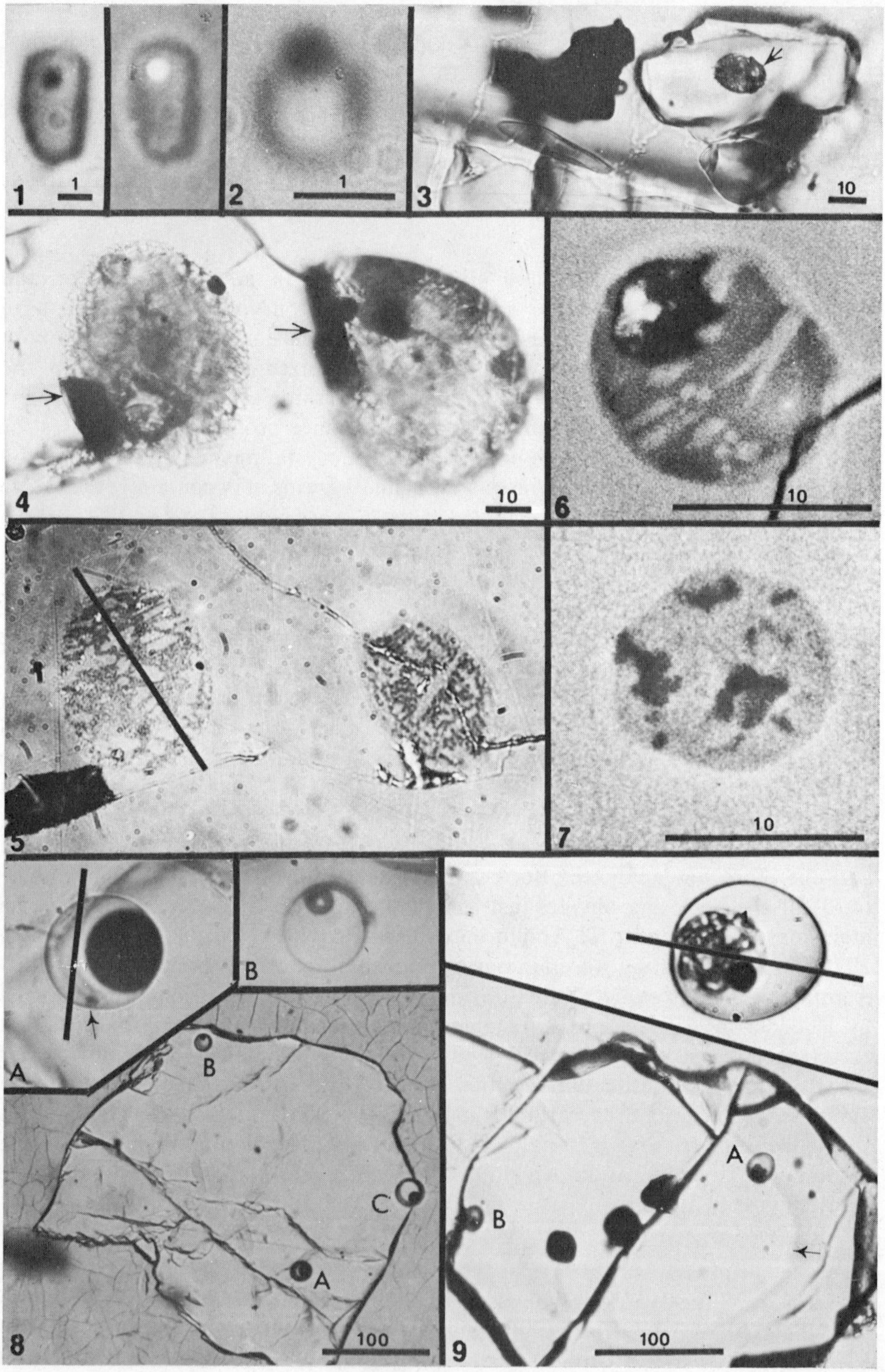
1
1
2
1
3
10
4
10
6
10
5
7
10
A
B
B
C
A
8
100
A
B
9
100

Figures 1–9 Inclusions in olivine of Luna 24. (Scale bar dimensions given in μm.)

Fig. 1. One of a group of five tiny melt inclusions in Fo_{64} olivine of grain 24149,50-26 (Table 1, Anal. 14), showing immiscible sulfide melt globules in glass. Left photo, transmitted plain light; right one in plain reflected light.

Fig. 2. Another tiny inclusion in same grain as Fig. 1, in transmitted plain light, showing relatively large globule of sulfide, presumably indicating presence of an immiscible liquid sulfide phase at the time of crystallization of this olivine.

Fig. 3. Small olivine grain in ferrobasalt fragment 24109,57-6, containing a relatively large crystalline melt inclusion (arrow), in transmitted plain light.

Figs. 4 and 5. Large, partly crystalline silicate melt inclusions in olivine of grain 24109,57-3, in transmitted plain light (Fig. 4), and plain reflected light (Fig. 5). The dark spinel plates (arrows) are tilted and hence appear more equant than they actually are. A small opaque globule of immiscible sulfide is visible in each. Neither spinel nor sulfide is cut by the polished surface (Fig. 5); this cuts only pyroxene laths in a darker glassy(?) matrix. An electron-microprobe traverse across the left inclusion (see line in Fig. 5) is shown in Fig. 10. (See also Table 1, Anals. 5–8).

Fig. 6. Crystalline inclusion in olivine of ferrobasalt fragment 24174,56-12, seen as an SEM seconary electron image. (See Table 1, Anals. 3 & 4).

Fig. 7. Crystalline inclusion in olivine of ferrobasalt fragment 24174,56-24, seen as an SEM secondary electron image. (See Table 1, Anals. 1 and 2).

Fig. 8. Single crystal, apparently unzoned Fo_{73} olivine grain 24109,50-2, containing three silicate glass inclusions, in transmitted plain light. The insets are enlarged views of inclusions A and B. The curved shadows in A are optical artifacts; the glass is optically featureless except for a small speck (sulfide?; arrow). Inclusion B is completely surrounded by the olivine crystal. A vertical electron-microprobe traverse across the glass of inclusion A (along the line) is shown in Fig. 11. (See Table 1, Anals. 9 and 10).

Fig. 9. Single-crystal, apparently unzoned Fo_{51} olivine grain 24174,50-93, containing two silicate glass inclusions, A and B, and three spinel crystals, in transmitted plain light. The inset is an enlarged view of inclusion A. Both glass inclusions contain sulfide globules (see insert). Only A is at the surface. An electron-microprobe traverse across A (along the line) is shown in Fig. 12. The curving line paralleling the grain edge (arrow) is an artifact from the bottom edge of this grain. (See Table 1, Anals. 11 and 12).

sulfide. The bubble/glass and glass/host crystal interfaces, as well as the glass itself, are featureless under optical (light) and scanning electron microscopes. Only two of the many glass inclusions were at the surface and available for analysis. Electron-microprobe analyses of these inclusions and of their optically clear and apparently unzoned host olivines are given in Table 1 (Anals. 9–12; Figs. 8,9). Both of these analyzed inclusions show a nearly opaque cluster of small grains within the bubble where it is cut by the polished surface; these are believed to be diamond grinding and polishing compound from sample prepara-

Table 1. Electron-microprobe analyses of inclusions and host phases in Luna 24 samples. P. W. Weiblen, analyst. Abbreviations used: adj.—adjacent; ilm.—ilmenite; inter.—interstitial; py.—pyroxene; DM—daughter mineral.

	Inclusions in olivine													
Anal. no.	1	2	3	4	5	6	7	8	9	10	11	12	13	14
Slide and grain no.	24174,56-24		24174,56-12		24109,57-3				24109,50-2		24174,50-93		24109, 57-1	24149, 50-26
Photo no.	P2	P2	P1	P1	3	1	2	4	P1	P2	P1	P1	P1	P5
Material	Inc.	Host	Inc.	Host	Inc.	Inc.	Py DM	Host	Inc.	Host	Inc.	Host	Host	Host
SiO_2	56.8	31.4	58.2	30.9	60.7	65.8	53.8	35.7	87.2	36.3	93.8	35.1	34.7	37.6
Al_2O_3	8.86		15.3		15.7	16.3	12.0		6.84	<0.05	1.51			<0.05
FeO	22.2	62.8	6.77	62.1	6.60	5.79	10.9	44.1	1.27	25.3	2.32	40.2	45.1	31.3
MgO	3.38	6.40	0.79	6.12	1.92	1.41	3.45	20.7	0.95	38.6	1.61	23.9	20.1	31.6
CaO	5.59	0.60	16.0		13.5	10.2	16.6		1.76	0.23	0.06	0.23	0.5	0.17
Na_2O	0.25		0.26		0.05	0.05	0.08		0.9		<0.5			
K_2O	0.05		0.08		0.05	0.04	0.07		0.02		0.11			
TiO_2	0.80		0.39		0.56	0.42	1.33		0.36					
MnO	0.21		0.25		0.10	0.11	0.16		0.15			0.25		0.30
Cr_2O_3					<0.05	<0.05	0.08		0.01			0.04		<0.05
Arith. total	98.14	101.20	98.04	99.12	99.18	100.12	98.47	100.5	99.46	100.43	99.41	99.72	100.4	100.97
Atoms Norm														
O Q	21.7	4.000	24.5	4.000	20.9	38.3	6.000	4.000	76.4	4.000	89.4	4.000	4.000	4.000
Si C		1.001		1.006			1.978	1.026	2.2	0.958	1.3	1.003	1.009	1.010
Ti or	0.3		0.5		0.4	0.2	0.037		0.1		0.7			
Al ab	2.2		2.2		0.6	0.4	0.521		7.7					
Cr an	23.4		41.3		35.4	44.2	0.002		8.8		0.3	0.001		
Fe wo	2.1	1.674	16.7	1.691	16.7	2.7	0.335	1.060		0.558		0.961	1.096	0.703
Mn en	8.6		2.0		7.8	3.5	0.005		2.4		4.0	0.006		0.007
Mg fs	40.1	0.304	12.0	0.297	16.0	9.9	0.189	0.887	1.7	1.518	4.3	1.018	0.871	1.265
Ca fo		0.020					0.654			0.006		0.007	0.016	0.005
Na fa							0.006							
K il	1.6		0.8		2.1	0.8	0.003		0.7					
cm					0.1									
Total	100.0		100.0		100.0	100.0			100.0		100.0			
ab wo							55.6							
or en							16.0							
an fs							28.4							
fo		15.4		14.9				45.6		73.1		51.4	44.3	64.3
fa		84.6		85.1				54.4		26.9		48.6	55.7	35.7

Table 1. (*cont.*)

	Inclusions in spinel					Inclusions in spinel				Low-Si inclusions in plagioclase					
Anal. no.	15	16	17	18	19	20	21	22	23	24	25	26	27	28	29
Slide and grain no.	24109,57-3			24109,57-4		24109,57-4				24109,57-2			24109,50-19		
Photo no.	P3-1	P3-2	P3-3	P4-1	P4-1	P4-2	P4-4	P4-3	P4-5	—	P2-1	P2-2	SEM-P-3	SEM-P-4	SEM-P-2
Material	Host	Host	Host	Inc.	Inc.	Host	Host	Host	Host	Inc.	Inc.	Inc.	Inc.	Inc.	Host
SiO_2				15.2	57.17	0.92	0.45	2.26	35.0	44.6	47.3	47.5	64.1	50.7	45.8
Al_2O_3	3.5	7.98	11.28	3.88	4.53	3.71	2.70	12.4		6.01	4.65	7.99	3.45	4.45	34.2
FeO	53.3	34.87	34.80	38.1	9.84	48.5	54.4	32.4	43.7	24.9	28.6	22.5	14.3	28.8	0.48
MgO				2.67	8.19	0.80	0.37	1.37	20.1	5.9	4.91	3.50	2.57	4.40	0.24
CaO				3.85	12.70	0.85	0.02	0.72	0.29	10.3	9.35	10.4	15.2	9.46	19.5
Na_2O				<0.05	<0.05					0.52			<0.5	<0.5	0.2
K_2O				0.05	0.20					0.04			0.11	0.10	0.01
TiO_2	17.2	6.24	1.92	16.9	7.09	20.5	24.5	3.10		4.57	3.30	3.69	0.36	1.35	<0.05
MnO				0.41	0.28	0.46	0.35	0.38	0.35	0.34	0.31	0.38	0.24	0.27	<0.05
Cr_2O_3	26.0	50.91	51.99	18.0	0.00	24.5	17.0	45.0		0.09	0.06	0.08	<0.05	<0.05	<0.05
Arith. total	100.0	100.00	99.99	99.06	100.00	100.24	99.79	97.63	99.44	97.38	98.48	96.04	100.33	99.53	100.43
Atoms Norm															
O Q	4.000	4.000	4.000		24.9	4.000	4.000	4.000	4.000	1.1	5.9	10.9	29.9	8.1	8.000
Si C						0.033	0.017	0.079	1.021						2.108
Ti or	0.481	0.166	0.051	0.3	1.2	0.554	0.676	0.081		0.3			0.6	0.6	
Al ab	0.154	0.333	0.468			0.157	0.117	0.510		4.5					1.856
Cr an	0.765	1.424	1.448	10.6	11.8	0.696	0.493	1.240		14.4	12.9	22.8	9.1	11.9	
Fe wo	1.659	1.032	1.025	3.7	21.4	1.457	1.669	0.945	1.066	15.9	14.3	13.0	27.7	14.8	0.018
Mn en				0.5	20.5	0.014	0.011	0.011	0.009	15.0	12.5	9.1	6.4	11.0	
Mg fs				2.1	6.7	0.043	0.020	0.071	0.874	39.8	47.9	36.8	25.6	51.0	0.017
Ca fo				4.4		0.033	0.001	0.027	0.009						0.962
Na fa				19.1											0.018
K il				32.4	13.5					8.9	6.4	7.3	0.7	2.6	
cm				26.9						0.1	0.1	0.1			
Total				100.0	100.0					100.0	100.0	100.0	100.0	100.0	
ab wo															1.82
or en															0.06
an fs															98.12
fo									45.0						
fa									55.0						

Table 1 (*cont.*)

	Low-Si inclusion in pyroxene		Late-stage inclusions and daughter minerals							
			Host = plagioclase			Host = pyroxene				
Anal. no.	30	31	32	33	34	35	36	37	38	39
Slide and grain no.	24170,4-6			24170,4-6		24149,50-98		24182,50-11		24174,56-25
Photo no.	P2Q-A	P2Q-B	P-4	P3-1	P3-F	P1-1	P1-2	P1-1	P1-2	P2-1
Material	Inc.	Host	Inc.	Inc.	Host	Inc.	Host	Inc.	Host	Inc.
SiO_2	48.9	49.1	50.2	47.1	43.5	76.4	48.8	70.6	47.8	82.2
Al_2O_3	14.9	0.96	29.3	5.35	36.5	11.3	0.84	10.0	0.53	4.15
FeO	10.2	28.9	3.2	32.0	0.36	4.66	37.0	12.1	31.7	6.28
MgO	7.08	7.16	<0.1	1.95	—	0.46	1.59	0.26	8.35	0.52
CaO	14.6	14.1	15.0	11.1	18.7	2.74	9.54	5.95	9.54	3.02
Na_2O	<1.0	<1.0	<0.1	0.2	1.25	0.03	<0.05	0.30	—	<0.05
K_2O	2.82	—	0.3	0.08	0.09	2.91	—	2.15	—	0.75
TiO_2	0.97	0.96	0.13	0.59	—	0.02	0.48	0.33	0.69	0.37
MnO	0.21	0.46	—	0.33	—	0.07	0.36	0.20	0.54	0.23
Cr_2O_3	0.04	0.07	—	<0.05	—	—	0.04	—	0.01	—
Arith. total	99.72	101.71	98.13	98.70	100.40	98.59	98.65	101.89	99.16	97.52
Atoms Norm										
O Q	—	6.000	14.6	6.000	8.000	55.4	6.000	39.1	6.000	70.3
Si C	—	1.941	1.7	1.927	2.011	3.2	2.038	—	1.949	—
Ti or	16.7	0.028	1.8	—	—	17.5	0.015	12.5	0.021	4.6
Al ab	—	0.045	—	0.253	1.989	0.3	0.042	2.5	0.025	—
Cr an	32.5	0.002	75.8	—	—	13.8	0.001	19.3	—	9.4
Fe wo	16.8	0.956	—	1.095	0.014	—	1.292	4.1	1.081	2.5
Mn en	12.2	0.015	—	0.018	—	1.2	0.013	0.6	0.019	1.3
Mg fs	11.8	0.422	5.8	0.119	—	8.6	0.099	21.3	0.507	11.2
Ca fo	3.9	0.597	—	0.487	0.926	—	0.427	—	0.417	—
Na fa	4.1	—	—	0.016	0.112	—	—	—	—	—
K il	1.9	—	0.3	0.004	0.005	—	—	0.6	—	0.7
cm	0.1	—	—	—	—	—	—	—	—	—
Total	100.0		100.0			100.0		100.0		100.0
ab wo		30.25		29	10.74		23.5		20.79	
or en		21.36		7	0.51		5.4		25.30	
an fs		48.39		64	88.75		71.1		53.91	

Table 1 (*cont.*)

	Host = pyroxene						Host = ilmenite			
Anal. no.	40	41	42	43	44	45	46	47	48	49
Slide and grain no.	24174,56-25		24109,50-47		24174,50-340		24174,56-7			
Photo no.	P2-2	P2-3	1	2	1	2	P1-1	P1-2	P1-3	P1-4
Material	Host	Adj.	Host	Adj.	Host	Adj.	Inc.	Py	Host	Inter.
SiO_2	48.0	48.8	46.9	50.7	46.9	48.5	74.0	44.1		78.9
Al_2O_3	0.72	0.71	1.28	1.29	0.23	0.99	9.41	0.84		8.42
FeO	35.4	32.6	31.2	13.8	47.0	18.3	6.21	37.0	48.9	2.73
MgO	4.67	8.14	8.18	15.6	0.44	13.6	0.17	1.85	0.31	0.11
CaO	10.1	7.69	10.7	17.3	4.87	17.4	4.43	13.7		0.94
Na_2O	<0.05		<.05		0.01	0.10	0.41			<0.05
K_2O							2.33			5.81
TiO_2	0.65	0.56	0.61	0.37	0.18	0.46	1.40	1.24	50.1	1.20
MnO	0.17	0.51	0.20	0.24	0.56	0.33	0.14	0.73		<0.1
Cr_2O_3	<0.05	0.05	0.14	0.56	0.02	0.51	<0.05	0.06		
Arith. total	99.71	99.06	99.21	99.86	100.21	100.19	98.50	99.52	99.31	98.11
Atoms Norm										
O Q	6.000	6.000	6.000	6.000	6.000	6.000	50.9	6.000	3.000	54.1
Si C	1.975	1.981	1.913	1.920	2.008	1.882		1.882		0.4
Ti or	0.020	0.017	0.019	0.011	0.006	0.013	14.0	0.040	0.968	35.0
Al ab	0.035	0.034	0.062	0.058	0.012	0.045	3.5	0.042		
Cr an		0.002	0.005	0.017	0.001	0.016	17.2	0.002		4.8
Fe wo	1.218	1.106	1.065	0.437	1.683	0.594	2.1	1.320	1.051	
Mn en	0.006	0.018	0.007	0.008	0.020	0.011	0.4	0.026		0.3
Mg fs	0.287	0.492	0.497	0.880	0.028	0.786	9.2	0.118	0.012	3.1
Ca fo	0.445	0.334	0.468	0.720	0.223	0.723		0.627		
Na fa					0.001	0.008				
K il							2.7			2.3
cm										
Total							100.0			100.0
ab wo	22.84	17.30	23.05	34.76	11.55	34.39		30.4		
or en	14.68	25.47	24.50	43.60	1.45	37.38		5.7		
an fs	62.48	57.24	52.45	21.64	87.00	28.23		63.9		

Table 1 (*concluded*)

	High-Fe/high-Si inclusions in plagioclase								Miscellaneous inclusions					
Anal. no.	50	51	52	53	54	55	56	57	58	59	60	61	62	63
Slide and grain no.	24077,41-10			24077,41-25			24109,57-5		24174,56-9		24182,50-9	24174,56-29	24174,50-311	
Photo no.	P3-2	P3-1	P3-3	P2-1	P2-2	P2-3	P1-1	P1-2	P1-1	P1-3	P1-1	P2-1	P1-1	P1-2
Material	Inc.	Inc.	Host	Inc.	Inc.	Host	Inc.	Host	Inc.	Host	Py	?	Inc.	Host
SiO_2	71.9	38.2	46.0	76.0	37.3	44.8	41.5	44.2	39.8	44.7	47.9	56.8	0.15	46.9
Al_2O_3	11.2	2.33	34.0	10.2	2.21	33.7	4.72	34.8	21.3	35.4	0.94	0.27		0.23
FeO	8.14	46.5	0.66	4.86	44.5	1.68	34.9	0.53	10.7	0.45	29.5	37.3	63.4	47.0
MgO	<0.5	0.73	<0.5	<0.5	<0.5	<0.5	1.63	0.32	18.5	0.48	13.1	0.32	36.4	0.44
CaO	5.47	6.36	19.4	5.22	5.66	19.2	12.3	19.3	9.60	18.5	7.97	5.05		4.87
Na_2O	<0.5	<0.5	<0.5	<0.5	<0.5	0.4	0.37	0.49		0.83		<0.05		<0.5
K_2O	3.81	0.27	0.11	1.73	0.27	0.12	<0.05	0.03	0.27	0.02		<0.05		
TiO_2	1.05	6.06	<0.05	0.57	9.32		3.40		0.09	0.03	0.61	0.41		0.18
MnO	0.05		<0.05	<0.05	0.72		0.39				0.45	0.38		0.56
Cr_2O_3		<0.05									0.40			0.02
Arith. total	101.62	100.45	100.17	98.58	99.98	99.90	99.56	99.67	100.26	100.41	100.87	100.53	99.95	100.20
Atoms Norm														
O Q	40.6		8.000	56.0		8.000	6.000	8.000		8.000	6.000	19.8		6.000
Si C			2.122			2.089	1.745	2.057	3.8	2.058	1.891			2.008
Ti or	22.2	1.6		10.4	1.6		0.108			0.001	0.018			0.006
Al ab			1.848			1.853	0.234	1.908		1.922	0.044			0.012
Cr an	19.0	5.5		23.1	5.3						0.012	0.7		0.001
Fe wo	3.2	10.8	0.026	1.3	9.6	0.065	1.227	0.021	47.7	0.017	0.974	10.1		1.683
Mn en		1.2					0.014		2.7		0.015	0.8		0.020
Mg fs	13.0	48.5		8.1	61.8		0.102	0.022	1.1	0.033	0.771	67.8		0.028
Ca fo		0.4	0.959			0.959	0.554	0.962	30.3	0.913	0.337			0.223
Na fa		20.5			3.9	0.036	0.030	0.044	13.9	0.074				
K il	2.0	11.5	0.006	1.1	17.8	0.007		0.001	0.5	0.001		0.8		
cm														
Total	100.0	100.0		100.0	100.0				100.0			100.0		
ab wo						3.61	29.4	4.38		7.5	16.19			11.55
or en			0.7			0.71	5.4	0.18		0.1	37.03			1.45
an fs			99.3			95.68	65.2	95.44		92.4	46.78			87.00

Notes on analyses

1. This inclusion appears to contain silica grains (Fig. 7). Defocused beam.
2. Host olivine for above.
3. This inclusion is rather coarsely crystalline (Fig. 6). Defocused beam.
4. Host olivine for above.
5. Ryder *et al.* (1978) reported a similar analysis for an inclusion in their grain 24109,57-10; this is probably the same grain. P_2O_5 <0.05% (See Figs. 4, 5, and 10). Defocused beam.
6. Area high in glass. P_2O_5 <0.05%.
7. This pyroxene daughter crystal is smaller than the area analyzed. P_2O_5 <0.05%.
8. Host olivine for above.
9. Glass inclusion (See Figs. 8 and 11). Na_2O based on 280 counts (100 seconds counting); ±0.5%.
10. Host olivine for above. NiO <0.05%.
11. Glass inclusion (See Figs. 9 and 12).
12. Host olivine for above. NiO <0.05%.
13. Host olivine for inclusion with 15 vol. % vapor phase.
14. Host olivine for tiny glass plus sulfide inclusions (Figs. 1 and 2). NiO <0.05%.

15–17. Three points on composite spinel grain host for silicate inclusion too small for analysis. Reconnaissance analyses normalized from sums of 105, 93, and 93%.

18. Tiny silicate inclusion (3.3 μm; Fig. 13) in spinel. The analysis is of a mixture that is not suitable for norm calculations.
19. Anal. 18 recalculated by subtraction of 73.47% spinel of composition Anal. 20.
20. Spinel host immediately adjacent to inclusion, in rim of grain.
21. Spinel host from another part of rim, opposite silicate inclusion.
22. Core of spinel grain.
23. Olivine host for spinel grain of Anals. 18–22. NiO <0.05%.
24. Analysis from Ryder *et al.* (1978), their grain 24109,57-51. P_2O_5 0.00%; BaO 0.11%.

25–26. Presumably same grain as in Anal. 24.

27. Wedge-shaped mass in inclusion in Fig. 15a.
28. Main part of same inclusion.
29. Host plagioclase for above.
30. Inclusion of glass (~3.5 μm) in pyroxene.
31. Host pyroxene for above.
32. Either impure An_{85} plagioclase daughter crystal(?) or glass associated with plagioclase, from inclusion in ilmenite (Roedder and Weiblen, 1977, Fig. 8).

33. Small impure pyroxene in crystallized melt inclusion in plagioclase (Fig. 17).
34. Host plagioclase for above.
35. Small (8 μm) glass inclusion in pyroxene.
36. Host pyroxene for above.
37. Small (5 μm) glass inclusion in pyroxene.
38. Host pyroxene for above.
39. Intermediate size inclusion (10 × 15 μm) in pyroxene.
40. Host pyroxene for above, crowded with inclusions.
41. Pyroxene ~100 μm from above, clear and free of inclusions.
42. Pyroxene crowded with inclusions.
43. Pyroxene ~100 μm from above, clear and free of inclusions (Fig. 28).
44. Pyroxene crowded with inclusions.
45. Pyroxene ~100 μm from above, clear and free of inclusions.
46. Inclusion trapped in ilmenite (Fig. 20).
47. Pyroxene crystal trapped with above.
48. Host ilmenite for above.
49. Interstitial inclusion 30 μm from above inclusion.
50. "High-silica" part of composite inclusion of immiscible melts (Roedder and Weiblen, 1977, Fig. 14).
51. "High-Fe" part of same composite inclusion.
52. Host plagioclase for Anals. 50 and 51.
53. "High-silica" part of composite inclusion of immiscible melts.
54. "High-Fe" part of same composite inclusion.
55. Host plagioclase for Anals. 53 and 54.
56. "High-Fe" (without adjacent "high-silica"; Fig. 19). P_2O_5-0.35%.
57. Host plagioclase for above.
58. Tiny solid inclusions in plagioclase (Roedder and Weiblen, 1977a, Figs. 15 and 16).
59. Host plagioclase for above.
60. Pyroxene crystal enclosed in plagioclase with inclusions similar to Anal. 58 (Roedder and Weiblen, 1977a, Fig. 15).
61. Average analysis (30-μm beam) of fine intergrowth (Fig. 21).
62. Sulfide inclusions in pyroxferroite (Fig. 23). Co, Cu, and Ni each $<0.1\%$.
63. Host pyroxferroite for above.

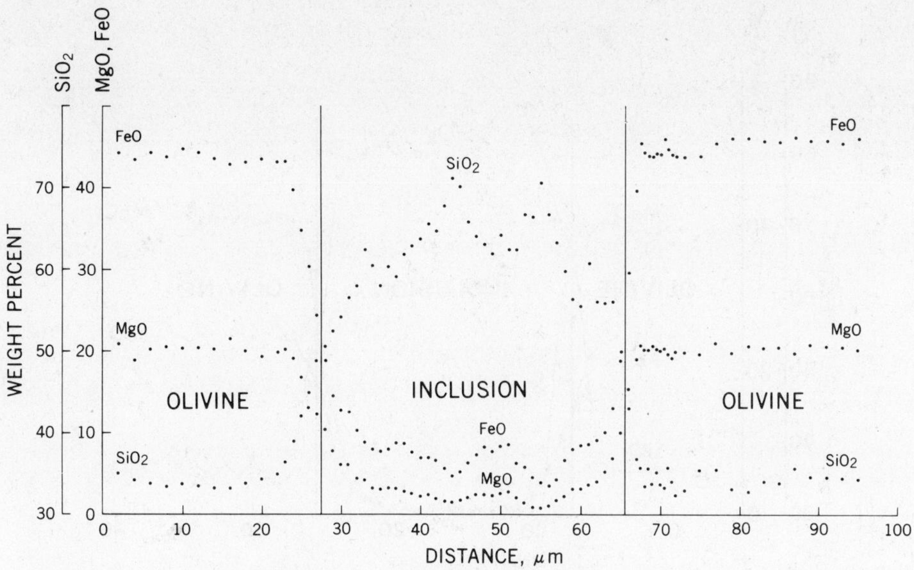

Fig. 10. Electron-microprobe traverse across left inclusion in Fig. 5, made with simultaneous 20-second counts at 1-μm intervals.

tion, since similar inclusions that are not cut by the polished surface have empty bubbles with clean walls.

The two analyzed inclusions are in relatively magnesian olivine (Fo_{51} and Fo_{73}). Fo_{73} is more magnesian than are most of the analyzed Luna 24 olivines but is similar to olivine found in a Luna 24 Mg-rich gabbro fragment (Bence *et al.*, 1977). Contrary to expectations, the glass of the inclusions in these more magnesian olivines is extremely SiO_2-rich (87–94%), far more so than the inclusions in the other more Fe-rich olivines and unlike any previously recorded inclusion in lunar or terrestrial olivines in several respects. Electron-microprobe traces (Figs. 11, 12) show these glasses to be reasonably uniform, discounting edge effects from host olivine and the bubble.

If spherical geometry is assumed for both inclusions and bubbles, calculations show that the bubbles make up 26, 4, and 12 vol. % of inclusions A, B, and C, respectively, in Fig. 8, and about 20 vol. % for the two in Fig. 9. These differences may be more apparent than real, however, as the true volume of glass in the inclusions can only be estimated at best. Both inclusions in Fig. 9 contain sulfide globules estimated at 1 vol. %, but only one of the three in Fig. 8 (A) appears to have any sulfide, and this is a rather small grain that is difficult to measure because of optical distortion (see inset, Fig. 8).

Early-stage inclusions in spinel

We analyzed several spinels in these samples which contained small silicate-

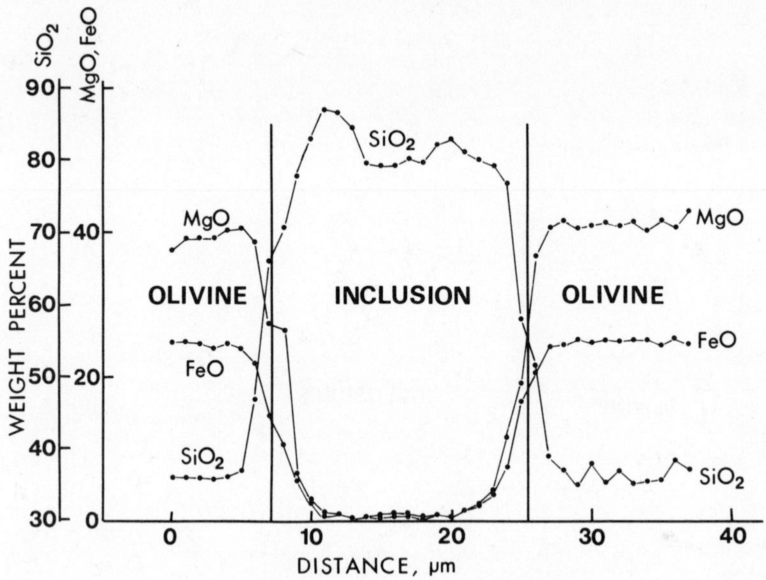

Fig. 11. Electron-microprobe traverse across the glass part of inclusion A in Fig. 8 made with simultaneous 20-second counts at 1-μm intervals.

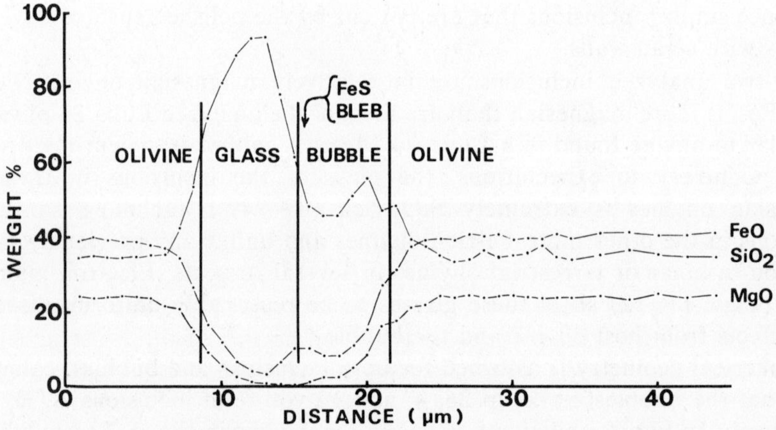

Fig. 12. Electron-microprobe traverse across inclusion A in Fig. 9, made with simultaneous 20-second counts at 1-μm intervals.

melt inclusions. The spinels range widely in composition, from aluminian chromite to chromian ulvöspinel. Haggerty (1977) found an even wider range of spinel compositions in Luna 24 samples, with a break which he attributed to a two-stage crystallization model under deep and then shallow conditions. We found silicate inclusions in only two spinel grains, 24109,57-3 and 24019,57-4

(Anals. 15–17, 20–22), but only one could be analyzed, and that analysis includes considerable host spinel due to its size (Anal. 18; Fig. 13). This analysis has been recalculated (Anal. 19) to remove ~74% spinel (based on an assumption of zero chromium in the inclusion). The inclusion occurs at the boundary between the aluminian chromite core (Anal. 22) and the chromian ulvöspinel rim (Anals. 20 and 21); this boundary is not visible in the photomicrograph (Fig. 13). This composite spinel grain, in turn, is apparently embedded in Fo_{45} olivine (Anal. 23).

Early-stage low-silica inclusions in plagioclase and pyroxene

We analyzed two large, relatively low-SiO_2 fine-grained crystalline inclusions in plagioclase grain 24109,57-2 (Fig. 16 and Anals. 25 and 26). These two are very similar to an inclusion analysis, reproduced here as Anal. 24, reported by Ryder *et al.* (1977) from their grain 24109,57-51; this is probably the same grain. Two higher SiO_2 but remarkably low Al_2O_3 analyses were obtained from an inclusion in grain 24109,50-19 (Figs. 14 and 15; Anals. 27 and 28), which shows that in addition to the homogeneous fine-grained crystalline texture, a larger scale chemical heterogeneity is present (Fig. 15a). Only one very small (~3.5 μm) inclusion in pyroxene was found that was also low in SiO_2 (Anal. 30).

Late-stage inclusions, in various hosts

Many of the Fe-rich pyroxene grains in these samples have two sharply defined zones, an optically clear and almost colorless core, and a pale- to strongly-pinkish and more Fe-rich and Mg-poor rim containing many inclusions (Anals. 40–45; see also Fig. 28 and Roedder and Weiblen, 1977, Figs. 5–7). Most of these melt inclusions have partly crystallized, particularly those in pyroxenes of the ferrogabbro, and the remaining glass is much higher in K_2O and Al_2O_3 than that in the inclusions in olivine. Smaller glass inclusions showed a similar range based on qualitative SEM analysis. Most of the minute inclusions in pyroxene are crystalline and consist of tridymite (?), in part as euhedral crystals (Fig. 18), vapor cavity, plagioclase, ilmenite, and minor sulfide (all by SEM). Similar crystalline inclusions in plagioclase consist mainly of Fe-rich pyroxene and silica (Fig. 17; Anals. 33–34), as well as FeS and phosphates (SEM).

One relatively large inclusion was found in the rare grains of ilmenite in these slides, in ferrogabbro fragment 24170,4-6 (Roedder and Weiblen, 1977a, Fig. 8). The cathodoluminescence in the electron probe shows an apparently homogeneous large central blue area, which was found on analysis to be either very impure low-Na, An_{85} plagioclase (on the basis of Ca alone), or glass plus plagioclase (Anal. 32), rimmed by a yellow band high in Ca and P (apatite or whitlockite?) on one side and a dark-blue, high-SiO_2 phase (cristobalite?) on the other.

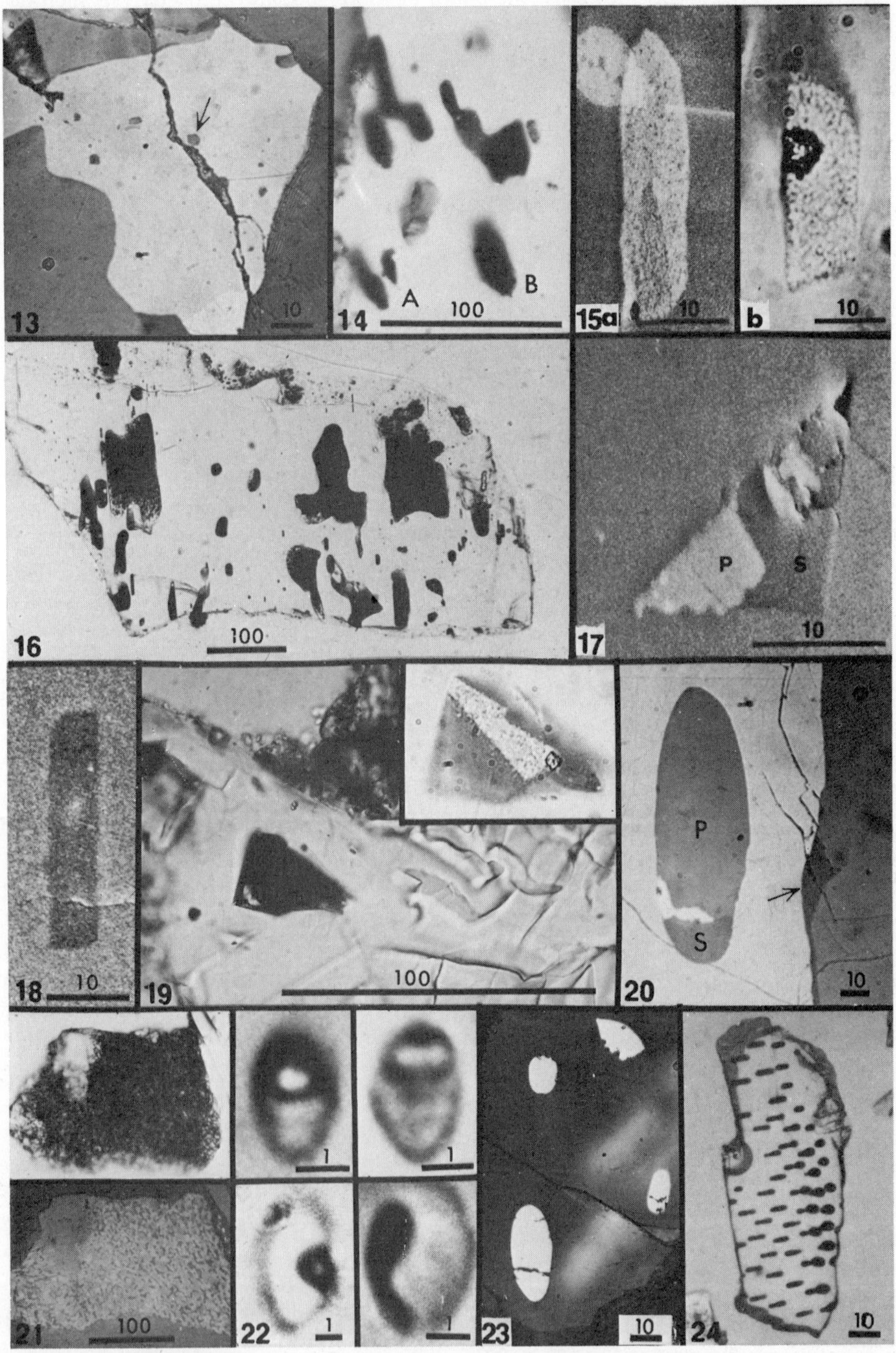
13
10
14
A
B
100
15a
10
b
10
16
100
17
P
S
10
18
10
19
100
20
P
S
10
21
100
22
1
1
1
1
23
10
24
10

Figs. 13–24. Inclusions in various minerals of Luna 24 (Scale bar dimensions given in μm.)

Fig. 13. Spinel grain in fragment 24109,57-4, showing 3.3 μm silicate melt inclusion (Anal. 18–23; arrow).

Fig. 14. Melt inclusions in plagioclase of agglutinate grain 24109,50-19, in transmitted plain light. Inclusions A and B are at surface (see Fig. 15).

Fig. 15. Detail on inclusions from Fig. 14. Fig. 15a is inclusion A (SEM secondary electron image), showing finely crystalline texture and two compositions (Anals. 27–28). The bright spot and line at top are artifacts. Fig. 15b is inclusion B, in transmitted plain light. Note irregular meniscus of bubble from crystallization of the melt.

Fig. 16. Large finely crystalline silicate melt inclusions, almost opaque, in plagioclase grain 24109,57-2 (Anals. 24-26).

Fig. 17. Crystallized interstitial inclusion in plagioclase of gabbro grain 24170,4-6 (SEM secondary electron image), showing major pyroxene (P; Anal. 33), silica (S), and other phases.

Fig. 18. Silica crystal (tridymite?) in pyroxene of gabbro grain 24170,4-1 (SEM secondary electron image).

Fig. 19. Immiscible "high-Fe" melt inclusion, in plagioclase of grain 24109,57-5, in transmitted plain light (Anal. 56). Inset is in reflected plain light.

Fig. 20. Inclusion of trapped pyroxene crystal (P) and "high-silica" melt (S) in ilmenite of grain 24174,56-7, in reflected plain light. Arrow points to adjacent interstitial "high-silica" melt inclusion. (Anals. 46–49).

Fig. 21. Fine-grained intergrowth of fayalite and silica in grain 24174,56-29 (anal. 61), in transmitted (upper) and reflected plain light (lower).

Fig. 22. Four individual microinclusions from plagioclase of grain 24182,50-19 in transmitted plain light. Each appears to consist of a crystal and a small vapor bubble. (Anal. 58).

Fig. 23. Large primary sulfide inclusions in single crystal pyroxferroite grain 24174,50-311 (Anal. 62).

Fig. 24. Primary sulfide inclusions in fayalite grain 24077,53-3, in transmitted plain light; in reflected light, these are seen to be circular in cross section.

Many glass inclusions are also present, either as ordinary melt inclusions in late-crystallizing parts of plagioclase, pyroxene, and ilmenite, or as interstitial inclusions between late-stage crystals (Fig. 20). Analyses of five such inclusions, all with high K_2O and >70% SiO_2, along with their host, daughter, and adjacent phases, are given in Anals. 35–49.

In addition to these high-SiO_2 melts, other obviously late-stage inclusions now consist of two components in various ratios, a high-SiO_2, high-K_2O glass, identical with that just described, and a high-Fe, low-SiO_2, fine-grained crystalline mixture, with a sharp meniscus between them (Roedder and Weiblen, 1977a, Fig. 14). Numerous small inclusions of this type were found within and interstitial to plagioclase crystals in the ferrobasaltic fragments of these samples, but only a few were large enough for analysis (Anals. 50–57). Two of these have both "high-silica" and "high-Fe" phases, but the third has just "high-Fe" (Anal. 56; Fig. 19).

Miscellaneous inclusion types

Fayalite-silica intergrowths. Several fairly large masses of an almost opaque, very fine grained crystalline intergrowth were found, the most obvious being grain 24174,56-29 (Fig. 21). These were examined because of their similarity to the crystallized "high-Fe" melt, but they apparently are not the same. They consist of two quite different phases, both of which are about 1–3 μm in diameter. The reflectivities are approximately those of fayalite and silica, and on moving the sample in the electron microprobe, the Si/Fe ratio was found to vary widely. The results on using a defocused beam on an apparently representative area are given in Anal. 61. Similar intergrowths of silica and fayalitic olivine were mentioned by Ryder *et al.* (1977, their Fig. 3b). We suggest that these intergrowths are the result of breakdown of pyroxferroite (Chao *et al.*, 1970, quoting work by Lindsley and coworkers). Anal. 61 is reasonably close to that of pyroxferroite, except for the Fe/Si ratio, and the nature of the sample causes that value to vary widely.

Microinclusions in plagioclase. This type of inclusion is best shown in grain 24182,50-19, which consists of a single crystal of plagioclase enclosing both a large single crystal of low-Ca, high-Fe pyroxene (Anal. 60) and many microinclusions. The pyroxene crystal is peculiar in that it in turn is cut by a 6-μm-thick diagonal plate of SiO_2 (Roedder and Weiblen, 1977a, Fig. 15), and both silica and pyroxene contain opaque blebs and other blebs rich in Ca, Al, and Si, possibly plagioclase (SEM data).

The microinclusions are randomly distributed in two zones in the host plagioclase, separated by an inclusion-free zone. Other grains of plagioclase sometimes have quite different concentrations of inclusions in adjacent twin zones. Most of the microinclusions are oval, 2–3 μm long, and appear to consist of two phases, a vapor bubble and a phase that has an index of refraction higher than plagioclase (Fig. 22). A crude attempt to estimate the volume of microinclusions in the inclusion-rich plagioclase, based on both areal extent (surface counts in reflected light) and inclusion counts in a specific volume of the thin section, yielded approximately 400 ppm by volume.

Attempts at SEM analysis of the microinclusions in grain 24182,50-19 were frustrated by problems of locating the inclusions, but another attempt was made, using the electron microprobe, on microinclusions in a similar An_{92} plagioclase grain, 24174,56-9. Several microinclusions in this sample were found using the microprobe; all showed higher Fe and Mg and lower Si, Al, and Ca than the adjacent plagioclase. The analyses of the largest of these inclusions, and the host plagioclase, are given as Anals. 58 and 59. Assuming that the Ca in the inclusion analysis represents solely a contribution to the analysis from the plagioclase enclosing these tiny inclusions, and if we subtract an appropriate amount of plagioclase composition (58.3% for Anal. 58), SiO_2, FeO, and MgO make up 93% of the residue, in the correct ratio for Fo_{75} olivine ($Mg_{1.572}$ $Fe_{0.513}$ $Si_{0.958}$ O_4).

Sulfide inclusions. In addition to the tiny globules of iron sulfide in the melt inclusions in olivine described above, inclusions of just sulfide also occur in these samples. Interstitial masses of sulfide are common, generally between crystals of late-stage minerals, as in the Apollo samples. Sulfide occurs also as obviously secondary inclusions in healed fractures, e.g., in plagioclase of grain 24174,56-35. In addition, however, some single-crystal fragments in the Luna 24 samples contain *primary* rounded inclusions of essentially pure sulfide, i.e., without visible metal phase. The largest of these that we found was in pyroxferroite crystal grain 24174,50-311 (Fig. 23 and Anals. 62–63), but many other smaller inclusions were found. Thus a rather regular series of round rods of sulfide occur in grain 24077,53-5 (Fig. 24), and similar rodlike arrays were found in grain 24149,49-4 and several other grains.

Interpretation

Sulfide liquid immiscibility

The presence of sulfide melt globules, constituting <1 vol. %, in most of the glassy inclusions in olivine (and they probably are present in the crystalline inclusions too, but are difficult to see) suggests that these melts, like those of the Apollo 11 and 12 samples (Roedder and Weiblen, 1970,1971), were close to saturation with respect to FeS and became saturated on cooling, after having been trapped in the host crystals. Some inclusions in olivine in the Luna 24 samples, however, show evidence of formation of this new immiscible liquid sulfide melt *before* entrapment by the enclosing olivine. Thus in Fig. 2, the sulfide globule constitutes >10 vol. % of this small inclusion (1.3 μm length). Presumably an immiscible globule of sulfide melt, suspended in the silicate melt, adhered to the growing surface of the olivine crystal and caused the trapping of some silicate melt at the same site. As the host olivine in this example is only Fo_{64}, it does not necessarily represent a very early phase, but to our knowledge, this is the only evidence of early sulfide saturation. The Luna 24 samples show many examples of the presence of an immiscible sulfide melt during several stages in their crystallization and may have been saturated with respect to sulfide throughout their crystallization. The primary sulfide globules in Fig. 23 and 24 almost certainly require the presence of sulfide melt during crystallization of the host silicate, and the secondary sulfide inclusions in fractured plagioclase of grain 24174,56-35 require the presence of a sulfide liquid or vapor during fracturing.

The regular arrays of sulfide, and particularly the rodlike inclusions shown in Fig. 24 may represent additional precipitation of liquid sulfide on an original array of tiny sulfide melt droplets, whose spacing was controlled by the original nucleation phenomena. Moore and Calk (1971) have shown that sulfide spherules, nucleated on the walls of vesicles in pillow basalts, are very regularly spaced; similar even spacing during carefully controlled nucleation is used today in a variety of industrial processes.

The apparent lack of a metal phase associated with the sulfide inclusions presents us with a problem (L. A. Taylor, pers. comm., 1978). Pure FeS melts at 1190°C (Jensen, 1942). The liquidus for the host ferrogabbros, determined by Grove and Vaniman (1977) on "brown glass" compositions, lies in the range 1140–1165°C, and pyroxene comes in at ~1130°C. Yet the shape of the sulfide inclusions in silicate (Fig. 23) implies entrapment of liquid globules by the growing crystal, and the rod-like inclusions (Fig. 24) almost require such entrapment of liquid. The melting point of FeS is greatly reduced by excess iron and much of the iron sulfide in the lunar samples had such excess iron, now visible as metal blebs with the sulfide, but the present inclusions do not. Furthermore, the analysis (Table 1, no. 62) shows that this sulfide is essentially free of other expected dilutants that might reduce the melting point. This dilemma cannot be resolved with existing data and further studies are planned.

Vapor bubbles in melt inclusions

Vapor bubbles in melt inclusions are a measure of the sum of 1) the difference in shrinkage of the trapped melt and the host crystal from the temperature of trapping to that at which the melt can no longer flow, and 2) the volume change from whatever crystallization took place, including that of olivine on the walls. The values obtained here for volume percent vapor bubble in glass inclusions, although crude, are much higher than those that could have been caused by simple shrinkage, so there must have been appreciable crystallization on the walls. The only other possible explanation, trapping of primary gas bubbles, seems unlikely for several reasons.

The small inclusions without bubbles have quenched to glass from a metastable stretched melt, probably under negative pressure (Roedder, 1967). This process is typical for small silicate-melt inclusions and results from failure to nucleate a new phase (vapor) in these small systems, either by heterogeneous or homogeneous nucleation (Roedder, 1971).

Differentiation trends

The analyses of the first three (crystalline) inclusions in olivine in Table 1, within the limitations inherent in their rather coarse heterogeneous nature, appear to represent essentially ferrobasaltic magma (Vaniman and Papike, 1977) depleted in Mg and Fe and enriched in everything else as a result of crystallization of various amounts of olivine, and possibly other phases, before the magma was trapped, or of olivine on the walls of the inclusion after the magma was trapped. The probable extent of crystallization of olivine is indicated by the very appreciable enrichments in normative silica in these inclusions (>20%); in contrast, Vaniman and Papike (1977) estimated that the Luna 24 ferrobasalts were undersaturated or near saturation.

The low-SiO_2 inclusions in plagioclase might also be considered to be a basaltic melt from which the plagioclase grew, in view of their relatively moderate-SiO_2 and low-K_2O values, but they are very low in Al_2O_3 (~4%), lower even than most of the inclusions in olivine. The wedge-shaped feature (Fig. 15a) is higher in SiO_2 and CaO and lower in FeO than is the other part of the inclusion. Although these two parts do not differ from each other as drastically as do most late-stage lunar immiscible melt pairs, the interface is sharp, possibly suggesting immiscibility. The parts differ greatly, however, from the "normal" late-stage immiscibility and from the inclusions in olivine.

When the values of K_D ($FeO_{xl.} \cdot MgO_{liq.}/MgO_{xl.} \cdot FeO_{liq.}$) are calculated for the olivine-inclusion and the pyroxene-inclusion pairs in Table 1, and compared with the laboratory equilibrium values of Roeder and Emslie (1970), Longhi *et al.* (1975), Walker *et al.* (1976), Lipin (1976), and Grove and Bence (1977), with one exception, the inclusion values are all found to be much higher. Thus, the six olivine-inclusion pairs from Table 1 give K_D values of 0.49 to 1.49, whereas the equilibrium values are ~0.3, and four pyroxene-inclusion pairs give K_D values of 0.35 to 2.8, rather than 0.17-0.28. Several *ad hoc* disequilibrium mechanisms can be visualized to explain these high values, invoking different diffusion rates for Fe and Mg, either in the main melt adjacent to the growing crystal prior to trapping of the inclusion, or in the crystal adjacent to the inclusion. The measured K_D values also do not necessarily reflect the local equilibrium values at the actual inclusion/host interface.

Despite the disequilibrium K_D values, when the data on all melt inclusions in Table 1 (except Anals. 9 and 11) are put on a plot of FeO/FeO + MgO vs. SiO_2 (Fig. 25), the results appear to define a more reasonable crystallization trend than do the Apollo 17 data (Roedder and Weiblen, 1977b). The trend on Fig. 25 is consistent with fractionation of olivine, pyroxene, plagioclase, and ilmenite. Such fractionation is also a permissible process, according to the experimental results on presumed parent compositions for Luna 24 ferrobasalt (Grove and Vaniman, 1977).

In contrast to the high-Ti Apollo 11 and 17 mare basalt melt-inclusion data, the Luna 24 data clearly show the effect of extensive olivine fractionation (Fig. 25). In this regard, the melt inclusion data suggest a cooling history for the Luna 24 samples that resembles that of the Apollo 12 olivine basalts more closely than that of any other Apollo basalt suite. This similarity between the two sites supports the observation made by Nyquist *et al.* (1977), on the basis of trace-element data, that there are affinities between the Apollo 12 (soils, basalts, and

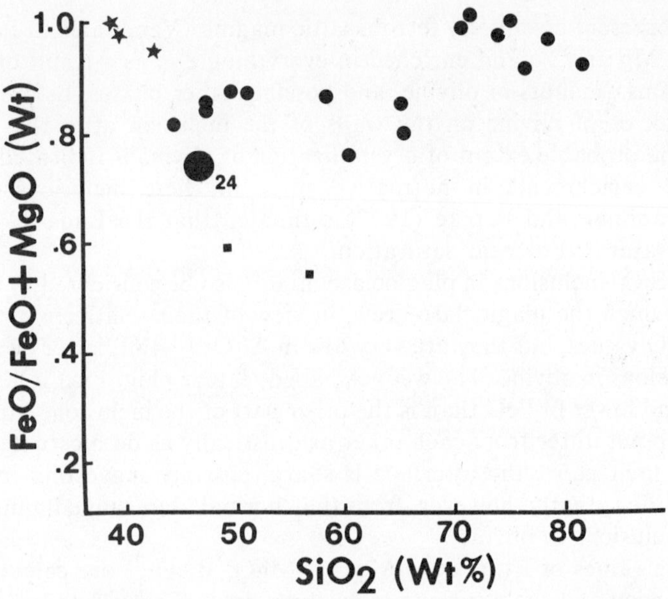

Fig. 25. Plot of FeO/FeO + MgO vs. SiO_2 for all melt inclusion compositions from table 1, except Anals. 9 and 11. The 3 stars are "high-iron" melts. The two small squares are both dubious analyses; the left one is of a 3.5-μm inclusion in spinel (Anal. 19) and the right is of a 3.5 μm inclusion in pyroxene (Anal. 30). The bulk Luna 24 composition is indicated (based on brown glass analyses by Vaniman and Papike, 1977).

possible source regions) and the Luna 24 sites. This fractionation also suggests that the Luna 24 samples came from disaggregation of a former lava pool or sill in which fractionation of olivine, pyroxene, plagioclase, and ilmenite took place. This suggestion is in agreement with the geological evidence of Head (1975) that the filling of Mare Crisium is more like that of a lava lake than lava sheets, as in Apollo 11 and 17.

Origin of the high-SiO_2 glass inclusions in olivine

Analyses 9 and 11, at 87.2 and 93.8% SiO_2, respectively, are extraordinary indeed, and as far as we know, completely without precedent in lunar or terrestrial samples. What is the origin of these high-SiO_2 melts; how did they become trapped in olivine; why is there a lack of reaction at the interface (Figs. 8,9,11,12)? These are perplexing and unanswered questions, and the fact that these melts are unique among all inclusions studied so far does not minimize these difficulties in any way. The distribution coefficients K_D for FeO and MgO between melt and crystal for these two inclusions are high, 0.49 and 1.17, just as are the rest of the values (see previous section). These two coefficients have particularly large errors because the concentrations in the inclusions are low, but they are well above the value of ~0.3 determined experimentally on more Fe-rich, Si-poor systems by Roeder and Emslie (1970), Walker *et al.* (1976), and Lipin (1976), and above the disequilibrium values found for inclusions in Apollo 12 olivines (Roedder and Weiblen, 1971). Several disequilibrium mechanisms for the origin of these melts were suggested but discarded (Roedder and Weiblen, 1977a). The low CaO and Al_2O_3 contents are the most difficult features to explain by any of these proposed mechanisms.

The inclusion and host compositions are suggestive, however, of the immiscible high-SiO_2 melt compositions found during laboratory work on the system MgO-FeO-SiO_2 (Bowen and Schairer, 1935). The olivine and inclusion compositions can be represented fairly well in this system (~1.7% extraneous material for inclusion Anal. 11)—the olivine at A and the inclusion at B (Fig. 26). A T-X section along AB is shown in Fig. 27. An *assumed* metastable subliquidus

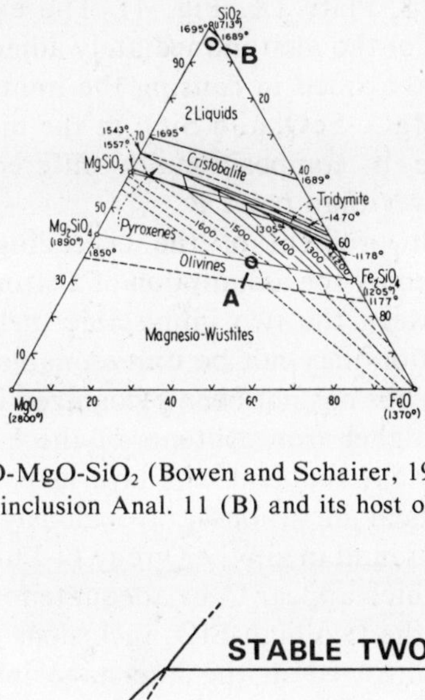

Fig. 26. The system FeO-MgO-SiO_2 (Bowen and Schairer, 1935), showing the compositions of inclusion Anal. 11 (B) and its host olivine (A).

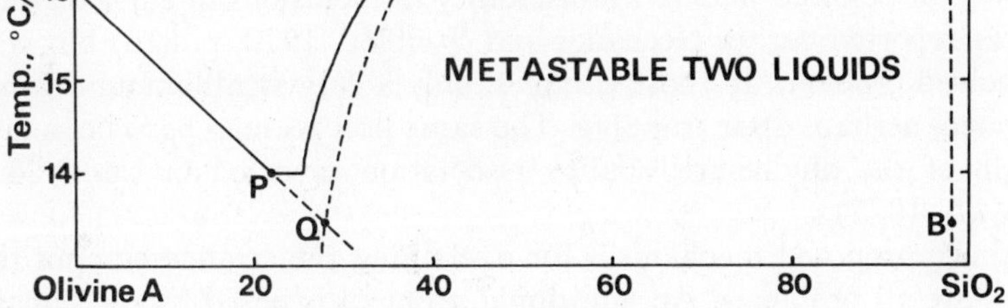

Fig. 27. T-X section A-B through Fig. 26, with assumed metastable extensions (See Fig. 26 for fields and text for discussion).

extension of the stable supraliquidus field of immiscibility has been sketched in. (Such subliquidus metastable immiscibility is accessible by proper thermal treatment and is widely used in the ceramics industry). *Under this specific assumption* (and note that no evidence precludes some entirely different assumptions, such as a metastable field of immiscibility that closes off at a lower consolute point at 1400-1600°C), the following scheme for the formation of composition 11 (and similarly for composition 9) can be envisaged: Metastable crystallization of Fo_{51} olivine from a ferrogabbroic liquid represented by composition P could carry the melt to point Q. (Such crystallization would not actually lie in this plane, as it would follow a curved path in the ternary system, but the analogy suffices.) Further crystallization of olivine would then cause separation of a globule of a metastable immiscible melt of composition B; such a globule could preferentially adhere to the growing olivine to the exclusion of the liquid Q and could become enclosed by the growth. Similar preferential wetting and trapping of only one of two immiscible phases is common in other types of inclusions (Roedder, 1972, Plate IX, Fig. 4). The expected local (nonequilibrium) enrichment in SiO_2 of the melt immediately adjacent to the growing olivine crystal face could also have aided in causing the immiscibility. The presence of constituents other than MgO, FeO, and SiO_2 in the melt from which the olivine crystallized would make it compositionally different, and the temperature probably considerably lower than point P.

The otherwise difficult problem of the exceedingly low CaO and Al_2O_3 contents could be explained by the assumption of a strongly disparate distribution of these components between the two immiscible melts; in view of the lack of alkalies, such an assumption may not be unreasonable. Why such metastability occurred in these samples but has not been recognized in the Apollo samples may be related either to the higher iron contents of the Luna 24 samples or to the more quiet crystallization environment of a lava lake, as postulated for Crisium by Head (1975). Additional inclusions are obviously needed, and 10 similar-appearing ones have been found in another group of Luna 24 slides recently made available to us, three of which appear to be adequate for analysis. The immiscible melts proposed here and the two high-SiO_2 inclusions in olivine differ greatly in composition from those involved in the late-stage immiscibility also found in these same samples.

Evidence of possible silicate immiscibility in inclusions in early olivine phenocrysts was reported earlier (Roedder and Weiblen, 1970, p. 816) but was left as an unresolved problem. In those samples, only a very small amount of high-SiO_2 melt formed, perhaps after trapping. The same process may have been involved in the origin of the olivine-cristobalite association reported in Luna 24 samples (Basu *et al.*, 1977).

The above proposed mechanism for explaining the strange compositions does not eliminate all problems. An additional problem is posed by the apparent lack of reaction at the glass/olivine interfaces; melt and olivine of these compositions should react to form pyroxene, as is also evident from Fig. 26, unless the activity of silica is lowered drastically by the presence of additional components, particularly Al_2O_3 (J.S. Huebner, pers. comm., 1977), but these melts are notably low in Al_2O_3. Electron-microprobe traverses across these interfaces give no evidence of reaction, dissolution or crystallization. Nucleation problems probably are not involved, as much smaller inclusions have nucleated pyroxene.

Late-stage inclusions

Most of the late-stage inclusions in these samples are characterized by high SiO_2 values (>70%), moderate Al_2O_3 (8–11%), and high K_2O values (1–6%, even though the Luna 24 bulk samples are particularly low in K_2O). The exceptions in Table 1 are probably mainly the result of unseen daughter minerals causing heterogeneity. We believe that the late-stage inclusions represent trapped and partially crystallized melt that contained immiscible blebs of high-K_2O, "high-silica" melt, just as was found in the Apollo mare basalts (Roedder and Weiblen, 1970, 1971). Presumably, the onset of immiscibility in the parent melt from which the host pyroxene crystals grew is marked by the sudden change from inclusion-free growth (from a homogeneous melt) to inclusion-laden growth (from a heterogeneous melt). This change does not occur at a fixed composition in the evolution of the composition of the host pyroxenes, but the positions on the pyroxene quadrilateral where it occurs in the Luna 24 samples (Figs. 28 and 29) are all very far from the positions reported by Walker and Taylor (1977) for similar abrupt changes in some Apollo pyroxenes. Walker and Taylor believe that the inclusion-laden parts have grown by a different mechanism than the inclusion-free parts, rather than invoking heterogeneity of the melt from immiscibility. Additional investigations are obviously in order.

Although some of these late-stage melts may have been as high in SiO_2 as the inclusions in olivine (before crystallization of silica), they are all much higher in Al_2O_3, FeO, and CaO, and some are two orders of magnitude higher in K_2O, so they are presumably not connected with the origin of the high-SiO_2 inclusions in olivine. The Luna 24 samples are characterized by Fe-rich compositions for both the bulk rocks and the minerals; hence, the fact that the observed Fe values in both the "high-silica" and the "high-Fe" melts from Luna 24 samples are higher than those in the corresponding melts from Apollo samples is not unexpected.

Microinclusions in plagioclase

We originally thought, before obtaining compositional data, that these microinclusions were simply small melt inclusions of essentially basaltic composition. Crystallization of the plagioclase from such a melt onto the walls of the inclusion would have left a pyroxene-rich residue. On the basis of this assumption and estimates of the optical properties of these tiny grains, we tentatively identified the main phase as pyroxene(?). Because of the present electron-microprobe evidence that these solids are probably olivine and not pyroxene, such a melt-inclusion origin seems more unlikely.

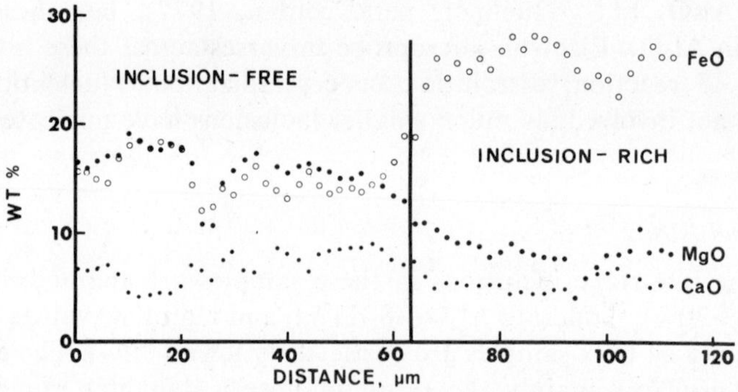

Fig. 28. Electron-microprobe traverse across inclusion-poor and inclusion-rich zones in grain 24109,50-47.

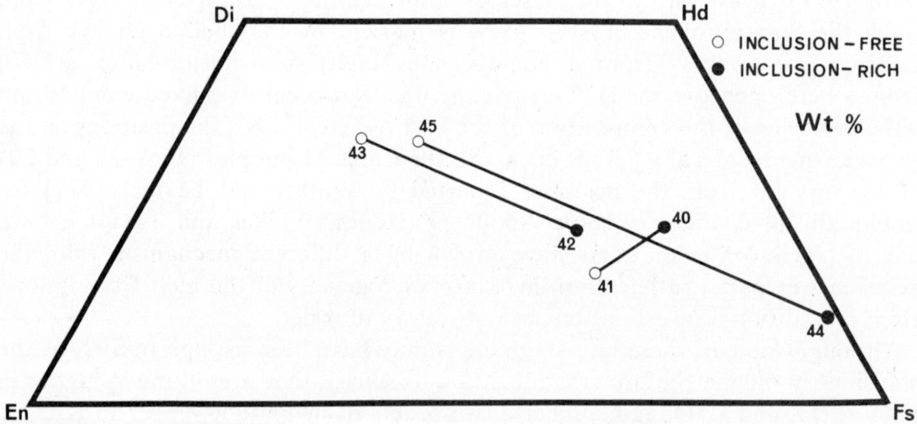

Fig. 29. Pyroxene quadrilateral showing plotted positions of the compositions of inclusion-poor and inclusion-rich pyroxene zones in three pyroxene grains, using the analyses indicated.

Microinclusions of this sort have been seen in many previous lunar plagioclase samples. We have noted them commonly in plagioclase clasts in Apollo 14 metamorphosed breccias and in the Luna 20 breccias (e.g., Roedder and Weiblen, 1973, Fig. 8). Lally *et al.* (1972) reported submicroscopic (<1 μm) precipitates of pyroxene in Apollo 15 sample 15415,25. The distribution of the microinclusions in Luna 24 plagioclase is also reminiscent of that of some symplectites formed by exsolution from olivine in the Apollo samples (Bell *et al.*, 1975); hence, we suggest that they might have formed as a result of exsolution of minor amounts (~400 ppm total) of FeO, MgO, and SiO_2 from the plagioclase structure. Why these should have come out in the ratio R_2SiO_4 is not at all clear, and a search for samples on which better analytical data may be obtained, possibly by transmission electron microscopy, is underway.

Conclusions

1. The Luna 24 magmas were saturated with respect to an apparently pure iron sulfide melt throughout much of their crystallization history. In apparent contrast with the Apollo magmas, the Luna 24 magmas underwent saturation with sulfide even while olivine was still crystallizing.
2. Considerable olivine has crystallized out on the walls of melt inclusions in olivine after the inclusions were trapped, forming relatively large vapor bubbles.
3. The silicate-melt inclusions in the Luna 24 sample are similar to those in Apollo samples, but are higher in Fe. They also show a rather well-defined crystallization trend on a plot of FeO/FeO + MgO vs SiO_2, even though the K_D values for FeO and MgO between crystal and melt inclusion (for both olivine and pyroxene) are much higher than equlibrium values.
4. The late-stage melt inclusions are highly enriched in SiO_2 and K_2O and have probably formed from an Fe-rich residual melt by immiscibility, as in Apollo, although both melts in Luna 24 are higher in Fe than are their Apollo counterparts.
5. Some of the melt inclusions have exceptionally and inexplicably low Al_2O_3 contents.
6. Two glass inclusions in olivine were found to have extremely high SiO_2 contents (87.2 and 93.8%) and low Al_2O_3 contents (6.84 and 1.51%). These inclusions may have originated by metastable subliquidus immiscibility, but they present other problems as well.
7. Abundant microinclusions (<3 μm) in some plagioclase crystals have an olivine composition; they may represent exsolution products.

Acknowledgments—This study was supported by National Aeronautics and Space Administration Grants T-1170-A and NSG-9050. We thank Gordon L. Nord, Jr., Rosalind Helz, Lawrence A. Taylor and T.M. Usselman for manuscript reviews, and Lawrence A. Taylor and Ursula Marvin for the opportunity to examine slides in their allotments.

References

Basu A., McKay D. S. and Fruland R. M. (1977) Petrography, mineralogy and source rocks of Luna-24 drill core soils (abstract.). In, *Papers Presented to the Conference on Luna 24*, p. 18–21. The Lunar Science Institute, Houston.

Bell P. M., Mao H. K., Roedder E. and Weiblen P. W. (1975) The problem of the origin of symplectites in olivine-bearing lunar rocks. *Proc. Lunar Sci. Conf. 6th*, p. 231–248.

Bence A. E., Grove T. L. and Scambos T. (1977) Gabbros from Mare Crisium: An analysis of Luna 24 soils. *Geophys. Res. Lett.*, **4**, 493–496.

Bowen N. L. and Schairer J. F. (1935) The system MgO-FeO-SiO_2, *Amer. J. Sci.*, 5th Ser., **29**, 151–217.

Chao E. C. T., Minkin J. A., Frondel C., Klein C. Jr., Drake J. C., Fuchs L., Tani B., Smith J. V., Anderson A. T., Moore P. B., Zechman G. R. Jr., Traill R. J., Plant A. G., Douglas J. A. V. and Dence M. R. (1970) Pyroxferroite, a new calcium-bearing iron silicate from Tranquillity Base. *Proc. Apollo 11 Lunar Sci. Conf.*, p. 65–79.

Grove T. L. and Bence A. E. (1977) Experimental study of pyroxene-liquid interaction in quartz-normative basalt 15597. *Proc. Lunar Sci. Conf. 8th*, p. 1549–1579.

Grove T. L. and Vaniman D. T. (1977) Experimental petrology of very low Ti basalts and origin of Luna 24 ferrobasalt (abstract). In *Papers Presented to the Conference of Luna 24*, p. 68–71. The Lunar Science Institute, Houston.

Haggerty S. E. (1977) Luna-24: Opaque mineral chemistry of gabbroic and basaltic fragments from Mare Crisium and the development of a new P-T spinel grid for lunar intrusive suites. *Geophys. Res. Lett.*, **4**, 489–492.

Head J. W. (1975) Mode of occurrence and style of emplacement of lunar mare deposits. In *Conference on Origins of Mare Basalts and Their Implications for Lunar Evolution*, Nov. 17, 1975, p. 61–65. The Lunar Science Institute, Houston.

Jensen E. (1942) Pyrrhotite; melting relations and composition. *Amer. J. Sci.*, **240**, 695–709.

Kohlstedt D. L. and Vander Sande J. B. (1975) An electron microscopy study of naturally occurring oxidation produced precipitates in iron-bearing olivines. *Contrib. Mineral. Petrol.*, **53**, 3–24.

Lally J. S., Fisher R. M., Christie J. M., Griggs D. T., Heuer A. H., Nord G. L., Jr. and Radcliffe S. V. (1972) Electron petrography of Apollo 14 and 15 rocks. *Proc. Lunar Sci. Conf. 3rd*, p. 401–422.

Lipin B. R. (1976) The origin of Fra Mauro basalts, (abstract). *Lunar Science VIII*, p. 495–497. The Lunar Science Institute, Houston.

Longhi J., Walker D. and Hays J. F. (1975) Fe-Mg distribution between olivine and lunar basaltic liquids (abstract). *EOS (Trans. Amer. Geophys. Union)* **56**, 471.

Moore J. G. and Calk Lewis (1971) Sulfide spherules in vesicles of dredged pillow basalt. *Amer. Mineral.*, **56**, 476–488.

Nyquist L., Weismann H., Bansal B., Wooden J., McKay G. and Hubbard N. (1977) Chemical and Sr-isotopic characteristics of the Luna 24 samples (abstract). In *Papers Presented to the Conference on Luna 24*, p. 139–142. The Lunar Science Institute, Houston.

Roedder E. (1965) Liquid CO_2 inclusions in olivine-bearing nodules and phenocrysts from basalt. *Amer. Mineral.*, **50**, 1740–1782.

Roedder E. (1967) Metastable superheated ice in liquid-water inclusions under high negative pressure. *Science* **155**, 1413–1417.

Roedder E. (1971) Metastability in fluid inclusions. *Soc. Mining Geol. Japan, Spec. Issue 3*, 327–334.

Roedder E. (1972) Composition of fluid inclusions. *U. S. Geol. Survey Prof. Paper* 440-JJ. 164 pp.

Roedder E. and Weiblen P. W. (1970) Lunar petrology of silicate melt inclusions, Apollo 11 rocks. *Proc. Apollo 11 Lunar Sci. Conf.*, p. 801–837.

Roedder E. and Weiblen P. W. (1971) Petrology of silicate melt inclusions, Apollo 11 and Apollo 12 and terrestrial equivalents. *Proc. Lunar Sci. Conf. 2nd*, p. 507–528.

Roedder E. and Weiblen P. W. (1973) Petrology of some lithic fragments for Luna 20. *Geochim. Cosmochim. Acta* **37**, 1031–1052.

Roedder E. and Weiblen P. W. (1977a) High-silica glass inclusions in olivine of Luna 24 samples. *Geophys. Res. Lett.* **4**, no. 10, 485–488.

Roedder E. and Weiblen P. W. (1977b) Compositional variation in late-stage differentiates in mare lavas, as indicated by silicate melt inclusions. *Proc. Lunar Sci. Conf. 8th*, p. 1767–1783.

Roeder P. L. and Emslie R. F. (1970) Olivine-liquid equilibrium. *Contrib. Mineral. Petrol.* **29**, no. 4, 275–289.

Ryder G., McSween H. Y. and Marvin U. B. (1977) Basalts from Mare Crisium. *The Moon* **17**, 263–287.

Vaniman D. T. and Papike J. J. (1977) Ferrobasalts from Mare Crisium: Luna 24. *Geophys. Res. Lett.*, **4**, no. 10, 497–500.

Walker B. M. and Taylor L. A. (1977) Crystallization histories based on pyroxene interface morphologies (abstract). In *Lunar Science VIII*, p. 961–963. The Lunar Science Institute, Houston.

Walker D., Kirkpatrick R. J., Longhi J. and Hays J. F. (1976) Crystallization history of lunar picritic basalt sample 12002: Phase-equilibria and cooling-rate studies. *Bull. Geol. Soc. Amer.* **87**, 646–656.

Merrill, R. B. and Papike, J. J., editors.
(1978) *Mare Crisium: The View from Luna 24*, Pergamon Press (New York), p. 523–535
Printed in the United States of America

Luna 24: Systematics in spinel mineral chemistry in the context of an intrusive petrogenetic grid

STEPHEN E. HAGGERTY

Department of Geology, University of Massachusetts, Amherst, Massachusetts 01003

Abstract—Although the opaque mineral oxides in the Luna 24 gabbro and basalts constitute <1% by volume of the crystalline fragments from the drill string, the mineral chemistry of these phases may provide important clues to their respective depths of origin. Spinels in the gabbroic fragments have a restricted bimodal distribution of aluminian chromite and ulvöspinel, whereas those in the basalts form a continuous sequence which defines a compositional variation from Al-Mg-chromites to ulvöspinels containing 1 wt.% Al_2O_3. A comparison of these spinel mineral chemistries with basaltic spinels from other mare regions, and with spinels from other lunar intrusive rocks suggest that the Luna 24 gabbroic spinels lie at the low pressure end of a P-T spinel grid, and that titanium enrichment trends on Mg-Al-Cr rich spinel cores are related to extrusion and formed at the terminal stages of magmatic crystallization. These overall conclusions are underscored by a detailed comparative study of spinels in terrestrial intrusive rocks. The effects of bulk composition and cooling rate on spinel compositions are unknown at present but if the compositional variations of the spinels are assigned to P-T effects, it is estimated that initial crystallization of the gabbro took place at approximately 100 km in the lunar interior (5 kb).

INTRODUCTION

The Soviet unmanned Luna 24 mission returned a 2 meter drill string of regolith from Mare Crisium in August, 1976 (Barsukov and Florensky, 1977; Barsukov *et al.*, 1977). The sample contains an abundance of basalt fragments, a gabbroic rich horizon, a small concentration of highland basalts, breccias and glasses. Mineral chemistry, bulk compositions and crystallization trends for the Luna 24 basalts and gabbros show that these rock types are iron enriched and that they are probably derived from a closely similar low titanium source region (Vaniman and Papike, 1977; Bence *et al.*, 1977). Basalts and gabbros from the drill core, therefore, provide a unique opportunity in comparative studies of cooling rates for an isochemical system.

A reflection microscopy and electron microprobe study of opaque mineral oxides in the FOCUS* aliquot was undertaken on 15 basalt fragments (0.09–0.50 mm) from core locations between 77 and 174 cms, and for spinels and ilmenites from the gabbroic horizon at 170 cms. Petrographic and mineral chemical data were previously reported (Haggerty, 1977) and the main conclusion reached in that study suggested that the Luna 24 gabbroic spinels are high

---

*FOCUS—Friends Of Crisium Unmanned Sampling. Membership—A. E. Bence, J. Friel, J. Goldstein, T. Grove, S. E. Haggerty, J. J. Papike, D. T. Vaniman, and P. Weiblen.

level crustal precipitates based on Fe-Mg and Cr-Al trends. This communication is an extension of the earlier study in that it includes a detailed discussion of the systematics in mineral chemistry for spinels from Luna 24. Spinels from other lunar regions are also discussed, and spinels from terrestrial intrusive suites are shown to have comparable crystallization trends to those established in equivalent lunar suites.

SPINEL MINERAL CHEMISTRY

A characteristic feature of spinels in the Luna 24 basalts and gabbro is their highly aluminous nature with spinels reaching a maximum of 19.8 wt.% Al_2O_3. Gabbroic spinels are of two types and form two distinct popluations: early crystallized aluminian-chromites (11.5–16.9 wt.% Al_2O_3) which form a tight cluster at the base of the multicomponent spinel prism (Fig. 1); and a second late stage generation which is ulvöspinel (1.2–1.5 wt.% Al_2O_3; 0.1–0.5 wt.% MgO) at the prism apex (Fig. 1). Chromites, which are typically enclosed in olivine, are optically and compositionally homogeneous. Ulvöspinels are fine-grained (50 μm maximum) and occur interstitially to plagioclase, pyroxene and olivine. These ulvöspinels contain rare ilmenite along {111} planes in association with metallic iron and this assemblage is interpreted to have resulted by subsolidus reduction (e.g. El Goresy *et al.*, 1972; Haggerty, 1972a).

Basaltic spinels are in marked contrast to spinels in the gabbro insofar as they exhibit continuous compositional zoning trends from crystal cores to mantles; this is expressed as a range in compositions from the base of the spinel prism to the prism apex (Fig. 1). These spinels, in accordance with the proposed classification by Haggerty (1972b), may be termed Al-Mg-chromites (7.0 wt.% MgO; 19.8 wt.% Al_2O_3); Al-chromites (10–13 wt.% Al_2O_3); Al-Ti-chromites (6 wt.% TiO_2, 13 wt.% Al_2O_3); Ti-Al-chromites (18 wt.% TiO_2, 6 wt.% Al_2O_3); Cr-Al-ulvöspinels (11 wt.% Cr_2O_3, 3 wt.% Al_2O_3); and ulvöspinels (1 wt.% Al_2O_3). In common with the Luna 24 gabbroic spinels and with spinels in other mare basalts the earliest crystallized compositions are Al and Mg enriched chromites, and the late stage (gabbroic) or mantle (basaltic) compositions are Fe and Ti enriched ulvöspinels. Although no single basaltic spinel assemblage was determined to have the entire range of compositions, there are three basalt fragments with spinels which span approximately 75% of the overall trend shown in Fig. 1.

To clarify the systematics of lunar spinel compositions from the data shown in Fig. 1, cations in octahedral coordination $[Cr^{3+}$ and $Al^{3+}]^{VI}$ and cations in tetrahedral coordination (Mg^{2+} and Fe^{2+} are in fourfold coordination in normal spinels, i.e. those on the base of the spinel prism $Mg^{IV}[Al_2]^{VI}O_4$, $Fe^{IV}[Al_2]^{VI}O_4$, $Fe^{IV}[Cr_2]^{VI}O_4$, $Mg^{IV}[Cr_2]^{VI}O_4$, whereas Fe^{2+} is in both tetrahedral and octahedral coordination in endmember ulvöspinel $Fe^{IV}[FeTi]^{VI}O_4$)* are plotted in Figs. 2 and 3 respectively as functions of Cr vs. Al and of Mg vs. Fe.

*For a complete discussion of spinel site occupancy distributions and of coordination variations for spinels of intermediate normal-inverse compositions see Haggerty (1972a) and references therein.

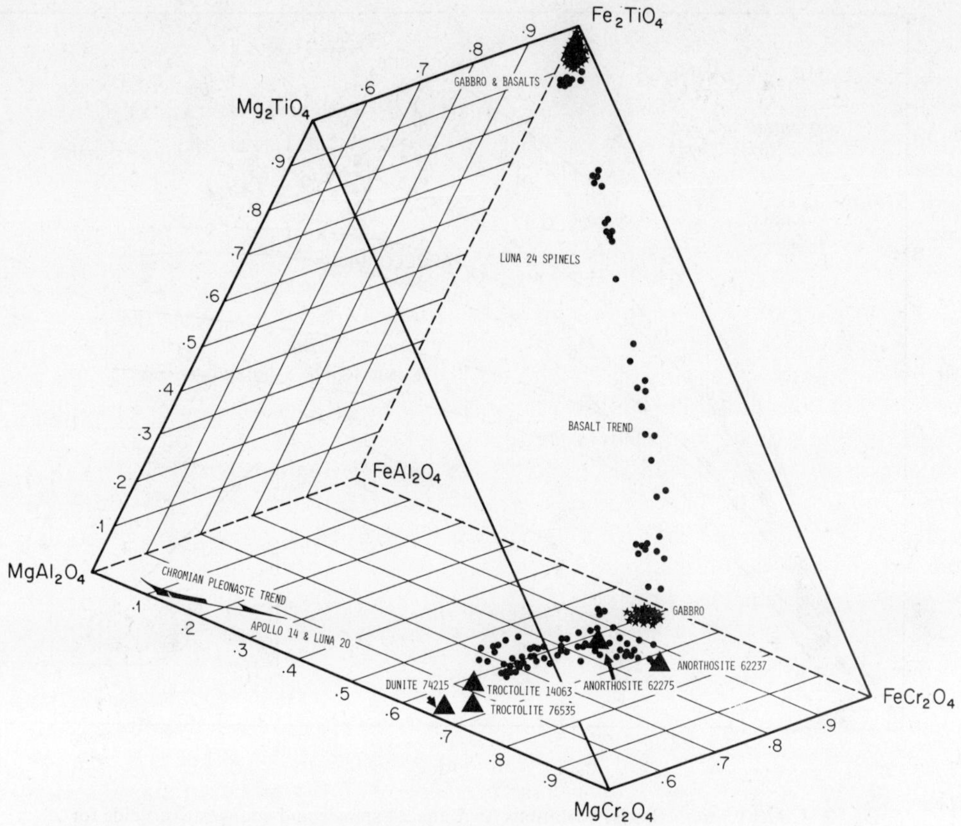

Fig. 1. Luna 24 electron microprobe data plotted within the spinel multicomponent prism. The gabbroic spinels in sample 24170 (170 cm down the drill core) are present on the base of the prism and at the ulvöspinel apex. The significance of the Apollo 14-Luna 20 chromian pleonaste trend, the anorthosites, the Apollo 17 troctolite and dunite compositions is discussed in the text.

Based on the relationship of Cr vs. Al, the basaltic and gabbroic spinels define two dominant populations: the first is high in chromium (Cr cations = 8.0–10.0/32 oxygens) and has a range in aluminum (Al cations = 4.0–7.0/32 oxygens, whereas the second is low in Cr (<1 Cr cation/32 oxygens) and low in Al (0.5–0.7 Al cations/32 oxygens). This bimodality is respectively equivalent to normal spinels on the base of the prism, and to spinels at the prism apex (Fig. 1). The basaltic spinels are strongly zoned, define a smooth compositional curve between these two extremes, and represent the range from Al-Mg-chromite to Al-ulvöspinel. Two noteworthy features of this diagram are: (1) that the most aluminous spinels are chromian depleted and have a trend towards a parabolic curve; and (2) that the Luna 24 spinels occupy a region of Cr and Al substitution

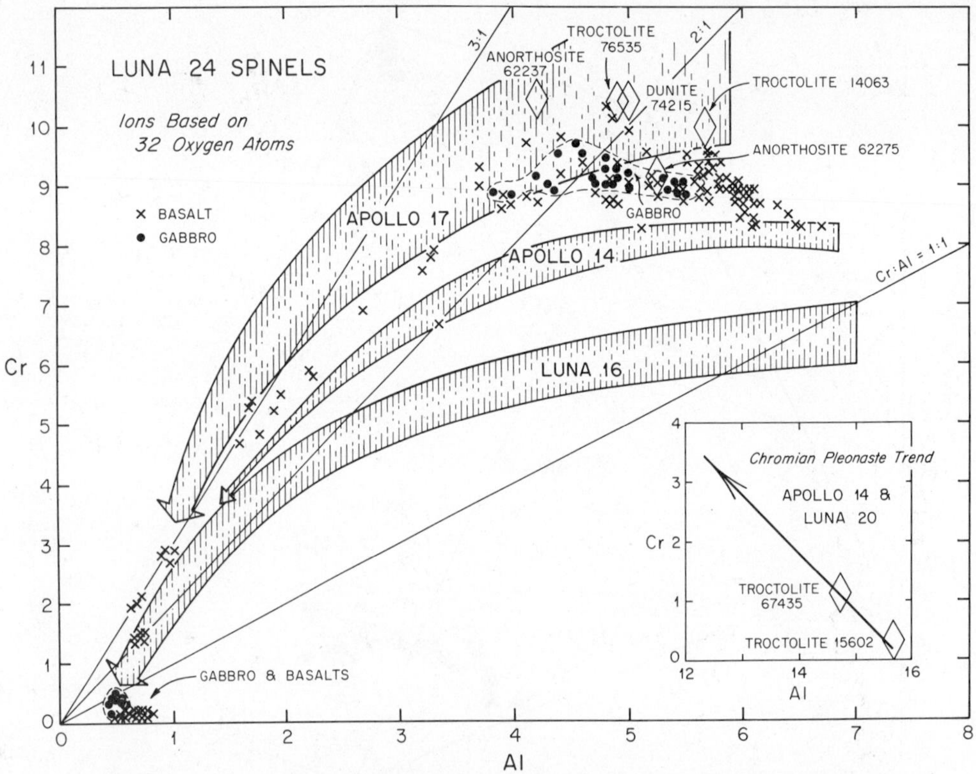

Fig. 2. Cr-Al (cationic basis) variations for Luna 24 spinels and comparative fields for Apollo 14, 17 and Luna 16 spinel fields. Spinels clustered between 8 and 9 Cr cats are chromian spinels; those between 0 and approximately 8 Cr cats are titanian spinels. Note the wide variations of Al contents for high, but constant Cr values and note also that one of the anorthosites, the dunite and a troctolite have virtually identical Cr values but a range in Al. The overall trend is parabolic: the negative limb is defined by chromian pleonastes, the crest by early formed basaltic spinels and spinels in a variety of intrusive suites, and the positive limb by late crystallized spinels.

which falls between the generalized envelopes of spinels from basalts at the Apollo 14 and the Apollo 17 sites.

The Fe vs. Mg variation diagram (Fig. 3) shows a progressive increase in Fe as a function of decreasing Mg, but two trends with pronounced differences in slope are apparent for low magnesium spinels which have a steep inclination between 10 and 16 Fe cations/32 oxygens, and for low iron spinels which have a gentle slope between 6 and 10 cations/32 oxygens. The coordinates of the pivotal region in which these changes in slope discontinuities occur are at Fe = 9.0 cations/32 oxygens, and Mg = 0.5 cations/32 oxygens; this region of the Fe-Mg diagram coincides precisely with the compositions of the Luna 24 gabbroic aluminian chromites. In common with the Cr-Al variation, the gabbroic spinels are

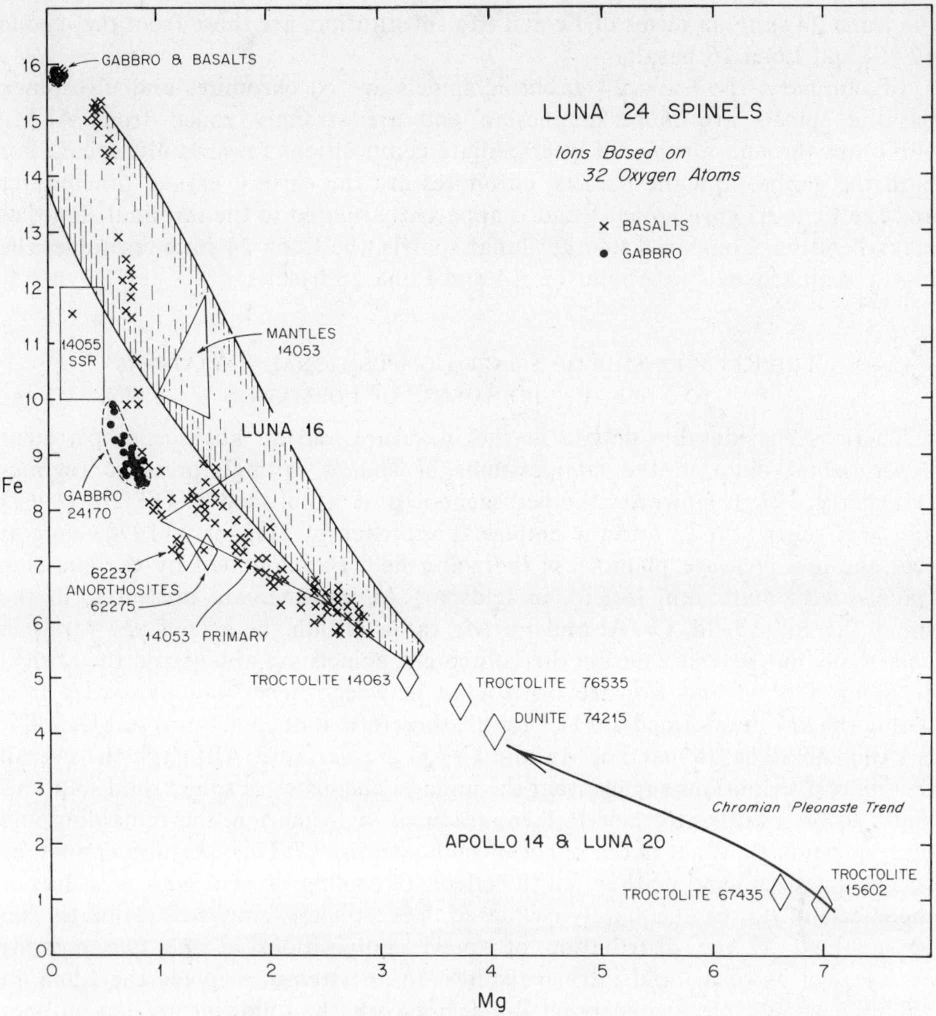

Fig. 3. Fe-Mg (cationic basis) variations for Luna 24 spinels. The intrusive Fe-Mg trend originates at 24170 (gabbro), trends down towards the anorthosites and continues to the dunite and troctolites. The abrupt changes in slope discontinuities and the two trends identified are: (a) a steep slope from ulvöspinel (Fe cats = 16.0) to Fe cats = 10; and (b) a more gentle slope originating at the gabbro and terminating at Fe cats = 6 and Mg cats = 3. The former are titanian spinels and the latter are chromian spinels.

bimodally distributed, whereas the basaltic spinels exhibit a range in Mg as a function of Fe. In addition, the more magnesian and iron depleted spinels, which are equivalent to those rich in Cr + Al (Fig. 2), are early crystal precipitates, whereas those enriched in Fe (and also Ti) and low in Mg, are late stage products. Spinels from other lunar sites which most closely approach spinels from the Luna 24 suite, in terms of Fe and Mg substitution, are those from the Apollo 12, 14 and Luna 16 basalts.

In summary, the Luna 24 gabbroic spinels are Al chromites and ulvöspinel; basaltic spinels are more magnesian and are strongly zoned from Al-Mg chromites through a range of intermediate compositions towards ulvöspinel. For both the gabbro and the basalts, chromites are the earliest crystal precipitates and the Fe + Ti enrichment trend is apparently related to the terminal stages of crystallization. Compared to other lunar spinels, the Luna 24 suite resembles the high alumina spinels in Apollo 12, 14 and Luna 16 basalts.

THE RELATIONSHIP OF SPINEL COMPOSITIONAL VARIATIONS TO THEIR ENVIRONMENTS OF FORMATION

There is considerable debate in the literature and an absence of pertinent experimental data on the compositions of spinels in high pressure regimes (Haggerty, 1978). However, the petrogenetic grid established by O'Hara (1967) and McGregor (1967), and the empirical approach by Haggerty (1978) suggest that the low pressure plagioclase lherzolite field is dominated by chrome rich spinels, with aluminum largely in feldspar. At intermediate pressures, in the spinel lherzolite field, Cr/Al and Fe/Mg ratios within the spinels are variable, and in the high pressure garnet lherzolite field, spinels are absent and the critical elements, Cr, Al and Mg are distributed between garnet and pyroxene. It is within the low to intermediate P-T range, therefore, that spinels are stable, and it is within these fields that Fe/Mg and Cr/Al are variable. Although the overall lherzolite distributions suggest that the mineral chemistry of spinel solid solutions ought to be sensitive to their P-T environment of formation, the remaining and open question is: What is the effect of bulk chemistry? This question cannot, as yet, be answered, and neither can the effects of cooling rates at high pressures or the effects of fO_2 be adequately evaluated. Nevertheless, empirical estimates can be obtained on the distribution of spinel compositions at the two possible extremes of extrusive and intrusive suites. In an attempt to place the Luna 24 gabbroic spinels into a conceptual P-T framework, the following section outlines examples of the known distributions in spinel mineral chemistry in lunar and terrestrial intrusive rocks. These data provide some insight into the environment of initial spinel crystallization in the Luna 24 samples.

In a survey of the compositional variations of lunar spinels, it is noted that there are two recurring properties of elemental systematics which have previously not been adequately addressed and which cannot, moreover, be entirely explained on the basis of crystallochemical behavior. These two properties are: (1) the partial parabolic trend in Cr-Al variations (Fig. 2); and (2) the abrupt changes in slope discontinuities in Fe-Mg variations (Fig. 3). Attempts to explain these relationships have relied upon arguments related to the complexities of cationic site occupancy distributions, and to spinels which are continuously zoned and hence range from normal to inverse structures (Haggerty, 1972a and 1972b). The true nature of cations distributed over sites in the spinel structure, and between these two extremes is conjectural, so that the explanation for the observed variations in Cr-Al and in Fe-Mg are unsatisfactory.

If it is assumed that the Luna 24 gabbro is indeed gabbroic, in the sense that the large grain size (mm rather than μm) reflects a process of slow cooling in the lunar interior, and if it is accepted, furthermore, that the unequivocal paragenesis of early formed chromites predates that of olivine, pyroxene or plagioclase crystallization, then a comparison of spinel compositions from other lunar intrusive rocks, and spinels from terrestrial intrusive suites appears valid in the context of attempting to establish the regime of initial crystallization for these minerals. The lunar data are sparse, and hence although the results are encouraging in this treatment, it is emphasized that interpretations based on assigning fundamental P-T significance to spinel variations is at present only qualitative.

Spinel compositions from a variety of lunar intrusive rocks are plotted in Figs. 1–3. These data include spinels from troctolites, anorthosites and spinels in a dunite as reported by Reid (1972), Prinz *et al.* (1973a), Gooley *et al.* (1974), Dymek *et al.* (1975), Haggerty (1975), and Steele and Smith (1976). Within the multicomponent spinel prism (Fig. 1), spinels from these rock types all lie in the compositional plane of normal spinels at the prism base. In terms of Cr and Al relationships (Fig. 2) there are two populations, one high in Cr (9–10.5 Cr cations/32 oxygens), and a second which is high in Al (14.7–15.8 Al cations /32 oxygens). The latter spinels, in troctolites 15602 and 67435, fall along a curve which is also occupied by spinels of pleonaste compositions from suites at the Apollo 14 and Luna 20 sites (Haggerty, 1972b, 1973; Roedder and Weiblen, 1972; Prinz *et al.*, 1973b; Reid, 1972; Brett *et al.*, 1973). Although there is a significant gap in compositions between those high in Cr and those high in Al, the tendency towards a parabolic relationship is nevertheless apparent. The progressive depletion in Cr, as Al increases, is particularly well displayed by the early formed spinels from the Luna 24 ferrobasalts (Fig. 2). For Fe vs. Mg (Fig. 3) there are once again gaps in the data, but a smooth curve is defined by spinels in the Luna 24 gabbro, spinels in the Apollo 16 anorthosites, the Apollo 17 dunite and the Apollo 14, 15, 16 and 17 troctolites. In addition, a large proportion of the earliest formed chromites in the Luna 24 basalts fall along a concentric curve which closely parallels the intrusive Fe-Mg curve (Fig. 3). The basaltic spinels, however, are both iron and magnesium enriched and are comparable to primary spinel compositions in the Apollo 14 basalt, 14053.

In a comparative test of the validity of the distributions displayed by spinels in lunar intrusive rocks, data for spinels in terrestrial suites are shown in Figs. 4 and 5 for Cr vs. Al and for Fe vs. Mg. The data base consists of 840 mineral analyses for spinels in layered intrusive suites, peridotites, podiform chromite deposits, kimberlites, and for spinels in xenoliths in volcanic rocks (Haggerty, 1976 and 1978). It can be assumed that, except for layered intrusive suites, these spinels are all derived from high pressure regimes and this is very clearly indicated in the Cr-Al and Fe-Mg variation plots. The inferred high pressure spinel trend is very nearly linear (Al = 16 cations/32 oxygens to Cr = 16 cations/32 oxygens) for

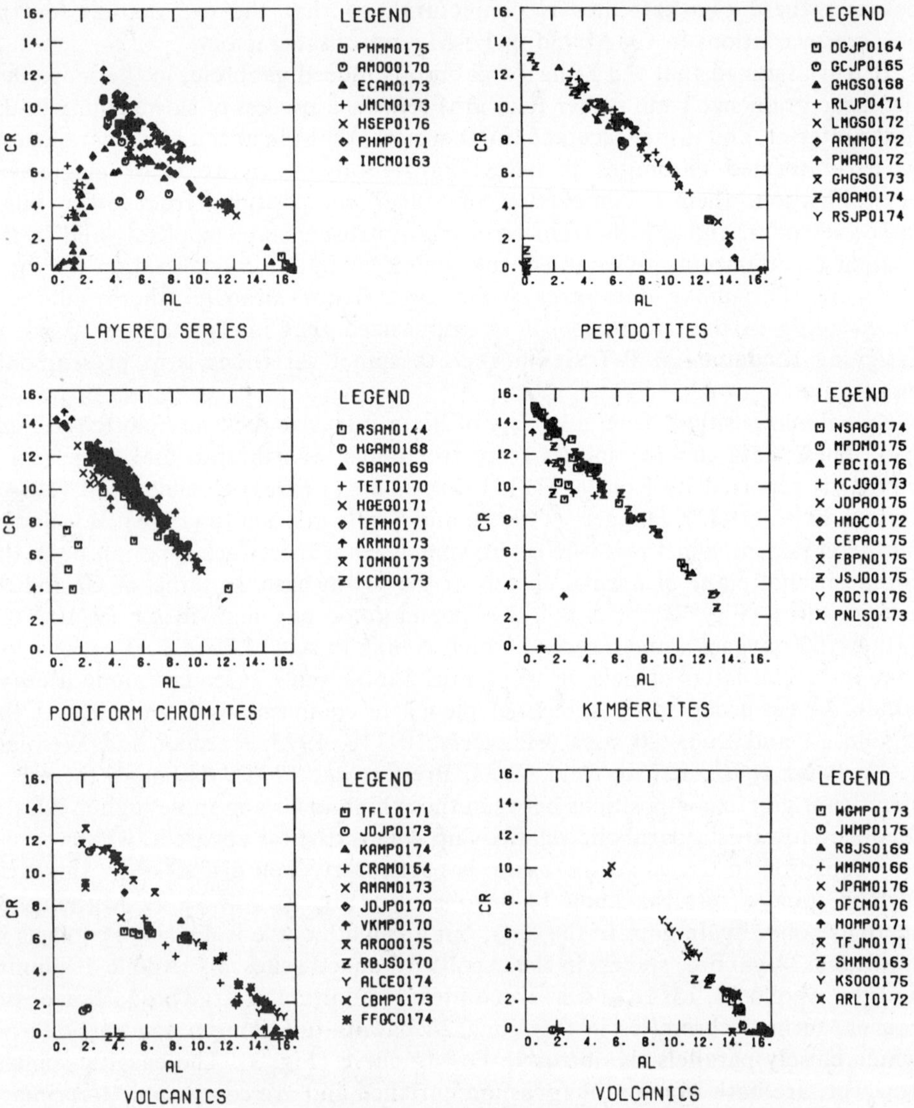

Fig. 4. Spinel data (cationic basis) for suites of terrestrial rocks having low to high pressure affinities plotted as functions of Cr vs. Al. Volcanic suites are for xenoliths in volcanic rocks. Legends are included to document the number of reference sources and are not intended for cross-correlation to the diagram. The data base is 840 mineral analyses (Haggerty, 1978).

all suites but the layered series spinels which exhibit a marked parabolic relationship. In a similar manner, the high pressure suites are all highly enriched in magnesium and depleted in iron (Fig. 5), which is a reflection of their

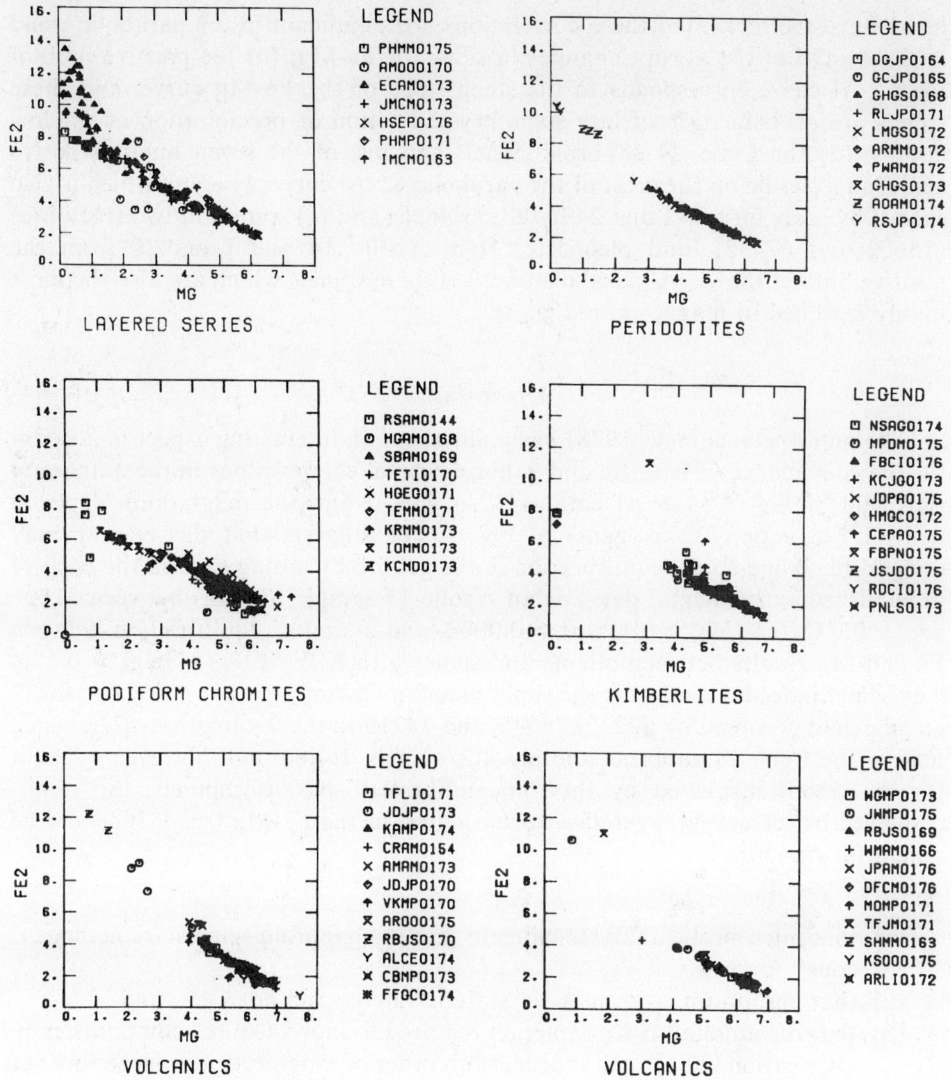

Fig. 5. Spinel data (cationic basis) for the same suite of terrestrial rocks as shown in Fig. 4 plotted as a function of Fe^{2+} vs. Mg. Comments to Fig. 4 are also applicable to these diagrams.

primitive source regions. On the other hand, spinels in the layered intrusive suites exhibit an iron enrichment trend and this is also matched by large increases in titanium which is virtually absent in high pressure spinels.

The following observations may be made from the two data sets shown in Figs. 2–5: (a) there is a remarkable correlation in compositional trends and in distribution profiles between terrestrial layered intrusive spinels and spinels in lunar intrusive rocks; (b) these correlations are significant in the parabolic trend of Cr-Al and of the abrupt changes in slope for Fe-Mg; (c) the positive limb of the Cr-Al curve corresponds to the steep slope of the Fe-Mg curve, and these trends are an indication of late stage crystallization or precipitation at shallow depths; (d) the Luna 24 gabbroic spinels and five of the seven lunar intrusive spinel analyses lie on the crest of the parabolic Cr-Al curve, a region which is also shared by early formed Luna 24 basaltic spinels; and (e) spinels in two troctolites (15602 and 67435), and pleonastes from Apollo 14 and Luna 20 form the positive limb of the Cr-Al parabola and it is these spinels which are also the most highly enriched in magnesium (Fig. 3).

DISCUSSION

Usselman (pers. comm., 1978) has pointed out an interesting aspect to account for the coincidence of basaltic and gabbroic spinel compositions in the aluminum rich field of Fig. 2 (4–6 Al cations/32 oxygens) and the magnesium depleted field (1 Mg cation/32 oxygens) of Fig. 3. He suggests that this overlap may arise from olivine-spinel equilibration in the gabbro by noting that on the basis of unpublished experimental data, on an Apollo 17 sample (74275) between 1120° and 1200°C, that $K_D^{Cr_2O_3}$ (01/Sp) $\simeq$ 0.0008, and a similar equilibration between Fe and Mg results between olivine and spinel with $K_D^{FeO/MgO}$ (01/Sp) = 0.170. This could indeed account for the suppression in Cr from the Cr = 10 cations/32 oxygen field occupied by 62237, 76535, and 74215 to the 9 Cr cations/32 oxygen field of the Luna 24 gabbroic and basaltic spinels. It does not, however, account for the trends displayed by the "unequilibrated" basaltic spinels, the trends displayed by terrestrial layered intrusive suites, or the Fe-Mg trend. It should be noted, furthermore:

(1) that microprobe analyses indicate that the gabbroic spinels are homogeneous;
(2) that chromium is in the Cr^{2+} state in olivine and not Cr^{3+}; and
(3) that the amount of Cr depletion required to move a spinel composition by 1 Cr cation/32 oxygen is at least an order of magnitude too large for that to be possible, given the substantial differences in K_D values between olivine and spinel.

By analogy, the correspondence between the terrestrial and lunar data sets implies that the lunar anorthosite, trocotolite, dunite and gabbroic spinels are the result of initial crystallization at depth. These data reflect the bulk compositional source regions of crystallization; hence, the more magnesian and more aluminous the spinel the more primitive and less evolved is the host rock. In this context, therefore, it may be argued that the Luna 24 gabbro has fractionated from a more mafic composition. However, the spinel data also suggest that initial crystallization would have occurred at relatively high crustal levels if the Fe-Mg and Cr-Al trends are meaningful pressure parameters as would seem to be indicated by the terrestrial examples. The Luna 24 gabbro, therefore, is considered to be at the low pressure end of the lunar intrusive Fe-Mg trend. With increasing depth, the trend continues to the Apollo 16 anorthosites, the Apollo 14 and Apollo 17 troctolites, the Apollo 17 dunite (which has re-equilibrated), and the Apollo 16 and Apollo 15 troctolites (Fig. 3). Included within the latter two rocks are the pleonaste crystal fragments from Luna 20 and Apollo 14. Because the Luna 24 basaltic spinels are magnesium enriched with respect to those in the gabbro, it would appear that they had nucleated at some depth greater than that of gabbroic crystallization. The close proximity of the basaltic spinel curve, however, suggests that the two magmas may originally have been generated from a closely similar source region, with the basalts being an early eruptive event and the gabbro a possible, but later congealed magma chamber. Although it is not possible to place absolute limits on these depths of initial crystallization, a value of approximately 100 km (5 kb) has been estimated, based on the compositions of spinels in the plagioclase lherzolite field (Haggerty, 1977).

The conclusions drawn from the data presented are dependent upon the parabolic relationship in Cr-Al substitution and the coupled slope digression in Fe-Mg substitution; the interpretation is that these data fingerprint those spinels that have evolved at depth with terminal crystallization at, or under, near surface conditions. These deep seated spinels have normal rather than inverse structures and are restricted to the base of the multicomponent spinel prism (Fig. 1).

Conclusions

The high magnesian and aluminous nature of spinels from the Luna 24 Mare Crisium site characterize this suite of rocks. The suite is best matched by Apollo 12 although there are also affinities with spinels from Apollo 14 and Luna 16 (Figs. 2–3). Relationships of high correlation are established for Cr-Al and Fe-Mg in a comparison of the compositional variations of Luna 24 gabbroic spinels with spinels from other lunar and terrestrial suites; these relationships may be interpreted to reflect magma evolution at various depths and it is proposed, therefore, that spinel compositions may be qualitatively employed in P-T grid applications. Luna 24 gabbroic spinels lie at the low pressure end of the spectrum and although discontinuous, increasing depths of origin are implied for lunar anorthosites and troctolites. Luna 24 basaltic spinels appear to have nucleated at depths greater than those of the gabbro but the former have reacted continuously at, or during, ascent to the lunar surface. The empirical distribution of elemental variations that has now been established on an extensively defined analytical data base, for the spinel mineral group, in lunar and terrestrial intrusive rocks, serves to document the need and usefulness of appropriate experiments on spinel-melt and on spinel-silicate systems, in a manner comparable to those of olivine-melt equilibria.

Acknowledgments—This research was supported by NASA under grant NGR-22-010-089 and by NSF (EAR76-23787). Both sources are gratefully acknowledged. The Soviet Academy of Science is also graciously acknowledged for the invaluable and successful return of samples from Mare Crisium. The microprobe facility in the Dept. of Geology is supported by the University of Massachusetts and by the National Science Foundation. Members of the FOCUS consortium have added to the excitement of Luna 24 through active discussion—and criticism. I am particularly indebted to Paul Weiblen and Tom Usselman who contributed to an unrefined manuscript through their meticulous reviews. The credit for skillful drafting goes to Marie Litterer.

REFERENCES

Barsukov V. L. and Florensky C. P. (1977) The lunar soil from Mare Crisium: preliminary data (abstract). In *Lunar Science VIII*, p. 61–63. The Lunar Science Institute, Houston.

Barsukov V. L., Ivanov A. V., Nazarov M. A., Rode O. D., Stakeev Yu. I., Tarasov L. S., Tobelko K. I. and Florensky C. P. (1977) Preliminary description of the regolith core from Mare Crisium (abstract). In *Lunar Science VIII*, p. 67–69. The Lunar Science Institute, Houston.

Bence A. E., Grove T. L. and Seambos T. (1977) Gabbros from Mare Crisium: An analysis of the Luna 24 soil. *Geophys. Res. Lett.* **4**, 493–496.

Brett R., Gooley R. C., Dowty E., Prinz M. and Keil K. (1973) Oxide minerals in lithic fragments from Luna 20 fines. *Geochim. Cosmochim. Acta* **37**, 761–773.

El Goresy A., Taylor L. A. and Ramdohr A. (1972) Fra Mauro crystalline rocks: Mineralogy, geochemistry and subsolidus reduction of the opaque minerals. *Proc. Lunar Sci. Conf. 3rd*, p. 339–349.

Dymek R. F., Albee A. L. and Chodos A. A. (1975) Comparative petrology of lunar cumulate rocks of possible primary origin: Dunite 72415, troctolite 76535, norite 78235, and anorthosite 62237. *Proc. Lunar Sci. Conf. 6th*, p. 302–341.

Gooley R. C., Brett R., Warner J. and Smyth J. R. (1974) A lunar rock of deep crustal origin: Sample 76535. *Geochim. Cosmochim. Acta* **38**, 1329–1340.

Haggerty S. E. (1972a) Apollo 14: Subsolidus reduction and compositional variations of spinels. *Proc. Lunar Sci. Conf. 3rd*, p. 305–332.

Haggerty S. E. (1972b) Luna 16: An opaque mineral study and a systematic examination of compositional variations of spinels from Mare Fecunditatis. *Earth Planet. Sci. Lett.* **13**, 328–352.

Haggerty S. E. (1973) Luna 20: Mineral chemistry of spinel, pleonaste, chromite, ulvöspinel, ilmenite and rutile. *Geochim. Cosmochim. Acta* **37**, 857–867.

Haggerty S. E. (1975) Geochemistry of opaque oxides in troctolites and basalts from Taurus Littrow (abstract). In *Lunar Science VI*, p. 321–323. The Lunar Science Institute, Houston.

Haggerty S. E. (1976) Opaque mineral oxides in terrestrial igneous rocks (short course notes). *Mineral. Soc. Amer.* **3**, 101–300.

Haggerty S. E. (1977) Luna 24: Opaque mineral chemistry of gabbroic and basaltic fragments from Mare Crisium. *Geophys. Res. Lett.* **4**, 489–492.

Haggerty S. E. (1978) Ilmenites and spinels in high pressure regimes. *Proc. 2nd Int. Kimberlite Conf.* In press.

MacGregor I. D. (1967) Mineralogy of model mantle compositions. In *Ultramafic and Related Rocks* (P. J. Wyllie, ed.), p. 382–393. Wyllie, New York.

O'Hara M. J. (1967) Mineral paragenesis in ultrabasic rocks. In *Ultramafic and Related Rocks* (P. J. Wyllie, ed.), p. 393–404. Wyllie, New York.

Prinz M., Dowty E., Keil K. and Bunch T. E. (1973a) Mineralogy, petrology and chemistry of lithic fragments from Luna 20 fines: Origin of the cumulate ANT suite and its relationship to high-alumina and mare basalts. *Geochim. Cosmochim. Acta.* **37**, 979–1006.

Prinz M., Dowty E., Keil K. and Bunch T. E. (1973b) Spinel troctolite and anorthosite in Apollo 16 samples. *Science* **179**, 74–76.

Reid J. B. (1972) Olivine-rich true spinel-bearing anorthosites from Apollo 15 and Luna 20 soils—possible fragments of the earliest formed crust. In *The Apollo 15 Lunar Samples* (J. W. Chamberlain and C. Watkins, eds.), p. 154–157. The Lunar Science Institute, Houston.

Roedder E. and Weiblen P. W. (1972) Occurence of chromian, hercynitic ("pleonaste") in Apollo 14 samples and its petrological implications. *Earth Planet. Sci. Lett.* **15**, 376–402.

Steele I. M. (1972) Chromian spinels from Apollo 14 rocks. *Earth Planet. Sci. Lett.* **14**, 190–194.

Steele I. M. and Smith J. V. (1976) Mineralogy and petrology of complex breccia 14063,14. *Proc. Lunar Sci. Conf. 7th*, p. 1949–1964.

Vaniman D. T. and Papike J. J. (1977) Ferrobasalts from Mare Crisium: Luna 24. *Geophys. Res. Lett.* **4**, 497–500.

Merrill, R. B. and Papike, J. J., editors.
(1978) *Mare Crisium: The View from Luna 24*, Pergamon Press (New York), p. 537–568
Printed in the United States of America

Chemistry, mineralogy and petrology of seven >1 mm fragments from Mare Crisium

J. C. Laul[1], D. T. Vaniman[2] and J. J. Papike[2]

[1]Physical Sciences Department, Battelle Northwest, Richland, Washington 99352
[2]Dept. of Earth and Space Sciences, State University of New York, Stony Brook, New York 11794

Abstract—We have studied seven >1 mm fragments (mostly ~2 mg each) from the Luna 24 site. In consortium, Schaeffer *et al.* (1978) have obtained an ^{39}Ar-^{40}Ar age of 3.24 B.Y. on one of these fragments (a metabasalt). The seven samples include an ophitic ferrobasalt (24174,7), remelted ferrobasalt vitrophyre (24182,12), recrystallized ferrobasalts (24077,62 & 24077,63), an Mg-rich vitrophyre (24077,4), an agglutinate (24210,50) and a plagioclase fragment (24182,8). The Luna 24 ferrobasalts are low in Ti content (1.1% TiO_2) with about 12% Al_2O_3, 22% FeO, 0.28% MnO, 6.8% MgO, 12% CaO, 0.26% Na_2O, 0.020% K_2O and 0.30% Cr_2O_3. This rock type has not been previously sampled on the moon. The low Ti content of Luna 24 ferrobasalt is similar to the very low Ti (VLT) basalts observed at the Apollo 17 site. The samples from Luna 24 show that VLT basalt may be an abundant mare rock type on the moon.

The Luna 24 ferrobasalts are quite low in refractory trace element contents (within a factor of ~2 of the moon's proposed primordial abundances). The chondritic normalized REE patterns of some ferrobasalts are of mare type with negative Eu anomalies. In contrast, however, all analyzed metabasalts and some of the ophitic ferrobasalts have positive Eu anomalies. The presence of either positive or negative Eu anomalies in otherwise isochemical ophitic basalts is perhaps due to REE fractionation on cooling, reduction/oxidation variations of Eu (Eu^{++}/Eu^{+++}), variations in composition of the source region, or variations in plagioclase concentration. The Mg-rich basaltic vitrophyre 24077,4 may represent a composition from which the ferrobasalt and an Mg, Cr-rich cumulate could be derived. The agglutinate 24210,50 is feldspathic and similar in composition to the fine fractions of the soil. The meteoritic component in this agglutinate is ~1% C1 equivalent. The Luna 24 basalts and soils follow established lunar trends for K/La, FeO/MnO and FeO/Sc.

Introduction

The Soviet program of unmanned sampling missions has returned our only samples from regions east of 40°E lunar longitude. Luna 16 returned samples of feldspathic, low-Ti mare basalt and subordinate highland rocks; Luna 20 returned additional highland samples. Luna 24, the mission to the Crisium mare basin, provided samples of a unique mare basalt which is iron-rich (atomic Mg/(Mg + Fe) ~0.3–0.45) and low in TiO_2 (~1.0 wt.%) and alkalis (Na_2O <0.3%, K_2O <0.5%: e.g., Barsukov (1977); Barsukov *et al.*, 1977; Tarasov *et al.*, 1977; Taylor *et al.*, 1977; Vaniman and Papike, 1977; Grove and Bence, 1977). The areal extent of this basalt type extends beyond the Luna 24 Crisium site, for the Crisium samples together with remote sensing studies demonstrate that early basin-filling mare basalt floods included large volumes of very low Ti (VLT) basalt (Papike *et al.*, 1978). The great variations in mare basalt types which have been predicted by spectral reflectance studies (e.g., Pieters, 1977; Johnson *et al.*, 1977) are now being confirmed by continuing sample studies.

This paper summarizes a consortium study of the chemistry, mineralogy and petrology of seven >1 mm fragments of ~2 mg each from the Luna 24 mission. As part of this consortium effort, Schaeffer *et al.* (1978) provide information on isotopic and age systematics for some of these samples. The fragments studied include 4 samples (24077,62; 24174,7; 24182,12; 24077,63) of mare ferrobasalt composition (the dominant lithic type at the Luna 24 site), one vitrophyre with the composition of an Mg-rich VLT mare basalt (24077,4), one agglutinate (24210,50) and one plagioclase fragment (24182,8). The study of these seven fragments, complemented by other chemical studies of Luna 24 fragments and soils (Blanchard *et al.*, 1977; Ma *et al.*, 1977; Tarasov *et al.*, 1977) provide important information about new lithic types from the eastern limb of the moon.

Our research objectives were: (1) to obtain complete instrumental neutron activation analysis (INAA) at the Battelle Northwest Laboratories for about 30 major and trace elements for each of six fragments. (The fragment 24077,63 was analyzed by Ma *et al.*, 1977); (2) to characterize each fragment by reflected light optical studies of polished surfaces; (3) to obtain microprobe analyses of minerals and/or glasses on each polished surface; and (4) to obtain where possible $^{39}Ar/^{40}Ar$ ages (ref. Schaeffer *et al.*, 1978). The objectives 2, 3 and 4 were carried out at Stony Brook.

EXPERIMENTAL METHODS

Instrumental Neutron Activation Analysis (INAA): Sample preparation was carried out in a clean room. The samples were weighed and transferred into sterilized 2/27 dram polyethylene vials and the vials were heat sealed. The 2/27 dram vial was placed in a 2/5 dram polyethylene vial and the 2/5 dram vial was then placed in a 2-dram polyethylene vial. The 2/5 and 2-dram vials were also heat sealed. Two Apollo 17 soils 71501 and 75081, and the U.S.G.S. Standards BCR-1, GSP-1, and PCC-1 were included as internal controls, to test for any systematic errors and to minimize any geometry problems in the counting procedures.

Silicon was determined by 14 Mev neutron activation at the University of Kentucky, following the procedure of Ali and Ehmann (1976). Following the Si determinations, the samples were irradiated at the Oregon State University (OSU) TRIGA reactor for the remaining elements. The INAA scheme

Neutron flux = 6×10^{12} n/cm²/sec
Irradiation I = 5 minutes
Irradiation II = 8 hours

Table 1. INAA scheme.

Decay interval	Isotope measured
I (1) 15 minutes	^{51}Ti, ^{27}Mg, ^{52}V, ^{28}Al, ^{49}Ca
(2) 3–5 hours	^{165}Dy, ^{56}Mn, ^{24}Na
II (3) 1–3 days	^{42}K
(4) 4–7 days	^{166}Ho, ^{153}Sm, ^{152}Eu, ^{177}Lu, $^{239}Np(U)$, ^{175}Yb, ^{198}Au, ^{131}Ba, ^{147}Nd, ^{140}La, ^{24}Na
(5) 40–50 days	^{182}Ta, ^{152}Eu, ^{141}Ce, ^{160}Tb, $^{233}Pa(Th)$, ^{51}Cr, ^{181}Hf, ^{85}Sr, ^{95}Zr, ^{134}Cs, $^{58}Co(Ni)$, ^{46}Sc, ^{86}Rb, ^{59}Fe, ^{65}Zn, ^{60}Co, ^{124}Sb

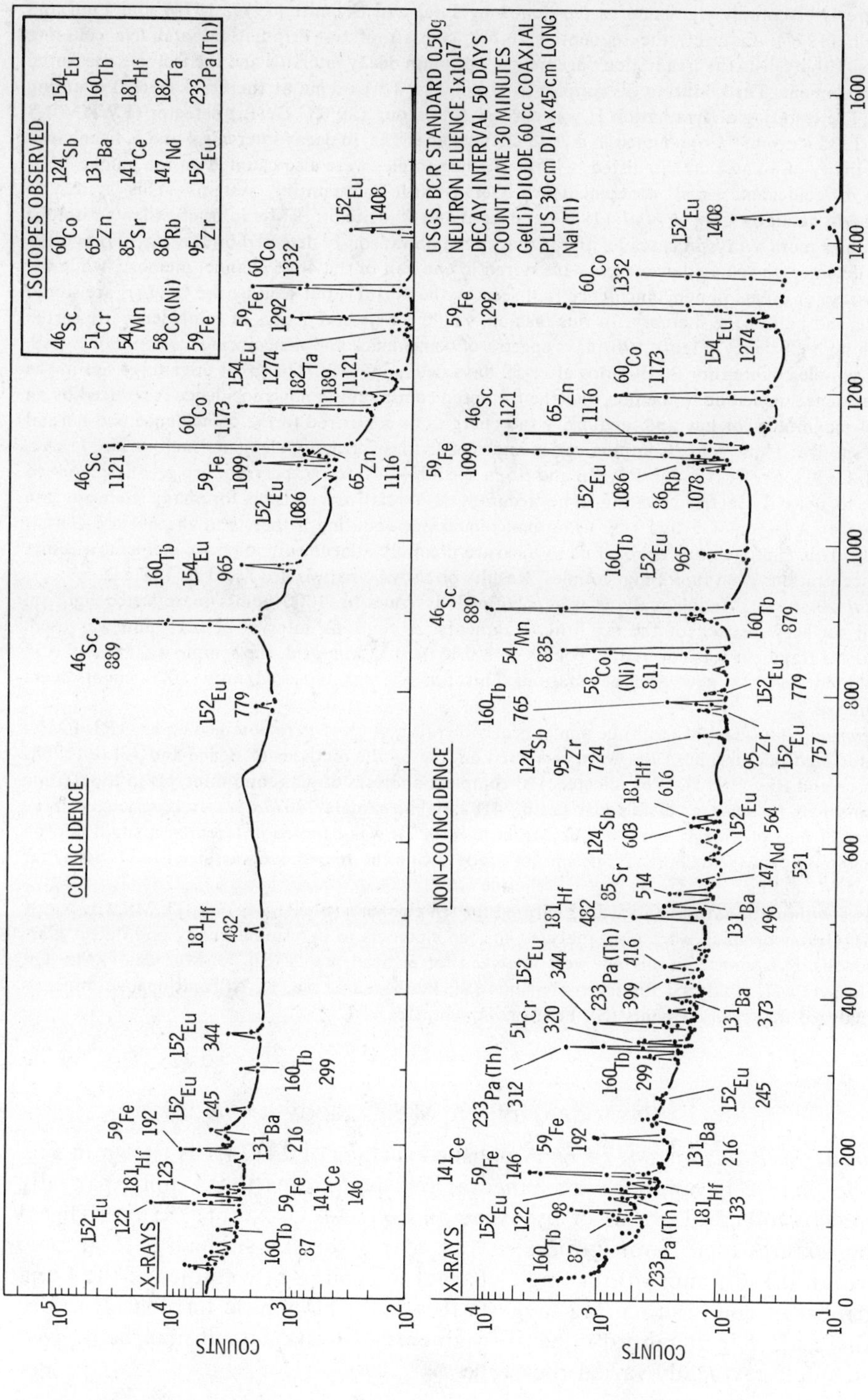

Fig. 1. Coincidence and non-coincidence gamma-ray spectra of neutron-irradiated BCR-1.

used here is essentially the same as developed by Laul and Schmitt (1973, 1974) and Laul and Rancitelli (1977). In brief, the sequential INAA consists of two irradiations and five counting intervals. Table 1 lists the irradiation parameters, optimum decay intervals and the isotopes measured for each element. The counting of samples in Irradiation I was done at the OSU Ge(Li) counting system. The counting of Irradiation II was carried out on our 130 CC Ge(Li) detector (FWHM 1.8 Kev for 1332 Kevγ of ^{60}Co) coupled to a 2048 channel analyzer. In decay intervals 4 and 5, to enhance the sensitivity of many elements listed in Table 1, the samples were also counted on the unique BNL facility of coincidence and noncoincidence Ge(Li)-NaI(Tl) counting systems. This system is described in detail by Cooper *et al.* (1970) and Laul and Rancitelli (1977). In brief, when an isotope emits two or more γ rays in cascade, and if they are simultaneously detected by the NaI(Tl) and the Ge(Li) detector, the coincidence events are stored in one half of the 4096 channel memory, while the single γ rays or events in noncoincidence [not seen by the NaI(Tl) but seen by the Ge(Li)] are stored in the second half of the memory. In this fashion, single γ rays and γ rays in coincidence are sorted out with high efficiency. Figure 1 displays spectra of coincidence and noncoincidence of the U.S.G.S. BCR-1 sample counted for 30 minutes after 50 days decay interval. The great advantage lies in the noncoincidence mode due to the fact that the Compton continuum in noncoincidence is reduced by an order of magnitude for low and medium γ ray energies as compared to the coincidence and normal Ge(Li) spectra. Thus, peak/Compton ratios and sensitivities for many elements such as Ho, U, Au, Ba, Nd, Ce, Cr, Sr, Zr, Cs, Ni, Rb, Zn and Sb in the noncoincidence spectrum are greatly increased relative to normal Ge(Li) counting. The accuracy and relative precision for many elements are improved by a factor of 5 to 10 by using noncoincidence counting rather than the normal Ge(Li) counting. This approach has enabled us to measure precisely a large suite of trace elements at quite low concentrations even in ~2 mg samples. Results of INAA analysis are listed in Table 2.

Modal Analyses: Modal analyses were obtained by counting 1000 points in reflected light on polished surfaces of each of the six lithic fragments. A counting interval of 0.15 mm was used. Whenever a fragment was covered by counts on a 0.15 × 0.15 mm grid, the sample was rotated ~5° and counted under a new grid orientation. This process was repeated until 1000 counts were accumulated.

Microprobe Analyses: Microprobe analyses of minerals and glass were obtained on an ARL EMX-SM automated microprobe. Data were corrected on-line by the methods of Bence and Albee (1968) and Albee and Ray (1970). We collected 420 complete analyses of glass and minerals in the 6 lithic fragments plus 18 analyses of feldspar grain 24182,8. The sample 24077,63, an exceptionally large fragment (7 mg) previously analyzed by Ma *et al.* (1977), was handled differently: a small portion was mounted in epoxy for optical/microprobe study, while the remainder was given to O. Schaeffer for ^{39}Ar/^{40}Ar dating.

Glasses within lithic fragments were analyzed for ten elements (Si, Fe, Mg, Ca, Ti, Mn, Cr, Na, K and Al). Grains of silica were also analyzed for ten elements to determine purity and detect glass intergrowths. Pyroxenes and olivines were analyzed for eight elements (Si, Fe, Mg, Ca, T, Mn, Cr, Al), feldspars were analyzed for seven elements (Si, Fe, Mg, Ca, Na, K, Al) and opaque minerals were analyzed for seven elements (Si, Fe, Mg, Ti, Mn, Cr, Al).

PETROGRAPHY AND MINERALOGY

Ophitic Ferrobasalt (24174,7): A polished surface of 24174,7 is shown in Fig. 2a. The texture is ophitic, with euhedral feldspar grains (~0.1 mm) partially enclosed by anhedral grains of pyroxene and olivine. Grove (1978) reproduced ophitic textures of 0.1 mm grain size and coarser blocky subhedral to graphic textures of 0.1–0.5 mm within single charges in cooling-rate studies of the Luna 24 ferrobasalt composition. He suggests that both the ophitic ferrobasalt lithics and the coarser ferrogabbro mineral fragments of Luna 24 soil may be derived from a single texturally varied rock type.

Table 2. Elemental abundances via INAA.*

	Luna 24							Controls		
								Apollo 17		U.S.G.S.
Sample No.	24077	24077	24077[a]	24174	24182	24182	24210	71501	75081	BCR-1
Sub. No.	4	62	63	7	8	12	50	18	21	—
Wt. (mg)	1.99	3.61	9.05	2.00	2.30	2.23	7.17	4.14	7.09	4.80
Description	Mg-Vitrophyre	Recryst. Ferro basalt	Recryst. Ferro basalt	Ferro basalt	Plag. Rich	Melt Rock	Agglutinate	Soil	Soil	Soil
(%) SiO_2	46.0	46.0	—	46.0	45.0	45.5	44.6	41.0	40.9	45.5
TiO_2	0.75	0.98	1.1	1.1	<0.10	1.1	0.65	8.6	9.0	2.20
Al_2O_3	10.1	12.5	12.8	12.1	34.7	12.5	15.4	11.3	11.0	13.7
FeO	22.4	21.5	21.1	22.1	0.97	22.2	16.5	17.5	17.4	12.2
MnO	0.260	0.278	0.247	0.280	0.015	0.280	0.235	0.232	0.229	0.180
MgO	10.5	7.1	8	6.0	0.5	7.4	10.3	9.0	9.6	3
CaO	10.8	12.2	13.5	11.6	19.4	12.1	12.3	10.7	10.9	6.90
Na_2O	0.210	0.260	0.240	0.265	0.693	0.260	0.250	0.350	0.386	3.20
K_2O	0.016	0.014	—	0.022	0.013	0.020	0.037	0.072	0.066	1.70
Cr_2O_3	0.424	0.296	0.240	0.300	0.0091	0.300	0.425	0.450	0.440	—
(ppm) Sc	43.6	46.0	43.0	57.0	1.8	47.0	42.0	67.0	67.0	32.0
V	150	160	180	140	<10	170	120	90	94	400
Co	59.3	43.9	38.0	43.3	1.9	43.4	40.0	30.1	28.4	36.0
Ba	40	40	—	50	10	40	60	100	100	670
Sr	90	100	—	110	210	90	90	150	180	330
La	1.80	1.62	1.1	2.87	0.18	1.80	3.60	7.47	7.60	25.5
Ce	5.5	5.0	—	8.6	0.5	6.0	9.3	27	28	54
Nd	5	4	—	7.0	<1	5	7.0	25	25	30
Sm	1.39	1.34	1.2	2.10	0.092	1.41	2.10	9.00	8.60	6.70
Eu	0.58	0.67	0.60	0.83	1.05	0.73	0.69	1.70	1.80	2.00
Tb	0.28	0.26	0.25	0.45	<0.02	0.31	0.50	2.2	2.1	1.0
Dy	1.7	1.6	1.5	2.9	<0.13	1.9	3.2	13	13	6.3
Ho	0.45	0.42	—	0.71	<0.04	0.48	0.78	3.2	3.0	1.2
Yb	1.22	1.10	1.0	2.00	0.045	1.24	2.10	7.70	7.40	3.40
Lu	0.18	0.16	0.15	0.31	0.007	0.18	0.33	1.10	1.03	0.52
Zr	<40	<40	—	50	<10	<40	60	220	200	180
Hf	1.1	1.0	0.90	1.4	0.03	1.1	1.7	7.3	7.3	4.7
Ta	0.13	0.13	0.10	0.24	<0.03	0.20	0.25	1.3	1.3	0.85
Th	0.20	0.20	—	0.20	<0.02	0.20	0.60	0.80	0.76	6.0
Ni	70	20	—	30	—	15	180	120	120	10
(ppb) Au	<1	<0.4	—	<0.4	—	<0.4	1.4	2.7	1.7	—

*Estimated errors based on counting statistics are: ±0.5–3% for SiO_2, Al_2O_3, FeO, MnO, Na_2O, Cr_2O_3, Sc, Co, La, Sm, Eu and Yb; ±5% for CaO, K_2O, V, Lu and Hf; ±5–10% for TiO_2, MgO, Ce,Tb, Dy, Ho, Ta, Th and Ni; ±10–15% for Ba, Nd, Sr and Zr. The upper limits are based on 3σ of the peak area.

[a]The date for 24077,63 are taken from Ma *et al.* (1977).

Pyroxene, olivine and opaque phase analyses in Table 3 are selected to represent the range of mineral zonation in 24174, 7. Figure 3a is a plot showing major-element pyroxene variation from an initial high-Ca composition ($Wo_{40}En_{35}Fs_{25}$) through a Ca-minimum ($Wo_{10}En_{50}Fs_{40}$) and rising to Ca, Fe enrichment with continued crystallization ($Wo_{37}En_5Fs_{58}$). Residual pyroxene grains ($Wo_{18}En_{82}$) approach pyroxferroite composition. Minor-element variations (Fig. 4a) show a high initial Al and Cr content; atomic ratios show that these minor elements represent approximately equal amounts of Al-CATS $CaAl(SiAl)O_6$ and Cr-CATS $CaCr(SiAl)O_6$ which account for >5 mol.% of the initial Ca-pyroxene composition. Residual pyroxenes are slightly enriched in

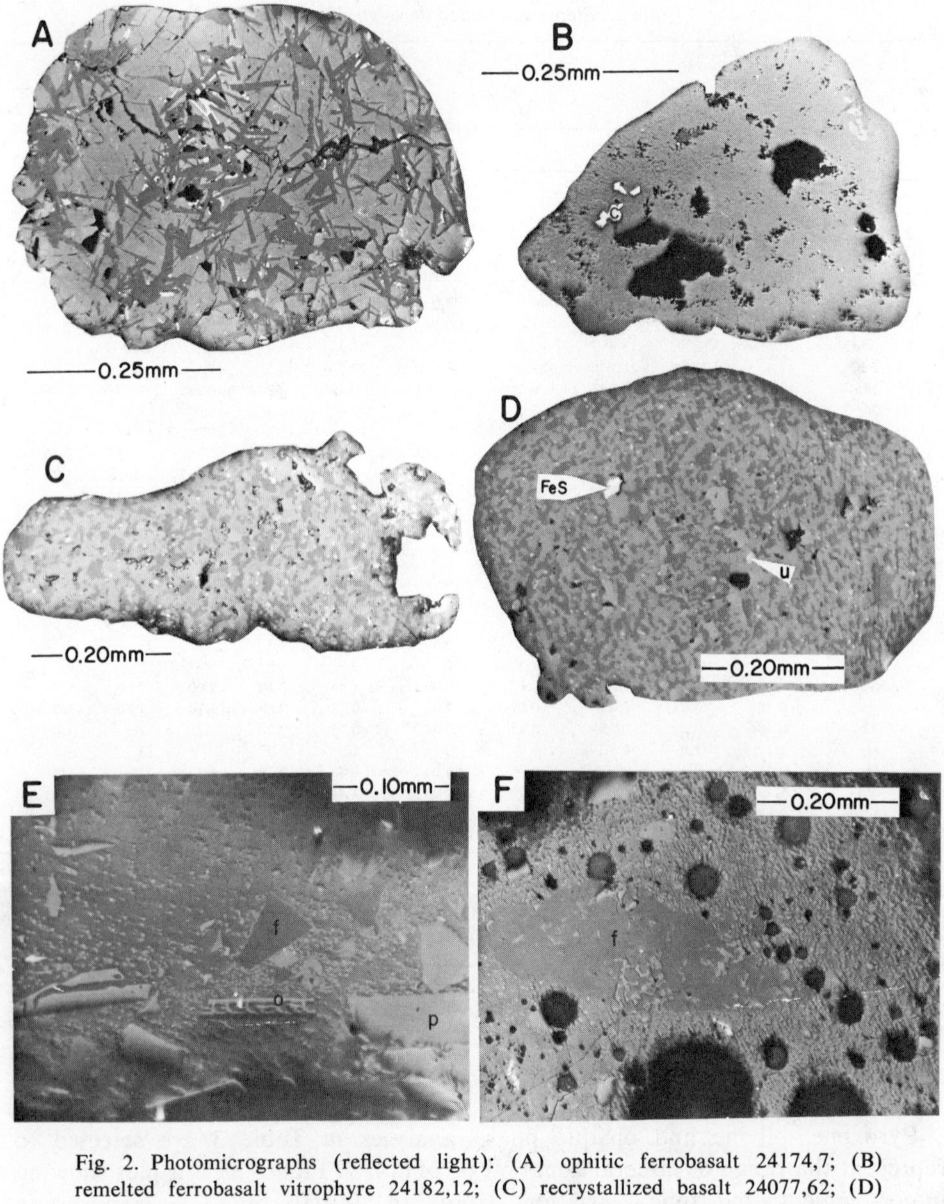

Fig. 2. Photomicrographs (reflected light): (A) ophitic ferrobasalt 24174,7; (B) remelted ferrobasalt vitrophyre 24182,12; (C) recrystallized basalt 24077,62; (D) recrystallized basalt 24077,63; (E) Mg-rich vitrophyre 24077,4; (F) agglutinate 24210,50. Symbols: (c) = chromite, (u) = ulvöspinel, (f) = feldspar, (o) = olivine.

$R^{2+}TiAl_2O_6$ (~2 mol.%). The high initial Al:Ti ratio (>>2:1) reflects the behavior of initial pyroxene crystallization in other VLT basalt types (Vaniman and Papike, 1977a).

Olivine compositions span a range from atomic $Mg_{41}Fe_{58}Mn_1$ to $Mg_1Fe_{96}Mn_3$.

The full range of olivine compositions in all ferrobasalt fragments which we have studied is $Mg_{56}Fe_{43}Mn_1$ to $Fe_{97}Mn_3$ (including data from six 0.5 mm fragments from depths of 77, 109, 182 cm in the core). Plagioclase compositions in 24174, 7 range from $An_{97}Ab_3$ (FeO ~0.5 wt.%) to $An_{88}Ab_{11}Or_1$ (FeO ~1.1%). the Mg-content of feldspar is minor or below detection limits.

Glassy Ferrobasalt (24182,12): This sample contains grains of chromian spinel, feldspar and olivine. The matrix (~97%) is homogeneous glass (Table 4). This rock is not a true monomict vitrophyre, for its spinel and olivine are anomalously Fe-rich: the spinel (atomic Fe/Mg = 8.0) should have an initial Fe/Mg ratio of 2.0, and the olivine (atomic Fe/Mg = 1.9) should have an initial composition of Fe/Mg ~0.67 (ref. Haggerty, 1977; and Grove and Vaniman, 1977). The anomalously iron-rich spinel and olivine may be xenocrysts, or relict crystals in an impact-remelted rock. The presence of small beads of FeS within the glass (Fig. 2b) suggests remelting. The minerals within this rock (with the possible exception of feldspar) do not have the compositions of near-liquidus phases; therefore, the fragment 24182,12 is polymict and is not a true ferrobasalt vitrophyre. Nevertheless, the bulk composition of the rock is very close to that of the ophitic ferrobasalt, suggesting that the remelted parent was essentially pristine ferrobasalt.

Recrystallized Ferrobasalts (Metaferrobasalts) 24077,62 and 24077,63: These rocks are very fine-grained (~0.01 mm) with an equigranular, hornfelsic texture. Ryder *et al.* (1977) and others have used the term "metabasalt" for this rock type, with the implication that the texture and mineralogy of the rock result from contact metamorphism between superimposed basalt flows. Schaeffer *et al.* (1978) have obtained ^{39}Ar-^{40}Ar ages of 3.33 ± 0.21 B.Y. for a ferrogabbro (equivalent to the ferrobasalt), and 3.24 ± 0.06 B.Y. for 24077,63. The similarity in these two ages is consistent with the interpretation of metamorphism coincident with emplacement.

Mineral compositions in the metabasalts have been re-equilibrated, and lack the zonation which is prominent in ophitic ferrobasalts. The restricted range of major-element composition in pyroxene and olivine is shown in Fig. 3b. Olivine composition is fixed at $\sim Mg_{26}Fe_{73}Mn_1$. Pyroxenes are exsolved or intergrown on a fine scale to produce analyzed compositions ranging along a line from $Wo_{33}En_{28}Fs_{39}$ to $Wo_{13}En_{32}Fs_{55}$. A Ti-Al_2^{IV} couple is the principal minor-element constituent of the pyroxenes ($R^{2+}TiAl_2O_6$ ~1 mol.%; Fig. 4b). Plagioclase composition is restricted to $Ab_{7\pm2}An_{93\pm2}Or_{\sim0}$, and the only opaque phases present are ulvöspinel and FeS. Mineral analyses are listed in Table 5.

The metabasalts are essentially isochemical with the ophitic ferrobasalts, but differ in modal composition. Table 6 and Fig. 5 summarize the modal differences: transformation of ophitic basalt to metabasalt requires the loss of chromian spinel, ilmenite and silica with a decrease in olivine, and results in the production of more ulvöspinel and pyroxene. The principal reaction is that of olivine + silica → pyroxene. Figure 6 compares the average pyroxene (an average of 530 analyses from 8 basalt fragments) and average olivine (average of 110 analyses from 7 fragments) of ophitic ferrobasalt with the average pyroxene and olivine of

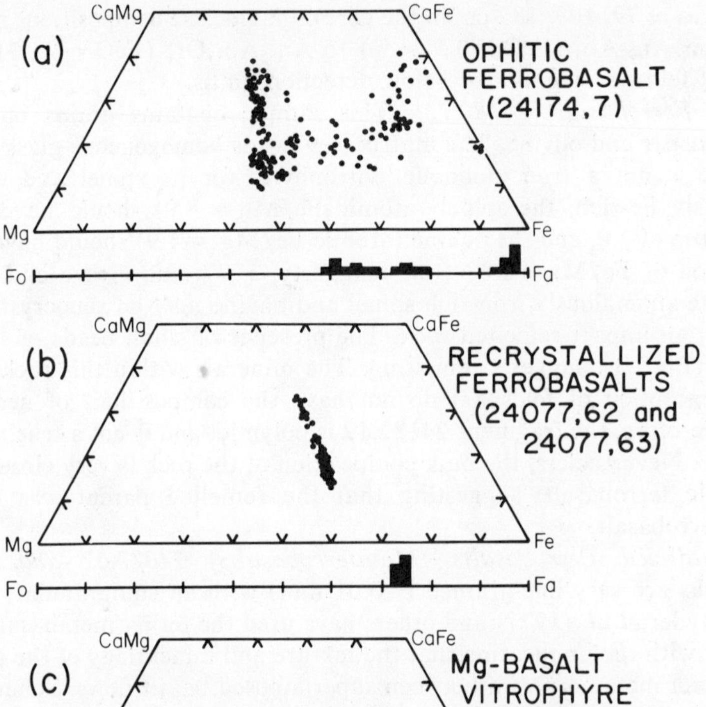

Fig. 3. Pyroxene and olivine compositions in five fragments: 24174,7; 24077,62; 24077,63; 24077,4 and 24210,50.

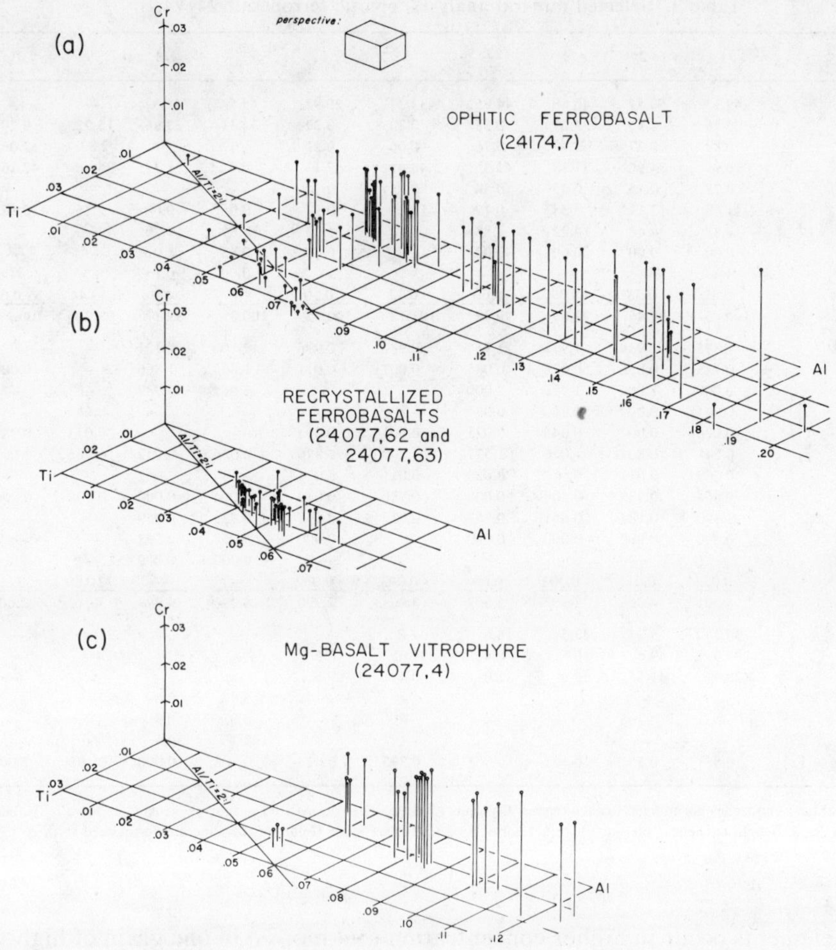

Fig. 4. Cr-Ti-Al compositional variation in pyroxenes of ophitic ferrobasalt (24174,7), recrystallized ferrobasalt (24077,62), and Mg-vitrophyre (24077,4).

metabasalt (data from 5 fragments including 24077,62 and 24077,63). The molar mass-balanced reaction is given in this figure. Major changes in bulk composition are not required for the transformation of ophitic basalt to metabasalt.

High-Mg Vitrophyre (24077,4): Representative mineral and glass analyses of 24077,4 are listed in Table 7. The range of major-element pyroxene and olivine composition is shown in Fig. 3c, and the variation of pyroxene minor-element composition is shown in Fig. 4c. Pyroxene compositions are characteristic of a magnesian VLT basalt. The most abundant high-Mg, low-Ca pyroxene ($Wo_{12}En_{60}Fs_{28}$) has a minor element component of ~2.5 mol.% (1% $R^{2+}Cr(SiAl)O_6$, 1% $R^{2+}Al(SiAl)O_6$, 0.5% $R^{2+}TiAl_2O_6$). The same ratios of

Table 3. Selected mineral analyses, ophitic ferrobasalt 24174,7.

	1	2	3	4	5	6	7	8	9	10
SiO_2	49.13	50.51	46.89	44.95	33.37	29.93	45.54	45.52	—	—
Al_2O_3	3.76	1.45	1.61	0.88	0.20	0.29	34.31	33.96	15.92	0.20
TiO_2	0.89	0.27	1.47	0.76	0.06	0.21	—	—	2.83	52.06
FeO	16.95	24.90	31.38	43.35	48.61	67.15	0.92	1.32	34.66	47.76
MnO	0.28	0.35	0.35	0.54	0.45	0.74	—	—	—	—
MgO	11.68	17.12	3.38	0.12	17.88	0.97	0.06	0.00	2.35	0.16
CaO	16.87	4.48	14.82	7.88	0.48	1.03	18.89	17.20	—	—
Na_2O	0.00	0.00	0.00	0.00	—	0.00	0.65	1.02	—	—
K_2O	—	—	—	—	—	—	0.00	0.15	—	—
Cr_2O_3	0.89	0.38	0.09	0.00	0.12	0.10	—	—	43.44	0.09
Σ	100.45	99.46	99.99	98.48	101.17	100.42	100.37	99.17	99.20	100.27
Si	1.875	1.944	1.921	1.955	0.985	0.996	2.100	2.121	—	—
Al^{IV}	0.125	0.056	0.078	0.045	0.007	0.011	1.865	1.866	—	0.006
Σtet	2.000	2.000	1.999	2.000	—	—	3.965	3.987	—	—
Al^{VI}	0.044	0.009	0.000	0.000	—	—	—	—	0.644	—
Ti	0.026	0.007	0.045	0.025	0.001	0.005	—	—	0.073	0.987
Fe	0.541	0.801	1.076	1.577	1.200	1.870	0.036	0.052	0.966	1.007
Mn	0.009	0.011	0.012	0.020	0.011	0.021	—	—	—	—
Mg	0.664	0.982	0.207	0.008	0.786	0.048	0.004	0.000	0.120	0.006
Ca	0.690	0.184	0.651	0.367	0.015	0.037	0.933	0.859	—	—
Na	0.000	0.000	0.000	0.000	—	0.000	0.058	0.092	—	—
K	—	—	—	—	—	—	0.000	0.009	—	—
Cr	0.027	0.011	0.003	0.000	0.003	0.002	—	—	1.180	0.002
Σcations	4.001	4.005	3.993	3.997	3.008	2.990	4.996	4.999	3.013	2.008
Wo	36.2	9.3	33.5	18.6	—	—	—	—	—	—
En/Fo	34.9	49.6	10.6	0.4	39.6	2.5	—	—	—	—
Fs/Fa	28.9	41.1	55.9	81.0	60.4	97.5	—	—	—	—
Ab	—	—	—	—	—	—	5.8	9.6	—	—
Or	—	—	—	—	—	—	0.0	0.9	—	—
An	—	—	—	—	—	—	94.2	89.5	—	—
Mg/(Mg + Fe)	0.551	0.551	0.161	0.005	0.396	0.025	0.108	0.000	0.108	0.006

Columns (1–4): pyroxenes normalized to six oxygens; Columns (5–6): olivines normalized to four oxygens; Columns (7–8): feldspars normalized to eight oxygens; Column (9): spinel normalized to four oxygens; Column (10): ilmenite normalized to three oxygens.

minor elements occur in higher concentration (~4 mol.%) in one grain of high-Ca pyroxene ($Wo_{29}En_{50}Fs_{21}$). The low-Ca magnesian pyroxenes are zoned to more Fe-rich compositions ($Wo_{26}En_{17}Fs_{57}$) with >1 mol.% $R^{2+}TiAl_2O_6$. The pyroxene zonation pattern is similar to the pyroxene trend in the Apollo 17 VLT basalt (Vaniman and Papike, 1977a), though 24077,4 has twice as much TiO_2 and one tenth less FeO and Al_2O_3.

High-Mg vitrophyres such as 24077,4 are rare in the Luna 24 soils. Fragment 1516, described by Tarasov *et al.* (1977), is very similar in composition to the vitrophyre described here. Marvin *et al.* (1977) discuss an olivine vitrophyre which is much more Mg-rich (16.9 wt.% MgO) than either the vitrophyre described by Tarasov *et al.* (1977) or the fragment described here (both with MgO ~10–10.5 wt.%). Taylor *et al.* (1977) and Grove and Vaniman (1977) also describe a variety of Mg-rich mare-type glasses with MgO contents ranging from 8.4 wt.% up to 17.7 wt.%. Grove and Vaniman (1978) speculate that some of these Mg-rich compositions may be the less fractionated components of a flow which fractionated to produce the ferrobasalt and an Mg, Cr-rich cumulate.

Table 4. Selected mineral analyses, remelted ferrobasalt vitrophyre 24182,12.

	1	2	3	4
SiO_2	33.31	45.48	—	46.48
Al_2O_3	0.14	34.66	21.77	12.59
TiO_2	0.11	—	1.54	0.98
FeO	51.21	0.42	35.04	21.54
MnO	0.45	—	0.51	0.25
MgO	15.49	0.10	2.19	6.62
CaO	0.31	18.80	—	11.84
Na_2O	—	0.64	—	0.16
K_2O	—	0.01	—	0.01
Cr_2O_3	0.07	—	39.86	0.12
Σ	101.09	100.11	100.91	100.59
Si	0.996	2.097	—	0.295
Al	0.005	1.884	0.848	0.094
Σ	1.001	3.981	—	—
Ti	0.003	—	0.038	0.005
Fe	1.281	0.016	0.968	0.114
Mn	0.011	—	0.014	0.001
Mg	0.690	0.007	0.108	0.062
Ca	0.010	0.929	—	0.080
Na	—	0.058	—	0.002
K	—	0.000	—	0.000
Cr	0.002	—	1.041	0.001
Σcations	2.998	4.991	3.017	0.654
Fo	34.8	—	—	—
Fa	65.2	—	—	—
Ab	—	5.9	—	—
Or	—	0.0	—	—
An	—	94.1	—	—
Mg/(Fe + Mg)	0.350	0.297	0.100	0.354

Column (1): olivine normalized to four oxygens; Column (2): feldspar normalized to eight oxygens; Column (3): spinel normalized to four oxygens; Column (4): glass normalized to one oxygen.

Agglutinate 24210,50: The composition of 24210,50 is similar to the average composition of soil fines (data of Blanchard *et al.*, 1977, for the <20 μ soil and data of Tarasov *et al.*, 1977, for the <75 μ soil). Compared to the bulk agglutinate, the glassy matrix (~60% of the fragment) is richer in Al_2O_3 and SiO_2, ~25% lower in FeO, ~20% lower in Cr_2O_3, and slightly lower in MgO (compare columns 6 and 7 in Table 8 to analysis of 24210,50 in Table 1). These differences between the bulk agglutinate and glassy matrix appear to be reconciled by the presence of mineral clasts which are relatively higher in Na (feldspar), Cr (spinel) and Mg (olivine).

Representative mineral analyses from 24210,50 are listed in Table 8. Pyroxene compositions (Fig. 3d) are similar to pyroxenes of the ophitic Luna 24 ferroba-

Table 5. Selected mineral analyses, recrystallized ferrobasalts (metabasalts) 24077,62 and 24077,63.

	1	2	3	4	5
SiO_2	49.93	49.40	32.05	45.12	0.20
Al_2O_3	1.20	0.70	0.31	34.95	1.83
TiO_2	0.79	0.46	0.18	—	30.49
FeO	22.48	31.67	56.48	1.11	58.53
MnO	0.27	0.44	0.51	—	—
MgO	10.01	11.40	11.20	0.00	0.97
CaO	15.16	5.75	0.26	18.14	—
Na_2O	0.00	0.00	—	0.78	—
K_2O	—	—	—	0.04	—
Cr_2O_3	0.20	0.11	0.12	—	8.46
Σ	100.04	99.93	101.11	100.14	100.48
Si	1.950	1.965	0.989	2.084	0.007
Al^{IV}	0.050	0.033	0.011	1.904	—
Σtet	2.000	1.998	1.000	3.988	—
Al^{VI}	0.005	0.000	—	—	0.079
Ti	0.023	0.014	0.004	—	0.834
Fe	0.735	1.053	1.457	0.043	1.781
Mn	0.009	0.015	0.013	—	—
Mg	0.583	0.675	0.515	0.000	0.053
Ca	0.635	0.245	0.008	0.898	—
Na	0.000	0.000	—	0.070	—
K	—	—	—	0.003	—
Cr	0.006	0.003	0.003	—	0.243
Σcations	3.996	4.003	3.000	5.002	2.997
Wo	32.4	12.3	—	—	—
En/Fo	29.7	34.0	25.9	—	—
Fs/Fa	37.9	53.7	74.1	—	—
Ab	—	—	—	7.2	—
Or	—	—	—	0.3	—
An	—	—	—	92.5	—
Mg/(Fe + Mg)	0.442	0.391	0.261	0.000	0.025

Columns (1,2): pyroxenes (24077,62) normalized to six oxygens; Column (3): olivine (24077,63) normalized to four oxygens; Column (4): feldspar (24077,62) normalized to eight oxygens, Column (5): ulvöspinel normalized to four oxygens.

salt. Minor-element content of pyroxene bears out this similarity (e.g., molecular Ti/Al <<0.5 at high Mg:Fe ratios). However, most of the olivine grains are too Mg-rich (molecular Fo >60%) to be derived from the ferrobasalt. Thus, these Mg-rich olivines (Fo_{67-60}) may represent the Mg-rich mineral component which accounts for the higher content of MgO in agglutinates and bulk soil than in the ferrobasalt. The single olivine crystal of Fo_{76} may have a highland origin: note that the CaO/Al_2O_3 ratio of the agglutinate and of the <75 μ soil (Fig. 12) is ~0.8, much less than the ratio for mare rocks (~1.0) and approaching the ratio for feldspathic highland rocks (~0.55).

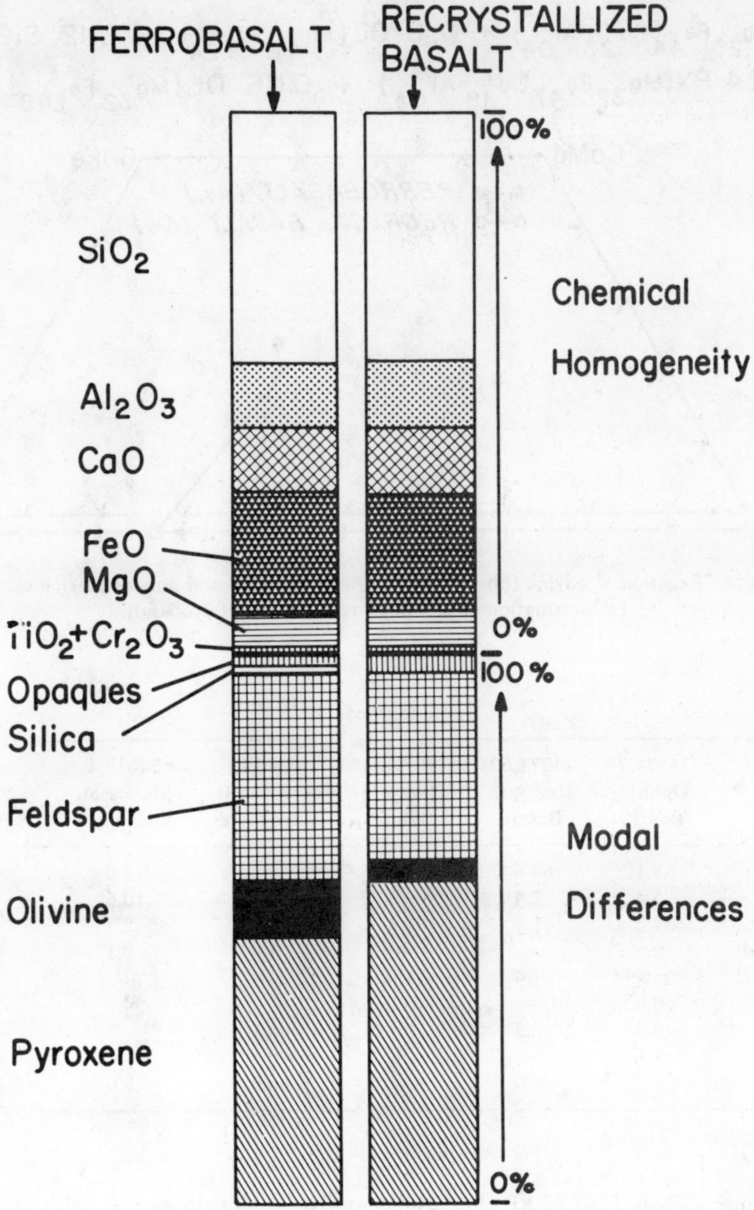

Fig. 5. Modal and bulk chemical composition: a comparison of 8 averaged ophitic ferrobasalts and 6 averaged recrystallized ferrobasalts.

Olivine + Silica → Pyroxene (mole proportions)

$$PX(Mg_{.29}Fe_{.44}Ca_{.23}Al_{.04}) + 0.18\ OL(Mg_{.76}Fe_{1.24}) + 0.12\ SiO_2$$
$$\rightarrow 1.24\ PX(Mg_{.32}Fe_{.47}Ca_{.18}Al_{.03}) + 0.06\ OL(Mg_{.52}Fe_{1.48})$$

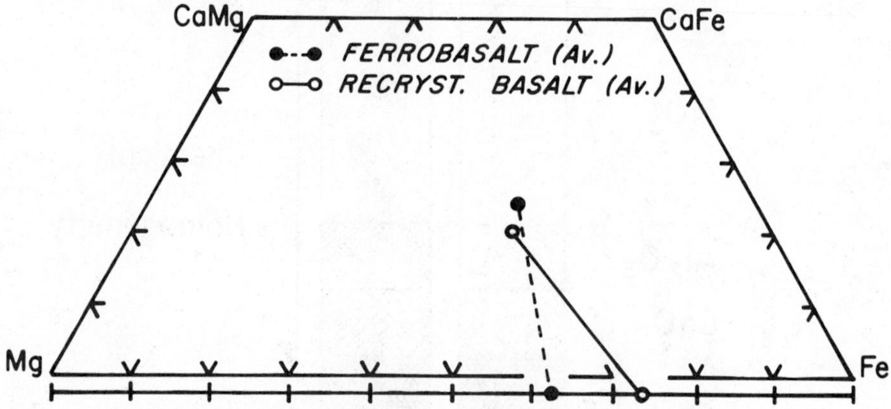

Fig. 6. Reaction of olivine (plus quartz) to form pyroxene and a more Fe-rich olivine: transformation of ophitic to recrystallized ferrobasalt.

Table 6. Modal data.

	24174,7 Ophitic Basalt	24077,62 Recryst. Basalt	24077,63 Recryst. Basalt	24182,12 Ferrobasalt Vitrophyre	24077,4 Mg-Basalt Vitrophyre	24210,50 Agglutinate
Pyroxene	58.1	64.6	61.5	—	8.0	13.5
Olivine	5.4	2.4	3.4	1.3	10.6	12.4
Feldspar	34.2	31.3	33.1	0.6	2.7	11.3
Cr-Spinel	0.1	—	—	1.4	0.1	2.3
Ulvöspinel	trace	1.4	1.9	—	—	—
Ilmenite	0.8	—	—	—	—	—
FeS	0.3	0.3	0.1	<0.1	—	<0.1
Silica	1.1	—	—	<0.1	—	—
Glass	—	—	—	96.6	78.6	60.2

Feldspar Grain (24182,8): The bulk analysis of this single feldspar crystal (Table 1) includes some minor attached ferromagnesian minerals (<1%). Microprobe analysis (Table 9) attached ferromagnesian minerals (<1%). Microprobe analysis (Table 9) shows that the grain is very slightly zoned ($An_{95.3}Ab_{4.7}$ to $An_{93.9}Ab_{6.1}$). The Fe and Mg content, Ca-Na range, and size of this crystal are characteristic of the Luna 24 "ferrogabbro" (Bence *et al.*, 1977).

Table 7. Selected mineral analyses and glass composition, magnesian vitrophyre 24077,4.

	1	2	3	4	5	6
SiO_2	51.32	52.37	47.09	37.19	45.18	47.23
Al_2O_3	2.92	1.93	1.46	0.39	34.01	10.68
TiO_2	0.60	0.46	0.77	0.00	—	0.85
FeO	12.64	17.04	32.09	27.11	0.99	19.86
MnO	0.17	0.22	0.37	0.28	—	0.19
MgO	17.87	21.33	5.35	34.93	0.09	8.86
CaO	13.75	6.23	12.28	0.50	19.02	11.25
Na_2O	0.00	0.00	0.00	—	0.51	0.13
K_2O	—	—	—	—	0.01	0.01
Cr_2O_3	0.82	0.77	0.13	0.41	—	0.40
Σ	100.09	100.35	99.23	100.81	99.81	99.46
Si	1.903	1.932	1.938	0.985	2.097	0.300
Al^{IV}	0.097	0.068	0.056	0.012	1.862	0.080
Σtet	2.000	2.000	1.994	0.997	3.959	—
Al^{VI}	0.031	0.016	0.000	—	—	—
Ti	0.017	0.012	0.024	0.000	—	0.004
Fe	0.392	0.525	1.104	0.600	0.039	0.106
Mn	0.005	0.007	0.013	0.006	—	0.001
Mg	0.988	1.172	0.328	1.378	0.006	0.084
Ca	0.546	0.246	0.541	0.014	0.946	0.077
Na	0.000	0.000	0.000	—	0.046	0.002
K	—	—	—	—	0.000	0.000
Cr	0.024	0.022	0.004	0.008	—	0.002
Σcations	4.003	4.000	4.008	3.003	4.996	0.656
Wo	28.3	12.6	27.2	—	—	—
En/Fo	51.1	60.1	16.5	69.4	—	—
Fs/Fa	20.6	27.3	56.3	30.6	—	—
Ab	—	—	—	—	4.7	—
Or	—	—	—	—	0.0	—
An	—	—	—	—	95.3	—
Mg/(Mg + Fe)	0.716	0.690	0.229	0.697	0.141	0.443

Columns (1–3): high-Ca, low-Ca, and Fe-rich pyroxenes normalized to six oxygens; Column (4): magnesian olivine normalized to four oxygens: Column (5): feldspar normalized to eight oxygens; Column (6): glass normalized to one oxygen.

Chemical Systematics

Table 2 summarizes the INAA data for 31 major and trace elements in our six Luna 24 fragments along with controls 71501, 75081 and (U.S.G.S.) BCR-1. The data for 24077,63 of Ma *et al.* (1977) are also included in Table 2 because we have included this sample in our petrographic, microprobe and age studies.

The element Ho (Table 2) is reported for the first time by INAA. Analysis of Ho is feasible by coincidence-noncoincidence counting. Our K_2O values for the Luna 24 rocks are reported with some certainty; other reported K_2O values are either upper limits or within the calculated error. The analyses of controls 71501,

Table 8. Selected mineral and glass analyses, agglutinate 24210,50.

	1	2	3	4	5	6	7
SiO_2	52.17	51.48	38.15	37.00	44.13	46.40	46.82
Al_2O_3	1.18	1.01	0.00	0.07	35.28	16.33	19.97
TiO_2	0.19	0.18	0.06	0.04	—	0.83	0.73
FeO	16.57	22.27	22.64	29.34	0.20	14.60	11.74
MnO	0.25	0.34	—	—	—	0.17	0.17
MgO	17.30	17.43	39.75	33.34	0.00	10.16	6.69
CaO	11.57	6.73	0.17	0.57	19.61	11.50	13.41
Na_2O	0.00	0.00	—	—	0.33	0.21	0.31
K_2O	—	—	—	—	0.01	0.03	0.03
Cr_2O_3	0.50	0.34	0.07	0.57	—	0.35	0.25
Σ	99.73	99.78	100.84	100.93	99.56	100.58	100.12
Si	1.962	1.960	0.985	0.987	2.051	0.284	0.284
Al^{IV}	0.038	0.040	0.000	0.002	1.933	0.118	0.143
Σ	2.000	2.000	0.985	0.989	3.984	—	—
Al^{VI}	0.014	0.005	—	—	—	—	—
Ti	0.006	0.005	0.001	0.001	—	0.004	0.003
Fe	0.521	0.709	0.488	0.655	0.008	0.075	0.060
Mn	0.008	0.011	—	—	—	0.001	0.001
Mg	0.970	0.989	1.529	1.325	0.000	0.093	0.060
Ca	0.466	0.274	0.004	0.016	0.976	0.075	0.087
Na	0.000	0.000	—	—	0.030	0.002	0.004
K	—	—	—	—	0.000	0.000	0.000
Cr	0.015	0.010	0.001	0.012	—	0.002	0.001
Σcations	4.000	4.003	3.008	2.998	4.998	0.654	0.643
Wo	23.7	13.8	—	—	—	—	—
En/Fo	49.4	49.9	75.6	66.7	—	—	—
Fs/Fa	26.9	36.3	24.4	33.3	—	—	—
Ab	—	—	—	—	3.0	—	—
Or	—	—	—	—	0.0	—	—
An	—	—	—	—	97.0	—	—
Mg/(Mg + Fe)	0.650	0.582	0.758	0.669	0.000	0.553	0.504

Columns (1–2): pyroxenes normalized to six oxygens; Columns (3–4): magnesian olivines (Mg: Fe >1.5) normalized to four oxygens; Column (5): feldspar normalized to eight oxygens; Columns (6–7): glasses normalized to one oxygen.

75081 and BCR-1 agree very well with the several replicate analyses reported by others (Laul and Schmitt, 1973, 1974; Ma *et al.*, 1977; Korotev, 1976). In comparing the INAA data with the microprobe data for major elements, the SiO_2 values determined by INAA were found to be systematically low by ~4.5% relative to the probe values. The INAA SiO_2 values were corrected by this factor.

Major-Element characteristics

There is a common major-element composition for all of the primary,

Table 9. Extremes of Ab-An variation, analyses of feldspar grain 24182,8.

	1	2
SiO_2	44.37	45.25
Al_2O_3	35.06	34.84
FeO	0.32	0.49
MgO	0.07	0.10
CaO	18.74	18.72
Na_2O	0.51	0.67
K_2O	0.00	0.00
Σ	99.07	100.07
Si	2.067	2.088
Al^{IV}	1.926	1.896
Σ	3.993	3.984
Fe	0.012	0.019
Mg	0.005	0.007
Ca	0.936	0.925
Na	0.046	0.060
K	0.000	0.000
Σcations	4.992	4.995
Ab	4.7	6.1
Or	0.0	0.0
An	95.3	93.9
Mg/(Mg + Fe)	0.277	0.269

recrystallized and remelted ferrobasalt representatives we have analyzed (24174,7; 24182,12; 24077,62) and for similar lithic fragments analyzed by Ma *et al.* (1977: 24109,78; 24077,63; 24109,76; 24149,29). In summary of these data the mean ferrobasalt composition can be characterized as ~46% SiO_2, 1.1% TiO_2, 12% Al_2O_3, 22% FeO, 0.28% MnO, 7% MgO, 12% CaO, 0.25% Na_2O, 0.02% K_2O, and 0.3% Cr_2O_3. This is a mare basalt composition which is highly fractionated (Mg/(Mg + Fe) = 0.36; common mare basalt range = 0.57 to 0.35), low in alkali content, and low in Ti content. These are characteristics of a fractionated type of very low-Ti (VLT) mare basalt. The Mg-rich vitrophyre 24077,4 represents a more magnesian VLT basalt composition at the Luna 24 site. The bulk soil (Blanchard *et al.*, 1977) is also representative of VLT mare basalt. This abundance of VLT basalt at the Crisium landing site provides direct evidence that the VLT basalt suite, which had previously been recognized by only a few fragments at the Apollo 17 site (Vaniman and Papike, 1977; Taylor *et al.*, 1977), is indeed widespread on the moon.

The only nonbasaltic lithic fragment we have studied is the agglutinate 24210,50. This rock is distinct from the basaltic fragments by low FeO and high Al_2O_3 content, a composition which does not appear to reflect the local VLT-type bulk soil. This composition is within the range of variation of other Luna 24 agglutinates (Simonds *et al.*, 1977: microprobe data). The <20 μm soil fines (Blanchard *et al.*, 1977) and <74 μm soil fines (Tarasov *et al.*, 1977) are similar to the agglutinates in Fe and Al content. Agglutinates such as 24210,50 may have been formed elsewhere and mixed with less mature soils at the Luna 24 site.

Trace-Element characteristics: Agglutinate 24210,50 and the ferrobasalts

Agglutinate (24210,50): The chondritic normalized REE pattern of this agglutinate (Fig. 7) is essentially flat, ~10X chondritic with a slight (~10%) negative Eu anomaly. Our sample (7.2 mg) is a subsample of the 25 mg sample of 24210,5 which has been described as a melt breccia with scattered gabbro clasts, small vesicles, and layered soil inclusions (Nagle and Walton, 1977). In many respects the agglutinate is similar in composition to the finest soil fraction. Agglutinate 24210,50 is similar to the soil fines <20 μm analyzed by Blanchard *et al.* (1977) in content of FeO (16.5 vs. 16.4%) and Cr_2O_3 (0.42 vs. 0.45%), and similar to the soil fines <74 μm analyzed by Barsukov *et al.* (1977) in Al_2O_3 (15.4 vs. 15.2–16%) and CaO (12.3 vs. 12.9–13.3%). However, the correspondence in Mg, Ti and alkali content of the agglutinate and the soil fines is not as good. Moreover the REE concentrations of the <20 μ soil fractions are ~14× ± 2× chondritic and show a slight depletion in Yb and Lu, unlike the flat REE pattern of 24210,50 at ~10× chondritic enrichment.

Since micrometeorites are the main agents in the formation of agglutinates, the <74 μm and <20 μm soil fractions and the agglutinates should be enriched in meteoritic indicator elements such as Ni and Au. Barsukov *et al.* (1977) noted high Ni (~130 ppm) in their <74 μm fraction, and Blanchard *et al.* (1977) observed quite high Ni content (~400 ppm) in their <20 μm fraction relative to the 90–150 μm and bulk soil. We observe 180 ppm of Ni and 1.4 ppb of Au in agglutinate 24210,50. The Ni content of ferrobasalts is about 20 ppm (Table 2). Based on the Ni and Au contents, the meteoritic contribution in agglutinate 24210,50 is estimated at ~1% Cl chondritic.

Ferrobasalts: The Luna 24 ferrobasalts are quite low in rare earth elements (REE) and other refractory trace elements (Table 2). The chondrite-normalized REE patterns are shown in Fig. 7. REE concentrations in the ferrobasalts lie in the range 5–10× chondritic, with patterns that are bow-shaped (concave downward). This pattern is typical of mare basalts, with the exception that Sm/La fractionation in the Luna 24 ferrobasalts is less than in other mare basalts. However, the most striking feature of the REE pattern is the occurrence of a positive Eu anomaly in some ferrobasalt compositions. This is contrary to the typically negative Eu anomaly in mare basalts. Prior to the analysis of these Luna 24 samples, the only analyzed mare rocks with positive Eu anomalies (Laul and Schmitt, 1972) were rake samples 15643 (olivine microgabbro) and 15388 (feldspathic microgabbro: Dowty *et al.*, 1973). The absolute REE concentrations in 15643 and 15388 also lie in the range of Luna 24 basalts.

The positive Eu anomaly seen in some Luna 24 basalts stands out against the

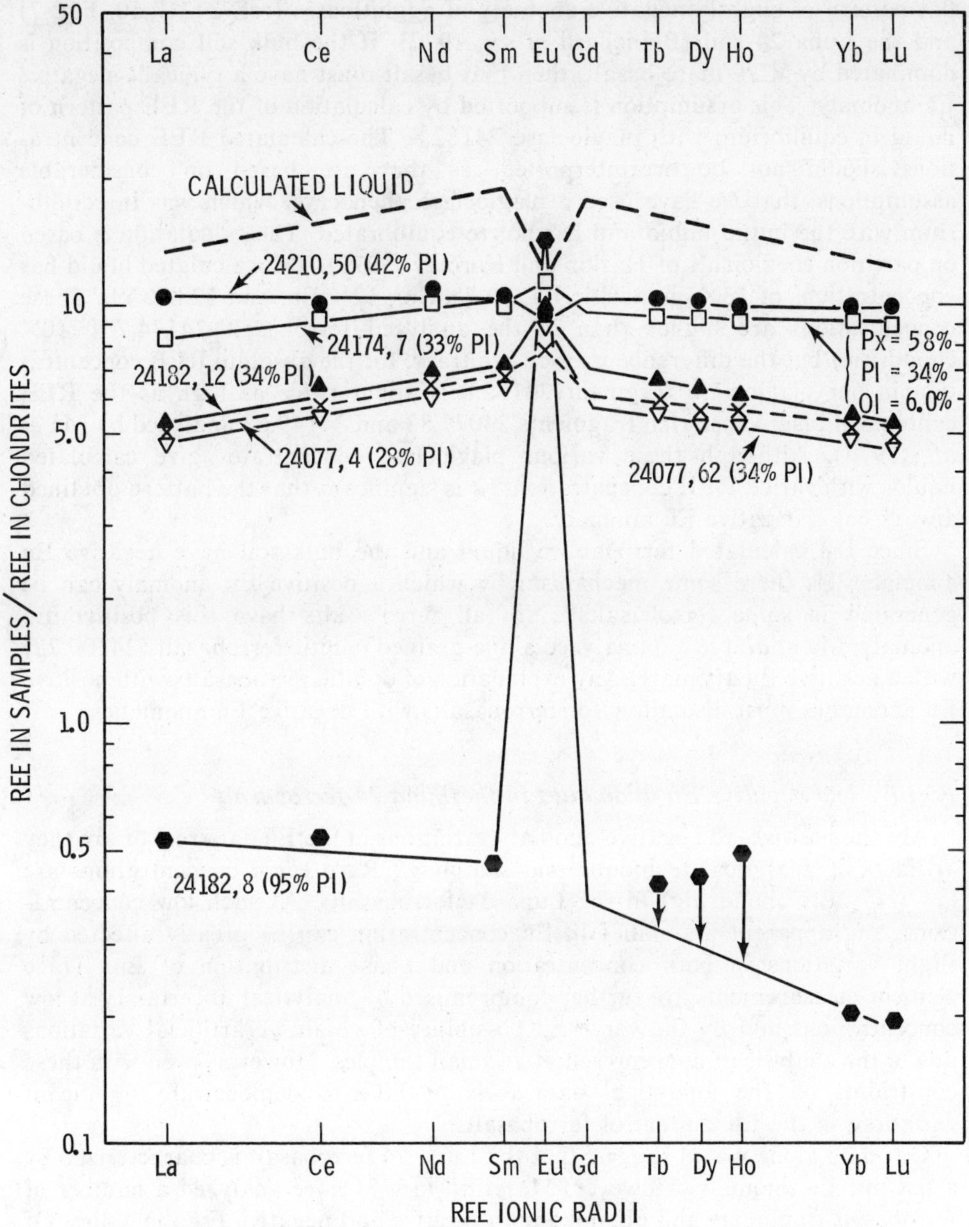

Fig. 7. Chondrite normalized REE patterns in Luna 24 samples: Mg-rich vitrophyre 24077,4; recrystallized ferrobasalt 24077,62; fine-grained ferrobasalt 24174, 7; plagioclase 24182,8; melt rock 24182,12; and agglutinate 24210,50. Plagioclase content for each sample is calculated based on Al_2O_3 values. See Laul and Schmitt (1973) for the chondritic values used for normalization. "Calculated liquid" is discussed in text.

flat pattern or slightly negative anomaly of agglutinates (ref. 24210,50, Fig. 7) and the Luna 24 soil (Blanchard *et al.*, 1977). If the bulk soil composition is dominated by VLT mare basalt, then that basalt must have a typically negative Eu anomaly. This assumption is supported by calculation of the REE pattern of liquid in equilibrium with plagioclase 24182,8. The calculated REE concentrations should not be overinterpreted, as they are based on considerable assumptions: that we have used a plagioclase phenocryst which was in equilibrium with the initial liquid and has not re-equilibrated. The calculation is based on partition coefficients of Haskin and Korotev (1977). The calculated liquid has concentrations of 13× chondritic La, 19.7× Sm, 12× Eu, and 12.8× Yb. These concentrations are higher than in the analyzed ferrobasalt 24174,7 (~10× chondritic) but the difference may be arbitrary, for the absolute REE concentration in our plagioclase fragment 24182,8 is about twice as high as the REE contents of plagioclase-rich fragments 24077,83 and 24149,63 analyzed by Ma *et al.* (1977). Although these various plagioclase concentrates give calculated liquids with varied REE concentrations, it is significant that the pattern obtained always has a negative Eu anomaly.

Since the calculated ferrogabbro liquid and the bulk soil have negative Eu anomalies, is there some mechanism by which a positive Eu anomaly can be generated in some ferrobasalts? Not all ferrobasalts have this positive Eu anomaly; Ma *et al.* (1977) analyzed a fine-grained ophitic ferrobasalt (24109,78) with a negative Eu anomaly. Any explanation of ophitic ferrobasalts with positive Eu anomalies must also allow for ferrobasalts with negative Eu anomalies.

Positive and negative Eu anomalies in the Luna 24 ferrobasalt

Are the positive- to negative-anomaly variations in the Eu data real or are they artifacts of analytical technique and sampling? Rare earth concentrations are low (~5–10× chondritic) in the Luna 24 ferrobasalts. At such low concentrations, any apparent anomalies in Eu concentration can be greatly affected by slight variations in both concentration and phase distribution of Eu. Trace element measurements are further compromised by analytical uncertainty at low concentrations and by the very real possibility of obtaining artificial variations due to the analysis of nonrepresentative small samples. However, even with these constraints on the analytical data it is possible to demonstrate significant variations in the Eu content of ferrobasalts.

Our data alone would suggest that the Luna 24 ferrobasalt is characterized by a positive Eu anomaly. However, Ma *et al.* (1977) have analyzed a number of ferrobasalt fragments and obtained both positive and negative Eu anomalies. Of these fragments, the best candidate for an ophitic ferrobasalt (24109,78) is very fine-grained (~0.03 mm) and has a negative Eu anomaly (Ma, *pers. comm.*), in contrast to the ophitic basalt in our sample suite (24174,7, ~0.1 mm grain size) which has a positive Eu anomaly. Note, however, that three metabasalt fragments analyzed by Ma *et al.* (24077,63; 24109,76; and 24149,29: ref. 1977) have positive Eu anomalies. It appears that all metabasalts have positive Eu anomalies, whereas the ophitic-textured ferrobasalts may have either positive or negative Eu anomalies.

There are several possible explanations for the occurrence of both positive and negative Eu anomalies in Luna 24 ferrobasalts. (1) Is there a bias due to the small size of the analyzed samples? Haskin and Korotev (1977) have reported both positive and negative Eu anomalies in 1–2 mm fragments of basalt 70135. However, their data were collected on 1–2 mm fragments of a rock with ~1–3 mm grain size, a situation in which the problems of small-sample analysis are greatly magnified. These problems are not so important in the Luna 24 ferrobasalt fragments. Ophitic Luna 24 ferrobasalts have grain sizes of 0.03–0.2 mm, and the recrystallized ferrobasalts have grain sizes of 0.01 mm. The samples studied are all >1 mm; therefore, the ophitic ferrobasalts have >125 grains/sample, and the recrystallized ferrobasalts have 10^5 to 10^6 grains/sample. The problem of sampling bias is minimal in the ophitic ferrobasalts and trivial in the recrystallized ferrobasalts. (2) Is there a higher feldspar concentration in ferrobasalts with positive Eu anomalies? There is a tendency for ferrobasalts with positive Eu anomalies to have lower CaO/Al_2O_3 ratios (CaO/Al_2O_3 decreases with increasing normative feldspar content). For instance, ophitic ferrobasalt 24174,7 (CaO/Al_2O_3 = 0.96) has a positive Eu anomaly whereas ophitic ferrobasalt 24109,78 (CaO/Al_2O_3 = 1.06; Ma *et al.*, 1977, analyst) has a negative Eu anomaly. McSween *et al.* (1977) have documented a variation in plagioclase content of recrystallized ferrobasalts, and Stolper (pers. comm.) suggests that heating of recrystallized basalts above the liquidus and extraction of a low-temperature liquid may result in the observed trend of plagioclase enrichment. Thus, a number of feldspar concentrating processes (accumulation, assimilation or remelting) may contribute to the variation of Eu/REE which is seen in the ferrobasalts. (3) Could the ferrobasalts with positive Eu anomalies contain Eu-rich phosphates? Eu concentrations in phosphate phases such as apatite might account for the positive Eu anomalies (A. Albee, pers. comm.). However, we have searched for phosphates in polished sections of one of the analyzed ophitic basalts (24174,7) and two recrystallized basalts (24077,62 and 24077,63) and found none. (4) Could the differing Eu anomalies be due to the degree of partial melting or variation in composition of the source? M. Drake (pers. comm.) has suggested that the varied Eu anomalies may be due to a variation in the amount of plagioclase in the source region. Grove and Vaniman (1978) suggest that the Luna 24 ferrobasalt is fractionated from a more mafic VLT composition represented by some Luna 24 green glasses. The results of high-pressure experiments on an Apollo 15 green glass composition (Grove and Lindsley, 1978) suggest a two pyroxene olivine-garnet source region for the Apollo 15 green glassmagma. There are no relevant experimental data on possible Luna 24 green glass-type magmas, and the available data on Luna 24 parental compositions is inadequate (Grove and Vaniman, 1978). The source region could lack plagioclase, but have variable proportions of clinopyroxene ± aluminous garnet to account for variations in the ratio of Eu^{++} to trivalent REE. At present there are not sufficient data to constrain the phase composition of the source region for magmas parental to the Luna 24 ferrobasalt; without a good model for the source region, calculations based on minor variations in partial melting or modal composition of the source are not warranted. (5) Could the variations in Eu anomaly be due to REE fractionation during cooling? Haskin and Korotev (1977) suggest that the trace-element concentrations in any particular grain depend on the environment in which the grain grew. First formed plagioclase ($D_{Eu} > 1$) will begin to deplete the melt in Eu, resulting in lower concentrations of Eu/trivalent REE in later plagioclase (disequilibrium crystallization). It is possible that the Luna 24 ferrobasalts with positive Eu anomalies may represent compositions with relatively more early formed plagioclase. (6) Could reduction and an increased Eu^{++}/Eu^{+++} ratio account for higher Eu concentration in crystallizing feldspar? The Luna 24 ferrobasalts with positive Eu anomalies may have crystallized from magma under more reducing conditions than did those ferrobasalts with negative Eu anomalies. El Goresy (1976) has cited mineralogic evidence for late-stage reduction in more Ti-rich mare basalts, but the behavior of fO_2 in the crystallization of Luna 24 ferrobasalts has not been determined.

We suggest one or more of the four possibilities listed above to explain the presence of both positive and negative Eu anomalies in the Luna 24 ferrobasalts. The variation in Eu anomaly may be caused by REE fractionation during cooling, by variation in reduction/oxidation during cooling, by variations in the melting history or composition of the source region (nature unknown), or by late-stage variations in plagioclase concentration.

Since there apparently is a real difference in Eu anomaly between different samples of Luna 24 ferrobasalt, why is it that such differences are not common in other mare basalts?

The positive and negative Eu anomalies can be quantified on a plot of Sm:Eu ratio vs. Sm (Fig. 8), where Sm is an indicator of trivalent REE concentration. Kesson and Ringwood (1976) used this plot to demonstrate that high- and low-Ti mare basalts appear to be a mixture of high-Ti magma types which are high in Sm/Eu (~10) and have large negative Eu anomalies, and low-Ti magma types with smaller Sm/Eu ratios (~5) and increasingly smaller negative Eu anomalies as they approach chondritic composition (zero anomaly). All VLT basalts have very small Eu anomalies (whether positive or negative) and small Sm/Eu ratios (≤3: ref. Warner *et al.*, 1978). In Fig. 8 the fields of positive and negative Eu anomalies are separated by a horizontal line passing through the chondritic Sm/Eu ratio. Although the ophitic basalt 24109,78 has a negative anomaly and does plot near the line characteristic of mare basalt types, small changes in Eu content (~0.35 ppm, a 50–60% change) can drive the Sm/Eu ratio into the field of positive Eu anomalies. It is not surprising that this process has not been seen in other mare basalts, for these would require a much larger modification of Eu content in order to acquire a positive Eu anomaly (~2.5 ppm Eu increase in low-Ti basalts, to ~8 ppm Eu increase in high-Ti basalts. These modifications would represent a 90% to 270% change in Eu content).

LEGEND

Ferrobasalts

○ = ophitic ferrobasalt 24174,7
∇ = ophitic ferrobasalt 24109,78
Π = medium-grained basalt 24143.4-006
● = remelted vitrophyric ferrobasalt 24182,12

Recrystallized Ferrobasalts

ꭙ = 24077,62
ꭙ = 24077,63
ꭙ = 24109,76
ꭙ = 24149,29

Mg-basalt Vitrophyre

⊖ = 24077,4

Agglutinates

A = 24210,50
a = average agglutinate of Simonds *et al.* (1977)

Soils

B = bulk soil
β = < 20μ soil
π = < 74μ soil

Averaged Compositions

$\bar{F}$ = average ferrobasalt
$\bar{B}$ = average bulk soil

Fig. 8a. Legend for Figs. 8b–12. Circles and "A" are analyses by J. C. Laul (1977); triangles are analyses by Ma *et al.* (1977); "B" and β are analyses by Blanchard *et al.* (1977); "II" and "π" are analyses by Tarasov *et al.* (1977) and Barsukov *et al.* (1977).

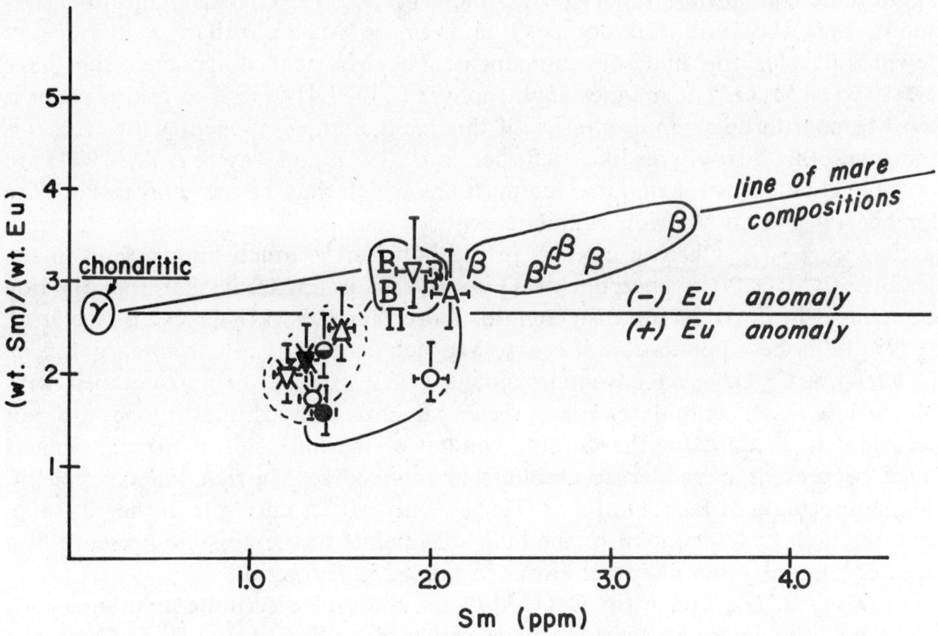

Fig. 8b. Sm/Eu (atomic) ratio plotted against Sm. Line of mare compositions is from Kesson and Ringwood (1976): See discussion in text. Error bars are plotted for Sm, and for maximum compounded error in Sm and Eu for extremes of Sm/Eu ratio. Symbols as in Fig. 8a.

Luna 24 lithic fragments in the context of Luna 24 soil

Much of the discussion at the Luna 24 Conference (see *Papers Presented to the Conference on Luna 24*) hosted by the Lunar Science Institute in Houston centered around the Mg, Cr-rich composition of Luna 24 soil relative to the Luna 24 ferrobasalt. There are no lithic fragments in the soil which are rich enough in Mg and Cr to account for this difference. However, Basu *et al.* (1977) and Papp *et al.* (1977) noted an abundance of Mg-rich olivine fragments (Mg:Fe >1.5 atomic) and Mg-rich pyroxene fragments (Mg:Fe >2 atomic) that cannot be derived from ferrobasalt or ferrogabbro. These Mg-rich fragments must account for a large portion of the high-Mg component in the bulk soil. When one notes that more than 60% of the olivine fragments analyzed by Basu *et al.* (1977) have atomic Mg:Fe ratios >1.5, it is evident that the Mg-rich component in the bulk soil is characterized by abundant coarse, magnesian olivine grains. The bivariate oxide plots in Figs. 9–11 place some restrictions on the composition of this Mg, Cr-rich component.

FeO vs. MgO: Figure 9 is a plot of FeO versus MgO in which six of the seven fragments in Table 2 are displayed (plagioclase 24182,8 is omitted). Additional data are plotted for three metabasalts and one ophitic ferrobasalt analyzed by Ma *et al.* (1977), for the six bulk soils analyzed by Blanchard *et al.* (1977), for four <74 μ soils analyzed by Barsukov *et al.* (1977), and for the average agglutinate composition reported by Simonds *et al.* (1977). This diagram clearly shows that the bulk soil composition is more Mg-rich than a mixture of ferrobasalt plus the fine soil component. The Mg-rich component must have >>10 wt.% MgO. The magnesian vitrophyre (24077,4) described in this paper is too Mg-poor to be a representative of this magnesian component. However, this rock and other vitrophyres like it (Ryder *et al.*, 1977, and Taylor *et al.*, 1977) are possible candidates for magma compositions which may be parental to both the ferrobasalt and an Mg-rich cumulate zone.

FeO vs. Cr_2O_3: Chrome content in the bulk soil is much higher than in the ferrobasalt or <20 μ components (Fig. 10). It is apparent that the Mg-rich component in Luna 24 soil also contains more than 0.5 wt.% Cr_2O_3. Bence *et al.* (1977) describe a population of coarse Mg-rich fragments of chromian pyroxene (>0.8 wt.% Cr_2O_3) and chromian olivine (~0.2 wt.% Cr_2O_3). Combined with plagioclase in a cumulate zone, these mineral compositions alone are not sufficient to account for the chrome content of the bulk soil. Chromian spinels must be present in moderate amounts in this coarse Mg-rich soil component. Though the data of Basu *et al.* (1977) show only ~0.2% chromite in the 20–250 μ soil, the high Cr_2O_3 content of the bulk soils points to the possible presence of a greater concentration of spinel grains in the >250 μ fraction.

CaO vs. Al_2O_3: The ratio CaO/Al_2O_3 is a useful discriminant of mare vs. highland lithologies. Mare rocks have ratios of wt.% CaO/wt.% Al_2O_3 in the range 0.9–1.1, whereas highland lithologies have a much lower ratio (0.55). Feldspathic "VHA" basalts, hybrid melt rocks and soil mixtures have intermediate ratios. Figure 11 shows that the calcium and aluminum contents of the ferrobasalt and bulk soil are very similar, and both fall near the mare ratio of

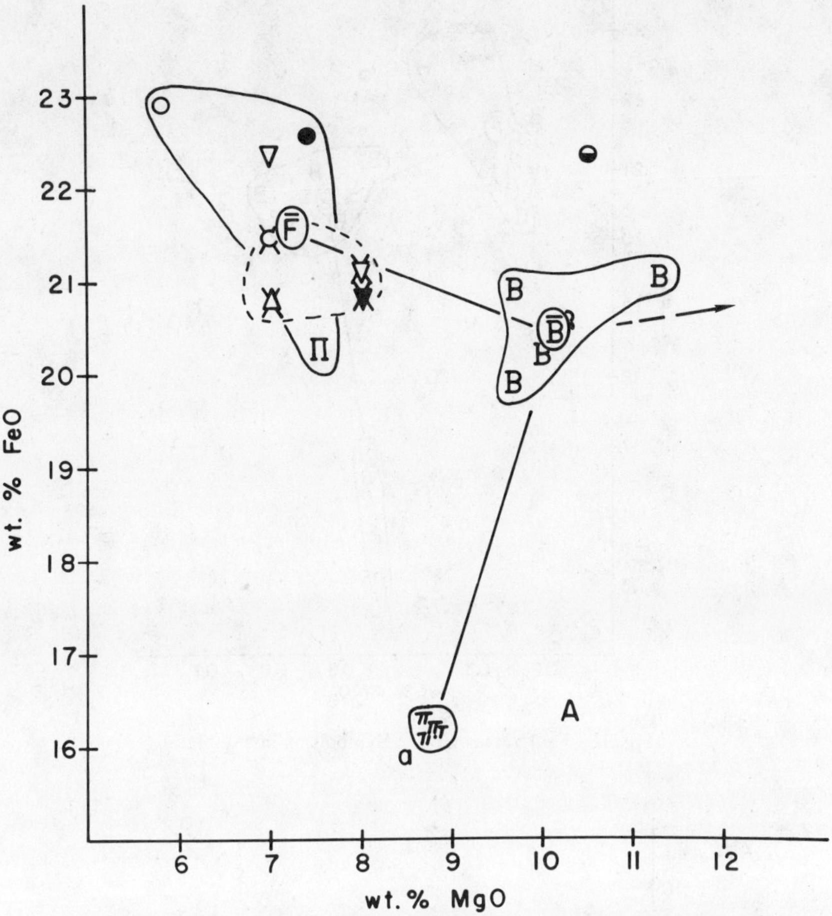

Fig. 9. Comparison of FeO and MgO, with fields outlining bulk soil, soil fines, igneous-textured ferrobasalts, and recrystallized ferrobasalts. Dashed arrow points in the general direction of the proposed "Mg, Cr-rich component." Symbols as in Figure 8a.

$CaO/Al_2O_3 = 1$. However, the <74 μ soils and agglutinates have a lower CaO/Al_2O_3 ratio (~0.84), which suggests that either a significant amount of highland material is included in the soil fines, or plagioclase is concentrated in the finer fractions that result from pulverization of the Luna 24 VLT basalts (a similar effect is noted by Haskin and Korotev, 1977, for finer size fractions of mare rock 70135).

We note that the Mg, Cr-rich component of the bulk soil is characterized by a mare CaO/Al_2O_3 ratio (≥1.0) to counterbalance the lower CaO/Al_2O_3 ratio in the <74 μ fraction. Since the <74 μ soil fraction and the ferrobasalt are both enriched in Ca and Al, the Mg, Cr-rich component must be depleted in these

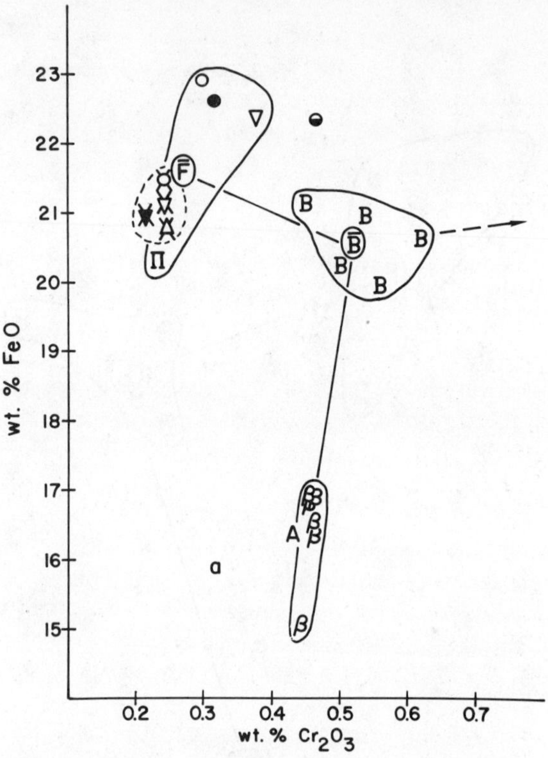

Fig. 10. FeO versus Cr_2O_3: Symbols as in Fig. 8a.

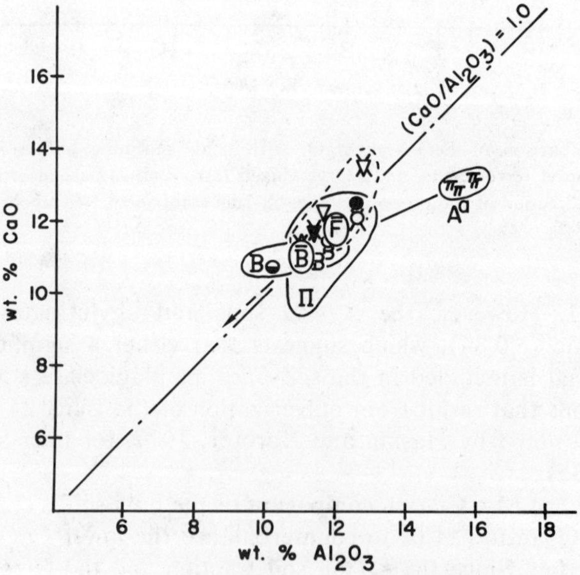

Fig. 11. CaO versus Al_2O_3: Symbols as in Fig. 8a.

elements relative to the bulk soil. The Mg, Cr-rich portion of the soil must have a content of $CaO \simeq Al_2O_3 < 11$ wt.%.

The combined constraints of high Fe and Mg content, high Cr content, and low Ca, Al content suggest that this Mg, Cr-rich component of the bulk soil may be dominated by either a less fractionated Luna 24 VLT basalt or a cumulate formed by crystal settling during the development of the ferrobasalt. Either of these rock types could be coarse-grained (100 μ to >1 mm), a grain-size constraint imposed by the sieve studies of McKay *et al.* (1977) and the microprobe studies of the >90 μ fraction by Basu *et al.* (1977) and Papp *et al.* (1977). The Mg, Cr-rich soil component may be derived from a Luna 24 VLT basalt which is less fractionated than the ferrobasalt, or a cumulate rock which is complementary to the Luna 24 ferrobasalt, or another VLT basalt type.

Luna 24 Lithic Fragments in the Context of Lunar Chemical Systematics

Interelement correlations are useful constraints to test any proposed model for magmatic processes and primordial abundances of the moon. Further, they serve as a criterion to establish main or subgroupings of rock types. Many of the volatile/non-volatile/refractory interelement correlations such as K/La, K/Hf, Th/U, FeO/MnO, FeO/Sc, K/Rb, Rb/Cs, etc., have been previously established for a wide variety of lunar rocks and soils (Laul and Schmitt, 1972, 1973, 1974; Laul *et al.*, 1974; Laul, 1977; Miller *et al.*, 1974; Wänke *et al.*, 1973, 1974; Anders *et al.*, 1971; *Proc. Lunar Sci. Conf.*, 1970–1977). The consistency of such correlations tends to favor the argument for homogeneous instead of heterogeneous accretion of the moon.

It is important to test the Luna 24 samples against the previously established correlations for K/La, FeO/MnO and FeO/Sc, since we have suggested that the very low K_2O and TiO_2 VLT basalts appear to be abundant on the moon. Figure 12 shows a plot of K versus La in lunar igneous and metaigneous rocks. Data are included from Luna 16, Luna 20 and all Apollo landings. The Luna 24 soil data are from Ma *et al.* (1977). A strong correlation between K and La is observed over a concentration range of two orders of magnitude for both elements. The KREEPY rocks from Apollo 14, 15 and 16 cluster in the top of the diagram. The KREEPY region is followed by Apollo 11 high-K rocks. Rocks from Apollo 14 and 16 and Luna 16 and 20 occupy the middle region of the diagram (medium-K). The rocks from Apollo 12, 15 and 17 cover a wide lower range of the figure (low-K). The Luna 24 rocks occupy the bottom of the diagram (very low-K). The Luna 24 ferrobasalts are the lowest in K and other trace elements of all rocks analyzed thus far. K and other LIL elements of the Luna 24 ferrobasalt occur in abundances about twice the moon's proposed primordial abundances. The ferrobasalts are unique in this respect.

The rocks 15643 (olivine microgabbro) and 15388 (feldspathic microgabbro), both with positive Eu anomalies, fall near the Luna 24 cluster. The Luna 24 soils lie slightly above the rocks and lie on a mixing line. This suggests that the soils contain other component(s), probably highland material as previously noted in

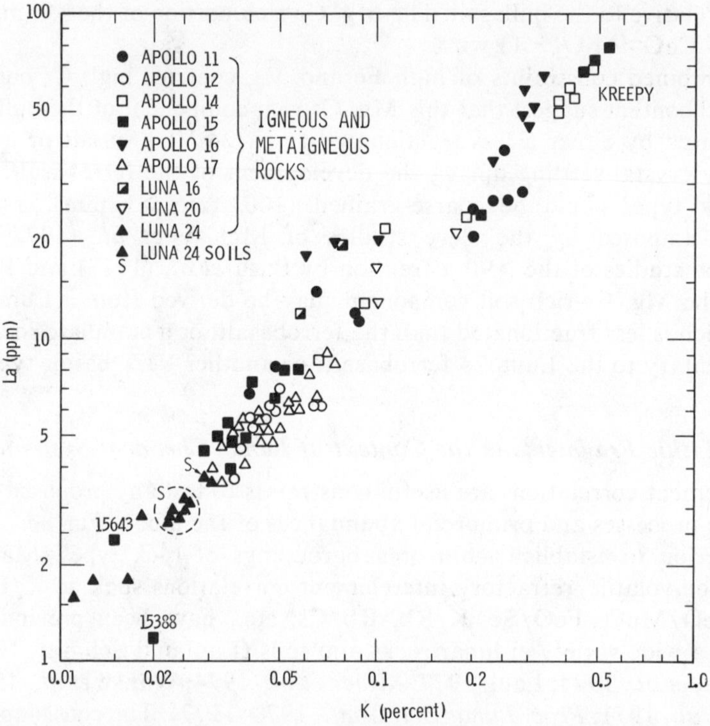

Fig. 12. The correlation of K versus La for igneous and metaigneous rocks from Apollo and Luna missions. K correlates well with La over two orders of magnitude. Luna 24 rocks lie in the bottom of the diagram. K versus La correlation was first shown by Wänke *et al.* (1973).

the <74 μm and <20 μm size fractions by Barsukov *et al.* (1977) and Blanchard *et al.* (1977).

Figure 13 shows correlations of FeO versus MnO and FeO versus Sc for Luna 24 rocks and soils. The solid lines are the established lunar ratios for FeO/MnO (80 ± 5) and FeO/Sc (5000 ± 400). The data for soils and other rocks are taken from Ma *et al.* (1977) and Blanchard *et al.* (1977). The correlations are not as strong as they should be. This is perhaps due to the fact that the Luna 24 samples are very small compared to the large samples analyzed in the Apollo missions. In small-size samples, there is a possibility of elemental fractionation due to nonrepresentative sampling or different grain sizes. The latter effect is seen in FeO/Sc ratios of different soil size fractions (Blanchard *et al.*, 1978) in Fig. 13: The <20 μm fraction has lower FeO/Sc ratios than the 90–150 μm fraction and the bulk soil. In view of the small sample size and analytical uncertainties in the data, these FeO/MnO and FeO/Sc correlations are valid for the Luna 24 rocks and soils and, in turn, strengthen the argument for homogeneous accretion of the moon.

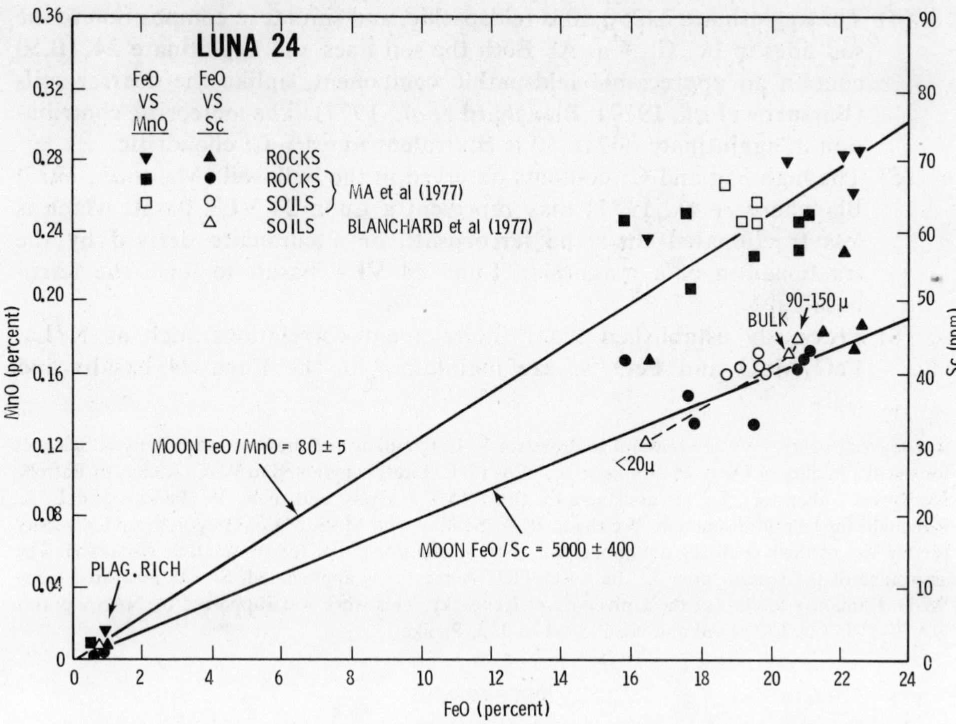

Fig. 13. The lunar correlations between FeO versus MnO, and FeO versus Sc are upheld in Luna 24 rocks and soils. The lunar ratios for FeO/MnO (80 ± 5) and FeO/Sc (5000 ± 400) are based on a wide variety of samples.

CONCLUSIONS

(1) The Luna 24 ferrobasalt is a highly fractionated mare rock very low in alkalis and TiO_2. The ferrobasalt is similar to the less fractionated Apollo 17 VLT basalts (Vaniman and Papike, 1977; Taylor *et al.*, 1977). Although Grove and Vaniman (1977) point out that the Apollo 17 and Luna 24 VLT basalts cannot be related by fractional crystallization, both of these rock types are characteristic of a family of VLT basalts which are now being recognized as an abundant mare rock type.

(2) The Luna 24 ferrobasalt is quite low in LIL trace-element content (~5–10X chondritic). This concentration is approximately twice the moon's proposed primordial abundances.

(3) The REE patterns of Luna 24 ferrobasalts are typical of VLT basalt with one striking irregularity: Some of the ophitic basalts have positive Eu anomalies, though others have the negative Eu anomaly typical of mare basalts. Variation in Eu anomaly might be due to REE fractionation on cooling, reduction/oxidation variations, variations in composition of the source region, or variations in plagioclase concentration.

(4) The agglutinate 24210,50 is feldspathic, and similar in composition to the soil fines in Fe, Cr, Ca, Al. Both the soil fines and agglutinate 24210,50 contain an appreciable feldspathic component, unlike the coarser soils (Barsukov *et al.*, 1977): Blanchard *et al.*, 1977). The meteoritic contribution in agglutinate 24210,50 is equivalent to ~1% Cl chondritic.

(5) The high Mg and Cr contents observed in the bulk soil (Ma *et al.*, 1977; Blanchard *et al.*, 1977) may represent a Luna 24 VLT basalt which is less fractionated than the ferrobasalt, or a cumulate derived by the fractionation of a magnesian Luna 24 VLT basalt to form the ferrobasalt.

(6) Previously established lunar interelement correlations such as K/La, FeO/MnO and FeO/Sc are maintained in the Luna 24 basalts and soils.

Acknowledgments—We are grateful to Professor V. L. Barsukov and the Soviet Academy of Sciences for the availability of Luna 24 samples. One of us (J. C. Laul) is indebted to W. C. Richey of Battelle Northwest Laboratory for his assistance in the INAA analysis, and to R. W. Perkins and L. A. Rancitelli for helpful discussion. We thank R. A. Schmitt and M.-S. Ma of Oregon State University for the use of their counting equipment for short irradiations, and for stimulating discussion. The assistance of the reactor crew at the OSU TRIGA reactor is appreciated. Si was measured using W. D. Ehmann's facility at the University of Kentucky. This work was supported by NASA grants NAS9–15357 to J. C. Laul and NSG-9044 to J. J. Papike.

REFERENCES

Albee A. L. and Ray L. (1970) Correction factors for electron probe microanalysis of silicates, carbonates, phosphates and sulfates. *Anal. Chem.* **42**, 1408–1414.

Ali M. Z. and Ehmann W. D. (1976) Chemical characterization of lunar core 60009. *Proc. Lunar Sci. Conf. 7th*, p. 241–258.

Anders E., Ganapathy R., Keays R. R., Laul J. C. and Morgan J. W. (1971) Volatile and siderophile elements in lunar rocks: Comparison with terrestrial and meteoritic basalt. *Proc. Lunar Sci. Conf. 2nd*, p. 1021–1036.

Barsukov V. L. (1977) Preliminary data for the regolith core brought to earth by automatic Lunar Station Luna 24. *Proc. Lunar Sci. Conf. 8th*, 3303–3318.

Barsukov V. L., Tarasov L. S., Dmitriev L. V., Kolesov G. M., Shevaleevsky I. D. and Garanin A. V. (1977) The geochemical and petrochemical features of regolith and rocks from Mare Crisium (preliminary data). *Proc. Lunar Sci. Conf. 8th*, p. 3319–3332.

Basu A., McKay D. S. and Fruland R. M. (1977) Petrography, mineralogy and source rocks of Luna 24 drill core soils (abstract). In *Papers Presented to the Conference on Luna 24*, p. 18–21. The Lunar Science Institute, Houston.

Bence A. E. and Albee A. L. (1968) Empirical correction factors for electron microanalysis of silicates and oxides. *J. Geol.* **76**, 382–403.

Bence A. E., Grove T. L. and Scambos T. (1977) Gabbros from Mare Crisium: An analysis of the Luna 24 soil. *Geophys. Res. Lett.* **4**, 493–496.

Blanchard D. P., Haskin L. A., Brannon J. C. and Aaboe E. (1977) Chemistry of soils and particles from Luna 24 (abstract). In *Papers Presented to the Conference on Luna 24*, p. 37–40. The Lunar Science Institute, Houston.

Cooper J. A., Rancitelli L. A. and Perkins R. W. (1970) An anticoincidence-shielded Ge(Li) gamma-ray spectrometer and its application to radioanalytical chemistry problems. *J. Rad. Anal. Chem.* **6**, 147.

Dowty E., Prinz M. and Keil K. (1973) Composition, mineralogy and petrology of 28 mare basalts from Apollo 15 rake samples. *Proc. Lunar Sci. Conf. 4th*, p. 423–444.

El Goresy A. (1976) Oxide minerals in lunar rocks. In *Oxide Minerals*. Miner. Soc. of America Short Course Notes 3, p. EG-1 to EG-46.

Grove T. L. (1978) Cooling histories of Luna 24 low-Ti ferrobasalts and ferrogabbros (abstract). In *Lunar and Planetary Science IX*, p. 424–426. Lunar and Planetary Institute, Houston.

Grove T. L. and Bence A. E. (1977) Petrogenesis of gabbros from Mare Crisium (abstract). In *Papers Presented to the Conference on Luna 24*, p. 64–67. The Lunar Science Institute, Houston.

Grove T. L. and Lindsey (1978) Compositional variation and origin of lunar ultramafic green glasses (abstract). In *Lunar and Planetary Science IX*, p. 430–432. Lunar and Planetary Institute, Houston.

Grove T. L. and Vaniman D. T. (1977) Experimental petrology of very low Ti basalts and origin of Luna 24 ferrobasalts (abstract). In *Papers Presented to the Conference on Luna 24*, p. 68–71. The Lunar Science Institute, Houston.

Grove T. L. and Vaniman D. T. (1978) Experimental petrology of very low Ti (VLT) basalts. In *Luna 24: The View from Mare Crisium* (R. B. Merrill and J. J. Papike, eds.). This volume.

Haggerty S. E. (1977) Luna 24: Opaque mineral chemistry of gabbroic and basaltic fragments from Mare Crisium. *Geophys. Res. Lett.* **4**, 489–492.

Haskin L. A. and Korotev R. L. (1977) Test of a model for trace element partition during closed system solidification of a silicate liquid. *Geochim. Cosmochim. Acta* **41**, 921–939.

Johnson T. V., Saunders R. S., Matson D. L. and Mosher J. A. (1977) A TiO_2 abundance map for the northern maria. *Proc. Lunar Sci. Conf. 8th*, p. 1029–1036.

Kesson S. E. and Ringwood A. E. (1976) A dynamic model for mare basalt petrogenesis. *Proc. Lunar Sci. Conf. 7th*, p. 1697–1722.

Korotev R. L. (1976) Geochemistry of grain-size fraction of soils from the Taurus-Littrow Valley Floor. *Proc. Lunar Sci. Conf. 7th*, p. 695–726.

Laul J. C. (1977) Chemical composition of Luna 24 melt rocks (24077,4; 24077,62; 24174,7; 24182,12; 24210,50) and gabbro (24182,8) (abstract). In *Papers Presented to the Conference on Luna 24*, p. 94–97. The Lunar Science Institute, Houston.

Laul J. C., Hill D. W. and Schmitt R. A. (1974) Chemical studies of Apollo 16 and 17 samples. *Proc. Lunar Sci. Conf. 5th*, p. 1047–1066.

Laul J. C. and Rancitelli L. A. (1977) Multielement analysis by sequential instrumental and radiochemical neutron activation. *J. Radioanal. Chem.* **38**, 461–475.

Laul J. C. and Schmitt R. A. (1972) The Apollo 15 Lunar Sample (abstract). In *Lunar Science III*, p. 225–227. The Lunar Science Institute, Houston.

Laul J. C. and Schmitt R. A. (1973a) Chemical composition of Luna 20 rocks and soil and Apollo 16 soils. *Geochim. Cosmochim. Acta* **37**, 927–942.

Laul J. C. and Schmitt R. A. (1973b) Chemical composition of Apollo 15, 16 and 17 samples. *Proc. Lunar Sci. Conf. 4th*, p. 1349–1367.

Laul J. C. and Schmitt R. A. (1974) Chemical composition of boulder-2 rocks and soils. *Earth Planet. Sci. Lett.* **23**, 206–219.

Ma M.–S., Schmitt R. A., Taylor G. J., Warner R. D., Lange D. E. and Keil K. (1977) Chemical and petrographic study of 18 Luna 24 lithic fragments. In *Papers Presented to the Conference on Luna 24*, p. 102–105. The Lunar Science Institute, Houston.

Marvin U. B., Ryder G. and McSween H. Y. (1977) Some exotic particles in the Luna 24 core sample. In *Papers Presented to the Conference on Luna 24*, p. 106–109. The Lunar Science Institute, Houston.

McKay D. S., Basu A., Waits G., Clanton U., Fuhrman R. and Fruland R. (1977) Grain size and evolution of Luna 24 soils (abstract). In *Papers Presented to the Conference on Luna 24*, p. 114–117. The Lunar Science Institute, Houston.

McSween H. Y., Taylor L. A. and Clark J. C. (1977) Luna 24 metabasalts: Possible evidence for assimilation of highlands material (abstract). In *Papers Presented to the Conference on Luna 24*, p. 118–120. The Lunar Science Institute, Houston.

Miller M. D., Pacer R. A., Ma M.-S., Hawke B. R., Lookhart G. L. and Ehmann W. D. (1974) Compositional studies of the lunar regolith at the Apollo 17 site. *Proc. Lunar Sci. Conf. 5th*, p. 1079–1086.

Nagle J. S. and Walton W. J. A. (1977) *Luna 24 Catalog and Preliminary Description.* Johnson Space Center, Houston. *Papers Presented to the Conference on Luna 24* (1977) The Lunar Science Institute, Houston, 213 pp.

Papike J. J. and Vaniman D. T. (1978) Luna 24 ferrobasalts and the mare basalt suite: Comparative chemistry, mineralogy and petrology. In *Luna 24: The View from Mare Crisium* (R. B. Merrill and J. J. Papike, eds.). This volume.

Papp H., Steel I. M. and Smith J. V. (1977) Luna 24 90–150 micron fraction: Implications for sampling of planetary and asteroidal regoliths (abstract). In *Papers Presented to the Conference on Luna 24*, p. 143–146. The Lunar Science Institute, Houston.

Pieters C. (1977) Characterization of lunar mare basalt types-II: Spectral classification of fresh mare craters. *Proc. Lunar Sci. Conf. 8th*, p. 1037–48.

Proceedings of the Apollo Lunar Science Conference (1970), vol. 1–3, Pergamon Press, New York.

Proceedings of the Lunar Science Conference (2nd–8th) (1971-1977), vol. 1–3, Pergamon Press, New York.

Ryder G., McSween H. Y. and Marvin U. B. (1977) Basalts from Mare Crisium. *The Moon* **17**, 263–287.

Schaeffer O. A., Bence A. E., Eichhorn G. and Vaniman D. T. (1978) ^{39}Ar-^{40}Ar and petrologic study of Mare Crisium: Age and petrology of Luna 24 samples 24077,13 and 24077,63 (abstract). In *Lunar and Planetary Science IX*, p. 1007–1009. Lunar and Planetary Institute, Houston.

Simonds C. H., Warner J. L., McGee P. E., Cochran A. and Brown R. W. (1977) Luna 24: Petrochemical potpourri (abstract). In *Papers Presented to the Conference on Luna 24*, p. 170–173. The Lunar Science Institute, Houston.

Tarasov L. S., Nazarov M. A., Shevaleevsky I. D., Kudryashova A. F., Gaverdovskaya A. S. and Korina M. I. (1977) Mineralogy and petrography of lunar rocks from Mare Crisium (preliminary data). *Proc. Lunar Science Conf. 8th*, p. 3333–3356.

Taylor G. J., Warner R., Wentworth S., Keil K. and Sayeed U. (1977a) Petrology and chemistry of Luna 24 mare basalts and basaltic glasses (abstract). In *Papers Presented to the Conference on Luna 24*, p. 189–92. The Lunar Science Institute, Houston.

Taylor G. J., Keil K. and Warner R. D. (1977b) Very low-Ti mare basalts. *Geophys. Res. Lett.* **4**, 207–210.

Vaniman D. T. and Papike J. J. (1977a) Very low Ti (VLT) basalts: A new mare rock type from the Apollo 17 drill core. *Proc. Lunar Sci. Conf. 8th*, p. 1443–71.

Vaniman D. T. and Papike J. J. (1977b) Ferrobasalts from Mare Crisium; Luna 24. *Geophys. Res. Lett.* **4**, 497–500.

Wänke H., Baddenhausen H., Dreibus G., Jagoutz E., Kruse H., Palme H., Spettel B. and Teschke F. (1973) Multielement analyses of Apollo 15, 16, and 17 samples and the bulk composition of the moon. *Proc. Lunar Sci. Conf. 4th*, p. 1461–1481.

Wänke H., Palme H., Baddenhausen H., Dreibus G., Jagoutz E., Kruse H., Spettel B., Teschke F. and Thacker R. (1974) Chemistry of Apollo 16 and 17 samples: Bulk composition, late-stage accumulation and early differentiation of the moon. *Proc. Lunar Sci. Conf. 5th*, p. 1307–1335.

Warner R. D., Taylor G. J., Keil K. and Nehru C. E. (1978) Green glassy rock 78526: An impact melt of very low-Ti mare basalts? (abstract). In *Lunar and Planetary Science IX*, p. 1225–1227. Lunar and Planetary Institute, Houston.

Merrill, R. B. and Papike, J. J., editors.
(1978) *Mare Crisium: The View from Luna 24*, Pergamon Press (New York), p. 569–592
Printed in the United States of America

Chemistry and petrology of Luna 24 lithic fragments and <250 μm soils: Constraints on the origin of VLT mare basalts

M.-S. MA AND R. A. SCHMITT

Department of Chemistry and the Radiation Center, Oregon State University
Corvallis, Oregon 97331

G. J. TAYLOR, R. D. WARNER, D. E. LANGE AND KLAUS KEIL

Department of Geology and Institute of Meteoritics, University of New Mexico
Albuquerque, New Mexico 87131

Abstract—We report results of a combined INAA-petrologic study of 17 small (0.2–1.5 mg) Luna 24 lithic and mineral fragments and INAA study of 5 bulk soils and mineral separates from gabbro 24170. We classify lithic and mineral fragments into VLT mare basalts (ferrobasalt and metabasalts), low-Ti, variolitic mare basalt, gabbros, melt rock and soil breccia. Data indicate 5 possible magma types, represented by: 1) VLT ferrobasalt and gabbro fragments, with low-TiO_2 (~1%), slightly bow-shaped REE pattern, and low REE concentrations (5-10X chondritic); 2) a ferrobasalt (Laul *et al.*, 1978) and metabasalt fragments with major and trace element contents similar to 1), but positive Eu anomalies; 3) one gabbro fragment with distinctive pyroxene compositional trend (increasing Ti with nearly constant Fe/Fe + Mg) and highest REE contents of any Luna 24 mare basaltic sample; 4) a gabbro fragment with considerably less V (46 vs. >150 ppm) and Cr_2O_3 (670 vs. 2000 ppm), than ferrobasalt and metabasalt fragments; and 5) variolitic basalt fragment with higher TiO_2 (2.3%) than other Luna 24 basalts and pyroxene that has increasing, then decreasing Ti with increasing Fe/Fe + Mg.

Trace element data place constraints on the nature of the source region and possible parent magmas for the Luna 24 VLT ferrobasalts: 1) The varying REE contents coupled with nearly constant Sc and Eu suggest that small amounts of clinopyroxene and plagioclase remained in the solid residue after partial melting produced the basaltic magmas. 2) Sm-Nd and Rb-Sr data indicate that there was an insignificant degree of fractionation between the Luna 24 VLT ferrobasaltic magma and its source material. 3) A magma with the composition of Apollo 15 green glass cannot evolve by olivine fractionation to yield Luna 24 VLT ferrobasalts.

INTRODUCTION

In this paper we report the results of a coordinated INAA (by OSU) and mineralogic-petrologic-electron microprobe (by UNM) study of 17 Luna 24 lithic fragments of sizes between 0.5–1 mm (Ma *et al.*, 1977b). The fragments were first analyzed by non-destructive INAA for major, minor and trace (including the REE) elements, and polished thin sections were then prepared of the same neutron-irradiated fragments for mineralogical, petrographic, and electron microprobe studies. Thus, close correlation is achieved between bulk compositions of the fragments and studies of their textures, mineral contents and mineral compositions. In addition, INAA was carried out on an additional metabasalt fragment (O. A. Schaeffer consortium), three mineral separates from gabbro 24170 and five soils (G. J. Wasserburg consortia).

The main purpose of the present study is to chemically and petrographically characterize the Luna 24 lithic fragments. These data, in turn, will be used to address questions such as the preponderance of Very-Low-Titanium (VLT) rock types at the Luna 24 landing site and the derivation of metabasalts from VLT ferrobasalts. Furthermore, we will show that coarse-grained (sometimes nearly monomineralic) lithic fragments represent a VLT mare gabbro type similar to gabbro 24170 and that this rock type was derived from a VLT magma similar to that which formed the fine-grained VLT ferrobasalts. In addition, we discuss the trace element distribution in the inferred magma and source material responsible for the origin of Luna 24 VLT basalt. We also show that the Luna 24 core, as represented by bulk soils and separate fragments, is quite homogeneous and of VLT composition but that other rock types are also represented in the Luna 24 collection that are different from the typical VLT ferrobasalt. These are low-Ti mare basalt and several different gabbro types. Finally, we discuss genetic relationships of Luna 24 VLT basaltic magmas to possible parental materials such as Apollo 15 green glass, Luna 24 olivine vitrophyre, etc.

Experimental

All samples were activated in the Oregon State University TRIGA reactor for the determination of elements that were activated to short-lived radionuclides. Subsequently, the 17 lithic fragments and metabasalt fragment 24077,63 were sealed in separate quartz vials and activated in the high flux ($1.0 \times 10^{14} n/cm^2 \bullet sec$) research reactor at the University of Missouri for 96 hours. Individual Fe wires were contiguous to each quartz vial for relative neutron flux monitoring. After irradiation, the samples were transferred carefully to new polyvials for gamma-ray counting. Microscopic examination during sample transfer ensured no loss of sample. Weights of samples were checked before and after transfer and no apparent losses were observed. Three countings by using two high efficiency Ge(Li) detectors were carried out. Due to the time delay during transportation of samples from Missouri to Oregon, K abundances were not determined in these samples. The long activation of the soil samples and the mineral separates of 24170 was performed in the Oregon State University TRIGA reactor. Soil samples and the mineral separates of 24170 were prepared at Cal Tech. No sample transfers were made at OSU after irradiations. Two to four blank polyvials were sealed, activated and counted in the same manner as were the sample containers.

USGS standard rocks BCR-1, GSP-1 and PCC-1 and Apollo soil samples previously analyzed by us were irradiated simultaneously with the Luna 24 samples to provide a measure for precision and accuracy. Procedures of INAA are essentially the same as reported by Wakita *et al.* (1970) and Laul and Schmitt (1973), and electron microprobe analyses were carried out following methods described by Keil (1967) and Bence and Albee (1968).

Results

Here we present the results of our mineralogical, petrologic, and chemical studies of Luna 24 particles and soils (Figs. 1–7, Tables 1–2). We classified the samples studied into five main categories. These are Very-Low-Titanium (VLT) mare basalt fragments, low-Ti mare basalt fragment, gabbro fragments, melt rock and soil breccia fragments, and soils. The VLT mare basalt fragments were subdivided into ferrobasalt and metabasalts, and the gabbro fragments, because of their coarse grain sizes that resulted in sampling of variable proportions of plagioclase and clinopyroxene, into plagioclase-rich, clinopyroxene-rich, and mixed mineral fragments. In addition, mineral separates of gabbroic fragments were also studied.

VLT mare basalt fragments

Ferrobasalt fragment 24109,78, although finer-grained than most Luna 24 ferrobasalt fragments, has an ophitic to subophitic texture (Fig. 1A). An olivine microphenocryst (Fo_{39}) occurs in one corner of the thin section, accompanied by two smaller olivine grains (Fo_{32} to Fo_{30}) and several subhedral chromite crystals (Fig. 1B). The chromites contain appreciable TiO_2 (7.7%) and Al_2O_3 (11.3%). The remainder of the fragment consists of plagioclase laths (most ~50 μm long), intergrown with subequant pyroxene (a modal analysis appears in Table 1). The plagioclase is low in alkalis (the average of 8 analyses is $An_{93.6}Ab_{6.2}Or_{0.2}$). Pyroxene compositions (Figs. 2A, 3A) encompass a more restricted range than those in most Luna 24 ferrobasalts; this is probably due to more rapid crystallization. Accessory amounts of ilmenite [Fe/(Fe + Mg) of 0.96] and troilite occur throughout the sample. Mesostasis areas are rare and very small (~1 μm). A small amount (<1%) of soil breccia adheres in thin patches along two edges of the thin section. This fragment resembles typical Luna 24 ferrobasalt fragments that we studied in thin sections (250–500 μm size range) of Luna 24 bulk soils (Taylor *et al.*, 1977b, 1978b), although it is finer-grained than most.

Compositionally, the fragment also resembles typical Luna 24 VLT mare basalt fragments (Table 2). Major element abundances, for example, are similar to those previously reported for Luna 24 VLT ferrobasalts (Taylor *et al.*, 1977a, b, 1978b; Vaniman and Papike, 1977a,b; Grove and Vaniman, 1977; Laul 1977), with Fe/Fe + Mg = 0.65 and CaO/Al_2O_3 = 1.06. It should be noted, however, that in this fragment, TiO_2, FeO, and Cr_2O_3 appear to be slightly higher and Al_2O_3 and CaO lower than in other fine-grained basalt fragments. The REE pattern is bow-shaped and shows a negative Eu anomaly (Fig. 4). The light REE are more depleted and fractionated relative to the heavy REE. The only other fine-grained ferrobasalt fragment analyzed for REE (24174,7; Laul, 1977) is similar in bulk composition to the fragment studied here, but has a positive Eu anomaly (see discussion).

Metabasalt fragments 24077,63, 24109,76, and 24149,29 are similar to other Luna 24 metabasalt fragments (Ryder *et al.*, 1978; Taylor *et al.*, 1977b, 1978b) and are characterized by microgranular textures and equilibrated mineral compositions. The petrology of sample 24077,63 is described by Laul *et al.* (1978); we summarize their results here. Fragments 24077,63 and 24109,76 have essentially identical textures, mineralogy, and mineral compositions: Both are microgranular with grain sizes of ~0.1 mm (Fig. 1C). They consist mainly of pyroxene (Fig. 2B), which are exsolved on a very fine scale to yield compositions

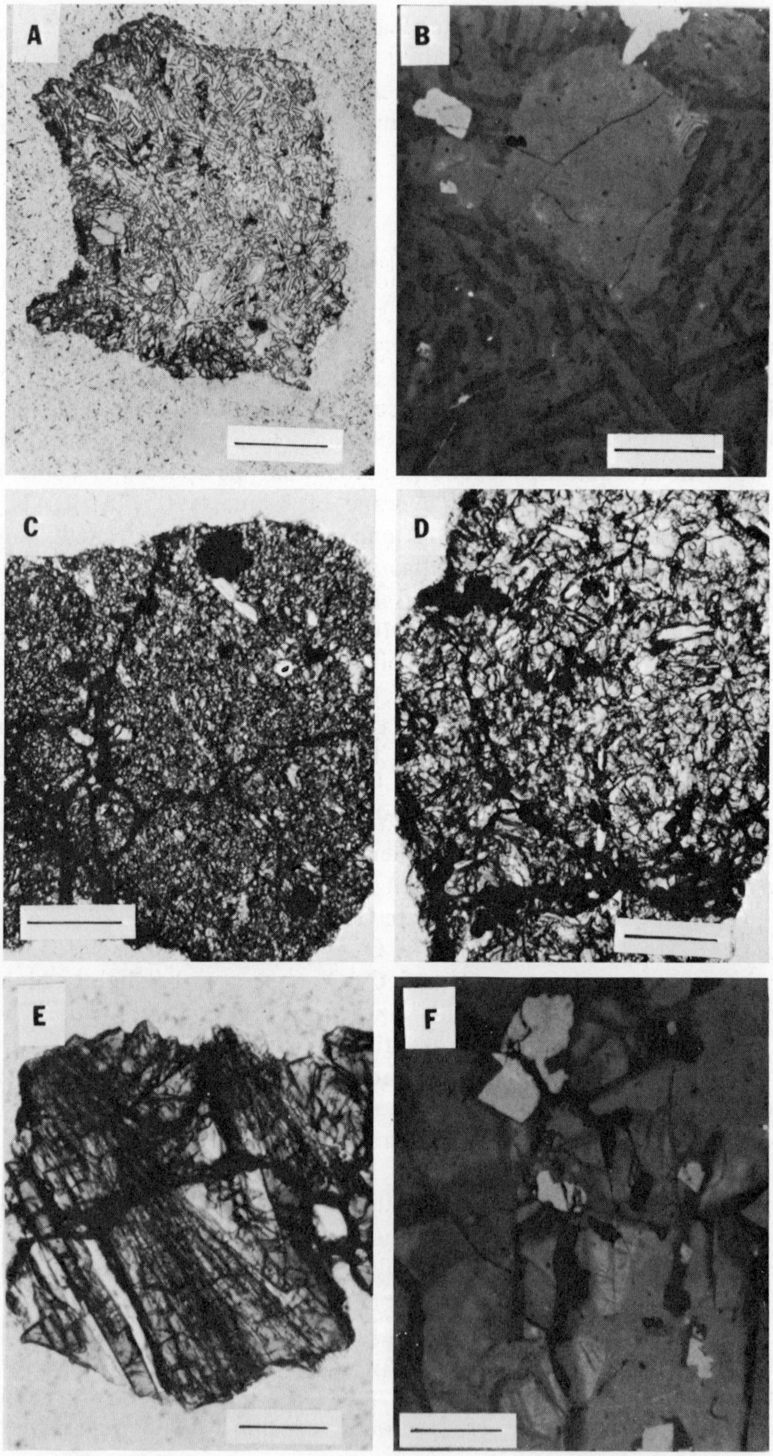
A
B
C
D
E
F

Fig. 1. Photomicrographs of some of the fragments studied. (A) Transmitted light view of ferrobasalt fragment 24109,78; scale bar is 200 μm. (B) Reflected light view of 24109,78 showing olivine microphenocryst surrounded by subophitic intergrowth of plagioclase and pyroxene; light grains are chromite. Scale bar is 50 μm. (C) Transmitted light view of metabasalt fragment 24109,76; scale bar is 200 μm. (D) Photograph of metabasalt fragment 24149,29 in transmitted light. Note the lath-shaped plagioclase crystals; scale bar is 200 μm. (E) Photograph in transmitted light of variolitic basalt fragment 24149,69; scale bar is 200 μm. (F) Reflected light view of gabbro fragment 24109,72. Note the euhedral chromite grains in the right half of the photograph, entirely enclosed by plagioclase. Larger ulvöspinel grains occur in top center. Scale bar is 100 μm.

Table 1. Modal compositions (vol.%) of one ferrobasalt and two metabasalt fragments. N is the number of points counted.

	A	B	C
N	407	321	443
plagioclase	35.8	33.3	35.2
pyroxene	61.0	62.0	58.7
olivine	2.2	2.5	2.9
chromite	0.4	—	—
ulvöspinel	—	2.2	2.0
ilmenite	0.4	—	0.7
metal-FeS	0.2	—	0.5

(A) Ferrobasalt fragment 24109,78.
(B) Metabasalt fragment 24109,76.
(C) Metabasalt fragment 24149,29.

that range along a well-defined line. The low-Ca end of the trend for 24077,63 is slightly lower in Ca than that for 24109,76. Plagioclase has a very restricted range in composition ($Ab_{5\text{-}7}An_{93\text{-}95}Or_{<1}$). Olivine is homogeneous (Fo_{26}). Ulvöspinel and troilite are the only opaque phases. Both 24077,63 and 24109,76 are typical Luna 24 metabasalt fragments. In contrast, sample 24149,29, although basically granular, is coarser-grained (0.3–0.4 mm average grain size), has a remnant subophitic ophitic texture (Fig. 1D), and contains both ulvöspinel and ilmenite. Most plagioclase crystals are lath-shaped and are up to 30 μm long. However, pyroxene (Fig. 2B), olivine, and plagioclase compositions are indistinguishable from those in 24077,63 and 24109,76. Thus, it appears that sample 24149,29 is less metamorphosed than most Luna 24 metabasalt fragments.

Metabasalt fragments have major (e.g. Fe/Fe + Mg, CaO/Al_2O_3) and trace element (Sc, V, Co) abundances similar to those of fine-grained ferrobasalt fragments (Table 2). However, the REE patterns consistently show positive Eu anomalies and both the light and heavy REE are fractionated and depleted relative to the middle REE. A similar REE pattern was also measured for another metabasalt fragment (Laul, 1977). Our bulk compositional data are very similar to those reported by Taylor *et al.* (1977b, 1978b), Vaniman and Papike

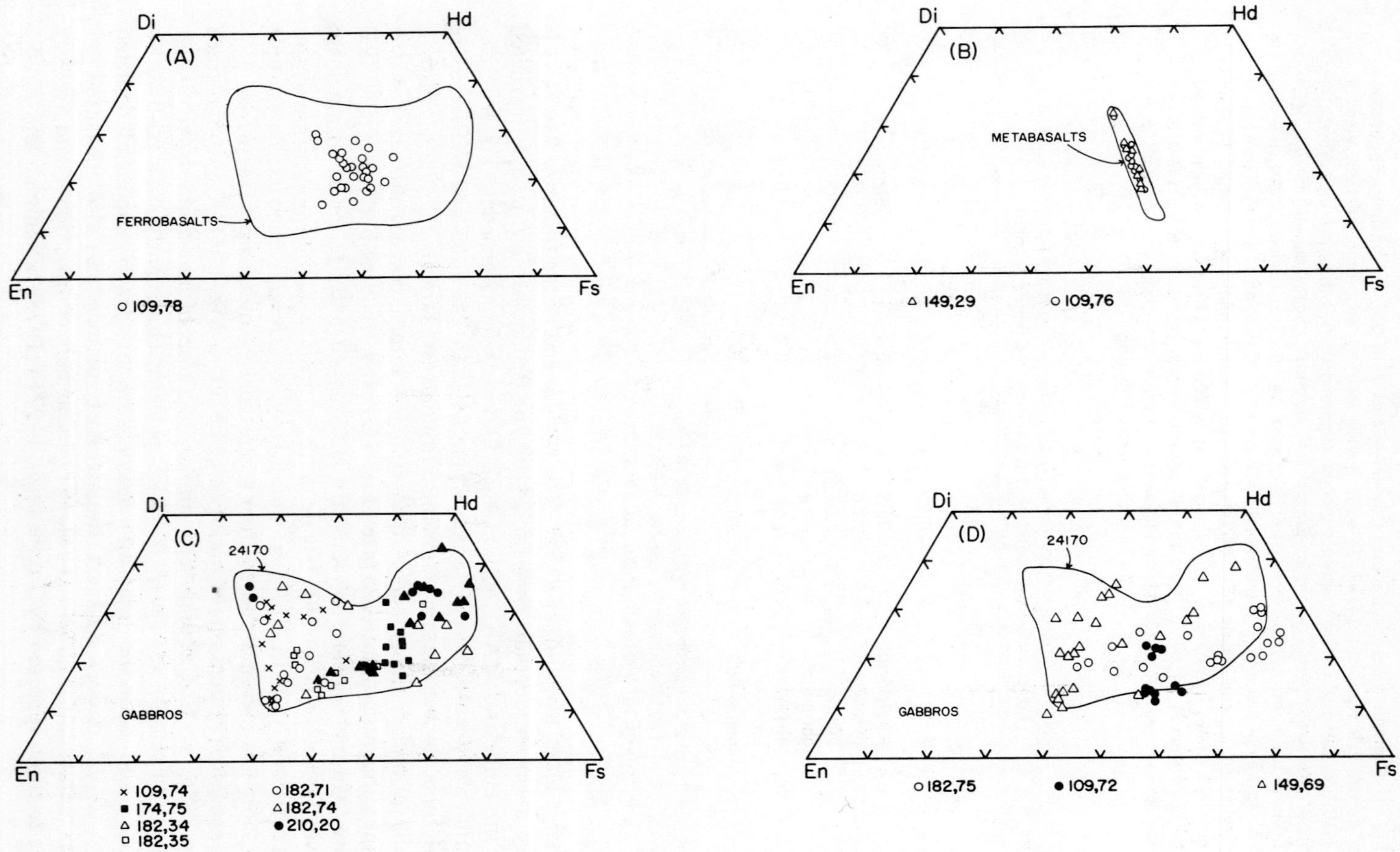

Fig. 2. Pyroxene compositions in fragments of ferrobasalt (A), metabasalts (B), gabbroic fragments similar to 24170 (C), and other gabbro or basaltic fragments (D). Outlined fields are from Taylor *et al.* (1977b).

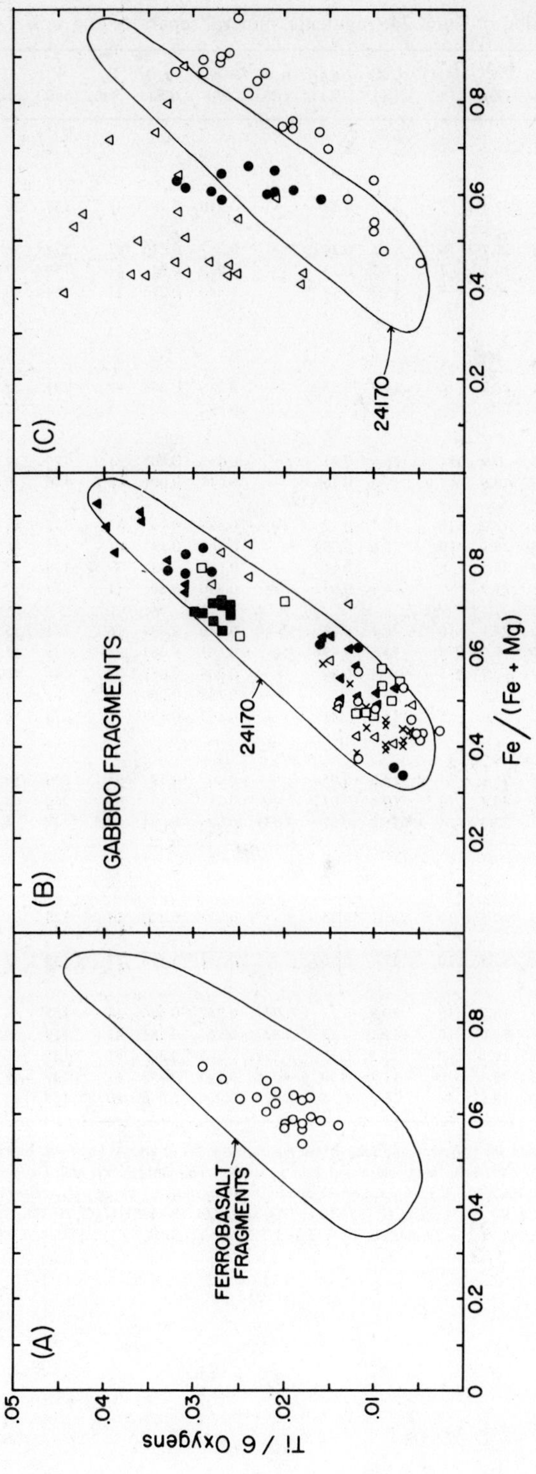

Fig. 3. Plot of the Ti content per 6 oxygen atoms vs the atomic Fe/Fe + Mg ratio in pyroxene in the fragments studied. (A) Ferrobasalt fragment 24109,78; note that the analyses conform to the trend (Taylor *et al.*, 1977b) established for ferrobasalts in Luna 24 soils. (B) Gabbro fragments that follow the trend established for gabbro 24170 (Taylor *et al.*, 1977b). Symbols same as in Fig. 2C. (C) Gabbro or basalt fragments with trends different from the 24170 trend; symbols same as Fig. 2D.

Table 2. Composition of Luna 24 fragments, mineral separates and soils.*

Sample	Weight mg	TiO_2 (%)	Al_2O_3 (%)	FeO (%)	MgO (%)	CaO (%)	Na_2O (%)	K_2O (%)	MnO (%)	Cr_2O_3 (%)	Sc (ppm)	V (ppm)	Co (ppm)	Zr (ppm)	Sr (ppm)
VLT mare basalt fragments															
Ferrobasalt															
24109,78	0.342	1.3	11.6	22.4	7	12.3	0.29	—	0.258	0.381	47	177	36	—	—
Metabasalt															
24077,63	9.055	1.1	12.8	21.1	8	13.5	0.24	—	0.247	0.240	43	178	38	—	—
24109,76	0.710	1.1	11.1	20.8	7	11.5	0.25	—	0.229	0.246	41	179	37	—	—
24149,29	0.934	1.3	11.4	20.9	8	11.7	0.24	—	0.246	0.216	42	154	40	—	—
Low-Ti mare basalt fragment															
Variolitic basalt															
24149,69	0.220	2.3	14.2	15.8	6	11.7	0.55	—	0.246	0.089	42	131	30	—	—
Gabbro fragments															
Plagioclase-rich															
24077,83	1.505	<0.3	31.6	0.9	~1	18.6	0.47	—	0.013	0.007	1.3	<20	2.2	—	—
24149,63	0.312	<0.7	32.4	0.6	<1	19.4	0.76	—	0.011	0.001	0.5	<40	0.9	—	—
Clinopyroxene-rich															
24182,35	0.210	0.8	2.3	27.0	10	11.1	0.12	—	0.390	0.153	81	172	46	—	—
24182,74	0.606	0.6	3.3	26.1	10	10.6	0.094	—	0.362	0.142	81	175	46	—	—
24182,75	0.550	0.6	2.4	28.5	8	11.2	0.11	—	0.381	0.125	79	139	43	—	—
24210,20	0.523	0.8	2.8	24.9	7	14.0	0.082	—	0.340	0.206	91	212	37	—	—
Mixed mineral															
24109,72	0.334	2.0	10.2	24.1	10	8.1	0.33	—	0.290	0.839	16	151	55	—	—
24109,74	0.207	0.6	21.5	9.0	4	14.5	0.49	—	0.161	0.087	28	79	17	—	—
24174,75	1.289	0.7	17.2	17.7	5	13.4	0.38	—	0.208	0.067	32	46	32	—	—
24182,34	0.310	0.5	21.7	3.9	6	16.6	0.53	—	0.139	0.060	7.1	81	4.9	—	—
24182,71	0.336	<0.7	24.9	7.8	5	15.9	0.41	—	0.103	0.139	16	74	16	—	—
Mineral separates from 24170															
Pyroxene-A	8.0	0.7	2.8	26.2	15.8	14.5	0.056	—	0.40	0.59	98	390	56	—	—
Pyroxene-B	2.3	0.6	3.2	30.9	12.3	10.9	0.075	—	0.38	0.43	85	260	62	—	—
Plagioclase	6.3	—	30.7	2.3	—	16.8	0.57	0.023	0.029	0.024	4.5	—	4.2	—	—
Melt rock and soil breccia fragments															
Melt rock															
24149,61	0.473	1.1	11.9	19.5	10	11.1	0.26	—	0.225	0.320	33	129	74	—	—
Soil breccia															
24077,81	0.577	1.0	13.4	17.6	10	12.7	0.30	—	0.251	0.316	37	149	47	—	—
Soils															
24109	23.8	1.0	12.1	18.7	10	10.8	0.27	0.027	0.267	0.380	40	155	46	70	81
24149	24.1	1.1	12.1	19.6	10	11.1	0.28	0.027	0.257	0.384	43	159	60	60	109
24174	25.1	0.8	12.1	19.1	9	11.3	0.28	0.029	0.245	0.388	41	139	46	50	108
24182	25.1	1.2	12.3	19.6	10	11.4	0.30	0.030	0.245	0.390	41	147	50	50	93
24210	14.1	0.1	12.0	19.8	10	10.8	0.28	0.028	0.254	0.446	40	159	42	50	103

*Estimated errors due to counting statistics in the pyroxene and plagioclase separates of 24170 are: Al_2O_3, FeO, MnO, Na_2O in plagioclase, MnO, Cr_2O_3, Sc, V and Co in pyroxenes, Sm and Eu in plagioclase and La, ~±5–10%, TiO_2, MgO, Yb and Ta, ~±10–20%; K_2O, Eu in pyroxenes, Tb, Dy and Lu, ~±20–30%; Hf and Ba, ~±50%. Estimated errors due to counting statistics in the soils are: Al_2O_3, FeO, Na_2O, MnO, Cr_2O_3, Sc, Co, La and Sm, ±1–5%; CaO, V, Ce, Eu, Yb, Lu and Ni, ±5–10%, TiO_2 and K_2O, ±8–15%; MgO, Hf, Th, Sr, Nd, Tb, Dy and Ta, ±10–25%; Ba and Zr, ±30–50%. The abundances of Ni, Zr, Sr, Ba, Ce, Nd and Th were obtained by Dr. J. C. Laul, using the coincidence-

Ba (ppm)	La (ppm)	Ce (ppm)	Nd (ppm)	Sm (ppm)	Eu (ppm)	Tb (ppm)	Dy (ppm)	Yb (ppm)	Lu (ppm)	Hf (ppm)	Ta (ppm)	Th (ppm)	Ni (ppm)	Ir (ppb)	Au (ppb)
—	2.0	—	—	1.9	0.58	0.44	2.8	1.9	0.29	1.1	0.1	—	80	—	—
—	1.1	—	—	1.2	0.63	0.25	1.5	1.0	0.15	0.9	0.1	—	—	—	—
—	1.9	—	—	1.5	0.58	0.33	2.1	1.3	0.17	1.0	0.1	—	<50	—	—
—	1.6	—	—	1.3	0.64	0.29	1.7	1.2	0.18	1.1	0.2	—	<50	—	—
—	2.1	—	—	2.4	0.75	0.51	3.6	2.1	0.30	1.3	0.3	—	<50	—	—
—	0.08	—	—	0.06	0.55	0.014	—	0.04	0.005	—	—	—	—	—	—
—	0.1	—	—	0.06	0.75	0.009	—	0.03	0.004	—	—	—	—	—	—
—	0.4	—	—	0.73	0.15	0.21	(1.8)	1.2	0.22	0.2	0.2	—	90	—	—
—	0.6	—	—	0.81	0.24	0.21	1.4	1.1	0.18	0.3	0.1	—	120	—	—
—	1.9	—	—	2.0	0.52	0.53	3.7	2.9	0.44	1.4	0.3	—	135	—	—
—	3.2	—	—	2.8	0.63	0.67	4.4	2.8	0.39	1.8	0.2	—	180	—	—
—	9.1	—	—	6.7	1.55	1.3	7.8	4.6	0.65	3.5	0.5	—	130	—	—
—	0.2	—	—	0.27	0.46	0.07	—	0.4	0.07	0.1	—	—	—	—	—
—	2.5	—	—	1.8	1.01	0.41	2.7	1.7	0.23	1.0	0.1	—	80	—	—
—	3.8	—	—	2.3	1.04	0.51	3.1	1.8	0.26	1.3	0.2	—	—	—	—
—	0.14	—	—	0.20	0.38	0.04	0.3	0.2	0.03	—	—	—	—	—	—
—	0.49	—	—	0.62	0.16	0.16	1.1	0.82	0.13	0.3	2.5	—	—	—	—
—	1.26	—	—	1.02	0.25	0.25	1.7	1.31	0.20	0.6	6.1	—	—	—	—
30	1.31	—	—	0.66	1.11	0.12	0.8	0.40	0.057	0.3	—	—	—	—	—
—	2.5	—	—	1.8	0.50	0.40	2.5	1.6	0.24	1.0	0.2	—	380	8 ± 3	7 ± 3
—	2.4	—	—	1.7	0.59	0.40	2.6	1.6	0.23	1.2	0.2	—	350	8 ± 3	8 ± 3
25	2.9	7.8	6.3	1.96	0.59	0.42	2.9	1.6	0.23	1.3	0.2	0.5	155	—	—
30	2.7	7.5	6.0	1.83	0.63	0.41	2.8	1.7	0.25	1.3	0.3	0.4	175	—	—
—	2.8	7.6	6.3	1.87	0.67	0.41	2.8	1.6	0.24	1.3	—	0.3	125	—	—
45	3.2	8.6	7.0	2.14	0.67	0.46	3.2	1.7	0.25	1.5	0.2	0.5	145	—	—
45	3.0	8.1	6.6	2.05	0.66	0.44	3.0	1.7	0.25	1.4	(0.5)	0.4	145	—	—

anticoincidence Ge(Li) counting facility of the Battelle Northwest Laboratory. Estimated errors due to counting statistics in the other fragments are: Al_2O_3, FeO, Na_2O, MnO (>0.1%), Cr_2O_3, Sc, Co, La (>1 ppm) and Sm, ±1–5%; CaO, MnO (<0.1%), Eu, Tb (>0.1 ppm) and Yb (>1 ppm), ±5–10%; TiO_2 (>1%), La (<1 ppm) and Lu (>0.1 ppm), ±10–15%; TiO_2 (<1%), MgO (<5%), Tb (<0.1 ppm), Dy (>1 ppm), Lu (<0.1 ppm), Hf and Ta, ±15–30%; MgO (<5%), Dy (<1 ppm) and Ni, ±30–50%.

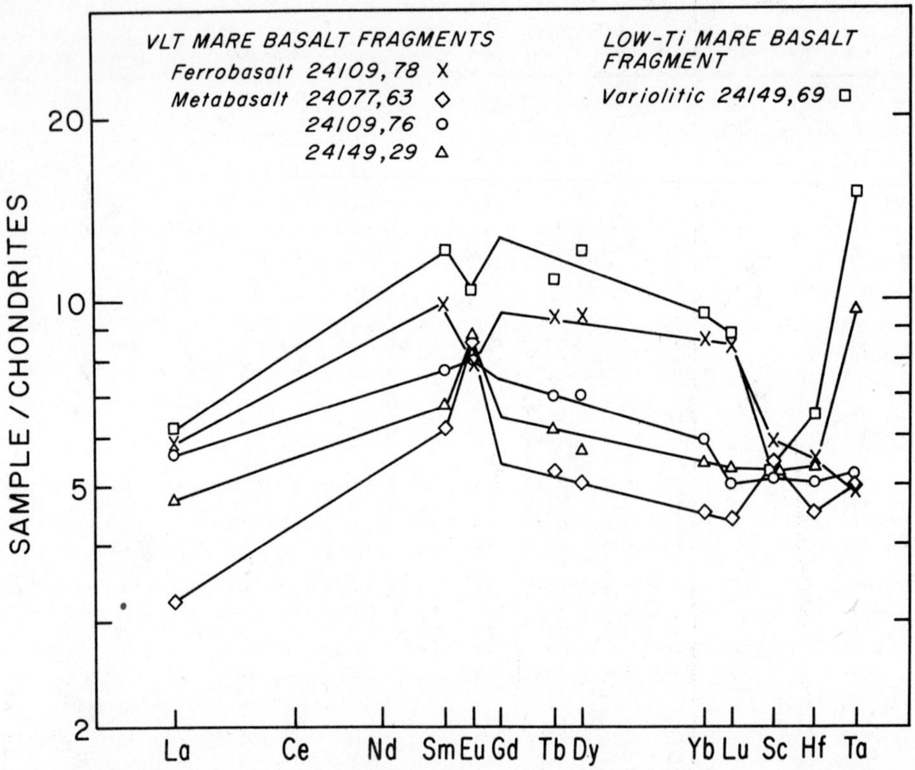

Fig. 4. The chondritic normalized REE, Sc, Hf and Ta patterns for the Luna 24 basalt fragments.

(1977b), and Laul (1977) for other Luna 24 metabasalt fragments. However, we did not observe the variations in composition from fragment to fragment reported by Ryder *et al.* (1977) and McSween *et al.* (1977). Also, their data for Al_2O_3, CaO, and Na_2O are generally higher and for FeO, MgO, TiO_2, and Cr_2O_3 generally lower than ours and those reported by Taylor *et al.* (1977b, 1978b), Vaniman and Papike (1977b), and Laul (1977).

Low-Ti mare basalt fragment

Variolitic basalt fragment 24149,69 is clearly distinct from other Luna 24 basalt fragments in texture, mineral and bulk composition. It is variolitic, with a grain size intermediate between that of the ferrobasalt and gabbro fragments (Fig. 1E). The thin section contains only pyroxene and plagioclase ($An_{89.5}Ab_{9.9}Or_{0.6}$), with two small (~10 μm) ilmenite crystals. Pyroxene compositional trends (Figs. 2D, 3C) are different from those of Luna 24 ferrobasalt and gabbro fragments: Ti increases as Fe/(Fe + Mg) increases (i.e., as crystallization proceeded), then begins to decrease. This pattern is unlike that of VLT mare basalt fragments, where Ti continuously increases as Fe/(Fe + Mg) increases (Vaniman and Papike, 1977c; Taylor *et al.*, 1977a). Such pyroxene compositions are more characteristic of basalts higher in Ti.

Bulk compositional data support the petrologic observations. Sample 24149,69 contains more TiO_2 (2.3 wt.%; Table 2) than any of the samples we studied. It also has high Na_2O content (0.55%, Table 2), resulting in plagioclase relatively high in Na_2O (~1.1%). Abundances of Eu and the heavy REE are also higher than in any of the Luna 24 ferrobasalt and metabasalt fragments. Thus, we assign this variolitic basalt fragment to a separate category distinct from the VLT mare basalt fragments and suggest referring to it as low-Ti mare basalt fragment.

Gabbro fragments

Coarse-grained rocks apparently make a significant contribution to the Luna 24 soils (Bence *et al.*, 1977; Ryder *et al.*, 1978; Taylor *et al.*, 1977b, 1978b). In fact, most of the fragments we studied are inferred to be from such coarse-grained, gabbroic rocks. Because of the generally small sizes of Luna 24 fragments collected, fragments of coarse-grained rocks are usually not representative samples but have highly variable proportions of the major minerals of the rock and, in fact, sometimes are fragments consisting of only one mineral, often in the form of a single crystal. However, we infer their gabbroic parentage on the basis of their compositions and grain sizes. Therefore, we classify our gabbroic samples into three categories: plagioclase-rich, clinopyroxene-rich, and mixed mineral fragments. In addition, we studied mineral separates from gabbro layer 24170.

Plagioclase-rich fragments 24077,83 and 24149,63 consist of a single plagioclase crystal ($An_{95.1}Ab_{4.7}Or_{0.2}$) and almost entirely of two plagioclase crystals ($An_{93.0}Ab_{6.8}Or_{0.2}$), respectively. 24149,63 also contains a few small (10–20 μm) patches of dark mesostasis. Both fragments have REE patterns characteristic of plagioclase, with large, positive Eu anomalies and depleted, heavy REE abundances (Fig. 5). Total REE concentrations are lower than those in other single Luna 24 plagioclase fragments analyzed by Blanchard *et al.* (1977) and Laul (1977) and also lower than those of plagioclase separated from gabbro 24170 (Fig. 6). In fact, REE abundances in 24077,83 and 24149,63 are among the lowest reported for lunar plagioclase (e.g. Laul *et al.*, 1974).

Clinopyroxene-rich fragments 24182,35, 24182,74, 24182,75, 24210,20. Our thin section of sample 24182,35 contains one pyroxene crystal, but we saw some plagioclase in the fragment when we examined it macroscopically before sectioning. Fragment 24182,74 consists of three roughly equant pyroxene crystals, with one plagioclase grain occurring between two of them. Sample 24182,75 contains one large (~1 mm) pyroxene crystal and a fine-grained mesostasis, composed of fayalite, silica, and Fe-rich pyroxene, occupies ~15% of the section. The section also contains one small olivine and two small plagioclase crystals.

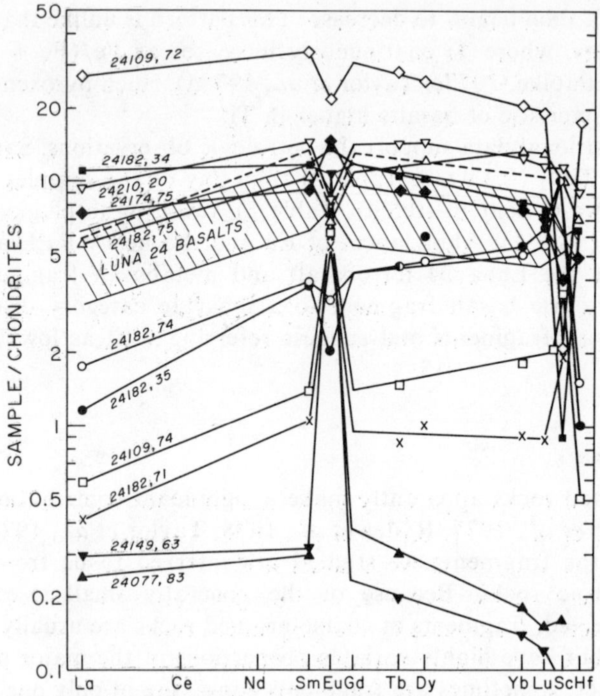

Fig. 5. The chondritic normalized REE, Sc and Hf patterns for the Luna 24 gabbro fragments: 24077,83 and 24149,63 are plagioclase-rich fragments; 24182,35, 24182,74, 24182,75 and 24210,20 are clinopyroxene-rich fragments; and 24109,72, 24109,74, 24174,75, 24182,34, and 24182,71 are mixed mineral fragments. The shaded field is for Luna 24 VLT Mare Basalt Fragments (from Fig. 4). Also shown is the calculated pattern of the magma from which the gabbros crystallized.

Sample 24210,20 is almost all pyroxene. Several crystals occur in a granular intergrowth. A long, narrow patch contains plagioclase and silica.

Except for 24182,75, these fragments have pyroxene compositions that plot in the same range and along the same trends as gabbro fragment 24170 (Figs. 2C, 3B). The only differences among the fragments is in the average pyroxene compositions and in the extent of zonation. Plagioclase compositions tend to be like those in 24170 as well ($An_{93\text{-}95}Ab_{4\text{-}7}Or_{0.1\text{-}0.3}$), although the plagioclase associated with silica in 24210,20 is quite sodic, $An_{73.0}Ab_{22.5}Or_{4.5}$ (average of 3 analyses). Sample 24182,75 differs slightly compared to the other pyroxene-rich samples. Its main distinguishing characteristic, other than the presence of an appreciable amount of mesostasis, is a pyroxene compositional trend to Fe-rich compositions that falls to the right of the 24170 field in Fig. 3C. Plagioclase compositions reflect the association with late-stage mesostasis ($An_{80.6}Ab_{16.8}Or_{2.5}$). One olivine crystal occurs outside of the mesostasis and has the composition Fo_8.

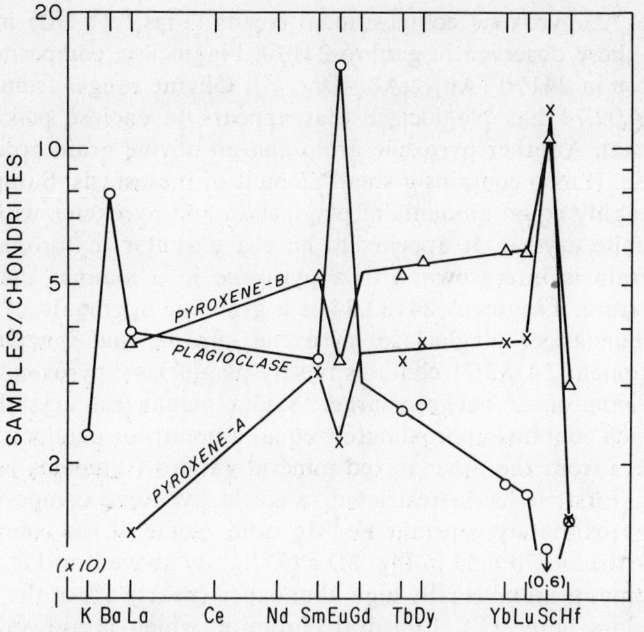

Fig. 6. The chondritic normalized K, Ba, REE, Sc and Hf patterns for the mineral separates of gabbro 24170.

Bulk compositions (Table 2) are similar between these four fragments, although some differences are noted and are probably largely due to sample heterogeneities as a result of coarse-grain sizes. All four fragments are characterized by low Al_2O_3 (2.3–3.3%), high FeO (24.9–28.5%) and Sc (79–91 ppm) (Table 2). TiO_2 abundances are comparable to those of pyroxenes in Luna 24 VLT ferrobasalt lithic fragments (e.g. Taylor *et al.*, 1977b). Compositions of fragments 24182,35 and 24182,74 are almost identical for major, minor, and trace elements (Table 2), and their REE, Sc and Hf patterns resemble that of the high-Ca clinopyroxene separate of 24170 (Fig. 6). Fragment 24182,75 has the highest bulk FeO content (28.5%; Table 2) of the four clinopyroxene-rich fragments, in keeping with its ferrosilite-rich clinopyroxene (Fig. 2D). It appears that this fragment represents a gabbro different from the typical Luna 24 gabbro, although its REE pattern nearly parallels those of 24182,35 and 24182,74, but shows higher REE concentrations (Fig. 5). Fragment 24210,20 has light REE abundances higher than 24182,75; however, the heavy REE abundances are slightly lower (Fig. 5). These differences in the REE patterns are attributed to differences in the mesostasis contents and the nature of the pyroxenes in the two fragments.

Mixed mineral fragments 24109,72, 24109,74, 24174,75, 24182,34, and 24182,71 are heterogeneous and have between 10 and 25 wt.% Al_2O_3. Except for sample 24109,72, pyroxene compositional trends (Figs. 2C, 3B) in these fragments follow those observed in gabbro 24170. Plagioclase compositions are also similar to those in 24170 ($An_{93\text{-}95}Ab_{4\text{-}7}Or_{0.1\text{-}0.3}$). Olivine ranges from Fo_{50} to Fo_3. Fragment 24109,74 has plagioclase that appears to enclose poikilitically one pyroxene crystal. Another pyroxene grain and an olivine grain occur outside of the plagioclase. It also contains a small amount of mesostasis. Sample 24174,75 consists of roughly equal amounts of plagioclase and pyroxene, with one olivine and one ilmenite crystal. It appears to have a granular texture, although one plagioclase grain is intergrown with a pyroxene in a manner that suggests a subophitic texture. Fragment 24182,34 is a granular aggregate of (in order of decreasing abundance) plagioclase, pyroxene, olivine, and ilmenite. Our thin section of fragment 24182,71 contains mostly plagioclase; pyroxene and a lesser amount of olivine occur between large, tabular plagioclase crystals. Fragment 24109,72, which contains approximately equal amounts of plagioclase and mafic silicates, differs from the other mixed-mineral gabbro fragments in two significant respects. First, it has a restricted range in pyroxene compositions, maintaining an approximately constant Fe/Mg ratio. Some of the compositions fall slightly below the 24170 field in Fig. 2D and slightly above it in Fig. 3C. Second, the sample contains an unusually high abundance (nearly 5% of the thin section) of opaque oxides (Fig. 1F). Euhedral chromite, which is always enclosed by plagioclase, has an Fe/(Fe + Mg) ratio of 0.60 to 0.76. Cr-Al ulvöspinel is subhedral and occurs between the larger silicate crystals; it has Fe/(Fe + Mg) of 0.94 to 0.96, 2.5 to 3.6% Al_2O_3, and 8.4 to 15.8% Cr_2O_3. Ulvöspinel is more abundant than chromite. Olivine (Fo_{43} to Fo_{32}) and plagioclase ($An_{93.5}Ab_{6.3}Or_{0.2}$) have similar compositions to those in the other mixed mineral fragments and in 24170.

Fragments 24109,74, 24182,34, and 24182,71 are characterized by high Al_2O_3 (21.5–24.9%) and CaO (14.5–16.6%) and low FeO (3.9–9.0%) contents. Thus, the bulk major element composition reflects the plagioclase-rich nature of these fragments. Similarly, REE abundances reflect the bulk composition for 24109,74 and 24182,71: They are low in REE and exhibit a large positive Eu anomaly (Fig. 5). In fact, bulk and trace element compositions of both fragments can be reproduced by mixing plagioclase- and clinopyroxene-rich gabbro fragments, for example, ~25% of clinopyroxene-rich fragments 24182,35 or 24182,74 with ~75% plagioclase-rich fragments 24077,83 or 24149,63. Thus, fragments 24109,74 and 24182,71 clearly are fragments from similar Luna 24 gabbros. Fragment 24182,34 has much higher REE abundances, a negative slope for the heavy REE (Dy-Lu) (Fig. 5) and low FeO and Sc abundances (Table 2). There is no petrographic evidence for KREEP-rich material in this sample that could explain the high REE abundances. Fragment 24174,75 has about equal amounts of plagioclase and clinopyroxene and, thus, its bulk major, minor, and trace element compositions resemble closely VLT ferrobasalt, of which the gabbros are a coarse-grained variety (Table 2, Fig. 5). The considerably lower V (46 ppm) and Cr_2O_3 (670 ppm) are noteworthy.

Although fragment 24109,72 consists of about equal proportions of plagioclase and mafic silicates and, thus, would also be expected to approach ferrobasalt in bulk composition, it is clearly different from that rock type (Table 2, Fig. 5). It is high in TiO_2, FeO and Cr_2O_3 and low in CaO and Al_2O_3, in keeping with the high abundance of Cr-Al ulvöspinel and chromite in this fragment. The REE and other LIL element abundances are the highest among all data reported for Luna 24 fragments. There is no petrographic evidence in this fragment for the presence of KREEP-rich material that could explain the high REE abundances and the negative heavy REE slope (Sm-Lu). Thus, mineralogic, petrographic, and bulk major, minor, and trace element data suggest that fragment 24109,72 represents a gabbro type different from the common Luna 24 VLT type.

Mineral separates from 24170. The mineralogy and petrology of gabbro layer 24170 has been described in detail by Albee *et al.* (1977), Bence *et al.* (1977), Taylor *et al.* (1978b) and Ryder *et al.* (1977). These authors agree that this coarse-grained rock crystallized from an Fe-rich, VLT mare basalt magma.

Clinopyroxene and plagioclase separates were prepared, weighed, and sealed at Cal. Tech. and are part of a consortium study by A. L. Albee (Albee *et al.*, 1977), G. J. Wasserburg (Papanastassiou *et al.*, 1977b), E. Anders and R. A. Schmitt. One plagioclase and two clinopyroxene separates were prepared by handpicking. INAA data (Table 2) for major and minor elements are systematically higher than the average electron microprobe analyses of clinopyroxene from 24170 reported by Taylor *et al.* (1977b, 1978b), resulting in a considerably lower calculated SiO_2 value in the INAA data. Since data on control samples (USGS standards BCR-1, PCC-1, GSP-1, and lunar soil) obtained simultaneously with those for the clinopyroxene separates during INAA agree well with the accepted literature values, we conclude that the differences between INAA and electron microprobe data are not in our analytical procedures. Furthermore, impurities cannot account for these differences: higher Al_2O_3 in the separates suggests only minor (<5%) plagioclase impurities, and the negative correlation of FeO with MgO in the clinopyroxene separates suggests that olivine is also not a significant contaminant (note that olivine from 24170 is very FeO-rich and MgO-poor; e.g. Taylor *et al.*, 1977b, 1978b). Thus, we conclude that the discrepancies in the INAA and microprobe data are due to weighing errors of the INAA samples. Comparison of the INAA data for the plagioclase separate (Table 2) with electron microprobe data for this phase (e.g. Taylor *et al.*, 1977b, 1978b), particularly the contents of FeO, MnO, and Cr_2O_3, indicate an impurity of ~7% clinopyroxene in the plagioclase separate.

The compositions of the two clinopyroxene separates are different (Table 2, Fig. 6). Since there is considerable zoning and grain to grain compositional variability in clinopyroxene from 24170 (e.g. Taylor *et al.*, 1977b, 1978b), it is likely that the compositional differences among the clinopyroxene separates are due to sampling of the coarse-grained, heterogeneous rock. The REE patterns for the two clinopyroxene separates are similar and nearly parallel to each other (Fig. 6). They are also similar to those of clinopyroxene-rich fragments 24182,35 and 24182,74 (Fig. 5). The higher REE and Hf abundances observed in separate B relative to A are consistent with the higher FeO content of the former. The higher Cr_2O_3 and V contents in the clinopyroxene separates compared to clinopyroxene-rich fragments 24182,35 and 24182,74 (Table 2) are attributed to accessory spinel in the former.

The REE pattern for the plagioclase separate is similar to that of low-K type anorthositic coarse fines 14161,35,2 (Hubbard *et al.*, 1971). Since the mineral compositions of the Luna 24 gabbro and ferrobasalt fragments are nearly identical (Grove and Bence, 1977) and since the REE distributions in the 24170 clinopyroxene separates are similar to those in the clinopyroxene of the gabbro fragments, it is difficult to interpret the high REE abundances in the plagioclase separate. Inclusion of ~10% of late-stage liquid derived by extensive fractional crystallization (~85-90%) of a ferrobasalt-like liquid could account for the observed REE abundances, but such large amounts of trapped liquid do not seem reasonable.

Melt rock and soil breccia fragments

Melt rock fragment 24149,61 is vesicular and consists of olivine and pyroxene clasts in a fine-grained, subophitic groundmass. Metallic nickel-iron is notably more abundant (~1%) than in Luna 24 ferrobasalt fragments. Major and minor element abundances (Table 2, Fig. 7) are very similar to those of Luna 24 soils. It

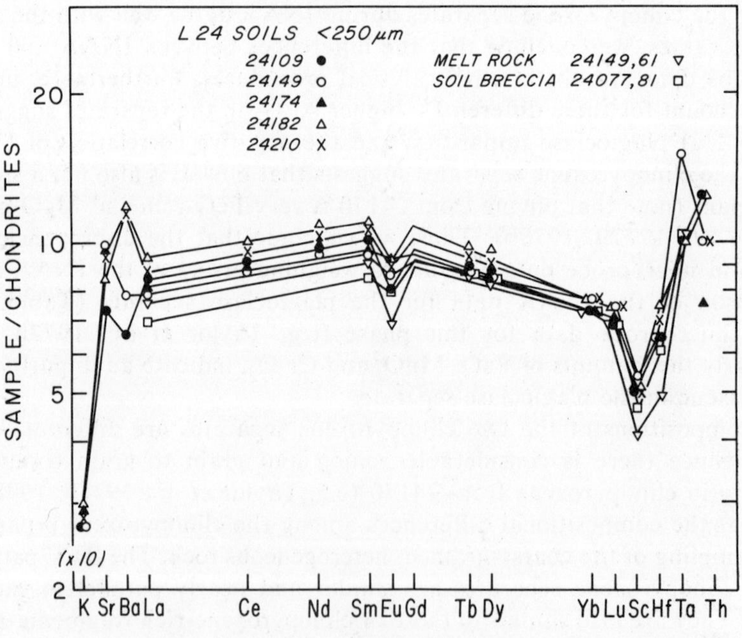

Fig. 7. Chondritic normalized K, Sr, Ba, REE, Sc, Hf, Ta and Th patterns for the Luna 24 soils.

appears that sample 24149,61 formed by shock melting of the local soil at the Luna 24 site.

Soil breccia fragment 24077,81 contains pyroxene, olivine and plagioclase clasts, set in a brown, glass-rich matrix. Its bulk composition (Table 2, Fig. 7) is essentially identical to those of Luna 24 soils. Thus, we interpret this fragment as a soil breccia.

Soils

Our joint min. pet. (UNM) - INAA (OSU) project did not include mineralogic-petrologic soil studies, although a separate allocation of 250–500 μm soil fractions was studied petrographically at UNM. Here we report the results of INAA studies of five Luna 24 soils in the <250 μm size range from 5 different stratigraphic levels of the core that were allocated to the OSU group as part of the Schmitt-Wasserburg consortium (Ma and Schmitt, 1977a).

The major, minor, and trace element abundances are surprisingly similar in these soils, regardless of their locations in the core (Table 2, Fig. 7). This is also observed for the <74 μm fines (Barsukov *et al.*, 1977) and for the bulk soils and the corresponding size fractions (Blanchard *et al.*, 1977). The REE patterns show a bow-shaped fractionation with a negative Eu anomaly, and the light REE are slightly higher than are the heavy REE (Fig. 7). Our REE abundances are similar to those obtained for the 90–150 μm fractions (Blanchard *et al.*, 1977) and are consistently lower with smaller Eu anomalies than those reported by Barsukov *et al.* (1977) for the <74 μm fines. The Luna 24 VLT soils are the lowest in LIL trace elements compared to those collected at the six Apollo and two other Luna landing sites. For example, REE abundances in Luna 16 soils (Gillum *et al.*, 1972) in Mare Fecunditatis, which is ~400 km south of the Luna 24 landing site, are ~4 times those found in Luna 24 soils.

Similarities in major, minor, and trace element abundances in the Luna 24 soils throughout the core, combined with similarities in compositions of basaltic fragments at different stratigraphic levels, suggest that the Luna 24 core (and site) is dominated by a single group of VLT mare basalts and mechanical derivatives thereof. However, a slightly higher MgO content is noted in the soils (~10%) in comparison to the VLT ferrobasaltic fragments (~7%) (Table 2). Monomineralic fragments of Mg-rich pyroxene and olivine (Basu *et al.*, 1977), olivine-rich lithic fragments, olivine vitrophyre and Mg-rich glasses (Taylor *et al.*, 1977b, 1978b) found in the soil may account for this difference.

DISCUSSION

REE pattern of Luna 24 gabbros

None of our gabbro fragments is large enough to be representative of the gabbroic lithologies present at Luna 24 and, thus, the measured REE patterns are biased by sampling. Consequently, we attempted to calculate the REE pattern of the magma from which plagioclase-rich fragments 24077,83 and 24149,63 crystallized. We used these plagioclase fragments instead of the clinopyroxene-rich fragments because the latter are not pure clinopyroxene and are compositionally zoned. The plagioclases, in contrast, are homogeneous in composition and essentially pure plagioclase. Using crystal-liquid partition coefficients (McKay and Weill, 1976; Haskin and Korotev, 1977), we calculated the parent magma composition by assuming equilibrium crystallization. (Calculations assuming closed system crystallization are not necessary because our samples were single crystals of plagioclase, not aggregates from crushed rock samples, and show no petrographic evidence of closed system crystallization such as adcumulus overgrowths or Na-rich rims).

The results of the calculations are shown in Fig. 5. The pattern is similar to that of Luna 24 VLT ferrobasalt fragments and, like ferrobasalt fragment 24109,78, the calculated liquid has a negative Eu anomaly. Laul *et al.* (1978) and Blanchard *et al.* (1977) also calculated REE patterns for the gabbroic magma from plagioclase crystals. Their results are similar to ours: The patterns are parallel to Luna 24 ferrobasalt patterns and have negative Eu anomalies.

Magma types at Luna 24

The data indicate that there are at least four, and possibly five, distinct magma types represented as rock fragments in the Luna 24 soil. The gabbro fragments and our fragment of ferrobasalt (24109,78) seem to be from the same magma type, characterized by a slightly bow-shaped REE pattern, with low concentrations of REE (5–10 × chondrites) and a negative Eu anomaly. Pyroxene compositions (Grove and Bence, 1977; Taylor *et al.*, 1977b) also indicate that most Luna 24 ferrobasalt and gabbro fragments are related.

A *possible* second magma type is represented by metabasalt and by ophitic ferrobasalt fragment 24174,7 (Laul *et al.*, 1978). Major and trace element fragment concentrations are quite similar to our ferrobasalt fragment 24109,78, except that fragments in this second group have *positive* Eu anomalies. It is critical to our understanding of the petrogenesis of VLT ferrobasalts that we determine whether this difference in Eu concentrations is inherent in the magmas or whether it is due to a process that took place after the magma had formed in the lunar interior.

There are a number of explanations for the existence of both positive and negative Eu anomalies. They could be due to analytical uncertainties. Since the Luna 24 samples have low (5–10 × chondrites) concentrations of the REE, and since the samples are small (~0.5 mg), analytical uncertainties are relatively high. However, the differences in Eu concentrations in, for example, ferrobasalt fragments 24109,78 (0.58 ppm) and 24174,7 (0.83 ppm) are well outside of experimental errors (±0.06 ppm for our data). It appears therefore, that the measured Eu anomalies are real. Is it possible, then, that they are due to sampling errors? For example, the samples with positive Eu anomalies might be preferentially enriched in plagioclase. This seems unlikely because all samples in question (ferrobasalt and metabasalt fragments) are very fine-grained and they have nearly identical Al_2O_3 contents. The same arguments also negate the idea that the fragments with positive Eu anomalies assimilated plagioclase.

Laul *et al.* (1978) propose that the positive Eu anomalies might arise during normal differentiation of a ferrobasalt magma. As crystallization proceeds, they argue, release of volatiles results in more reducing conditions, hence higher Eu^{2+}/Eu^{3+}. Since plagioclase readily incorporates Eu^{2+}, the bulk fragment analysis will show an increase in Eu content with increasing degree of differentiation. To support the idea, Laul *et al.* (1978) show a plot of Fe/(Fe + Mg) vs Eu for ferrobasalt and metabasalt whole-fragment analyses, in which the two parameters are directly correlated. However, the correlation is entirely dependent on one sample, ferrobasalt fragment 24174,7, which has the highest Eu and Fe/(Fe + Mg); the other samples have Eu contents within experimental uncertainties of each other. We conclude, therefore, that the differences in Eu contents are just as likely due to differences in magma composition as to reduction.

Gabbro fragment 24174,75 seems to represent another magma type. Its petrography and mineral composition are not notably distinctive compared to other gabbro fragments, but it contains considerably less V (46 ppm vs. >150 ppm) and Cr_2O_3 (670 ppm vs. >2000 ppm) than ferrobasalt and metabasalt fragments. Sc and Co contents are only slightly lower. These differences cannot be due to sampling errors. We suggest that sample 24174,75 represents a distinctive VLT mare basalt magma.

Gabbro fragment 24109,72 is also likely to be from a different magma than the ones from which most Luna 24 ferrobasalts crystallized. Two parameters distinguish it from other Luna 24 basaltic lithologies. First, it has a unique pyroxene compositional trend (Fig. 3C), characterized by increasing TiO_2 with nearly constant Fe/(Fe + Mg). Second, it has the highest REE contents of any Luna 24 mare basaltic sample (Fig. 5).

Finally, sample 24149,69 is variolitic, has a unique pyroxene compositional trend (Fig. 3C), higher bulk TiO_2 (2.3 wt.%) than any other fragment, and the most bow-shaped REE pattern of any basaltic sample (Fig. 4). It appears that it represents a low-Ti mare basalt type, distinct from the VLT basalts.

Petrogenesis of VLT mare basalts

VLT mare basalts are present at Apollo 17 (Vaniman and Papike, 1977c; Taylor *et al.*, 1977a) and Luna 24. REE and other trace element data are available for both suites, although the data for Apollo 17 VLT mare basalts are limited to only one sample, vitrophyre 78526 (Laul and Schmitt, 1975; Murali *et al.*, 1977; Warner *et al.*, 1978). In addition, trace element data have been reported for Apollo 15 green glass (Taylor *et al.*, 1973; Ma and Schmitt, unpublished data), which has been suggested to be related to VLT mare basalts. Fig. 8 shows the REE patterns for Luna 24 ferrobasalt (both positive and negative Eu anomalies), rock 78526, and green glass 15426.

Grove and Vaniman (1977) and Norman *et al.* (1977) suggest that Luna 24

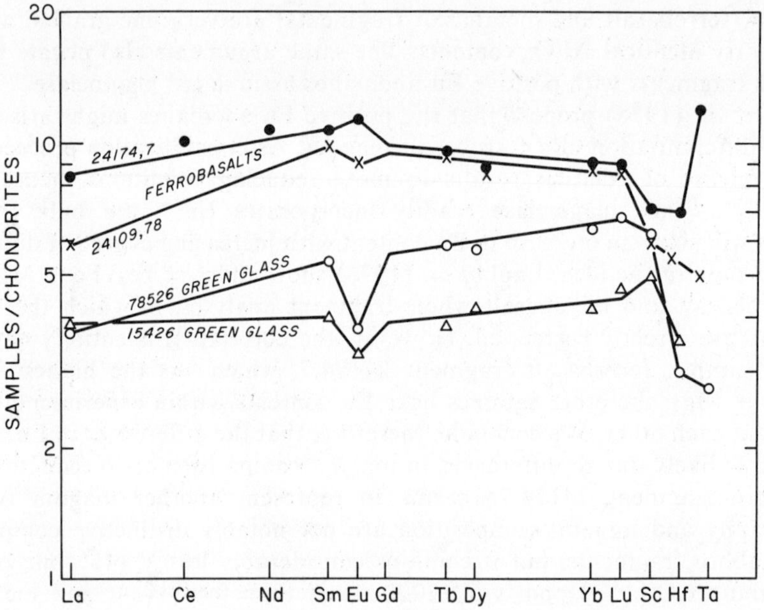

Fig. 8. Chondritic-normalized REE patterns for Luna 24 ferrobasalt lithic fragments 24109,78 (this paper) and 24174,7 (Laul *et al.*, 1978), VLT basalt 78526 (Warner *et al.*, 1978), and green glass 15426 (our unpubl. data).

ferrobasalts could have been derived from Apollo 15 green glass by fractional crystallization. However, the data in Fig. 8 prove that a magma identical in composition to Apollo 15 green glass could not fractionate olivine or low-Ca pyroxene to yield Luna 24 ferrobasalts. Such a fractionation scheme would not change the REE pattern appreciably, except to raise the concentrations of the light REE relative to the heavy REE, and would not match the bow-shaped Luna 24 REE pattern. Since Apollo 15 green glass and Luna 24 ferrobasalts have distinctly different REE patterns, no relationship via fractional crystallization exists. This interpretation is also supported by experimental results in which the liquid line of descent from Apollo 15 green glass does not pass through the Luna 24 (or Apollo 17 VLT) compositions, particularly on plots involving CaO/Al_2O_3 (Grove and Vaniman, oral presentation at the Conference on Luna 24). The hypothesis is also not supported by theoretical calculations involving the removal of liquidus olivine (Taylor *et al.*, 1978a). Apollo 15 green glass and VLT mare basalts are almost certainly broadly related, but there is no direct genetic relationship between the samples studied to date.

Taylor *et al.* (1977a) proposed that Apollo 17 VLT mare basalts, such as vitrophyre 78526, could have fractionated olivine and low-Ca pyroxene to yield magmas like Luna 24 ferrobasalts. However, detailed mixing models (Taylor *et al.*, 1977c) rule this out. Differences in the REE patterns of 78526 and Luna 24 ferrobasalts (Fig. 8) also rule out the notion that 78526 represents a parent magma for the Luna 24 ferrobasalts. Other Apollo 17 VLT mare basalt compositions may be suitable (Taylor *et al.*, 1978a), but, unfortunately, the appropriate compositions are derived from analyses of glass fragments in Apollo 17 soils and soil breccias for which no REE data are available.

We can use trace element data to infer the mineralogical character of the source regions of the Luna 24 VLT ferrobasalts. As Fig. 4 shows, the ferrobasalts vary in REE contents (e.g., La varies by a factor of ~3), yet they have nearly constant Sc concentrations. The most straightforward interpretation of these data is that the Luna 24 ferrobasalts formed by varying degrees (up to ~50%) of batch equilibrium partial melting of a source that contained clinopyroxene in the solid residue. We reason as follows: Assuming the Luna 24 ferrobasalts derived from a common source, the range in REE contents indicates that the percentage of partial melting must have varied because mineral-liquid partition coefficients for the REE are usually <1 (e.g. D_{La}(cpx) is ~0.1). Sc contents would also have varied, however, unless clinopyroxene remained in the residue. This is a consequence of Sc having a clinopyroxene-melt partition coefficient greater than 1 (D_{Sc}(cpx) is ~1.6; Leeman, 1974); once clinopyroxene is exhausted from the solid, Sc is partitioned readily into the melt and its concentration (like those of the REE) will depend on the fraction of melt relative to solid. By similar reasoning, the nearly constant Eu abundances in Luna 24 ferrobasalts (Fig. 4, ~8X chondrites) suggests that a small amount of plagioclase remained in the residue during the batch equilibrium partial melting episodes that formed these basalts. It appears, therefore, that the source region of the the Luna 24 VLT ferrobasalts contained both clinopyroxene and plagioclase in the residue.

The REE distribution in the Luna 24 source material that generated gabbroic rock 24170 may be estimated by applying Rb/Sr and Sm/Nd constraints. Using data reported by Papanastassiou *et al.* (1977b) and the equations given by Shih *et al.* (1975) and Papanastassiou *et al.* (1977a), we calculated the relative Rb-Sr and Sm-Nd fractionations (F) between the basalt and its source at crystallization time. (If F = 1, no fractionation took place). We estimated the whole rock ^{147}Sm/^{144}Nd by assuming that the rock contains 70% pyroxene and 30% plagioclase (which was estimated by D. Papanastassiou from his mineral separations of 24170). The calculated F is slightly greater than one for the relative Sm-Nd fractionation and slightly less than one for the relative Rb-Sr fractionation. It is difficult to explain these F's by a crystal-liquid fractionation process involving major mineral phases such as clinopyroxene, orthopyroxene, and olivine (Shih *et al.*, 1975) because for each mineral $D_{Rb}(s/1) < D_{Sr}(s/1$, and $D_{Nd}(s/1) < D_{Sm}(s/1)$ (Shih *et al.*, 1975; Shih and Schonfield, 1976). However, the calculated F is extremely sensitive to the uncertainties in the measurements used to estimate the fractionation between basalt and its source; F = ~2 can be obtained by using the extremes for the Rb-Sr fractionation. The precise modal mineralogy of rock 24170 is difficult to determine (Albee *et al.*, 1977); F is slightly less than one for Sm-Nd fractionation if we assume a different modal mineralogy for the whole rock (e.g., assuming 70% plagioclase and 30% pyroxene).

From these estimations and a consideration of the ranges in fractionation allowed by these isotopic constraints, we suggest that an insignificant degree of fractionation for the REE and Rb-Sr between the basaltic magma and its source occurred during the formation of the Luna 24 VLT ferrobasaltic magma. Therefore, we suggest that the Luna 24 VLT ferrobasalts and gabbros formed by large degrees of melting of a fractionated source material (probably a cumulate) in which possibly olivine and orthopyroxene, with small amounts of clinopyroxene and plagioclase (see argument above), remained in the residue.

Acknowledgments—This work was supported by NASA grants NGL-32-002-039 and NGL-32-004-063. We thank G. J. Wasserburg and D. A. Papanastassiou for providing mineral separates and acknowledge the assistance of S. Gunn and A. Meyer at the University of Missouri research reactor, and T. V. Anderson, W. T. Carpenter and S. L. Bennett at the Oregon State University TRIGA reactor for neutron activations, and J. R. Budahn for help in the INAA work.

REFERENCES

Albee A. L., Chodos A. A. and Dymek R. F. (1977) Petrography of Luna 24 sample 24170 (abstract). In *Papers Presented to the Conference on Luna 24*, p. 5–7. The Lunar Science Institute, Houston.

Barsukov V. L., Tarasov L. S., Dmitriev L. V., Kolesov G. M., Shevaleevsky I. D. and Garanin A. V. (1977) The geochemical and petrographic features of regolith and rocks from Mare Crisium (preliminary data). *Proc. Lunar Sci. Conf. 8th*, p. 3319–3332.

Basu, A., McKay D. S. and Fruland R. M. (1977) Petrography, mineralogy and source rocks of Luna 24 drill core soil (abstract). In *Papers Presented to the Conference on Luna 24*, p. 18-21. The Lunar Science Institute, Houston.

Bence A. E. and Albee A. L. (1968) Empirical correction factors for the electron microanalysis of silicates and oxides. *J. Geol.* **76**, 382–403.

Bence, A. E., Grove T. L. and Scambos T. (1977) Gabbros from Mare Crisium: An analysis of the Luna 24 soil. *Geophys. Res. Lett.* **4**, 493–496.

Blanchard D. P., Haskin L. A., Brannon J. C. and Aaboe E. (1977) Chemistry of soils and particles from Luna 24 (abstract). In *Papers Presented to the Conference on Luna 24*, p. 37–40. The Lunar Science Institute, Houston.

Gillum D. E., Ehmann W. D., Wakita H. and Schmitt R. A. (1972) Bulk and rare earth abundances in the Luna-16 soil levels A and D. *Earth Planet. Sci. Lett.* **13**, 444–449.

Grove T. L. and Bence A. E. (1977) Petrogenesis of gabbros from Mare Crisium (abstract). In *Papers Presented to the Conference on Luna 24*, p. 64–67. The Lunar Science Institute, Houston.

Grove T. L. and Vaniman D. T. (1977) Experimental petrology of very low Ti basalts and origin of Luna 24 ferrobasalt (abstract). In *Papers Presented to the Conference on Luna 24*, p. 68–71. The Lunar Science Institute, Houston.

Haskin L. A. and Korotev R. L. (1977) Test of a model for trace element partition during closed-system solidification of a silicate liquid. *Geochim. Cosmochim. Acta.* **41**, 921–939.

Hubbard N. J., Gast P. W., Meyer C., Nyquist L. E. and Shih C. (1971) Chemical composition of lunar anorthosites and their parent liquid. *Earth Planet. Sci. Lett.* **13**, 71–75.

Keil K. (1967) The electron microprobe X-ray analyzer and its application in mineralogy. *Fortschr. Mineral.* **44**, 4–66.

Laul J. C. (1977) Chemical composition of Luna 24 melt rocks. (24077,4; 24077,62; 24174,7; 24182,12; 24210,50) and gabbro (24182,8) (abstract). In *Papers Presented to the Conference on Luna 24*, p. 94–97. The Lunar Science Institute, Houston.

Laul J. C., Hill D. W. and Schmitt R. A. (1974) Chemical studies of Apollo 16 and 17 samples. *Proc. Lunar Sci. Conf. 5th*, p. 1047–1066.

Laul J. C. and Schmitt R. A. (1973) Chemical composition of Luna 20 rocks and soil and Apollo 16 soils. *Geochim. Cosmochim. Acta* **37**, 927–942.

Laul J. C. and Schmitt R. A. (1975) Chemical composition of Apollo 17 samples: Boulder breccias (2), rake breccias (8), and others (abstract). In *Lunar Science VI*, p. 489–491. The Lunar Science Institute, Houston.

Laul J. C., Vaniman D. T. and Papike J. J. (1978) Chemistry, mineralogy and petrology of seven > 1mm fragments from Mare Crisium. In *Mare Crisium: The View from Luna 24* (R. B. Merrill and J. J. Papike, eds.). This volume.

Leeman W. P. (1974) Petrology of basaltic lavas from the Snake River plain, Idaho (Part I); Experimental determination of partitioning of divalent cations between olivine and basaltic liquid (Part II). Ph.D. dissertation, Univ. of Oregon, Eugene, Oregon.

Ma M.-S. and Schmitt R. A. (1977a) Luna 24 soils: a chemical study (abstract). In *Papers Presented to the Conference on Luna 24*, p. 98–101. The Lunar Science Institute, Houston.

Ma M.-S., Schmitt R. A., Taylor G. J., Warner R. D., Lange D. E., and Keil K. (1977b) Chemical and petrographic studies of 18 Luna 24 lithic fragments (abstract). In *Papers Presented to the Conference on Luna 24*, p. 102–105. The Lunar Science Institute, Houston.

McKay G. A. and Weill D. F. (1976) Petrogenesis of KREEP. *Proc. Lunar Sci. Conf. 7th*, p. 2427–2447.

McSween H. Y. Jr., Taylor L. A. and Clark J. C. (1977) Luna 24 metabasalts: possible evidence for assimilation of highland material? (abstract) In *Papers Presented to the Conference on Luna 24*, p. 118–120. The Lunar Science Institute, Houston.

Murali A. V., Ma M.-S., Laul J. C., and Schmitt R. A. (1977) Chemical composition of breccias, feldspathic basalt and anorthosites from Apollo 15 (15308, 15359, 15382, and 15362), Apollo 16 (60618 and 65785), Apollo 17 (72435, 72536, 72559, 72735, 72738, 78526 and 78527) and Luna 20 (22012 and 22013) (abstract). In *Lunar Science VIII*, p. 700–702. The Lunar Science Institute, Houston.

Norman M., Coish R. A., and Taylor L. A. (1977) Glass chemistry and magma evolution at Mare Crisium (abstract). In *Papers Presented to the Conference on Luna 24*, p. 136–138. The Lunar Science Institute, Houston.

Papanastassiou D. A., DePaolo D. J. and Wasserburg G. J. (1977a) Rb-Sr and Sm-Nd chronology and genealogy of mare basalts from the Sea of Tranquillity. *Proc. Lunar Sci. Conf. 8th*, p. 1639–1672.

Papanastassiou D. A., McCulloch M. T., DePaolo D. J., Huneke J. C. and Wasserburg G. J. (1977b) Ages and evolution of Luna 24 samples (abstract). In *Papers Presented to the Conference on Luna 24*. The Lunar Science Institute, Houston.

Ryder G., McSween H. Y. Jr. and Marvin U. B. (1977) Luna 24 VLT basalt: character and origin (abstract). In *Papers Presented to the Conference on Luna 24*, p. 160–163. The Lunar Science Institute, Houston.

Ryder G., McSween H. Y. Jr. and Marvin U. B. (1977) Basalts from Mare Crisium. *The Moon*, **17**, 263–287.

Shih C.-Y., Haskin L. A., Wiesmann H., Bansal B. M. and Brannon J. C. (1975) On the origin of high-Ti mare basalt. *Proc. Lunar Sci. Conf. 6th*, p. 1255–1285.

Shih C.-Y. and Schonfeld E. (1976) Mare basalt genesis: A cumulate-remelting model. *Proc. Lunar Sci. Conf. 7th*, p. 1757–1792.

Taylor G. J., Keil K. and Warner R. D. (1977a) Very low-Ti mare basalts. *Geophys. Res. Lett.* **4**, 207–210.

Taylor G. J., Warner R. D., Wentworth S., Keil K. and Sayeed U. (1977b) Petrology and chemistry of Luna 24 mare basalts and basaltic glasses (abstract). In *Papers Presented to the Conference on Luna 24*, p. 189–192.

Taylor G. J., Warner R. D., and Keil K. (1977c) Chemical distinctions among very low-Ti mare basalts (abstract). In *Papers Presented to the Conference on Luna 24*, p. 183–185. The Lunar Science Institute, Houston.

Taylor G. J., Warner R. D., and Keil K. (1978a) VLT mare basalts: Impact mixing, parent magma types, and petrogenesis. In *Mare Crisium: The View from Luna 24* (R. B. Merrill and J. J. Papike, eds.). This volume.

Taylor G. J., Warner R. D., Wentworth S., Keil K., and Sayeed U. (1978b) Lunar 24 lithologies: Petrochemical relationships among lithic fragments, mineral fragments, and glasses. In *Mare Crisium: The View from Luna 24* (R. B. Merrill and J. J. Papike, eds.). This volume.

Taylor S. R., Gorton M. P., Muir P., Nance W., Rudowski R. and Ware N. (1973) Lunar highlands composition: Apennine Front. *Proc. Lunar Sci. Conf. 4th*, p. 1445–1459.

Vaniman D. T. and Papike J. J. (1977a) Ferrobasalts from Mare Crisium: Luna 24. *Geophys. Res. Lett.* **4**, 497–500.

Vaniman D. T. and Papike J. J. (1977b) Luna 24 ferrobasalt: petrology and comparison with other mare basalt types (abstract). In *Papers Presented to the Conference on Luna 24*, p. 196–199. The Lunar Science Institute, Houston.

Vaniman D. T. and Papike J. J. (1977c) Very low Ti (VLT) basalts: A new mare rock type from the Apollo 17 drill core. *Proc. Lunar Sci. Conf. 8th,* 1443–1471.

Wakita H., Schmitt R. A. and Rey P. (1970) Elemental abundances of major, minor, and trace elements in Apollo 11 lunar rocks, soil and core samples. *Proc. Apollo 11 Lunar Sci. Conf.*, p. 1685–1717.

Warner R. D., Taylor G. J., Keil K., Planner H. N., Nehru C. E., Ma M.-S. and Schmitt R. A. (1978) Green glass vitrophyre 78526: An impact melt of very low-Ti mare basalt composition. *Proc. Lunar Planet. Sci. Conf. 9th.* In press.

Merrill, R. B. and Papike, J. J., editors.
(1978) *Mare Crisium: The View from Luna 24*, Pergamon Press (New York), p. 593–611
Printed in the United States of America

Trace element composition of Luna 24 Crisium VLT basalt

LARRY A. HASKIN
Dept. of Earth and Planetary Sciences and McDonnell Center for Space Sciences,
Washington University, St. Louis, Missouri 63130.

Abstract—Thirty-four fragments of Luna 24 material weighing between 0.2 and 9.1 mg have been analyzed for trace elements by four laboratories. From these analyses, data from 15 particles were used to estimate concentrations in Mare Crisium VLT basalt for REE, Co, Sc, FeO, and Na_2O. Only one fragment is clearly of non-mare origin; five fragments are soil breccias or agglutinates, with compositions similar to those of Luna 24 <250 μm fines. All other particles appear to be lithic fragments whose compositional characteristics are consistent with an origin from Crisium VLT magma. They all have recognizable counterparts in fragments and mineral separates from crushed 70135 mare basalt and also in 1–2 mm basalt fragments from Apollo 17 soils. The estimated REE concentrations (except that of Eu) in Crisium VLT basalt are lower and the light REE are relatively depleted, compared with the Luna 24 fines. This is consistent with known effects of differential comminution of basalts and with selective incorporation of non-mare material into mare-derived soils. There is weak and ambiguous evidence for a small (~10%) positive Eu anomaly relative to chondrites in Crisium VLT basalt; uncertainties in sampling and analysis are sufficient that there might be no Eu anomaly or even a slight negative one. The fragment data do not require more than one parent basalt trace element composition.

INTRODUCTION

The purpose of this paper is to consider the origins of the individual particles analyzed from the Luna 24 core and the information they provide on the trace element composition of Mare Crisium basalt. Acquisition of lunar soil from southeastern Mare Crisium by the Luna 24 mission has provided mare basalt from a previously unsampled, major basin (e.g., Barsukov, 1977; Butler and Morrison, 1977). From petrographic and chemical study it has been inferred that soils from the Luna 24 core consist overwhelmingly of mare basalt fragments (e.g., Barsukov, 1977; Barsukov *et al.*, 1977; Tarasov *et al.*, 1977; Taylor *et al.*, 1977; Vaniman and Papike, 1977; Bence *et al.*, 1977; Ryder *et al.*, 1977; Basu *et al.*, 1977; Coish and Taylor, 1977; Simonds *et al.*, 1977). Blanchard *et al.* (1977) discuss the chemical composition of Luna 24 soil and review a variety of other experimental information. They conclude that, in contrast to most other mare sites, Luna 24 soil compositions mainly reflect local mare basalt composition, consistent with the petrographic evidence cited above. Blanchard *et al.* (1977) also indicate clearly that the soils are mixtures and not representative of bulk parent basalt. Bence and Grove (1977) state that the total abundance of fragments of highlands origin in the samples they examined petrographically did not exceed ~2%.

This paper will refer to the parent of the Luna 24 soil fragments as Crisium

VLT (very low Ti) basalt, after the nomenclature of Vaniman and Papike (1977). Most lithic fragments in the Luna 24 core are classified as ferrogabbro, with TiO_2 contents lower than observed for the principal mare basalts sampled at Apollo landing sites. On the basis of published spectral evidence (Pieters *et al.*, 1976; Pieters and McCord, 1977) it is unclear how representative of Crisium basalts the Luna 24 material is.

To what extent can the concentrations and relative abundances of these elements in the basalts be inferred from studies on sized soil fractions and individual small fragments from the Luna 24 core? Do the trace element data require more than a single parent basalt composition? Can contributions of trace elements to the soil from the local mare basalt be separated from those from non-mare materials or meteorites?

Available data for trace elements in Luna 24 fragments

Previously we reported preliminary trace-element data for 5 fragments of Luna 24 soil (Blanchard *et al.*, 1977). A companion paper to the present one (Blanchard *et al.*, 1978) presents final trace element data for 6 Luna 24 fragments ranging in weight from 0.5 to 7.6 mg, analyzed by instrumental neutron activation. (That paper also presents major and trace element data for sized fractions of 6 Luna 24 soils, bulk <250 μm, 90–150 μm, <20 μm.) Figure 1 shows the REE abundances of the 6 fragments in comparison with an average for chondrites. Note the considerable deviation of several of the samples from the soil average (Blanchard *et al.*, 1978).

Particle 24077,17 is identical in trace element composition to the matrix of typical Apollo 17 boulders, e.g., 73215 (Blanchard *et al.*, 1976). In thin section it is clearly recognizable as non-mare material (G. Taylor, pers. comm.). It is the only fragment among the 34 for which there are trace element data published that is clearly of non-mare origin.

Particle 24149,5 is a soil breccia (Nagle and Walton, 1977) and its REE and other trace element abundances closely resemble those of average Luna 24 soil. Particle 24077,21 is an agglutinate and particle 24077,12 a recrystallized breccia (Nagle and Walton). Particle 24077,21 is low compared to the range for 6 bulk soils in FeO, Sc, Co, Hf, and heavy REE and slightly enriched in Na_2O. 24077,12 is enriched relative to the range for the bulk soils in FeO and Sc, and depleted in Cr, Co, Hf, and REE (except Eu). Particle 24174,5 is a typical Crisium VLT basalt fragment (ferro-gabbro, G. Taylor, pers. comm.). Its REE distribution reflects the presence of plagioclase and pyroxene. Compared with the bulk soils it is depleted in FeO, Cr, Sc, Co, Ni, Hf, and REE (except Eu) and is enriched in Na_2O. Particle 24077,19 is a VLT gabbro fragment consisting of nearly pure plagioclase. Relative to the bulk soils it is depleted in all elements determined except Na_2O and Eu. The characteristic plagioclase REE distribution and the FeO and Sc contents of only 0.77% and 1.8 ppm limit the amount of pyroxene to $\lesssim 2\%$.

Ma and Schmitt (1977) reported compositional data for 5 soils and Ma *et al.*

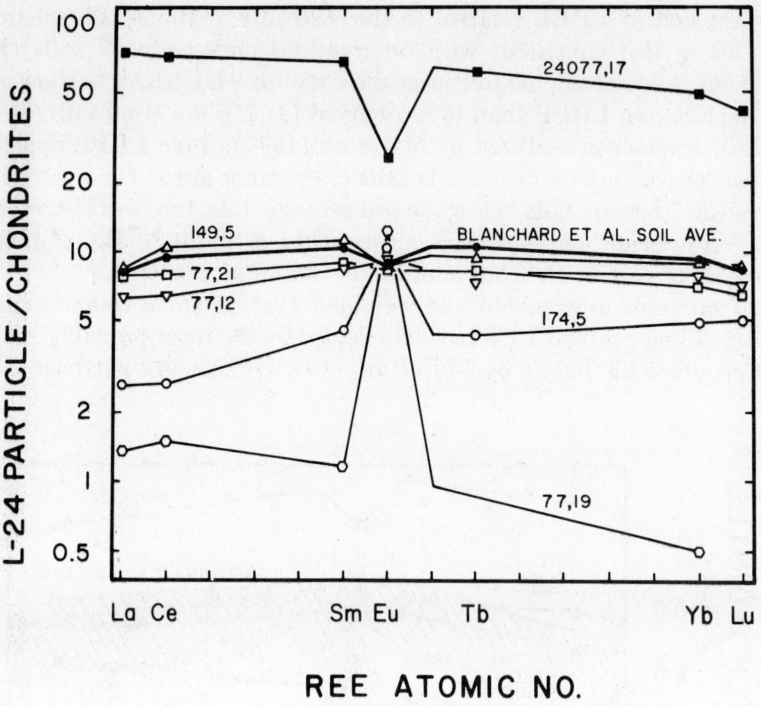

Fig. 1. REE comparison diagram for 6 Luna 24 fragments reported in this work. (Analytical data are in Blanchard *et al.*, 1978.) Soil average is from Blanchard *et al.* (1978).

(1977) reported compositions for 18 fragments weighing from 0.2 to 9.1 mg. Laul (1977) reported compositional data for 6 fragments weighing from 2.0 to 7.2 mg. Nyquist *et al.* (1977a) reported trace-element data for 5 soils. Barsukov *et al.* (1977) reported compositional data including trace elements for 8 soils and 4 fragments. REE distributions for a number of these fragments are shown in Fig. 2. Ma *et al.* (1977) on the basis of petrography and chemical composition designated 8 fragments as basalts. One particle among the 8 has since turned out to be a soil breccia. The REE average and range for the remaining 7 fragments is shown in Fig. 2. Ma *et al.* (pers. comm.) have selected 4 of the fragments which are fine-grained VLT basalts as the most representative. The REE compositional range for the 7 fragments is similar to that observed for soils, except the LREE are more depleted in the basalts and 4 of the fragments have small positive Eu anomalies. The LREE relative abundances are higher in the fine-grained soil material than in the 7 basalt fragments. A similar relationship between REE distributions in 1–2 mm mare basalt fragments and host <1 mm fines was found for Apollo 17 soils (Blanchard *et al.*, 1975). This difference can result either from inclusion of fine-grained non-mare material in the fines or from differential comminution of basalt (Korotev, 1976a). The 90–150 μm size fractions of L-24 soil are depleted in LREE relative to the <20 μm fractions (Blanchard *et al.*, 1978). This is also consistent with observations on Apollo 17 soils (Korotev, 1976a). Thus, it is reasonable to expect the Crisium VLT basalt to show a steeper relative depletion in LREE than that observed for 250 μm sized soils, as seen for the 7 basalt fragments analyzed by Ma *et al.* (1977). Such LREE depletion is a well-known characteristic of mare basalts from other sites. The average HREE content of the 7 basalts falls below the soil average; this, too, was observed for the Apollo 17 fragments and can be a consequence of differential comminution of parent basalt or can result from addition of non-mare material.

Table 1 contains unweighted and weighted average trace element concentrations of the 4 fine-grained VLT basalts selected as the most probably representative of Crisium VLT basalt by Ma *et al.*, (1977); these are particles 24077,63;

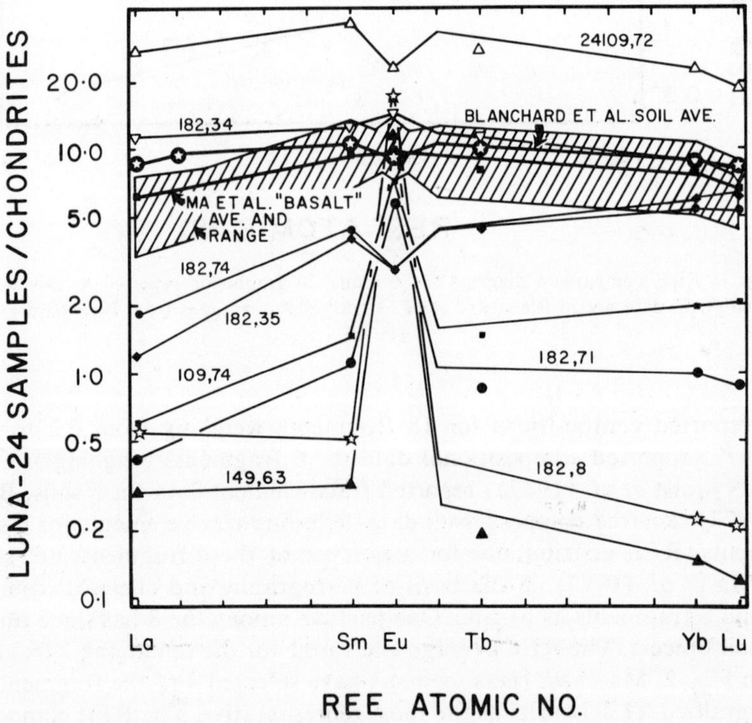

Fig. 2. REE comparison diagram showing range and average for 7 Luna 24 fragments initially selected as representing Crisium VLT basalt by Ma *et al.* (1977). Luna 24 soil average is from Blanchard *et al.* (1978). Individual particle data are from Ma *et al.* (1977) and Laul (1977). See text for details. Data for some particles discussed in the text are not shown, but would be similar to particles from the same grouping as indicated in the text. Because diagram is interpolated over several values of atomic number for which there are no data, the shape of the REE distribution may be slightly distorted. Lines are not always drawn through values for Tb, which sometimes have large analytical errors.

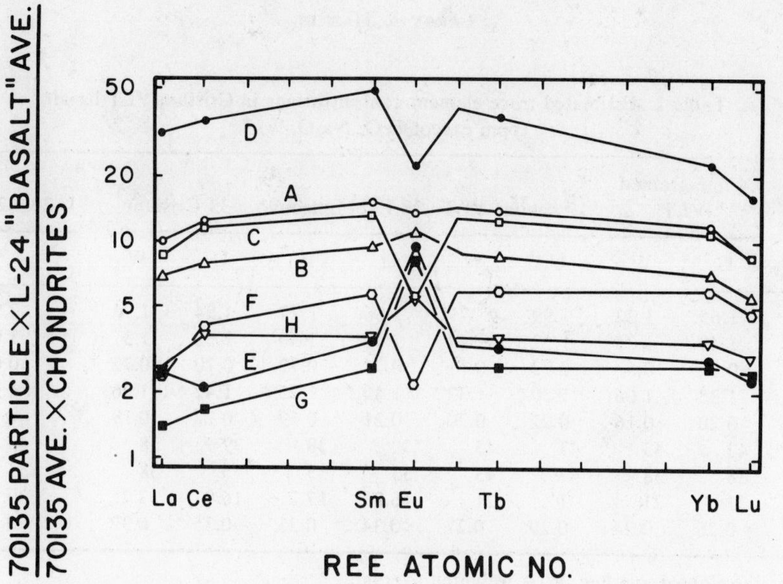

Fig. 3. REE comparison diagram for 8, 1–2 mm fragments from a crushing experiment on high-Ti mare basalt 70135,27 (Haskin and Korotev, 1977). REE concentrations have been normalized as discussed in the text to remove the compositional differences between 70135,27 and average Crisium VLT basalt.

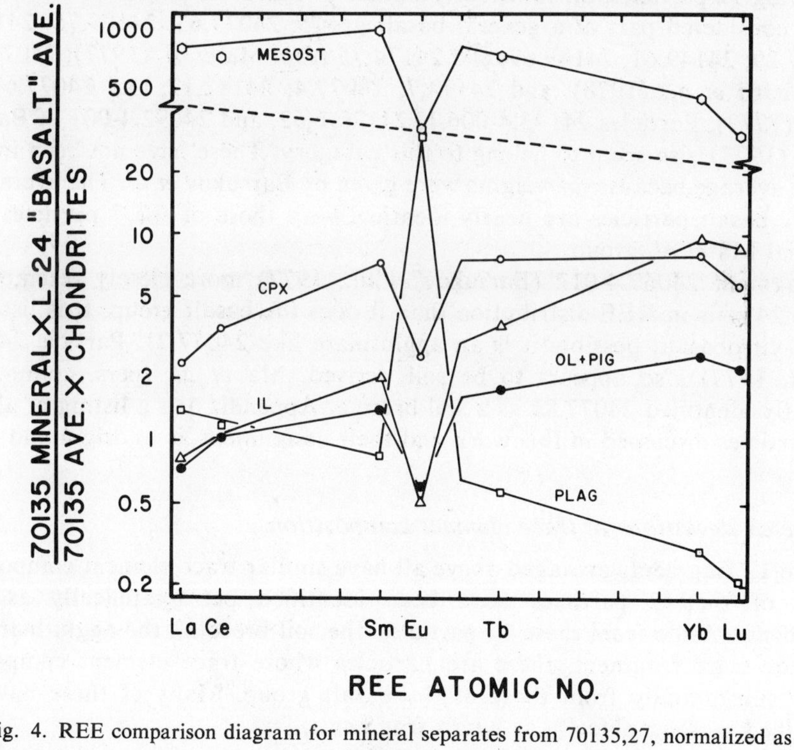

Fig. 4. REE comparison diagram for mineral separates from 70135,27, normalized as in Fig. 3.

Table 1. Estimated trace element concentrations in Crisium VLT basalt (ppm except FeO, Na_2O, %).*

	4 fine-grained VLT		12 similar VLT		28 L-24 igneous		33 Crisium		L-24 <250 μm
	U	W	U	W	U	W	U	W	
La	1.65	1.22	1.99	1.79	1.86	1.72	1.94	1.75	2.9
Sm	1.48	1.25	1.52	1.44	1.49	1.30	1.52	1.31	1.93
Eu	0.6	0.6	0.72	0.71	0.71	0.72	0.70	0.72	0.63
Yb	1.35	1.06	1.50	1.32	1.40	1.25	1.42	1.26	1.80
Lu	0.20	0.16	0.22	0.20	0.21	0.19	0.22	0.19	0.28
Sc	43	43	43	45	39	38	39	38	43
Co	38	38	43	43	33	35	35	38	51
FeO	21	21	21	21	16.8	17.2	16.9	17.2	20.5
Na_2O	0.26	0.24	0.29	0.27	0.34	0.33	0.35	0.32	0.28

*U = unweighted average, W = weighted average.

24109,76; 24109,78; 24149,29. Table 1 also contains the averages for the following 12 particles, all sufficiently similar in character to the Ma *et al.* basalts to be considered part of a general basalt group: 24077,63; 24109,76; 24109,78; 24149,29; 24149,61; 24149,69; and 24174,75 from Ma *et al.* (1977); 24077,12 of Blanchard *et al.* (1978); and 24174,7; 24077,4; 24182,12; and 24077,62 from Laul (1977). Particles 24143,4-006.1, 24170.1-62, and 24092.4-001 of Barsukov *et al.* (1977) also seem to belong to this category. These have not been included in the average because no weights were given by Barsukov *et al.* The averages for the 12 basalt particles are nearly identical with those of the 7 particles of the original Ma *et al.* group.

Fragment 24067.4-012 (Barsukov *et al.*, 1977) more closely resembles the Luna 24 soils in REE distribution than it does the basalt group. It is listed as a black vitrobasalt; possibly it is an agglutinate like 24077,21. Particle 24210,50 (Laul, 1977) also appears to be soil derived. Ma *et al.* (pers. comm.) have recently identified 24077,81 as a soil breccia. Appendix 1 is a listing of all Luna 24 particles discussed in this work and their assignment as to origin and type.

Expected deviations in trace element composition

The 12 fragments averaged above all have similar trace element compositions. Most of the 12 particles have been identified petrographically as basalt fragments. Aside from these 12 particles, the soil breccias, the agglutinates, and the non-mare fragment, there are particles whose trace element compositions differ substantially from those of the basalt group. Many of these have been identified petrographically as basalt fragments.

Mixing model calculations for the <250 μm soil compositions indicate the need for a high-MgO component in addition to the Crisium VLT basalt (Nyquist *et al.*, 1977a; Blanchard *et al.*, 1978). Do the trace element characteristics of the individual particles require or support the presence of more than one basaltic component? Alternatively, could the variations in trace element composition arise mainly because such small fragments of basalt as those analyzed from the Luna 24 core cannot all be representative of their parent basalt compositions? The trace element compositions of the Crisium VLT particles should reflect, besides possible additions or losses of components during the soil forming processes, the trace-element concentrations of the parent basalt and the relative amounts of the different basalt minerals in the fragments (Haskin and Korotev, 1977).

Probably the soundest approach to interpreting the compositional variations among the individual Luna 24 fragments is comparison with sets of analogous fragments whose relationships to their parent basalts are better known. One such set consists of 8 fragments weighing 4.4 to 7.0 mg that were analyzed in the basalt crushing experiment of Haskin and Korotev (1977) on 70135,27; the mineral separates from that experiment are also informative. Another set is the 1–2 mm basalt particles from Apollo 17 soils (Blanchard *et al.*, 1975) for which the parent basalt compositions are reasonably well known (Shih *et al.*, 1975; Rhodes *et al.*, 1976). In order to facilitate comparison of these Apollo 17 materials with the Luna 24 fragments, the difference in parent basalt trace element compositions has been nearly removed, for purposes of illustration, as follows. The compositions of the 8 fragments and the separated minerals from 70135 were divided by the composition of their parent (taken to be the <500 m weighted average, Haskin and Korotev, 1977), then multiplied by the average composition of a Crisium VLT parent (taken to be the average of 12 basalt fragments, Table 1.) The resulting REE concentrations, normalized to chondrites, are shown in Figs. 3 and 4 for comparison with the data of Figs. 1 and 2.

Note the rather broad spread of values (Fig. 3). The spreading is enhanced somewhat in that 70135 is relatively coarse grained for a lunar basalt, but not even fine-grained lunar basalts are always homogeneous on a size scale of a few cm (Lindstrom and Haskin, unpublished results). While these variations can be understood to a first approximation in terms of mineral and mesostasis composition, they cannot yet be modelled in detail. Haskin and Korotev found that a mixing model using mineral separate and mesostasis compositions could account very well for sized portions of crushed 70135,27 (except the <20 μm fraction) but could not account well for compositions of the 8 individual fragments (see Appendix 2). The fragment compositions reflected the vagaries of highly localized crystallization and effects of crushing. Thus, it is not possible to derive from fragment compositions accurate trace element compositions for parent melt by the use of the trace element models and distribution coefficients, even when fairly defensible values for those coefficients are available (e.g., Korotev and Haskin). The 70135 fragments and minerals are very useful, however, in understanding qualitatively the nature of the Luna 24 basalt fragments.

Crisium VLT basalt and Eu anomalies

Of the 15 fragments selected for the VLT basalt group, 11 have positive Eu anomalies relative to the REE distribution of the 9 chondrite composite (Haskin *et al.*, 1968) and 4 have negative ones. This suggests that the Crisium VLT parent would have a positive Eu anomaly, which would be an apparent deviation from the usual character of mare basalts, ie., negative Eu anomalies that decrease in size with decreasing Sm contents. Might the 4 fragments with negative Eu anomalies indicate a second type of Crisium VLT basalt? How compelling is the evidence that a Crisium VLT basalt has a positive Eu anomaly?

For one thing, the nature of the Eu anomaly, if any, depends on the REE distribution to which the samples are normalized (in this case, the average for 9 chondrites). Thus, the presence of a small positive Eu anomaly in Crisium VLT basalts, as suggested by the above data, need not mean a departure of these basalts from the usual character of mare basalt, as mentioned above. The proper reference distribution might not be chondritic (e.g., Nyquist et al., 1977b; Hubbard and Minear, 1975).

From Fig. 3 it is seen that 4 of the 70135 particles have positive Eu anomalies and 4 have negative Eu ones, relative to their parent basalt. (This is not an artifact of the procedure used to normalize these fragments to a Crisium VLT-like parent. A direct comparison of the 70135 particles to the average for the crushed chip from which these particles came (Haskin and Korotev, 1977, weighted average for <500 μm size fractions) gives exactly the same result.) A direct comparison to the average for the 8 particles themselves gives 5 positive and three negative Eu anomalies. Thus, the presence of a positive or negative Eu anomaly in a particle does not necessarily indicate that the parent basalt also had one.

Three separate chips from 70135 weighing 0.23 mg, 1.4 gm, and 1.6 gm have been analyzed for REE (see Haskin and Korotev, 1977, for discussion) and show a range of REE concentration of a factor of more than 3 for La, 2 for Lu. This example exceeds the range found for all chips from all Apollo 17 basalts by Shih *et al.* (1975). Relative to the unweighted average of the concentrations for all 3 chips, one chip shows a positive Eu anomaly, one chip a negative one, and one chip no anomaly. Relative to the weighted average for those chips, two show negative Eu anomalies, one a positive one. When the 8 small fragments of 70135 discussed above are compared to the weighted average for the three separate chips of 70135 instead of to the concentrations found for the chip of 70135,27 from which they came, 5 are found to have positive Eu anomalies and 3 to have negative ones, as opposed to an equal number of each. In short, there is no simple pattern of deviation of Eu anomalies in very small fragments from a mean value that can readily be used to determine the nature of the Eu anomaly in the parent.

When values for 28 petrographically identified mare basalt fragments from Apollo 17 soils 70182,1; 71502,3; and 78222,2 (Blanchard *et al.*, 1975) are compared with the average concentration for Apollo 17 mare basalt (Shih *et al.*,

1975), 12 are found to have positive Eu anomalies, and 16 to have negative ones. When the REE abundances in these fragments are normalized to a hypothetical Crisium VLT parent composition as described for the 8 70135 fragments (Fig. 3), only 14 of them fall in the general range for Luna 24 Crisium VLT fragments (as defined earlier on the basis of 15 Luna 24 particles). Considering only these 14 Apollo 17 soil fragments in a straight comparison to average Apollo 17 basalt (i.e., without normalization to a Crisium VLT parent composition), 8 have positive Eu anomalies and 6 have negative ones. The average size of the positive Eu anomalies (~16%) is nearly twice that of the negative ones (~8%). It is uncertain, of course, whether the average Apollo 17 basalt as determined from the values of Shih *et al.* (1975) is really the proper set of values for comparison with the Apollo 17 soil fragments.

However, there is some reason to expect a preponderance of fragments that yield positive Eu anomalies relative to their parent basalt compositions. The bulk of the REE in a mare basalt is found in the mesostasis (Haskin and Korotev, 1977) except for Eu and the heaviest REE, which are preferentially in plagioclase (Eu) and clinopyroxene and ilmenite (HREE). This last mineral is relatively lacking in Crisium VLT basalt. Fragments without an average share of mesostasis but with an average balance of other minerals would have positive Eu anomalies relative to their parents. As mesostasis is easily crushed and concentrates into the finest fractions during basalt comminution, more mesostasis-poor than mesostasis-rich 1–2 mm particles might be produced.

Depending on the REE values selected for comparison, this anticipated preponderance of positive Eu anomalies was or was not found among the 8 particles in the 70315 experiment. Eight particles is scarcely enough for a good statistical distribution. Nevertheless, neither the experiment with the 8 70135 fragments or the Apollo 17 soil fragments yields as high a ratio of positive to negative Eu anomalies relative to assumed parent basalt as is found for the 15 Luna 24 fragments (relative to the chondrite values of Haskin *et al.*, 1968). For the 8 70135 fragments, there are 50 to 63% positive anomalies; for the 14 Apollo 17 soil fragments there are 57% positive anomalies; for the 15 Luna 24 fragments (relative to chondrites) there are 73% positive anomalies.

This constitutes weak evidence for a positive Eu anomaly in Crisium VLT basalt relative to chondrites. Uncertainties considered, it is consistent with no Eu anomaly in Crisium VLT basalt. Nyquist *et al.* (1977a and pers. comm.) offer isotopic evidence favoring a positive anomaly in Crisium VLT basalt, but do not consider that their evidence unambiguously requires such an anomaly.

Fragments very different from the Crisium VLT average

In this section those fragments are considered that have trace element abundances substantially different from those of the 15 fragments selected as approximately representative of Crisium VLT basalt. Also excluded from this discussion are the soil breccias, agglutinates, and non-mare fragment. The purpose of this section is to consider whether any of these fragments indicates the presence of another basalt type or igneous rock type, or whether all may reasonably be considered products of crushing of average Crisium VLT basalt. The particles are considered in light of our general knowledge of lunar igneous materials as well as in comparison to the 8 fragments and separated minerals from the 70135 basalt crushing experiment.

Ma *et al.* (1977) provided qualitative interpretation of trace element distributions of several fragments in terms of their mineralogy. That discussion is expanded here and extended to the rest of the Luna 24 fragments considered in this work but not already classified above; it happens that all of these remaining particles have counterparts in the 8 fragments or the mineral separates from the 70135 basalt crushing experiment. The discussion centers mainly on trace elements, but the conclusions are consistent with, and take into account, major elements, in cases where they have been determined.

Some analyzed Luna 24 fragments consist almost entirely of plagioclase (24077,19 of Blanchard *et al.*, 1978; 24182,8 of Laul, 1977; 24077,83 and 24149,63 of Ma *et al.*, 1977). The size of the positive Eu anomaly in each of these particles is less than or equal to that of separated 70135 plagioclase (Haskin and Korotev, 1977). The size of the Eu anomaly in the 70135 plagioclase is suppressed somewhat by closed-system crystallization in comparison to the value expected for pure plagioclase grown in equilibrium with a liquid of the composition of the basalt whole rock, as explained by Haskin and Korotev (1977). The sizes of the Eu anomalies in the 4 Luna 24 plagioclase-rich particles are consistent with the derivation of those particles from Crisium VLT basalt. The REE concentrations in particles 24077,83; 24149,63; and 24182,8 are lower and the relative LREE abundances more horizontal (Fig. 2) than for the normalized 70135 plagioclase. These abundances are not lower than the product of the distribution coefficients estimated for plagioclase (Haskin and Korotev, 1977) and the average for the 7 Crisium VLT fragments of Ma *et al.* (1977) (except for La in 24077,83, which is perhaps different just outside of analytical uncertainty). The difference between these plagioclase fragments and those of 70135 can be accounted for if these fragments completed their growth at an earlier stage of solidification of the Crisium VLT liquid than did the average 70135 plagioclase from its parent liquid. Alternatively, they may have formed in a region of a basalt flow in which a higher proportion of the residual liquid from fractional crystallization chilled to form mesostasis than in 70135. However, the reason for the horizontal LREE distributions in the 3 fragments is not known.

Concentrations of other trace elements determined in the plagioclase-rich Luna 24 fragments (Sc, Co, Ni, Hf, Cr, FeO) are consistent with the derivation of those fragments from Crisium VLT liquid. Those for 24149,63 (Ma *et al.*, 1977) are nearly identical with those of normalized 70135 plagioclase. Those for the other plagioclase-rich fragments are higher than for normalized 70135 plagioclase, consistent with inclusions of small amounts of pyroxene in those fragments. The concentrations of the REE and other trace elements cannot be used to distinguish plagioclase of mare basalt origin from that of highlands origin (e.g., anorthosites 15415, Hubbard *et al.*, 1971; 60015, 69955, 60025, Laul and Schmitt 1973; 60025, 67075, 65702, Haskin *et al.*, 1973).

Fragments 24182,35 and 24182,74 (Ma *et al.*, 1977) consist mainly of pyroxene. This is reflected in their REE distributions (Fig.2) as compared with that of normalized 70135 pyroxene (Fig. 4). There are some differences. The shallower Eu anomaly in the fragments may reflect inclusion of small amounts of plagioclase or mesostasis. The higher contents of HREE, Co, and Sc probably result because there was much more ilmenite competing for those elements in 70135 than in Crisium VLT basalts. The augite of 70135 (e.g., Warner *et al.*, 1975) is more Ca-rich and Fe-poor (11.4% FeO) than the Crisium VLT pyroxene fragments (26% FeO, Ma *et al.*, 1977). This can substantially effect the distribution coefficients for the HREE, Co, and Sc.

Particles 24109,74 and 24182,71 (Ma *et al.*, 1977) and 24174,5 (Fig.1) correspond to mesostasis-poor mixtures of plagioclase and pyroxene. Their counterparts are particles E and G from 70135 (Fig. 3). The modal analyses for the 8 particles from 70135, taken from Korotev (1976b), are given in Appendix 2. They are shown on the basis of point counting of thin sections made from the particles after their analysis and on the basis of best-fit to a mixing model for minerals and mesostasis. Note from the X^2/ν values the poor mixing model fits to most of the fragments. (See Haskin and Korotev, 1977, for discussion.) The low LREE concentrations of Crisium VLT plagioclase are apparent in plagioclase-rich particles 24109,74 and 24182,71. Concentrations of HREE, Sc, Co, Cr, Na_2O, and Hf fall within the range for mixtures of normalized 70135 plagioclase clinopyroxene except for an enrichment of Co and Hf in 24174,5. This probably reflects the higher mesostasis concentration in that fragment.

Particles 24210,20 and 24182,75 (Ma *et al.*, 1977) are equivalent to particles A and C of 70135. The particles are pyroxene-rich, with Crisium pyroxene absorbing much of the role of ilmenite in 70135 for Co and Sc. Particle 24182,34 (Ma *et al.*) corresponds to particle B of 70135. This particle has enough plagioclase to produce a positive Eu anomaly. From the low Sc, Co, and FeO contents, 24182,34 is low in pyroxene. All of the above particles are more mesostasis-rich mixtures of plagioclase and pyroxene than were 24109,74; 24182,71; and 24174,5. This interpretation, based on REE and other trace element data, is consistent with what is known about their petrography. The FeO concentration in 24182,34 of only 3.9% is difficult to reconcile with late-stage, Fe-rich pyroxene being responsible for the high REE concentrations, as suggested by Ma *et al.*, (1977).

Fragment 24109,72 (Ma *et al.*, 1977) bears the same relationship to average Crisium VLT basalt as particle D does to its parent 70135. The REE content is dominated by mesostasis. The high FeO, MgO, and Co concentrations and low Sc content suggest the presence of olivine in this particle. No petrographic data have been published for it.

In summary, all the remaining Luna 24 particles discussed in this section have characteristics expected of fragments of normal Crisium VLT basalt.

Luna 24 particle averages and Crisium VLT trace element composition

Various compositional averages can be constructed from the Luna 24 particle data to serve as estimates of typical Crisium VLT basalt trace-element composition. Four such averages are discussed in this section. For each subset of particles considered, two extreme types of average were computed. One type is an unweighted average (sum of the concentrations for a given element in each particle divided by the number of particles). The other type is a weighted average (the concentration of the trace element in a particle multiplied by the weight of the particle, the resulting products for all particles summed, then that sum divided by the sum of the weights of all the particles). If the individual particles were somehow properly representative of Crisium VLT basalt in proportion to their masses, the latter type of average would be the correct one. Since there is no good reason to presume that an individual particle is properly representative of any larger-scale material, the former type of average may be just as good. The differences in results from the two types of averaging help to remind us of the uncertainties with which the Crisium VLT basalt composition can be determined.

The first pair of averages in Table 1 is that for the 4 fine-grained VLT basalts selected by Ma and Schmitt (1977, pers. comm.). The weighted average is substantially lower than the unweighted one for REE (except Eu), indicative of a substantial range of values among the four particles. The weighted average is dominated by particle 24077,63, which weighs 9.1 mg and constitutes 82% of the combined masses of all 4. Concentrations of FeO, Na_2O, Co, and Sc are constant among the 4 fragments; thus, their concentrations are the same in both averages.

The averages for the 12 of 15 similar VLT particles for which weights were available are given in the second pair of columns in Table 1. Inclusion of the 3 samples of Barsukov *et al.*, (1977) would not change the unweighted average. The values for REE are higher than those of the 4 fine-grained fragments. Values for Na_2O and Co are also higher; values for FeO and Sc are unchanged. Again, the values for REE in the unweighted average exceed those of the weighted average, although not by as much as in the case of the 4 particles.

The next pair of columns contains the averages for all 28 Luna 24 igneous fragments. Values for REE are fairly similar to those for the 12 VLT basalt particles. Those for Sc and FeO are noticeably lower and that for Na_2O is higher, a consequence of including significant amounts of plagioclase and plagioclase-rich particles in the average. The final pair of columns gives the averages for 33 fragments, all of those analyzed except the one clearly non-mare fragment. Values are, of course, quite similar to those for the 28 igneous fragments, but concentrations of La and Sm are noticeably higher in the unweighted average. The presence of excess light REE in mare soils as compared with basalt chips or fragments from those soils at Apollo 17 mare sampling sites (Blanchard *et al.*, 1975) was discussed previously. The non-igneous particles added into this average of 33 fragments were all soil derived.

The final column of Table 1 gives the average compositon of Luna 24 fines (bulk <250 μm) as determined by Blanchard *et al.* (1977, 1978). The fines are enriched in REE (except Eu) compared to all types of particle averages, and are preferentially enriched in the LREE, as discussed in an earlier section. The bulk soil (<250 μm) composition clearly differs from any reasonable Luna 24 basalt average. Blanchard *et al.* (1978) consider in detail the origin of the materials producing the soil.

Which average, if any, best describes the Crisium VLT basalt? The 33 particle averages can be eliminated, because Luna 24 soils, which contribute to that average, are known to contain non-mare material. The 28 particle average is also eliminated on the basis that the number of particles is small for obtaining a statistical average; thus, the distribution of particles with compositions deviating strongly from the average is surely non-representative. This is reflected in the unreasonably low concentration of FeO, a major constituent.

It might be argued that the average of 4 fine-grained VLT fragments best represents typical Crisium VLT basalt on the basis that finer-grained fragments contain more grains and should be more representative of their parent compositions than coarser grained fragments. Fig. 5 is a histogram of La and Sm concentrations of 23 Luna 24 polymineralic igneous fragments (the 29 igneous fragments minus the 6 monomineralic ones), the Apollo 17 soil fragments, and the 8 70135 fragments. (Values for the Apollo materials were normalized as described before to a hypothetical Crisium basalt parent to facilitate comparison.) The spread for the 4 fine-grained Luna 24 VLT fragments is substantial, covering most of the range for the particles in the 12 similar VLT average. Peaks for the histograms of 23 Luna 24 polymineralic fragments are broad, but peak regions are dominated by the 12 similar VLT basalts. There is no obvious reason to select the 4 fine-grained particles to represent the Crisium VLT average. The distributions for the 22 Luna 24 multimineralic igneous fragments are very similar to those for the Apollo 17 soil mare basalt fragments, which are in a general sense analogous in origin and acquisition to the Luna 24 particles. The 8 70135 particles are fairly consistent in compositional distribution to the Apollo 17 soil and Luna 24 particle groups; they are possibly more dispersed, as might be expected for fragments from a relatively coarse-grained rock.

It is concluded that the average for 12 similar VLT particles (Table 1) is as good an average as can presently be provided for the composition of typical Crisium VLT basalt. Uncertainties are in the range of ±30% for REE (except Eu), but probably only a few % for Eu, Sc, Co, FeO, and Na_2O. The relative abundances of the REE (except Eu) are probably known to about ±20% for the La/Yb ratio. The positive Eu anomaly, if any, probably does not exceed about 15% and may be entirely an artifact of the sampling. The average value of Sm/Eu for the 15 similar VLT particles is 2.36, or 90% of the chondritic value of 2.62. Eight of the 15 values for Eu used in obtaining these numbers were given only to one significant figure, which reduces even further the potential significance of the 10% apparent anomaly.

On the basis of these established properties of Crisium VLT basalt it is clearly not necessary that fragments need contain highland plagioclase to account for

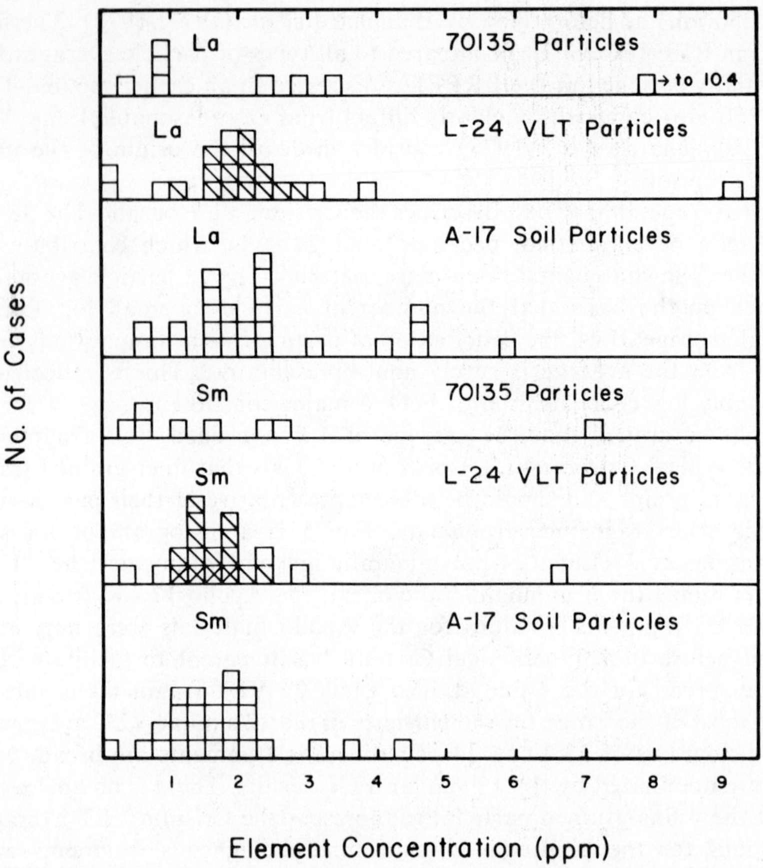

Fig. 5. Histogram of La and Sm concentrations. Boxes with X are the 4 fine-grained VLT fragments selected by Ma *et al.* as representing Crisium basalt (R. Schmitt, *pers. comm.*). Boxes with / or X are those for 15 particles selected in this work as best examples of Crisium VLT basalt. Associated empty boxes are remaining analyzed Luna 24 multimineralic igneous fragments. Values for Apollo 17 soils and 8 70135 fragments were normalized to a hypothetical Crisium VLT basalt parent, as described in the text, to facilitate comparison.

their positive Eu anomalies (Laul, 1977). The Eu anomaly in 24182,8 does not relate that sample to highland anorthosite (Laul, 1977). There is no definite requirement for a positive Eu anomaly in the basalts and exhaustion of plagioclase from the basalt source region cannot be strongly argued on that basis (Nyquist *et al.*, 1977a).

CONCLUSIONS

The following conclusions can be drawn concerning Crisium VLT basalts:

1. The average trace-element concentrations for 12 VLT basalt fragments

(columns 2 and 3, Table 1) are the best available estimates of concentrations for bulk samples of Crisium VLT basalt. Concentrations of the REE (except Eu) are known to approximately ±30%; concentrations of Eu, Sc, Co, FeO, and Na_2O are known to within several percent. Relative abundances of La, Sm, Yb, and Lu are known to within approximately 20% (based on the range observed in La/Yb ratios).

2. There is weak evidence that average Crisium basalt might have a small positive Eu anomaly relative to chondritic matter as found for the composite of 9 chondrites (Haskin *et al.*, 1968). The average size of the observed Eu anomaly among 15 Luna 24 VLT fragments relative to chondrites is +10%. However, positive Eu anomalies are expected for a majority of surviving basalt fragments relative to crushed parent basalt. Data for Apollo 17 basalt fragments and fragments of basalt 70135 produced in the laboratory support this expectation, but not quite to the extent observed for Luna 24. The maximum possible size of a positive Eu anomaly for Crisium basalt is about 15%. Uncertainties in REE concentrations are large enough that the apparent positive Eu anomaly estimated for Crisium VLT basalts could actually be no anomaly, or could even be a slightly negative anomaly. Thus, there is no compelling evidence that the abundance of Eu relative to those of the other REE is a departure from the trend of decreasing negative Eu anomaly with decreasing Sm concentration found for mare basalts from other regions. On a graph of Sm/Eu ratio versus Sm, useful as a measure of this trend, a straight line drawn from the value for chondrites and passing among the points for basalts from all Apollo missions crosses well within the error bar for Crisium VLT basalt.
3. Concentrations of trace elements in bulk Luna 24 soil do not give a satisfactory estimate for those in bulk Crisium VLT basalt; the soils contain components from sources other than that basalt (Blanchard *et al.*, 1978).
4. Average concentrations of trace elements in all particles or in all igneous fragments analyzed from the Luna 24 soils do not provide a satisfactory estimate for those in bulk Crisium VLT basalt. Mineralogical and trace-element compositions of fragments such as those analyzed are highly variable, and these are strong indications that the selection of fragments analyzed whose compositions were strongly different from the best estimated average Crisium VLT basalt composition was not statistically representative. Furthermore, surviving fragments are expected to retain less than their proportionate share of REE-rich mesostasis, relative to the parent basalt.
5. Of the 34 particles analyzed, only one (24077,17) is clearly foreign and cannot be derived from Crisium VLT basalt or from soil fines at the Crisium site by agglutination or compaction. Four particles are agglutinates or soil breccias. All other soil particles have compositional counterparts in fragments or minerals analyzed and studied petrographically in the 70135 basalt crushing experiment of Haskin and Korotev (1977) and

also in the 1-2 mm fragments of mare basalt from Apollo 17 soils analyzed and studied petrographically by Blanchard *et al.* (1975).

6. Whereas studies of fines from the Luna 24 core require more than one mare-related component (e.g., Blanchard *et al.*, 1978) there is no convincing information in concentrations of REE, Co, Sc, FeO, or Na_2O among the analyzed fragments to indicate more than one parent basalt. Presumably then, the trace element characteristics of the different components are similar, if the different components are basaltic.

Acknowledgments—Thanks to D. Blanchard for helpful discussions, to L. Nyquist for a particularly thorough and helpful review of the original manuscript, to J. Brannon and M. Reilly for assistance with calculations, and to M. Reilly, A. Tolin, and W. Day for assistance in preparing the manuscript. The availability of Luna 24 samples is gratefully acknowledged. This work was supported in part by NASA grant NSG-9055.

REFERENCES

Barsukov V. L. (1977) Preliminary data for the regolith core brought to Earth by the automatic lunar station Luna 24. *Proc. Lunar Sci. Conf. 8th*, p. 3303–3318.

Barsukov V. L., Tarasov L. S., Dmitriev L. V., Kolesov G. M., Shevaleevsky I. D. and Garanin A. V. (1977) The geochemical and petrochemical features of regolith and rocks from Mare Crisium (preliminary data). *Proc. Lunar Sci. Conf. 8th*, p. 3319–3332.

Basu A., McKay D. S. and Fruland R. M. (1977) Petrography, mineralogy and source rocks of Luna 24 drill core soils (abstract). In *Papers Presented to the Conference on Luna 24*, p. 18–21. The Lunar Science Institute, Houston.

Bence A. E. and Grove T. L. (1977) The highland component in the Luna 24 core (abstract). In *Papers Presented to the Conference on Luna 24*, p. 22–24. The Lunar Science Institute, Houston.

Bence A. E., Grove T. L. and Scambos T. (1977) Gabbros from Mare Crisium: An analysis of the Luna 24 soil. *Geophys. Res. Lett.* **4**, 493–496.

Blanchard D. P., Brannon J. C., Aaboe E. and Budahn J. R. (1978) Major and trace element chemistry of Luna 24 samples from Mare Crisium. In *Mare Crisum: The View From Luna 24* (R. B. Merrill and J. J. Papike, eds.). This volume.

Blanchard D. P., Haskin L. A., Brannon J. C. and Aaboe E. (1977) Chemistry of soils and particles from Luna 24 (abstract). In *Papers Presented to the Conference on Luna 24*, p. 37–40. The Lunar Science Institute, Houston.

Blanchard D. P., Jacobs J. W., Brannon J. C., and Haskin L. A. (1976) Major and trace element compositions of matrix and aphanitic clasts from consortium breccia 73215. *Proc. Lunar Sci. Conf. 7th*, p. 2179–2187.

Blanchard D. P., Korotev R. L., Brannon J. C., Jacobs J. W., Haskin L. A., Reid A. M., Donaldson C. H. and Brown R. W. (1975) A geochemical and petrographic study of 1–2 mm fines from Apollo 17. *Proc. Lunar Sci. Conf. 6th*, p. 2321–2341.

Butler P. Jr. and Morrison D. A. (1977) Geology of the Luna 24 landing site. *Proc. Lunar Sci. Conf. 8th*, p. 3281–3301.

Coish R. A. and Taylor L. A. (1977) Igneous rocks from Mare Crisium: Mineralogy and petrology (abstract). In *Papers Presented to the Conference on Luna 24*, p. 48–50. The Lunar Science Institute, Houston.

Haskin L. A., Haskin M. A., Frey F. A. and Wildemen T. R. (1968) Relative and absolute terrestrial abundances of the rare earths. In *Origin and Distribution of the Elements* (L. H. Ahrens, ed.), p. 889–912. Pergamon, New York.

Haskin L. A., Helmke P. A., Blanchard D. P., Jacobs J. W. and Telander K. (1973) Major and trace element abundances in samples from the lunar highlands. *Proc. Lunar Sci. Conf. 4th*, p. 1275–1296.

Haskin L. A. and Korotev R. L. (1977) Test of a model for trace element partition during closed-system solidification of a silicate liquid. *Geochim. Cosmochim. Acta* **41**, 921–939.

Hubbard N. J., Gast P. W., Meyer C., Nyquist L. E., Shih C.-Y, and Wiesmann H. (1971) Chemical composition of lunar anorthosites and their parent liquids. *Earth Planet. Sci. Lett.* **13**, 71–75.

Hubbard N. J. and Minear J. W. (1975) A physical and chemical model of early lunar history. *Proc. Lunar Sci. Conf. 6th*, p. 1057–1085.

Korotev R. L. (1976a) Geochemistry of grain size fractions of soils from the Taurus-Littrow valley floor. *Proc. Lunar Sci. Conf. 7th*, p. 695–726.

Korotev R. L. (1976b) Geochemical modelling of the distribution of rare-earth and other elements in a basalt and grain-size fractions of soils from the Apollo 17 valley floor and a well-tested procedure for accurate instrumental neutron activation analysis of geologic materials. Ph.D. dissertation, Univ. of Wisconsin, Madison.

Laul J. C. (1977) Chemical composition of Luna 24 melt rocks and gabbro (abstract). In *Papers Presented to the Conference on Luna 24*, p. 94–97. The Lunar Science Institute, Houston.

Laul J. C. and Schmitt R. A. (1973) Chemical composition of Apollo 15, 16, and 17 samples. *Proc. Lunar Sci. Conf. 4th*, p. 1349–1367.

Ma M.-S. and Schmitt R. A. (1977) Luna 24 soils: A chemical study (abstract). In *Papers Presented to the Conference on Luna 24*, p. 98–101. The Lunar Science Institute, Houston.

Ma M.-S., Schmitt R. A., Taylor G. J., Warner R. D., Lange D. E. and Keil K. (1977) Chemical and petrographic studies of 18 Luna 24 lithic fragments (abstract). In *Papers Presented to the Conference on Luna 24*, p. 102–105. The Lunar Science Institute, Houston.

Nagle J. S. and Walton J. A. (1977) Luna 24: Catalog and preliminary description of samples presented to the U.S.A. Curatorial Office, NASA Johnson Space Center, Houston. 86 pp.

Nyquist L. E., Bansal B. M., Wooden J. L. and Wiesmann H. (1977b) Sr-isotopic constraints on the petrogenesis of Apollo 12 mare basalts. *Proc. Lunar Sci. Conf. 8th*, p. 1383–1415.

Nyquist L. E., Wiesmann H., Bansal B., Wooden J. L., McKay G. and Hubbard N. J. (1977a) Chemical and Sr-isotopic characteristics of the Luna 24 samples (abstract). In *Papers Presented to the Conference on Luna 24*, p. 139–142. The Lunar Science Institute, Houston.

Pieters C. and McCord T. B. (1977) Mare Crisium: A second look at regional spectral properties (abstract). In *Papers Presented to the Conference on Luna 24*, p. 147. The Lunar Science Institute, Houston.

Pieters C., McCord T. B. and Adams J. B. (1976) Regional basalt types in the Luna 24 landing area as derived from remote observations. *Geophys. Res. Lett.* **3**, 697–700.

Rhodes J. M., Hubbard N. J., Wiesmann H., Rodgers K. V., Brannon J. C. and Bansal B. M. (1976) Chemistry, classification, and petrogenesis of Apollo 17 mare basalts. *Proc. Lunar Sci. Conf. 7th*, p. 1467–1489.

Ryder G., McSween H.Y., Jr. and Marvin U.B. (1977) Luna 24 VLT basalt: Character and origin (abstract). In *Papers Presented to the Conference on Luna 24*, p. 160–163. The Lunar Science Institute, Houston.

Shih C.-Y., Haskin L. A., Wiesmann H., Bansal B. and Brannon J. C. (1975) On the origin of high-Ti mare basalts. *Proc. Lunar Sci. Conf. 6th*, p. 1255–1285.

Simonds C. H., Warner J. L., McGee P. E., Cochran A. and Brown R. W. (1977) Luna 24: Petrochemical potpourri (abstract). In *Papers Presented to the Conference on Luna 24*, p. 170–173. The Lunar Science Institute, Houston.

Tarasov L. S., Nazarov M. A., Shevaleevsky I. D., Kudryashova A. F., Gaverdovskaya A. S. and Korina M. I. (1977) Mineralogy and petrography of lunar rocks from Mare Crisium (preliminary data). *Proc. Lunar Sci. Conf. 8th*, p. 3333–3356.

Taylor G. J., Warner R. D. and Keil K. (1977) Chemical distinctions among very low-Ti mare basalts (abstract). In *Papers Presented to the Conference on Luna 24*, p. 183–185. The Lunar Science Institute, Houston.

Vaniman D. T. and Papike J. J. (1977) Ferrobasalts from Mare Crisium: Luna 24. *Geophys. Res. Lett.* **4**, 497–500.

Warner R. D., Keil K., Prinz M., Laul J. C., Murali A. V. and Schmitt R. A. (1975) Mineralogy, petrology, and chemistry of mare basalts from Apollo 17 rake samples. *Proc. Lunar Sci. Conf. 6th*, p. 193–220.

Appendix 1. Luna-24 particles considered in this work.*

Particle No.	Mass (mg)	Type	Reference
24149,29	0.934	fine-gr. VLT	Ma *et al.* (1977)
24109,78	0.342	"	"
24109,76	0.710	"	"
24077,63	9.055	"	"
24174,75	1.289	coarse gr. VLT	"
24149,69	0.220	medium gr. VLT	"
24149,61	0.473	shock melt (VLT)	"
24077,12	7.59	(VLT)	Blanchard *et al.* (1977)
24077,62	3.61	(VLT)	Laul (1977)
24077,4	1.99	melt breccia (VLT)	"
24182,12	2.23	"	"
24174,7	2.00	recryst. breccia VLT	"
24143,2006.1	—	(VLT)	Barsukov *et al.* (1977)
24170.1-6.2	—	"	"
24092.4-001	—	"	"
24077,19	4.03	Plagioclase	Blanchard *et al.* (1977, 1978)
24182,8	2.30	"	Laul (1977)
24077,83	1.505	"	Ma *et al.* (1977)
24149,63	0.312	"	"
24182,35	0.210	pyroxene	"
24182,74	0.606	"	"
24109,74	0.27	mesostasis-poor VLT	"
24182,71	0.336	"	"
24174,5	5.89	"	Blanchard *et al.* (1977)
24210,20	0.523	pyroxene-rich VLT	Ma *et al.* (1977)
24182,75	0.550	"	"
24182,34	0.310	plagioclase-rich VLT	"
24109,72	0.334	mesostasis-rich VLT	Ma *et al.* (1977)
24077,21	0.52	agglutinate	Blanchard *et al.* (1977, 1978)
24210,50	7.17	plagioclase-rich VLT	Laul (1977)
24067.4-012	—	black vitrobasalt (agglut.)	Barsukov *et al.* (1977) "
24149,5	0.695	soil breccia	Blanchard *et al.* (1977, 1978)
24077,81	0.577	"	Ma *et al.* (1977)
24077,17	3.72	non-mare	Blanchard *et al.* (1977, 1978)

*Parentheses under particle type indicate assignments made in this paper and not confirmed by petrographic analysis.

Appendix 2. Modal analyses and mixing-model results for individual 1–2 mm particles from 70135. Fractions of phases in percent by volume (modal) and by weight (mixing-model). Whole rock mode is for the 90–150 μm size fraction of basalt sample crushed to give 1–2 mm particles. (See Haskin and Korotev, 1977 for details.)

Petrographic mode	A	B	C	D	E	F	G	H	Whole Rock
cpx	43	32	69	35	44	80	41	48	56.5*
ilm	32	14	26	20	5	20	20	36	18.3
plg	24	54	5	41	51	0	39	15	24.2
meso/inclus.	0.5	0	0	4	0	0	0	0	0.8
points	201	148	186	309	101	243	184	201	2071
points/mg	33.4	33.6	32.2	56.6	20.9	34.9	37.4	40.0	—

Mixing-model mode	A	B	C	D	E	F	G	H	Whole Rock
aug	31	18	54	19	16	66	19	17	53.8*
"pig"	−2	23	7	13	43	14	42	49	
ilm	44	29	34	33	6	19	8	15	25.2
plg	26	29	4	31	35	1	30	18	20.3
meso	1.01	0.63	0.81	3.72	0.21	0.09	0.09	0.18	0.56
metal	0.19	0.18	0.02	0.02	0.05	0.56	0.02	0.18	0.12
χ^2/ν	4.59	4.73	0.36	7.40	31.2	18.8	20.0	48.3	0.84

*Total pyroxene and olivine

Merrill, R. B. and Papike, J. J., editors.
(1978) *Mare Crisium: The View from Luna 24*, Pergamon Press (New York), p. 613–630
Printed in the United States of America

Major and trace element chemistry of Luna 24 samples from Mare Crisium

DOUGLAS P. BLANCHARD

NASA Johnson Space Center, Houston, Texas

JOYCE C. BRANNON

Lockheed Electronics Co., Inc., Houston, Texas and Washington University, St. Louis, Missouri

ERIK AABOE

Yale University, New Haven, CT and The Lunar Science Institute, Houston, Texas

JAMES R. BUDAHN

Lockheed Electronics Co., Inc., Houston, Texas

Abstract—We have analyzed 6 Luna 24 soils for major and trace elements by atomic absorption spectrophotometry (AAS) and instrumental neutron activation analysis (INAA). We have also analyzed size fractions (90–150 μm and <20 μm) of these same soils and 6 fragments (>250 μm) for trace elements by INAA. We have used this data and those of other workers to identify several compositional components in the Luna 24 soils and have tested the reasonableness of these components by using mixing calculations to reconstruct the soils.

The Luna 24 regolith contains compositional evidence for a history of mixing of several components. The soils are well mixed (i.e., have nearly homogeneous bulk compositions) throughout the length of the core tube yet the size fractions of each soil have quite dissimilar compositions. On this and other grounds we conclude that, since emplacement, the regolith has not been extensively reworked which would tend to homogenize the elemental distribution among grain size fractions; most of the properties of the regolith which are indicative of reworking (surface maturity) have been inherited from one or more of the components in the soil.

The observed elemental distributions as a function of grain-size are due in part to admixture of noritic breccia component preferentially to the finest fractions and in part to differential comminution of one or more of the other components of the soil.

Other than the small percentage of noritic breccia in the Luna 24 regolith we see no evidence for any other component that is clearly foreign to the Luna 24 site. The small percentage of noritic breccia does contribute significantly to the REE of the bulk soil, but has only insignificant effect on the major element composition. Among the other probable components, we have identified the ferrobasalt composition and one or more components with higher MgO contents. Mixing results do not definitively identify the high-MgO component. There is evidence from the petrographic analysis of mineral grains that there is probably a very coarse-grained component with very forsteritic olivine contributing to the regolith. Whether this coarse-grained component, which is not represented among the lithic fragments in the soil, can wholly account for the "excess" MgO in the soils or whether there is an additional high-MgO basalt is problematical.

INTRODUCTION

In August, 1976, the automated Soviet space probe Luna 24 returned to the earth a core tube of lunar material from the previously unsampled Mare Crisium area of the moon. Through the efforts of the scientists at the Vernadsky Institute and the Government of the U.S.S.R., highly valuable samples from this core were made available to investigators throughout the world. We have studied samples of six soils (<250 μm) from various depths in the Luna 24 core and a total of six fragments (>250 μm) from three of those soils. The work reported here was done in conjunction with a working group which has been conducting multidisciplined studies of lunar regolith processes. The results of the others in this working group are reported elsewhere in this volume (Blanford and Wood, 1978; Bogard and Hirsch, 1978; McKay *et al.*, 1978; and Morris, 1978).

Our interests in Luna 24 samples were:

1. To gain insight into the detailed history of the regolith at the Luna 24 site by studying relationships between particle sizes and compositions;
2. To identify compositions of materials which have combined to form the soils; and
3. To suggest sources on the moon which might supply these components.

In a companion paper to this one, Haskin (1978) discusses the data on the fragments, the question of multiplicity of basalts at the Luna 24 site, and constraints on their source regions.

SAMPLES AND METHODS

We have analyzed samples of "bulk" soils (<250 μm) as we received them with no further treatment. We received most soils directly from the Curator; soil 24210,11 was obtained from D. Bogard. Bulk soils were analyzed for the elements listed in Table 2 by nondestructive instrumental neutron activation analysis (INAA) (Jacobs *et al.*, 1977). They were then dissolved in HF in a heated (130°C), pressurized teflon bomb, stabilized with boric acid, diluted, and analyzed for SiO_2, TiO_2, FeO, MgO, Ca_2O and K_2O by atomic absorption spectrophotometry (AAS) (Blanchard, 1973). Based on counting statistics and results of repetitive analyses of well known U.S.G.S. standard rocks, we estimate that the uncertainties associated with the INAA technique are 2–3% for all elements except Yb (6%), Ta, Ce, Tb (15%) and Ni, Hf, Th (20–25%). Uncertainties associated with AAS are 2–3% except for TiO_2 (15–20%). Uncertainties of AAS analysis of 24210,11 are slightly higher due to the small sample size. Values for FeO concentrations determined by the two methods agreed very well; the FeO values in Table 1 are the INAA data which have slightly smaller uncertainties.

Because of the unusually valuable nature of the Luna 24 samples, we took the additional analytical precaution of analyzing a well-analyzed lunar soil 71501 in conjunction with the Luna 24 samples. Our results for 71501 were in close agreement with previous analyses in this laboratory (Korotev, 1976) and with results on the same soil by Rhodes *et al.* (1974). There is some uncertainty about the cause of rather large disagreement among analysts of the Luna 24 bulk soils; it appears to be, in most cases, much larger than the analytical uncertainties. Most notably, values for FeO in the average bulk soil ranges from 17.51% reported by Simonds *et al.* (1977) to 20.5 reported here. Barsukov *et al.* (1977) report a range of FeO values from 18.8% to 20.9% for bulk soils; on fractions <75 μm in grain size their analyses range from 16.1% to 16.3%. Our results indicate the same dichotomy in FeO between bulk and fine fractions; our average bulk soil contains 20.5% FeO, our average <20 μm soil fraction contains 16.4% FeO. We believe that some of the range in reported values for FeO in the bulk soils may be a result of nonrepresentative sampling, but the very low values for FeO on bulk soils probably reflect systematic analytical errors.

Size fractions of the six soils were separated by sieving in the Johnson Space Center Scanning Electron Microscope Laboratory (McKay *et al.*, 1978). The aliquants of soils which were sieved were not the same aliquants as those used for AAS and INAA of "bulk soils" reported here. We analyzed the 90–150 μm and <20 μm size fractions by INAA only (Table 2). Other aliquants of these same

Table 1. Major element compositions of Luna 24 soils (>250 μm).

	Mass (mg)	SiO_2	TiO_2	Al_2O_3	FeO	MgO	CaO	Na_2O	K_2O	Cr_2O_3	Total
24077,7	9.72	45.9	0.9	11.8	19.9	9.7	11.2	0.29	0.03	0.56	100.3
24109,14	13.56	44.1	1.0	11.3	20.6	10.3	11.0	0.27	0.02	0.62	99.2
24149,16	11.07	44.6	1.3	11.0	20.3	10.1	10.8	0.28	0.03	0.48	98.9
24174,12	12.83	45.5	0.9	11.1	20.9	9.7	10.8	0.29	0.03	0.49	99.7
24182,17	11.28	44.5	1.3	11.5	20.2	10.0	11.1	0.29	0.04	0.50	99.4
24210,11	3.51	47.5	1.0	9.7	21.1	11.3	10.8	0.25	—	0.45	102.2
Average		45.4	1.1	11.1	20.5	10.2	11.0	0.28	0.03	0.52	100.1
± s		±1.3	±0.2	±0.7	±0.5	±0.6	±0.2	±0.02	±0.01	±0.06	

size fractions have been studied petrographically by Basu *et al.* (1977), for solar and cosmogenic gases by Bogard and Hirsch (1978), and for particle track densities by Blanford and Wood (1978). Some of the aliquants used for INAA have been transferred to the JSC-MSID and isotope laboratory for further chemical and isotopic characterization.

The six Luna 24 particles (>250 μm) were analyzed by INAA (Table 3) as received from the Lunar Sample Curator. Visual examination under binocular microscope verified that each particle corresponded to its description in the Luna 24 catalog except that 24149,5, described as a soil breccia, was received as a disaggregated soil. After INAA, the particles were transferred to other investigators as follows:

24077,12—Split and sent to Keil (1 mg) and Nyquist (8 mg).
24077,17—Split and sent to Keil (1 mg) and Nyquist (3 mg).
24077,19—Sent to Nyquist (4 mg).
24077,21—Thin sectioned and sent to L. Taylor (0.52 mg).
24149,5—Sent to Nyquist (7 mg).
24174,5—Split and sent to Keil (1 mg) and Nyquist (3 mg).

Discussion

Four broad observations can be made from our data:

1. Bulk soils are nearly identical to each other in composition.
2. The compositions of the <20 μm fractions are quite different from the bulk soil, nevertheless, the <20 μm fractions are very similar to each other. Thus while the elements are not homogeneously distributed within any one soil, the soils themselves are well mixed with one another, each having the same elemental distribution among grain sizes.
3. The bulk soils are distinctly basaltic in composition and generally similar to the average of the basaltic fragments from the >250 portion of the soil samples. Individually, however, the basalt chips have a wide range of compositions.
4. Most of the individually analyzed fragments have compositions consistent with their being basaltic and are probably locally derived. Only 24077,17 is distinctly foreign to the Luna 24 site; only a small percentage of a component with its composition can be tolerated in the soils without noticably perturbing their composition.

Table 2. Results of INAA analyses of size fractions of Luna 24 soils.

	Mass (mg)	FeO	Na_2O	Cr_2O_3	Sc	Co	Ni	Hf	Th	Ta	La	Ce	Sm	Eu	Tb	Yb	Lu
Bulk (<250 μm)																	
24077,7	9.72	19.9	0.290	0.56	41.1	50.9	160	1.5	0.41	—	3.10	8.9	2.00	0.66	0.49	1.76	0.273
24109,14	13.56	20.6	0.272	0.62	43.4	52.5	230	1.36	0.25	—	3.22	9.7	2.04	0.66	0.49	1.78	0.272
24149,16	11.07	20.3	0.282	0.48	43.0	50.7	100	1.9	0.26	—	2.32	7.3	1.73	0.60	0.40	1.63	0.263
24174,12	12.83	20.9	0.287	0.49	43.9	50.4	130	1.47	0.24	—	2.74	8.5	1.95	0.63	0.46	1.89	0.299
24182,17	11.28	20.2	0.295	0.50	42.3	48.4	90	1.6	0.40	—	3.24	9.9	2.09	0.70	0.56	1.93	0.298
24210,11	3.51	21.1	0.252	0.45	44.6	54.0	—	1.5	0.30	—	2.64	6.8	1.74	0.54	0.51	1.81	0.27
Average		20.5	0.28	0.52	43.1	51.0	140	1.6	0.31	—	2.88	8.53	1.93	0.63	0.49	1.80	0.28
± s		±0.5	±0.02	±0.06	±1.2	±1.6	±60	±0.2	±0.08	—	±0.37	±1.26	±0.15	±0.06	±0.05	±0.11	±0.02
(90–150 μm)																	
24077,9	1.48	20.1	0.321	0.52	43.2	61.0	230	1.0	—	—	2.51	7.0	1.66	0.61	—	1.8	0.25
24109,13	4.44	20.7	0.234	0.41	45.8	51.0	—	1.0	—	—	2.38	6.9	1.69	0.54	0.36	1.55	0.22
24149,15	4.99	19.5	0.250	0.39	42.4	47.0	—	1.0	—	—	2.05	6.4	1.45	0.53	0.60	1.61	0.23
24174,10	4.00	20.9	0.255	0.39	45.4	48.0	—	1.2	—	—	2.10	6.5	1.49	0.52	0.40	1.56	0.26
24182,15	4.88	20.8	0.262	0.38	45.8	50.0	90	1.2	—	—	2.07	5.0	1.58	0.55	0.45	1.80	0.28
24210,9	3.45	22.2	0.237	0.44	44.6	56.0	140	1.3	—	—	2.14	5.9	1.48	0.54	0.36	1.56	0.24
Average		20.7	0.26	0.42	44.5	52.1	150	1.1	—	—	2.21	6.30	1.56	0.55	0.43	1.65	0.25
± s		±0.9	±0.03	±0.05	±1.4	±5.3	±70	±0.1	—	—	±0.19	±0.75	±0.10	±0.03	±0.10	±0.12	±0.02
(<20 μm)																	
24077	5.21	16.3	0.57	0.46	30.5	47.0	430	2.1	0.66	—	4.35	10.3	2.57	0.83	0.64	2.04	0.31
24109	5.31	16.5	0.341	0.46	31.5	46.7	400	1.8	0.53	—	5.20	11.9	2.73	0.81	0.65	2.24	0.31
24149	3.24	16.8	0.359	0.45	30.6	44.2	380	2.2	0.57	—	4.43	12.0	2.64	0.81	0.63	2.21	0.31
24174	4.22	16.8	0.352	0.46	31.4	45.6	430	2.1	0.70	0.29	5.0	13.0	2.90	0.89	0.65	2.26	0.35
24182	7.69	16.7	0.377	0.45	30.4	46.0	390	2.3	0.67	0.27	5.8	17.3	3.35	0.92	0.75	2.55	0.36
24210	1.98	15.0	0.310	0.44	27.3	42.3	270	1.7	—	—	3.99	10.3	2.23	0.69	0.58	1.85	0.29
Average		16.4	0.38	0.45	30.3	45.3	380	2.0	0.63	—	4.80	12.5	2.74	0.85	0.65	2.19	0.32
± s		±0.7	±0.09	±0.01	±4.2	±1.8	±60	±0.2	±0.07	—	±0.66	±2.6	±0.37	±0.08	±0.06	±0.23	±0.03

Table 3. Results of INAA analyses of fragments from the Luna 24 core.

	Mass (mg)	FeO	Na_2O	Cr_2O_3	Sc	Co	Ni	Hf	Th	Ta	La	Ce	Sm	Eu	Tb	Yb	Lu
24077,12 Basalt	7.59	22.1	0.29	0.42	47.4	46.0	—	0.9	—	—	2.08	5.7	1.53	0.63	0.34	1.55	0.24
24077,17 Recrystallize Breccia	3.72	7.60	0.66	0.184	13.3	20.2	180	9.9	14	4.8	24.9	68	12.1	1.74	2.9	9.7	1.40
24077,19 Plag-rich Gabbro	4.03	0.77	0.63	0.018	1.76	2.04	20	0.79	—	—	0.45	1.30	0.21	0.84	—	0.10	<0.009
24077,21 Agglutinate	0.52	16.3	0.31	0.46	36.4	46	180	1.0	—	—	2.6	7.0	1.6	0.58	0.38	1.4	0.22
24149,5 Soil Breccia	0.695	20.3	0.27	0.51	47	50.7	150	1.3	—	0.3	2.8	9.2	2.0	0.59	0.43	1.75	0.29
24174,5 Px-rich Gabbro	5.89	16.6	0.38	0.20	35.8	29.7	—	0.88	—	—	0.86	2.33	0.80	0.70	0.20	0.96	0.17

The major questions we will address here are: (1) What is the cause of the inhomogeneous elemental distribution among grain sizes and what does it imply about the regolith and its history at the Luna 24 site? (2) How many components are represented in the soils and what are their compositions?

Relationships of Composition and Particle Size

Most of the Luna 24 soils we have analyzed have bimodal size distributions (McKay *et al.*, 1978). The Luna 24 "bulk" soils are, in fact, the <250 μm sieve fractions of the soil samples received from the Soviets. Figure 1 shows the size distributions (McKay *et al.*, 1978) of the six Luna 24 soils we have analyzed. The open areas on the left of each histogram are the >250 μm fractions. It can be seen that for several soils the split at 250 μm has excluded nearly the entire larger particle size mode. The cross hatched areas denote the 90–150 μm and the <20 μm fractions. While the 90–150 μm fraction is a relatively small fraction it is generally near the peak of the finer mode, and thus is probably a reasonable sample of materials making up that mode. For Luna 24 soils the <20 μm fractions comprise a rather substantial fraction of the <250 μm soils (16 to 26%).

Above the histograms in Fig. 1 are plotted the concentrations of selected

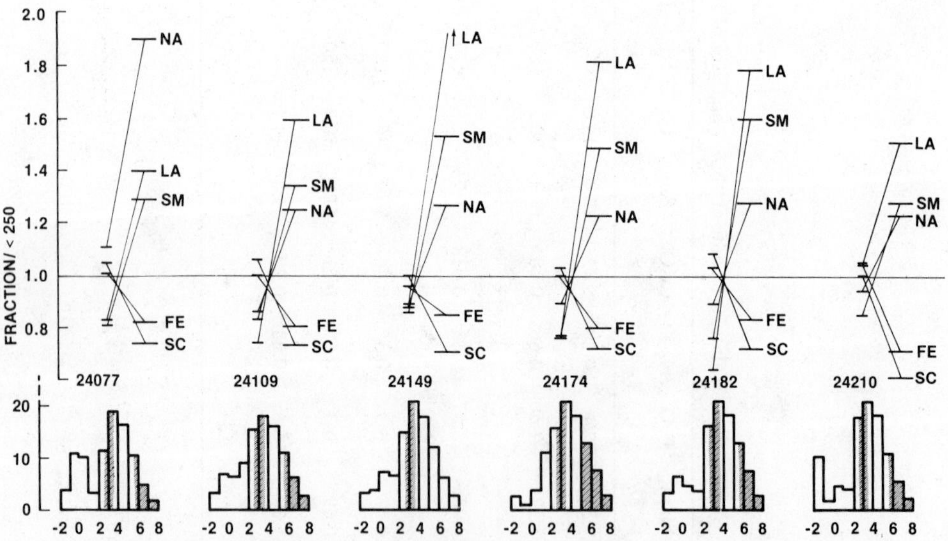

Fig. 1. Size distributions and elemental distributions between 90–150 μm and <20 μm size fractions for Luna 24 soils. Size distributions are from McKay *et al.* (1978); size units are ∅ units. The area on the left of each histogram which is outlined only is the >250 μm fraction. Shaded areas are 90–150 μm and <20 μm size fractions which were analyzed in this work. Directly above each of the shaded size fractions is plotted its concentration for the specified element, normalized to the concentration in the bulk soil. See text for discussion.

elements in the 90–150 μm and <20 μm fractions; each is normalized to the concentrations in the bulk <250 μm soils. The distribution of elements among the size fractions is very similar in all the Luna 24 soils; only in the uppermost soil 24077 is there a change in the relative order of enrichments with Na_2O concentrated in the finest fraction. In all cases Na, La and Sm are enriched in the <20 μm fraction and the La/Sm ratio is higher in the <20 μm fraction than in the bulk soils. Again, in every case the Sc and Fe are somewhat enriched in the 90–150 μm and depleted in the <20 μm fractions relative to the bulk soils.

We have compared these size-composition patterns with those of other mare soils (Korotev, 1976) and with results of a basalt-crushing experiment reported by Korotev and Haskin (1976). Figure 2 shows the histograms and fractionation patterns associated with the size fractions from the crushed basalt 70135 and Apollo 17 mare soils.

Evidence for Differential Comminution

The elemental distribution due to differential comminution during the crushing of lunar basalt 70135 is nearly identical, both in direction and magnitude to that observed for Luna 24 soils. The size distribution of basalt debris is heavily skewed to the coarse particles making the fractionation in the

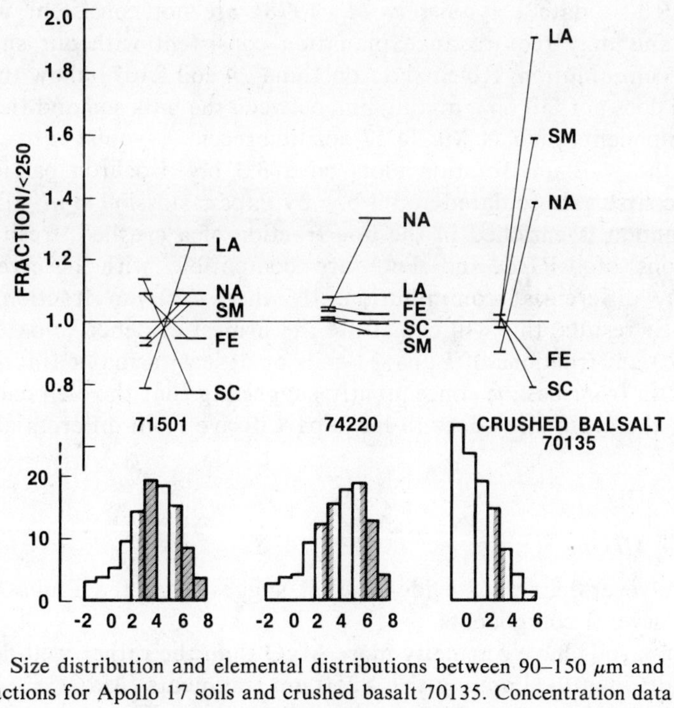

Fig. 2. Size distribution and elemental distributions between 90–150 μm and 120 μm size fractions for Apollo 17 soils and crushed basalt 70135. Concentration data plotted as in Fig. 1.

<20 μm size fraction of the basalt somewhat less important than that in the Luna 24 soils from a total mass balance standpoint. The compositional range of 1–2 mm particles produced by the crushing experiment of Korotev and Haskin (1976) is only slightly exceeded by the compositional range of the >250 μm particles analyzed from these soils. This observation is discussed further by Haskin (1978).

Elemental distribution patterns such as these are not unique to cases of differential comminution. The results of analyses of a typical mare soil from Apollo 17 (Korotev, 1976) also show similar fractionation patterns, although they are somewhat less extreme. Using detailed mixing calculations utilizing well characterized materials at the Apollo 17 site, Korotev (1976) accounted for the compositional differences among the size fractions of the Taurus-Littrow valley soils as the result of the admixture to the soils of fine-grained material with compositions typical of the surrounding highland province.

We have been unable to definitively rule out differential comminution as a mechanism to produce the elemental distributions in the Luna 24 soils. We lack the detailed data on compositions of materials present in the vicinity of the Luna 24 site to evaluate the relative importance of differential comminution vs. mixing of components of differing composition and size distribution ala Korotev (1976). The analogous compositional data on the Luna 24 soils and crushed basalt at this point only imply the possibility of compositional effects due to differential comminution.

Limited Rb-Sr data (Nyquist *et al.*, 1978) are not consistent with mixing arguments and may require an explanation consistent with our suggestion of differential comminution. Rb-Sr data on Luna 24 soil 24077 show that the <20 μm fraction does not fall on a mixing line between the bulk soil and the REE- and Rb-rich components such as Apollo 17 noritic breccia (Nyquist *et al.*, 1978). The datum for the <20 μm fraction plots on a 3.3 b.y. isochron passing through 24170, the coarse gabbro dated at 3.3 b.y. by Papanastassiou *et al.* (1977). If the <20 μm fraction is enriched in the fine fraction of a crushed, fresh basalt, the concentrations of LREE and FeO are compatible with observed patterns produced by differential comminution. If other <20 μm fractions also give similar Rb-Sr results, this will constitute the firmest evidence to date to suggest that the <20 μm fractions of Luna 24 soils or at least a major fraction of them derive directly from basalt; consequently suggesting that the elemental distribution among the grain sizes does in large part derive from differential comminution.

Evidence for Mixing

There are several lines of evidence that suggest that the Luna 24 soils are mixtures of several components.

1. The bulk soils have distinctly more MgO than the rather well-defined suite of ferrobasalts identified among the >250 μm fragments (Ma *et al.*, 1977; Laul, 1977). As can be seen on Fig. 3, the soils are among the most magnesian

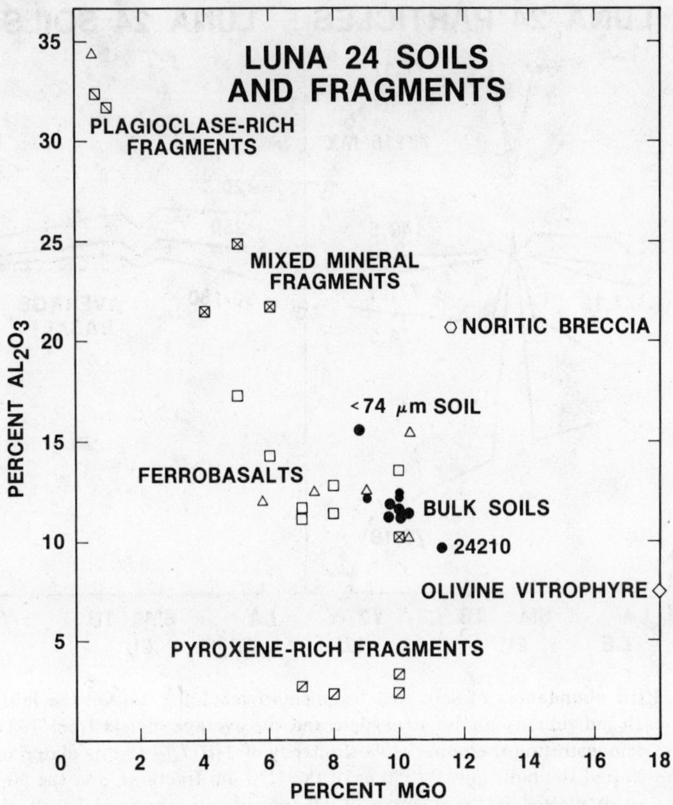

Fig. 3. Al_2O_3 vs. MgO diagram for Luna 24 soils and fragments. Large filled circles are bulk soil analyses from this work and from Ma *et al.* (1977) except for the <74 μm soils which is from Barsukov *et al.* (1977). Open squares are fragments of basalts, plagioclase-rich, pyroxene-rich, and mixed mineral fragments from Ma *et al.* (1977). Triangular symbols are data from Laul (1977). The olivine vitrophyre composition is from Taylor *et al.* (1977). The noritic breccia is plotted as the composition of average matrix of 73215 (Blanchard *et al.*, 1976).

materials at the Luna 24 site. Only the rare olivine vitrophyres (Taylor *et al.*, 1977) are distinctly more magnesian. We discuss the identities of the other components in the soils later, but for now we point out the high MgO content of the soils seems to require at least one component in addition to the ferrobasalt which has MgO content of 10% or more.

2. Bulk soils have somewhat enriched total REE and a more LREE-enriched pattern than the average of the basalt chips analyzed from Luna 24 (Fig. 4). The bulk soils have a net negative Eu anomaly while the average of the basalt chips has a slight positive Eu anomaly. These differences are in the opposite sense of what we expect from differential comminution based on analogy to Korotev's (1976) results; however, with the limited data available, the absolute significance

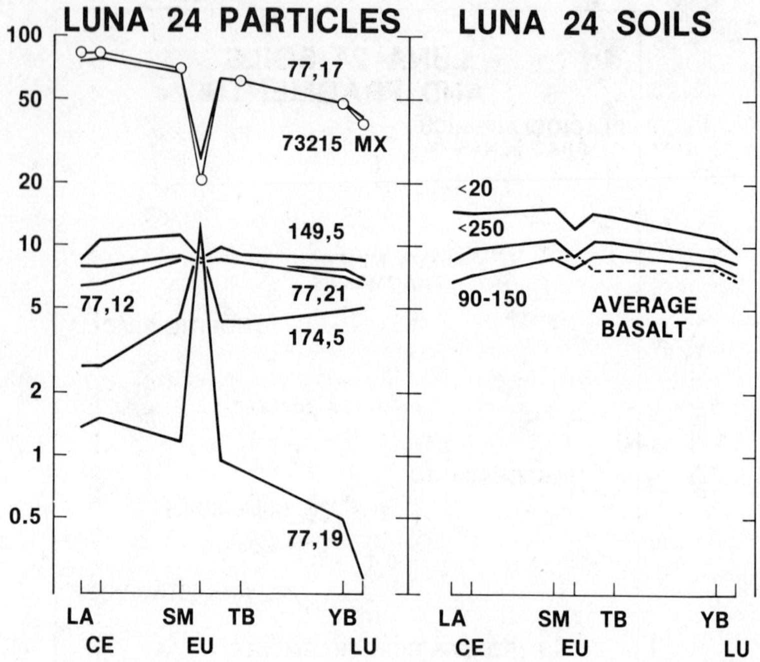

Fig. 4. REE abundances of soils and fragments from Luna 24. On the left side are plotted each individually analyzed particle and the average matrix from 73215 (open symbols) demonstrating the remarkable similarity of 24077,17 to this material. On the right are plotted the bulk soils (<250 μm), the <20 μm fractions, and the 90–150 μm fractions. Also plotted is the average of all individually analyzed basalt chips; the composition was estimated by taking the average of all basaltic particles listed in the footnote to Table 4.

of these differences is questionable. The relationships among the average basalts and bulk and size fraction data on soils is consistent with addition of small amounts of a REE-rich component to the finest fraction.

3. Among the particles we analyzed was a recrystallized breccia nearly identical in composition to the aphanitic matrix of 73215 (Blanchard, *et al.*, 1976). This particle is clearly foreign to the Luna 24 site; its presence in the particle suite makes it not unreasonable to expect to find material of similar composition mixed into the soils.

4. Both the Co concentrations and the Co/Ni ratios in the <20 μm fractions are clearly lower than the bulk soil. This suggests that the <20 μm fraction did not derive from the bulk soil by comminution by micrometeorite impact which tends to increase both Ni and Co concentrations in the <20 μm fraction (Korotev, 1976). There is no Ni data for Korotev's crushed basalt but the concentration of Co in the <20 μm fraction is greater than the bulk rock, thus the trend observed in the Luna 24 soils is in the opposite direction to that caused by differential comminution. We have recently found that the 74001/2 core samples

have a small but significant Co depletion in the <20 μm fraction of the orange and black glass that make up this core (Blanchard and Budahn, 1978). Korotev noted a similar Co depletion in the <20 μm fraction of orange soil 74220 (Korotev, 1976). 74001/2 has had very little surface exposure and has a totally homogeneous composition throughout the core, suggesting that the soil is a single component soil at least in the regions below about 5 cm. We have pointed out the possibility that differential comminution of glass and olivine crystals in the 74001/2 core could explain the Co effect. However, the 74001/2 is a core of very special, unique soil which in many ways is not analogous to the Luna 24 core materials. Consequently, we conclude that the Ni-Co relationships in the size fractions of Luna 24 soils are best explained by addition of a fine-grained component with distinctly higher Ni and lower Co content than the bulk soil.

5. The Luna 24 soils have bimodal size distributions (McKay *et al.*, 1978) strongly suggesting at least two components—a fresh component with coarse mean grain size and a component with a more mature particle size distribution. However, these components do not necessarily have different compositions.

6. Particle track densities of the plagioclase fragments show a bimodal distribution (Blanford and Wood, 1978). The component with which the high track density plagioclases are associated must have received extensive surface exposure, but the measurements of I_s/FeO exposure index for the bulk soils are relatively low (19–39), indicative of a submature to immature soil (Morris, 1978). These observations are consistent if a mature component with high track densities and high I_s/FeO were mixed with a fresh, immature component with low track densities and low I_s/FeO.

7. There is petrographic evidence that some highland material is mixed with the otherwise predominantly basaltic materials which make up the major portion of the Luna 24 soils. Bence and Grove (1977) report finding less than 2% highland material. Basu *et al.* (1977) identified fragments of highland anorthosite and gabbroic anorthosite amounting to ~1.5% of the total soil. Both groups report total absence of feldspathic basalts and KREEP basalts.

8. The population of monomineralic fragments in the Luna 24 soils includes olivines that are far too Mg-rich to have been in equilibrium with the ferrobasalt fragments (e.g., McKay *et al.*, 1978; Taylor *et al.*, 1977). McKay *et al.* (1978) argue that the distribution of monomineralic olivines among the grain size fractions suggests that the component which contributed these forsteritic olivines was coarse-grained; so coarse-grained that it is not represented as fragments in the Luna 24 fragment suite.

There is evidence, therefore, that suggests that both mixing and differential comminution processes have been operative in the Luna 24 regolith, resulting in the observed compositional differences among size fractions. We believe that the evidence for mixing is undeniable and that mixing is primarily responsible for the observed compositions of the soils. However, differential comminution effects cannot be ruled out. Perhaps as more isotopic data become available, especially on the <20 μm size fractions, limits can be placed on the extents to which the two processes have participated to produce the observed effects.

on the <20 μm size fractions, limits can be placed on the extents to which the two processes have participated to produce the observed effects.

Components for Mixing Calculations

We have determined that it is necessary to have at least the following components to fulfill the requirements of the lines of evidence for mixing detailed above.

1. A ferrobasalt/gabbro: This is the dominant rock type identified among the <250 μm fragments; its composition parallels that of the bulk soils more nearly than any of the other fragment types (except soil breccias and agglutinates, of course) and thus it is an obvious soil component.

2. Noritic breccia: This component fulfills the requirements of having high REE abundances with LREE enrichment. It is probably predominantly mixed in as fine material, although we have identified a fragment of this composition among the >250 μm fragments. It is a well identified composition on the moon, especially at the Apollo 17 site. It is a highland material and probably accounts for most of the highland materials found in the Luna 24 regolith. With 180 ppm Ni and 20 ppm Co this component fulfills the requirement of a component with lower Co than the bulk soil, but does not have enough Ni to completely explain the Ni differences.

3. Meteoritic component: Most assuredly, as in almost every lunar soil, some of this component exists. The noritic breccia also probably contains some meteoritic component, thus this component might more aptly be called additional meteoritic component.

4. High-Mg mafic component: The need for this component is based largely on the anomalously high abundances of coarse monomineralic fragments of forsteritic olivine in the soils. These olivines are not equilibrium olivines with the ferrobasalt (McKay *et al.*, 1978). We have attempted to estimate the major element composition of this very coarse-grained component by using the analysis of olivine vitrophyre by Taylor *et al.* (1977). Basu *et al.* (1977) have suggested that the olivine vitrophyre may be a volcanic or impact melted, nearly isochemical, equivalent of the inferred, very coarse-grained cumulate from which the anomalously high abundances of MgO-rich mineral fragments derived. Having no trace element data, we have arbitrarily set the trace element abundances equal to those of the average soil, thus in effect using only the major elements to calculate the proportions of this component in the soils.

5. High-MgO basalt: The necessity for this component is probably the most weakly defined. We based our choice on two factors: (1) There are basalt fragments reported with 10% MgO suggesting, though not unequivocably (see Haskin, 1978), that there may be a high-MgO basalt present at the Luna 24 site; and (2) the relatively few samples of olivine vitrophyre that have been identified (e.g., Basu *et al.*, 1977; Taylor *et al.*, 1977) suggests that a more plentiful, high magnesium component may also be needed.

Results of Mixing Calculations

We have calculated mixtures of the above components whose compositions are in Table 4 for the average <250 μm soil, and each of the 6 individual bulk soils. Fits were calculated using the elements and weighting factors listed in Table 4 except for the size fractions where only the INAA elements were fit. The weighting factors were arbitrarily chosen to emphasize major and trace elements more or less evenly. The mixing program is an adaptation of a program used by

Table 4. Compositions of components used in mixing calculations.

	High-Mg Basalt	Ferro-Basalt	Noritic Breccia	High-Mg Mafic	Meteorite	Relative Weight
SiO_2	42.9	45.2	46.4	44.2	23.1	0.001
TiO_2	0.93	1.03	0.70	0.62	0.08	0.2
Al_2O_3	12.7	12.4	20.6	7.7	1.78	1.0
FeO	19.7	21.0	7.6	21.6	22.0	1.0
MgO	10.2	7.1	11.6	18.2	15.6	1.0
CaO	11.7	12.3	11.6	7.0	1.51	1.0
Na_2O	0.26	0.25	0.66	0.15	0.69	0.2
K_2O	0.027	0.016	0.2	0.05	0.07	0.2
Cr_2O_3	0.39	0.28	0.18	0.53	0.36	0.2
Sc	41	46	13.3	43.1	5.1	1.0
Co	53	40	20.2	51	520	0.5
Ni	365	60	180	140	10,000	0.5
La	2.48	1.6	24.9	2.88	0.33	0.5
Sm	1.70	1.35	12.1	1.93	0.18	1.0
Eu	0.61	0.69	1.74	0.63	0.069	0.5
Tb	0.38	0.29	2.9	0.49	0.047	0.001
Yb	1.63	1.17	9.7	1.8	0.20	0.5

Ferrobasalt—Major element composition is average of:
24077,62; 24174,7; and 24182,12 (Laul, 1977)
24077,63; 24109,76; 24109,78; and 24149,29 (Ma *et al.*, 1977)
24109,51; 24109,52; 24077,41; and 24182,25 (Vaniman and Papike, 1977)
—Trace element composition is average of:
24077,62 and 24182,12 (Laul, 1977)
24077,63; 24109,76; and 24149,29 (Ma *et al.*, 1977)
High Mg Basalt—Major and trace element data is average of:
24077,4 and 24210,50 (Laul, 1977)
24077,81 and 24149,61 (Ma *et al.*, 1977)
24077,12 (this work)
Noritic Breccia—
Major element data is taken from the average matrix of 73215 (Blanchard *et al.*, 1976).
Trace element data from 24077,17 (this work).
High-Mg Mafic—
Major element composition from Taylor *et al.* (1977).
Trace element data arbitrarily set equal to average Luna 24 soil (Table 1) see text.
Meteorite—
Major elements from Goles *et al.* (1971) and Schonfeld and Meyer (1972).
Trace elements from Haskin *et al.* (1968).

the JSC Earth Resources Observation Lab.; the program constrains the components to remain positive.

Our intent was not to calculate *the* definitive combination of components for each soil, but rather to test the reasonableness of the components which, on other grounds, we have defined as necessary. The results of the calculations are given in Table 5. The calculated vs. observed composition of the average bulk soil is given in Table 6 as an example of the typical match between observed and calculated compositions; generally, the calculated compositions closely matched the observed.

Since the proportions of ferrogabbro to high-Mg basalt were hypersensitive to small changes in the weighting parameters or small differences in observed compositions, we conclude there is little or no significance to the ratio of the two

Table 5. Results of mixing calculations on bulk (<250 μm) samples of soil.

	Ferro-Basalt	High-Mg Basalt	Noritic Breccia	High-Mg Mafic	Meteorite
24077	50	29	3	18	0
24109	30	51	2	17	0
24149	64	7	1	28	0
24174	57	17	2	24	0
24182	66	3	4	27	0
24210	No Suitable Fit Calculated				
Avg. Soil	52	20	2	25	

Table 6. Typical result of mixing calculations for average soil.

	Calculated	Observed
SiO_2	45.0	45.4
TiO_2	0.9	1.1
Al_2O_3	11.5	11.1
FeO	20.7	20.5
MgO	10.7	10.2
CaO	11.0	10.9
Na_2O	0.24	0.28
K_2O	0.025	0.030
Cr_2O_3	0.37	0.52
Sc	43.8	43.1
Co	45.2	51.0
Ni	145	140
La	2.63	2.88
Sm	1.82	1.93
Eu	0.64	0.63
Yb	1.65	1.80

basaltic components. Probably the more significant observation about the mixing results is that the totals of the basaltic components (Ferrobasalt + High Mg Basalt) are more or less the same (69–81%) for all the soils. Mixing calculations using only the INAA suite of elements indicate that, if there are two basaltic components as we have suggested, the finest fraction of the soils is much enriched in the High Mg Basalt. The significance of this observation based on a limited number of elements will be more easily assessed when MgO data for the $<20\ \mu m$ fraction becomes available.

The proportion of noritic breccia component ranges from 1 to 4% in the bulk soils. The proportion of this component was not nearly as sensitive to parameter changes in the mixing model as were the basalt components. Clearly the high REE abundances of this component determine its proportion rather firmly. Mixing calculations on size fractions not surprisingly suggest that noritic breccia was concentrated in the $<20\ \mu m$ fraction. Arguments about differential comminution aside, this suggests that the noritic breccia entered these soil samples predominantly as very fine material.

The proportion of our high-Mg mafic component remains relatively constant among the soils. Remembering that the trace element composition is arbitrarily assigned so as to minimize its effect in the calculations, the fit most certainly depends primarily on the MgO difference between the basalts and the soils. Basu *et al.* (1977) has observed that monomineralic fragments account for 50% or more of the soils. Thus both petrographic and mixing model evidence would predict a rather sizable proportion of this coarse-grained component in the soils.

Interestingly there was no apparent requirement for additional meteoritic component in any of the mixes. This is consistent with the argument that the Luna 24 regolith has not been substantially reworked since its emplacement. Whatever meteoritic component may be present in the soils was most probably carried in as part of the noritic breccia. Mixing data on the size fractions require no meteoritic component either, even for the $<20\ \mu m$ fractions, thus re-enforcing our argument for mixing given above.

No mixing calculation adequately described the composition of 24210. This soil has higher FeO and MgO and lower Al_2O_3 than the other soils. As noted previously, the major element analysis by AAS of this soil may be somewhat less precise than the others because of the small sample size (3.51 mg). INAA results indicate that 24210 may be very slightly enriched in mafic component (slightly higher FeO, Sc and Co); however, these enrichments are so slight that we ascribe the inability to obtain a reasonable fit for this soil to the uncertainties in the major element analysis, especially for Al_2O_3 and SiO_2.

Concluding Discussion

The Luna 24 regolith contains compositional evidence for a history of mixing of several components. The soils are well mixed (i.e., have nearly the homogeneous bulk compositions) throughout the length of the core tube yet the size fractions of each of the soils have quite dissimilar compositions. On this and other grounds we conclude that, since emplacement the regolith has not been extensively reworked; reworking would tend to homogenize the elemental distribution among grain size fractions. Most of the properties of the regolith which are indicative of reworking (surface maturity) have been inherited from one or more of the components in the soil.

The observed elemental distributions as a function of grain-size are due in part to admixture of noritic breccia component preferentially to the finest fractions and in part to differential comminution of one or more of the other components of the soil. We note that either the very coarse-grained rock component (whose composition is estimated in mixing calculations by the high-Mg mafic component) and/or the observed ferrobasalt rock type could be a candidate for the object of this inferred differential comminution. At present there is too little data on size fractions and exact compositions of local basalts to estimate the extent to which each of these effects has contributed to the elemental distribution among grain sizes.

Other than the small percentage of noritic breccia in the Luna 24 regolith we see no evidence for any other component that is clearly foreign to the Luna 24 site. The small percentage of noritic breccia does contribute significantly to the REE of the bulk soil, but has only insignificant effect on the major element composition. Among the other probable components, we have identified the ferrobasalt composition and one or more components with higher MgO contents. Mixing results do not definitively identify the high-MgO component. There is evidence from the petrographic analysis of mineral grains that there is probably a very coarse-grained component with very foresterictic olivine contributing to the regolith. Whether this coarse-grained component, which is not represented among the fragments in the soil, can wholly account for the "excess" MgO in the soils or whether there is an additional high-MgO basalt is problematical.

Acknowledgments—We thank Bob Browning and the Texas A & M reactor crew for irradiating the samples. We thank the JSC-SEM lab for their careful preparation of our size fraction samples. We have had very helpful discussions with L. Nyquist regarding appropriateness of mixing model components. We thank C. Hardy for preparing this manuscript. This is Lunar and Planetary Institute Contribution #321.

REFERENCES

Barsukov V. L., Tarasov L. S., Dmitriev L. V., Kalesov G. M., Shevaleevski I. D. and Garanin A. V. (1977) The geochemical and petrochemical features of regolith and rocks from Mare Crisium (preliminary data). *Proc. Lunar Sci. Conf. 8th*, p. 3319–3332.

Basu A., McKay D. S. and Fruland R. M. (1977) Petrography, mineralogy, and source rocks of Luna 24 drill core soils (abstract). In *Papers Presented to the Conference on Luna 24*, p. 18–21. The Lunar Science Institute, Houston.

Bence A. E. and Grove T. L. (1977) The highland component in the Luna 24 core (abstract). In *Papers Presented to the Conference on Luna 24*, p. 22–24. The Lunar Science Institute, Houston.

Blanchard D. P. (1973) Analysis of Apollo 15 rake sample walnuts and lunar breccias 15465,3 and 15565,7 for major and trace elements by atomic absorption spectrophotometry and neutron activation analysis. Ph.D. Thesis, University of Wisconsin.

Blanchard D. P. and Budahn J. R. (1978) Compositional variations among size fractions in a homogeneous, unmatured regolith: Drive core 74001/2 (abstract). In *Lunar and Planetary Science IX*, p. 100–102. Lunar and Planetary Institute, Houston.

Blanchard D. P., Jacobs J. W., Brannon J. C. and Haskin L. A. (1976) Major and trace element compositions of matrix and aphanite clasts from consortium breccia 73215. *Proc. Lunar Sci. Conf. 7th*, p. 2179–2187.

Blanford G. E. and Wood G. C. (1978) Particle track densities in the Luna 24 core. In *Mare Crisium: The View from Luna 24* (R. B. Merrill and J. J. Papike, eds.). This volume.

Bogard D. D. and Hirsch W. C. (1978) Noble gases in Luna 24 core soils. In *Mare Crisium: The View from Luna 24* (R. B. Merrill and J. J. Papike, eds.). This volume.

Goles G. G., Duncan A. R., Lindstrom D. J., Martin M. R., Beyer R. L., Osawa M., Randle K., Meek L. T., Steinborn T. L. and McKay S. M. (1971) Analysis of Apollo 12 specimens: Compositional variations, differentiation processes, and lunar soil mixing models. *Proc. Lunar Sci. Conf. 2nd*, p. 1063–1081.

Haskin L. A. (1978) Trace element composition of Luna 24 Crisium VLT basalt. In *Mare Crisium: The View from Luna 24* (R. B. Merrill and J. J. Papike, eds.). This volume.

Haskin L. A., Wildeman T. R. and Haskin M. A. (1968) An accurate procedure for the determination of the rare earths by neutron activation. *J Radioanal. Chem.* **1**, 337–348.

Jacobs J. W., Korotev R. L., Blanchard D. P. and Haskin L. A. (1977) A well tested procedure for instrumental neutron activation analysis of rocks and minerals. *J. Radioanal. Chem.* **40**, 93–114.

Korotev R. L. (1976) Geochemistry of grain size fractions of soils from the Taurus-Littrow valley floor. *Proc. Lunar Sci. Conf. 7th*, p. 695–726.

Korotev R. L. and Haskin L. A. (1976) Inhomogeneity of trace element distributions from studies of the rare earths and other elements in size fractions of crushed basalt 70135 (abstract). In *Papers Presented to the Conference on Origins of Mare Basalts and Their Implications for Lunar Evolution*, p. 86–90. The Lunar Science Institute, Houston.

Laul J. C. (1977) Chemical composition of Luna 24 melt rocks (24077,4; 24077,62; 24174,7; 24182,12; 24210,50) and gabbro (24182,2) (abstract). In *Papers Presented to the Conference on Luna 24*, p. 94–97. The Lunar Science Institute, Houston.

Ma M.-S., Schmitt R. A., Taylor G. J., Warner R. D., Lange D. E. and Keil K. (1977) Chemical and petrographic studies of 18 Luna 24 lithic fragments (abstract). In *Papers Presented to the Conference on Luna 24*, p. 102–105. The Lunar Science Institute, Houston.

McKay D. S., Basu A. and Waits G. (1978) Grain size and the evolution of Luna 24 soils. In *Mare Crisium: The View from Luna 24* (R. B. Merrill and J. J. Papike, eds.). This volume.

Morris R. V. (1978) FMR and magnetic properties of Luna 24 soils and >1 mm soil particles. In *Mare Crisium: The View from Luna 24* (R. B. Merrill and J. J. Papike, eds.). This volume.

Nyquist L., Wiesmann H., Bansal B., Wooden J. and McKay G. (1978) Chemical and Sr-isotopic characteristics of the Luna 24 samples. In *Mare Crisium: The View from Luna 24* (R. B. Merrill and J. J. Papike, eds.). This volume.

Papanastassiou D. A., McCullock M. T., DePaola D. J., Huneke J. C. and Wasserburg G. J. (1977) Ages and evolution of Luna 24 samples. Unpublished abstract.

Rhodes J. M., Rodgers K. V., Shih C., Bansal B. M., Nyquist L. E., Wiesmann H. and Hubbard N. J. (1974) The relationships between geology and soil chemistry at the Apollo 17 landing site. *Proc. Lunar Sci. Conf. 5th*, p. 1097–1117.

Schonfeld E. and Meyer C., Jr. (1972) The abundances of components of the lunar soils by a least squares mixing-model and the formation age of KREEP. *Proc. Lunar Sci. Conf.* 3rd, p. 1397–1420.

Simonds C. H., Warner J. L., McGee P. E., Cochran A. and Brown R. W. (1977) Luna 24: Petrochemical potpourri (abstract). In *Papers Presented to the Conference on Luna 24*, p. 170–173. The Lunar Science Institute, Houston.

Taylor G. J., Warner R., Wentworth S., Keil K. and Sayeed U. (1977) Petrology and chemistry of Luna 24 mare basalts and basaltic glasses (abstract). In *Papers Presented to the Conference on Luna 24*, p. 189–192. The Lunar Science Institute, Houston.

Vaniman D. T. and Papike J. J. (1977) Luna 24 ferrobasalt: Petrology and comparisons with other mare basalt types (abstract). In *Papers Presented to the Conference on Luna 24*, p. 196–199. The Lunar Science Institute, Houston.

Merrill, R. B. and Papike, J. J., editors.
(1978) *Mare Crisium: The View from Luna 24*, Pergamon Press (New York), p. 631–656
Printed in the United States of America

Chemical and Sr-isotopic characteristics of the Luna 24 samples

L. E. NYQUIST
NASA-Johnson Space Center, Houston, TX 77058

H. WIESMANN, B. BANSAL, J. WOODEN
Lockheed Electronics Co., Inc., Houston, TX 77058

and G. MCKAY
National Research Council, NASA-Johnson Space Center, Houston, TX 77058

Abstract—Analytical data are reported for selected major elements (Mg, Ca, Al, Fe, Ti), trace elements (alkalis, alkaline earths, rare earth elements, U, and Zr) and $^{87}Sr/^{86}Sr$ for five Luna 24 bulk soils. Li, K, Rb, Sr, and $^{87}Sr/^{86}Sr$ are reported for a <20 μm size fraction of soil 24077. Rb, Sr, and $^{87}Sr/^{86}Sr$ are reported for a plagioclase fragment 24077,19 and for plagioclase from the "crushed basalt" 24170.

The major and trace element composition of the bulk soil can be matched by a mixing model which uses "lo-Mg basalt" (52%), "hi-Mg basalt" (23%), "olivine vitrophyre" (20%), LKFM (4%), and anorthositic gabbro (1%) compositions as end member components. The resulting fit to the average bulk soil composition is good with maximum deviation of 3% for major elements and 7% for trace elements between measured and calculated values. The relative proportions of the basaltic components are not well constrained, however.

$^{87}Sr/^{86}Sr$ measured for the plagioclase samples imply an initial $^{87}Sr/^{86}Sr$ for "lo-Mg basalt" which is lower than any we have measured for other mare basalts. Applying the results of the mixing calculation to the Rb-Sr systematics of the soils leads to the conclusion that one or both of the high Mg basaltic components present in the soils probably has a higher Rb/Sr ratio and a more evolved initial $^{87}Sr/^{86}Sr$ than the widely sampled lo-Mg ferrobasalt.

It is possible to fit the Rb-Sr and REE systematics of at least the lo-Mg basalts with a cumulate remelting model. Good fits to the trivalent REE pattern of a lithic fragment whose REE abundances approximate those of average lo-Mg basalt was obtained for ~3–10% partial melting of cumulates containing opx and ol in proportion 1:1, 7–14% of cpx, and small (0–7%) amounts of plagioclase. Features of this model believed to be generally valid include: (i) the presence of pyroxene in the source, implied by light REE fractionation for Luna 24 basalts, (ii) the presence in the source of opx, which discriminates in favor of heavy REE, suggested by the higher relative abundances of heavy REE as compared to other lunar basalts, (iii) the presence of plagioclase in the source, suggested by low Rb/Sr ratios, low initial $^{87}Sr/^{86}Sr$ ratios and positive Eu anomalies (if indeed characteristic). If the Luna 24 basalt source cumulates formed from a global magma ocean of uniform composition it is inferred that they formed after crustal formation but before or coincident with formation of cumulates parental to Apollo 15 basalts.

INTRODUCTION

The Soviet Luna 24 mission to Mare Crisium has proven important for several reasons. It has extended the region of the moon which has been directly sampled. This is particularly significant because Mare Crisium is one of the easternmost maria and is closer to the backside highlands than are the maria sampled by the Apollo missions. Thus, if the lunar backside-frontside assymetry is due in part to primary compositional heterogeneities arising during accretion, study of the Luna 24 samples may provide clues to the nature of these heterogeneities.

The Luna 24 samples have extended the geochemical range of mare basalt types, in the sense that they provide samples of very low TiO_2 content which are unlike most other lunar basalts (Barsukov *et al.*, 1977). They have also focused attention on other very low TiO_2 basalts ("VLT" basalts, Vaniman and Papike, 1977a, Taylor *et al.*, 1977a). One of the questions which must now be answered is whether this extension of lunar basalt types reflects primary regional lunar heterogeneities. Conversely, can the Luna 24 basalts be understood within the framework of earlier hypotheses of lunar evolution?

We have attempted to contribute to answering these questions by (i) determining the chemical and Sr-isotopic characteristics of Luna 24 samples, and (ii) by trying to interpret these characteristics within the framework of the scenario of trace element and Sr-isotopic evolution of lunar mare basalts presented by Nyquist *et al.* (1977).

It was obvious that the study of Luna 24 samples would be hindered by sampling problems. On one hand there is the need to sample all lithologies present at the Luna 24 site in order to get a realistic picture of the make-up of the site. On the other hand, there is a need to determine representative analyses of the lithologies which are sampled. Our analytical techniques require a minimum sample size of ~3 mg, hindering us in sampling individual lithologies because of the scarcity of comparatively large lithic fragments among the Luna 24 samples. In light of these considerations and the realities of sample availability we adopted a strategy of first analysing bulk soils in order to get an overview of regolith composition. These bulk soil analyses were followed by analyses of individual fragments and soil fractions to characterize end member components. These latter analyses are as yet incomplete. In them we are cooperating with independent investigations conducted under the leadership of D. Blanchard, L. Haskin and K. Keil. Because our general objectives are similar to those of several other investigators this paper will also make extensive use of data from the literature.

Analytical Procedures

During our soil analyses we determined a large suite of major and trace elements in addition to Rb, Sr, and $^{87}Sr/^{86}Sr$ to maximize the data obtained from each sample and to correlate our analyses to those of other investigators. Figure 1 shows the analytical scheme. The sample was first dissolved in conc. HF + conc. $HC10_4$ and "spiked" with enriched isotopes for trace element concentration measurements by the mass spectrometric isotope dilution technique (MSID). Because contamination from major element spikes is significant in comparison to the amounts of trace elements, an ~5% aliquant of the sample solution was weighed out and separately spiked for Ti, Ca, and Mg analysis by MSID. The aliquant was diluted to 100 ml., 20 ml of this solution was passed through a cation exchange (AG 50 × 12, 200–400 mesh) column and Ca and Mg separated out prior to MSID analysis. Another 20 ml portion was passed through an anion exchange (AG 1 × 8) column to separate Fe and Al. An additional 10 ml portion from the major element aliquant was first passed through the anion column to remove Fe. The effluent and washings from this column were passed

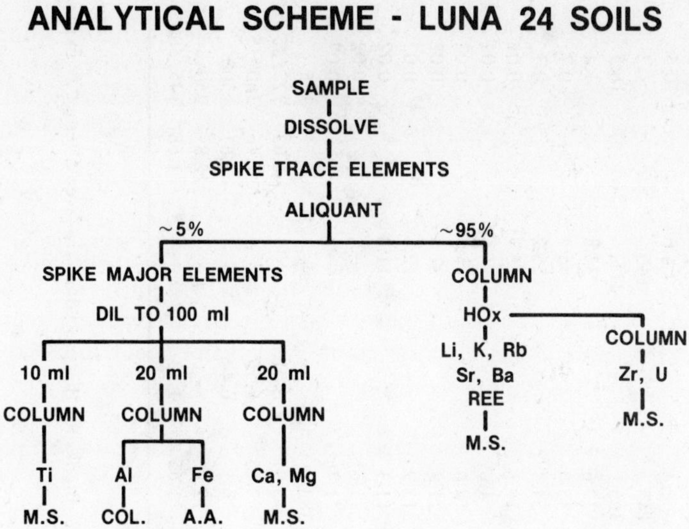

Fig. 1. Analytical scheme for Luna 24 bulk soils. HOx = oxalic acid. An abbreviated scheme is used for lithic fragments and soil fractions (see text).

through another cation exchange (AG 50 × 8) column at high ionic strength (8M $HC1O_4$ + 1M HCl mix) and Ti was eluted using a (3.6M $HC1O_4$ + 5.4M HCl) mix. Al was determined by colorimetry, Fe by atomic absorption spectrometry, and Ti by MSID.

The larger sample aliquant (~95%) was processed through a cation exchange column (same as for Ca, Mg) using an elution procedure similar to that of Gast *et al.* (1970). In this procedure, due to the loading of the sample in 0.1M oxalic acid, U and Zr are non-adsorbed elements and consequently come in the oxalic acid effluent. This oxalic acid fraction is further separated into its constituents by a series of anion and cation exchange columns to obtain single element fractions for U and Zr prior to mass spectrometric analysis. After passing 0.1M oxalic acid Li, K, Rb, Sr, Ba, Ce and La are separated into single element fractions by sequential elution with 0.6N HBr, 1.2N HBr, 2.0N HCL, 2.5N, 3.0N, and 4.0N HNO_3. The REE elements Nd-Lu are divided into three fractions in the 2.5N HNO_3 effluent. Heavy and light REE are enriched in the first and last fractions, respectively. The stable isotope spikes are chosen to minimize concentration uncertainties arising from isobaric interference during mass spectrometric analysis. Nevertheless, some troublesome interferences arise during REE analysis and must be "burnt off" in the mass spectrometer. This problem is most acute for the heavy REE because of isobaric interference from the light REE oxides. Analysis of two REE fractions for the same element serves as a check that interferences have not affected the result.

The concentration of spike solutions is determined by calibration with standard solutions. NBS supplied salts were used for Li, K, Rb, and Sr standard solutions. Commercially available ultra high purity salts (99.99%) were used for REE + Ba standard solutions. These solutions were checked by titration against EDTA.

The blanks for the multi-element chemical procedure used for the soils are given in Table 1. Chemical procedures for other samples are simplified from this basic procedure depending on the goals of the analysis and prior sample history. Samples 24077,19 and 24077,9004 had been previously analysed by INAA (Blanchard *et al.*, 1977). Sample 24077,19 was a nearly pure plagioclase grain and was analysed only for Rb,Sr, and $^{87}Sr/^{86}Sr$ to obtain an estimate of initial $^{87}Sr/^{86}Sr$ for the parent basalt. Sample 24077,9004 was a <20 μm soil sieve fraction. Li, K, Rb, Sr, Ba, Mg, Ca, Ti, and Al, as well as $^{87}Sr/^{86}Sr$, were measured to complement the INAA analyses. Because of the small sample size it was not possible to determine the major elements on a sample aliquant. Because of cross

Table 1. Major and trace element analyses of Luna 24 soils.

Sample #, wt.	24077,36 26mg	24210,12 19mg	24182,20 36mg	24109,17 27mg	24149,9 18mg	24077,<20 μm[a] 5mg	12070,50[b] 1977a	12070,50[c] 1977b	12070,50[d] 1974	12070[e] XRF	Blank ng
Li ppm	5.6	6.0	6.2	5.7	6.0	9.2	17.4		17.2		1.4
K ppm	252	246	294	256	226	537	2000		2077		120
Rb ppm	0.545	0.581	0.699	0.549	0.468	1.17	6.48		6.64		0.5
Sr ppm	94.2	86.9	96.9	90.1	90.6	117.4	145		146		1.2
Ba ppm	37.2	36.3	45.5	37.0	34.8	69.8	365		375		16.5
La ppm	3.09	2.94	3.72	3.11	3.01	—	33.6		33.8		1.4
Ce ppm	7.99	7.50	9.55	7.64	7.38	—	85.0		86.7		0.7
Nd ppm	5.74	5.44	7.02	5.60	5.59	—	54.2		55.2		0.3
Sm ppm	1.87	1.78	2.23	1.83	1.86	—	16.1		15.9		0.02
Eu ppm	0.687	0.637	0.738	0.662	0.671	—	1.69		1.68		0.02
Gd ppm	2.47	2.37	2.90	2.44	2.43	—	18.6		19.4		0.02
Dy ppm	2.95	2.88	3.53	2.95	3.02	—	21.8		21.5		0.02
Er ppm	1.85	1.80	2.14	1.86	1.91	—	12.9		12.8		0.02
Yb ppm	1.66	1.65	1.92	1.68	1.69	—	11.3		11.8		0.02
Lu ppm	0.243	0.239	0.273	0.250	0.249	—	1.61		—		0.02
U ppm	0.12	0.12	—	0.08	0.13	—	1.76		—		0.07
Zr ppm	44	36	49	38	44	—	383		—		365
Mg %	5.67	6.18	5.83	5.83	5.83	4.69	5.91	6.01	—	6.03	725/35
Ca %	8.15	7.83	8.10	8.09	8.12	8.13	7.45	7.56	—	7.46	3600/5
Al %	7.20	5.91	6.52	6.36	6.25	7.44	6.5	6.44	—	6.61	1500
Fe %	14.5	14.9	14.34	14.5	14.8	—	12.7	—	—	12.8	250
Ti %	0.59	0.58	0.65	0.64	0.66	0.63	1.65	1.53	—	1.68	145/500

[a]See Blanchard *et al.* (1977) for INAA.

[b]Control sample analysed using the aliquanting procedure.

[c]Control sample analysed by spiking Mg, Ca, and Ti after elution from cation exchange column. Blanks for this procedure are given after a "/" in the last column.

[d]Wiesmann and Hubbard (1975).

[e]XRF analysis by Rhodes (1972).

contamination from the major element spikes it was necessary to first spike for trace elements and to collect the major element fractions as they were eluted from the cation exchange column. Ca, Mg, and Ti spikes were then added as before and concentration determined by MSID, while Al was determined by colorimetry. This post-spiking procedure has an inherent risk because of the possibility that the elements being analysed will not be quantitatively removed from the column. However, careful column calibration and analysis of the 12070 control sample (Table 1) showed the procedure to be reliable to within at least a few percent. Rb and Sr blanks were ~0.04 ng and ~0.8 ng, respectively, for 24077,19 and 24077,9004.

Sr-isotope mass spectrometry followed the standard procedures of this laboratory (cf, Nyquist *et al.*, 1977). An aliquant of the Sr fraction of 24170 plagioclase was transferred to us for Sr-isotopic analysis by G. J. Wasserburg. Sr-isotopic measurements made in the Cal. Tech. and JSC laboratories can be directly compared for this sample (see Lunatic Asylum, 1978). The two analyses differ by 0.00007 ± 0.00007, within the previously established interlaboratory bias of 0.00012 ± 0.00005 (Nyquist *et al.*, 1972).

Results

Chemical data

Major and trace element data for Luna 24 soils are reported in Table 1. Because the soils are mixtures, interpretation of our data for them requires knowledge of the spectrum of rock types which contribute to the regolith at the Luna 24 site. In this section we identify candidates for important chemically defined rock types at the site. We then discuss the results of a least squares mixing calculation undertaken to test the hypothesis that these rock types comprise the Luna 24 regolith.

Major and trace element data for lithic fragments have been reported by Barsukov *et al.* (1977), Blanchard *et al.* (1977), Laul (1977) and Ma *et al.* (1977). A more extensive major element data base exists, but in order to best perform the mixing calculation, we need to know both trace and major element abundances, and thus we relied most heavily on the work of the above authors.

Based on chemical composition, a number of distinct rock types can be identified among lithic fragments whose compositions are reported in the literature. Differences between these rock types are well illustrated on a plot of Al_2O_3 vs. MgO. Figure 2 shows abundances of these elements taken from the literature for 19 lithic fragments of mare composition as well as our own abundances for bulk soils. The fragment data define three groups. The average compositions of each of the groups is given in Table 2 and plotted in Fig. 2.

The mare basalt types and corresponding compositional groups are: Group 1. A basalt/gabbro with very low TiO_2 (~1%) and low MgO (~6.5%). This is the most widely sampled basalt lithology comprising 14 of the 19 fragments plotted in Fig. 2. Fragments of this general composition are referred to as "cristobalite gabbro," "dolerite," and "medium grained basalt" by Tarasov *et al.* (1977), "Luna 24 ferrobasalt" (Vaniman and Papike, 1977b,c) "subophitic to ophitic VLT low-Mg basalts" (Ryder *et al.*, 1977) and "fine grained basalt" (Taylor *et al.*, 1977c). They are isochemical to the "low-Mg mare" glass cluster of Simonds *et al.* (1977) and to Luna 24 brown glass (Vaniman and Papike, 1977c). This composition is referred to here as the Luna 24 low-Mg basalt composition. Group

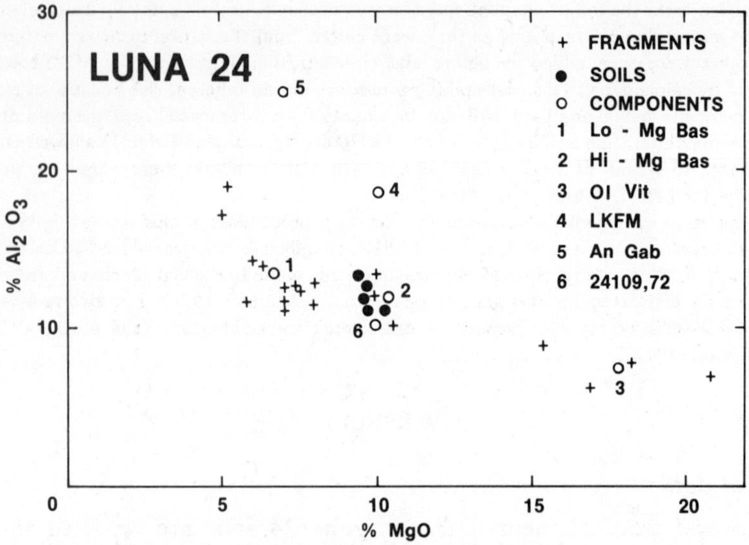

Fig. 2. Al_2O_3 vs. MgO for bulk soils (present investigation), lithic fragments, and average lithic fragment compositions (references in Table 2).

2. VLT basalt with ~10% MgO. Only four of the fragments plotted in Fig. 2 have MgO ~10%. One of these (24109,72) has TiO_2 = 2% and trace element abundances three-fold higher than the others (Ma *et al.*, 1977). This fragment may belong to a distinct lithology or may reflect a sampling bias. The average composition of the other three fragments is given in Table 3 as "hi-Mg basalt." This composition is isochemical, or nearly so, to the "hi-Mg mare" glass cluster of Simonds *et al.* (1977) and to Luna 24 green glass (Norman *et al.*, 1977; Vaniman and Papike, 1977b). Group 3. Olivine vitrophyre. Four fragments of this composition have been reported by Ryder *et al.* (1977) and Taylor *et al.* (1977a) and are plotted in Fig. 2 at MgO values between 15–20%.

The compositions of a number of Luna 24 non-mare lithic fragments have been reported in the literature. These fragments tend to fall within two distinct compositional groups which are shown in Fig. 2. The groups are: Group 4. Low-K Fra Mauro (LKFM) basalt. Taylor *et al.* (1977b) report data for two "aphanites" with this composition. The average composition (Table 2 and Fig. 2) is similar to that of the LKFM glass cluster of Simonds *et al.* (1977). Group 5. Anorthositic gabbro or "highland basalt." Taylor *et al.* (1977b) report seven non-mare lithic fragments of this composition. The average composition (Table 2 and Fig. 2) is similar to that of the "highland basalt" glass cluster of Simonds *et al.* (1977).

We have used a least squares mixing calculation to test the hypothesis that the above rock types comprise the regolith at the Luna 24 site. Sample 24109,72 was included in these calculations as a representative of "Group 6", possibly a

Table 2. Mixing model components and results.

Element	Lo-Mg[a] Basalt	Hi-Mg[b] Basalt	Olivine[c] Vit.	LKFM[d]	Anor.[e] Gab.	24109[f] ,72	Avg.[g] Soil	Calc. Soil
SiO_2 %	46.0	43.9	44.4	45.8	44.4	44.1	45.4	45.1
TiO_2 %	1.15	0.95	0.91	1.44	0.29	2.0	1.04	1.06
Al_2O_3 %	13.4	11.8	7.4	18.6	24.9	10.2	12.2	12.1
FeO %	19.8	19.8	20.0	10.8	6.6	24.1	18.8	19.3
MgO %	6.7	10.2	17.8	10.1	7.0	10.0	9.74	9.81
CaO %	12.5	11.5	7.7	12.0	14.8	8.1	11.3	11.3
Na_2O %	0.33	0.26	0.12	0.60	0.43	0.33	0.28	0.28
La ppm	2.00	2.20	(2.20)	24.5	4.07	9.1	3.17	3.01
Sm ppm	1.56	1.63	(1.63)	12.1	1.81	6.7	1.91	2.02
Eu ppm	0.72	0.57	(0.57)	1.7	0.89	1.6	0.68	0.70
Dy ppm	2.29	2.27	(2.27)	16.8	2.65	7.8	3.07	2.87
Yb ppm	1.49	1.47	(1.47)	9.7	1.70	4.6	1.72	1.81
Sc ppm	44.7	38.0	(38.0)	13.0	10.0	16.0	42.9	40.1
Component %	52.5	23.1	19.5	4.1	0.9	0		

[a]Majors: 24170.1-0.2; 24143.4-006.1; 24067.4-012 Barsukov *et al.* (1977) 24077,14; 24174,7; 24182,12 Laul (1977) 24174,5 Blanchard *et al.* (1977); 24077,63; 24109,76; 24109,78; 24149,29; 24149,69; 24174,75 Ma *et al.* (1977) SiO_2 by difference when not measured.
Traces: Above plus 24092.4-001 Barsukov *et al.* (1977).

[b]Majors and traces: 24077,4 Laul (1977) and 24077,81; 24149,61 Ma *et al.* (1977). SiO_2 by difference when not measured.

[c]Majors: 24149,42-18; 24174,65-7; 24174,59–12 Taylor *et al.* (1977c). "Olivine vitrophyre" of Marvin *et al.* (1977).
Traces: Assumed same as for Hi-Mg basalt.

[d]Majors: Aphanites 24077,43-30; 24077,43-31 Taylor *et al.* (1977b).
Traces: 24077,17 Blanchard *et al.* (1977).

[e]Majors: 24109,60-45; 24174,58-41; 24210,43-45; 24174,58-44; 24174,60-65; 24210,43-20; 24174,65-19 Taylor *et al.* (1977b).
Traces: 15445,17; 67955,56; 76315,62; 76230.4; 77017,2; 78155,2 Hubbard *et al.* (1974).

[f]Ma *et al.* (1977).

[g]24077,36; 24109,17; 24149,9; 24182,20; 24210,12. This investigation except SiO_2, Na_2O, and Sc from Blanchard *et al.* (1977).

separate rock type. The abundance of some of the component lithologies (especially LKFM) may be calculated with more precision if trace elements are included in the mixing calculation. Trace element abundances are directly available for the low- and hi-Mg basalts and 24109,72 (the reason for our choice of the data base) but not for the other components. Blanchard *et al.* (1977) have found REE and other trace element abundances in 24077,17 to be similar to those of Apollo 17 noritic breccias. It is inferred that this fragment is of LKFM composition and trace element values measured for it are assumed representative of LKFM at the Luna 24 site. Trace element data are not yet available for anorthositic gabbros from the Luna 24 site. We have chosen an average value characteristic of Apollo 15, 16, and 17 anorthositic gabbros (Hubbard *et al.*, 1974) for our mixing calculation.

We have no reliable way of estimating trace element abundances for the olivine vitrophyre component. The high MgO content of the soils, compared to that of the widely sampled low-Mg basalt, requires the presence of a component with MgO content even higher than that of the soils. Since olivine vitrophyre is the only observed lithic type with sufficiently high MgO content, we include it as a component in the mixing calculation despite the lack of reliable trace element data. As a first approximation, we estimate the trace element content of olivine vitrophyre to be the same as that of the hi-Mg basalt. This estimate is supported by the apparent similarity of TiO_2 contents for these two basalt types, since TiO_2 often behaves as an incompatible element.

The results of the mixing calculation, including best-fit component abundances and the resulting calculated soil composition, are given in Table 2. These results indicate that the soil is dominated by lo-Mg and hi-Mg basalts and olivine vitrophyre. Agreement between calculated soil and average soil is better than 3% for major elements and 7% for trace elements. Hence a good fit to the average soil composition can be obtained by using the compositions given in Table 2 as end members.

There is considerable uncertainty in the relative abundances of the mare components. This uncertainty arises primarily because the three mare components are nearly colinear on correlation diagrams for any two major elements and because the hi-Mg basalt component is so close to the bulk soil in composition. Thus the major element composition of the soil can be closely matched for any combination of the above three mare components satisfying the requirement that the ratio lo-Mg basalt:olivine vitrophyre $\approx$2:1 regardless of the percentage of hi-Mg basalt component.

The 52% lo-Mg basalt component given by the mixing calculation seems reasonable based on its frequent occurrence among the lithic fragments. In contrast, the 20% abundance of "olivine vitrophyre" yielded by the mixing calculation contradicts the apparent rarity of this rock type among the fragments (Marvin *et al.*, 1977). Monomineralic olivine fragments of $\sim Fo_{75}$, which are abundant in the Luna 24 soils, have been interpreted by Basu *et al.* (1977) to be comminution products of a coarse-grained ultramafic cumulate. These authors propose that olivine vitrophyre is the volcanic isochemical equivalent of this cumulate. Hence it is possible that the "olivine vitrophyre" chemical component used in the mixing calculation is present in the soil primarily in a petrographically unrecognizable form. The 23% abundance calculated for the hi-Mg basalt component suggests that this component may be under-represented among the fragments sampled. However, as discussed above, the computed abundance for this component is extremely sensitive to the input parameters of the mixing calculation.

The $\sim$1% contribution calculated for the anorthositic gabbro component is very sensitive to the Al_2O_3 and FeO concentrations used for end members and soil. However, none of our trials of various input parameters yielded abundances of this component in excess of 1% and several trials yielded 0%. The 0.9% value given in Table 2 is within the range of 0–2% "highland-derived ANT" found by

McKay *et al.* (1977) for the various size fractions of the Luna 24 soil and in agreement with the estimate of Bence and Grove (1977) of <2% of highland lithologies in the 0.15–0.5 mm size fraction of the soil.

The mixing model given in Table 2 does not require any contribution from fragments such as 24109,72 (TiO_2 = 2%). This result derives in large measure from the comparatively high TiO_2 content used for the lo-Mg basalt. If a low TiO_2 value such as the 0.69% given by Taylor *et al.* (1977c) for their "fine grained basalt" is used for the lo-Mg component, abundances of ~10% are calculated for the 24109,72-type component. In this case the increase occurs primarily at the expenses of the LKFM and hi-Mg basalt components.

The ~4% contribution calculated for the LKFM component is strongly controlled by REE abundances. Figure 3 shows chondrite normalized REE abundances for average bulk soil and average lithic fragments. The calculated soil values obtained from the mixing model are also shown and closely match the

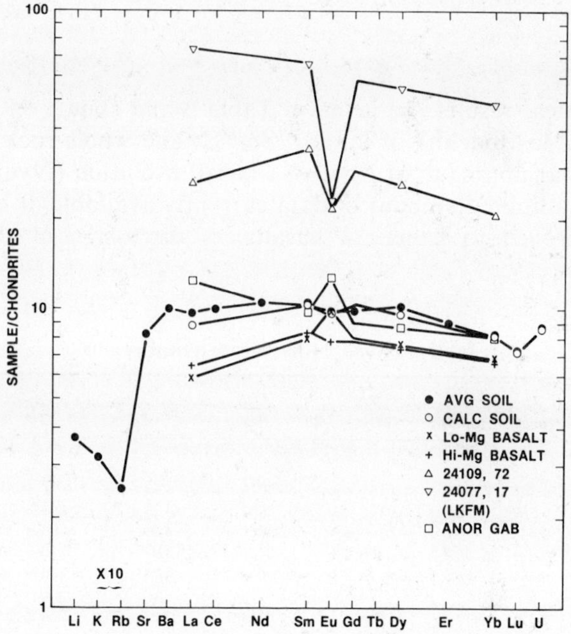

Fig. 3. Chondrite-normalized trace element abundances. "Average soil" is the average for the five Luna 24 bulk soils listed in Table 1. "Calc soil" gives the values from the mixing calculation (Table 2). Lo-Mg and hi-Mg basalts are average compositions from Table 2. Data for 24109,72 are from Ma *et al.* (1977), those for 24077,17 are from Blanchard *et al.* (1977). "Anor Gab" values are the average for four anorthositic gabbros analysed by Hubbard *et al.* (1974) (see footnote to Table 2). Data from this laboratory normalized to the "chondrite composite" values of Hubbard and Gast (1971). Data of Blanchard *et al.* (1977) are normalized to the chondritic values of Haskin *et al.* (1968). All other data are normalized to the chondritic values quoted by Laul *et al.* (1974).

average measured values. It should be noted that we have chosen to normalize the fragment data and calculated soil values to the chondritic abundances used by the Oregon State Univ. group (Laul *et al.* 1974), because data from this group dominate the lithic fragment averages. However, because of what we believe to be systematic interlaboratory differences we have chosen to normalize data from other investigators to their own chondrite values. We believe that this produces the most self-consistent chondrite normalized abundance patterns.

In summary, the available major and trace element chemical data for Luna 24 samples indicate the presence of as many as three basaltic compositions in the regolith. These components together dominate the soil (~95%) but their relative proportions are not well determined. REE abundances indicate the probable presence of ~4% of LKFM in the soil. This conclusion is weakened by lack of REE data for the "olivine vitrophyre" component. The low MgO values of the apparently dominant lo-Mg basalt and the significantly higher MgO values of the bulk soils seem to require a significant proportion of a high MgO component chemically similar to the olivine vitrophyres.

Rb-Sr Data

Rb-Sr analytical results are given in Table 3 and shown on the Sr-evolution diagram in Fig. 4. Modeling of initial $^{87}Sr/^{86}Sr$ and whole rock $^{87}Rb/^{86}Sr$ ratios was central to our approach to Apollo 12 basalt evolution (Nyquist *et al.* 1977). Despite the very limited amount of data currently available, it is still possible to apply this approach to Luna 24 basalts. A discussion of the data base is appropriate first.

Table 3. Luna 24 Rb-Sr analytical results.

Sample	wt (mg)	Rb (ppm)	Sr (ppm)	$^{87}Rb/^{86}Sr$ [a]	$^{87}Sr/^{86}Sr$ [b]	T_B [c]	T_L [d]	I(3.3) [e]	I(3.65) [f]
I. Soils									
24077,36	26	0.545	94.2	0.0167 ± 3	0.70026 ± 6	4.83 ± .33	5.11 ± .33	0.69949	0.69939
24109,17	26	0.549	90.1	0.0176 ± 3	0.70020 ± 5	4.36 ± .27	4.63 ± .27	0.69939	0.69928
24149,19	18	0.468	90.6	0.0149 ± 3	0.70009 ± 5	4.67 ± .33	4.99 ± .33	0.69941	0.69932
24182,20	36	0.699	96.9	0.0209 ± 3	0.70041 ± 7	4.37 ± .30	4.60 ± .30	0.69945	0.69932
24210,12	19	0.581	86.9	0.0193 ± 4	0.70043 ± 6	4.78 ± .33	5.04 ± .33	0.69954	0.69942
24077,9004[g] (<20μ)	5	1.17	117.4	0.0288 ± 3	0.70054 ± 8	3.51 ± .23	3.68 ± .23	0.69919	0.69904
II. Basalt Fragments									
24077,19	4.1	0.145	164.4	0.00255 ± 8	0.69921 ± 10	—	—	0.69909	0.69908
24170PL[h]	19	0.179	208.5	0.00248	0.69926 ± 6			0.69914	0.69913

[a]Uncertainties correspond to last figures.
[b]Uncertainties correspond to last figures and are $2\sigma_m$. Normalized to $^{88}Sr/^{86}Sr = 8.37521$.
[c]Model age assuming I = 0.69910 (BABI plus our bias).
[d]Model age assuming I = 0.69903 (Apollo 16 anorthosites for T = 4.6 AE).
[e]Initial $^{87}Sr/^{86}Sr$ assuming T = 3.3 AE.
[f]Initial $^{87}Sr/^{86}Sr$ assuming T = 3.65 AE.
[g]Sample analyzed by INAA by Blanchard *et al.* (1977).
[h]Sr-measured on an aliquot of the Sr-fraction from 24170 plagioclase (Lunatic Asylum, 1978). Rb as reported by Lunatic Asylum, (1978).

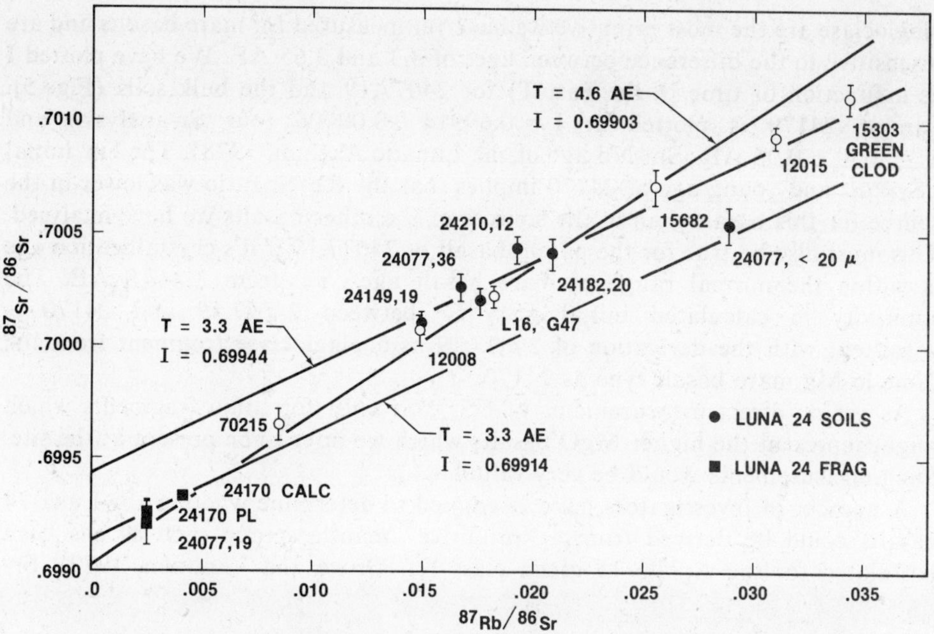

Fig. 4. $^{87}Sr/^{86}Sr$ vs. $^{87}Rb/^{86}Sr$ for data from Table 3. Data from our laboratory for selected mare basalts, Luna 16 soil (L16, G47) and green clod 15303 are shown for reference (open circles). Reference isochrons for an age of 3.3 AE (Sm-Nd age of 24170, Lunatic Asylum, 1978) are shown for 24170 plagioclase and for bulk soils. A reference isochron of age 4.6 AE is also shown through LUNI (Lunar Initial from Apollo 16 anorthosites) = 0.69903. The 24170 "whole rock" datum was calculated from the mineral separate data and estimated mode of the Lunatic Asylum (1978).

The currently available Sr-isotopic data for Luna 24 lithic fragments are limited to only three investigations and two basalts. The samples studied were: 24170 (Lunatic Asylum, 1978 and this investigation), actually a coarse layer in the Luna 24 core believed to be a crushed rock; 24171 (Birck *et al.*, 1977), the same material as 24170; and 24077,19 (this investigation) a single 4 mg plagioclase chip believed to be from a mare basalt because of its visual appearance and FeO content of 0.8%.

Sample 24170 is a lo-Mg basalt (Taylor *et al.*, 1977c) and a Sm-Nd age of 3.30 ± 0.05 AE has been determined for it (Lunatic Asylum, 1978). There is some age ambiguity arising from seemingly older but much less precise Rb-Sr ages for 24170 and 24171 of 3.7 ± 0.6 AE (Lunatic Asylum, 1978; Birck *et al.*, 1977) and the older $^{39}Ar/^{40}Ar$ age of 3.65 ± 0.12 AE reported by Stettler and Albarede (1977) for lithic fragment L24A from core sample 24096. The chemical data reported by Stettler and Albarede (1977) suggest that L24A is also a lo-Mg basalt.

We have formally calculated initial $^{87}Sr/^{86}Sr$ (I) for our samples for ages of 3.3 and 3.65 AE (Table 3). The initial $^{87}Sr/^{86}Sr$ values for 24077,19 and 24170 plagioclase are the most primitive we have yet measured for mare basalts and are insensitive to the difference between ages of 3.3 and 3.65 AE. We have plotted I as a function of time in the past(T) for 24077,19 and the bulk soils (Fig. 5). Sample 24170 is plotted at I = 0.69914 ± 0.00006 (our Sr-analysis) and T = 3.30 ± 0.05 AE (Sm-Nd age of the Lunatic Asylum, 1978). The low initial $^{87}Sr/^{86}Sr$ and young age of 24170 implies that the Rb/Sr ratio was lower in the source for this basalt than in the sources of the other basalts we have analysed. This must also be true for the parent basalt of 24077,19 if it's crystallization age is within the normal range of mare basalt ages, i.e. from 3.2–3.9 AE. The similarity in calculated initial $^{87}Sr/^{86}Sr$ between 24077,19 and 24170 is consistent with the derivation of 24077,19 as a plagioclase fragment from the same lo-Mg mare basalt type as 24170.

As yet no direct measurements of $^{87}Sr/^{86}Sr$ exist for lithic fragments which might represent the higher MgO basalts which we infer to be present at the site. Such measurements would be very valuable.

A number of investigators have attempted to determine whether the Luna 24 basalts could be derived from a "primitive" mantle source such as has been postulated for the Apollo 15 green glass (cf. Grove and Vaniman, 1977). Sr-

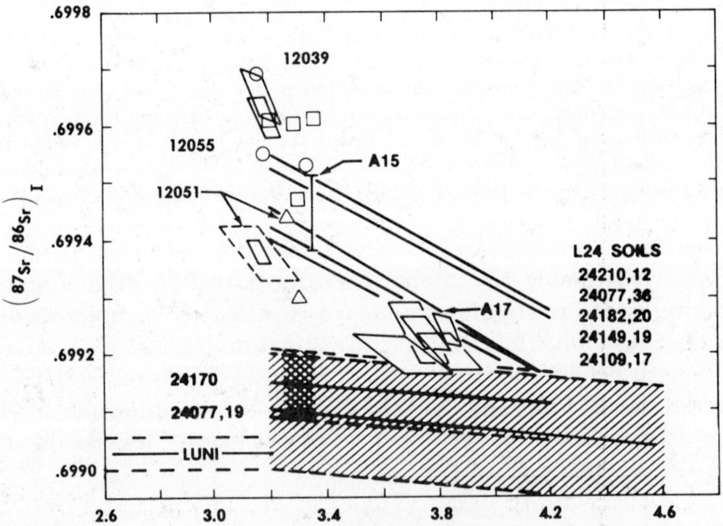

Fig. 5. Initial $^{87}Sr/^{86}Sr$ vs. time before present. Parallelograms show ages (T) and initial $^{87}Sr/^{86}Sr$ values (I) for Apollo 12 and Apollo 17 basalts as determined in our laboratory (Nyquist *et al.*, 1976, 1977). Circles, squares and triangles are (T,I) values for Apollo 12 basalts as determined in the Cal Tech laboratory (Papanastassiou and Wasserburg, 1971) adjusted for interlaboratory bias. The range in I calculated from whole rock Rb-Sr data for Apollo 15 basalts (Nyquist *et al.*, 1973) is shown by the vertical bar. (T,I) parameters for 24170 (cross-hatched area) are from our plag analysis and the Sm-Nd age of the Lunatic Asylum (1978). $^{87}Sr/^{86}Sr$ is shown as a function of time in the past for plag fragment 24077,19 and bulk Luna 24 soils.

isotopic data for a green glass clod (Fig. 4 and Nyquist *et al.*, 1973) combined with the $^{39}Ar/^{40}Ar$ age of 3.38 ± .06 AE (Podosek and Huneke, 1973) imply an initial $^{87}Sr/^{86}Sr$ for Apollo 15 green glass of 0.6994–0.6995, well within the range for normal Apollo 15 mare basalts (Fig. 5). Thus the Sr-isotopic data are inconsistent with derivation of the lo-Mg Luna 24 basalt from the source of Apollo 15 green glass. It would be of considerable interest to know whether the other Luna 24 basalt-types with higher MgO contents could be derived from the green glass source. This is particularly true of the olivine vitrophyres, which are very close to Apollo 15 green glass in bulk composition.

We can apply the mixing relations discussed above to the bulk soil Rb-Sr data to place limits on the initial $^{87}Sr/^{86}Sr$ of the high MgO rocks. To do so requires knowledge of Rb and Sr for all components except the hi-Mg basalt and olivine vitrophyres which must be considered together in this calculation.

The Lunatic Asylum (1978) report Rb = 0.114 ppm, Sr = 104 ppm, $^{87}Rb/^{86}Sr$ = 0.0032 and $^{87}Sr/^{86}Sr$ = 0.69923 ± 0.00007 for a 2.5 mg chip of 24170. Because of the small sample size and coarse grain size of 24170 it is probable that this chip does not truly represent the whole rock. Whole rock Rb-Sr data can be estimated from the Rb-Sr mineral data and the rock mode of 30% plagioclase, 65% pyroxene, and ~5% olivine as estimated by these authors from their mineral yields. We obtain Rb = 0.106 ppm and Sr = 74.4 ppm if Rb and Sr concentrations in olivine are assumed equal to those in pyroxene (the olivine contributions must be small in any case). From the calculated $^{87}Rb/^{86}Sr$ = 0.0041 and a 3.3 AE isochron through our 24170 plagioclase analysis in Fig. 4 we obtain a whole rock $^{87}Sr/^{86}Sr$ ratio of 0.69934 ("24170 calc" in Fig. 4). We prefer these values for 24170 "whole rock" and use them throughout the remainder of the paper. In particular we adopt these values as characteristic of lo-Mg basalt.

Application of the mixing model to the Rb-Sr data also requires knowledge of the Rb-Sr systematics of LKFM and anorthositic gabbro. We assume these components to be the same at the Luna 24 site as elsewhere on the moon. We adopt values of Rb = 6.6 ppm and Sr = 175 ppm for LKFM based on analyses of Apollo 17 noritic breccias (Nyquist *et al.*, 1974). For anorthositic gabbro, Rb = 1.41 ppm and Sr = 148 ppm is obtained by averaging data for the six samples used as the data base for the mixing calculation (see footnote to Table 2).

With the above Rb and Sr values for lo-Mg basalt, LKFM, and anorthositic gabbro and the component abundances given in Table 2 we calculate Rb = 0.54 ppm and Sr = 104 ppm for the combined hi-Mg basalt and olivine vitrophyre components. We have demonstrated above that the component abundances used for this calculation give a good fit to major element and other trace element data.

The $^{87}Rb/^{86}Sr$ ratio of 0.015 for the combined high-Mg component is not very different from an average $^{87}Rb/^{86}Sr$ = 0.0179 for the bulk soils. Assuming that addition of the non-mare component causes displacement approximately along the 4.6 AE reference isochron leads to the conclusion that the combined hi-Mg components probably have whole rock Rb-Sr systematics similar to Apollo 12 ilmenite basalts such as 12008. Thus at least one of the high Mg components probably has an initial $^{87}Sr/^{86}Sr$ which is significantly more evolved than that of the lo-Mg basalt.

Our analysis of the <20 μm fraction of 24077 was an attempt to directly check the assumptions behind the mixing calculation. Blanchard *et al.* (1977) found trace element abundances in the <20 μm soil fractions to be enriched over bulk soil values by ~25–50%. If this enrichment is due to a larger proportion of LKFM in this soil fraction one would expect the Rb-Sr data for the <20 μm fractions to be displaced to higher Rb/Sr values approximately along the 4.6 AE reference isochron in Fig. 4. Our analysis of 24077 (<20 μm) plots instead on a 3.3 AE reference isochron through 24170 plagioclase. Taken at face value this is strong indication that the trace element enrichments in this size fraction are not due to the addition of more LKFM but are due instead to differential comminution of the mare basaltic component. One must hypothesize that this differential comminution resulted in enrichment of trace element rich residual phases in this fine size range. This possibility was also suggested by Blanchard *et al.* (1977). The residual phases might be associated with cristobalite which was reported in basalts of the lo-Mg type by Tarasov *et al.* (1977) but which did not appear to be present in the coarsely crushed 24170 layer (Lunatic Asylum, 1978).

McKay *et al.* (1977) reported that highland derived ANT rocks are absent from the 20-45 μm soil size fractions. It is reasonable to assume that this is also true of the <20 μm fraction. The petrographic data do not otherwise provide strong evidence for or against the comminution hypothesis.

We consider the 24077 (<20 μm) data as preliminary although extremely interesting. It is noteworthy that, like Rb, K and Na (Blanchard *et al.*, 1977) are enriched two-fold relative to the bulk soil. Li enrichment is somewhat less. These general alkali enrichments seem to exclude the possibility that the location of the 24077 (<20 μm) datum to the right of the reference 4.6 AE isochron is due to an analytical problem such as an unusually high Rb blank. However, because of the small grain size and extensive handling of this material it is prone to acquire surface contamination. This, plus some unanswered questions of mass balance prevent us from overemphasizing these data until they can be confirmed by further analyses. Taken at face value, however, they indicate that the Rb-Sr results for 24077 (<20 μm) are dominated by the presence of alkali-rich phases from basalts of similar age and initial ratio as 24170. We note, moreover, that the situation we have apparently found for Luna 24 soils is analogous to that found by Evensen *et al.* (1973) in their analyses of grain size fractions of soil 15531. In that instance the finest grain size fractions also asymptotically approached the mare basalt mineral isochron when plotted on a Rb-Sr isochron diagram.

Trace Element and Sr-isotopic Evolution of Luna 24 Basalts

Our study of Apollo 12 basalts led to a general approach to modeling the Sr-isotopic and REE evolution of lunar basalts (Nyquist *et al.*, 1977). Here we inquire whether this approach can be extended to Luna 24 basalts.

Following our earlier approach we assume: (i) mare basalts are partial melts of cumulates which formed from an "early liquid" possibly the widely hypothesized lunar magma ocean, (ii) relative abundances of the REE in the "early liquid" were non-chondritic during cumulate formation with chondrite-normalized values progressively decreasing for heavier REE, (iii) $^{87}Rb/^{86}Sr = 0.05$ in the early liquid, (iv) the bulk solid/liquid distribution coefficient $D(Rb) = 0.01$ for the cumulate mineralogy. The reader is referred to the earlier work (Nyquist *et al.* 1977) for a discussion of these assumptions.

It was assumed in the earlier work that $^{87}Sr/^{86}Sr = 0.69917$ in the "early liquid" and that the source cumulates formed at 4.4 AE ago. This $^{87}Sr/^{86}Sr$ ratio appears slightly too high to be consistent with the initial $^{87}Sr/^{86}Sr$ ratio of lo-Mg basalt of 0.69914 ± 0.00006 at 3.3 AE ago. As a first approximation we can assume that the Rb/Sr ratio in the source is the same as that in the basalt. For a whole rock $^{87}Rb/^{86}Sr = 0.0041$ as calculated for 24170 the evolution in $^{87}Sr/^{86}Sr$ between 3.3 and 4.4 AE ago would be only 0.00006 suggesting that 0.69908 ± 0.00006 would be a more appropriate value for the "early liquid" at the time of cumulate formation. The model age for the calculated 24170 whole rock data is 4.4 AE for this initial ratio and is consistent with the assumption that Rb/Sr in the source is equal to Rb/Sr in the basalt. However, evolution of $^{87}Sr/^{86}Sr$ by 0.00006 should be considered an upper limit because it is probable that the Rb/Sr ratio of the source was less than that of the basalt. As the expected evolution in $^{87}Sr/^{86}Sr$ in the low Mg source between 3.3 and 4.4 AE ago is comparable to the analytical precision of the data we will not attempt to model it in more detail.

The low Rb/Sr ratio of 24170 is consistent with its low initial $^{87}Sr/^{86}Sr$ and provides a useful constraint on its petrogenesis. Thus $\overline{D}(Rb) = 0.01$ and $^{87}Rb/^{86}Sr \leq 0.0041$ for the 24170 source implies $\overline{D}(Sr) \geq 0.12$. Such a high bulk Sr distribution coefficient for a source composed only of mafic minerals would imply a dominant proportion of high-Ca cpx, as was used in modeling the evolution of Apollo 17 basalts (Nyquist *et al.*, 1977). However, Luna 24 basalts are so different from Apollo 17 basalts in other geochemical aspects that this conclusion seems untenable. Thus we seek an alternate explanation for the comparatively high $\overline{D}(Sr)$. Even a small addition of plagioclase to an otherwise mafic source can increase the bulk solid/liquid Sr-distribution coefficient significantly because the plag/liquid distribution coefficient for Sr is ten- to one hundred-fold greater than px/liquid or ol/liquid distribution coefficients. Thus the presence of plagioclase in the source region provides a ready explanation for the low initial $^{87}Sr/^{86}Sr$ and the low $^{87}Rb/^{86}Sr$. On the other hand, plag/liquid distribution coefficients for the trivalent REE are comparable to px/liquid and ol/liquid distribution coefficients and the addition of a small amount of plagioclase to an otherwise mafic source has little effect on the bulk solid/liquid distribution coefficients for the trivalent REE. It is thus possible to model the REE and Rb-Sr data by: (i) determining the cumulate mineralogy and degree of partial melting required to match the relative abundances of the trivalent REE, (ii) determining the percent plagioclase required to obtain $\overline{D}(Sr) = 0.12$, (iii) iterating if necessary, (iv) determining the absolute REE abundances in the "early liquid" required to give the proper absolute REE abundances for the basalts.

Fig. 6 illustrates application of this approach. The REE data chosen to be matched by the model calculations are those of lithic fragments 24109,78 and 24174,5. (Data from Ma *et al.*, 1977 and Blanchard *et al.*, 1977.) Fragment 24109,78 was chosen because it is a well identified fine-grained basalt with a negative Eu-anomaly. Fragment 24174,5 was chosen because it has one of the largest positive Eu-anomalies among the fragments which have been studied and

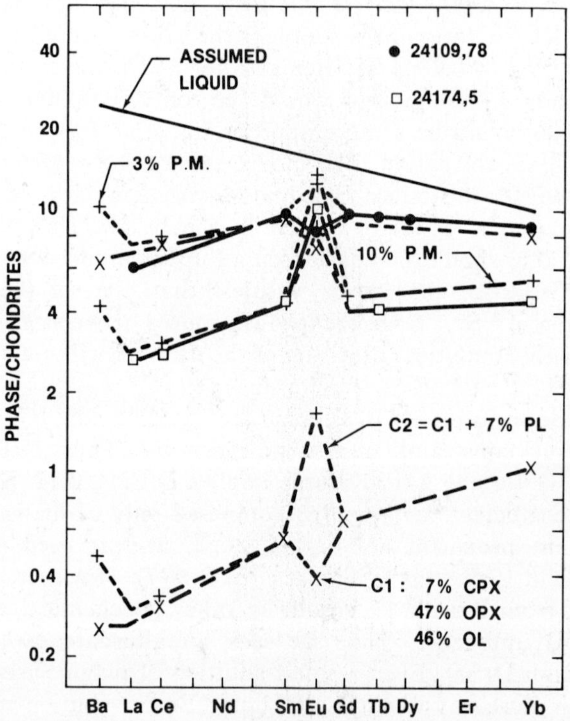

Fig. 6. Calculated chondrite-normalized Ba and REE abundance patterns for cumulates and melts from an assumed liquid of the composition shown. Bulk distribution coefficients are given in Table 4. Eu-abundances calculated for (i) modal melting (i.e. 7% plag left in residue) and (ii) non-modal melting for which plagioclase is exhausted from the source are shown by the lower and upper Eu values, respectively, for the 10% melting case. Data for 24109,78 are from Ma *et al.* (1977) normalized to chondritic values of Laul *et al.* (1974). Data for 24174,5 are from Blanchard *et al.* (1977) normalized to the chondrite composite values of Hubbard and Gast (1971). A good fit is obtained for the 24109,78 pattern for 3% melting of C1 and to the 24174,5 pattern for 10% melting of C2.

because we plan to acquire Rb-Sr data for this fragment. Unfortunately, both REE and Rb-Sr data are not yet available on the same lithic fragments.

Compared to average lo-Mg basalt (Table 2), sample 24109,78 has the same La abundance, 20–30% higher Sm, Dy, and Yb abundances and a 17% lower Eu abundance. Sample 24174,5 has the same Eu abundance as average lo-Mg basalt and La, Sm, and Yb abundances which are 40–60% lower. A case could be made for applying the model to the average fragment abundances, but modeling these two fragments seemed preferable because of the danger of erroneously defining average values and because these two samples illustrate several aspects of the calculation particularly well.

A cumulate containing 7% cpx, 47% opx and 47% ol (cumulate C2 for Apollo 12 basalts, Nyquist *et al.*, 1977) yields a calculated partial melt whose chondrite-normalized abundance pattern is nearly flat for the heavy REE. Because a flat chondrite-normalized heavy REE pattern is characteristic of Luna 24 basalts, we chose the above cumulate mineralogy as our starting point for modeling these basalts, and refer to it here as C1(24). Bulk solid/liquid distribution coefficients for this cumulate are given in Table 4. The REE abundances calculated for this cumulate in the manner of Nyquist *et al.* (1977) are shown in Fig. 6.

The same uniformly fractionated relative abundance pattern was used for the "early liquid" as in the earlier work. Using the calculational procedure described by Nyquist *et al.* (1977) it was found that the shape of the trivalent REE pattern of 24109,78 was well matched for 3% partial melting. In order to match the absolute REE abundances of 24109,78 for Cl(24) and 3% melting, the absolute REE abundances in the early liquid must be ~20× chondritic at La.

After the relative proportions among the mafic components were determined the percent plagioclase needed to match the low Rb/Sr ratio of 24170 (i.e., to make $\overline{D}(Sr) = 0.12$) was calculated. Using the plagioclase distribution coeffi-

Table 4. Bulk distribution coefficients for model cumulate sources.

	C1(24)	C2(24)	C3(24)	C4(24)
K	.011	.012	.009	.010
Rb	.009	.011	.010	.011
Sr	.018	.122	.025	.129
Ba	.011	.020	.012	.021
Ce	.014	.016	.018	.020
Sm	.034	.034	.051	.049
Eu	.026	.109	.035	.117
Gd	.047	.045	.067	.064
Yb	.111	.104	.124	.116

[a]Calculated for cumulate mineralogies shown in Figs. 6 and 7. Mineral/liquid distribution coefficients used in the calculations are the same as given by Nyquist *et al.* (1977) for ol, opx, and cpx, and by Shih *et al.* (1975) for plag.

cients preferred by Shih *et al.* (1975), about 7% plagioclase must be added to C1(24) to change $^{87}Rb/^{86}Sr$ from 0.05 in the early liquid to 0.0041 as estimated for 24170. Addition of 7% plagioclase to the C1(24) mineralogy produced a calculated abundance pattern for the source which has a distinct positive Eu-anomaly and chondrite-normalized Ba abundance greater than the chondrite-normalized La abundance. This cumulate mineralogy (7% plag, 6% cpx, 44% opx, and 43% ol) is referred to here as C2(24). Bulk distribution coefficients calculated for C2(24) are given in Table 4 and calculated REE abundances shown in Fig. 6. The inclusion of plagioclase in the source mineralogy has little effect on the calculated abundances of trivalent REE in the partial melt but causes the calculated Eu-anomaly to be positive. In this respect the measured REE abundance pattern of 24109,78 is no longer matched. However, there is reason to believe that 24109,78 is unrepresentative in having a negative Eu-anomaly because it belongs to the lo-Mg basalt group for which the average Eu-anomaly is positive by an amount comparable to that calculated for 3% melting of C2(24).

A good match can be obtained between the measured abundance pattern of 24174,5 and that calculated for 10% melting of C2(24). Sample 24174,5 is perhaps also unrepresentative as evidenced by the pronounced positive Eu-anomaly and low FeO = 16.6% (Blanchard *et al.* 1977). However, matching its pattern illustrates the effects of simply varying the degree of partial melting used in the calculation.

Two variations on the model were tried. In one, it was assumed that the plagioclase was completely consumed during melting rather than remaining as 7% of the residue. This variation results in a partial melt whose Rb/Sr ratio is nearly the same as that of the source, and whose positive Eu anomaly is greater than that calculated assuming 7% plagioclase remains in the residue. For example, for the case in 10% melting, if plagioclase is totally consumed the calculated positive Eu anomaly of the partial melt is 20% greater than that observed for 24174,5, while if 7% plagioclase remains in the residue, the calculated Eu anomaly is 20% less than that observed (Fig. 6). It is reasonable that the true situation would be intermediate to these extremes.

In a second variation on the calculation it was assumed that $^{87}Rb/^{86}Sr = 0.03$ in the "early liquid." This value is the present day ratio for Apollo 12 olivine pigeonite basalts and is considered to be a lower limit if the Apollo 12 and Luna 24 sources formed from the same early liquid. In this case the calculated plagioclase content of the source is only 4%, and the calculated Eu content of the source is 32% lower than that of C2(24) (Fig. 6). For 10% and 3% partial melting, the resulting Eu contents in the melts are 20% and 10% lower, respectively, than the corresponding partial melts of C2(24). The qualitative features of the results are unchanged relative to the 7% plag cumulate.

The calculations illustrated in Fig. 6 were based on the assumption that $^{87}Rb/^{86}Sr$ in the basalt source was equal to that estimated for 24170 whole rock. Thus for total self-consistency it should be required that the Rb/Sr ratio not change on melting. This can only be accomplished in the present model by requiring that plag be exhausted from the source. Alternatively it could be postulated that $^{87}Rb/^{86}Sr$ in the source was two-fold or more lower than that measured for the basalt. If so, the initial $^{87}Sr/^{86}Sr$ for the basalts at their genesis was only trivially (~0.00003) greater than was $^{87}Sr/^{86}Sr$ in the "early liquid" at the time of formation of the basalt source region. This is attractive because it would make $^{87}Sr/^{86}Sr$ in the "early liquid" closer to the value of 0.69917 deduced for Apollo 17 basalts. We have not pursued this point further because the variations in calculated values are within the analytical uncertainties of the isotopic data.

If it is hypothesized that plagioclase is exhausted from the source region by partial melting then the degree of melting must be greater than 4–7%, the percentage of plagioclase which is apparently required by the Rb-Sr systematics. Figure 7 shows a model calculation which satisfied this requirement. In this calculation the percentage of cpx was increased to 14% to produce a flatter heavy REE pattern for the source region so that the flat pattern of Luna 24 basalts is

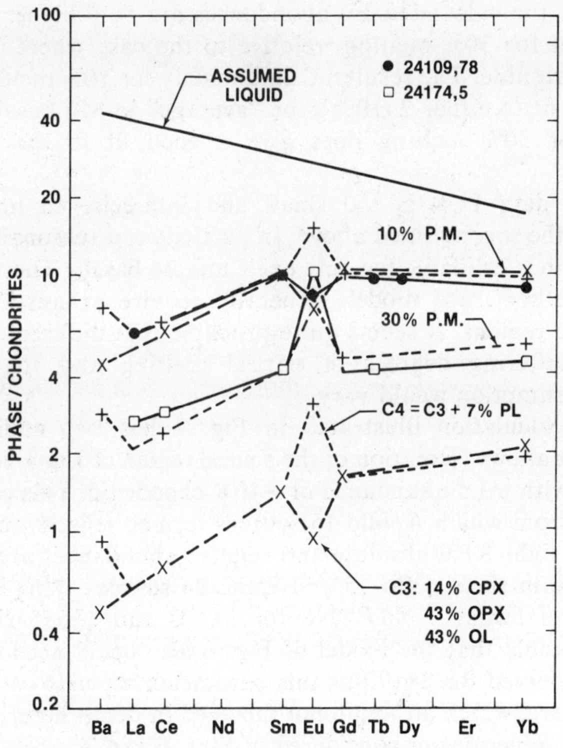

Fig. 7. Calculated chondrite-normalized Ba and REE abundance patterns for cumulates and melts from an assumed liquid of the composition shown. Bulk distribution coefficients are given in Table 4. Data as in Fig. 6. A good fit to the abundance pattern of 24109,78 is obtained for 10% melting of C1 and to the abundance pattern of 24174,5 for 30% melting of C2.

achieved at greater degrees of melting. The mafic cumulate of Fig. 7 is designated C3(24) and has the composition 14% cpx, 43% opx, and 43% ol. Addition of 7% plag results in a composition C4(24) with 7% plag, 13% cpx, 40% opx, and 40% ol. Bulk distribution coefficients for both of these compositions are given in Table 4 and the calculated REE patterns for the source regions are shown in Fig. 7. The trivalent REE patterns of 24109,78 and 24174,5 are quite well matched by 10% and 30% partial melting, respectively, of either C3 or C4. The calculated slope of the light REE pattern is slightly too steep, whereas in Fig. 6 it is slightly too shallow suggesting that the best fit to the trivalent REE would be obtained for intermediate values of partial melting and cpx content. We consider either of the fits displayed in Fig. 6 or 7 to be acceptable, however.

Partial melting of C3 produces a negative Eu-anomaly which (perhaps fortuitously) matches that of 24109,78 quite well. Partial melting of C4 produces a positive Eu-anomaly, the magnitude of which depends on the amount of plag which is assumed to remain in the source. For clarity Fig. 7 illustrates only the case where the modal mineralogy is assumed to remain unchanged during melting, i.e. the residue retains 7% plag. If plag is assumed to be consumed during melting, the calculated Eu-abundances are 56% higher for 10% melting and 18% higher for 30% melting, relative to the case where 7% plagioclase is retained in the residue. The resultant Eu-anomaly for 10% melting is too large to provide a good fit to either 24109,78 or "average" lo-Mg basalts. The resultant Eu-anomaly for 30% melting does give a good fit to the REE pattern of 24174,5.

The present data base is too small and imprecise to uniquely constrain calculations of the sort outlined above. In particular, a reasonably reliable whole rock Rb/Sr ratio is available for only one Luna 24 basalt. Thus, we refrain from concluding whether these models generally require exhaustion of plagioclase from the source regions. It seems quite probable that different basalts would be produced by differing degrees of partial melting and that the degree of plagioclase consumption would vary.

The model calculation illustrated in Fig. 7 has two additional attractive aspects. First, it allows formation of the source region of Luna 24 basalts from an "early liquid" with a La abundance of ~40× chondritic. This could be the same "early liquid" from which Apollo 15 sources formed (Fig. 8 and Nyquist *et al.*, 1977). Second, light REE absolute and relative abundances are calculated to be nearly the same in the Apollo 15 and Luna 24 sources. This is consistent with similar values of initial $^{143}Nd/^{144}Nd$ for 24170 and 15555 (Lunatic Asylum, 1978). It is possible that the model of Fig. 6 also could account for the initial $^{143}Nd/^{144}Nd$ observed for 24170 as this parameter depends only on the Sm/Nd ratio in the source, which although not modeled in detail here, should be similar for both source mineralogies considered in Figs. 6 and 7.

Conclusions

The spectrum of rock types present at the Luna 24 site is being defined as a

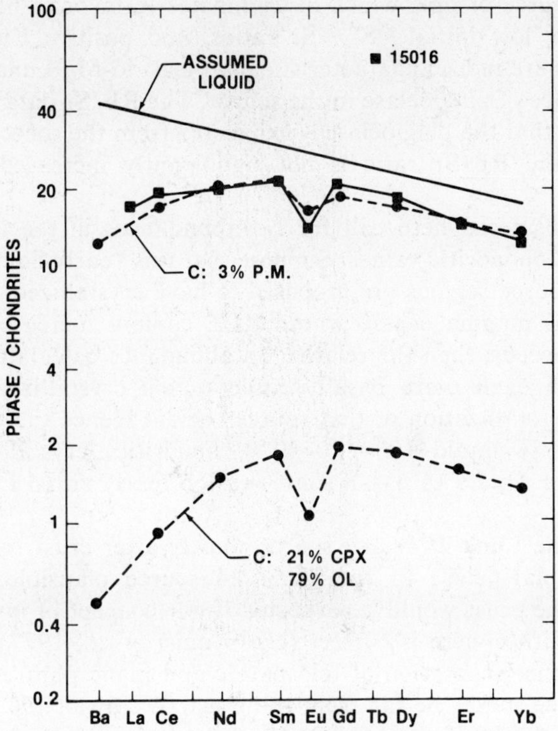

Fig. 8. Calculated chondrite-normalized Ba and REE abundance pattern for cumulate C and a 3% partial melt for an assumed liquid with the same REE abundances as in Fig. 7. Data for 15016 from Helmke *et al.* (1973) normalized to the chondrite composite (Hubbard and Gast, 1971).

result of the efforts of lunar scientists throughout the world. However, there is an obvious need for more high quality chemical and isotopic data on individual lithic fragments. Such data on high-Mg mare components would be particularly important. Until this larger data base can be produced it will be important to use analyses of lunar soils to infer the probable characteristics of individual rock types which may not have been directly sampled. Our study of Luna 24 soils leads us to the conclusion that one or more of the hi-Mg rock types represented in the soils is of mare composition and probably had a higher Rb/Sr ratio, more evolved initial $^{87}Sr/^{86}Sr$, and highter Sm/Eu ratio than the widely sampled lo-Mg ferrobasalt.

We have been able to satisfactorily model the REE and Rb-Sr characteristics of at least lo-Mg Luna 24 basalts using the procedures previously applied by us to Apollo 12 basalts. The model calculations are admittedly non-unique but have the following features of general validity: (i) light REE fractionation for Luna 24 basalts implies the presence of pyroxene in the source, (ii) the higher relative abundances of heavy REE as compared to other lunar basalts suggest the presence in the source of opx, which discriminates in favor of heavy REE, (iii) low Rb/Sr ratios, low initial $^{87}Sr/^{86}Sr$ ratios, and positive Eu anomalies (if positive anomalies are indeed characteristic of at least lo-Mg Luna 24 basalts) all suggest the presence of plagioclase in the source. The Rb/Sr data are most easily fit if it is assumed that the plagioclase is exhausted from the source region during melting so that the Rb/Sr ratio is not significantly increased during basalt genesis.

The models considered here call for La abundances in the assumed "early liquid" of 20–40× chondritic values. Suppose that this "early liquid" from which the mare basalt source regions are proposed to have crystallized is equated with an evolving global magma ocean, whose REE content increases with time as crystallization proceeds; then the relative La abundance calculated for the early liquid from which each mare basalt source region crystallized indicates the relative time of crystallization of that source region. Hence the Luna 24 basalt source region (early liquid La = 20–40 × chondritic) crystallized before or coincident with the Apollo 15 basalt source region (early liquid La $\simeq$40 × chondritic).

Formation of the Luna 24 basalt source shortly after crust formation makes the presence of plagioclase in the Luna 24 source plausible. Close spatial association with the crust would be expected if our concept of inward solidification of the lunar lithosphere is correct (cf. Nyquist *et al.*, 1977). On the other hand, if gravitational separation of feldspathic and mafic phrases predominates the Luna 24 source might be far from the crust. We await the results of high pressure experimental studies to help clarify the question of the depth of origin of the Luna 24 basalts.

Hubbard and Vilas (1977) have suggested the possibility of laterally varying source materials for mare basalts from orbital geochemical data. Nyquist *et al.* (1977) suggested small scale lateral heterogeneities might exist in the lunar mantle because Apollo 12 basalts, all with crystallization ages of ~3.2 AE, apparently derived from two or more different sources. Lateral heterogeneities in the mantle may also be implied by similar ages but different source compositions for Apollo 15 and Luna 24 basalts.

As shown here the trace element and isotopic characteristics of Luna 24 basalts are compatible with genesis via small to moderate (3%–30%) fractional melts of predominantly mafic cumulates. In this scenario comparatively low abundances of REE (and Ti as well because it is an "incompatible element" for the source mineralogies considered) are due to low source abundances and/or relatively large melt fractions. We do not consider the hypothesis of genesis via melting of an undifferentiated homogeneous lunar mantle to be tenable because of the very low initial $^{87}Sr/^{86}Sr$ of at least the lo-Mg basalts. Whereas one might postulate a primitive source of unique composition due to heterogeneous lunar accretion, this is not attractive as an explanation for low Rb/Sr ratios because K/U, another ratio of a volatile to a refractory element, is the same for Luna 24 soils as the lunar surface average (K/U = 2100; Metzger *et al.*, 1977).

It is not our purpose here to attempt a complete discussion of basalt genesis.

Thus we have not considered assimilation models for the petrogenesis of the Luna 24 basalts. Models involving crustal assimilation may yet prove attractive because of the aluminous nature of the Luna 24 basalts. However, the low absolute REE abundances and depleted light REE pattern of these basalts seem at odds with assimilation of crustal material by late stage (3.3 AE) evolved magmas. Assimilation of crustal material by primitive magmas from an undifferentiated lunar interior might be postulated. Note, however, that the light REE abundance pattern of anorthositic gabbros complements that of Luna 24 basalts, rather than resembling it as might be expected in this case. Hence, we feel that the currently available trace element and isotopic data can be most easily explained within the context of the cumulate remelting hypothesis. It is not necessary to postulate primary accretional heterogeneities in order to understand these data.

Acknowledgments—The mixing program used was adapted for geochemical use by D. P. Blanchard and S. Su from a program used by the JSC Earth Resources Observation Laboratory. Discussions with several JSC colleagues, especially D. P. Blanchard, were extremely helpful.

References

Barsukov V. L., Tarasov L. S., Dmitriev L. V., Kolesov G. M., Shevaleevsky I. D., and Garanin A. U. (1977) The geochemical and petrochemical features of regolith and rocks from Mare Crisium (preliminary data). *Proc. Lunar Sci. Conf. 8th*, p. 3319–3332.

Basu A., McKay D. S., Fruland R. M. (1977) Petrography, mineralogy, and source rocks of Luna 24 drill core soils (abstract). In *Papers Presented to the Conference on Luna 24*, p. 18–21. The Lunar Science Institute, Houston.

Bence A. E. and Grove T. L. (1977) The highland component in the Lunar 24 core (abstract). In *Papers Presented to the Conference on Luna 24*, p. 22–24. The Lunar Science Institute, Houston.

Birck J. L., Manhes G., Richard P., Joron J. L., Treuit M., and Allegre C. J. (1977) ^{87}Rb/^{87}Sr age of Luna 24 microgabbros, and isotopic and trace element study of soil 24096 (abstract). In *Papers Presented to the Conference on Luna 24*, p. 34–36. The Lunar Science Institute, Houston.

Blanchard D. P., Haskin L. A., Brannon J. C., and Aaboe E. (1977) Chemistry of soils and particles from Luna 24 (abstract). In *Papers Presented to the Conference on Luna 24*, p. 37–40. The Lunar Science Institute, Houston.

Evensen N. M., Murthy V. R., and Coscio M. R. Jr. (1973) Rb-Sr ages of some mare basalts and the isotopic and trace element systematics in lunar fines. *Proc. Lunar Sci. Conf. 4th*, p. 1707–1724.

Gast P. W., Hubbard N. J., and Wiesmann H. (1970) Chemical composition and petrogenesis of basalts from Tranquility Base. *Proc. Apollo 11 Lunar Sci. Conf.*, p. 1143–1163.

Grove T. L. and Vaniman D. T. (1977) Experimental petrology of very low Ti basalts and origin of Luna 24 ferrobasalt (abstract). In *Papers Presented to the Conference on Luna 24*, p. 68–71. The Lunar Science Institute, Houston.

Haskin L. A., Wildeman T. R., and Haskin M. A. (1968) An accurate procedure for the determination of the rare earths by neutron activation. *J. Radioanal. Chem.* **1,** 337–348.

Helmke P. A., Blanchard D. P., Haskin L. A., Telander K., Weiss C., and Jacobs J. W. (1973) Major and trace elements in igneous rocks from Apollo 15. *The Moon* **8,** 129–148.

Hubbard N. J. and Gast P. W. (1971) Chemical composition and origin of nonmare lunar basalts. *Proc. Lunar Sci. Conf. 2nd*, p. 999–1020.

Hubbard N. J., Rhodes J. M., Wiesmann H., Shih C.-Y., and Bansal B. M. (1974) The chemical definition and interpretation of rock types returned from the non-mare regions of the moon. *Proc. Lunar Sci. Conf. 5th*, p. 1227–1246.

Hubbard N. J. and Vilas F. (1977) From Serenity to Langemak: A regional chemical setting for Crisium (abstract). In *Papers Presented to the Conference on Luna 24*, p. 85–88. The Lunar Science Institute, Houston.

Laul J. C. (1977) Chemical composition of Luna 24 melt rocks (24077,4; 24077,62; 24174,7; 24182,12; 24210,50) and gabbro (24182,8) (abstract). In *Papers Presented to the Conference on Luna 24*, p. 94–97. The Lunar Science Institute, Houston.

Laul J. C., Hill D. W., and Schmitt R. A. (1974) Chemical studies of Apollo 16 and 17 samples. *Proc. Lunar Sci. Conf. 5th*, p. 1047–1066.

LSPET (Lunar Sample Preliminary Examination Team) (1972) Preliminary Examination of Lunar Samples. *Apollo 15 Prelim. Sci. Rep.* NASA SP-289, p. 6-1 to 6-25.

Lunatic Asylum, Contributing inmates: Wasserburg G. J., Papanastassiou D. A., McCulloch M. T., Dymek R. F., De Paolo D. J., Chodos A. A., and Albee A. L. (1978) Petrology, chemistry, age and irradiation history of Luna 24 samples. In *Mare Crisium: The View from Luna 24* (R. B. Merrill and J. J. Papike, eds.). This volume.

Ma M. S., Schmitt R. A., Taylor G. J., Warner R. D., Lange D. E., and Keil K. (1977) Chemical and petrographic studies of 18 Luna 24 lithic fragments (abstract). In *Papers Presented to the Conference on Luna 24*, p. 102–105. The Lunar Science Institute, Houston.

Marvin U. B., Ryder G., and McSween H. Y. Jr. (1977) Some exotic particles in the Luna 24 core sample (abstract). In *Papers Presented to the Conference on Luna 24*, p. 106–108. The Lunar Science Institute, Houston.

McKay D. S., Basu A., Waits G., Clanton U., Fuhrman R., and Fruland R. (1977) Grain size and evolution of Luna 24 soils (abstract). In *Papers Presented to the Conference on Luna 24*, p. 114–117. The Lunar Science Institute, Houston.

Metzger A. E., Haines E. L., Parker R. E., and Radocinski R. G. (1977) Thorium concentrations in the lunar surface, I: Regional values and crustal content. *Proc. Lunar Sci. Conf. 8th*, p. 949–999.

Norman M., Coish R. A., and Taylor L. A. (1977) Glass chemistry and magma evolution at Mare Crisium (abstract). In *Papers Presented to the Conference on Luna 24*, p. 136–138. The Lunar Science Institute, Houston.

Nyquist L. E., Bansal B. M., Wiesmann H. (1976) Sr isotopic constraints on the petrogenesis of Apollo 17 mare basalts. *Proc. Lunar Sci. Conf. 7th*, p. 1507–1528.

Nyquist L. E., Bansal B. M., Wooden J. L., and Wiesmann H. (1977) Sr. isotopic constraints on the petrogenesis of Apollo 12 mare basalts. *Proc. Lunar Sci. Conf. 8th*, p. 1383–1415.

Nyquist L. E., Hubbard N. J., Gast P. W., Bansal B. M., and Wiesmann H. (1973) Rb-Sr systematics for chemically defined Apollo 15 and 16 materials. *Proc. Lunar Sci. Conf. 4th*, p. 1823–1846.

Nyquist L. E., Hubbard N. J., Gast P. W., Church S. E., Bansal B. M., and Wiesmann H. (1972) Rb-Sr systematics for chemically defined Apollo 14 breccias. *Proc. Lunar Sci. Conf. 3rd*, p. 1515–1530.

Papanastassiou D. A. and Wasserburg G. J. (1971) Lunar chronology and evolution from Rb-Sr studies of Apollo 11 and 12 samples. *Earth Planet. Sci. Lett.* **11**, 37–62.

Podosek F. A. and Huneke J. C. (1973) Argon in Apollo 15 green glass spherules (15426): ^{40}Ar-^{39}Ar age and trapped argon. *Earth Planet. Sci. Lett.* **19**, 413–421.

Ryder G., McSween H. Y., Jr., and Marvin U. B. (1977) Basalts from Mare Crisium. *The Moon* **17**, 263–287.

Shih C.-Y., Haskin L. A., Wiesmann H., Bansal B. M., and Brannon J. C. (1975) On the origin of high-Ti mare basalts. *Proc. Lunar Sci. Conf. 6th*, p. 1255–1285.

Simonds C. H., Warner J. L., McGee P. E., Cochran A., and Brown R. W. (1977) Luna 24: Petrochemical potpourri (abstract). In *Papers Presented to the Conference on Luna 24*, p. 170–173. The Lunar Science Institute, Houston.

Stettler A. and Albarede F. (1977) Ar^{39}-Ar^{40} pattern and light noble gas systematics of two mm-sized rock fragments from Mare Crisium (abstract). In *Papers Presented to the Conference on Luna 24*, p. 175–178. The Lunar Science Institute, Houston.

Tarasov L. S., Nazarov M. A., Shevaleevsky I. D., Kudryashova A. F., Gaverdovskaya A. S., and Korina M. I. (1977) Mineralogy and petrography of lunar rocks from Mare Crisium (preliminary data). *Proc. Lunar Sci. Conf. 8th*, p. 3333–3356.

Taylor G. J., Keil K., and Warner R. D. (1977a) Very low-Ti mare basalts. *Geophys. Res. Lett.* **4**, 207–210.

Taylor G. J., Warner R., Wentworth S., and Keil K. (1977b) The Luna 24 regolith: lithologic abundances in the 250–500 μm size fraction, and compositions of agglutinates and nonmare lithic fragments and glasses (abstract). In *Papers Presented to the Conference on Luna 24*, p. 186–188. The Lunar Science Institute, Houston.

Taylor G. J., Warner R., Wentworth S., Keil K., and Sayeed U. (1977c) Petrology and chemistry of Luna 24 mare basalts and basaltic glasses (abstract). In *Papers Presented to the Conference on Luna 24*, p. 189–192. The Lunar Science Institute, Houston.

Vaniman D. T. and Papike J. J. (1977a) Very low Ti (VLT) basalts: A new mare rock type from the Apollo 17 drill core. *Proc. Lunar Sci. Conf. 8th*, p. 1443–1471.

Vaniman D. T. and Papike J. J. (1977b) Ferrobasalts from Mare Crisium: Luna 24. *Geophys. Res. Lett.* **4**, 497–500.

Vaniman D. T. and Papike J. J. (1977c) Luna 24 ferrobasalt: Petrology and comparisons with other mare basalt types (abstract). In *Papers Presented to the Conference on Luna 24*, p. 196–199. The Lunar Science Institute, Houston.

Wiesmann H. and Hubbard N. J. (1975) A compilation of the lunar sample data generated by the Gast, Nyquist and Hubbard lunar sample PI-ships. NASA Johnson Space Center. Unpublished report.

Merrill, R. B. and Papike, J. J., editors.
(1978) *Mare Crisium: The View from Luna 24*, Pergamon Press (New York), p. 657–678
Printed in the United States of America

Petrology, chemistry, age and irradiation history of Luna 24 samples

THE LUNATIC ASYLUM*
Division of Geological and Planetary Sciences**
California Institute of Technology, Pasadena, California 91125

Abstract—Sample 24170 is comprised of fragments from a single igneous rock characterized by high Fe in the pyroxene and olivine, extensive zoning in pyroxene, high Ca in olivine and low trace element concentrations. These characteristics indicate that 24170 is a mare basalt. The crystallization age of 24170 is 3.30 ± 0.05 AE as determined by concordant Sm-Nd and ^{40}Ar-^{39}Ar ages. This shows the presence of relatively young flows at Mare Crisium. This sample was derived from a parent reservoir with nearly chondritic Sm/Nd ratio but with extremely low Rb/Sr (Rb/Sr $\approx$ 0.002). All soil samples show low trace element contents and indicate minimum contamination by KREEPUTh-rich materials. The isotopic relations for Rb-Sr and Sm-Nd reflect the absence of significant chemical fractionation for the soil components at times younger than ~4.5 AE.

One soil sample from the Luna 24 core shows extremely large neutron capture effects. This measurement places a new lower limit to the neutron production rate in the regolith which is closer to the calculated value of Lingenfelter *et al.* (1972). This new lower limit and the comparatively low fluence values for most lunar soils require the addition of unirradiated materials from depth to the lunar surface, presumably as a result of relatively infrequent deep craters which spread out thin debris blankets on a wide scale.

INTRODUCTION

Sample 24170 is a light-colored, 2 cm-thick layer of coarse particles from the 170 cm depth of the regolith core returned from Mare Crisium by the Luna 24 mission. The sample "appears to represent a rather-large fragment of coarse-grained gabbro broken up by drilling" (Florensky, *et al.* 1977), and is the only Luna 24 sample large enough to permit integrated and detailed isotopic, chemical and petrologic studies.

The work reported here is part of a consortium effort with R. A. Schmitt, E. Anders and their coworkers. The experiment design for 24170 includes: a) mineralogy and petrology on micro-thin sections and on grain mounts of mineral separates; b) chemical analyses by INAA, RNAA and isotope dilution for selected elements; c) Rb-Sr and Sm-Nd crystallization ages; d) ^{40}Ar-^{39}Ar gas

* Contributing Inmates: G. J. Wasserburg, F. Radicati di Brozolo,‡ D. A. Papanastassiou, M. T. McCulloch, J. C. Huneke, R. F. Dymek†, D. J. DePaolo, A. A. Chodos, A. L. Albee

** Contribution Number 3033

† Present address: Department of Geological Sciences, Harvard University, Cambridge, Massachusetts 02138

‡ Permanent address: Laboratorio di Geoncronologia e Geochimica Isotopica Consiglio Nazionale delle Ricerche, Pisa, Italy.

retention age and exposure age; and e) neutron fluence and stratigraphy of the Luna 24 core utilizing data on several other samples as well as on 24170. Due to the complexity of the investigations and the limited sample size, some of these experiments are still in progress.

The eight largest lithic fragments (24170,1) totaling 96 mg, were initially allocated for the consortium study and the concentrations of K, Rb, Sr, Sm, Nd were measured on 0.5 mg samples of plagioclase and pyroxene to establish whether it was feasible to determine an internal isochron. Rb/Sr ratios were extremely low and showed insufficient spread for a precise age. The Sm and Nd abundances were low, but an internal isochron seemed possible if additional material could be obtained and if the amount of Nd required for a precise determination could be greatly reduced from that used in our previous studies. The sample was augmented by allocation of 24170,0, the residual fine-grained material from which 24170,1 had been picked. About one percent of exotic fragments were removed by magnetic separation. The remaining mass contained no exotic materials (less than 0.1%). The combined sample used for the mineral separations weighed 130 mg.

Before initiating the mineral separations we established that precise Nd isotopic data could be obtained on 20 ng of Nd with no loss of precision. This constitutes a reduction by a factor of four in the amount of Nd used per analysis over last year's work. The sample size was then sufficient for a single mass spectrometric measurement on the plagioclase from 24170.

PETROLOGY AND MINERALOGY OF 24170

Sample 24170 consists of fragments from a relatively coarse-grained, subophitic basalt (or microgabbro) comprised of plagioclase, pyroxene, and olivine, together with minor amounts of ilmenite, cristobalite, and troilite. Inspection of the fragments showed no zap pits or other textures suggestive of surfaces exposed at the lunar exterior. Petrographic and electron microprobe studies have been performed on two polished micro-thin sections (FQM-648, FQM-650), which were prepared from separate lithic fragments (Fig. 1), and on grain mounts, which were prepared to monitor the mineral separations for isotopic studies. The thin sections have a combined area of $\sim 1.5\ mm^2$ and are too small for reliable modal estimates. However, the plagioclase-pyroxene-olivine contents of the two thin sections are about 50-40-10% and 60-30-10%. Tarasov *et al.* (1977) calculated a composition of 50 mole per cent plagioclase, 27% pyroxene, 19% olivine, 1.2% ilmenite, 0.7% chromite and 0.3% apatite from the chemical analysis of a powdered portion of 24170. However, the relative yields in this laboratory of mineral separates obtained from fragments totaling 130 mg indicate a much lower plagioclase content—plagioclase ~30 wt.%, pyroxene ~65 wt.% and olivine ~5 wt.%. This estimate is in accord with the reported mode for a sieved and handpicked 0.15–0.50 mm fraction—34% plagioclase, 52% pyroxene, 10% olivine, 3% cristobalite, and 0.6% ilmenite (Bence *et al.*, 1977). A very similar modal estimate on another sample from the Luna 24 site was

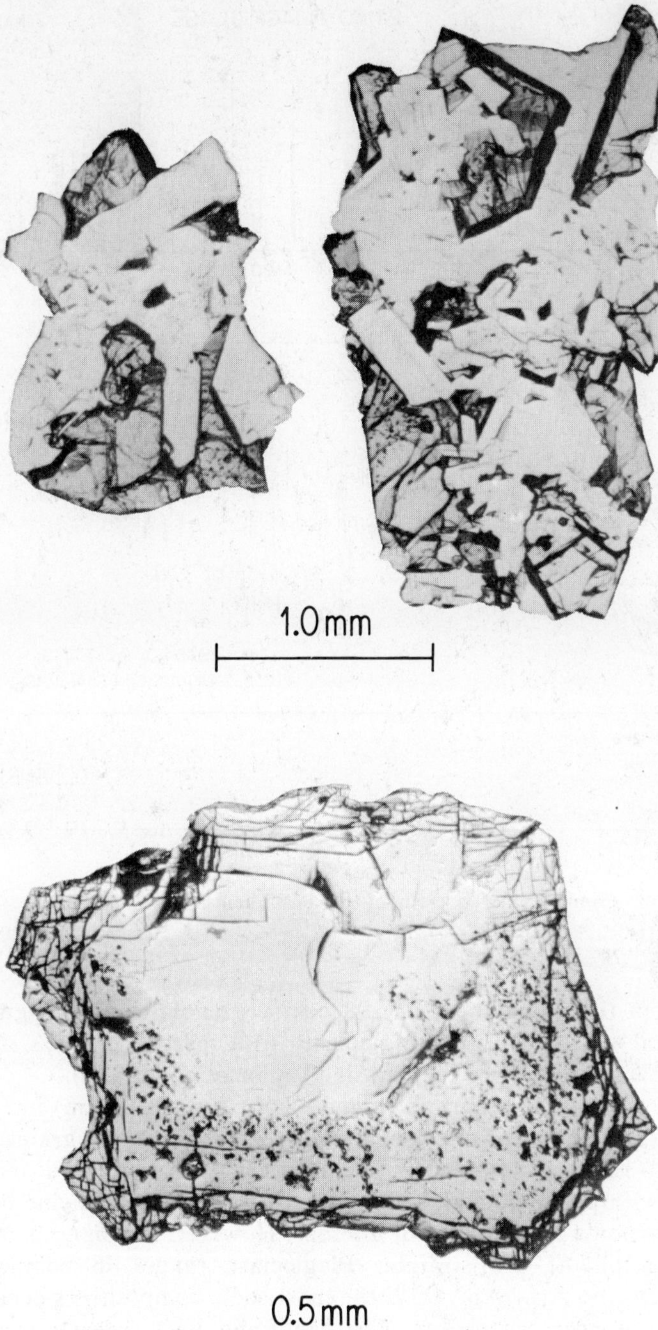

Fig. 1. Photomicrographics of two lithic fragments (FQM-648 and FQM-640) and of a zoned pyroxene grain (FQM-649) from sample 24170,1.

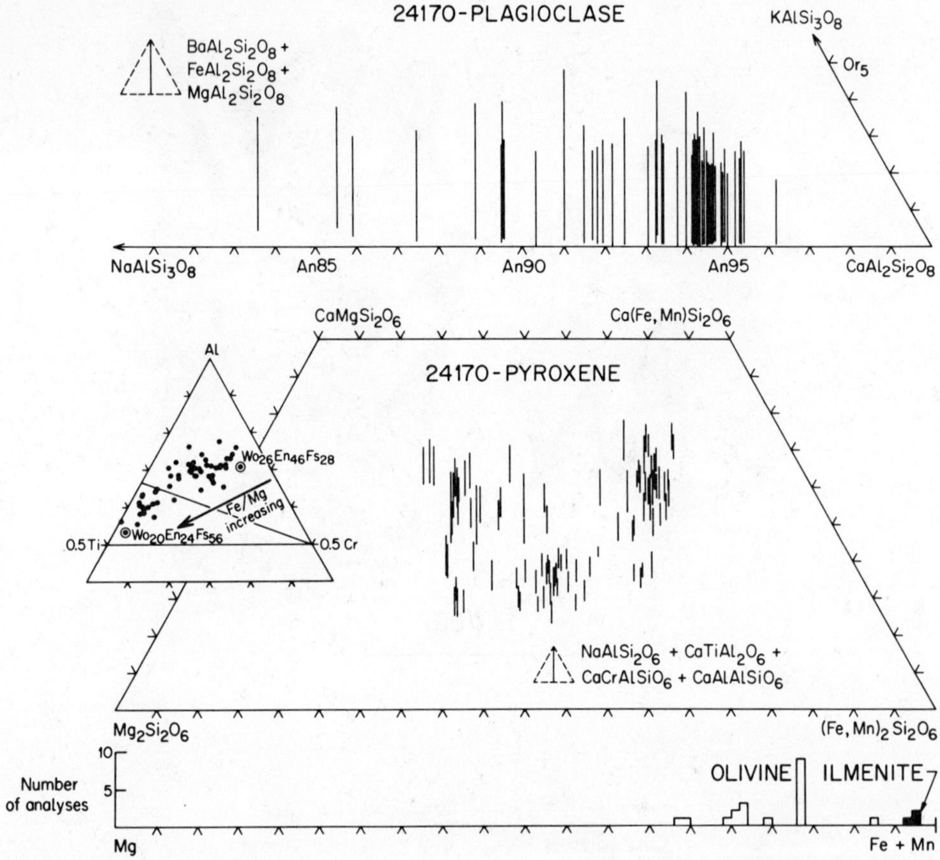

Fig. 2. Compositional variation of the constitutent minerals in sample 24170.

obtained from the abundances of 298 coarse monomineralic fragments in five Luna 24 soil samples—24% plagioclase, 61% pyroxene, 13% olivine, 0.3% opaques, 0.3% silica and 1% mesostasis (Taylor *et al.*, 1977b).

In 24170, plagioclase occurs as elongate laths (up to 500 μm) intergrown with pyroxene, in aggregates of interlocking (25–100 μm) tabular grains, and as tiny (10–50 μm) rounded to tabular inclusions in pyroxene and olivine. Grain boundaries with pyroxene are straight whereas plagioclase-olivine contacts tend to be undulose on a scale of tens of microns. However, the margins of plagioclase inclusions in olivine are straight. Plagioclase ranges in composition from $An_{84.4}Ab_{15.1}Or_{0.5}$ to $An_{97.0}Ab_{3.0}$ with the most sodic compositions occurring at the edges of grains near cristobalite. The MgO and FeO contents are low with a maximum of ~0.1 wt.% and ~0.7 wt.% respectively in the more sodic compositions.

Pyroxene ranges from pale pink to dark pink to yellow-brown in color and

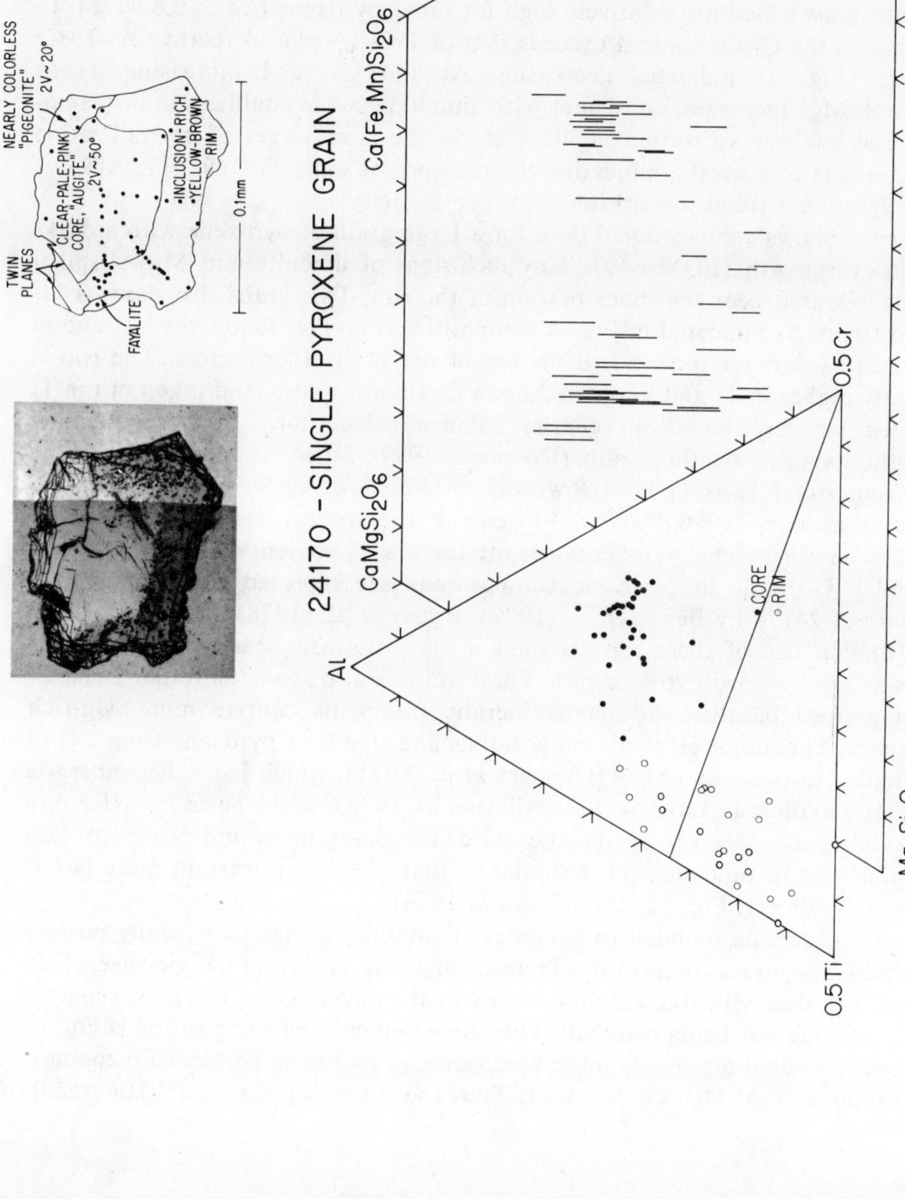

Fig. 3. Compositional variation in a single pyroxene grain from sample 24170.

occurs as anhedral, angular grains intergrown with plagioclase. Analyses of the pyroxene define a U-shaped trend on the pyroxene quadrilateral (Fig. 2). Pyroxene is zoned continuously from $\sim Wo_{26}En_{46}Fs_{28}$ to less-calcic and thence to more iron-rich compositions. The rims of larger grains, tiny (5–10 μm) grains intergrown with ilmenite, and tiny (5–10 μm) angular grains in the interstices between plagioclase laths fall in the range $Wo_{15}En_{43}Fs_{42}$ to $Wo_{32}En_{18}Fs_{50}$, the dominant trend being an increase in Ca-content. One pyroxene grain optically displays thin, poorly-defined lamellae. High resolution beam scans suggest that these may be platelets of high-Ca pyroxene. The contents of Al_2O_3 (0.7–1.9 wt.%) and TiO_2 (0.2–1.5 wt.%) are low, but the Cr_2O_3 contents range up to concentrations which are relatively high for lunar pyroxene (up to 0.8 wt.%). In some grains the Cr_2O_3 content exceeds that of TiO_2. A plot of relative Al-Ti-Cr contents (Fig. 2) indicates decreasing Al and Cr, and increasing Ti as Fe/(Fe + Mg) increases, consistent with simultaneous crystallization of plagioclase, and late crystallization of ilmenite. In the later stages of crystallization Ti/(Al + Cr) exceeds 0.5 indicating the presence of some Ti^{3+} ($R^{2+}Ti^{3+}AlSiO_6$) and suggesting a trend toward lower oxygen fugacity.

Figure 3 shows compositional data for a large grain of pyroxene with a dark-colored overgrowth (FQM-649). Tiny inclusions of ilmenite and Mg-Al spinel are concentrated near the inner margin of the rim. This grain also shows a U-shaped trend. As indicated in Fig. 3, compositional profiles follow the U-trend in some sectors, but jump across to the top of the U in other sectors. The rim is distinctly higher in Ti and lower in Al and Cr than the core, and much of the Ti must be trivalent based on charge balance calculations. However, optical absorption spectra on this grain (Rossman, 1977) failed to detect Ti^{3+} in the outer zone (total Ti as TiO_2 ~0.9 wt.%), but did show the presence of Cr in the central zone (Cr_2O_3 ~0.8 wt.%). In general, the spectra are similar to those obtained on other lunar pyroxenes despite the low Ti content of 24170.

Similar U-trends in pyroxene compositions are reported for finer-grained fractions of 24170 by Bence *et al.* (1977), Ryder *et al.* (1978) and Taylor *et al.* (1977b), but two of these reports show a "tail" toward hedenbergite and one shows a "tail" toward pyroxferroite. These authors also show that other Luna 24 coarse-grained basaltic and monomineralic fragments contain more Mg-rich pyroxenes. The major element compositions and trends in pyroxene from 24170 are similar to those in 14053 (Gancarz *et al.* 1971), which has a hedenbergite "tail" and to those in Luna 16 B-1 (Albee *et al.* 1972), which has a pyroxferroite tail. The trend toward Ti on the Al-Ti-Cr diagram is indicative of late crystallization of ilmenite and is similar to that reported in certain mare basalt clasts in 15205 (cf. Fig. 12, Dymek *et al.*, 1974).

Olivine occurs as rounded to irregular 10–400 μm grains that locally possess amoeboid apophyses that protrude into adjacent grains of plagioclase. This feature, together with the undulose nature of the olivine grain margins, suggests that the olivine was being resorbed. The range in measured composition is Fo_{31} to Fo_9, but individual grains are quite homogeneous (<2 mole per cent Fo zoning). The contents of Al_2O_3 (<0.20 wt.%), TiO_2 (<0.14 wt.%), Cr_2O_3 (<0.06 wt.%), and NiO (<0.09 wt.%) are low, but the concentration of CaO (0.4–0.6 wt.%) is high. No obvious reaction relation occurs between pyroxene and olivine, and fayalite (Fo_9) is in contact with hedenbergite-rich pyroxene ($\sim Wo_{36}En_{14}Fs_{50}$) at the margin of a large pyroxene grain (Fig. 3).

Tiny (<10 μm), angular grains of ilmenite [Fe/(Fe + Mg) = 0.97], together with micron-size blebs of troilite, occur in the interstices between plagioclase laths. Thin plates of ilmenite border euhedral surfaces of both plagioclase and pyroxene. No areas of K-rich mesostasis were found, but electron beam scans and optical examination suggest that a thin film of such mesostasis is associated with ilmenite on some grain surfaces. Electron beam scans also indicate the presence of Ca-phosphate in grains too small to analyze. A single grain of cristobalite ($\sim$50 μm) occurs between grains of plagioclase near the edge of FQM-650. No reaction relationships of SiO_2 with Fe-rich olivine or pyroxene such as described in other Luna 24 samples by Tarasov *et al.* (1977) or Ryder *et al.* (1978) were observed.

The detailed mineral chemistry supports the conclusion (Florensky *et al.*, 1977) that the fragments that comprise sample 24170 represent a rock that was broken up by the drilling. The mineral composition data obtained on fine-grained portions of 24170 (Bence *et al.*, 1977; Ryder *et al.*, 1977, 1978; Taylor *et al.*, 1977a,b) correspond in detail to those which we obtained on the larger lithic fragments.

The textural and composition features in 24170 are all consistent with crystallization from a melt with the following sequence: 1) plagioclase; 2) pyroxene; 3) olivine; 4) ilmenite, cristobalite and troilite. The high Fe-content of the pyroxene and olivine, the extensive zoning in the pyroxene, the high Ca-content of the olivine, and the low K-content of the plagioclase, taken together, suggest that Luna 24 sample 24170 is similar to other lunar mare basalts. The extensive early crystallization of plagioclase, the near absence of ilmenite, the low minor element content of the pyroxene, the trend in the relative amounts of Al-Ti-Cr in pyroxene, and the olivine-pyroxene relations are distinctive.

The late crystallization of only small amounts of ilmenite serves both to distinguish microgabbro 24170 from nearly all other lunar mare basalts and to emphasize its similarity to the fragments of finer-grained low-Ti ferrobasalt which are abundant in other Luna 24 samples. The obvious similarities (and subtle differences) between Luna 24 ferrobasalt and Apollo 17 very-low Ti basalts have been summarized by Vaniman and Papike (1977) and Taylor *et al.*, (1977a), and will not be discussed here. However, it is appropriate to compare 24170 with other rock types from the Luna 24 samples.

Coish and Taylor (1977), Ryder *et al.* (1978) and Tarasov *et al.* (1977) describe fragments of both vitrophyric basalt and ophitic basalt in which olivine was the first phase to crystallize, followed by plagioclase and pyroxene. In 24170, plagioclase was the first phase to crystallize. Consequently, this sequence of crystallization contrasts 24170 with most other Luna 24 ferrobasalts.

Since 24170 is relatively coarse-grained and probably cooled more slowly than either the vitrophyres or ophitic basalts, it could be argued that the differences in the order of phase appearance were controlled by variations in cooling rate. For example, the suppression of plagioclase crystallization by rapid cooling has been documented experimentally by Grove and Bence (1977) for Apollo 15 quartz normative basalts. However, the major and minor element chemistry of pyroxene in 24170 is virtually identical to that of the Luna 24 ferrobasalts (cf., Vaniman and Papike, 1977; Ryder *et al.*, 1977) suggesting that cooling rate was not the chief cause of differences in crystallization sequence. The pyroxene in the finer-grained Luna 24 basalts does, however, contain more Al than that in 24170, consistent with the late crystallization of plagioclase (Vaniman and Papike, 1977).

The differences in crystallization sequence can be explained in another way. In the pseudoternary system olivine-anorthite-silica, the pyroxene-olivine-plagioclase and olivine-spinel-plagioclase peritectic points shift toward higher plagioclase and silica contents respectively with higher pressure (D. Walker, pers. comm.). Thus a magma formed by partial melting of an olivine and pyroxene-bearing source region could evolve at high pressure by removal of olivine driving the residual liquids toward the olivine-plagioclase cotectic. The early portion of this fractionation trend is in the olivine primary field at both high pressure and low pressure, but the later portion of this trend would move from the olivine primary field at high pressure to the plagioclase primary field upon eruption. Eruption of a magma from the early part of the trend would result in a basalt with the crystallization sequence olivine→plagioclase→pyroxene. However, olivine would no longer be in equilibrium with a magma erupted from the latter portion of this trend and it would tend to react out. Such a model can explain the peculiar texture of the olivine that may be indicative of resorption. Thus, the resultant magma could show early crystallization of plagioclase with no evidence for a prior history that involved substantial olivine separation.

Age and Source of 24170

Samples of plagioclase and pyroxene (containing less than 5% olivine) were dissolved and analyzed for K, Rb, Sr, Ba, Sm and Nd. Samples of the plagioclase and pyroxene mineral separates have been provided to R. A. Schmitt at the Oregon State University for INAA. The analytical procedures used are described by Papanastassiou *et al.* (1977). Using these procedures we ascertained that high quality isotopic data could be obtained for samples containing only 20 ng of Nd. Chemistry blanks for Sm and Nd were 5 and 20 pg, respectively, and are negligible for the samples analyzed despite the small sample size.

The Sm-Nd data on 24170 are shown in Table 1 and Fig. 4. The errors shown for the data are two standard deviations (2σ). We obtain an age of 3.30 ± 0.05 AE and an initial $^{143}Nd/^{144}Nd = 0.507716 \mp 73$. The uncertainties of the age and the initial Nd represent our best estimates and correspond to the maximum change in the slope of the isochron allowed by the 1σ uncertainties of the two data points. From consideration of the 2σ analytical uncertainties, the maximum uncertainty in age for the two data points shown corresponds to 0.10 AE. The range in $^{143}Nd/^{144}Nd$ for the data is 22 parts in 10^4 and permits this precise age determination. The Rb-Sr data are shown in Table 1 and Fig. 5. The data show extremely low Rb/Sr for all samples and define an imprecise age of 3.7 ± 0.6 AE (2σ) and a very low initial $^{87}Sr/^{86}Sr$ of 0.69906 ∓ 4 (2σ). If we consider the precise Sm-Nd age and the plagioclase and chip data, we obtain a slightly higher initial $^{87}Sr/^{86}Sr$ of 0.69907 ∓ 4. Within analytical errors the Rb-Sr age is compatible with the much more precise Sm-Nd age.

The initial $^{87}Sr/^{86}Sr$ in the plagioclase shows that 24170 evolved in an extremely low Rb/Sr parent reservoir with Rb/Sr ≈ 0.002. We note that the initial $^{87}Sr/^{86}Sr$ of 24170 is significantly greater than BABI and shows that this sample must be significantly younger than 4.5 AE, despite the large uncertainty in the Rb-Sr age.

The mineralogic arguments presented earlier indicate that the microgabbro layer is a single rock, possibly crushed during drilling on the moon (Barsukov, 1977). Therefore, we assume that the isotopic data on 24170 yield meaningful age information. The ^{40}Ar-^{39}Ar ages of two individual fragments from 24170, which are presented later in this paper, are concordant with the Sm-Nd age and, therefore, the age obtained by both techniques defines the crystallization age of the microgabbro. This age indicates a relatively young lava flow at Mare Crisium and represents the first clear evidence that relatively young lavas were emplaced on the eastern part of the earth-facing lunar hemisphere. This datum and the evidence for young volcanism at the Ocean of Storms and at the Sea of Fertility (Papanastassiou and Wasserburg, 1971; 1972) show that relatively young mare flows erupted over widely separated areas on the moon very near the time believed to mark the cessation of mare volcanism. The age and low Ti-content of

Table 1. Analytical results, 24170.

Sample	Weight mg	Sm^a (10⁻⁸ mole/g)	$^{144}Nd^a$ (10⁻⁸ mole/g)	$^{147}Sm/^{144}Nd^b$ ($\pm 2\sigma$)	$^{143}Nd/^{144}Nd^c$ ($\pm 2\sigma$)
Plagioclase	19	0.4615	0.3738	0.18522 ± 10	0.511751 ± 16
Pyroxene	19	0.5830	0.3698	0.23645 ± 15	0.512867 ± 20
		Rb (10^{-8} mole/g)	^{88}Sr (10^{-8} mole/g)	$^{87}Rb/^{86}Sr^d$ $\times 10^2$	$^{87}Sr/^{86}Sr^e$ ($\pm 2\sigma$)
Chip	2.5	0.1339	98.3	0.318	0.69923 ± 7
Plagioclase	19	0.209	196.9	0.247	0.69919 ± 4
Pyroxene	19	0.0869	15.73	1.286	0.69974 ± 8

a. Isotopic composition of Sm from Russ (1974); isotopic composition of Nd from DePaolo and Wasserburg (1976).

b. Uncertainties do not include uncertainty in tracer concentration of 0.1%.

c. Normalized to $^{146}Nd/^{142}Nd = 0.636155$ to correct for mass discrimination.

d. Uncertainty ±0.4%.

e. Normalized to $^{86}Sr/^{88}Sr = 0.1194$ and $^{84}Sr/^{88}Sr = 0.006748$.

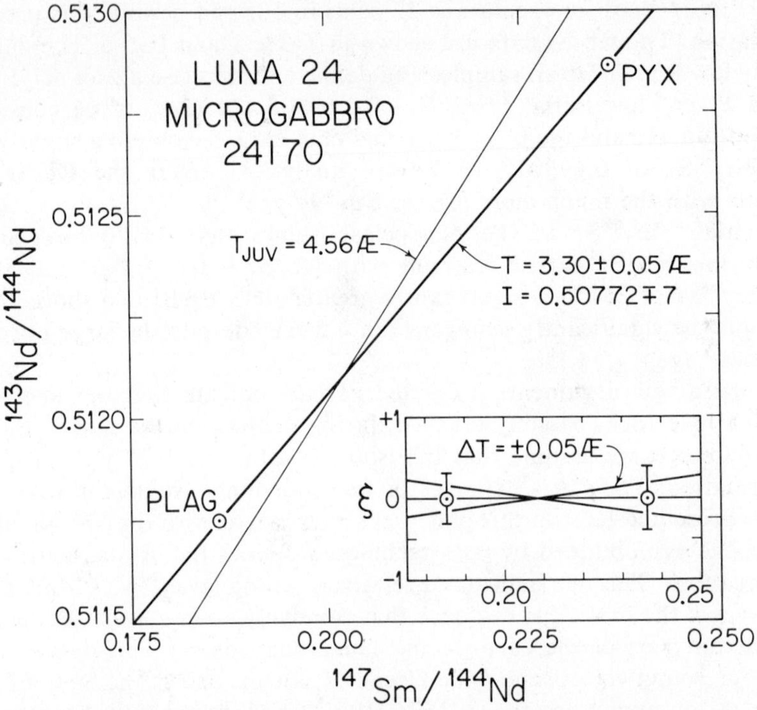

Fig. 4. Sm-Nd evolution diagram for microgabbro 24170. The inset shows the experimental uncertainties in the data (in parts in 10^4) and the uncertainty of ±0.05 AE in the age. The maximum uncertainty in the age can be seen to be ±0.10 AE.

24170 are in agreement with the general observation that low-Ti mare basalt samples have younger ages than high-Ti basalts (Papike *et al.*, 1974).

Birck *et al.* (1977) have reported a Rb-Sr age of 3.74 ± 0.56 AE (2σ) on microgabbro fragments from sample 24171. This sample is probably identical to sample 24170 as both are part of the 2-cm thick layer of crushed microgabbro in the Luna 24 core. Birck *et al.* (1977) did not present their analytical Rb-Sr data on 24171; however, we presume that the large uncertainty in their reported Rb-Sr age parallels our experience and reflects the low Rb/Sr characteristic of the Luna 24 samples. The Rb-Sr age uncertainties obtained by both groups permit an age for 24170 and 24171 ranging from 3.2 AE to 4.3 AE. Since the time scale for post-Imbrium mare volcanism is relatively narrow (3.0–4.0 AE), the assignment of a particular sample as young or old requires a precise age determination. The Rb-Sr age uncertainties as reported here and by Birck *et al.* (1977) are too large to permit the identification of rocks of older (or younger) ages at this site.

The low initial $^{87}Sr/^{86}Sr$ for 24170 indicates a very low Rb/Sr for the parent material of this sample similar to but distinctly lower than the sources of most

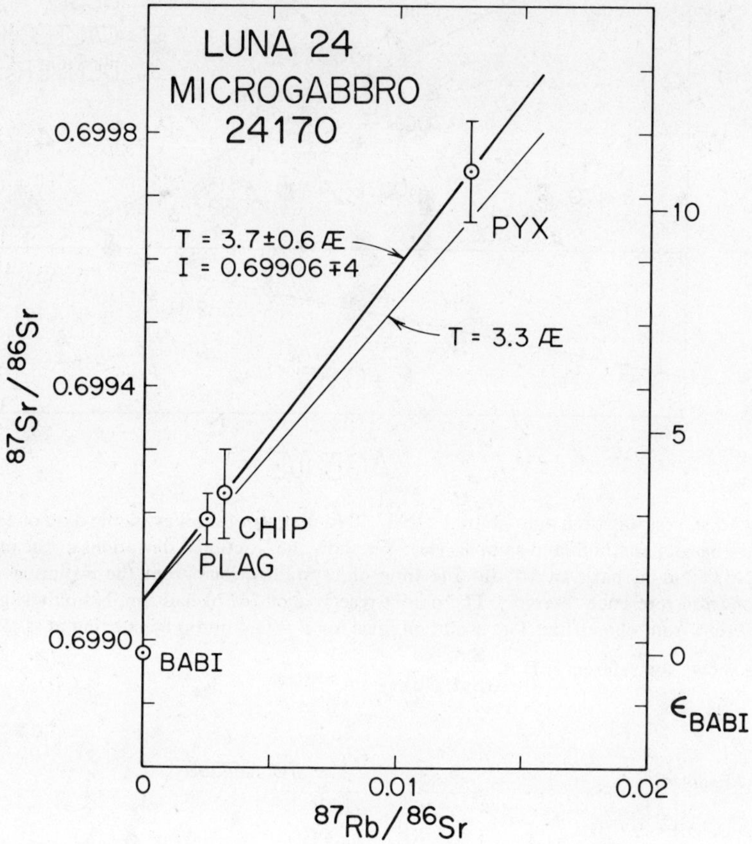

Fig. 5. Rb-Sr evolution diagram for microgabbro 24170. The extremely low Rb/Sr in the sample chip and the mineral separates prevent the determination of a precise age. On the right-hand ordinate we show the small relative enrichment in the measured $^{87}Sr/^{86}Sr$ with respect to BABI (in parts in 10^4).

other mare basalts (Fig. 7). From the general display of the Rb-Sr data on mare basalts in Fig. 7 and the T_{BABI} model ages of mare basalts, it is not possible to identify any particular mare basalts as being produced near their time of crystallization from an undifferentiated part of the lunar mantle. In particular, it is not possible to ascertain whether the extremely low Rb/Sr for the parent materials of 24170 is indicative of production of 24170 at 3.30 AE from an undifferentiated part of the lunar interior.

The initial $^{143}Nd/^{144}Nd$ ($\epsilon_I^{CHUR} = +1.9 \pm 0.6$) indicates that the 24170 magma was derived from a reservoir in the lunar mantle which had Sm/Nd about 6 ± 2% higher than chondritic Sm/Nd over the time from 4.56 AE to 3.3 AE (Fig. 6). This reservoir therefore appears to have had a REE pattern fractionated slightly from chondritic, but much less fractionated than the source reservoirs of

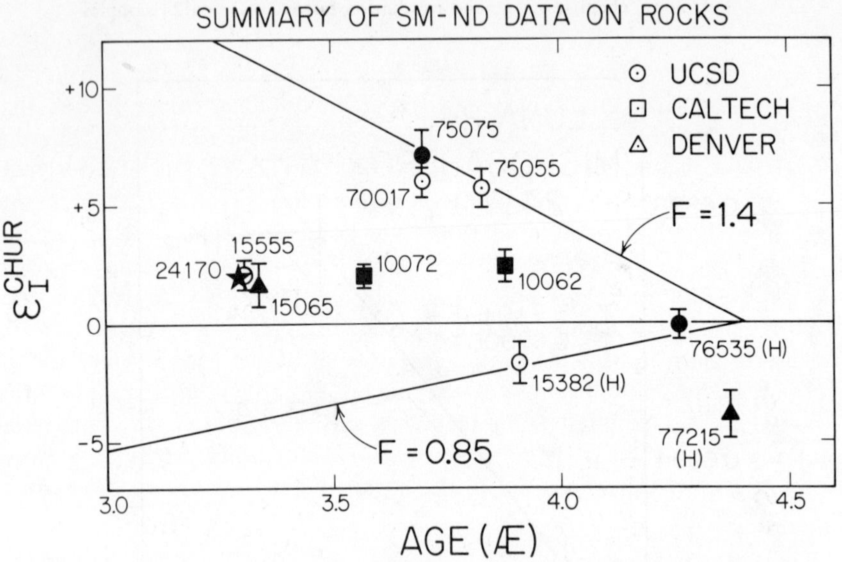

Fig. 6. Crystallization age—Initial $^{143}Nd/^{144}Nd$ diagram for all available data on lunar mare basalts and highland samples (H). We show the fractional deviations of the initial $^{143}Nd/^{144}Nd$ in parts in 10^4 (at the time of crystallization) from the evolution of a chondritic reference reservoir. The parent reservoir of 24170 had Sm/Nd only slightly different from chondritic. The evolution lines for F = 1.4 and 0.85 starting at 4.45 AE are shown for reference; $F = \dfrac{Sm/Nd}{(Sm/Nd)_{CHUR}}$

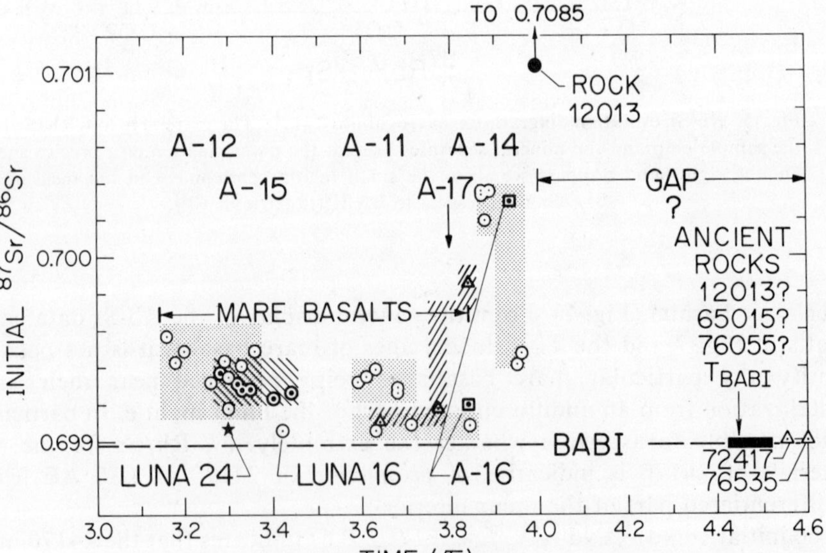

Fig. 7. Age—Initial $^{87}Sr/^{86}Sr$ for lunar samples. Mare basalt volcanism is restricted in time between 3.1 and 3.9 AE. Sample 24170 is relatively young and characterized by an extremely low initial $^{87}Sr/^{86}Sr$ similar to Luna 16 and low-K Apollo 11 basalts.

some Apollo 17 high-Ti basalts (Lugmair *et al.*, 1976). The data are consistent with derivation of the 24170 magma from a source richer in plagioclase than the source of low-K, high-Ti basalts (cf. Papanastassiou *et al.*, 1977). The data on 24170 and on Apollo 15 basalts suggest that low-Ti basalts are derived from near-chondritic reservoirs in contrast to *some* high-Ti basalts which may be derived from highly fractionated reservoirs (Lugmair *et al.*, 1976; Papanastassiou *et al.*, 1977). However, this similarity between the parent reservoirs of Apollo 15 basalts and the Luna 24 sample does not hold for Rb/Sr since the inferred Rb/Sr of the Luna 24 gabbro parent reservoir is approximately a factor of four lower than that of the parent reservoir of basalt 15555. Since the precise Sm/Nd of the bulk moon is not known, it is not possible to determine whether the 24170 magma was melted from primitive undifferentiated lunar material. At present there is no clear evidence from Sm/Nd studies on mare basalts that undifferentiated material existed in the moon to act as a source for mare basalt magmas.

Soil Samples

We have analyzed three soil samples from the Luna 24 core for K, Rb, Sr, Ba and one for Sm and Nd (Table 2). The data show very low levels of K, Rb, Ba, Sr, Sm and Nd, and demonstrate the near absence of significant KREEPUTh-rich components. The Luna 24 soils analyzed are extremely uniform in elemental and Sr isotopic composition, indicating the absence of significant chemical layering in the sediment column sampled by this core. The Rb-Sr data yield model ages T_{BABI} of 4.84 ± 0.26 to 4.43 ± 0.21 AE, which are not well defined but are consistent with the observation of model ages of ~4.5 AE for most lunar soils. The Sm-Nd data on the soils indicate that the soils have had a nearly chondritic Sm/Nd evolution. This requires the absence from these soils of significant components with large Sm/Nd fractionation or the essentially complete remixing of materials with complementary Sm/Nd fractionation

Table 2. Luna 24 soils.

Sample	K ppm	Rb ppm	Sr ppm	Ba ppm	$^{87}Sr/^{86}Sr$ (2σ)	T_{BABI} AE
24109	224	0.484	86.3	37	0.70011 ± 6	4.84 ± 0.26
24149	222	0.473	88.0	36	0.70002 ± 9	4.65 ± 0.40
24174	239	0.525	91.1	37	0.70004 ± 5	4.43 ± 0.21

Sample	Sm ppm	Nd ppm	$^{143}Nd/^{144}Nd$[a] (2σ)	T_{JUV} AE
24109	1.87	5.73	0.511952 ± 16	4.56 ± 0.02
24174	1.86	5.67	0.511970 ± 33	4.55 ± 0.03

a. Corrected for neutron capture effects on $^{143}Nd/^{144}Nd$ (4.5 parts in 10^5).

Table 3. Neutron effects in sm.

Sample	$^{149}Sm/^{154}Sm$	a ϵ_{Sm}	b Σ_{eff}	c $\Psi_{E\ <1.0\ ev}$	c,d $\Psi^{N}_{E\ <1.0\ ev}$	c,e $\Psi^{N}_{E\ <0.18\ ev}$
24109	0.60162 ± 32	97 ± 5	0.69	33.4	24.9	8.7
24170 plag	0.60580 ± 21	28 ± 3.5	0.69	9.7	7.2	2.5
24170 pyx	0.60575 ± 42	29 ± 7	0.69	10.0	7.5	2.6
10072 pyx	0.60636 ± 20	19 ± 3	1.03	7.9	8.1	2.8
10072 Quint[f]	0.60633 ± 13	19 ± 2	1.03	7.9	8.1	2.8
Luna 16, G2[g]	0.60485 ± 6	44 ± 1	0.83	16.5	14.8	4.6
60001,8[h]	0.60322 ± 13	71 ± 2	0.46	21.4	12.6	4.0
Terrestrial[i]	0.60750 ± 2	0	—	0	0	0
Reagent Sm[j]	0.60747 ± 7	0	—	0	0	0

a. Parts in 10^4.
b. Units 10^{-2} cm^2/g. Σ_{eff} for Luna 24 samples was calculated using major element composition data given by Bence *et al.* (1977) and trace element concentrations measured in this laboratory. The procedure for calculating this parameter is summarized by Curtis and Wasserburg (1975).
c. Fluence; units: 10^{16} neutron/cm^2.
d. Fluence normalized to an arbitrary, standard chemical composition, Σ_{eff} = 0.00925 cm^2/g (cf. Curtis and Wasserburg, 1975).
e. Equivalent normalized Gd fluence using estimate of $\epsilon_{Gd}/\epsilon_{Sm}$ = 1.4.
f. Quintessence-rich sample, cf. Papanastassiou *et al.* (1977).
g. Russ (1972); This sample previously yielded the highest normalized neutron fluence of any lunar soil.
h. Russ (1973); This sample had the largest ϵ_{Sm} previously measured.
i. Russ *et al.* (1971).
j. Measured on sample spiked with ^{147}Sm tracer.

produced at possibly diverse times throughout the duration of lunar magmatic activity.

AR MEASUREMENTS

^{40}Ar-^{39}Ar gas retention ages and $^{38}Ar/Ca$ cosmic ray exposure ages have been measured on two individual small fragments from 24170,1. These fragments weighed 0.9 mg and 0.5 mg and the measurements required use of newly established microtechniques to obtain detailed ^{40}Ar-^{39}Ar thermal release age spectra on very small fragments. High quality analyses were obtained for the individual fragments containing 10^{-7} g K (100 ppm K) and 3×10^{-9} ccSTP ^{40}Ar. This constitutes an improvement in sensitivity of one to two orders of magnitude over previous techniques and permits the determination of a high resolution age spectrum with small samples. The microtechniques are discussed in detail by Huneke (1978).

The Ar blank of the system was carefully monitored throughout the experiments and was typically 2.5×10^{-11} ccSTP air Ar. A simulation extraction using an empty filament assembly showed no increase over the cold system blank. The

Ar was measured mass spectrometrically with an electron multiplier at a gain of $\sim 10^4$ and with a sensitivity of 4.6×10^{-9} ccSTP/volt using a $10^9 \Omega$ feedback resistor on the electrometer. Discrimination was 0.05% per mass unit favoring lighter masses. The mass 39 background in the mass spectrometer was constant and equivalent to $\sim 6 \times 10^{-13}$ ccSTP Ar. Interferences at mass 35–38 are due to Cl^+ and HCl^+ with a $^{35}Cl^+$ signal equivalent to $\sim 3 \times 10^{-11}$ ccSTP Ar and $HCl^+/Cl^+ = 0.015 \pm 0.005$.

The samples were irradiated in the University of Missouri Research Reactor at Columbia, Missouri. The conversion coefficients for the irradiation (designated LAC 1) determined from the hb3gr hornblende monitor (Turner *et al.*, 1971) and a CaF_2 monitor are:

$$C_{39}(K) = 1.65 \times 10^{-3} \text{ ccSTP } ^{39}Ar/g \text{ K}$$
$$C_{37}(Ca) = 2.39 \times 10^{-4} \text{ ccSTP } ^{37}Ar/g \text{ Ca}$$

on February 1, 1978. Spatial variations in neutron fluence were monitored by Ni fluxwires between samples. Ar contributions from neutron reactions on Ca were $^{36}Ar/^{37}Ar/^{38}Ar/^{39}Ar = 0.00078/\equiv 1/0.00026/0.0021$ measured on CaF_2. $^{38}Ar_{Ca}$ is assumed to be 1/3 $^{36}Ar_{Ca}$. Since we did not measure the Ar in a simultaneously irradiated K salt, contributions from neutron reactions on K are taken as $^{38}Ar/^{39}Ar/^{40}Ar = 0.01/\equiv 1/0.01$ (Turner *et al.*, 1971). ^{38}Ar contributions from ^{37}Cl (n,γ) are not subtracted.

The Ar amounts and isotopic compositions are given in Table 4. For the calculation of exposure and gas retention ages, the Ar is divided between trapped, spallogenic and radiogenic Ar using $(^{36}Ar/^{38}Ar/^{40}Ar)_s = 0.6/1/0.15 \pm 0.15$ and $(^{36}Ar/^{38}Ar/^{40}Ar)_t = 1/0.187/0.5 \pm 0.5$ (see below).

Trapped solar wind ^{36}Ar and lunar parentless ^{40}Ar are much less prominent in the 24170 samples than has been the case for other K/Ar analyses of Luna 24 fragments (Stettler and Albarede, 1977). In the Ar release steps which define the ^{40}Ar-^{39}Ar age of 24170, $^{36}Ar_t$ is totally absent. The small amounts of trapped solar wind ^{36}Ar present most likely derive from residual dust adhering to the sample. The low temperature release steps containing $^{36}Ar_t$ are corrected for $^{40}Ar_t$ using $(^{40}Ar/^{36}Ar)_t = 0.5 \pm 0.5$ estimated from a graph of $^{36}Ar_t/^{40}Ar$ versus $^{39}Ar^*/^{40}Ar$ (*denotes K-derived Ar). This ratio is compatible with the range of 0.3–1.0 previously observed in Luna 24 soils and soil fragments (Bogard and Hirsch, 1977; Stettler and Albarede, 1977). We emphasize that the ^{40}Ar-^{39}Ar ages for the 24170 fragments are not sensitive to the subtraction of parentless lunar ^{40}Ar contributions, which are at most several per cent in low temperature steps. This source of uncertainty, which is present in the ^{40}Ar-^{39}Ar ages previously reported, is absent in the present samples. Using the conversion coefficients $C_{39}(K)$ and $C_{37}(Ca)$, K and Ca contents of 120 ppm K and 8.4% Ca for fragment R-2 and 130 ppm K and 10.6% Ca for fragment R-5 are calculated from the ^{39}Ar and ^{37}Ar contents.

The cosmic ray exposure age is calculated from spallogenic ^{38}Ar produced from Ca and Fe and ^{37}Ar produced from Ca. Using a production rate of 1.4×10^{-8} ccSTP/g Ca/m.y. and allowing for an additional 15% contribution to

Table 4. Ar amounts and isotopic composition in neutron irradiated fragments of Luna 24 24170,1 microgabbro.[f]

Filament Temp °C	$^{40}Ar^{a}$ $\times 10^{-10}$ ccSTP	$^{36}Ar^{b}/^{40}Ar$	$^{37}Ar^{c}/^{40}Ar$	$^{38}Ar^{d}/^{40}Ar$	$^{39}Ar^{*e}/^{40}Ar^{*}$	Age AE
24170,1R-2(0.9 mg)		$\times 10^{-4}$	$\times 10^{-2}$	$\times 10^{-4}$	$\times 10^{-4}$	
835	0.29	542 ± 205	500 ± 187	48 ± 60	888 ± 344	2.33 ± 0.51
965	1.48	588 ± 47	394 ± 29	351 ± 29	611 ± 49	2.85 ± 0.12
1060	2.64	841 ± 36	341 ± 14	456 ± 20	489 ± 27	3.17 ± 0.08
1140	3.50	738 ± 25	316 ± 10	454 ± 16	456 ± 20	3.27 ± 0.07
1150	2.32	469 ± 24	368 ± 18	447 ± 23	453 ± 24	3.29 ± 0.08
1200	2.71	419 ± 19	455 ± 19	536 ± 23	459 ± 20	3.27 ± 0.07
1210	2.18	403 ± 22	547 ± 27	631 ± 34	456 ± 24	3.28 ± 0.08
1250	4.01	444 ± 13	625 ± 17	741 ± 22	451 ± 14	3.30 ± 0.05
1275	3.89	412 ± 13	589 ± 17	710 ± 21	458 ± 15	3.27 ± 0.05
1320	3.47	333 ± 12	477 ± 16	565 ± 20	452 ± 16	3.29 ± 0.05
1400	3.48	294 ± 10	382 ± 12	470 ± 17	440 ± 15	3.33 ± 0.05
1545	4.81	616 ± 15	621 ± 14	903 ± 22	443 ± 13	3.32 ± 0.05
1665	0.89	343 ± 48	489 ± 63	467 ± 64	275 ± 38	4.07 ± 0.22
Total	35.67	488 ± 7	483 ± 6	597 ± 8	460 ± 6	
24170,1R-5(0.5 mg)		$\times 10^{-4}$	$\times 10^{-2}$	$\times 10^{-4}$	$\times 10^{-4}$	
850	0.17	347 ± 308	364 ± 316	283 ± 246	648 ± 570	2.77 ± 1.25
900	0.36	394 ± 222	363 ± 202	348 ± 195	659 ± 373	2.74 ± 0,80
965	0.82	441 ± 78	302 ± 53	348 ± 61	526 ± 94	3.07 ± 0.27
1020	1.31	537 ± 62	273 ± 31	371 ± 42	484 ± 57	3.19 ± 0.18
1090	2.32	623 ± 40	293 ± 19	430 ± 27	485 ± 34	3.19 ± 0.11
1165	2.86	475 ± 25	422 ± 21	564 ± 29	478 ± 26	3.21 ± 0.08
1250	7.76	480 ± 11	683 ± 13	852 ± 17	455 ± 11	3.28 ± 0.04
1310	2.08	345 ± 25	451 ± 31	578 ± 41	459 ± 33	3.27 ± 0.11
1400	0.95	381 ± 61	294 ± 45	500 ± 78	419 ± 66	3.41 ± 0.24
1545	2.54	675 ± 55	785 ± 63	1059 ± 85	407 ± 34	3.45 ± 0.13
1665	0.34	392 ± 304	526 ± 304	658 ± 381	152 ± 90	5.05 ± 1.00
Total	21.52	499 ± 17	529 ± 20	686 ± 28	459 ± 13	

a. ^{40}Ar corrected for blanks and contributions from neutron reactions on K.

b. ^{36}Ar corrected for spectrometer background, blank and contributions from neutron reactions on Ca.

c. ^{37}Ar corrected for spectrometer background, radioactive decay between 1st of February 1978 and time of analysis, and neutron fluence relative to the monitor (0.9908).

d. ^{38}Ar corrected for spectrometer background, blank and contributions from neutron reactions on K and Ca.

e. $^{39}Ar^{*}$ corrected for spectrometer backgound, radioactive decay between 1st of February 1978 and time of analysis, neutron fluence relative to the monitor (0.9908) and contributions from neutron reactions on Ca. $^{40}Ar^{*}$ corrected for trapped and spallogenic ^{40}Ar contributions (see text).

f. Errors include standard deviation in the extrapolation of the ratios to the inlet time into the spectrometer, errors in mass spectrometer and irradiation interferences, and a conservative error of ±50% in the amount of blank subtraction. The uncertainty in measured ^{40}Ar amount is 5%.

$^{38}Ar_s$ from Fe in the sample, the exposure age is 165 m.y. and is essentially identical for the two fragments. The identical exposure ages strengthen the argument that the 24170 microgabbro layer was derived from a single rock.

The K/Ar age of each step is listed in the last column of Table 4 and shown in Fig. 8 as a function of ^{39}Ar released. The ages are calculated using the newly

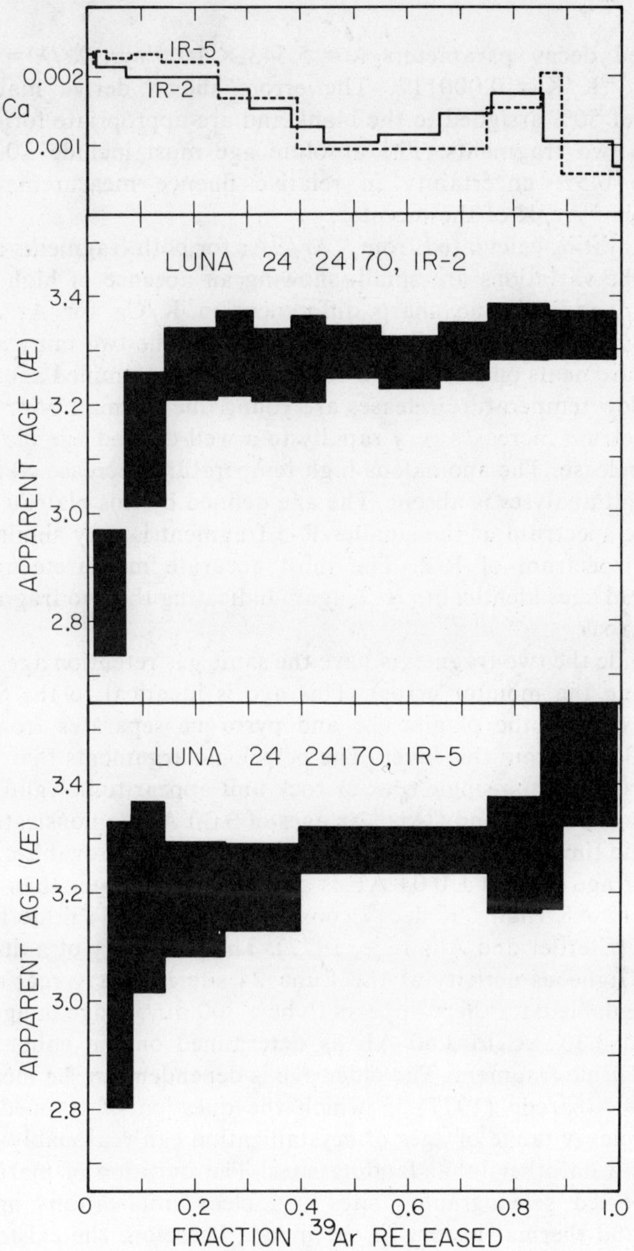

Fig. 8. Apparent age and K/Ca variations as a function of ^{39}Ar released from two individual fragments of the microgabbro layer 24170. A relatively high resolution Ar age spectrum was obtained. The ages are initially young due to previous small ^{40}Ar losses and increase very rapidly to a well-defined age plateau at 3.30 AE over the last 85% of ^{39}Ar release. K/Ca shows only small variations with no evidence for Ar release from high K/Ca phases. The age spectra show no anomalous features. The gas retention ages of the two individual fragments are identical at 3.30 ± 0.04 AE and are the same as the Sm-Nd crystallization age.

recommended decay parameters $\lambda = 5.543 \times 10^{-10}$/yr, $\lambda_\epsilon/\lambda = 0.105$ and K composition ^{40}K/K = 0.000117. The errors shown derive mainly from the uncertainty of 50% assigned to the blank and are appropriate for comparing the data on the two fragments. The absolute age must include ±0.03 AE corresponding to 0.5% uncertainty in relative fluence measurements and 1.5% uncertainty in ^{40}Ar/K of the monitor.

The K/Ca ratios calculated from ^{39}Ar/^{37}Ar for both fragments are also shown in Fig. 8. The variations are small, showing an absence of high K mesostasis; there are, in addition, no sharp differences in K/Ca for Ar released from plagioclase and pyroxene. K/Ca is comparable for the two minerals.

The measurements on R-2 yield a very precisely determined age. The apparent ages of the low temperature releases are young due to small prior losses of ^{40}Ar. The age spectrum increases very rapidly to a well-defined age plateau over 85% of the ^{39}Ar release. The anomalous high temperature decrease which is common in lunar basalt analyses is absent. The age defined by this plateau is 3.30 ± 0.02 AE. The age spectrum of the smaller R-5 fragment is very similar to the more precise age spectrum of R-2. The most accurate measurements of the R-5 spectrum yield ages identical to R-2, again indicating the two fragments are parts of the same rock.

We conclude the two fragments have the same gas retention age of 3.30 ± 0.04 AE (including the monitor error). This age is identical to the Sm-Nd age of 3.30 ± 0.05 AE on the plagioclase and pyroxene separates from a combined sample of 130 mg from this layer. The petrologic arguments that the 130 mg of material derives from a single rock or rock unit appear to be valid. In any event, the concordant Sm-Nd and ^{40}Ar-^{39}Ar ages of 3.30 AE demonstrate that this age represents the time of crystallization of the Luna 24 microgabbro.

The K/Ar age of 3.30 ± 0.04 AE is significantly younger than the K/Ar age of 3.60 ± 0.12 AE (new ^{40}K decay constants) of a low-Ti lithic fragment from layer 24096 (Stettler and Albarede, 1977). The possibility of a distinct range in the times of igneous activity at the Luna 24 site is a very real one. From the presently available data there appears to be a 300 m.y. range in ages at the Luna 24 site from 3.30 AE to 3.60 AE as determined on the gabbro and another undescribed lithic fragment. The older age is dependent on the measurements by Stettler and Albarede (1977) in which the question of trapped ^{40}Ar plays a significant role. A range of ages of crystallization can reasonably be anticipated from the data on other lunar landing sites. The duration of mare volcanism at widely dispersed selenographic sites has clear implications on the magma generation and thermal history of the moon. Therefore the existence of a wide range in ages at the Luna 24 site is an important question which must receive additional attention in future measurements on adequately described igneous lithic fragments with low trapped Ar content.

IRRADIATION HISTORY

We show in Table 3 neutron capture effects in Sm for lunar samples. The data on 24109, 24170 and 10072 were obtained on samples spiked for Sm and Nd and used for the Sm-Nd studies. Some of the previous observations on Sm by Russ (1972, 1973) are given in Table 3 for comparison. Analysis of spiked terrestrial Sm yields results in excellent agreement with the results of unspiked Sm and indicates the absence of artifacts. We show the isotopic shifts in $^{149}Sm/^{154}Sm$ and the neutron dosage calculated for neutron energies E <1.0ev (applicable to Sm) and the dosage calculated for E <0.18ev (applicable to Gd). Soil 24109 shows the largest depletion ever observed in $^{149}Sm/^{154}Sm$ which corresponds to a low energy neutron fluence two times larger than previously observed. Sample 24170 has a much lower neutron fluence than soil 24109 indicating that the age of deposition of the regolith at this site is much less than 3.3 AE.

The high fluence calculated for 24109 is important because it provides a lower limit to the ^{157}Gd and ^{149}Sm neutron capture rates in lunar rocks, which have been poorly constrained. The capture rates for these isotopes were calculated by Lingenfelter *et al.*, (1972, referred to as LCH). Based on the results of the Lunar Neutron Probe Experiment, Woolum *et al.* (1975) suggested that the capture rate of ^{157}Gd ($\equiv P^{157}$) was only about 50% of that estimated by LCH. Neutron capture rates are important because inferences about sedimentation and mixing rates in the lunar regolith based on isotopic data are critically dependent on them. For instance, if the LCH estimates of P^{157} are correct, then for simple vertical mixing models (Curtis and Wasserburg, 1975) the existing data indicate that the regolith is mixed and homogenized to depths of tens of meters as often as it is mixed to depths of one meter. This is in disagreement with cratering models (cf. Gault *et al.*, 1974) which predict that the upper few meters should experience much more frequent mixing than deeper levels. As a possible resolution to this conflict, Curtis and Wasserburg (1975) noted that if the capture rate P^{157} were as low as 30% of the LCH value, then the requirement for extensive deep mixing would disappear. The largest previously measured isotopic effects allowed the ^{157}Gd production rate to be as low as ~20% of the LCH value.

A *minimum* average capture rate (P^{157}_{min}) can be calculated from the ϵ_{Gd} and the age of a site (T) by the relation $P^{157}_{min} = \epsilon_{Gd}/10^4$ T, where ϵ_{Gd} is given in parts in 10^4. We calculate $P^{157}_{min} = 0.0041$ AE^{-1} for 24109 using T = 3.3 AE, and assuming $\epsilon_{Gd} = 1.4\ \epsilon_{Sm}$ (Russ *et al.*, 1971). This number is one-half of the P^{157} estimated by LCH for a soil with the chemical composition of 24109 at a depth of ~150g/cm^2, where the capture rate is maximum. Therefore, P^{157} must be greater than 50% of the LCH value. From an analogous calculation, the production rate of ^{149}Sm (P^{149}) must be greater than 60% of the LCH value. These minimum values apply for the case that 24109 material has remained at a depth of 150g/cm^2 for the entire 3.3 AE history of this site. However, 24170, which was at about twice the depth in the regolith as 24109, and closer to the maximum production zone, has been exposed to only about 30% as many neutrons as 24109. This indicates that this soil column has been disturbed more recently than 3.3 AE and thus that sample 24109 probably did not spend the entire 3.3 AE at a depth of 150g/cm^2. This is in agreement with the relatively low ^{38}Ar exposure age of 24170. Therefore, it is most likely that P^{157} must be significantly greater than 50% of the LCH value. This new minimum means that the low fluences found in soils from other sites require that deep mixing occur in the lunar regolith.

CONCLUSIONS

Sample 24170 appears from petrographic and isotopic evidence to be comprised of fragments from a single igneous rock. The characteristics of high Fe in the pyroxene and olivine, extensive zoning in the pyroxene, high Ca in the olivine and low trace element concentrations indicate that 24170 is a mare basalt. The crystallization age of 24170 is 3.30 ± 0.05 AE as determined by concordant Sm-Nd and ^{40}Ar-^{39}Ar measurements. Insufficient time resolution is obtained from Rb-Sr. The data establish the presence of relatively young flows at Mare Crisium. This sample was produced from a parent reservoir with Sm/Nd ratio fractionated only slightly (6%) from chondritic but with extremely low Rb/Sr (Rb/Sr $\approx$ 0.002).

All soil samples show low trace element contents and indicate minimum contamination by KREEPUTh-rich materials. The isotopic relations for Rb-Sr and Sm-Nd reflect the absence of significant chemical fractionation for the soil components at times younger than ~4.5 AE.

One soil sample from the Luna 24 core shows extremely large neutron capture effects. This measurement places a new lower limit to the neutron production rate in the regolith which is close to the calculated value by Lingenfelter *et al.* (1972). This new lower limit and the comparatively low fluence values of most lunar soils require the addition of unirradiated materials from depth to the lunar surface, presumably as a result of relatively infrequent deep craters which spread out thin debris blankets on a wide scale.

Acknowledgments—This work was carried out on samples returned from the moon by the Soviet Luna 24 mission and made available to us for scientific investigations through the sample exchange agreement between the U.S.A. and the U.S.S.R. We wish to thank the Academy of Sciences of the U.S.S.R. and our colleagues at the Vernadsky Institute for their aid in selecting and preparing the Luna 24 materials and for the opportunity to work on these unique lunar samples. The research was supported by the National Aeronautics and Space Administration under grants NGL 05-002-188 and NGL 05-002-038. G. Lugmair and L. E. Nyquist provided constructive reviews of the paper.

REFERENCES

Albee A. L., Chodos A. A., Gancarz A. J., Haines E. L., Papanastassiou D. A., Ray L., Tera F., Wasserburg G. J. and Wen T. (1972) Mineralogy, petrology, and chemistry of a Luna 16 basaltic fragment, Sample B-1. *Earth Planet. Sci. Lett.* **13**, 353–367.

Barsukov V. L. (1977) Preliminary data for the regolith core brought to earth by the automatic lunar station Luna 24. *Proc. Lunar Sci. Conf. 8th*, p. 3303–3318.

Bence A. E., Grove T. L. and Scambos T. (1977) Gabbros from Mare Crisium: an analysis of the Luna 24 soil. *Geophys. Res. Lett.* **4**, 493–496.

Birck J. L., Manhes G., Richard P., Joron J. L., Treuil M. and Allegre C. J. (1977) ^{87}Rb/^{87}Sr age of Luna 24 microgabbros, and isotopic and trace element study of soil 24096 (abstract). In *Papers Presented to the Conference on Luna 24*, p. 34–36. The Lunar Science Institute, Houston.

Bogard D. D. and Hirsch W. C. (1977) Noble gases in Luna 24 core soils (abstract). In *Papers Presented to the Conference on Luna 24*, p. 44–47. The Lunar Science Institute, Houston.

Coish R. A. and Taylor L. A. (1977) Igneous rocks from Mare Crisium: mineralogy and petrology (abstract). In *Papers Presented to the Conference on Luna 24*, p. 48–50. The Lunar Science Institute, Houston.

Curtis D. B. and Wasserburg G. J. (1975) Apollo 17 neutron stratigraphy: sedimentation and mixing in the lunar regolith. *The Moon* **13**, 185–227.

DePaolo D. J. and Wasserburg G. J. (1976) Nd isotopic variations and petrogenetic models. *Geophys. Res. Lett.* **3**, 249–252.

Dymek R. F., Albee A. L. and Chodos A. A. (1974) Glass-coated soil breccia 15205: selenologic history and petrologic constraints on the nature of its source region. *Proc. Lunar Sci. Conf. 5th*, p. 235–260.

Florensky C. P., Basilevsky A. T., Ivanov A. V., Pronin A. A. and Rode O. D. (1977) Luna 24: Geologic setting of landing site and characteristics of sample core (preliminary data). *Proc. Lunar Sci. Conf. 8th*, p. 3257–3279.

Gancarz A. J., Albee A. L. and Chodos A. A. (1971) Petrologic and mineralogic investigation of some crystalline rocks returned by the Apollo 14 mission. *Earth Planet. Sci. Lett.* **12**, 1–18.

Gault D. E., Hörz F., Brownlee D. E. and Hartung J. B. (1974) Mixing of the lunar regolith. *Proc. Lunar Sci. Conf. 5th*, p. 2365–2386.

Grove T. L. and Bence A. E. (1977) Experimental study of pyroxene-liquid interaction in quartz-normative basalt 15597. *Proc. Lunar Sci. Conf. 8th*, p. 1549–1579.

Huneke J. C. (1978) ^{40}Ar-^{39}Ar microanalysis of single 74220 glass balls and 72435 breccia clasts. *Proc. Lunar and Planet. Sci. Conf. 9th.* In press.

Lingenfelter R. E., Canfield E. H. and Hampel V. E. (1972) The lunar neutron flux revisited. *Earth Planet. Sci. Lett.* **16**, 355–369.

Lugmair G. W., Marti K., Kurtz J. P. and Scheinin N. B. (1976) History and genesis of lunar troctolite 76535 or: How old is old? *Proc. Lunar Sci. Conf. 7th.* p. 2009–2033.

Papanastassiou D. A., DePaolo D. J. and Wasserburg G. J. (1977) Rb-Sr and Sm-Nd chronology and genealogy of mare basalts from the Sea of Tranquility. *Proc. Lunar Sci. Conf. 8th*, p. 1639–1672.

Papanastassiou D. A. and Wasserburg G. J. (1971) Lunar chronology and evolution from Rb-Sr studies of Apollo 11 and 12 samples. *Earth Planet. Sci. Lett.* **11**, 37–62.

Papanastassiou D. A. and Wasserburg G. J. (1972) Rb-Sr age of a Luna 16 basalt and the model age of lunar soils. *Earth Planet. Sci. Lett.* **13**, 368–374.

Papike J. J., Bence A. E. and Lindsley D. H. (1974) Mare basalts from the Taurus-Littrow region of the moon. *Proc. Lunar Sci. Conf. 5th*, p. 471–504.

Rossman G. R. (1977) Optical absorption spectra of major minerals in Luna 24 sample 24170 (abstract). In *Papers Presented to the Conference on Luna 24*, p. 156–159. The Lunar Science Institute, Houston.

Russ G. P. (1972) Neutron capture on Gd and Sm in the Luna 16, G-2 soil. *Earth Planet. Sci. Lett.* **13**, 384–386.

Russ G. P. (1973) Apollo 16 neutron stratigraphy. *Earth Planet. Sci. Lett.* **19**, 275–289.

Russ G. P. (1974) Neutron stratigraphy in the lunar regolith. Ph.D Thesis, California Institute of Technology, Pasadena.

Russ G. P., Burnett D. S., Lingenfelter R. E. and Wasserburg G. J. (1971) Neutron capture on ^{149}Sm in lunar samples. *Earth Planet. Sci. Lett.* **13**, 53–60.

Russ G. P., Burnett D. S. and Wasserburg G. J. (1972) Lunar neutron stratigraphy. *Earth Planet. Sci. Lett.* **15**, 172–186.

Ryder G., McSween H. Y., Jr., and Marvin U. B. (1977) Luna 24 VLT basalt: Character and origin (abstract). In *Papers Presented to the Conference on Luna 24*, p. 160–163. The Lunar Science Institute, Houston.

Ryder G., McSween H. Y., Jr., and Marvin U. B. (1978) Basalts from Mare Crisium. *The Moon.* **17**, 263–287.

Stettler A. and Albarede F. (1977) ^{39}Ar-^{40}Ar pattern and light noble gas systematics of two mm-sized rock fragments from Mare Crisium (abstract). In *Papers Presented to the Conference on Luna 24*, p. 175–178. The Lunar Science Institute, Houston.

Tarasov L. S., Nazarov M. A., Shevaleevsky I. D., Kudryashova A. F., Gaverdovskaya A. S. and Korina M. I. (1977) Mineralogy and petrography of lunar rocks from Mare Crisium (preliminary data). *Proc. Lunar Sci. Conf. 8th*, p. 3333–3356.

Taylor G. J., Warner R. D. and Keil K. (1977a) Chemical distinctions among very low-Ti mare basalts (abstract). In *Papers Presented to the Conference on Luna 24*, p. 183–188. The Lunar Science Institute, Houston.

Taylor G. J., Warner R., Wentworth S., Keil K. and Sayeed U. (1977b) Petrology and chemistry of Luna 24 mare basalts and basaltic glasses (abstract). In *Papers Presented to the Conference on Luna 24*, p. 189–192. The Lunar Science Institute, Houston.

Turner G., Huneke J. C., Podosek F. A. and Wasserburg G. J. (1971) ^{40}Ar-^{39}Ar ages and cosmic ray exposure ages of Apollo 14 samples. *Earth Planet. Sci. Lett.* **12**, 19–35.

Vaniman D. T. and Papike J. J. (1977) Ferrobasalts from Mare Crisium: Luna 24. *Geophys. Res. Lett.* **4**, 497–500.

Woolum D. S., Burnett D. S., Furst M., and Weiss J. R. (1975) Measurement of the lunar neutron density profile. *The Moon* **12**, 231–250.

Merrill, R. B. and Papike, J. J., editors.
(1978) *Mare Crisium: The View from Luna 24*, Pergamon Press (New York), p. 679–694
Printed in the United States of America

Implications from Luna 24 sample 24170 to U-Pb evolution in the lunar mantle

DANIEL M. UNRUH and MITSUNOBU TATSUMOTO

U.S. Geological Survey, Mail Stop 963, Box 25046, Denver, Colorado, U.S.A. 80225

Abstract—U, Th and Pb analyses have been performed on mineral separates from a 12 mg aliquot of Luna 24 sample 24170 in an attempt to determine the age of the basalt and U-Pb evolution in the Mare Crisium area. The Pb in the Luna 24 separates is quite nonradiogenic ($^{206}Pb/^{204}Pb \sim 22 - 50$) compared to other lunar basalts ($^{206}Pb/^{204}Pb \gtrsim 300$). These data may reflect unaccounted for terrestrial contamination, however terrestrial contamination alone will not account for all of the trends suggested by the data. The Luna 24 basalt has had a rather complex post-crystallizational history. A 0.5 ± 0.5 B.y. disturbance to the U-Pb system is suggested by the U-Pb data (if the disturbance represents a single event). Some separates also appear to be "contaminated" by Luna 24 soil-type Pb. Three-stage U-Pb evolution model calculations indicate that the source cumulates of the Luna 24 basalt evolved under a very low $^{238}U/^{204}Pb$ environment (~12–15) relative to Apollo mare basalts. The calculations may reflect a laterally heterogeneous lunar magma ocean, U-Pb fractionation during cumulate formation (primary differentiation) and/or lack of KREEP addition to the Luna 24 basalt. In spite of the calculated low initial $^{238}U/^{204}Pb$ ratios the data are nevertheless consistent with the hypothesis that the moon originally accreted as a volatile-depleted body. The low observed $^{238}U/^{204}Pb$ ratios are roughly compatible with a crude relationship between the K content and the $^{238}U/^{204}Pb$ ratio observed in Apollo mare basalts.

INTRODUCTION

U, Th and Pb concentrations and Pb isotopic compositions have been determined on mineral separates and "whole-rocks" from a meager 12 mg aliquot of Luna 24 microgabbro 24170 in an attempt to determine the age and U-Th-Pb evolution history of the Mare Crisium Area. Leaching experiments using 2N HCl have also been performed in order to determine the nature and extent of apparent Pb isotopic disequilibrium in lunar rocks, which we previously observed in mare basalt 15085 (Unruh and Tatsumoto, 1977).

Sample 24170 represents a portion of a layer of rather coarse particles (≳1 mm) from the upper portion of Unit IV of the Luna 24 core, and probably represents a part of a single rock that was crushed during the coring process (Barsukov *et al.*, 1977). The rock has been classified as a very low titanium (VLT) coarse-grained ophitic basalt or gabbro (Ryder *et al.*, 1977). The Luna 24 basalts are distinct among mare basalts in their low TiO_2 content, yet low Mg/Fe ratios (Tarasov *et al.*, 1977; Ryder *et al.*, 1977; Vaniman and Papike, 1978). VLT basalts have been observed at the Apollo 17 site (Vaniman and Papike, 1977), although these basalts generally are more magnesian.

Age data for the Luna 24 samples have been obtained at several different laboratories. An Ar-Ar age of 3.65 ± 0.12 B.y. (10^9 yr) has been obtained on a lithic fragment from sample 24096 (Stettler and Albarede, 1977) and Rb-Sr ages of 3.74 ± 0.56 B.y. (2σ, Birck *et al.*, 1977) and 3.7 ± 0.6 B.y. (2σ, Papanastassiou *et al.*, 1977) have been reported for Luna 24 gabbros 24170 and 24171, respectively. The large uncertainties in the Rb-Sr ages are a result of small ranges in Rb/Sr ratios and Sr isotopic compositions among the various minerals. Sm-Nd data yield an age of 3.30 ± 0.05 B.y. (2 points only, Papanastassiou *et al.*, 1977) which is not statistically different from the Rb-Sr ages, yet is distinctly younger than the Ar-Ar age. Preliminary Pb data (Tatsumoto and Unruh, 1977) were consistent with an ~3.7 B.y. age, although subsequent analyses (this work) have shown the agreement between the apparent Pb-Pb age and the Rb-Sr and Ar-Ar ages to be fortuitous.

Although the data are not as precise as those for 15085 (Unruh and Tatsumoto, 1977) owing to the extremely limited sample size, the general features of the U-Pb systematics data are similar to those of 15085 and the data are compatible with the U-Pb open system concept developed in our previous paper, (e.g., Unruh and Tatsumoto, 1977).

EXPERIMENTAL PROCEDURES

Twelve mg of 24170 (Sub. No. 2), which was partially crushed, either by natural processes on the moon, by coring (Barsukov *et al.*, 1977), or during sample processing and allocation, were used for our analyses. The sample consisted primarily of mostly-monomineralic fragments of pyroxene, plagioclase, olivine, and a few opaque grains. These were hand-picked on a laminar flow bench without further crushing to avoid sample loss and possible contamination. As a result our mineral separates are not as pure as those usually obtained. A few polymineralic grains were also observed. These were saved for "whole-rock" analyses. A few exotic particles were also observed. These included 5 small soil breccia (?) fragments, 2 pyroxene grains coated with either impact melt or fused soil, and ~20 μgm (10^{-6}gm) of what appeared to be orange glass spherules and Fe-oxides. One opaque grain coated with this material was also observed. The "orange-glass" has not been observed in other Luna 24 samples studied by other groups (except possibly the Ti-rich devitrified glass observed by Ryder *et al.*, 1977) and hence may not represent lunar material, but oxidation of the stainless steel sample container.

All samples were washed with double distilled acetone and rinsed with H_2O prior to leaching or decomposition in order to remove any terrestrial Pb contamination adsorbed on the grain surfaces. Silicate samples were decomposed using a mixture of ~16N HNO_3 and 60% HF (~0.01 ml of each/mg of sample) in a sealed teflon bomb at ~120°C for ~10 hrs. Opaque oxides were decomposed using 9N HCl and 16N HNO_3 in the same manner. A few small grains, presumably chromite, were not decomposed, however. All samples were spiked with a ^{205}Pb-^{235}U-^{230}Th-enriched tracer prior to dissolution. After dissolution, samples were evaporated and converted first to chloride and then to bromide forms, using 9N HCl and 9N HBr, respectively, and Pb was separated in 1.5N HBr medium using a 35 μl anion exchange resin (AGl-X8, 200–400 mesh) column (ϕ 0.15 cm × H 2.0 cm). The column is made of heat-shrinkable teflon FEP and a porous teflon (40 micron) frit (Tatsumoto and Unruh, 1976). The column was connected to a vacuum pump to maintain a constant flow rate through the column of ~100 μl/min. Pb was eluted from the column using 1.0 N HNO_3. The Pb recovery from the procedure has been checked using an in-house basalt standard and was found to be rather consistent at 93–99% for 10–70 mg-sized samples, but somewhat less (≳75%) for samples larger than 100 mg.

Reagents were purified using either bottle-to-bottle sub-boiling distillation (HF, HCl) or sub-boiling distillation in a French-made quartz still (H_2O, HNO_3, HBr). Total chemistry blanks for the Luna 24 samples ranged from 0.04 to 0.06 ng (10^{-9}g). These blanks are somewhat higher than those predicted from the sum of the reagent blanks (<0.03 ng) and the discrepancies are attributed to memory from the decomposition bombs and to a rather impure silica-gel source.

Samples were loaded on an M-shaped mini-Re filament using phosphoric acid and silica-gel (Tatsumoto and Unruh, 1976) and were run on an NBS-type, 90°-sector two-stage (tandem) mass spectrometer, with two stages joined in an off-axis "S" configuration (Shields, 1967, 1968). The mass spectrometer is equipped with a removable Faraday cage collector on the first stage and a pulse counting system, consisting of an electron multiplier and pulse-height discriminator, on the second stage (for details, see Tatsumoto and Unruh, 1976).

U and Th were separated by anion exchange in 7N HNO_3 medium and the mass spectrometry was performed using the conventional triple-Re-filament technique. U and Th total chemistry blanks were less than 0.2 and 0.5 picograms, respectively.

Prior to analysis, some samples were leached with 2N HCl for ~10 minutes. Both the leach and residue fractions were analyzed in order to determine (1) the extent and nature of any pre-analysis Pb contamination and (2) to determine how U, Th and Pb are distributed within a rock. Work on unfractured terrestrial mineral grains (Shimizu, 1974) indicates that the trace elements residing within the crystal lattice are not removed by the leaching, but those residing on grain boundaries or in microfractures may be easily leached. The results of the leaching experiments are discussed in a later section.

Results and Discussion

The U, Th and Pb concentrations obtained from the various mineral fractions, leaches (−L) and leach-residues (−R) are shown in Table 1. The data are compared to those obtained from Apollo mare basalts in Table 2. On the whole, the Luna 24 sample (Σ 24170 in Table 1) contains about one-half the U and Th of mare basalts sampled by the Apollo missions, although these concentrations are quite variable among the various separates. Pb concentrations, except for the dark pyroxene fractions (PXD), are also lower than those generally observed in Apollo mare basalts. The low U, Th and Pb concentrations observed in this study are in agreement with the low trace-element abundances observed by other workers (e.g., Nyquist *et al.*, 1977; Papanastassiou *et al.*, 1977) in 24170 and in Luna 24 soils. The $^{238}U/^{204}Pb$ ratios ($\equiv\mu$) are also quite low compared to most lunar basalts (Table 2).

The Pb isotopic compositions of the various mineral separates are shown in Table 3, and the $^{206}Pb/^{204}Pb$ vs. $^{207}Pb/^{204}Pb$ data are shown in Fig. 1. The Pb data are very nonradiogenic compared to those from Apollo mare basalts ($^{206}Pb/^{204}Pb \sim 20$–49 vs. $^{206}Pb/^{204}Pb > 300$). The Pb in the Luna 24 soils analyzed by Birck *et al.* (1977) is also quite nonradiogenic ($^{206}Pb/^{204}Pb \sim 75$–120) compared to Apollo soils but is similar to that of Luna 16 ($^{206}Pb/^{204}Pb \sim 100$; Tera and Wasserburg, 1972) and Luna 20 ($^{206}Pb/^{204}Pb \sim 90$; Tatsumoto, 1973). The Luna soil data, however, are more radiogenic than our analyses of 24170.

Pb Contamination

One possible mechanism to explain the low μ values and nonradiogenic character of our analyses is terrestrial Pb contamination, either prior to or during

Table 1. U, Th and Pb concentrations in 24170.

Sample[1]	WT mg	Pb Blank ng	PPM U	PPM Th	PPM Pb	$^{232}Th/^{238}U$	$^{238}U/^{204}Pb$	% ^{204}Pb Blank[2]	% Leached U	% Leached Th	% Leached Pb
WR1-L		0.021	0.121 ± 0.016	0.420 ± 0.037	0.095 ± 0.002	3.60 ± 0.57	169. ± 26	40.5			
WR1-R		0.052	0.057 ± 0.007	0.054 ± 0.001	0.108 ± 0.015	0.98 ± 0.12	44. ± 13	47.6			
ΣWR1	0.6962	0.073	0.178 ± 0.023	0.474 ± 0.038	0.203 ± 0.017	2.75 ± 0.42	88 ± 21	45.3	68	89	47
WR2-L		0.051	0.005 ± 0.005	—	0.025 ± 0.008	—	(23)	90.5			
WR-2R		0.051	0.037 ± 0.020	—	0.137 ± 0.028	—	(33)	63.1			
ΣWR2	0.4167	0.102	0.042 ± 0.025	—	0.162 ± 0.036	—	(32)	75.0	(12)	—	15
WR3	0.9122	—	0.061 ± 0.004	0.287 ± 0.003	—	4.84 ± 0.32	—	—			
PXL1	1.8556	0.040	0.005 ± 0.003	0.014 ± 0.001	0.051 ± 0.007	(2.9)	(7)	32.4			
PXL2	0.9854	0.053	0.028 ± 0.001	—	0.29 ± 0.02	—	7. ± 3	17.1			
OP	0.7793	0.052	0.220 ± 0.008	0.584 ± 0.002	0.739 ± 0.013	2.75 ± 0.10	31.5 ± 1.9	13.0			
PXD1	0.9733	0.040	0.052 ± 0.004	0.217 ± 0.002	1.19 ± 0.01	4.30 ± 0.33	3.07 ± 0.24	3.7			
PXD2-L		0.053	0.015 ± 0.004	0.040 ± 0.002	0.094 ± 0.01	2.8	10.6 ± 3.3	34.2			
PXD2-R		0.053	0.011 ± 0.007	0.033 ± 0.009	0.956 ± 0.013	(3.1)	0.77 ± 0.51	5.0			
ΣPXD2	1.1762	0.106	0.026 ± 0.011	0.073 ± 0.011	1.04 ± 0.023	(2.9)	1.7 ± 1.0	8.6	58	55	10
PL-L		0.053	0.063 ± 0.010	0.101 ± 0.005	0.633 ± 0.008	1.65 ± 0.27	7.3 ± 1.0	16.5			
PL-R		0.053	0.020 ± 0.011	0.059 ± 0.013	0.328 ± 0.003	(3.1)	4.3 ± 2.3	26.7			
ΣPL	0.4917	0.106	0.083 ± 0.021	0.170 ± 0.018	0.961 ± 0.011	1.99 ± 0.54	6.3 ± 1.2	20.4	76	59	66
Σ24170[3]			0.064 ± 0.008	0.211 ± 0.008	0.545 ± 0.015	3.4 ± 0.4	8.8				

[1]Samples designated by "-L" are leaches by "-R" are leach residues and by "Σ" are recombined leach and residue fractions. "WR" = whole rock, "PXL" = Light-colored pyroxene, "PXD" = dark-brown pyroxene, "OP" = opaque, "PL" = plagioclase. Data in parentheses are very uncertain.

[2]$\% \; ^{204}Pb \text{ Blank} = \frac{^{204}Pb_{Blank \times 100}}{^{204}Pb_{Blank} + ^{204}Pb_{Sample}}$

[3]Σ24170 Combined weighted average of all Luna 24 data.

Table 2. Average U, Th, and Pb concentrations; Th/U and U/Pb ratios of mare basalts.

Sample	PPM						$^{238}U/^{204}Pb$		$^{232}Th/^{238}U$		Ref.[1]
	U		Th		Pb						
	Ave.	Range	Ave.	Range	Ave.	Range	Ave.	Range	Ave.	Range	
Apollo 11, Low K	0.215	0.156–0.259	0.762	0.531–0.869	0.40	0.29–0.50	322	160 450	3.66	3.47–3.96	1, 2
Apollo 11, High K	0.866	0.854 0.884	3.37	3.35–3.41	1.63	1.56–1.69	855	490 1400	4.03	3.90–4.08	1, 2
Apollo 12	0.243	0.191–0.372	0.872	0.658–1.31	0.422	0.283–0.624	677	460–990	3.72	3.56–4.05	3
Apollo 15	0.144	0.118–0.192	0.550	0.445–0.733	0.253	0.200–0.354	472	380–611	3.93	3.77–4.01	4
Apollo 17	0.130	0.118–0.151	0.432	0.388–0.488	0.278	0.242–0.326	411	250–450	3.43	3.35–3.54	5
Σ24170	0.064		0.211		0.545		8.8		3.4		

[1]References

1. Tatsumoto, 1970.
2. Silver, 1970.
3. Tatsumoto *et al.*, 1971.
4. Tatsumoto *et al.*, 1972b.
5. Nunes *et al.*, 1974.

analysis. From Table 1 it is evident that up to ~90% of the ^{204}Pb analyzed in some samples is from the analytical blank, as a result of the small sample sizes (≲1 mg) and low Pb concentrations. Therefore, if a systematic bias exists in our blank correction, or if the blanks are really much more variable than the data indicate, the observed μ and $^{206}Pb/^{204}Pb$ ratios might really be much higher than those shown in Tables 1 and 3. In order to further investigate the Pb contamination possibility, we have performed leaching experiments and duplicate analyses on some separates. The following important observations are made from the data in Tables 1 and 3: (1) The very nonradiogenic character of some samples is independent of the Pb concentration (compare PXL1 and PXD1 in Tables 1 and 3); (2) The leach analyses are not less radiogenic than the residues, and the extensions of tie-lines connecting the respective leach and residue fractions do not pass through the value for modern terrestrial (blank) Pb (Fig. 1); (3) The high Pb content observed for the dark pyroxene (PXD) is reproducible; (4) The Pb isotopic composition of the dark pyroxene appears to be reproducible; and (5) The percentage of leachable Pb appears to be quite variable among the various separates, but roughly corresponds to the amounts of U and Th leached (Table 1). These observations suggest that there was not a great deal of pre-analysis Pb contamination remaining on this sample after the acetone washing. If there was significant undetected terrestrial contamination there would be a negative correlation between the $^{206}Pb/^{204}Pb$ ratio and the Pb concentration (for the same separate), and pre-analysis contamination would also cause the leach and residue

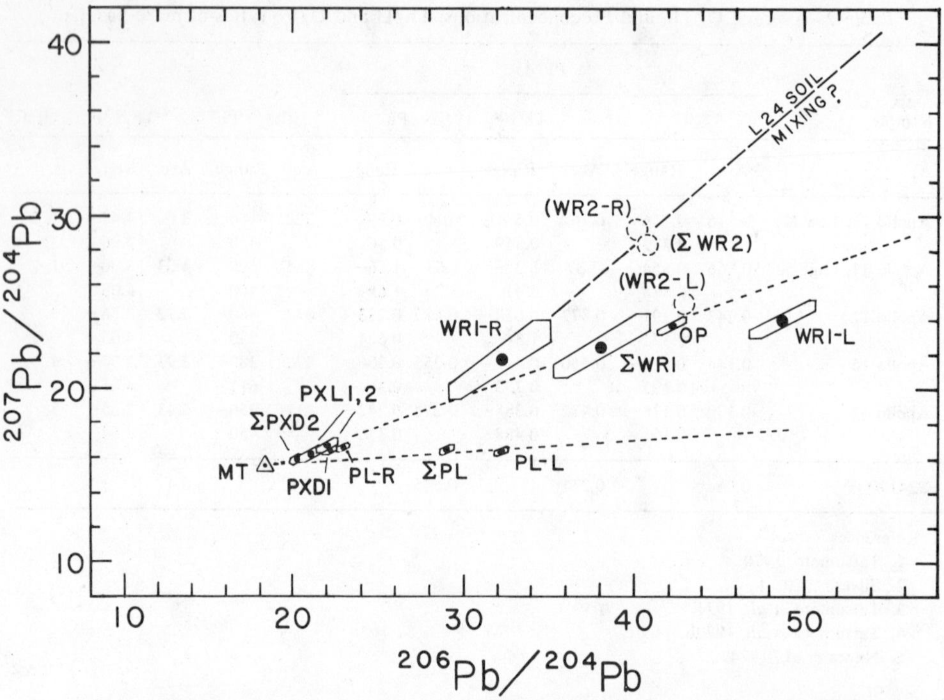

Fig. 1. $^{207}Pb/^{204}Pb$ vs. $^{206}Pb/^{204}Pb$ plot for mineral separates from 24170. The data are scattered and do not yield an isochron. A tie-line (broken line) connects WR1-R, WR2-R and ΣWR2 data with the 24096 soil data (not shown; Birck *et al.*, 1977), and suggests possible L24 soil-Pb contamination" to some 24170 data. The dotted lines represent possible modern terrestrial Pb (MT) mixing lines. That is, increasing the amount of analytical blank causes the data points to move directly away from MT along or between the dotted lines. Symbols are the same as in Table 1. The WR2 data are very uncertain (see Table 1).

Pb isotopic data (Fig. 1) to define a mixing line which passes through modern terrestrial Pb (MT in Fig. 1). The reproducibility of the pyroxene data indicates that these data are probably real and not artifacts of analytical contamination. The various WR data cannot be compared because the variations in the U and Th concentrations suggest real heterogeneities. Thus it appears that the nonradiogenic character of this sample is real, although the small sample-size and resulting large Pb blank/sample ratios may make the data somewhat less quantitative than the calculated uncertainties (assuming ±0.02 ng blank uncertainty) in Tables 1 and 3 indicate. Hence, the ensuing discussions will be of a qualitative rather than quantitative nature whenever possible.

Leaching Experiments

In a previous work (Unruh and Tatsumoto, 1977) we determined that large

Table 3. Pb isotopic compositions and model ages for mineral separates from 24170.

	Observed	Corrected for analytical blank			Model ages in 10^9yr[1]			
Sample	$^{206}Pb/^{204}Pb$	$^{206}Pb/^{204}Pb$	$^{207}Pb/^{206}Pb$	$^{208}Pb/^{206}Pb$	$^{206}Pb/^{238}U$	$^{207}Pb/^{235}U$	$u^{207}Pb/^{206}Pb$	$^{208}Pb/^{232}Th$
WR1-L	36.1 ± 0.2	48.7 ± 2.0	0.4922 ± 0.0091	1.636 ± 0.012	1.35 ± 0.17	2.54 ± 0.16	3.70 ± 0.06	1.6 ± 0.2
WR1-R	25.7 ± 0.5	32.2 ± 3.2	0.672 ± 0.027	1.28 ± 0.10	2.7 ± 0.6	3.7 ± 0.5	4.22 ± 0.13	4.9 ± 1.5
ΣWR1	—	38.1 ± 2.8	0.590 ± 0.017	1.44 ± 0.06	(1.82)	(3.04)	(4.00)	(1.99)
WR2-L	20.3 ± 0.1	43. ± 15.	0.58 ± 0.06	1.69 ± 0.08	—	—	4.0 ± 0.6	—
WR2-R	26.2 ± 0.1	40. ± 10.	0.72 ± 0.08	1.77 ± 0.08	4.2 ± 1.2	4.43 ± 0.8	4.53 ± 0.27	—
ΣWR2	—	41. ± 11.	0.70 ± 0.08	1.75 ± 0.08	(4.44)	(4.45)	(4.46)	—
PXL1	20.9 ± 0.1	22.0 ± 0.5	0.760 ± 0.013	1.89 ± 0.03	6.5 ± 3.1	4.9 ± 0.9	4.27 ± 0.07	9.3 ± 5.7
PXL2	20.6 ± 0.3	21.1 ± 0.7	0.767 ± 0.017	1.97 ± 0.04	6.4 ± 2.0	4.8 ± 0.6	4.24 ± 0.08	—
OP	39.1 ± 0.1	42.2 ± 0.9	0.5619 ± 0.0055	1.324 ± 0.014	4.61 ± 0.15	4.15 ± 0.07	3.94 ± 0.03	5.4 ± 0.4
PXD1	22.0 ± 0.1	22.2 ± 0.1	0.7453 ± 0.0021	1.856 ± 0.005	10.6 ± 0.4	5.7 ± 0.1	4.20 ± 0.01	12.8 ± 0.6
PXD2-L	20.0 ± 0.5	21.0 ± 0.7	0.767 ± 0.012	1.93 ± 0.02	4.8 ± 1.3	4.4 ± 0.6	4.23 ± 0.08	6.7 ± 2.1
PXD2-R	20.1 ± 0.1	20.2 ± 0.2	0.789 ± 0.005	1.967 ± 0.013	17. ± 4.	6.9 ± 0.7	4.29 ± 0.03	32. ± 5.
ΣPXD2	—	20.3 ± 0.3	0.787 ± 0.006	1.96 ± 0.02	(13)	(6.2)	(4.29)	(23)
PL-L	30. ± 0.2	32.3 ± 0.3	0.5069 ± 0.0078	1.072 ± 0.013	9.2 ± 0.8	4.8 ± 0.2	3.27 ± 0.10	7. ± 3.
PL-R	21.7 ± 0.2	23.0 ± 0.2	0.7214 ± 0.0033	1.749 ± 0.010	9. ± 3.	5.4 ± 0.5	4.12 ± 0.03	12. ± 4.
ΣPL	—	29.1 ± 0.3	0.5670 ± 0.0062	1.260 ± 0.012	(9.2)	(5.0)	(3.54)	(9.2)

[1] Corrected for Canyon Diablo Troilite Pb as the initial Pb.

amounts of U, Th and Pb were easily leached (2N HCl) from mare basalt 15085. Tera *et al.* (1972) and Tera and Wasserburg (1975) have also found relatively large abundances of leachable Pb in anorthosite 15415 and mare basalt 75055. The Luna 24 data exhibit a similar trend. As shown in Table 1, up to ~66% of the Pb, 76% of the U and 90% of the Th can be leached in cold 2N HCl. These values compare with up to 71%, 50%, and 41% of leachable Pb, U and Th, respectively, in 15085 (Unruh and Tatsumoto, 1977). These data suggest that U, Th and Pb (and other incompatible trace elements) are (1) located on grain boundaries or in microfractures, (2) concentrated in late-stage (semisoluble) mesostases as Lovering and Wark (1975) have observed for K and U, and/or (3) concentrated in semisoluble mineral phases or inclusions such as oxides or sulfides. The leaching should not remove elements residing within the silicate crystal lattice (Shimizu, 1974), unless the crystal is altered or highly fractured. There is, of course, no guarantee that all of a particular element has been leached from a particular phase, or even that the U, Th, and Pb have been leached from the same phase.

In our 15085 study we observed that the leach fractions always gave younger $^{207}Pb/^{206}Pb$ model ages—that is, on a $^{207}Pb/^{204}Pb$ vs. $^{206}Pb/^{204}Pb$ plot the leach fractions plotted distinctly lower or to the right of the corresponding residue fractions. We observe the same trend for the 24170 data. The leach fractions always give younger model Pb-Pb ages (Table 3 and Fig. 1). The differences are real regardless of the possible analytical problems discussed in the previous section, as increasing the amount of Pb contamination causes the respective leach and residue fractions to further diverge with respect to the $^{207}Pb/^{206}Pb$ ratio (MT mixing lines in Fig. 1). Ludwig and Silver (1977) have also observed apparent Pb isotopic disequilibrium in terrestrial K-feldspars.

Several mechanisms may be called on to account for the observed isotopic disequilibrium (for a thorough discussion see Unruh and Tatsumoto, 1977): (1) A thermal disturbance, such as a meteorite impact or burial in a hot ejecta blanket could cause partial reequilibration of the U-Pb system on Pb loss from the system; (2) Both the ^{235}U-^{207}Pb and ^{238}U-^{206}Pb systems evolve through complex decay schemes. Intermediate daughter diffusion (Rn; Frisen and Heymann, 1972) or Pb or U diffusion (Tilton, 1960; Wetherill, 1963; Wasserburg, 1963) could also account for the apparent Pb isotopic disequilibrium; (3) Conceivably, the disequilibrium could be artificially produced. Due to differences in the decay schemes and in α-recoil energies, it might be possible to preferentially leach certain Pb isotopes. At present, however, we consider this mechanism unlikely. Ultimately some combination of the three mechanisms may be required to explain the data, but at present the data are neither precise enough nor abundant enough to provide a unique interpretation.

U-Pb Evolution of 24170

Ages for the Luna 24 samples have been determined by several different methods at three different laboratories. Stettler and Albarede (1977) have obtained a 3.65 ± 0.12 B.y. (10^9yr) Ar-Ar age for two lithic fragments from 24096. Birck *et al.* (1977) obtained a 3.74 ± 0.56 B.y. (2σ) age for sample 24171, a portion of the same sample as 24170. Papanastassiou *et al.* (1977) obtained a Rb-Sr age of 3.7 ± 0.6 B.y. (2σ) and an initial $^{87}Sr/^{86}Sr$ ratio of 0.69907 ± 0.00004 (2σ) for sample 24170. The large uncertainties in the Rb-Sr ages result from low Rb and Sr abundances, low Rb/Sr ratios and a small spread in the Sr isotopic compositions among the various mineral separates (Papanastassiou *et al.*, 1977). A 3.30 ± 0.05 B.y. Sm-Nd age has been reported for 24170 (Papanastassiou *et al.*, 1977) although due to the relatively small amounts of sample available, only two points were used to calculate the isochron. The Ar-Ar and Rb-Sr ages are in excellent agreement whereas the Sm-Nd age is somewhat younger than the Ar-Ar ages, yet not statistically younger than the Rb-Sr ages.

The Pb-Pb data are quite scattered (Fig. 1) and do not help to more precisely pin down the age of the Luna 24 basalts. The scatter in the Pb-Pb data in Fig. 1 suggests that there has been some open system behavior of the U-Pb system since crystallization. The disturbance may have been caused by partial re-equilibration of U-Pb during meteorite impact or burial in hot ejecta as evidenced by coatings of apparent impact melt which we have observed on a few grains. Alternatively, Pb may have been added to the rock from an outside source—terrestrial, lunar, or meteoritic. Nearly all lunar soils show excess Pb relative to U. The excess Pb probably originated from Pb (either lunar or meteoritic) that was volatilized during meteorite impact and which later condensed and was adsorbed by lunar surface-soils (broken line in Fig. 1). As we stated in a previous section, terrestrial contamination may be important, but terrestrial contamination alone will not account for the scatter observed in Fig. 1.

The U-Pb data are plotted on a "modified" concordia plot in Fig. 2, a and b (Nunes *et al.*, 1975). No initial Pb correction has been made to the data in Fig. 2a. Samples containing excess Pb relative to U plot to the left of the concordia curve. Shown for comparison are soils from Luna 24 (L24S: Birck *et al.*, 1977). A tie line connecting ΣWR1, ΣWR2, and the cluster of points to the lower left (PXL1, 2; PXD1, 2) intersects concordia at ~4.7 and 0.7 B.y. The older intercept has no significance since no initial Pb correction has been made. Although the lower intercept is very uncertain, it does reflect a relatively recent disturbance to the U-Pb system. The ΣWR1 data reflect overall depletion of Pb relative to U. If the excess Pb is of terrestrial origin, then it must have a significantly higher $^{207}Pb/^{206}Pb$ ratio than that of our blank Pb, since the MT (= modern terrestrial Pb) data do not plot on the line defined by the other data. This observation is more clearly illustrated in Fig. 2b. Here the data are corrected for blank Pb as the "initial" Pb isotopic composition (we assume that all of the ^{204}Pb is from MT contamination). The data become much more scattered, and the PXD and PL data still indicate excess Pb relative to U. The data (except PL-L and ΣPL) are bounded by ~2.9 − 0 B.y. and ~4.5 − 0.7 B.y. discordia lines (solid lines, Fig. 2b). The scatter in the data and the young lower concordia intercepts again reflect a post-crystallization disturbance to the U-Pb system. The ~4.5 − 0.7 B.y. discordia line suggested by ΣWR1 and ΣWR2 passes through the points

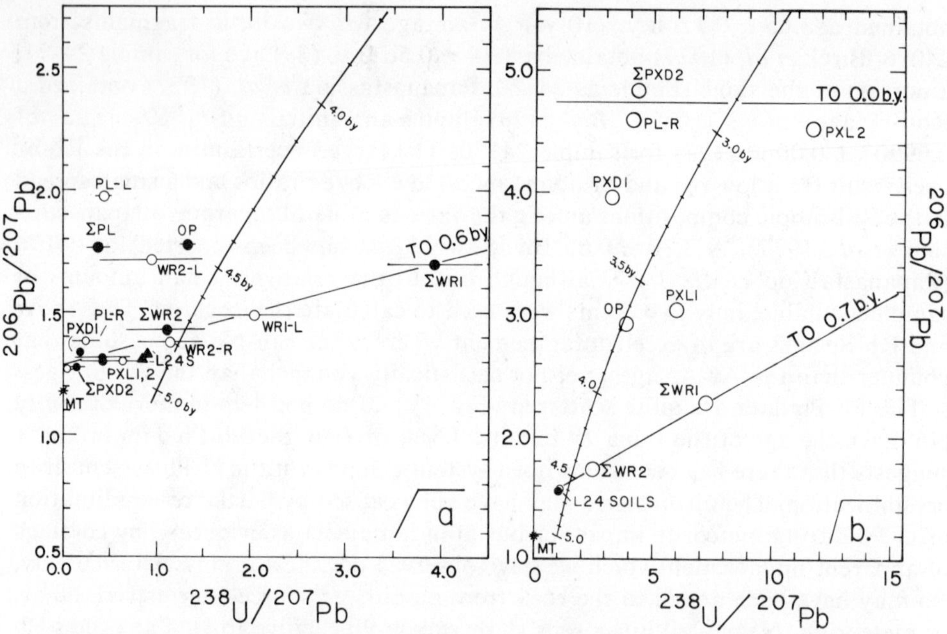

Fig. 2. $^{206}Pb/^{207}Pb$ vs. $^{238}U/^{207}Pb$ modified U-Pb evolution diagram for 24170. (a.) The Pb data are not corrected for initial Pb. As a result the ~4.7 B.y. older concordia intercept suggested by much of the data has no significance. A recent disturbance (0.6 B.y.) to the U-Pb system is suggested by the data (except PL and OP). If the 4.7–0.6 B.y. line is interpreted as a mixing line between 24170 Pb and Pb contamination, then the contaminant must have a higher $^{206}Pb/^{207}Pb$ ratio than that of our blank (MT) because the MT point does not lie on the mixing line. The Luna 24 soils (L24S; Birck *et al.*, 1977) have about the right $^{206}Pb/^{207}Pb$ ratios to be possible contaminants. (b.) The same data are corrected for modern terrestrial Pb as the "initial" Pb (it is assumed that all of the ^{204}Pb is from MT contamination). The data are quite scattered and, except for PL, are bounded by 4.55–0.7 B.y. and 2.9–0.0 B.y. discordia lines (solid lines). The ΣP1 data lie far above the diagram ($^{206}Pb/^{207}Pb$ ~12), which suggests that the data are over-corrected. Soil-Pb contamination is suggested by the tie-line connecting the WR data and the L24 soils.

defined by the L24 soil data (Birck *et al.*, 1977). There is therefore the suggestion that at least some samples may have been contaminated with L24 soil-type Pb (see also Fig. 1).

Two other attempts have been made to determine a proper "initial" Pb isotopic composition (not shown). When the data are corrected for Canyon Diablo troilite Pb (Tatsumoto *et al.*, 1973b), the data are still scattered and younger concordia intercepts of ~0.1–0.8 B.y. are suggested by the data. If we assume that the least radiogenic sample (PXD-2) has the composition of the contamination (nearly all of the Pb in this sample is from contamination—source unknown) and correct the other data for this Pb composition, the data remain scattered and again a young event(s) of ~0–0.5 B.y. is suggested.

Regardless of the initial Pb correction, the data consistently exhibit two features: (1) a young event (or events) at ~0–1 B.y. which caused partial re-equilibration of the U-Pb system is suggested, and (2) tie-lines connecting ΣWR1 and ΣWR2 pass through the points defined by the Luna 24 soils. These observations and those from Fig. 1 indicate that U-Pb evolution in 24170 may best be interpreted by a two-stage *post-crystallization* history involving (1) U-Pb re-equilibration either by diffusion processes or a discrete event (or events) at ~0.5 ± 0.5 B.y., and (2) addition of Luna 24 soil-type Pb (± modern terrestrial Pb) to 24170.

The Low μ ($^{238}U/^{204}Pb$)

The U-Pb data thus far obtained from the Luna 16 (Tera and Wasserburg, 1972), Luna 20 (Tatsumoto, 1973), and Luna 24 (Birck *et al.*, 1977; this work) indicate that the Eastern portion of the moon (Crisium-Fecunditatis area) has evolved under a lower μ environment than that of the near side (South Imbrium area; Nunes *et al.*, 1974). We have made an attempt to semiquantitatively determine the μ ratio of the L24 basalt source using a 3-stage model for ΣWR1 and OP (which are our most reliable data). In this model we assume $T_0 = 4.55$ B.y. = primary differentiation age of the moon (the age of the parent cumulate formation), $T_1 = 3.3$ B.y. = crystallization age of 24170, and $T_2 = 0.5$ B.y. = age of the post-crystallization disturbance to the U-Pb system. We assume that the initial Pb isotopic composition at T_0 was that of Canyon Diablo troilite Pb. Using these assumptions we can then calculate μ_0 (μ from T_0 to T_1) and μ_1 (μ from T_1 to T_2) for both OP and ΣWR1. We calculate μ_0 ~15 and ~12 and μ_1 ~28 and 44 for ΣWR1 and OP, respectively. In view of the previous discussion, these values probably represent lower limits. Although the calculated values are model-dependent and represent only an estimate of the real value, it is quite significant that some areas of the moon may have evolved under drastically lower μ environments than those of the Apollo areas. The overall trend of an approximate two–threefold increase in the μ ratio through time is not as model-dependent, however, and is identical to trends calculated for Apollo mare basalts (e.g. Tatsumoto, 1970; Tatsumoto *et al.*, 1971, 1972b; Nunes *et al.*, 1974; Tera and Wasserburg, 1974).

Three mechanisms may be called on to account for the apparent low initial μ of the L24 basalt source: (1) lunar lateral heterogeneity; (2) U-Pb fractionation during cumulate formation; and (3) lack of significant KREEP mixing. If the initial magma ocean from which the mare basalt sources presumably crystallized (e.g., Drake and Consolmagno, 1976) were heterogeneous (e.g., Jovanovic and Reed, 1976), then of course we would expect these heterogeneities to be reflected in the mare basalts. However, the magma ocean may have been approximately homogenized via convection and diffusion, unless the time interval from the formation of the magma ocean to crystallization of the mare basalt source cumulates was very short, and there was virtually no mixing between convection cells (Jovanovic and Reed, 1976).

In our previous work (Unruh and Tatsumoto, 1977) we suggested that the μ values of the mare basalt source cumulates were probably significantly lower than those of the initial magma ocean. However, if the bulk D value for Pb is ~20 times that of U (e.g., Tatsumoto, 1978) and is significantly large ($\gtrsim$0.1), then it is possible that the U/Pb ratio of the mare basalt source cumulates progressively increased as fractional crystallization proceeded. The apparent low initial μ values and low trace element abundances would suggest that the L24 basalts were derived from cumulates that crystallized earlier (at greater depth) than those from the Apollo sites. However the low Mg/Fe ratios of the L24 samples (e.g., Albee *et al.*, 1977; Vaniman and Papike, 1978) suggest that the L24 source cumulate may have been formed from a magma which was more fractionated with respect to Mg/Fe than the Apollo samples, and thus should have formed later.

We have previously proposed a KREEP mixing model to explain the U-Pb systematics of Apollo mare basalts (cf. Nunes *et al.*, 1974; Unruh and Tatsumoto, 1977). If we assume a simple fractional crystallization model for lunar primary differentiation, then most trace elements would have been progressively concentrated in the residual liquid. Ringwood and Kesson (1976) have suggested that the residual liquid would not have solidified as a thin layer below the lunar crust, but would have segregated into "pods" with dimensions of 5–20 km. If these "pods" were more concentrated toward the near side of the moon, then KREEP material may have been concentrated in specific areas (South Imbrium?; Nunes *et al.*, 1974). These KREEP "zones" could then act as contaminants to mare basalt magmas passing through them and could cause trace element enrichment and an increase in the Rb/Sr and U/Pb ratios of the mare magma.

If we consider the Luna 24 basalt to represent uncontaminated material, then up to ~50–80% of the U and Th in Apollo mare basalts may actually be KREEP contamination (Tables 1 and 2). If we consider the KREEP contaminant to have had U abundances similar to that of rock 14310 (~3 ppm, Tatsumoto *et al.*, 1972a), then only ~4% of bulk KREEP material is required to double the U concentration in the mare magma. If we assume a higher U concentration for the contaminant such as that observed in rock 12013 (~6 ppm; Tatsumoto *et al.*, 1971) or if the mare magma came into partial equilibrium with the surrounding solid KREEP material, then the amount of admixed bulk KREEP material required to double the U concentration in the mare magma may be less than 1%. This model is rather speculative, but it does appear to satisfy many of the trace element characteristics of mare basalts, and to account for the apparent low μ ratios of the Luna 24 basalts.

CONCLUSIONS

Due to the small sample sizes available, and low U, Th and Pb abundances, the data are not nearly as precise as those usually obtained. However, some conclusions regarding the data can still be made. Low initial μ ratios

($\equiv^{238}U/^{204}Pb$; refractory/volatile indicator) for the Fecunditatis-Crisium area are suggested by this work and soil analyses from Luna 16, 20 and 24 (Tera and Wasserburg, 1972; Tatsumoto, 1973; Birck *et al.*, 1977). Three-stage model calculations for ΣWR1 and OP suggest that the μ ratio of the cumulate source could have been as low ~12–15. These initial μ's are about one-tenth to one-twentieth of those calculated for Apollo mare basalts and are about one-half that of the initial μ of orange soil 74220 (e.g., Tatsumoto *et al.*, 1973a; Tera and Wasserburg, 1976). An initial μ of ~12–15 is however still ~1.5–2 times higher than that inferred for the earth's mantle (~8).

The data may simply reflect a laterally heterogeneous magma ocean, but could also reflect U-Pb fractionation during cumulate formation (vertical heterogeneities) and/or a lesser amount of KREEP contamination in the Luna 24 samples relative to Apollo (near-side) mare basalts. If we assume that the Luna 24 basalt does not contain significant KREEP "contamination," then these low initial μ's may be more representative of the μ ratios of the mare basalt cumulate sources than are the calculated first-stage μ's of Apollo mare basalts. If the bulk crystal-liquid partition for Pb is ~20 times that of U (but still significantly less than 1; e.g., Unruh and Tatsumoto, 1977) then the initial μ of the undifferentiated lunar magma ocean must still have been quite high, ~50–150, relative to the earth. Thus, in spite of the calculated low μ for the cumulate source of this sample, the data are still compatible with the hypothesis that the moon originally accreted as a volatile-depleted body (e.g., Ganapathy and Anders, 1974).

The Pb isotopic data (Fig. 1) and the U-Pb data (Fig. 2) indicate that sample 24170 has had a rather complex U-Pb evolution history. The U-Pb system was disturbed at ~0.5 ± 0.5 B.y. ago, if the disturbance represents a single event, and the data indicate that some samples (WR2 and WR1) have been contaminated with Luna 24 soil-type Pb. There is also the possibility that some samples (PXD) may contain significant amounts of undetected terrestrial contamination. The apparent $^{207}Pb/^{206}Pb$ isotopic disequilibrium observed between the leach and residue fractions supports the hypothesis that the U-Pb systems of mare basalt mineral fractions may not have remained closed after crystallization. The leach residue trends observed in this study are nearly identical to those observed for 15085 (Unruh and Tatsumoto, 1977). Chen *et al.* (1978) have shown that mare basalt 75075 has also apparently experienced post-crystallizational U-Pb open system behavior. We would speculate that further detailed analyses of other lunar basalts may prove this open system behavior to be the rule rather than the exception.

The large percentages of easily leachable U, Th, and Pb found both in this study and our 15085 study indicate that significant portions of U, Th and Pb reside on grain boundaries, in semi-soluble mesostases, or in semi-soluble nonsilicate mineral phases or inclusions. This distribution of U, Th and Pb within the rock and the relatively volatile nature of Pb can probably account for the observed isotopic disequilibrium between the leach and residue fractions, but at present the relationship is not understood.

Lunar K and U data for mare basalts exhibit a rather well defined incompatible element correlation (K/U = 3 ± 2 × 10^3; Lovering and Wark, 1975 and references therein) and there is also a rather crude positive correlation between the K content and the μ ratio. Apollo 11 high-K (~2500 ppm K) mare basalts show the highest μ ratios (Table 2); however Apollo 11 low-K and Apollos 12, 15 and 17 mare basalts all have similar K contents and μ ratios. Hence, the "correlation" consists of essentially only 2 points. K contents for Luna 24 soils, glasses and rock fragments generally range from ~80–250 ppm (e.g., Vaniman and Papike, 1978; Nyquist *et al.*, 1977) as compared to ~300–600 ppm K for most Apollo mare basalts (except Apollo 11 high-K; e.g., Gast *et al.*, 1970; Christian *et al.*, 1972; Compston *et al.*, 1971; Eldridge *et al.*, 1974). The low U concentration observed in 24170 is roughly consistent with the low K content (K/U ~ 1.6 − 3.9 × 10^3 for K = 100–250 ppm and U = 0.064 ppm; Table 1), and the apparent low μ ratio is consistent with the suggested correlation between the K content and the μ ratio. One should note however, that the K content in the Luna 24 rock fragments appears to be quite variable, and as much as ~500 ppm K has been observed in some fragments (Barsukov *et al.*, 1977).

The low $^{87}Sr/^{86}Sr$ initial ratio of ~0.69908 (Papanastassiou *et al.*, 1977; Birck *et al.*, 1977) relative to Apollo mare basalts of similar age (e.g., ~0.6993 at Apollo 15; Papanastassiou and Wasserburg, 1973) suggest that the source region for the Luna 24 basalts evolved in a lower Rb/Sr environment than did those of Apollo mare basalts. The low initial $^{87}Sr/^{86}Sr$ ratio, the low Rb/Sr ratio, and low Rb, Sr and REE abundances (e.g., Papanastassiou *et al.*, 1977; Nyquist *et al.*, 1977) in the Luna 24 basalt correlate well with the low μ ratios and low U and Th abundances reported here, and support the hypothesis that the lunar mantle is laterally heterogeneous with respect to trace elements.

Acknowledgments—We thank the Soviet Academy of Science and the U.S. National Aeronautics and Space Administration for providing the Luna 24 sample. This research was primarily supported by NASA Interagency Transfer Order T-2407A. We benefitted from valuable comments by R. E. Zartman and K. R. Ludwig, and from reviews by G. R. Tilton and F. Tera. We are also indebted to Karen Fenicle for preparing the manuscript in time to meet the numerous deadlines.

REFERENCES

Albee A. L., Chodos A. P. and Dymek R. F. (1977) Petrography of Luna 24 sample 24170 (abstract). In *Papers Presented to the Conference on Luna 24*, p. 5–7. The Lunar Science Institute, Houston.

Barsukov V. L., Tarasov L. S., Dmitriev L. V., Kalesov G. M., Shevaleevsky I. D. and Garanin A. V. (1977) The geochemical and petrochemical features of regolith and rocks from Mare Crisium (preliminary data). *Proc. Lunar Sci. Conf. 8th*, p. 3319–3332.

Birck J. L., Manhes G., Richard P., Joron J. L., Treuil M. and Allegre C. J. (1977) $^{87}Rb/^{87}Sr$ age of Luna 24 microgabbros and isotopic and trace element study of soil 24096 (abstract). In *Papers Presented to the Conference on Luna 24*, p. 34–36. The Lunar Science Institute, Houston.

Chen J. H., Mattinson J. M., Tilton G. R. and Vidal Ph. (1978) Lead isotope systematics of mare basalt 75075 (abstract). In *Lunar and Planetary Science IX*, p. 160–162. The Lunar and Planetary Institute, Houston.

Christian R. P., Annell C. S., Carron M. K., Cuttitta F., Dwornik E. J., Ligon D. T. Jr. and Rose H. J. (1972) Chemical composition of some Apollo 15 igneous rocks. In *The Apollo 15 Lunar Samples* (J. W. Chamberlain and C. Watkins, eds.) p. 206–209. The Lunar Science Institute, Houston.

Compston W., Berry H., Vernon M. J., Chappell B. W. and Kage M. J. (1971) Rubidium-strontium chronology and chemistry of lunar material from the Ocean of Storms. *Proc. Lunar Sci. Conf. 2nd.*, p. 1471–1485.

Drake M. J. and Consolmagno G. J. (1976) Critical review of models for the evolution of high-Ti mare basalts. *Proc. Lunar Sci. Conf. 7th.*, p. 1633–1657.

Eldridge J. S., O'Kelley C. D. and Northcutt K. J. (1974) Primordial radioelement concentrations in rocks and soils from Taurus-Littrow. *Proc. Lunar Sci. Conf. 5th.*, p. 1025–1033.

Frisen L. J. and Heymann D. (1972) Model for Radon diffusion through the lunar regolith. *The Moon* **3**, 461–471.

Ganapathy R. and Anders E. (1974) Bulk compositions of the moon and earth, estimated from meteorites. *Proc. Lunar Sci. Conf. 5th*, p. 1181–1206.

Gast P. W., Hubbard N. J. and Wiesmann H. (1970) Chemical composition and petrogenesis of basalts from Tranquility Base. *Apollo 11 Lunar Sci. Conf.*, p. 1143–1163.

Jovanovic S. and Reed G. W. Jr. (1976) Convection cells in the early lunar magma ocean: Trace-element evidence. *Proc. Lunar Sci. Conf. 7th*, p. 3447–3459.

Lovering J. F. and Wark D. A. (1975) The lunar crust—chemically defined rock groups and their potassium-uranium fractionation. *Proc. Lunar Sci. Conf. 6th*, p. 1203–1217.

Ludwig K. R. and Silver L. T. (1977) Lead-isotope inhomogeneity in Precambrian igneous K-feldspars. *Geochim. Cosmochim. Acta* **41**, 1457–1471.

Nunes P. D., Tatsumoto M. and Unruh D. M. (1974) U-Th-Pb systematics of Apollo 17 lunar samples and implications for a lunar basin excavation chronology. *Proc. Lunar Sci. Conf. 5th.*, p. 1487–1514.

Nunes P. D., Tatsumoto M. and Unruh D. M. (1975) U-Th-Pb systematics of anorthositic gabbros 78155 and 77017-Implications for early lunar evolution. *Proc. Lunar Sci. Conf. 6th.*, p. 1431–1444.

Nyquist L., Wiesmann H., Bansal B., Wooden J., McKay G. and Hubbard N. (1977) Chemical and Sr-isotopic characteristics of the Luna 24 samples (abstract). In *Papers Presented to the Conference on Luna 24*, p. 139–142. The Lunar Science Institute, Houston.

Papanastassiou D. A., McCulloch M. T., De Paolo D. J., Huneke J. C. and Wasserburg G. J. (1977) Ages and evolution of Luna 24 samples. Paper presented at the Conference on Luna 24, Dec. 1–3, 1977, Houston (unpaginated).

Papanastassiou D. A. and Wasserburg G. S. (1973) Rb-Sr ages and initial strontium in basalts from Apollo 15. *Earth Planet. Sci. Lett.* **17**, 324–337.

Ringwood A. E. and Kesson S. E. (1976) A dynamic model for mare basalt petrogenesis. *Proc. Lunar Sci. Conf. 7th.*, p. 1697–1722.

Ryder G., McSween H. Y. Jr. and Marvin U. B. (1977) Basalts from Mare Crisium. *The Moon* **17,** 263-287.

Shields W. R., ed. (1967) Analytical Mass Spectrometry Section: Summary of Activities July 1966 to June 1967. National Bureau of Standards, Technical Note 426, Washington, D. C.

Shields W. R., ed. (1968) Analytical Mass Spectrometry Section: Summary of Activities July 1967 to June 1968. National Bureau of Standards, Technical Note 456, Washington, D. C.

Shimizu N. (1974) An experimental study of the partitioning of K, Rb, Cs, Sr, and Ba between clinopyroxene and liquid at high pressures. *Geochim. Cosmochim. Acta* **38**, 1789–1798.

Silver L. T. (1970) Uranium-thorium-lead isotopes in some Tranquility Base samples and their implications for lunar history. *Proc. Apollo 11 Lunar Sci. Conf.*, p. 1533–1574.

Stettler A. and Albarede F. (1977) Ar^{39}-Ar^{40} pattern and light noble gas systematics of two mm-sized rock fragments from Mare Crisium (abstract). In *Papers Presented to the Conference on Luna 24*, p. 175–178. The Lunar Science Institute, Houston.

Tarasov L. S., Nazarov M. A., Shevaleeusky I. D., Kudryashova A. F., Gaverkovskaya A. S. and Korina M. I. (1977) Mineralogy and petrography of lunar rocks from Mare Crisium (preliminary data). *Proc. Lunar Sci. Conf. 8th.*, p. 3333–3356.

Tatsumoto M. (1970) Age of the moon: An isotopic study of U-Th-Pb systematics of Apollo 11 lunar samples-II. *Proc. Apollo 11 Lunar Sci. Conf.*, p. 1595–1612.

Tatsumoto M. (1973) U-Th-Pb measurements of Luna 20 soil. *Geochim. Cosmochim. Acta.* **37**, 1079–1086.

Tatsumoto M. (1978) Isotopic composition of Pb in oceanic basalt and its implications to mantle evolution. In *Trace Elements and Their Role in Igneous Petrology* (C. J. Allegre and S. R. Hart, eds.). Elsevier, Amsterdam. In press.

Tatsumoto M., Hedge C. E., Doe B. R. and Unruh D. M. (1972a) U-Th-Pb and Rb-Sr measurements on some Apollo 14 lunar samples. *Proc. Lunar Sci. Conf. 3rd*, p. 1531–1555.

Tatsumoto M., Hedge C. E., Knight R. J., Unruh D. M. and Doe B. R. (1972b) U-Th-Pb, Rb-Sr, and K measurements on some Apollo 15 and Apollo 16 samples (abstract). In *The Apollo 15 Lunar Samples*, p. 391–395. The Lunar Science Institute, Houston.

Tatsumoto M., Knight R. J. and Allegre C. J. (1973b) Time differences in the formation of meteorites as determined by $^{207}Pb/^{206}Pb$. *Science* **180**, 1279–1283.

Tatsumoto M., Knight R. J. and Doe B. R. (1971) U-Th-Pb systematics of Apollo 12 lunar samples. *Proc. Lunar Sci. Conf. 2nd*, p. 1521–1546.

Tatsumoto M., Nunes P. D., Knight R. J., Hedge C. E. and Unruh D. M. (1973a) U-Th-Pb, Rb-Sr, and K measurements of two Apollo 17 samples (abstract). *EOS (Trans. Amer. Geophys. Union)* **54**, 614–615.

Tatsumoto M. and Unruh D. M. (1976) KREEP basalt age: Grain by grain U-Th-Pb systematics study of the quartz monzodiorite clast 15045,88. *Proc. Lunar Sci. Conf. 7th.*, p. 2107–2129.

Tatsumoto M. and Unruh D. M. (1977) Primitive Pb in Luna 24 microgabbro 24170 (abstract). In *Papers Presented to the Conference on Luna 24*, p. 179–181. The Lunar Science Institute, Houston.

Tera F., Ray L. A. and Wasserburg G. J. (1972) Distribution of Pb-U-Th in lunar anorthosite 15415 and inferences about its age. In *The Apollo 15 Lunar Samples* (J. W. Chamberlain and C. Watkins, eds.), p. 396–401. The Lunar Science Institute, Houston.

Tera F. and Wasserburg G. J. (1972) U-Th-Pb analyses of soil from the Sea of Fertility. *Earth Planet. Sci. Lett.* **13**, 457–466.

Tera F. and Wasserburg G. J. (1974) U-Th-Pb systematics on lunar rocks and inferences about lunar evolution and the age of the moon. *Proc. Lunar Sci. Conf. 5th.*, p. 1571–1599.

Tera F. and Wasserburg G. J. (1975) The evolution and history of mare basalts as inferred from U-Th-Pb systematics (abstract). In *Lunar Science VI*, p. 807–809. The Lunar Science Institute, Houston.

Tera F. and Wasserburg G. J. (1976) Lunar ballgames and other sports (abstract). In *Lunar Science VII*, p. 858–860. The Lunar Science Institute, Houston.

Tilton G. R. (1960) Volume diffusion as a mechanism for discordant lead ages. *J. Geophys. Res.* **65**, 2933–2945.

Unruh D. M. and Tatsumoto M. (1977) Evolution of mare basalts: The complexity of the U-Th-Pb system. *Proc. Lunar Sci. Conf. 8th.*, p. 1673–1696.

Vaniman D. T. and Papike J. J. (1977) Very low Ti (VLT) basalts: A new mare rock type from the Apollo 17 drill core. *Proc. Lunar Sci. Conf. 8th*, p. 1443–1471.

Vaniman D. T. and Papike J. J. (1978) Ferrobasalts from Mare Crisium: Luna 24. *Geophys. Res. Lett.* In press.

Wasserburg G. J. (1963) Diffusion processes in lead-uranium systems. *J. Geophys. Res.* **68**, 4823–4846.

Wetherill G. W. (1963) Discordant uranium-lead ages, part 2: Discordant age resulting from diffusion of lead and uranium. *J. Geophys. Res.* **68**, 2957–2965.

Merrill, R. B. and Papike, J. J., editors.
(1978) *Mare Crisium: The View from Luna 24*, Pergamon Press (New York), p. 695–700
Printed in the United States of America

Luna 24 origins: Some trace element constraints*

S. JOVANOVIC, K. J. JENSEN and G. W. REED, JR.

Chemistry Division, Argonne National Laboratory, Argonne, Illinois 60439

Abstract—Luna 24 basaltic material originated in the particular volume of the lunar crust that produced Apollo 11 and Luna 16 basaltic material; it has the same Cl_r/P_2O_5 signature. Ru-Os contents (Barsukov *et al.*, 1977) and a number of radiogenic age measurements place Luna 24 basalt origins stratigraphically in a layer between those from which Apollo 15 and Apollo 11 and 17 basalts were derived.

INTRODUCTION

The 160 cm long core returned from Mare Crisium by the Soviet Union Luna 24 mission provides samples of regolith from the mare farthest east on the near side of the moon. Other mare material from near this region was returned by Luna 16 from Mare Fecunditatis. Samples were requested in the expectation that we might further test our proposal that lunar samples which cluster into one of the three major *r*esidual Cl_r/P_2O_5 ratio groups are related by the differentiation of an isolated magma, possibly in a convection cell, in the early lunar surface ocean (Jovanovic and Reed, 1976).

We have measured Cl, Br, I, and U by radiochemical activation analysis and, in the case of the halogens, have separated a labile hot water-leachable and a residual fraction. In addition Fe, Cr, Na, Co, Sc, Sm, and Eu were measured in order to relate our samples to those measured by other groups. P_2O_5 was determined by colorimetry. The water soluble fractions are considered to be deposits from vapors. Most lunar samples have *r*esidual Cl_r/P_2O_5 ratios of either 0.004 ± 0.001, 0.009 ± 0.002 or 0.022 ± 0.005. Samples ranging from anorthositic to basaltic are found in each ratio group. An interpretation of the data is that Cl_r and P_2O_5 occur almost exclusively in late residual liquids and if the original magma was well mixed they will retain the same element ratios during crystallization. At least three such isolated magma bodies were postulated to have existed in the early lunar magma ocean. We have labeled these magma bodies cell I, II and III (Jovanovic and Reed, 1976). In this discussion we will assign our L-24 samples to a cell and examine the implications with respect to a laterally inhomogeneous moon. We will also use results on Ru and Os by Barsukov *et al.* (1977) to examine the stratigraphic origins of L-24 basaltic material relative to other basalts.

---

*Work supported by the National Aeronautics and Space Administration.

RESULTS AND DISCUSSION

The data are given in Tables 1 and 2. Luna 24, <0.250 mm fines have Cl contents comparable to the low concentrations in Apollo 12 and 15 basalts. About one half of the Cl and Br are present as surface deposits. U contents are among the lowest found in mare soils. Our *r*esidual Cl_r results are near the soil values reported by Barsukov *et al.* (1977) and our total Cl results are near those reported for basalts. Our *r*esidual Br_r is similar in amount to their total Br; and U and P_2O_5 results are lower. Within the uncertainties noted in the footnotes to Table 2, our data for the other elements listed above overlap results reported for a variety of samples, Table 2. That the data from basalts, melts and soil are comparable supports arguments published by Ryder *et al.*, 1977; Butler and Morrison, 1977; Florensky *et al.*, (1977) and supported by others at the Luna 24 conference (1977) that the sampled material is almost exclusively local. This suggests that our samples are also representative of local basaltic material.

The relative amounts of hot water leachable Cl, Br and I on Luna 24 soils are similar to those found on other lunar samples (Jovanovic and Reed, 1975). This suggests that the sources of and conditions causing these surface deposits were similar to those operating elsewhere on the moon.

Regional origins: The Cl_r/P_2O_5 ratios for the three samples, 24077, 24174 and 24186 are 0.0097, ≥0.0062, and 0.0078, respectively, placing Luna 24 in the cell with a mean Cl_r/P_2O_5 ratio of 0.009 ± 0.002 (Jovanovic and Reed, 1976), which includes Apollo 11 basalts, many soils and breccias from the Apennine front, Fra Mauro and the Taurus-Littrow massifs. Many of these samples are KREEP rich in contrast to Luna 24. Sampling of this cell is unique in that now very low titanium (VLT), KREEPy and high Ti basaltic materials have been returned.

No consensus was reached on the petrogenesis of Luna 24 basaltic fragments at the Luna 24 conference (1977). Our results on the Cl_r/P_2O_5 ratios do not directly address this question. They do identify Luna 24 material with a source region from which Apollo 15 green glass, 15427, Apollo 17 green glassy basaltic rock 78526 (Taylor *et al.*, 1977a), Apollo 11 basalts, Luna 16 basaltic and Apollo 14 and 15 KREEP materials were derived. This region does not include Apollo 12 and 15 basalts nor some Apollo 17 basaltic material. Some support for a relationship between Luna 24, Luna 16 and Apollo 11 basaltic materials is found in their $^{87}Rb/^{86}Sr$ ratios. In a preliminary report we noted that these ratios form different groups which contain the same samples as the Cl_r/P_2O_5 groups (Jovanovic *et al.*, 1978). Other support may be inferred by the observation that on the MgO-FeO diagram of Kurat *et al.* (1976) the field of the Luna 24 samples [data from Barsukov *et al.*, (1977), Vaniman and Papike (1977) and Taylor *et al.*, (1977a)] overlaps that of Luna 16 basalts and includes both low and high K Apollo 11 basalts. Neither Apollo 15 green glass 15427 nor Apollo 17 green glassy basaltic rock 78526, both VLT, fall near this trend, although Cl_r/P_2O_5 places them in the same cell.

We have used basalts as material which originated below the depths in the moon perturbed by impacts to delineate approximate areal boundaries for the

Table 1. Halogens, U and P_2O_5 in Lunar Samples Returned by the Luna 24 Mission. Samples are <0.250 mm bulk fines.*

Sample	wt. mg	Cl r^{+} ppm	Cl l^{+} ppm	Br* r ppb	Br* l ppb	I ppb	U ppb	P_2O_5 wt.%	P_2O_5 (mg)$^{++}$
24077,6	28.2	3.2	4.2	39	≤38	4.2	89	0.033	(5.0)
24174,11	21.4	3.6	2.6	24	41	1.7	56 ± 10	≤0.058	(7.7)
24182,16	17.7	3.8	4.1	≤20	≤36	1.8	140 ± 20	0.049	(6.9)

*Counting statistical errors are 10% or less, or as indicated; except 20% or less for Br. We reported Li of 1.7 ppm and 3.6 ppm for 24174 and 24182 in the Luna 24 conference abstract (Jovanovic *et al.*, 1977) but now question these numbers since the monitor was found to be about 2 times higher than the average from a number of other irradiations.

$^{+}$r = residue after leach; l = 10 min. hot water leach; I observed in leach only; $^{++}$ = weight of sample used for P_2O_5 measurement.

Table 2. Major, Minor and Trace Element Data for Characterization of Luna 24 <0.250 mm Soil Samples.*

Sample	FeO wt.%	Cr_2O_3 wt.%	Na_2O wt.%	Co ppm	Sc ppm	Sm ppm	Eu* ppm
24077,6	19.3	0.496	0.22	54.1	53.2	2.8	0.97
24174,11	17.0	0.345	0.15	39.9	37.2	1.8	0.60
24182,16	18.3	0.448	0.20	57.1	44.1	2.7	0.92
Literature	19[a],20.9[b],18.9[c] 18.5[d],20.55[e], 20.46[f]	0.35[a],0.36[b], 0.37[c],0.18[d], 0.42[e],0.16[f], 0.3639[g]	0.35[a],0.26[b], 0.34[c],0.26[d], 0.09[e],0.10[f]	43[a] 44[b] 52.9[g]	35[a] 37[b] 40.2[g]	2.5[a] 2.7[b]	0.58[a] 0.65[b]

*Counting statistical errors are 4% or less, except 10% for Eu. Absolute errors may be greater due to weighing and to pipetting of aliquants. These elements are determined in aliquants of solutions of irradiated samples. W-1 is used as the monitor in these determinations.

[a,b]Barsukov *et al.* (1977): a = 24175,1-0, b = 24183,1-0 basalts. [c,d]Ryder *et al.* (1977): c,d = 24077,42-35,12 impact melt. [e,f]Vaniman and Papike (1977): e = 24182 green glass, f = 24182 brown glass. [g]Birck *et al.* (1977): soil 24096.

cells. One such boundary was based on Apollo 12 and 15 basalts which have the same Cl_r/P_2O_5 signature. Another was based on Apollo 17 and Luna 16 basaltic material. The Luna 24 samples provide, possibly, a better eastern anchor for this boundary. All three cells are represented among the Apollo 17 basalts. We consider some to be exotic. Recently Jovanovic *et al.* (1978) discussed the possible close proximity of magmas from two cells. Luna 24 may require that the boundary of the cell be moved north eastward. If so, the proposal of Coish and Taylor (1977) for a direct vertical ascent of magma to minimize any changes prior to extrusion may be relevant. However, a more distant magma supply could be considered, namely, via a subsurface channel originating in the Fecunditatis or Tranquillitatis basins.

Stratigraphic origins: The basalt Ru-Os data of Barsukov *et al.* (1977) can be used to extend our proposed stratigraphic trends based on comparison of Ru/Os ratios with radiogenic ages (Jovanovic and Reed, 1977). Radiogenic ages are generally interpreted as related to the basalt magma-source layers as melting occurred at progressively greater depths with time. The Ru/Os ratios of Apollo 11 and 17 ($<$0.2), 15 ($\sim$2.5) and 12 ($\sim$25) basalts are correlated with the age. The Luna 24 Ru/Os ratio of $\gtrsim$0.5 lies between those of Apollo 11 and 17 and Apollo 15 basalts. The limit on the ratio is due to the Os data. The relatively high limits for Os (10–15 ppb) reported for 3 of 4 samples would have to be reduced by a factor of 3 or greater to be consistent with the 0.3–3.0 ppb Os that we have found in other basalts. With a lowered Os content the Luna 24 Ru/Os ratio might approximate the Apollo 15 ratio. The Luna 24 ages of 3.5 to 3.75 B.y. (Boyce *et al.*, 1977; Birck *et al.*, 1977; Stettler and Albarede, 1977; Tatsumoto and Unruh, 1977; Papanastassiou *et al.*, 1977) lie between the $\sim$3.8 (Apollo 11 and 17) and $\sim$3.2 (Apollo 12) B.y. ages of the other basalts. This provides further support for the proposal that Ru/Os and age are somehow related. A scenario that would compromise this relationship would be the magma holding chambers proposed by Vaniman and Papike (1977) as a means of relating the chronology of various VLT basalts. For that matter, any requirement for a near surface chamber or surface magma flow in which fractional crystallization could have occurred, to account for the highly evolved Luna 24 basalts would probably perturb the relationship. This assumption is based on the following. Holding chambers suggest a two stage magma transfer. The variations of Ru-Os ratios in basalts suggest that these elements tended to fractionate during either cumulus formation or the partial melting of cumulus to provide basalt parental magmas or during both processes. The ages only record the final act, i.e., crystallization of the specimen in hand. The Ru-Os ratio may be effected during each igneous event leading to the final crystallization of the sample. The correlation suggested by the data would seem to be extremely fortuitous if the processes leading to basalt parental magma extrusion involved more than a simple single stage production and transfer of liquid.

SUMMARY

Luna 24 basalts are from the same cell as Apollo 11 basalts and from a greater depth. The two basalt types could be derivatives of primary cumulates formed in the original lunar surface ocean in which trace elements (Ti, as well) tended to concentrate in the near surface layers, Apollo 11, and to be relatively depleted at depth, Luna 24. This interpretation is consistent with but extends the statement of Ryder *et al.* (1977) that Luna 24 VLT basalts are the products of an independently derived liquid from a source material which was itself VLT. Taylor *et al.* (1977b) extend the idea of similar source regions to include VLT basalts other than Luna 24.

The various VLT basalts (our data include only Luna 24, 15427 and 78526 material of basaltic affinity) may be derived from different strata in a VLT cumulate layer. Other cumulate layers may have provided the high Ti Apollo 11 and low Ti Apollo 14 and 15 KREEPy basaltic material in this particular cell. On the basis of the apparent age-depth correlation, tentatively supported by Ru-Os basalt data, the Luna 24 source lies just below the Apollo 11 source and above that for Apollo 15 green glass. The ≥3.9 B.y. age for KREEP materials (See Nyquist, 1977 for summary of data) places their low-Ti source above the high-Ti basalt source. All of these overlie another low-Ti layer, not yet identified in the samples from this cell, which has been recognized in another cell that contains Apollo 12 and 15 basalts. These basalts have the same Cl_r/P_2O_5 signature but both their ages and their Ru/Os ratios suggest that they originated in different source regions of this other cell (Jovanovic and Reed, 1977).

Acknowledgments—We thank the operating staff of the High Flux Reactor, Brookhaven National Laboratory for their assistance and the Chemistry Department for the use of their facilities. The association of G. W. R. with and the use of facilities of the Enrico Fermi Institute, University of Chicago is gratefully acknowledged. We thank Karen Knoerr for so expeditiously typing this manuscript.

REFERENCES

Barsukov V. L., Tarasov L. S., Dmitriev L. V., Kolosev G. M., Shevaleevsky J. D. and Garanin A. V. (1977) The geochemical and petrochemical features of regolith and locus from Mare Crisium (preliminary data). *Proc. Lunar Sci. Conf. 8th*, p. 3319–3332.

Birck J. L., Manlens G., Richard P., Joron J. L., Trenil M. and Allegre C. J. (1977) $^{87}Rb/^{87}Sr$ age of Luna 24 microgabbros, and isotopic and trace element study of soil 24096 (abstract). In *Papers Presented to the Conference on Luna 24*, p. 34–36. The Lunar Science Institute, Houston.

Boyce J. M., Schaber G. G. and Dial A. L. Jr. (1977) Age of Luna 24 mare basalts based on crater studies. *Nature* **265**, 38–39.

Butler P. Jr. and Morrison D. A. (1977) Geology of Luna 24 landing site. *Proc. Lunar Sci. Conf. 8th*, p. 3281–3301.

Coish R. A. and Taylor L. A. (1977) Igneous rocks from Mare Crisium: Mineralogy and petrology (abstract). In *Papers Presented to the Conference on Luna 24*, p. 48–50. The Lunar Science Institute, Houston.

Florensky C. P., Basilevsky A. T., Ivanov A. V., Pronin A. A. and Rode O. D. (1977) Luna 24: Geological setting of landing site and characteristics of sample core (preliminary data). *Proc. Lunar Sci. Conf. 8th*, p. 3257–3279.

Jovanovic S., Jensen K. J. and Reed G. W. Jr. (1977) Origins of Luna 24 basalts: A geochemical perspective (abstract). In *Papers Presented to the Conference on Luna 24*, p. 90–93. The Lunar Science Institute, Houston.

Jovanovic S., Jensen K. J. and Reed G. W. Jr. (1978) A laterally inhomogeneous moon: Cl-P_2O_5, Rb-Sr evidence (abstract). In *Lunar and Planetary Science IX*, p. 600–602. Lunar and Planetary Institute, Houston.

Jovanovic S. and Reed G. W. Jr. (1975) Soil breccia relationships and vapor deposits on the moon. *Proc. Lunar Sci. Conf. 6th*, p. 1753–1759.

Jovanovic S. and Reed G. W. Jr. (1976) Convection cells in the early lunar magma ocean: Trace-element evidence. *Proc. Lunar Sci. Conf. 7th*, p. 3447–3459.

Jovanovic S. and Reed G. W. Jr. (1977) Trace element geochemistry and early lunar differentiation. *Proc. Lunar Sci. Conf. 8th*, p. 623–632.

Kurat G., Kracher A., Keil K., Warner R. and Prinz M. (1976) Composition and origin of Luna 16 aluminous mare basalts. *Proc. Lunar Sci. Conf. 7th*, p. 1301–1321.

Nyquist L. E. (1977) Lunar Rb-Sr chronology. *Phys. Chem. Earth* **10**, 103–142.

Papanastassiou D. A., McCulloch M. T., DePaulo D. J., Huneck J. C. and Wasserburg G. J. (1977) Ages and evolution at Luna 24 samples. Preprint.

Ryder G., McSween H. Y. Jr. and Marvin U. B. (1977) Basalts from Mare Crisium. *The Moon* **17**, 263-287.

Stettler A. and Albarede F. (1977) Ar^{39}–Ar^{40} pattern and light noble gas systematics of two mm-sized rock fragments from Mare Crisium (abstract). In *Papers Presented to the Conference on Luna 24*, p. 175–178. The Lunar Science Institute, Houston.

Tatsumoto M. and Unruh D. M. (1977) Primitive Pb in Luna 24 microgabbro 24170 (abstract). In *Papers Presented to the Conference on Luna 24*, p. 179–182. The Lunar Science Institute, Houston.

Taylor G. J., Keil K. and Warner R. D. (1977a) Very low-Ti mare basalts. *Geophys. Res. Lett.* **4**, 207–210.

Taylor G. J., Warner R. D. and Keil K. (1977b) Chemical distinction among very low-Ti mare basalts (abstract). In *Papers Presented to the Conference on Luna 24*, p. 183–185. The Lunar Science Institute, Houston.

Vaniman D. T. and Papike J. J. (1977) Ferrobasalts from Mare Crisium: Luna 24. *Geophys. Res. Lett.* **4**, 497–500.

Merrill, R. B. and Papike, J. J., editors.
(1978) *Mare Crisium: The View from Luna 24*, Pergamon Press (New York), p. 701–709
Printed in the United States of America

Rock magnetic and palaeomagnetic studies on Luna 24 samples

A. STEPHENSON, D. W. COLLINSON and S. K. RUNCORN

Institute of Lunar and Planetary Science, School of Physics,
The University, Newcastle upon Tyne, NE1 7RU.

Abstract—Magnetic measurements on forty seven Luna 24 samples indicate that at least one sample acquired a magnetization in a field of about 0.3 Oe. Most samples possessed an NRM below the detection limit of the magnetometer but a typical upper limit to the field in these cases is 0.5 Oe. In agreement with other work done on lunar samples, the ratio IRM_s/ARM_s was found to be constant for a wide variety of sample types.

INTRODUCTION

Forty seven Luna 24 samples, most of which were of mass about 1 mg, were examined magnetically. The measurements fall into two categories: (a) measurement of NRM and estimation of palaeointensity by an ARM method; (b) rock magnetic measurements. The fragments included basalts, anorthosites and agglutinates, and their preliminary description obtained by the use of a binocular microscope by Dr. S.O. Agrell is given in Table 1. The samples came from three different depths below the surface, namely 90, 125 and 196 cm. These depths are incorporated in the sample numbers of Table 1.

NRM MEASUREMENTS

All but one of the samples had a mass in the range of 0.1 to 4 mg and were sealed in pyrex tubes. The remaining sample (number 47 in Table 1) had a mass of 17.6 mg and was not sealed. Because most of the samples were to be used by other investigators for further measurements, it was considered impracticable to attempt to remove each sample from its tube for measurement. To estimate the NRM, therefore, the tube was rotated about its vertical long axis in a Digico magnetometer with a specially adapted sensor of small dimensions to increase the sensitivity. The usual six position method which improves the sensitivity could not be used. Using this arrangement the minimum detectable moment was about 5×10^{-7} G cm^3. Each sample was shaken into an arbitrary orientation in its tube, and because of its irregularities and the constrictions at the sealed ends of the tube, the samples did not move relative to their tubes when these were rotated about their vertical axes of symmetry. Thus only the horizontal components of magnetization of each sample were measured. In unfavourable cases, therefore, when any magnetization happened by chance to be vertically orientated, no magnetization would be detected.

Only 3 of the 46 samples possessed a detectable NRM by this method. These

Table 1. Stephenson—magnetic studies on Luna 24.

N	Sample Number	Description	NRM	ARM_s	h_p	IRM_s	χ
1	24090,1,9001		<0.4	<0.4	—	2.71	<0.04
2	9002		<0.5	<0.5	—	0.77	<0.04
3	9003	basalts	<1.1	<1.1	—	1.64	<0.09
4	9004		<1.5	<1.5	—	2.68	<0.14
5	9005		<2.8	<2.8	—	7.70	<0.25
6	9006		<1.5	<1.5	—	<1.5	0.21
7	9007		<1.5	4.35	<0.5	102	2.00
8	9008		<1.2	1.50	<1	40.5	0.39
9	9009	agglutinates	<1.2	5.34	<0.3	81.4	1.66
10	9010		<1.6	4.25	<0.5	78.9	1.10
11	9011		<0.7	3.50	<0.3	74.1	0.94
12	9012		<0.7	1.29	<0.7	43.8	0.82
13	9013		<0.3	4.18	<0.1	93.4	1.75
14	9014		0.89	3.87	0.3	78.5	1.64
15	9015		<1.3	0.67	<3	15.6	0.29
16	9016	weakly sintered clods	<0.5	9.17	<0.07	381	3.66
17	9017		<0.6	<0.6	—	5.46	0.17
18	9018		<1.0	0.59	<2	16.1	0.41
19	9019		<2.9	6.54	<0.6	107	2.15
20	9020		<6.9	<6.9	—	67.1	1.98
21	9021	poikilitic	<1.3	6.05	<0.3	200	3.32
22	9022	melt rocks	<0.2	<0.2	—	5.09	0.12
23	9023	shock	5.4	3.52	2	88.6	1.48
24	9024	anorthosites	<1.1	<1.1	—	3.44	0.45
25	9025	noritic anorthosite	<0.4	<0.4	—	2.94	0.09
26	24125,1,9001		<0.5	<0.5	—	2.35	0.10
27	9002	basalts	<0.6	<0.6	—	1.10	0.04
28	9003		<0.8	<0.8	—	2.70	0.08
29	9004		<0.3	3.95	<0.1	143	1.77
30	9005		<0.8	3.32	<0.3	102	1.68
31	9006		<0.7	2.58	<0.4	89.6	1.75
32	9007	agglutinates	<1.2	4.38	<0.4	77.9	1.30
33	9008		<1.6	7.86	<0.3	79.9	2.49
34	9009		<1.0	2.60	<0.5	58.5	1.05
35	9010		<1.1	1.68	<0.9	53.5	1.26
36	9011		2.9	8.58	0.45	305	0.63
37	24196,1,9001	basaltic glass	<0.1	<0.1	—	0.59	0.05
38	9002		<0.3	<0.3	—	0.71	0.11
39	9003		<0.3	<0.3	—	1.96	0.11
40	9004	anorthositic norites	<0.8	<0.8	—	2.93	0.15
41	9005		<1.1	<1.1	—	12.0	0.20
42	9006		<0.1	0.23	<0.6	9.49	0.32
43	9007		<0.3	<0.3	—	5.05	0.11
44	9008	weakly	<0.5	2.43	<0.3	58.8	0.93
45	9009	sintered	<0.9	<0.88	—	4.93	0.19
46	9010	material	<1.5	<1.5	—	20.1	0.95
47	9011	granulitic anorthosite norite	0.02	0.10	0.3	1.77	0.11

Summary of data obtained on all 47 samples.
Units of NRM, ARM_s and IRM_s are 10^{-3} G cm^3 g^{-1}.
Units of h_p are Oe and χ is in units of 10^{-3} G cm^3 g^{-1} Oe^{-1}.

were numbers 14, 23 and 36 and were described as a weakly sintered clod, a shock anorthosite and an agglutinate, respectively. The agglutinate sample, however, had inadvertently been fused into the end of the tube during the sealing process so that its magnetization (2.9×10^{-3} G cm^3 g^{-1}) was undoubtedly a TRM acquired on cooling in the earth's field. The temperature reached during this heating was probably in excess of the Curie point of iron (770°C).

The NRM of the anorthosite was strong (5.4×10^{-3} G cm^3 g^{-1}) but was rather soft, being halved by the application of a peak alternating field of 30 Oe. The sintered clod, however, had an NRM which was much weaker (0.89×10^{-3} G cm^3 g^{-1}) but much more stable, there being no detectable change in intensity on application of a 30 Oe demagnetizing field. The presence of high coercivity material in this sample was further confirmed since an alternating peak field of 360 Oe was required to halve the saturated IRM.

All three samples were then given an ARM in a direct field of 1.8 Oe on which was superimposed an alternating field of peak value 1000 Oe along the same axis as the 1.8 Oe field. (The usual tumbling could not be employed because the sample orientation in the tube would be disturbed.) The ARM values are shown in Table 1. Assuming that the original NRM is a TRM acquired in a field h_p, and that TRM is a factor of 1.34 larger than the ARM acquired in the same steady field (Stephenson and Collinson, 1974), the palaeointensity for the sintered clod is estimated to be 0.3 Oe. It should be stressed that the error in this estimate is uncertain and could be very large.

As regards the anorthosite sample, if its NRM can be regarded as being predominantly an original TRM, and this must be regarded as uncertain, application of the above method gives an order of magnitude estimate for h_p as 2 Oe. In the case of the inadvertently heated agglutinate sample, applying the same argument gives a "palaeointensity" of 0.45 Oe which is similar to the earth's field at the place where the samples were sealed and therefore provides an unexpected but convenient check that the above method is reasonable.

Sample 47, in spite of its much larger mass (17.6 mg), had an NRM which was just about on the detection limit of the magnetometer. In this case the detection limit was lower because of the increased mass and because the sample could be orientated in the six positions required for a measurement. The NRM was estimated to be about 0.02×10^{-3} G cm^3 g^{-1}. When the ARM was given, however, there was some difficulty in measuring it because of the magnetic moment due to the sample holder. This was allowed for in two ways: by subtracting the ARM acquired by the holder from the combined ARM of the sample and holder, and also by demagnetizing the holder in 1000 Oe and remeasuring sample and holder together. The two methods produced somewhat inconsistent results (although the reason was not apparent), and after repeat measurements using the same methods, the average ARM, based on the 4 measurements was $(0.104 \pm 0.022) \times 10^{-3}$ G cm^3 g^{-1}. Using the same palaeointensity method as above gives a value of 0.3 Oe for this sample which is similar to the value for sample 14.

It should be emphasised that the above estimates are very uncertain, especially for the magnetically soft anorthosite sample. It is interesting but rather surprising to note that the general magnetic characteristics of samples 14 and 23 are very similar. The susceptibility, saturated ARM and saturated IRM (see Table 1) are all within about 10% of each other even though the NRM is very different both in magnitude and ease of demagnetization. It seems likely that the anorthosite (sample 23) may have been incorrectly classified since its magnetic characteristics are much closer to the agglutinates than to the other anorthosites (see Table 1). The above method is also heavily biased towards detecting those samples with a very high remanence. It is likely that many of the other samples, especially the basalts, possessed an NRM which was below the detection limit of the apparatus since relatively few lunar samples have been found to have an NRM as high as 10^{-3} G cm^3 g^{-1} (see Pearce *et al.*, 1973) which is the approximate detectable lower limit in these measurements. Of 50 lunar samples included in a histogram of NRM by Pearce *et al.*, only 2 exceed 10^{-3} G cm^3 g^{-1} so that the number detected here is very similar to that expected from their histogram.

Although an NRM was not detected in most cases it is nevertheless possible to estimate an upper limit to the field using the above approximate palaeointensity method, i.e.

$$h_p = \frac{\text{NRM}}{\text{ARM}} \times \frac{1.8}{1.34}$$

Upper limits shown in Table 1 are typically about 0.5 Oe and so the method is unfortunately just not sensitive enough to contribute to the possible relationship between field and age given by Stephenson *et al.*, (1975). It should be stressed that these upper limits are based on the assumption that any NRM below the detection limit is an unchanged original magnetization acquired billions of years ago.

ROCK MAGNETIC MEASUREMENTS

The susceptibilities of all the samples were measured in a peak field of 7 Oe at a frequency of 10 KHz. Because the samples were so small, the diamagnetism of the pyrex tubes in which they were placed had to be allowed for. This was done by measuring the diamagnetic susceptibility of several empty tubes and taking an average.

All the samples (except number 47 which was tumbled) were also given an ARM without tumbling in a 1.8 Oe steady field on which was superimposed a decreasing alternating field of maximum value 1000 peak Oe. The samples were then given an IRM in a field of 2.6 KOe which would be expected to produce near saturation in most samples since typical remanence coercive forces rarely exceed 500 Oe (Nagata *et al.*, 1973; Pearce *et al.*, 1973). In all cases the ARM

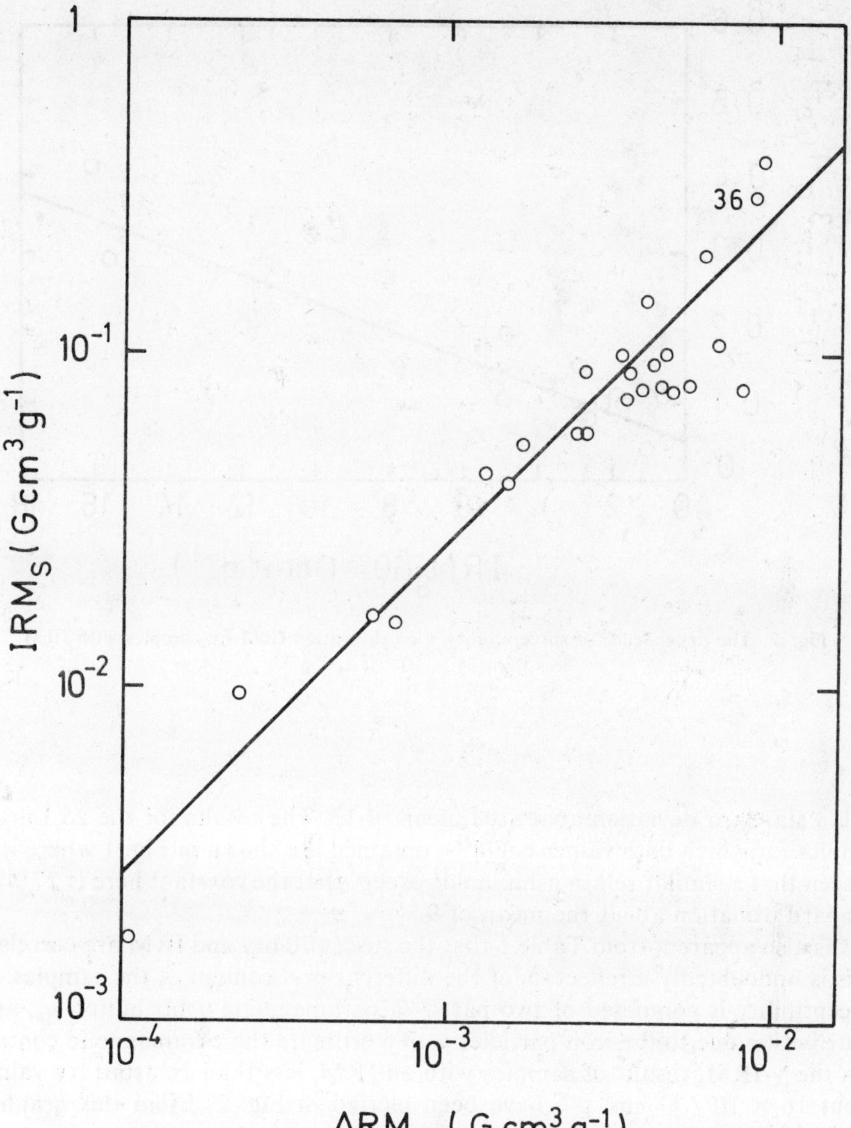

Fig. 1. The relationship between IRM_s and saturated ARM. Sample 36 (chemically altered) is indicated.

produced in the basalts was undetectable whereas apart from one sample, a detectable IRM was produced. All these rock magnetic results are shown in Table 1.

It has been found by Stephenson *et al.* (1977b) that from 8 lunar samples the ratio IRM_s/ARM_s in 1.8 Oe is approximately constant and equal to about 33

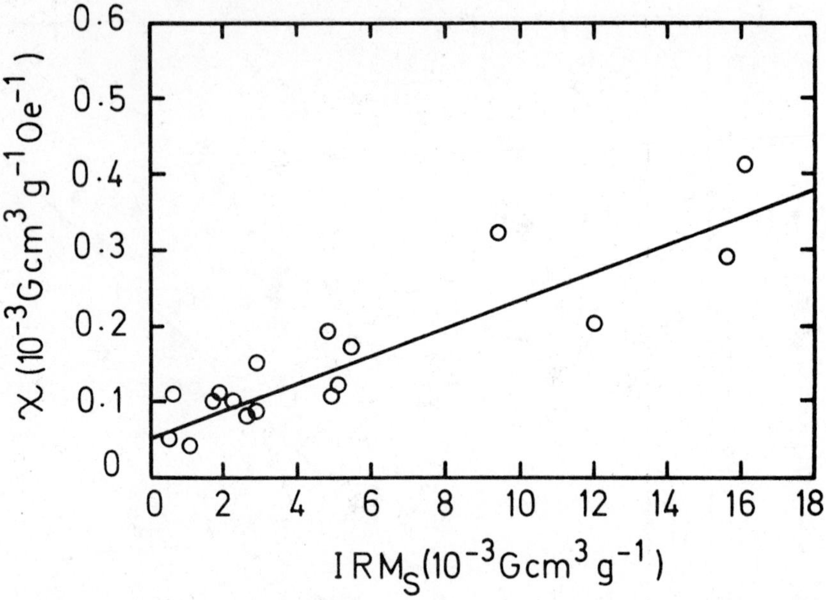

Fig. 2. The dependence of susceptibility χ on saturation IRM for samples with IRM_s less than 16×10^{-3} G cm^3 g^{-1}.

with a standard deviation about the mean of 13. The results for the 25 Luna 24 samples for which both values could be obtained are shown in Fig. 1 where it can be seen that a similar relationship holds except that the constant here is 27 with a standard deviation about the mean of 9.

It is also apparent from Table 1 that the susceptibility and IRM are correlated. This is undoubtedly a reflection of the different iron content of the samples. The susceptibility is composed of two parts—a paramagnetic contribution χ_p, and a contribution due to the iron particles χ_f. To estimate the paramagnetic contribution the χ-IRM_s results of samples with an IRM_s less than an arbitrary value of about 16×10^{-3} G cm^3 g^{-1} have been plotted in Fig. 2. From this graph the intercept is 0.057×10^{-3} G cm^3 g^{-1} Oe^{-1} and this has been taken as an average value of χ_p of all samples. This is similar to values of χ_p found for Apollo samples (Pearce *et al.*, 1974). Samples with IRM_s larger than 16×10^{-3} G cm^3 g^{-1} have not been included in Fig. 2 because their contribution χ_f would swamp the paramagnetic component χ_p. When the value χ_p is subtracted from the measurements in Table 1 to give the component χ_f, the graph of Fig. 3 results. The slope shown in this figure is of value unity and hence χ_f is proportional to IRM_s, the average value of the ratio χ_f/IRM_s being about 20×10^{-3} Oe^{-1}. This compares with the value 25×10^{-3} Oe^{-1} obtained from Luna 16 and 20 soil fractions (Stephenson *et al.*, 1977a).

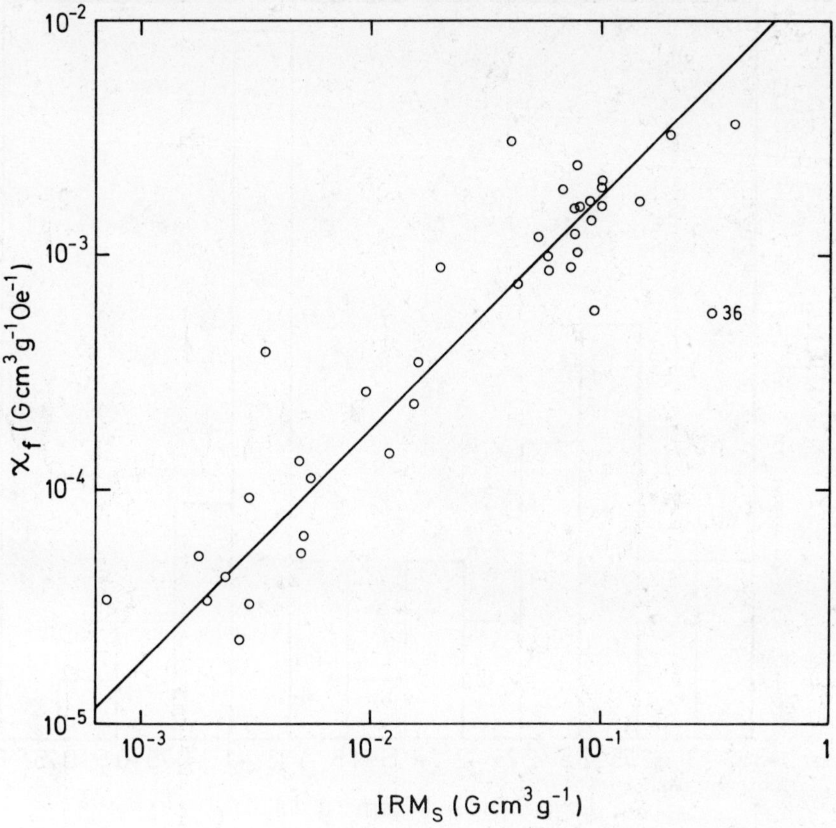

Fig. 3. The dependence of the ferromagnetic susceptibility χ_f (paramagnetic component subtracted) on saturation IRM. Sample 36 (chemically altered) is indicated.

Discussion

Figure 4 shows a histogram of the saturated IRM where it is clear that there are two populations. The group with lower IRM_s consists mainly of the basalts and anorthosites and presumably has properties dominated by multi-domain iron grains. The agglutinates which form the higher IRM group have properties dominated by single domain grains many of which are superparamagnetic. In view of the large variation in iron particle size it is somewhat surprising that χ_f and IRM_s should exhibit such a good correlation since it might be expected that the ratio χ_f/IRM_s would be very different for the two extremes of iron particle size. It seems likely that this correlation may be fortuitous rather than being of any fundamental significance since the ratio for the single domain bearing samples will be dependent on the fraction of iron particles larger than about 130 Å (Stephenson, 1971). It is interesting to note that sample 36 (the heated sample)

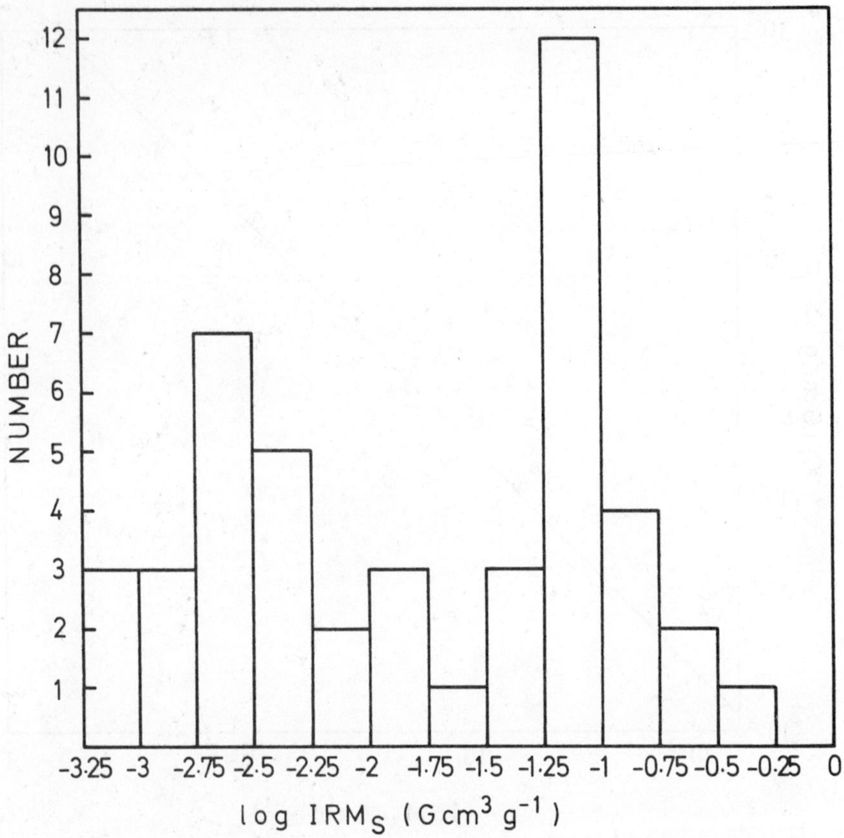

Fig. 4. Histogram of saturated IRM values for all samples.

shows the greatest deviation from the trend in Fig. 3, presumably as a result of chemical change, but does not deviate significantly from the average $IRM_s - ARM_s$ slope in Fig. 1. This suggests that the IRM_s/ARM_s ratio may be independent of composition. The observation that this ratio is similar for iron and magnetite (Stephenson *et al.*, 1977) supports this conclusion.

The most reliable palaeofield estimate from these samples is considered to be that for sample 14 which gave a value of 0.3 Oe. For various reasons such as ease of demagnetization of NRM and inconsistent readings, the other values can not be relied upon. Even that of sample 14 is of lower reliability than most of the other samples shown in the results of Stephenson *et al.* (1977) and, because of the approximate method used, may have a considerable error associated with it. Nevertheless this sample possessed a relatively stable NRM and thus supports previous conclusions that surface fields were once present on the moon and were of the same order of magnitude as the present earth's field.

Acknowledgments—The help of Dr. L. Molyneux in increasing the sensitivity of the measuring instruments is gratefully acknowledged.

References

Nagata T., Fisher R. M., Schwerer F. C., Fuller M. D. and Dunn J. R. (1973) Magnetic properties and natural remanent magnetization of Apollo 15 and 16 lunar materials. *Proc. Lunar Sci. Conf. 4th*, p. 3019–3043.

Pearce G. W., Gose W. A. and Strangway D. W. (1973) Magnetic studies on Apollo 15 and 16 lunar samples. *Proc. Lunar Sci. Conf. 4th*, p. 3045–3076.

Stephenson A. (1971) Single domain grain distributions. II The distribution of single domain iron grains in Apollo lunar dust. *Phys. Earth Planet. Inter.* **4**, 361–369.

Stephenson A. and Collinson D. W. (1974) Lunar magnetic field palaeointensities determined by an anhysteretic remanent magnetization method. *Earth Planet. Sci. Lett.* **23**, 220–228.

Stephenson A., Collinson D. W. and Runcorn S. K. (1977a) Magnetic characteristics of Luna 16 and 20 samples. *Philos. Trans.. R. Soc.* **284**, 151–156.

Stephenson A., Runcorn S. K. and Collinson D. W. (1975) On changes in the intensity of the ancient lunar magnetic field. *Proc. Lunar Sci. Conf. 6th*, p. 3049–3062.

Stephenson A., Runcorn S. K. and Collinson D. W. (1977b) Palaeointensity estimates from lunar samples 10017 and 10020. *Proc. Lunar Sci. Conf. 8th*, p. 679–687.

Sample Index

Sample Index

Subject Index

Author Index